Molecular Beams in Physics and Chemistry

Bretislav Friedrich · Horst Schmidt-Böcking
Editors

Molecular Beams in Physics and Chemistry

From Otto Stern's Pioneering Exploits
to Present-Day Feats

 Springer

Editors
Bretislav Friedrich
Fritz-Haber-Institut der
Max-Planck-Gesellschaft
Berlin, Germany

Horst Schmidt-Böcking
Institut für Kernphysik
Universität Frankfurt
Frankfurt am Main, Hessen, Germany

ISBN 978-3-030-63965-5   ISBN 978-3-030-63963-1   (eBook)
https://doi.org/10.1007/978-3-030-63963-1

This Springer imprint is published by the registered company Springer Nature Switzerland AG
The registered company address is: Gewerbestrasse 11, 6330 Cham, Switzerland

Preface

In 1919, at the Physics Department of the University of Frankfurt, Otto Stern carried out the first quantitative experiment using molecular beams. With this experiment, aimed at measuring the velocity distribution of molecules in a gas, Stern launched the molecular beam method. Stern would be awarded the 1943 Nobel Prize in Physics "for his contribution to the development of the molecular ray [beam] method and his discovery of the magnetic moment of the proton." In his Nobel lecture, delivered in 1946, Stern extolled the virtues of molecular beams thus: "The most distinctive characteristic property of the molecular ray method is its simplicity and directness. It enables us to make measurements on isolated neutral atoms or molecules with macroscopic tools. For this reason, it is especially valuable for testing and demonstrating directly fundamental assumptions of the theory."

On September 1–5, 2019, a symposium was held at Frankfurt to mark the centennial of Stern's pioneering experiment and to show that many key areas of modern science, in particular of physics and chemistry, originated in the seminal molecular beam work of Otto Stern and his school.

Of special significance was the Stern–Gerlach experiment, carried out at Frankfurt in 1920–1922, which introduced the key concept of sorting quantum states via space quantization of angular momentum. Among its descendants are the prototypes for nuclear magnetic resonance, optical pumping, the laser, and atomic clocks, as well as incisive discoveries such as the Lamb shift and the anomalous increment in the magnetic moment of the electron, which launched quantum electrodynamics. In the 1960s, the molecular beam technique made inroads into chemistry as well, by enabling the study of elementary chemical reactions as single binary collisions of chemically well-defined reagents. The ensuing study of chemical reaction dynamics has remained one of the chief preoccupations of chemical/molecular physics to date. In the 1990s, a renaissance began in atomic physics, nurtured by the development of techniques to cool and trap atoms. Based on a combination of molecular beams with laser cooling, these techniques enabled the realization of quantum degeneracy in atomic gases, launched condensed-matter physics with tunable interactions, as well as transforming metrology.

At the Otto Stern Symposium, forty-eight talks and thirty-five posters were presented to an international audience of one hundred fourteen attendees. The symposium was chaired by Dudley Herschbach (Harvard University) and J. Peter Toennies (Max-Planck-Institut für Dynamik und Selsbstorganisation, Göttingen) and presented as the 702th Heraeus-Seminar.

The Physics Department at Frankfurt has been recognized by the European Physical Society (EPS) as an "EPS Historic Site." A plaque marking the site was unveiled during the Otto Stern Symposium by the President of the EPS Petra Rudolf, the President of the University of Frankfurt Birgitta Wolff, and the President of the German Physical Society Dieter Meschede. It honors the work of Max Born, Otto Stern, Walther Gerlach, Elisabeth Bormann, and Alfred Landé performed in Frankfurt during the period 1919–1922.

In order to make the content and insights of the Otto Stern Symposium more enduring and, at the same time, accessible to a larger audience, we invited the symposium speakers and others to contribute chapters to the present volume, which is being published both as a print book and online with open access.

The volume consists of a total of twenty-seven contributions, the first two serving as a prelude to the following parts: I. Historical Perspectives; II. Foundations of Quantum Physics and Precision Measurements; III. Femto- and Atto-Science; IV. Cold and Controlled Molecules; V. Matter Waves; and VI. Exotic Beams.

We trust this volume will help readers to keep abreast of current developments in molecular beam research as well as to appreciate the history and evolution of this powerful method and the knowledge it reveals.

The symposium was funded by grants from the Wilhelm and Else Heraeus Foundation https://www.we-heraeus-stiftung.de/english/, the Deutsche Forschungsgemeinschaft https://www.dfg.de/, Vereinigung von Freunden und Förderern der Johann Wolfgang Goethe-Universität http://www.uni-frankfurt.de/34841010/ueber_vff and Stiftung zur Förderung der internationalen wissenschaftlichen Beziehungen der Johann Wolfgang Goethe-Universität https://www.uni-frankfurt.de/38294561, and the Community Fund of Frontiers Media https://www.frontiersin.org. We thank them all for their generous support.

We also thank Roentdek GmbH, Kelkheim, for generously funding the boat trip on the Rhine that concluded the symposium as well as for contributing to the cost of the symposium dinner.

We are also grateful to the Senckenberg-Stiftung and the Physikalischer Verein Frankfurt and their Presidents Volker Mosbrugger und Wolfgang Grünbein as well as to Professor Andreas Mulch of the Senckenberg-Stiftung for kindly making the facilities of the historic Arthur-Weinberg Haus (Alte Physik, from 1919) available for the symposium. We also thank their colleagues Dr. Tobias Schneck and Professor Bruno Deiss for their kind help. Our special thanks go to Dr. Sebastian Eckardt, Dr. Markus Schöffler, Dr. Christian Janke, and Marianne Frey as well as to others from the Institut für Kernphysik of the University of Frankfurt for ensuring a perfect execution of our organizational plans both during the sessions and the breaks.

We also thank to Joachim Weinert and Sandra Schwab for the realization of the Historic Site plaque.

Last but not least, we thank Dr. Angela Lahee, Executive Editor at Springer Nature, for her dedicated support of the project that resulted in this book.

Berlin, Germany   Bretislav Friedrich
Frankfurt am Main, Germany   Horst Schmidt-Böcking

Contents

Chapter 1
An Homage to Otto Stern

Dudley Herschbach

Abstract This chapter outlines an International Symposium held at Frankfurt on 1–5 September 2019. It marked the centennial of quantitative experiments with molecular beams, pioneered by Otto Stern. The European Physical Society declared Stern's original laboratory a Historic Site, the fifth in Germany. As a graduate student in 1955, I learned about Otto Stern (1888–1969) and the impact of his molecular beams on quantum physics. I was intrigued and undertook crossed-beam experiments at Berkeley. In 1960 Otto came to a seminar that I gave. Later I met him, and heard some of his stories. The rest of the chapter describes his Nobel Prize and other Fests. In 1958 his long-term colleague, Immanuel Estermann, organized a celebration and Festschrift for Otto's 70th birthday. In 1988, as a guest editor, I organized a Festschift for the centennial of Otto's birth. That year, the German Physical Society established the Stern-Gerlach Prize as its highest award for experimental physics. Bretislav Friedrich and I wrote three papers about Stern. Since 2000, Horst Schmidt-Böcking at Frankfurt and colleagues have produced historical articles, along with a book about Otto, edited and bound all of his research papers into books, and diligently pursued letters to and from Otto, collecting them into large volumes.

1 The Frankfurt Conference

The joyful voluntary for trumpet [1] and organ made for a wonderful start for the Otto Stern Conference on 2 September 2019 in Alte Aula at the University of Frankfurt [2]. Professors Horst Schmidt-Böcking (Frankfurt) and Bretislav Friedrich (Berlin) were the Organizers; they developed a festive conference with a hefty booklet. As elders, J. Peter Toennies (Göttingen) and I (Harvard) were glad to be Honorary Chairs. About 140 participants were engaged in talks and discussions over three

D. Herschbach (✉)
Department of Chemistry and Chemical Biology, Harvard University, 12 Oxford St., Cambridge, MA 02138, USA
e-mail: dherschbach@gmail.com

975 Memorial Drive, Apt 712, Cambridge, MA 02138-5754, USA

© The Author(s) 2021   1
B. Friedrich and H. Schmidt-Böcking (eds.), *Molecular Beams in Physics and Chemistry*,
https://doi.org/10.1007/978-3-030-63963-1_1

Fig. 1 Photo of Otto Stern during his Frankfurt time, circa 1920; courtesy of Alan Templeton, grandnephew of Otto Stern

days. The first session of the Conference focused on history, marking the centennial of experiments with molecular beams launched by Otto Stern (Fig. 1). A dozen other sessions highlighted current areas of modern physics and chemistry. On the second day, Stern's original laboratory was declared a European Physical Society Historic Site, the fifth in Germany. The ceremony included a keynote lecture, along with superb music [3], and unveiling of a plaque (Fig. 2) honoring the key discoveries made during 1919–1922 at Frankfurt. Most iconic was the experiment by Stern and Walther Gerlach that proved the reality of space quantization, thereby contributing decisively to the development of quantum mechanics.

The Conference booklet [4] had two historical articles. One is titled "*Stern and Gerlach: How a Bad Cigar Helped Reorient Atomic Physics*," by B. Friedrich and D. Herschbach (Physics Today, 2003 [5]). The second article, extensive and titled "*Otto Stern (1888–1969): The founding father of experimental atomic physics*," by J. P. Toennies, H. Schmidt-Böcking, B. Friedrich, and J. C. A. Lower (Ann. Phys., 2011 [6]). The booklet articles had some festive aspects, suited for Otto. Along with his cigar, he liked amusements, movies, music, dancing, dining and travel by ship. At the Conference dinner, held in the Dorint Oberursel, Professor Ludger Wöste (Berlin) exhibited (Fig. 3) some of his *Physical Amusements*, fascinating and charming toys [7]. More fun came with a post-conference event, on September 5. A bus from Frankfurt took us to Geisenheim, for a boat ride on the Rhein to Braubach and back, accompanied with a wind ensemble, lively hornblowers!

European Physical Society – EPS Historic Site

This building housed Max Born's Institute for Theoretical Physics where key discoveries were made during the period 1919-1922 that contributed decisively to the development of quantum mechanics. The Institute launched experiments in 1919 via the molecular beam technique by Otto Stern, for which he was awarded the 1943 Nobel Prize in Physics. Experiments done in 1920 by Max Born and Elisabeth Bormann sent a beam of silver atoms measuring the free-path length in gases and probing various gases to estimate sizes of molecules. An iconic experiment in 1922 by Otto Stern and Walther Gerlach demonstrated space quantization of atomic magnetic moments and thereby also, for the first time, of the quantization of atomic angular momenta. In 1921, Alfred Landé postulated here the coupling of angular momenta as the basis of the electron dynamics within atoms. This building is the seat of the Physical Society of Frankfurt (the oldest in Germany, founded in 1824).

European Physical Society – EPS Historic Site

In diesem Gebäude wurden in den Jahren 1919 bis 1922 im Institut von Max Born bahnbrechende physikalische Entdeckungen gemacht, die entscheidend zur Entwicklung der Quantenmechanik beigetragen haben. Das sind die Entwicklung der Molekularstrahlmethode im Jahre 1919 durch Otto Stern, für die er den Nobelpreis für Physik des Jahres 1943 erhielt, sowie der im Jahre 1922 erbrachte experimentelle Nachweis der Richtungsquantelung atomarer magnetischer Momente durch Otto Stern und Walther Gerlach, die damit auch erstmals die Drehimpulsquantelung in Atomen nachgewiesen haben. Max Born zusammen mit Elisabeth Bormann haben hier 1920 erstmals die freie Weglänge von Atomen in Gasen und die Größe von Molekülen gemessen. Alfred Landé hat hier 1921 erstmals die Drehimpulskopplung als die Grundlage der inneratomaren Elektronendynamik postuliert. In diesem Gebäude ist der Physikalische Verein Frankfurt (der älteste Deutschlands, gegründet 1824), zu Hause.

Fig. 2 Plaque that marks the European Physical Society Historic Site honoring the Physics Department at Frankfurt, dedicated on 3 September 2019, the fifth such in Germany

Fig. 3 Photo of Ludger Wöste, exhibiting one of his *Physical Amusements*

The Conference booklet also mentions that Otto Stern had a heyday period at the University of Hamburg (1923–1933), but Stern was forced by the Nazi regime to emigrate. He settled in the United States, first in Pittsburg at the Carnegie Institute (1933–1945) and then in Berkeley (1946–1969). He became a U.S. citizen in 1939 which enabled him to serve as a consultant in some military research projects. After the Second World War, Stern was generously helping many of his friends and colleagues with CARE packages. And he would not miss an opportunity to visit Europe to see his friends at conferences and meetings, in particular in Copenhagen, London, and foremost, in Zurich.

2   Learning About Otto Stern and Molecular Beams

In the spring of 1955, as a student at Stanford University, I took a course on statistical thermodynamics taught by a physics professor, Walter Meyerhof (1922–2006). In a brief digression, less than 5 min, he described Stern's first beam experiment done in 1919 at Frankfurt to test the Maxwell-Boltzmann velocity distribution. Meyerhof had to emigrate (Fig. 4) in the Nazi era, and barely escaped the Gestapo [8]. Otherwise, it is likely he would not have been in a Stanford classroom, captivating a susceptible student. For me, learning about molecular beams was love at first sight. I remember a

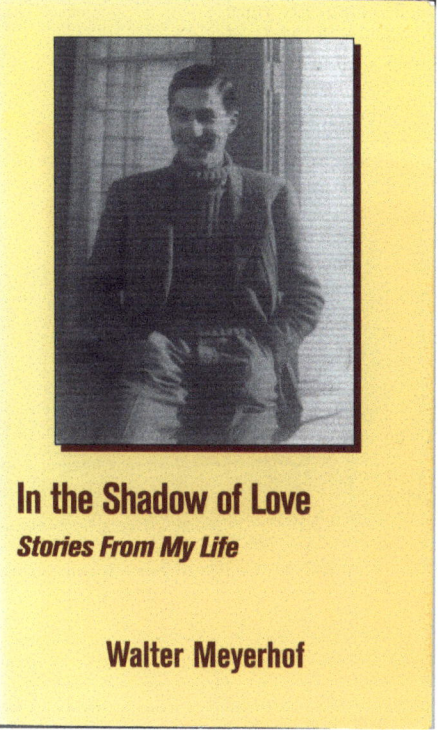

Fig. 4 Photo of Walter Meyerhof, on face of his book *In the Shadow of Love–Stories from My Life* (Fithian Press, 2002)

flush of excitement at the thought that this was *the way* to study elementary chemical reactions. Only five years later my own first beam apparatus was functioning at Berkeley and I had met Otto Stern himself.

Meanwhile, my mentor at Stanford, Harold Johnston (1920–2012), had imbued me with his passion for chemical kinetics. It seemed to me a fundamental thing to try to understand how reactions occur at the molecular level. I wanted to find out what molecules are really doing, making and breaking bonds, instead of the gross macroscopic way that chemists were limited to before, trying to unravel many elementary steps at the same time. Hearing about Otto Stern, I thought by using molecular beams, you can really find out whether or not a reaction occurs as an elementary step. I immediately contacted Hal Johnston, with the naïve notion that chemical reactions could be studied by crossing two such molecular beams in a vacuum to isolate single collisions and directly detect the products. Hal laughed and said, "Well, sure, of course, but there's not enough intensity." It looked difficult. Molecular beam methods had found many applications in physics, but as of the early 1950s, very little had been done in chemistry.

In the fall of 1955 I learned more about Otto, when I moved to Harvard as a graduate student, aiming to obtain a Ph.D. in chemical physics. By golly, Norman Ramsey had just completed his book, titled *Molecular Beams* (Fig. 5). Ramsey gave a sparkling course, handing out the galley proofs. His excellent book reviewed the

MOLECULAR
BEAMS

BY

NORMAN F. RAMSEY
PROFESSOR OF PHYSICS AND
JOHN SIMON GUGGENHEIM FELLOW
HARVARD UNIVERSITY

OXFORD
AT THE CLARENDON PRESS
1956

essence of Stern's work and covered a wealth of further experimental and theoretical methods that produced many important discoveries. Early in Ramsey's course, he discussed Stern's velocity analysis study and actually announced, in his booming voice: "This would be a wonderful way to do chemistry!"

Ramsey also described the career of his mentor, Isidor Rabi, who made epochal molecular beam contributions to physics. Rabi had worked in Stern's lab at Hamburg in 1927–1929 as a postdoctoral fellow before joining the physics faculty at Columbia. There he gladly displayed in his office a photo of Stern (Fig. 6) that he took in the early 1960s. In 1938, Rabi invented a versatile new beam instrument, delivering radiofrequency spectroscopy with extremely high resolving power. In October of 1955, Rabi was invited to give a special lecture at Harvard Physics. His title was "Science and the Humanities." I was intrigued and still am. A friend, John Rigden, wrote a superb book: *Rabi, Scientist and Citizen* (Fig. 7).

In the chemistry department, an ebullient young instructor, William Klemperer, invited me to help build a high-temperature microwave spectrometer. This led us to study ionization of alkali atoms as a function of the surface temperature. Ramsey kindly lent us one of his beam machines over the Christmas vacation in 1956. This was a key episode for both Bill and me. He too fell in love with molecular beams, and immediately undertook to build an electric resonance beam apparatus. Bill and his

Fig. 6 Photo of Otto Stern
in his early 70s that I.I. Rabi
kept on display in his office
at Columbia University

students developed that into a cornucopia for molecular spectroscopy, unprecedented
in resolution and chemical scope [9].

3   Meeting Otto Stern and Hearing Stories from Him

In the summer of 1959, I joined the chemistry faculty at the University of California at
Berkeley as an assistant professor. With two graduate students, George Kwei and Jim
Norris, we built a rudimentary crossed-beam apparatus that enabled us to measure
the angular distributions for reactants and products. Our first reaction was $K + CH_3I$
$\rightarrow KI + CH_3$. In the fall of 1960, the physics department invited me to give a seminar
about our work. In presenting the seminar, I naturally began with homage to Otto
Stern, writing his name on the blackboard and sketching his velocity analysis and
magnetic deflection experiments. During my seminar, I was surprised that two of
the professors in the first row were engaged in animated conversation and swiveling
around to look back at the audience. After the seminar, one of them asked me, "Did
you know Otto Stern was in the audience?" Actually, I had noticed a fellow seated by

Fig. 7 Photo of book of
Rabi, Scientist and Citizen,
by author John S. Rigden
(Basic Books, New York,
1987)

himself, many rows up and back at left. In size and dark attire, he resembled Charlie
Chaplin.

A meeting was arranged so that researchers using molecular beams at Berkeley
could meet him. That was a week or so after the seminar. Professors Howard Shugart
and William Nierenberg gathered a group of more than a dozen graduate students
and postdoctoral fellows, systematically measuring spins and magnetic moments
of radioactive nuclei using the Rabi molecular beam magnetic resonance method.
George Kwei and Jim Norris came along with me. At the meeting, supplied with
coffee, tea, and cookies, Stern at first seemed very shy. Soon, however, in response
to questions, he began telling stories with gleeful verve. Six of them I have retold
often.

1. In his velocity analysis experiment, the results were in approximate agreement
   with the Maxwell-Boltzmann distribution, as anticipated, but deviated from it in
   a systematic way. After sending off a paper, Stern received a letter pointing out
   that he should have included an additional factor of v, the velocity, that enters
   because the detected atoms must pass through a slit. That amendment improved
   the agreement with theory. After explaining this, Stern laughed heartily as he
   added: "That letter came from Albert Einstein!"
2. He spoke happily about his gratitude to Max Born, who was renowned as a fine
   speaker and raised money to build Stern's apparatus at Frankfurt by giving public
   lectures.

3. With wry humor, Stern recalled that when he began teaching a physics course, he found it necessary to work late into the night preparing his lectures. He got into the habit of drinking strong black coffee to stay awake. Since then, he had found he could not fall asleep unless he first had a cup of such coffee.

4. The birth of the celebrated Stern-Gerlach experiment was told by Stern this way [10]: "The question whether a gas might be magnetically birefringent (in the words we used in those days) was raised at a seminar. The next morning I woke early, too early to go to the lab. As it was too cold to get out of bed, I lay there thinking about the seminar question and had the idea for the experiment."

5. Stern said when he got to the lab, "I recruited Gerlach as a collaborator. He was a skillful experimentalist, and I was not. In fact, each part of the apparatus that I constructed had to be remade by Gerlach." Cheerfully, Stern also said: "We were never able to get the apparatus to work before midnight."

6. Stern's "cigar story" was my favorite. As I remember, he told it with relish: "When finally all seemed to function properly, we had a strange experience. After venting to release the vacuum, Gerlach removed the detector flange. But he could see no trace of the silver atom beam and handed the flange to me. With Gerlach looking over my shoulder as I peered closely at the plate, we were surprised to see gradually emerge two distinct traces of the beam. Several times we repeated the experiment, with the same mysterious results. Finally we realized what it was. I smoked cheap cigars. These had a lot of sulfur in them, so my breath on the plate turned the silver into silver sulfide, which is jet black so easily visible. It was like developing a photographic film."

This meeting with Stern lasted about two hours, whereas his cigar episode happened about four decades earlier. Another four decades came ahead: a new Center for Experimental Physics at the University of Frankfurt was dedicated in February 2002 to be named in honor of Stern and Gerlach. At the dedication, I expected to tell Stern's cigar story, having told it many times over forty years. However, historical sleuthing by Bretislav Friedrich showed that two major aspects of my version of the cigar story were wrong. The cigar episode must have occurred at an earlier stage, because Stern was away in Rostock. When Gerlach had finally resolved a pair of distinct traces and by then he was using a photographic development process. The occasion of the Frankfurt dedication prompted Bretislav and me to carry out an experimental test. We found that bad breath did not suffice, although when cigar smoke is exhaled directly onto the deposition plate, the silver traces did rapidly become visible.

I had hoped to meet Stern again at a seminar. But I didn't have the sense to ask Shugart to invite Stern again. In 1963 my group and lab moved to Harvard; alas, I failed to invite him there.

4 Fests with Otto Stern Present

With the Stern-Gerlach experiment, Stern had acquired fame and liked to visit other countries. In 1930 he lectured for some weeks at the University of California at Berkeley and was awarded an honorary degree of L.L.D. On the way there, during December 1929, he met Ernest Lawrence on coincident visits to Harvard. Unaccustomed to Prohibition, Stern asked Lawrence to take him to a speak-easy. While contemplating the circular rings left by their wine glasses, Lawrence diagrammed an idea he had been mulling over for months, a means to accelerate ions in a magnetic field. Stern urged him to stop talking about it, get back to his lab at Berkeley, and work on the idea. Lawrence took the advice and soon developed his cyclotron [11]. As early as 1931, Stern reported in Europe with great enthusiasm on the future of the cyclotron. However, when Stern was forced to emigrate in 1933, he did not receive an offer from Berkeley.

Otto likely enjoyed a fine cigar on December 10, 1944. The Nobel Prizes broke the five-year respite owing the Second World War; no prizes were awarded from 1940 until 1944. The Swedish Academy made up part of the loss by naming the 1943 winners along with those for 1944. The 1943 prize for physics went to Otto Stern and the 1944 prize to Isidor Rabi [12]. They couldn't go to Sweden—the war was still on—so the ceremony was held in New York, at the Waldorf-Astoria (Figs. 8 and 9). Rabi said: "It was an enormous pleasure and an excuse for many parties ..." At the parties, a little ditty was sung with the refrain: "Twinkle, twinkle Otto Stern/How did

Fig. 8 Otto Stern's *Nobel Document. Photo, courtesy* Diana Templeton Killen

Fig. 9 The Swedish ambassador Eric Boström presents the Nobel awards in physics to Stern (left) and Rabi (middle) at the New York Waldorf Astoria Hotel on December 10, 1944. *Courtesy* Diana Templeton Killen

Rabi so much learn?" Otto did come to Stockholm for the 1946 Nobel celebration, and he delivered his Les Prix Nobel lecture, only 7 pages [13].

In 1958, a Festschrift was held for Stern's 70th birthday, organized by Immanuel Estermann (1900–1973). A long-term colleague, Estermann obtained his doctorate in 1921 at Hamburg, and began working with Otto, first at Rostock, then at Hamburg. When forced to emigrate in 1933, he was hired by the Carnegie Institute (now Carnegie-Mellon University) at Pittsburgh alongside with Otto. During the Second World War, Immanuel worked first on Radar and then transferred to the Manhattan Project. After Otto retired to Berkeley in 1945, Estermann left Pittsburgh in 1950 to join the Office of Naval Research. He also became editor of the series of Advances in Atomic and Molecular Physics.

Estermann edited a book: *Recent Research in Molecular Beams* (Fig. 10), a collection of ten chapters dedicated to Otto Stern (Academic Press, 1959) [14]. Estermann wrote the first chapter about the historic work in Hamburg (1922–1933). The other chapters describe fresh research among seven institutions. Only one dealt with chemistry. Sheldon Datz and Ellison Taylor, at Oak Ridge National Laboratory, in 1955 had published a crossed molecular beam reaction, $K + HBr \rightarrow KBr + H$. It made an impact on eager physical chemists. By 1965, a Gordon Research Conference in New Hampshire was accepted. A lively group of 60 graduate students and mentors were discussing theory and experiments for reactions with molecular beams. When I mentioned Otto Stern, a shout came from Sheldon Datz: "For all of us, he is our Father." Of course, I responded: "Otto is a bachelor." There was a roar: "We are all

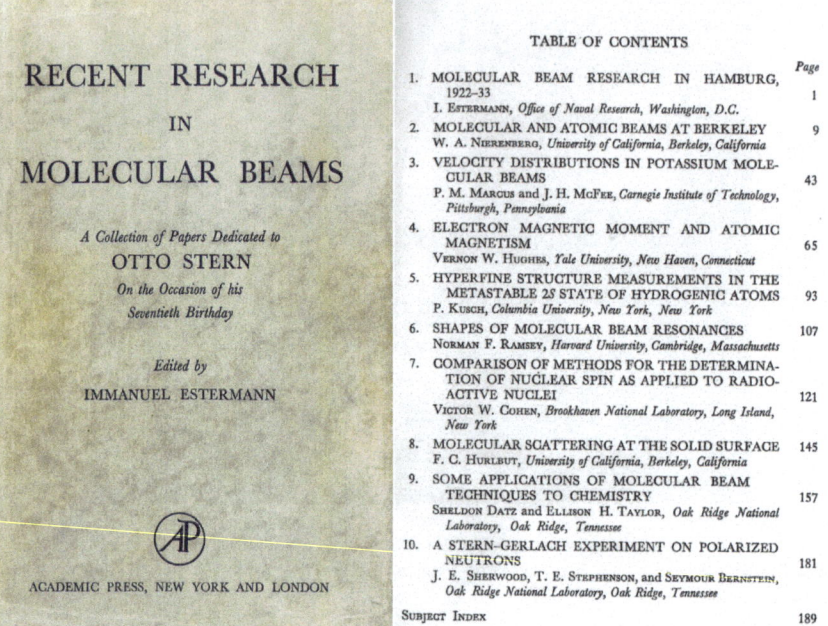

Fig. 10 Photo of book *Recent Research in Molecular Beams, A collection of papers dedicated to Otto Stern on the occasion of his 70th birthday*; edited by Immanuel Estermann (Academic Press, 1959). Table of Contents displayed

bastards!" Since then, dynamics of molecular reactive collisions has flourished, with conferences every two years or so for more than 50 years.

In 1961, Otto Stern had an oral interview, by Res Jost [15]. Also, in 1962, Immanuel Estermann had an extensive oral history interview by John L. Heilbron [16]. Immanuel was engaged in writing a book on the History of the Molecular Beam Method when he died in 1973. A paper in 1975 was published in Am. J. Phys. [17] covering the essence of the first two chapters (edited by S. N. Foner) on the important evolutionary period, 1919–1933. It contains some amusing historical sidelights on the research personalities that dominated that period.

In 1973 Emilio Segrè (1905–1982) delivered a biographical memoir of Otto Stern for the National Academy of Sciences [18]. Segrè had worked with Otto Stern and Otto Frisch (1904–1979) during 1931–1933 at Hamburg on space quantization. When Otto Stern retired to Berkeley, Emilo was on the faculty, so they often met. During his last years, Otto remained interested in discoveries in particle physics and astrophysics. A few days before his death, Otto argued vehemently about enormous energy output of quasars and was dissatisfied that astrophysicists rejected his interpretation! Emilio and many others count Otto Stern among the greatest physicists of the twentieth century.

5   Centennial of Otto Stern and Beyond

In 1987, after writing a long article, *Molecular Dynamics of Elementary Chemical Reactions* [19], I felt attention was deserved in 1988, to have a Festschrift for the centennial of Stern's birth. In his Hamburg era, 1923–1933, Stern had inaugurated a series of papers which he called Untersuchungen zur Molekularstrahlmethode (U.z.M.) published in Zeitschrift für Physik. The series reached 30 papers. That journal fifty years later had grown to four categories. So I urged the Editor in Chief, Ingolf V. Hertel, to produce a centennial issue. He asked me to do it as a Guest Editor for *Z. Phys. D Atoms, Molecules and Clusters* (Fig. 11). Here are parts of the Preface, An homage to Otto Stern:

> His legacy abides in many domains of physics, but especially in vigorous progeny exempli-
> fying his favorite theme: "the characteristic simplicity and directness of the molecular ray
> method." Concepts and techniques developed by Stern have proved remarkably durable and
> versatile, yet still more vital for science is his exemplary pursuit of insight and beauty.

Next comes a reprint of Otto's 1921 paper (plus an English translation); it proposed "an experiment which, if successful, will decide unequivocally between the quantum and classical views." A list of his publications follows—only 60 (Stern's total was 71, including publications in nonscientific venues). Then come reminiscences of Stern by I. I. Rabi as told to John Rigden, some in the last days before Rabi's death (11 January

Fig. 11 Festschrift in memoriam Otto Stern on the 100th anniversary of his birth: Zeitschrift für Physik D, Atoms, Molecules and Clusters **10**, 109–392, June 1988 (Springer International); with six samples among the 31 articles

1988). A review of Stern's development of molecular beams was given by Norman Ramsey, from his lecture presented at a convocation in Hamburg commemorating Stern (4 February, 1988). Ramsey provided a list [20] of 32 major "advances that contributed to physics from the field of molecular beams … during the past seventy years."

The Festschrift indeed had 31 exceptional papers, largely from Stern's kindred spirits. Here are six samples (Fig. 11). Among them are a "continuous Stern-Gerlach effect" that glimpses the primordial Big Bang. Or is "spin coherence like Humpty-Dumpty?" Also, a liquid jet. Or an Otto Stern double bank shot. Or using an electrospray source that generates molecular beams of huge proteins.

Also in 1988 the German Physical Society established the Stern-Gerlach Prize. In 1993 the Prize became the Stern-Gerlach Medal. It is awarded for excellence in experimental physics, in parallel with the existing Max Planck Medal for excellence in theory.

Hamburg also had in 1988 an Otto Stern Symposium, as noted, with Norman Ramsey. A two-day Stern event was held in 2013 with many speakers. This is available on *YouTube*. A single-day Stern event was held in 2018.

In 1998 Bretislav Friedrich and I contributed to an unusual event: *Science in Culture*, held in Proceedings of the American Academy of Arts and Science, Cambridge, Massachusetts [21]. The event was dedicated to Gerald Holton, an outstanding historian of science, for his studies of Einstein. Bretislav and I delivered a sizeable paper titled: *Space Quantization: Otto Stern's Lucky Star* [21]. We hoped to make it accessible to anyone with only vague memories of high-school science, and to induce chuckles rather than growls.

During December 11–14, 2000 there was held in Berlin a Quantum Theory Centenary, celebrating the famous talk of Max Planck. Fifty scientists were invited to present reviews of their fields to a large international audience. The proceedings were collected as a Festschrift in the "Annalen der Physik". I was asked to talk about Otto Stern and molecular beams, before 1935. That led to five decisive episodes: discovery of space quantization; de Broglie matter waves; anomalous magnetic moments of the proton and neutron; recoil of an atom of emission of a photon; and the limitation of scattering cross-sections for molecular collisions imposed by the uncertainty principle [22]. The Centenary Symposium was splendid, having quantum entanglement and teleportation, discovery of quarks, quantum cosmology and more!

In 2002, when the Stern-Gerlach Center for Experimental Physics at Frankfurt was named, a memorial plaque (Fig. 12) was mounted near the entrance of the building where the Stern-Gerlach experiment took place. Horst Schmidt-Böcking had a major role in the installation of the SGE plaque and much more. At the 2019 Conference, the plaque was moved near the room where the SGE was done. The inscription, in translation reads: "*In February 1922 … was made the fundamental discovery of space quantization of the magnetic moments of atoms. The Stern-Gerlach Experiment is the basis of important scientific and technological developments in the 20th century, such as nuclear magnetic resonance, atomic clocks, or lasers …*"

Frankfurt was busy well before the 2019 Conference. In 2005, Wolfgang Trageser collected papers [23] to form a *Stern-Stunden* book (Fig. 13). In 2011, Horst produced

Fig. 12 A memorial plaque honoring Otto Stern and Walther Gerlach was mounted in February 2002 next to the entrance of the building where the S-G experiment took place 80 years earlier

with Karin Reich [24] an Otto Stern book (Fig. 14). He wrote historical articles [25] with others (2011, 2016) and edited all of Otto's research papers [26] into books (Fig. 15). Moreover, Horst with Alan Templeton and Wolfgang Trageser were extraordinarily diligent in pursuing letters to and from Otto, organizing and collecting them into large volumes [27] (Fig. 16).

When I visited Berkeley again, to give a Commencement address in 2012, Alan took me to the Chemistry Library to see Otto's magnificent desk that he had donated to the library (Fig. 17).

The 2019 Conference aimed to show that many key areas of modern physics and chemistry originated in the seminal molecular beam work of Otto Stern and his colleagues. The sessions highlighted the state of the art: foundations of quantum mechanics, as well the problems of quantum measurement; magnetic and electronic resonance spectroscopy, including magnetic resonance imaging and its medical applications; high-precision measurements; cold atoms and molecules; reaction dynamics; matter-wave scattering; magneto-optical traps and optical lattices; and exotic beams, among microdroplet chemistry, liquid beams, and helium droplet beams.

Beyond the history session, memories of Otto and his colleagues endure. Alan Templeton gave a festive talk: *My uncle Otto Stern*. Other presenters were Peter Toennies: *Otto Stern and Wave-Particle Duality*; Dan Kleppner: *Our Patrimony from*

Fig. 13 Photo of book by W. Trageser, ed., *Stern-Stunden Höhepunkte Frankfurter Physik*, comprised of collected articles. A sampling was made from [5] and [21], pp. 149–170

Otto Stern and My Memories of Otto Frisch; Karl von Meyenn: *Stern's Friendship with Wolfgang Pauli*; and Horst: *Stern's Relation to Gerlach*.

Concluding my introduction to the Conference, I offered a song by Cole Porter, "*Experiment*," more than 80 years old [28].

6   Epilogue

This is the closing paragraph of Otto Stern's Nobel Lecture [13]:

> The most distinctive characteristic property of the molecular ray method is its simplicity and directness. It enables us to make measurements on isolated neutral atoms or molecules with macroscopic tools. For this reason, it is especially valuable for testing and demonstrating directly fundamental assumptions of the theory.

Fig. 14 Photo of book by H. Schmidt-Böcking and K. Reich, *Otto Stern*: *Physiker, Querdenker, Nobelpreistrager* (Frankfurt/Main: Societäts-Verlag, 2011)

Fig. 15 H. Schmidt-Böcking, K. Reich, A. Templeton, W. Trageser, V. Vill, eds., *Otto Sterns Veröffentlichungen—Band 1, Sterns Veröffentlichungen 1912 bis 1916* (Springer Spektrum, 2016)

Fig. 16 H. Schmidt-Böcking, A. Templeton, W. Trageser, eds., *Otto Sterns gesammelte Briefe—Band 1, Hochschullaufbahn und die Zeit des Nationalsozialismus* (Springer Spektrum, 2018)

Fig. 17 Dudley with Alan Templeton, visiting Otto Stern's desk, now in the Chemistry Library at the University of California, Berkeley

Acknowledgements For helpful information, I am grateful to Horst, to Bretislav, to Peter, to Alan, to Diana Templeton-Killen, to my wife Georgene, and to Cole Porter.

Appendix: A Historical Puzzle

Recently, I learned from Eugene Wigner (1902–1995) that in the 1920s Michael Polanyi (1891–1976) had an original idea to use molecular beams [29]. At Haber's Institute in Berlin-Dahlem, he was famed for chemical kinetics, using a diffusion flame method involving sodium vapor, halogens, and organic halides. He must have known about the celebrated molecular beams used by Otto Stern at Frankfurt and Hamburg, not so far from Dahlem. Searches in the archives of correspondence of both turned up only one letter from Stern to Polanyi. It is dated 10 October 1928, but with questions unrelated to beams or reactions.

I first met Michael Polanyi in 1962 when he came to Berkeley to deliver a series of lectures on the philosophy of science. He also visited my lab and observed a molecular beam experiment. On other occasions, especially at a Faraday Discussion in London in 1973, Michael heard about many beam results, but didn't mention that he had once intended to try beams. In 1962, I missed an opportunity to arrange for Michael Polanyi to meet and exchange stories with Otto Stern.

Appendix: Lyrics of Cole Porter's "Experiment"

Before you leave these portals
To meet less fortunate mortals
There's just one final message
I would give to you
You all have learned reliance
On the sacred teachings of science
So I hope, through life you never will decline
In spite of philistine
Defiance
To do what all good scientists do

Experiment
Make it your motto day and night
Experiment
And it will lead you to the light
The apple on the top of the tree
Is never too high to achieve
So take an example from Eve
Experiment
Be curious
Though interfering friends may frown
Get furious
At each attempt to hold you down
If this advice you'll only employ
The future can offer you infinite joy
And merriment
Experiment
And you'll see

References

1. Department of Chemistry and Chemical Biology, Harvard University, Cambridge, MA 02138-5754, U.S.A. https://chemistry.harvard.edu/people/dudley-herschbach-dherschbach@gmail.com
2. Musikalische Umrahmung at the Opening of the Otto Stern Fest; the musicians were: Wolfgang Huhn, trumpet; Karsten Schwind, organ
3. Musikalische Umrahmung at the European Physical Society Historic Site Ceremony; the musician was: Roman Kuperschmidt, clarinet

4. Conference Booklet, *Otto Stern's Molecular Beam Research and Its Impact on Science*, ed. by B. Friedrich, D. Herschbach, H. Slchmidt-Böcking, P. Toennies. *An International Symposium*, Frontiers Phys., vol 7, (2019), p. 208. Website: https://indico.fhi-berlin.mpg.de/event/35/
5. B. Friedrich, D. Herschbach, Stern and Gerlach: How a Bad Cigar Helped Reorient Atomic Physics. Phys. Today **56**, 53–59 (2003)
6. J.P. Toennies, H. Schmidt-Böcking, B. Friedrich, J.C.A. Lower, Otto Stern (1888–1969): The Founding Father of Experimental Atomic Theory. Ann. Phys. (Berlin) **523**, 1045–1070 (2011)
7. Ludger Wöste (Berlin), *Physical Amusements*, Part in [4]
8. W.E. Meyerhof, *In the Shadow of Love—Stories from My Life* (Fithian Press, Santa Barbara, 2002), pp. 38–67
9. D. Herschbach, Obituary: William Klemperer (1927–2017). Nat. Astron. **2**, 24–25 (2018)
10. These are not Stern's exact words, but I have presented his stories in first person as an attempt to capture his way of telling them
11. N.P. Davis, *Lawrence & Oppenheimer* (Simon and Schuster, New York, 1968), pp. 27–28
12. J.S. Rigden, *Rabi, Scientist and Citizen* (Basic Books, New York, 1987), pp. 169–170
13. O. Stern, *The Method of Molecular Rays*, Les Prix Nobel en. (M.P.A.L Hallstrom, Stockholm, 1946), pp. 123–130. Available at https://www.nobelprize.org/prizes/physics/1943/stern/lecture/
14. I. Estermann (ed.), *Recent Research in Molecular Beams* (Academic Press, New York, 1959)
15. Interview of Otto Stern by Res Jost, 2 December 1961, tape recording, ETH-Bibliothek Zürich (CH-001807-7 Hs 1008:8). https://search.library.ethz.ch/primo-explore/fulldisplay?docid= cmistar5eecbf09e08143568a2a3046acf875c5&context=L&vid=DADS&lang=en_US&sea rch_scope=default_scope&adaptor=Local%20Search%20Engine&tab=default_tab&query= any,contains,Hs%201008:8&mode=Basic
16. Interview of Immanuel Estermann by John L. Heilbron, 13 December 1962, transcript, 24 pages, available from American Institute of Physics (OH 4593), College Park, Maryland, 20740, USA at https://www.aip.org/history-programs/niels-bohr-library/oral-histories/4593
17. I. Estermann, *History of molecular beam research: personal reminiscences of the important evolutionary period 1919–1933*. Am. J. Phys. **43**, 611–680 (1975) (Edited by S.N. Foner)
18. E. Segrè, Otto Stern. Biogr. Mem. Nat. Acad. Sci. **43**, 215–236 (1973)
19. D.R. Herschbach, *Molecular Dynamics of Elementary Chemical Reactions*, in Les Prix Nobel 1986 (Almquist & Wiksell Int'l, Stockholm, 1987), pp. 117–166. Also, Angew. Chem. Int. Ed. **26**, 1221–1243 (1987)
20. N.F. Ramsey, Molecular beams: our legacy from Otto Stern. Z. Phys. D **10**, 121–125 (1988)
21. B. Friedrich, D. Herschbach, Space quantization: Otto Stern's lucky star. Daedalus Sci. Culture **127**, 165–191 (1998)
22. D. Herschbach, Molecular beams entwined with quantum theory: a bouquet for Max Planck. Ann. Phys. (Leipzig) **10**(1–2), 163–176 (2001)
23. W. Trageser, ed., *Stern-Stunden* book, comprised of collected articles. Part formed from (5) and (21)
24. H. Schmidt-Böcking, K. Reich, *Otto Stern: Physiker, Querdenker, Nobelpreistrager* (Societäts-Verlag, Frankfurt/Main, 2011)
25. H. Schmidt-Böcking, L. Schmidt, H.J. Ludde, W. Trageser A. Templeton T. Sauer. The Stern-Gerlach experiment revisited. Eur. Phys. J. H. **41**, 327–364 (2016). Also see [6]
26. H. Schmidt-Böcking, K. Reich, A. Templeton, W. Trageser, V. Vill (eds.), *Otto Sterns Veröffentlichungen—Band 1-5*, vol. 1–5 (Springer Spektrum, Berlin, Heidelberg, 2016)
27. H. Schmidt-Böcking, A. Templeton, W. Trageser (eds.), *Otto Sterns gesammelte Briefe—Band 1-2*. (Springer Spektrum, Berlin, 2018). In [25] is a list of Archival locations where Stern's letters have been found
28. Cole Porter (1891–1964) song lyrics of "Experiment" (1931) are available at http://www.son glyrics.com/cole-porter/experiment-lyrics/
29. D.R. Herschbach, Michael Polanyi: Patriarch of chemical dynamics and Tacit Knowing. Angew. Chem. Int. Ed. **56**, 3434–3444 (2017)

Chapter 2
A Greeting from Hamburg to the Otto Stern Symposium

Peter E. Toschek

Dear Chairmen Profs. Herschbach and Toennies, dear organizers Profs. Schmidt-Böcking and Friedrich, dear Colleagues, dear Ladies and Gentlemen.

The occasion for our gathering today in this location is a truly historic one: the centenary of the first quantitative experiment with molecular beams, a technique that would enable, in 1922, the first convincing proof of quantum mechanics—the fundamental theory of light and matter—and to give due honor to the pioneer physicist who developed this technique: Otto Stern.

It is my pleasant duty to congratulate, in the name of the *Akademie der Wissenschaften in Hamburg* and in my own name, our Frankfurt colleagues and the *Johann-Wolfgang-Goethe-Universität* to this most appropriate and promising event.

Otto Stern's and Walther Gerlach's ground-breaking experiment—the detection of the "*Richtungsquantelung*" (space quantization) of atoms in a magnetic field—was spectacularly performed here in Frankfurt. Fortunately enough, the founding fathers of Hamburg University—that celebrates its centenary this year as well—recognized Stern's brilliance and awarded him the Chair of Physical Chemistry in 1923. Highlights of his and his team's activities in the following decade included such fundamental discoveries as the verification of de Broglie's matter waves, the measurement of proton's magnetic moment, which is the first measurement of a nuclear quantity, and the demonstration of recoil of atoms upon emission and absorption of light—the first requirement for the much later ubiquitous laser cooling of atoms. The key to these magnificent achievements was the further exploitation of Stern's molecular beam technique that allows to control two atomic translational degrees of freedom. The step towards a 3-D control of atomic motion, forming what could be called a "zero-dimensional atomic beam," is a precondition for making the atom available for

---

The author died on 25 June 2020.

P. E. Toschek (✉)
Institut für Laserphysik, Universität Hamburg, Hamburg, Germany

© The Author(s) 2021
B. Friedrich and H. Schmidt-Böcking (eds.), *Molecular Beams in Physics and Chemistry*,
https://doi.org/10.1007/978-3-030-63963-1_2

manipulation over long time durations. It took half a century for this advancement to be achieved.

In 1933, at the zenith of his career, Otto Stern had to face callous rejection and horrible injustice. Both centers of his prolific scientific activity—Frankfurt and Hamburg—have many good reasons to remember and to highlight this giant of physics, his multi-facetted achievements, and the wide-ranging consequences of his ideas.

The Board and the Members of the *Akademie der Wissenschaften in Hamburg* extend their most sincere wishes to speakers, guests, and organizers for a most successful and inspiring Symposium, commensurate with the innovative and communicative spirit of Otto Stern.

Frankfurt, 1 September 2019

Part I
Historical Perspectives

Chapter 3
My Uncle Otto Stern

Lieselotte K. Templeton

It was only since 1946 [1945] when my uncle moved to Berkeley that I got to know him well. Before this time we had never lived in the same town, and I had only seen him rarely. Otto Stern moved into a house he had bought several years earlier in the Berkeley hills not far from my parents' house. Because he was a bachelor, he hired a housekeeper who came in six days a week for a few hours to cook and keep house. He loved good food and good wine. The housekeeper for the last years did not keep the house as clean as he would have liked, but her cooking met with his approval, so she stayed for many years. On Sundays, he would have dinner with his sister, Berta (my mother) and family, or he would go into town for dinner in a restaurant and then to a movie. He loved movies, and Shirley MacLaine was one of his favorite actresses.

Mother was a good talker and he used to kid her that he was kind of deaf in his right ear, because she sat on that side at dinner when they were children. He could be quite talkative himself, especially if you got him to reminisce.

During World War I he was drafted into the German Army, made a weatherman and sent to a small town, Lomsha, in Russian Poland. There he used to go up in an airplane and make meteorological measurements. When the airplane crashed, luckily not hurting him, it was decided that he would just use balloons. He said that it was not very difficult to predict the weather; it was always terrible, very cold. Anyway, he had plenty of free time and used it to calculate a very large determinant which he always called the Lomsha determinant and he published a paper about it [1]. Another of his papers is also dated from Lomsha [2].

After World War II when one could travel again to Europe, he would go to Zurich every year and a half or two years and stay there for about six months. I think one of Zurich's attractions for him was the fact that he could talk German there, especially to the physicists at the ETH. He had had a classical education in the Gymnasium

---

The author died on 10 October 2009.

L. K. Templeton (✉)
Department of Chemistry, University of California, Berkeley, CA 94720, USA

© The Author(s) 2021   27
B. Friedrich and H. Schmidt-Böcking (eds.), *Molecular Beams in Physics and Chemistry*,
https://doi.org/10.1007/978-3-030-63963-1_3

(high school) which included instruction in Latin and Greek, but no modern foreign language. He only learned English as an adult and never felt that comfortable speaking it. He had friends in Zurich, but I believe that there was also a sentimental reason why he went there. It probably brought back memories of his days with [Albert] Einstein and [Max] von Laue before World War I and the long walks and discussions they had. He developed a friendship with von Laue in Zurich which lasted a lifetime.

On those trips to Europe he used to take the train across the U.S. to New York, stopping in Chicago to visit with his friend [James] Franck. From New York he took a boat to Europe. He liked to take Dutch ships because their food was very good and he felt they were just the right size so one would not feel the vibrations so much.

On one of these early trips, he stopped in Copenhagen where he stayed with Niels Bohr. At the end of the visit he asked Mrs. Bohr where was the maid, because he wanted to give her a tip for the excellent service he had received. To his embarrassment he found out that it was Mrs. Bohr who had made up his room and that there was no maid.

On one of the later trips he was questioned by the customs officials for about 2 h. Why was he taking so many trips to Europe, why was he going to Amsterdam and so on and on. It turned out that they had been given a tip that a diamond smuggler by the name of Stern was coming back from Amsterdam and they suspected my uncle of being that person. After a time he was able to convince them that they had stopped the wrong man.

He took the train, because he felt the airplanes were not safe. He contended they lacked the instruments at that time to tell how far above the ground they were, and flying across the United States there were a lot of high mountains. My husband and I knew trains were on the way out when he began take airplanes in his last years.

His stops in New York always included a visit, both to his dentist and to his doctor there, Rudi Stern, who was a cousin. After Rudi's death in 1962 he had to find a doctor here in the Bay Area.

On the whole he seemed to be in rather good health, but he had arthritis in his hands which bothered him. One day he was trying to boil an egg. He had grandfather's—his fathers—gold pocket watch in his hand and dropped the watch into the water instead of the egg. I believe that this was the reason he bought the inexpensive "dollar" pocket watches. They seemed so out of character with his habit of having custom-made shirts, etc. After my mother's death in 1963 he came quite often to our house for dinner. He told me once, "Lilo, please don't use your good crystal glasses when I come to dinner, I might drop my glass."

After World War II he was entitled to a pension from Germany as a former professor. He refused to accept it, because he wanted no official connection with Germany. He had an unwritten rule not to go there, but broke it on two occasions [in fact, at least on eight, see Chap. 5] for which he made a lot of excuses. In the first instance, he went to East Berlin to visit his old friend Max Volmer. Volmer, as a sick, old man, had been released by the Russians to his old villa in East Berlin in the 1960s. Since it was difficult for Volmer to travel, my uncle went to him. The other occasion was a meeting arranged by the Nobel Foundation in Lindau at Lake

Fig. 1 Portrait of Albert Einstein by John Philipp with Einstein's signature. The inscription reads: Albert Einstein d'apres nature John Philipp 1929. Courtesy of Diana Templeton-Killen

Constance. It was about a year or two before his death. He used to say it was really a Nobel meeting and it was only a fluke that it happened to be in Germany.

One of the nicest things I inherited from my uncle is a portrait of Einstein, Fig. 1. It used to hang in his office in Hamburg. When the Nazis came into power in 1933, he was told one day that they were going to come the next day to take and destroy it. He took the picture home with him, and it was the only picture hanging in his study in Berkeley. He always looked up to Einstein, who was a role model for him.

His study was a spacious room (originally the living room) but full of books and papers. If one wished to sit down, it was first necessary to move several copies of Physical Review or the Neue Zürcher Zeitung from one of the chairs. The dining room walls were lined with bookcases full of his old journals.

One reason he took an early retirement in 1946 [1945] was that he felt teaching took too much of his time; he wished to devote more time to some of his ideas. He wanted to derive a correlation between thermodynamics and quantum theory. His conviction was that the third law is fundamental, and that if it is postulated correctly, it should be possible co derive the wave mechanics as a consequence. He used to grumble that he did not have anybody to talk to about it in Berkeley. I think he did not fare much better in Zurich. Anyway this project did not progress too well. He did publish one paper, his last [3], on this subject. Unfortunately he was not able to accomplish quite his goal.

References

1. O. Stern, Über eine Methode zur Berechnung der Entropie von Systemen elastisch gekoppelter Massenpunkte. Ann. Phys. **51**, 237–260 (1916)
2. O. Stern, Die Entropie fester Lösungen. Ann. Phys. **49**, 823–841 (1916)
3. O. Stern, On a proposal to base wave mechanics on Nernst's Theorem. Helv. Phys. Acta **35**, 367–368 (1962)

Chapter 4
My Great Uncle

Alan Templeton

When I was a child, my favorite relative without a doubt was my great uncle, Otto Stern, because he nearly always did exactly what he wanted, and he did very little else. Otto lived just 3 km away from us in a beautiful part of North Berkeley that is known for its fine views of San Francisco Bay, its pleasant prewar houses, and its many appealing gardens. I loved exploring Otto's backyard because he left it completely untended. It gave me the feeling of walking into a fairy tale, far removed from the everyday world of rules and order.

One day I asked him, "Uncle Otto, why do you let the garden grow wild?"

And he said to me in a completely matter-of-fact manner, "I don't like to garden, so I don't."

Many people claim not to care what other people think, but Otto was the only person I have ever known who seemed genuinely immune to such concerns. He cared deeply about the family, his friends, his sincere and trusted colleagues, but not about impressing the neighbors. He was very polite and unassuming, and he nearly always wore a three-piece suit, but otherwise he was wonderfully unconventional. If you want to understand how he became such a clever experimentalist, an innovative thinker, a *Querdenker*, I think it is tied to several things: he was quite possibly the most brilliant representative of a family that was and is full of smart people, he was affluent enough that money was rarely a major concern, he was highly independent with a natural curiosity, and he seldom followed the crowd. As far as I can tell, Otto never had a car, never learned to cook, avoided flying like the plague, and enjoyed life immensely. Also very telling, he never bragged about any of his accomplishments, not in the slightest. Showing off is certainly frowned upon in our family, and I was always taught that it is best to teach by example. Otto excelled at this.

Conversing with Otto, his wit and humor were immediately evident, and his intelligence shone through, yet he could also be rather humble with the occasional

A. Templeton (✉)
360 Grand Avenue, Ste. 395, Oakland, CA 94610, USA
e-mail: alantempca@yahoo.com

© The Author(s) 2021
B. Friedrich and H. Schmidt-Böcking (eds.), *Molecular Beams in Physics and Chemistry*,
https://doi.org/10.1007/978-3-030-63963-1_4

self-deprecating remark. But underlying all of this was a quiet confidence which left a lasting impression. He was very much his own man, unconcerned with current fashion in science or any other field. He had the experience of seeing himself become a rather famous scientist, and then become somewhat forgotten. I do not think it bothered him. He knew what he had accomplished was of lasting value. He had no need to be in the limelight. During his Berkeley years, he often visited the campus to see friends and colleagues, and for a long time, he attended the physics seminar. At the latter, he often sat quietly in the back rows, drawing little attention to himself. But his favorite person at Berkeley was his niece, my mother, Lilo, who was a physical chemist (as was my father, David). Lilo spent her entire adult life in science, the first woman in the family to ever do so. This was a daring choice for a woman born in Breslau in 1918. She was determined to have a life in chemistry, and it was made all the more feasible because she had the unwavering support and encouragement of Otto. The two of them always remained close, and I think they understood each other quite well.

But whereas Lilo and David's house was relatively orderly, light and airy, Otto's house felt completely different. The first thing one noticed was the pervasive odor of cigar smoke. He really did love to smoke them, often rather inexpensive ones, much to the chagrin of various members of the family. The interior tended to be fairly dark with lots of wooden furniture, most of it brought over from Europe. It was immediately obvious which room was the most important: the highly cluttered office, filled with books and papers everywhere. Otto always employed a housekeeper to clean and prepare meals for him. I always had the impression she was very good at her job, but it was clear she was not allowed to touch anything in the office, which remained perpetually messy, though the piles of paper made sense to Otto. At the center of it was the exquisitely crafted desk designed by Li (Elise Stern), Otto's younger sister. She was always Otto's favorite within the family, and probably his favorite person in the whole world. By all accounts, Li was a lively, free-spirited, highly independent woman who loved the arts, design, travel, fashion, and good conversation. During Otto's highly productive years in Hamburg, they lived just several blocks apart from each other in the Uhlenhorst district, then as now a rather chic neighborhood with attractive apartment buildings and small houses, lots of shops and restaurants, and a favorable location near the waters of the Außenalster.

In his later years, we often had Sunday lunch with Otto, usually at one of the nice restaurants with a view of San Francisco Bay, and always somewhere with attentive service and a certain air of elegance. His favorite of these was the Spinnaker, a locally famous eatery on the Sausalito waterfront which has a spectacular view of San Francisco and the water. In retrospect, I think Otto enjoyed it so much because it reminded him of happier days spent in Hamburg. If you want to savor the Otto Stern lifestyle for yourself, there is no better way than having lunch or dinner at the Jahreszeiten Grill Hamburg inside the Vier Jahreszeiten Hotel, still one of the finest addresses in that thriving city. The Art Deco interior, the superb cuisine, the extensive wine list, the well-heeled crowd, it has again recaptured much of its vibrancy and elegance from an earlier time. But to truly honor Otto, there is an even better way: reward one or more younger colleagues who have been working hard by treating them to a long, leisurely meal at a fine restaurant in your own part of the world. Take them somewhere refined where they could not easily afford to dine on their own, and during the course of this pleasant indulgence, have a wide-ranging conversation in which you discuss many different subjects, not just science, and discover what they truly care about, exploring their hopes and aspirations for the future. This is what Otto would have done.

That highly productive period, late 1918 to early 1933, spent primarily in Frankfurt and Hamburg, was a golden age for Otto as a scientist, and I suspect that it also included the happiest years of his life. It was bookended by two much more difficult times. It is my understanding that Otto volunteered to serve Germany in the First World War. This would be completely plausible. It is not that he had any desire to wage war, far from it. Rather, Otto would have seen it as an obligation of citizenship, and many members of the family served in that devastating conflict. But what does one do with a young, promising scientist in wartime? The German command made him a weatherman along the Eastern Front. His main responsibility was to fly a biplane once a day near the front lines in order to take weather readings. This worked fine until one day the Russians shot down his plane. His rather flimsy biplane crashed into the ground. Amazingly, Otto was not seriously hurt, and he managed to rush back to safety without being taken prisoner, but it was a very traumatic experience which marked him for life.

In late 1968, shortly before my first flight, I asked Otto what flying was like. He looked at me and said, "The physics of flying is mostly well founded, though not always!" He said this in a cheerful tone with his characteristic smile. I can still see him in my mind's eye. He then explained to me his earlier experiences with biplanes which seemed absolutely incredible to me. I suspect his tremendous distaste for commercial air travel stemmed from those memories. Throughout the postwar years, he traveled to Europe nearly every year. Each journey started by taking the train to New York City where he would visit friends, see his doctor, and enjoy the city life before boarding one of the magnificent ocean liners of the day to travel to Europe in style. It really is a superior form of travel. Having done it myself in recent years, I highly recommend it. Otto really did know how to live well.

In those postwar years, Otto observed a general boycott of Germany. The crimes of the Nazi regime were unforgivable, and the sense of betrayal was profound and indelible. But he nonetheless visited Germany a number of times after the war, though each stay tended to be quite brief. Two of these episodes were related to me. The first of these, in the mid 1950s, was to the still war-ravaged and divided city of Berlin. He knew the city well, and his father, Oskar, stepmother, Paula, and younger sister, Li, among other relatives, had all lived in the stylish Charlottenburg district of Berlin for many years well before the war. While many members of our family made it safely to the United States or Britain in the 1930s, not all were so lucky. Paula Stern played a very important role in the family, principally raising Li, and always staying in close contact with Otto and his siblings. Based on her letters, I can tell you she had a bright and lively mind. Once she was widowed, she spent her later years living in Wiesbaden with her two sisters. All three of them would later starve to death at the Theresienstadt Concentration Camp, victims of the Holocaust. Otto was painfully aware of this and so many other tragic deaths. So why did he travel to Berlin in the 1950s? To visit Max Vollmer, his dear friend and colleague, who had recently returned to East Berlin after being forced to work in the Soviet Union for many years. Vollmer was in declining health, and Otto wanted to see his old friend one last time. For Otto, friendship was more important than politics, and rightly so. It is my understanding that nearly all of his postwar visits to Germany focused on seeing specific friends and colleagues who remained important to him. Otto was a very loyal friend. The other trip to Germany that was often metioned, in the mid 1960s, was a brief jaunt to Lindau on Lake Constance (Bodensee) for a Nobel-sponsored event. He made a point of telling Lilo, his niece, that he was only going because it was a Nobel event, not a German event. Otto wanted to make it clear that his overall boycott of Germany was still essentially in effect. After the conference, he immediately went back to Zurich.

Otto nearly always spent time in Zurich during those postwar annual trips because it allowed him the pleasure of being in a sophisticated German-speaking city without going to Germany or Austria. Although Otto spoke very good English and was grateful to be an American citizen, and wanted to be considered a U.S. scientist, culturally he always remained central European, and I suspect he was nearly always thinking in German. He was certainly most at home speaking his native tongue. But politically, he was thoroughly American, and that goes back to events in 1933.

In late March or early April 1933, Otto's older sister, Berta (my grandmother), was tipped off by a family friend who worked at Breslau City Hall: her name was on a confidential list compiled by the local Nazi authorities of persons to be arrested for political reasons. The friend advised her to leave Germany, the sooner the better. Let me assure you that any government that perceived my grandmother as a threat was a very bad regime. In April 1933, Berta, her husband, and her children left Germany, eventually living in the town of Versailles, France, for three years before emigrating to the United States. They had really wanted to live in either Austria or Switzerland, but both those countries refused to accept them. Otto would have been well aware of this difficult drama unfolding for his older sister in the spring of 1933. I therefore believe it is likely that Otto started considering his own departure from Hamburg as early as April 1933.

In any event, Otto understood early on that the Nazi regime now in control of Hamburg University and the nation would not make life easy for him, or for the rest of the family. In late spring, the new authorities refused to renew the positions for the majority of his laboratory staff for the coming year. In the summer of 1933, when he actively sought a position in America, he soon received a generous offer from the Carnegie Institute of Technology in Pittsburgh (now Carnegie-Mellon). He said he would happily accept, provided they also offered a job to his favorite assistant, Emmanuel Estermann, newly unemployed. They graciously agreed, saving Estermann and his family as well. With a new position secured, Otto resigned his post in Hamburg. He was not expelled from Hamburg, he quit. The distinction was important to Otto. He turned his back on the new Nazi administration of the University before it could formally dismiss him. To the degree it was possible, Otto left Germany on his own terms. Li left at the same time or soon after, embracing life in America.

While Li lived mainly in New York City, Otto lived in Pittsburgh, though they certainly visited each other on a regular basis. He was well supported by Carnegie, and he was very appreciative for this fine job, but he never warmed to the city of

Pittsburgh. Keep in mind, this was not the renovated Pittsburgh of today. In the 1930s and 1940s, Pittsburgh was a much grimier place. Otto reported to the family that if he left a window open at his home, within several hours there would be a layer of soot on the sill from all of the steel mills of greater Pittsburgh. He was also underwhelmed by the cultural life of Depression-era Pittsburgh, and the local cuisine was found to be wanting. The hot summers were another unwanted surprise. And yet, professionally he did land in a good place, and I think it is fitting that he would become within ten years the first resident of Pittsburgh to ever be awarded the Nobel prize. The award ceremony took place in early 1944 in New York City, as the war was still raging. This was an ideal location, because it meant he could celebrate this triumph with Li and various friends in New York.

Tragically, Li would die the following year from medical problems, her life cut short at age 46. It was this painful loss which likely persuaded Otto to resign his position at Carnegie and relocate to Berkeley in 1945. He bought a house a short walk from his sister Berta and her husband Walter, and a short bus ride away from UC Berkeley. Berta and Walter would both predecease Otto. By the end of 1963, we were his closest surviving relatives.

Despite all the upheaval and misfortunes Otto witnessed, he never lost his wit or humor. I will leave you with one more example of it. One day he telephoned our house, and asked for my father: "David, I want to see you alone, can you come over?" This was an unusual request, as usually Otto so enjoyed speaking with Lilo, his "favorite niece" as he often called her. This was clearly true as she was his only niece.

"Yes, of course," replied David, "I'll be there in a few minutes."

After sitting down on opposite sides of Otto's magnificent desk, Otto said to him, "David, after careful thought, I have decided to make you the executor of my estate, because I trust you to do a good job, and I am not leaving you anything!" I guarantee you Otto said this with his ready smile, confident that David would understand the essence of the proposal. While it is true Otto left nothing specifically to David, he was quite generous to the rest of us, which was no surprise to anyone who knew him.

Chapter 5
Otto Stern's Molecular Beam Method and Its Impact on Quantum Physics

Bretislav Friedrich and Horst Schmidt-Böcking

Abstract Motivated by his interest in thermodynamics and the emerging quantum mechanics, Otto Stern (1888–1969) launched in 1919 his molecular beam method to examine the fundamental assumptions of theory that transpire in atomic, molecular, optical, and nuclear physics. Stern's experimental endeavors at Frankfurt (1919–1922), Hamburg (1923–1933), and Pittsburgh (1933–1945) provided insights into the quantum world that were independent of spectroscopy and that concerned well-defined isolated systems, hitherto accessible only to *Gedanken* experiments. In this chapter we look at how Stern's molecular beam research came about and review six of his seminal experiments along with their context and reception by the physics community: the Stern-Gerlach experiment; the three-stage Stern-Gerlach experiment; experimental evidence for de Broglie's matter waves; measurements of the magnetic dipole moment of the proton and the deuteron; experimental demonstration of momentum transfer upon absorption or emission of a photon; the experimental verification of the Maxwell-Boltzmann velocity distribution via deflection of a molecular beam by gravity. Regarded as paragons of thoroughness and ingenuity, these experiments entail accurate transversal momentum measurements with resolution better than 0.1 atomic units. Some of these experiments would be taken up by others where Stern left off only decades later (matter-wave scattering or photon momentum transfer). We conclude by highlighting aspects of Stern's legacy as reflected by the honors that have been bestowed upon him to date.

1 Prolog

Otto Stern (1888–1969) is primarily known for developing the molecular beam method into a powerful tool of experimental quantum physics. His seminal molecular beam experiments, carried out during the period 1919–1945 in Frankfurt, Ham-

B. Friedrich (✉)
Fritz-Haber-Institut der Max-Planck-Gesellschaft, Faradayweg 4-6, 14195 Berlin, Germany
e-mail: bretislav.friedrich@fhi-berlin.mpg.de

H. Schmidt-Böcking
Institut für Kernphysik, Universität Frankfurt, 60348 Frankfurt, Germany

© The Author(s) 2021
B. Friedrich and H. Schmidt-Böcking (eds.), *Molecular Beams in Physics and Chemistry*,
https://doi.org/10.1007/978-3-030-63963-1_5

burg, and Pittsburgh, were conceived as "questions posed to nature." The relentless answers nature provided were often at odds with expectations based either on contemporary theory or on intuition, including Stern's own. Prime examples of Stern's experiments with unexpected—and far-reaching outcomes—include those on space quantization[1] and the magnetic moment of the proton and deuteron. In 1944, Otto Stern was awarded the 1943 Nobel prize in Physics (unshared) "for his contribution to the development of the molecular ray [beam] method and his discovery of the magnetic moment of the proton." In his Nobel lecture, delivered in 1946, Stern extolled the virtues of molecular beams: "The most distinctive characteristic property of the molecular ray method is its simplicity and directness. It enables us to make measurements on isolated neutral atoms or molecules with macroscopic tools. For this reason it is especially valuable for testing and demonstrating directly fundamental assumptions of the theory."

The majority of Stern's publications, fifty-seven out of a total of seventy-two, deal with atomic, molecular, optical, and nuclear physics problems. Besides, Stern maintained an abiding interest in thermodynamics in general and the concept of entropy in particular. Although only fifteen of Stern's publications tackle topics from physical chemistry, among them his acclaimed paper on the electric double-layer (Stern 1924), and thermodynamics, including his last paper (Stern 1962), Stern's recently published correspondence reveals that he exchanged many more letters about these subjects than about his pursuits in atomic and molecular physics—roughly in the inverse ratio of his published works on these two subjects. Stern's correspondents included his friend and mentor Albert Einstein (1879–1955) as well as his friends and colleagues Wolfgang Pauli (1900–1958) and Niels Bohr (1885–1962). What especially preoccupied Stern was the relationship between entropy (degree of order) and quantum mechanics (Stern 1962) and the issue of the reversibility of measurements (Schmidt-Böcking et al. 2019). Stern's principal correspondent on the topic of molecular beams was his former assistant, Isidor Rabi (1898–1988).

Stern's deep interest in thermodynamics dates back to his apprenticeship, Fig. 1, at the University of Breslau with the pioneer of quantum statistical mechanics Otto Sackur (1880–1914). Sackur, Fig. 2, was one of the first to apply quantum ideas to a non-periodic motion, namely to the translation of atoms/molecules in a gas (Sackur 1911; Badino and Friedrich 2013). He recognized that Planck's quantum of action, h, enters the treatment of a gaseous system by quantizing its phase space and, based on this insight, derived a quantum expression for the entropy of a monoatomic gas, known as the Sackur-Tetrode equation (Tetrode 1912).

After completing his doctoral thesis, Stern joined Einstein in April 1912 at the German University in Prague[2] and moved on with him the same year to the ETH Zurich, where, under Einstein's auspices, he became *Privatdozent* for theoretical physics, in 1913 (Fig. 3). On invitation from Max von Laue (1879–1960), Stern

---

[1] Space quantization is a commonly accepted translation of the original German term *Richtungsquantelung*.

[2] Stern's contact to Einstein was mediated by Sackur via Sackur's and Einstein's common colleague and friend Fritz Haber (1868–1934).

Fig. 1 Otto Stern
(1888–1969), about 1912
(OSC)

would serve in the same capacity at the Royal University of Frankfurt, established in August 1914, two weeks after the outbreak of World War One.

During his time in Prague and Zurich, Stern would attend Einstein's lectures on theoretical physics and, as Einstein's sparring partner, develop a penchant for unconventional, out-of-the-box thinking.[3] Attracted to Einstein mainly because of his work on quantum physics, Stern and Einstein would co-author a paper on the zero-point energy of gaseous systems as exemplified by molecular hydrogen (Einstein and Stern 1913). This paper was written in response to an experiment by Arnold Eucken (1884–1950), which revealed anomalous behavior of hydrogen's heat capacity at low temperatures (Eucken 1912). During the war years, Stern continued mulling over, corresponding (Schulmann et al. 1998; Docs. 191, 192, 198, 201, 205), (Schmidt-Böcking et al. 2019; pp. 32–38), and publishing on related themes (Stern 1916a,b).

After a post-war interlude in the laboratory of Walther Nernst (1865–1941) at the Berlin University, where he worked with Max Volmer (1885–1965) on the kinetics of fluorescence, Stern returned in 1919 to his post at the Institute for Theoretical Physics at Frankfurt. Max von Laue swapped meanwhile his position with his Berlin colleague Max Born (1882–1970), who became the new head of the Frankfurt Institute. The institute occupied just two rooms in the Arthur von Weinberg-Haus on Robert-Meyer-Strasse 2, and consisted, apart from Born, of two assistants—Elisabeth Bormann and

---

[3] In German, one would call this ability *Querdenken*.

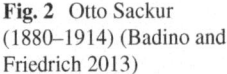

Fig. 2 Otto Sackur
(1880–1914) (Badino and
Friedrich 2013)

Otto Stern, and a technician, Mr. Adolf Schmidt. It was in this setting that Otto Stern
launched his epochal molecular beam research. But why did he?

Stern revealed his motive in his paper on the thermal molecular velocities (Stern
1920b): as a follow-up to his 1913 paper with Einstein, he set out to examine whether
the classical Maxwell-Boltzmann distribution of molecular velocities is the whole
story or whether zero-point energy plays a role and manifests itself in distorting the
classical thermal velocity distribution. Based on his ingenious experiment, described
in detail, along with its reproduction, in Chap. 9, Stern concluded that the velocity
distribution of a gas was Maxwell-Boltzmannian, with a root-mean-square velocity
$\sqrt{3kT/m}$, with k the Boltzmann constant, T the absolute temperature, and m the
atomic/molecular mass. Curiously, the evaluation of the experiment had to undergo
an amendment—due to Einstein, who noticed that Stern had not used the correct
root-mean-square velocity formula, $\sqrt{4kT/m}$, to compare his experimental result
with (Stern 1920c), (Buchwald et al. 2012; p. 355). What remains puzzling is how
Stern could have inferred from measuring just the root-mean-square velocity (with an
error of about 5%) of a hot silver beam that the velocity distribution was undistorted
by the zero-point energy or any other effects. However, at the end of his career as

Fig. 3 1913 in Pierre Weiss' laboratory in Zürich. From left: Karl Herzfeld (1892–1978), Otto Stern, Albert Einstein, and Auguste Piccard (1884–1962) (OSC)

an experimentalist, Stern would undertake a measurement of the complete velocity distribution and find a distorted Maxwell-Boltzmann distribution—due to scattering of slow molecules, see Sect. 2.6.

Hence it was Stern's dual interest in thermodynamics and quantum theory that motivated his work with molecular beams, whose rudimentary form was first implemented by Louis Dunoyer (1880–1963) in 1911 (Dunoyer 1911).

We note that the anomalous behavior of the heat capacity of hydrogen (Eucken 1912) that Stern and Einstein sought to explain in terms of the zero-point energy was in fact due to the existence of hydrogen's ortho and para allotropic modifications, see Sect. 2.4.

Whereas Stern's first molecular beam experiment did not answer his fundamental question about the manifestation of a quantum effect in the affirmative, his second did: the Stern-Gerlach experiment surprisingly confirmed the existence of space quantization, a concept developed, independently, by Arnold Sommerfeld (1868–1951) (Sommerfeld 1916) and Peter Debye (1884–1966) (Debye 1916), with two major corollaries: the quantization of electronic angular momenta in an atom in units of $\hbar \equiv h/(2\pi)$, as predicted by Bohr's model of the atom, and the existence of an elementary atomic magnetic dipole moment, of the size of a Bohr magneton, $\mu_B = e\hbar/(2m_e)$, with e the magnitude of the electron charge and m_e the electron mass. In order to carry out the extremely difficult experiment, Stern teamed up with an able experimentalist, Walther Gerlach (1889–1979), from Frankfurt's Institute for Experimental Physics located in the same building. When Gerlach, trained by Friedrich Paschen in Tübingen, appeared on the scene, Born exclaimed: "Thank God, now we have someone who knows how to do experiments. Come on, man, give us a hand" (Gerlach 1963a; p. 3).

Completed in February 1922, the Stern-Gerlach experiment (SGE) (Stern 1921; Gerlach and Stern 1922a,b, 1924; Gerlach 1925) caused a stir in the community, as everything about it appeared novel and non-classical. Einstein together with Paul

Ehrenfest (1880–1933) rushed to find a physical explanation for the process of space quantization (Einstein and Ehrenfest 1922), but without success (Unna and Sauer 2013). Although Gerlach ended up doing the experiment essentially alone, Einstein and Ehrenfest coined the term Stern-Gerlach experiment rather than Gerlach-Stern experiment, in recognition of the fact that it was Stern who conceived the idea for the experiment that was to "decide unequivocally between quantum-theoretical and classical views" (Stern 1921).

Incredulous about the outcome of the SGE, Stern left Frankfurt in October 1921 for the University of Rostock, where he assumed a Professorship in Theoretical Physics. He would visit Frankfurt and consult with Gerlach regularly until the completion of the SGE. While in Rostock, a place without much experimental infrastructure, Stern received an offer from the University of Hamburg for a Professorship in Physical Chemistry, which he accepted as of 1 January 1923. Founded in 1919, the University of Hamburg created decent conditions for Stern's work, which became excellent from 1929 on as a way of countering an offer that Stern then received from the University of Frankfurt. Stern's Hamburg laboratory had a slow start, with Immanuel Estermann (1900–1973) and Friedrich Knauer (1897–1979) as Stern's assistants, but began flourishing in about 1926. During the heyday period that lasted until the summer of 1933, they were joined by Thomas Erwin Phipps (1895–1990), Otto Robert Frisch (1904–1979), Robert Schnurmann (1904–1955), Otto Brill (1881–1954), Ronald Fraser (1899–1985), Isidor Isaac Rabi, John Bellamy Taylor (1875–1963), and Emilio Segrè (1905–1989) among others see Fig. 4. There were also graduate students around but, as noted by Estermann (1962):

> Stern very rarely put his name on the papers that were published by his more advanced graduate students, as a matter of fact. Practically all the theses were published by the student alone, just with a note somewhere acknowledging the assistance or inspiration or what-not of Stern. The papers or work that was done jointly with some of the grown-up people was published then as a joint paper.

It was in 1926 that Stern wrote programmatic papers (Stern 1926; Stern and Knauer 1926) on the molecular beam method and launched an eponymous series of publications in Zeitschrift für Physik, *Untersuchungen zur Molekulkarstrahlmethode aus dem Institut für physikalische Chemie der Hamburgischen Universität*—U.z.M. The series was cut short at Number 30 by the rise of the Nazis to power in Germany and Stern's subsequent emigration, in September 1933. The programmatic papers discussed improvements of the beam intensity, beam collimation, and the sensitivity of beam detection, as well as projects that such improvements would make feasible. The determination of the de Broglie wavelength of a matter wave and the measurements of the magnetic dipole moment of the proton and of the photon recoil were featured prominently on the list.

In 1928–1929, Stern and Estermann carried out the first matter-wave diffraction experiments in which they scattered a helium-atom or hydrogen-molecule beam off the surface of a LiF or NaCl crystal (Estermann and Stern 1930). The diffraction pattern they observed allowed them to determine the de Broglie wavelength, λ, of the beams. In follow-up experiments, Estermann, Frisch, and Stern made use of velocity selection to define and control the velocity and thereby the momentum, p, of the He atoms or H_2 molecules and corroborated the validity of Louis de Broglie's wavelength formula, $\lambda = h/p$, within an accuracy of 1% (Estermann, Frisch, and Stern 1932). Throughout his life, Otto Stern regarded the experimental confirmation of the wave-particle duality as his most important contribution to physics (Stern 1961).

Still in Hamburg, Stern and Frisch succeeded in measuring the magnetic dipole moment of the proton in an SGE-type deflection experiment that made use of the ortho and para allotropic modifications of molecular hydrogen (Frisch and Stern 1933). They found that proton's magnetic moment, μ_p, was by about a factor of 2.5 larger than the nuclear magneton, $\mu_n = e\hbar/(2m_p)$, with e the magnitude of the elementary charge and m_p the proton mass. The theory of Paul Dirac (1902–1984), until then undisputed, treated the proton as a positively charged point-like particle similar to the electron, but with a different mass (and opposite charge) (Dirac 1928, 1930). The true value of proton's magnetic moment therefore indicated that the proton must have an internal structure and cannot be an elementary particle like the electron. Otto Stern thus became a pioneer of elementary particle physics.

"With the sword of Nazism hanging over [their] heads" (Estermann 1975), Otto Robert Frisch, with Stern's support, demonstrated the existence of a momentum kick atoms receive upon the absorption or emission of a photon (Frisch 1933a), a process predicted by Einstein (1917). We note that such momentum kick is the basis for laser cooling of atoms—and, recently, also of molecules—and thus a key to achieving quantum degeneracy in gases and much more.

Also under the Nazi threat, Thomas Phipps, Otto Robert Frisch, and Emilio Segrè carried out a three-stage SGE (Frisch and Segrè 1933), inspired by a letter to Stern from Einstein (Schmidt-Böcking et al. 2019; p. 129). This experiment made use of two Stern-Gerlach magnets with an additional inhomogeneous magnetic field between them. The three-stage SGE allowed to probe spin-flips of silver atoms due to the intermediate field – and thus anticipated Rabi's resonance method.

On 7 April 1933, the "Law for the Restoration of the Professional Civil Service"— designed to exclude Jews and political opponents from civil service positions in Nazi Germany—was promulgated and Stern's assistants of Jewish descent—Immanuel Estermann, Otto Robert Frisch, and Robert Schnurmann (only Friedrich Knauer was "Aryan"—and a Nazi (Stern 1961))—were dismissed in the summer of 1933 as a result. Stern was exempted from the law because of his military service in World War One, but resigned at the end of September 1933 and emigrated to the United States, where he took up a professorship at the Carnegie Institute of Technology in Pittsburgh. Here is how Immanuel Estermann described Stern's—and his own— emigration to the U.S. (Estermann 1962):

[In] 1933 it became pretty obvious that [our] days [in Germany] were numbered ...I would have gotten out even earlier; I sent my family out as early as April or May 1933 – to England. I had a brother [there and an offer for a temporary job] ...Now, Stern didn't want to go; he thought that, well, he could survive Nazism in Germany. But he became convinced in June, or so, that it wouldn't work either. So he turned in his resignation. But we were then right in the middle of the proton-deuteron experiment, and decided as long as we would be left

Fig. 4 Stern's group in Hamburg 1928. From left: Friedrich Knauer, Otto Brill, Otto Stern, Ronald Fraser, Isidor Isaac Rabi, John B. Taylor, and Immanuel Estermann. Courtesy of Fritz Thieme, Universität Hamburg

alone we would continue this work. And we worked until August; then we finally quit. Several months before …the then President of Carnegie Tech …made a trip to Germany to try to find some good scientists who might be induced to come to Carnegie. So he made arrangements with [Stern and myself] to come to Carnegie …I had no thought of ever going back to Germany …and we actually took a considerable part of the equipment with us …We got authorization from Carnegie to buy the same kind of a magnet and pumps and so forth, and they were shipped to Pittsburgh by the manufacturer who duplicated the ones that we had had in Hamburg so that we could reestablish the apparatus. And the parts that were made specifically for the purpose in the local shop I think we got permission from the [Hamburg] University authorities to take along.

Stern, together with Estermann, would thus restore and even improve some of their scientific apparatus in Pittsburgh, but not their leadership role in experimental quantum physics. That role fell to Stern's former affiliate, Isidor Rabi. Stern and Rabi would share the stage at the Nobel ceremony at the Waldorf-Astoria, New York City in 1944, where Rabi received the 1944 Nobel Prize in Physics.

However, at Pittsburgh, Otto Stern with his collaborators carried out additional key experiments, confirming the value of proton's magnetic moment and continuing the measurements of the magnetic dipole moment of the deuteron (Estermann and Stern 1934), begun in Hamburg (Estermann Stern 1933b). Stern and coworkers also verified the Maxwell-Boltzmann velocity distribution in an (effusive) beam of Cs and K atoms by observing the atoms' free fall. "The measurement of the intensity distribution in a beam deflected by gravity represents the velocity distribution of the beam atoms and permits an accurate determination of this distribution" (Estermann, Simpson, and Stern 1947a).

The environment at the Carnegie Institute and Stern's attitude towards it was described by Estermann as follows (Estermann 1962):

[After the retirement of Carnegie's president because of his illness] there was no support from the top after the first year anymore. Stern was something of a prima donna, as you have probably noticed. If things didn't come his way he would retire into his (corner), and pick up his marbles and go home, so to speak; which made life even more difficult. His whole personality is not suited to an American University ...There was probably nobody in the physics department at Carnegie Tech who had ever heard of him before, or heard of anything of modern physics before.

In 1945, Stern retired to Berkeley, where his sister lived, and became a private citizen.

Between 1924 and 1944, Otto Stern received eighty-three nominations for a Nobel Prize in Physics,[4] more than Planck (nominated seventy-four times) and Einstein (nominated sixty-two times) or any other physicists of his time. The attitude of Stern's nominators was aptly expressed by Max Born in his nomination (Schmidt-Böcking et al. 2019; p. 299):

It seemed to me that Stern's achievements exceed those of all other experimenters so much, both by the boldness of the thoughts and by masterfully overcoming the experimental difficulties, that I do not want to name any other physicist as a candidate for the Nobel Prize besides him.

In 1944, Stern was awarded the Nobel in Physics for 1943.

2   Otto Stern's Seminal Experiments

In what follows, we review briefly six seminal experiments proposed by Otto Stern and/or carried out in his laboratories at Frankfurt, Hamburg, and Pittsburgh during the period 1920–1945.

- The Stern-Gerlach experiment, carried out with Walther Gerlach at Frankfurt in 1920–1922
- The three-stage SGE experiment, carried out together with Thomas Phipps, Otto Robert Frisch, and Emilio Segrè at Hamburg in 1933

---

[4]The official number of nominations provided by the Nobel Archives (The Nobel Population 1901–1950, A census 2002, The Royal Swedish Academy, Produced by Universal Academy Press, Inc.) for Otto Stern is eighty-two. Thirty nominations were for the Stern-Gerlach experiment, fifty-two for Stern's other molecular beam work. Einstein nominated Stern twice (in 1924 and in 1940), but the first nomination, of Stern and Gerlach for a shared prize, (Buchwald et al. 2015; Doc. 132), was not counted, because of Einstein's parallel nomination of other scientists that year (James Franck and Gustav Hertz). The rules applicable in 1924 admitted only one set of nominees by a given nominator. We note that Viktor Hess claimed in a letter to Otto Stern, dated 11 November 1944, that he had nominated Stern in 1937 and 1938 for the Nobel Prize in Physics (Schmidt-Böcking et al. 2019; p. 372). The curator of the Nobel Archives, Karl Grandin, determined that Hess's claim was incorrect.

- The experimental verification of de Broglie's relation for the wavelength of matter waves, performed with Friedrich Knauer, Immanuel Estermann, and Otto Robert Frisch at Hamburg in 1929–1933
- The measurement of the magnetic dipole moment of the proton and deuteron, with Otto Robert Frisch, Immanuel Estermann, and Oliver Simpson at Hamburg and Pittsburgh in 1933–1937
- Experimental demonstration of momentum transfer upon absorption or emission of a photon by Otto Robert Frisch, at Hamburg in 1933
- The experimental verification of the Maxwell-Boltzmann velocity distribution via deflection of a molecular beam by gravity, with Immanuel Estermann and Oliver Simpson at Pittsburgh in 1938–1945

2.1   The Stern-Gerlach Experiment

On 26 August 1921, Otto Stern submitted a paper to the *Zeitschrift für Physik*, in which he proposed "a way to examine experimentally space quantization in a magnetic field," i.e., investigate whether "the component of the angular momentum [of an atom] in the direction of the magnetic field can only have values that are integer multiples of [$\hbar$]" (Stern 1921). Stern realized that such a behavior would contrast sharply with a classical one, as classical mechanics did not impose any restriction on the projection of the angular momentum on the field. Stern thus saw the experiment as a way to "decide unequivocally between quantum-theoretical and classical views." All that was needed was "to observe the deflection of a beam of atoms in a suitable inhomogeneous magnetic field." The perception of space quantization as "otherworldly" transpired in Stern's remark that

> one cannot envision at all how the atoms of a gas, whose angular momenta [in the absence] of a magnetic field point in all possible directions, would acquire the preordained directions upon entry into the magnetic field.

In addition, Stern realized that space quantization of orbital angular momentum of atoms would lead to magnetic birefringence, which he would attempt to observe—in vain—in later experiments with Gerlach in Rostock.

By his own admission, Stern was prompted to publish his proposal when he came across the page proofs of a paper by Hartmut Kallmann (1896–1978) and Fritz Reiche (1883–1969) on the analogous deflection of polar molecules in an inhomogeneous electric field (Kallmann and Reiche 1921). According to Gerlach, upon learning about the work of Kallmann and Reiche, Stern exclaimed: "For God's sake, now they are going to start and take space quantization away from us. I'd better publish it fast" (Gerlach 1963b).

Stern's "prophetic paper" (Stern 1921) exemplifies the meticulous preparations of Stern's experiments that invariably entailed detailed feasibility calculations as well as quantitative assessments of the expected outcomes.

Fig. 5 Members of the Frankfurt Physics faculty in 1920. From right: sitting Otto Stern, Max Born, and Richard Wachsmuth (1868–1941), standing: 3rd from right Alfred Landé (1888–1976), and 4th Walther Gerlach. Standing left of Gerlach is likely Elisabeth Bormann (1895–1986) (OSC)

Stern's calculations suggested that the experiment to "decide unequivocally between quantum-theoretical and classical views" will be very difficult to carry out. Therefore, as noted, Stern invited Walther Gerlach, an assistant to Richard Wachsmuth (1868–1941), the director of Frankfurt's Institute for Experimental Physics, Fig. 5. Gerlach was regarded as an excellent experimentalist and had even attempted his own molecular beam experiment to study dia- and para-magnetism, see Chap. 8.

The actual Stern-Gerlach apparatus, which comprised an oven to produce an effusive beam of silver atoms, beam stops, the deflection region, and the beam collecting plate, was small, not much larger than a fountain pen, Fig. 6. The high vacuum needed to produce and sustain the atomic beam was produced by two glass mercury diffusion pumps, one for the source chamber and one for the detector chamber. The deflection region was squeezed between the pole pieces—edge and groove, a design proposed by Erwin Madelung (1881–1972) (Stern 1961)—of an electromagnet. The required transverse-momentum resolution was about 0.1 a.u. (an electron with a kinetic energy of 13.6 eV has a momentum of 1 a.u.). The expected angular deflection of the beam (just a few mrads) required high mechanical precision, on the order of a μm. For its operation, the apparatus required a delicate balance between heated (oven) and cooled (detector plate) components. A more detailed description of the apparatus and its operation is given in Chap. 8 by Gerlach's student Wilhelm Schütz.

The apparatus was constructed and operated during the hyperinflation period that beset Germany in the aftermath of World War One. Support for the experiment came from several sources, most notably the *Physikalischer Verein Frankfurt*, founded in 1824. The *Verein's* long-time chairman was Wilhelm Eugen Hartmann (1853–1915), founder of the *Hartmann & Braun* company that provided Stern and Gerlach with

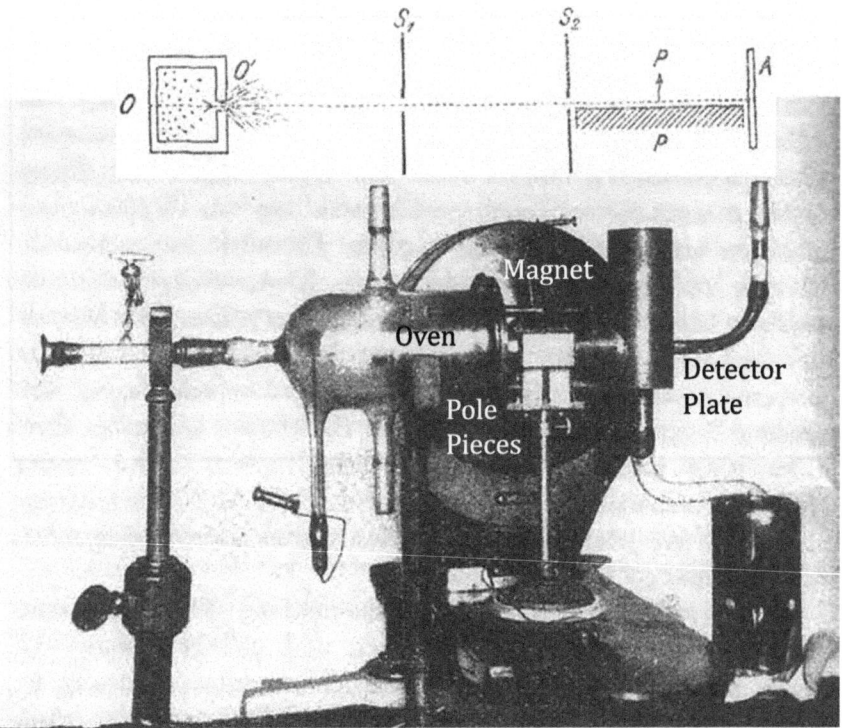

Fig. 6 The Stern-Gerlach apparatus of 1922, with improvements of 1924 and 1925. The schematic in the inset shows the silver beam effusing from an oven (O) and passing through a pinhole (S_1) and a rectangular slit (S_2) before entering the magnetic field (whose direction is indicated by the arrow) between the pole pieces (P) and finally reaching the detector plate (A) (Gerlach and Stern 1924; Gerlach 1925)

a small Dubois magnet. The *Messer* company donated some liquid air (Gerlach and Stern 1922a). Einstein, then director of the Kaiser Wilhelm Institute for Physics in Berlin, provided 20,000 Marks for the purchase of an electromagnet from *Hartmann & Braun* (Buchwald et al. 2012; p. 802), 813 (AEA 77681, 77355). Additional funding came from the *Association of Friends and Sponsors of the University of Frankfurt* as well as from the entrance fee to Max Born's popular lectures on general relativity (Stern 1961). Silver of high purity was acquired from *Heraeus*.

Unfortunately, original documents and drawings related to the SGE are no longer available. Gerlach took the documents with him to Tübingen and then to Munich where he kept them at the Physics Institute of the *Ludwig-Maximilians-Universität*. But in March 1943, almost everything was destroyed by fire following a bombing raid (Huber 2014).

On the night of 5 November 1921, Gerlach—with Stern absent—scored his first major success by observing a broadening of the silver beam consistent with a magnetic moment of 1 to 2 Bohr magnetons (Gerlach 1969; Huber 2014; Schmidt-

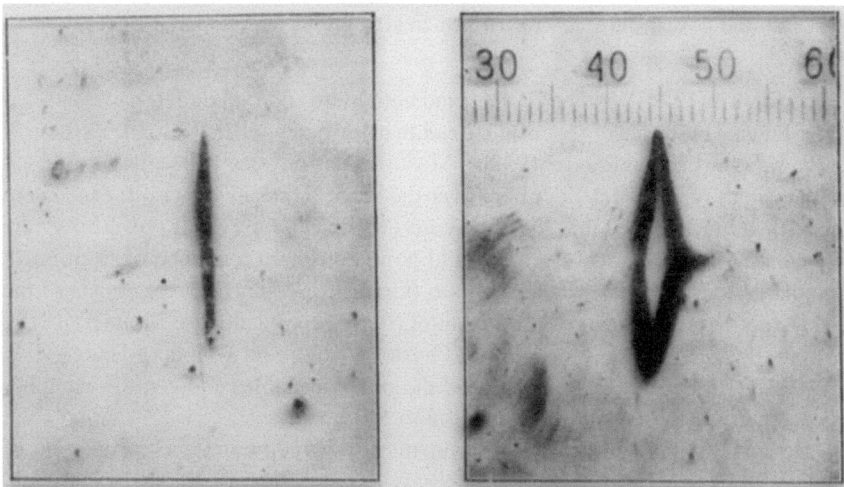

Fig. 7 Silver sulfide (Ag$_2$S) deposits obtained in the SGE. The microphotographs are from Otto Stern's personal collection, published images were included in (Gerlach and Stern 1922b). Left side: Ag beam deposit obtained in the absence of the magnetic field (deposit length about 1.1 mm, width about 0.06 to 0.1 mm). Right: Beam deposit with the magnetic field switched on; the deposit is split into two components broadened due to the beam velocity distribution. The asymmetry of the magnetic field strength between the two magnetic pole pieces is reflected by the shape of the deposit as atoms passing near the tip of the S pole are more strongly deflected

Böcking and Reich 2011). However, the low angular resolution of the apparatus left the key question about the existence of space quantization unanswered.

In early February 1922, Gerlach and Stern met at a physics conference in Göttingen and discussed further improvements of the apparatus, especially the arrangement and the shape of the apertures. An invitation letter to Stern from David Hilbert (1862–1943) to come over for a cup of coffee corroborates that Stern was indeed in Göttingen at the time (Schmidt-Böcking et al. 2019; p. 115). Like most beam experiments, the SGE suffered from a low beam intensity which was, in this case, partly due to beam scattering off the tiny platinum apertures, needed, in turn, for achieving sufficient angular resolution. With some more time on their hands—thanks to a railroad strike (Friedrich and Herschbach 1998, 2003)—Gerlach and Stern finally decided to replace the circular aperture in front of the magnetic field with a rectangular slit (0.8 mm × 30 µm). Upon his return to Frankfurt, Gerlach implemented the slit, which led quickly to a breakthrough: During the night of 7 February 1922, Gerlach was able to observe, for the first time, the splitting of the silver beam into two components, with nothing in between, Fig. 7.

Wilhelm Schütz (1900–1972), Gerlach's PhD student at the time, described in 1969 the toil of the Stern-Gerlach experiment in detail (Schütz 1969). For an extended quote, see Chap. 8 on Gerlach. After the successful completion of the experiment

[Schütz] was tasked with sending a telegram to Professor Stern in Rostock, with the text: "Bohr is right after all!"

On March 1, 1922, Walther Gerlach and Otto Stern submitted their paper entitled "Experimental evidence of space quantization in the magnetic field" to the *Zeitschrift für Physik* (Gerlach and Stern 1922b). Most of their physics colleagues expressed surprise about or even bewilderment over the reported result. After all, even Stern himself had not believed that the "quantum-theoretical view" will prevail over the classical one. However, as Gerlach would point out, Stern remained open-minded: "The dissection will tell" was their motto (Gerlach 1969). The protagonists of the SGE are shown together in the company of Stern's confidant Lise Meitner (1878–1968) in Fig. 8, Fig. 9 shows Frankfurt Physics (Arthur von Weinberg-Haus) while Fig. 10 shows the emblematic splitting of the silver beam once more with an angular scale added.

Here is a sampling of the responses from the physics community to the outcome of the SGE: Wolfgang Pauli wrote on 17 February 1922 a postcard to Gerlach (Hermann, von Meyenn, and Weisskopf 1979; p. 55):

My heartfelt congratulations on a successful experiment! Hopefully it will convert even the nonbeliever Stern. I would just like to mention one detail. It is not easy to explain that one side is stronger than the other. Shouldn't it be some secondary perturbation? You mentioned me in your letter to Franck. However, the paramagnetic effect that I calculated at the time (based on Langevin) is far too small and is out of the question here. So I'm innocent on this matter. Best regards to you, and to Prof. Madelung and to Landé.

In his 1922 letter to Max Born, Einstein emphasized (Buchwald et al. 2012; Doc.191):

The most interesting achievement at this point is the experiment of Stern and Gerlach. The alignment of the atoms without collisions via radiative [exchange] is not comprehensible based on the current [theoretical] methods; it should take more than 100 years for the atoms to align. I have done a little calculation about this with Ehrenfest. [Heinrich] Rubens considers the experimental result to be absolutely certain.

Niels Bohr wrote to Gerlach (Gerlach 1969):

I would be very grateful if you or Stern could let me know, in a few lines, whether you interpret your experimental results in this way that the atoms are oriented only parallel or opposed, but not normal to the field, as one could provide theoretical reasons for the latter assertion.

James Franck wrote to Gerlach (Gerlach 1969):

More important is whether this proves the existence of space quantization. Please add a few words of explanation to your puzzle, such as what's really going on.

Friedrich Paschen stated (Gerlach 1969):

Your experiment proves for the first time the reality of Bohr's [stationary] states.

Arnold Sommerfeld noted (Sommerfeld 1924):

Through their clever experimental arrangement, Stern and Gerlach not only demonstrated ad oculos [for the eyes] the space quantization of atoms in a magnetic field, but they also proved the quantum origin of electricity and its connection with atomic structure.

But even after the SGE was completed, Stern remained incredulous—contrary to the hope that Pauli expressed in his postcard to Gerlach. In his Zurich interview with Res Jost, Stern said (Stern 1961):

What was really interesting was the experiment that I did together with Gerlach on space quantization. I had thought that [quantum theory] couldn't be right ... I was still very skeptical about quantum theory and thought that a hydrogen or alkali atom must exhibit birefringence in a magnetic field ... At that time I had thought about [space quantization] and realized that one could test it experimentally. I was attuned to molecular beams through the measurement of molecular velocities and so I tried the experiment. I did it jointly with Gerlach, because it was a difficult matter, and so I wanted to have a real experimental physicist working with me. It went quite nicely ... for instance, I would build a little torsional balance to measure the electric [magnetic] field that worked but not very well. Then Gerlach would build a very fine one that worked much better. Incidentally, I'd like to emphasize one thing on this occasion, [namely] that we did not cite [acknowledge] sufficiently at the time the help that we received from Madelung. Born was already gone then [moved to his new post at Göttingen] and his successor was Madelung. Madelung essentially suggested to us the [realization of the inhomogeneous] magnetic field [by making use] of an edge [and groove combination]. But the way the experiment turned out, I didn't understand at all. [How could there be] the discrete beams—and yet, [there was] no birefringence. We [even] made some additional experiments about it. It was absolutely impossible to understand. This is also quite clear, one needed not only the new quantum theory, but also the magnetic electron. These two things weren't there yet at the time. ... I still do have objections against the idea of beauty of quantum mechanics. But she is correct.

As has been noted elsewhere (Friedrich and Herschbach 1998, 2003), the splitting of the beam of ground-state silver atoms $Ag(^2S)$ into two components as well as the apparent magnitude of the magnetic dipole moment involved was the result of a

Fig. 8 From left: Walther Gerlach, Lise Meitner, and Otto Stern in Zürich, about 1927. Photo: Ruth Speiser and Bruno Lüthi, private communication

Fig. 9 Building where the Stern-Gerlach experiment was carried out. Left: photo from 1910, Archiv der Universität Frankfurt, Johann Wolfgang Goethe-Universität Frankfurt am Main, Senckenberganlage 31–33, 60325 Frankfurt. Right: photo from 2020 by Horst Schmidt-Böcking

"kind conspiracy of nature:" Firstly, it was not the orbital angular momentum (which is zero for a ^2S state and not 1 $\hbar$ as assumed by Bohr) that was space-quantized, but rather the spin angular momentum of the electron with quantum number $s = 1/2$ and projections $m_s = \pm 1/2$, which would be discovered only in 1925 (Uhlenbeck and Goudsmit 1925). Secondly, it was electron's anomalous gyromagnetic ratio, $g_e \approx 2.002319$, combined with the half-integral quantum number $s = 1/2$ that created the impression that the magnitude of the observed magnetic dipole moment $\mu = g_e \mu_B m_s$ was that of a Bohr magneton.

Interestingly, a similar "duplicity of nature" played a role in the treatment of the anomalous Zeeman effect by Alfred Landé (1888–1976), then also at Max Born's Frankfurt Institute for Theoretical Physics. Based on the available Zeeman spectra and the recognition of the role of the coupling of electronic angular momenta in determining atomic structure, Landé found a formula for the atomic magnetic dipole moment (Landé 1921a,b). Landé's empirical formula also rendered correctly the double-splitting of the silver atom beam as observed in the SGE, with $k = 1/2$ the angular momentum of the atom's "interior" and a g-factor of 2 (Landé 1923), cf. also (Tomonaga 1997). Thus Landé's insight presaged the role of half-integral quantum numbers and thus of electron spin in shaping the electronic structure of atoms. Even Born, who shared an office with Landé, had underestimated the significance of Landé's formula.

The SGE has raised a number of interpretative questions (Ribeiro 2010; Wennerström and Westlund 2012; Devereux 2015; Utz et al. 2015; Griffiths 2015; Sauer 2016) that inspired a large body of experimental work, some of it still ongoing. Among them are: What is the role, if any, of diffraction of the molecular beam off the apertures? Is there spin relaxation? Do the atoms on their way from the source to the detector have to be treated as quantum mechanical waves or as classical particles? Is there interference between the two spin states of the silver atoms? The last two questions have been answered in the affirmative (Machluf et al. 2013; Margalit et al. 2015; Zhou et al. 2018; Margalit et al. 2018; Amit et al. 2019; Zhou et al.

Fig. 10 The SGE result plotted in scattering angles (milli rad). Otto Stern's private slide collection. Senckenberg Bibliothek der Universität Frankfurt, Johann Wolfgang Goethe-Universität Frankfurt am Main, 60325 Frankfurt. Calculation of the deflection angles by Horst Schmidt-Böcking

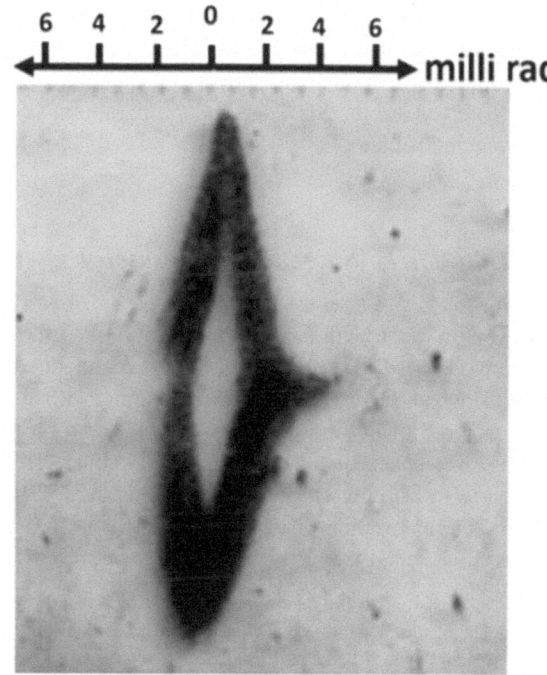

2020). These questions and more are addressed in separate chapters in this volume, especially in Chaps. 11, 12, 14, and 15.

There seems to be a consensus that the following questions have been answered by the SGE definitively:

1. The SGE has determined that each silver atom has a magnetic dipole moment of about one Bohr magneton.
2. The SGE presented the first direct experimental evidence that angular momentum is quantized in units of $\hbar$.
3. The SGE confirmed Sommerfeld's and Debye's hypothesis of "Richtungs-Quantelung" (space quantization) of angular momenta in magnetic (and electric) fields.
4. The SGE was the first measurement that examined the ground-state of an atom—without involvement of higher states, as is the case in spectroscopy.
5. The SGE produced the first fully spin-polarized atomic beam.
6. The SGE produces population inversion—a crucial ingredient for the development of the maser and laser (Friedrich and Herschbach 1998, 2003).
7. Deflecting atoms in a well-defined momentum state by an external field makes it possible to study their internal properties (electronic and nuclear). Measuring the kinematics of particles with high momentum resolution (0.1 a.u.) amounts to a new kind of microscopy, similar to mass spectrometry (Aston 1919; Downard 2007).

8. The SGE demonstrated that angular momentum "collapses" into a classically inexplicable projection on the direction of the external magnetic field, only accounted for upon the discovery of quantum mechanics, see, e.g., (Utz et al. 2015). To date, the SGE serves as a paradigm for the notorious quantum measurement problem.

2.2   The Three-Stage Stern-Gerlach Experiment

Stern kept in touch with Einstein throughout the time they both lived and worked in Germany (1914–1932) not only via correspondence but also by visiting him every now and then in Berlin (Stern 1961). In keeping with his quip that "On quantum theory I use up more of my brains [Hirnschmalz] than on relativity", Einstein continued mulling over space quantization. On 21 January 1928, he wrote a letter to Stern (as well as to Ehrenfest) (Schmidt-Böcking et al. 2019; pp. 128–131), in which he described a far-reaching idea for an experiment to explore further aspects of space quantization, see also Fig. 11:

> On the occasion of our quantum seminar, two questions have come up that concern the behavior of a molecular beam in a magnetic field, so they just fall within your work area. Perhaps you have already made equivalent experiments and if not then this suggestion could be of some use.
>
> I. Assume that an atom is oriented this ↑ or this ↓ way in a vertical magnet[ic field]. Assume the magnetic field is slowly changing its direction. Does the orientation of each individual atom follow [the direction of] the field?
>
> Test: An atomic beam passes consecutively through two oppositely oriented inhomogeneous magnetic fields. Assume that an atom is oriented in such a way as to be deflected upward in the first field. If [the atom] flips its orientation [in the region between the two fields], then, because of the reversal [of the orientation] of both the [second] field and the dipole, the beam must [be deflected by the second field] as if the two magnetic fields were oriented in the same direction.
>
> This is all the more paradoxical given that the deflection increases linearly with field strength.
>
> II. It is a part of our current understanding that the field determines the orientation of the atom and the field gradient the magnitude of the deflection. The field and the field gradient

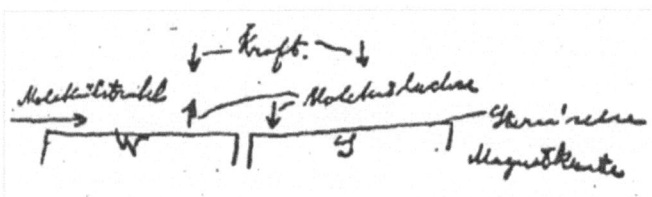

Fig. 11 Einstein's sketch of a three-stage SGE (letter to Ehrenfest) (Schmidt-Böcking et al. 2019; p. 131)

can be varied independently of one another. Let us consider that the field gradient is fixed and the field is varied; in which case only the direction of the [field] but not its magnitude should matter. The field can be arbitrarily weak, without affecting the deflection. It should therefore be possible to entirely change the [sense of the] deflections by a mere change of the direction of the arbitrarily weak magnetic field. This is surely paradoxical, but consistent with our current view. Perhaps it would be convenient to generate the inhomogeneous field by running [electric] current through a water-cooled pipe.

If you already have data available that answer the two questions, please communicate these to me. Should this not be the case, it would be worthwhile to answer these questions experimentally.

Hence Einstein recognized that if reorientation of the dipoles (i.e., spin flip) took place in the intermediate region between the two oppositely oriented Stern-Gerlach fields, the second Stern-Gerlach field would have pushed the atoms further away from the original beam direction. But this also meant that in the absence of reorientation of the atoms' magnetic dipoles (without a spin flip), the atomic beam could be refocused by the second Stern-Gerlach field on the same spot that the beam would have hit in the absence of the deflecting fields (i.e., along the original beam direction). Reorientation (spin flip) would then result in a dip in the beam intensity along the original direction. This idea, whose variant was implemented by Stern and his coworkers, would later resonate with Isidor Rabi, see below.

The possibility of a spin flip was considered by a number of workers, including Charles Galton Darwin (Darwin 1928), Landé (Landé 1929), Werner Heisenberg, as noted in (Phipps and Stern 1932), and P. Güttinger (Güttinger 1932), who concluded that the magnetic dipoles would flip if their interaction with the intermediate magnetic field were non-adiabatic. Heisenberg formulated a criterion for a non-adiabatic interaction, which was subsequently refined by Güttinger: What matters is the ratio of the Larmor period of the dipole to the dipole's interaction time with the field. Should this ratio be large, the interaction will tend to be non-adiabatic and hence the spin flip likely.

Otto Stern together with Guggenheim Fellow Thomas Phipps took it from there. On 9 September 1931 they submitted a paper that described their attempt to observe spin flips in a beam of potassium atoms (Phipps and Stern 1932). In their experiment, they implemented the intermediate field by placing three tiny spatially separated electromagnets in series and letting the spin-selected beam run between their pole pieces. Adjacent magnets were rotated by 120° with respect to one another, effecting a 360° overall rotation of the magnetic field direction. The spatially varying magnetic field became a time-varying magnetic field once the atoms flew through it. The time variation of the field was such that the above non-adiabaticity condition needed for spin flips was fulfilled. The triple-magnet contraption was placed in a magnetic shield [*Panzerkugel*] fashioned with apertures to let the beam through. The magnetic shield was supposed to keep the magnetic fields generated by the two Stern-Gerlach magnets (selector and analyser) out of the region where the small magnets interacted with the spin-selected potassium beam. Otherwise the field of the triple-magnets would have been overshadowed by that of the Stern-Gerlach fields and there would be no spatial/time variation of the intermediate triple-magnet field. The potassium

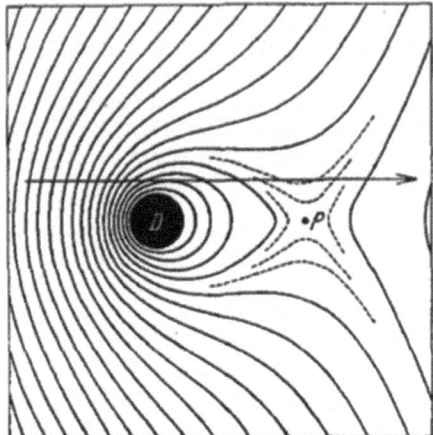

Fig. 12 Magnetic field lines in the intermediate region of the Frisch-Segrè apparatus. D is the current-carrying wire, P is where the magnetic field vanishes, and the arrow shows the path of the potassium beam (Frisch and Segrè 1933)

beam was sensitively detected with excellent angular resolution using a Langmuir-Taylor (hot tungsten wire) detector. Unfortunately, the outcome of the Phipps-Stern experiment was negative—no spin flips had been observed—likely due to insufficient shielding of the intermediate region.

Upon Phipps's return to America, the experiment was continued by Otto Robert Frisch and Rockefeller Fellow Emilio Segrè, who made use of the Phipps-Stern apparatus, but designed the intermediate flipping field quite differently: As Segrè recollected (Segrè 1973)

> I inherited [Phipps's] apparatus, but could not make much headway until on reading Maxwell's Electricity I found a trick by which one could achieve a certain magnetic field configuration essential to the success of the experiment.

Incidentally, this configuration was the same as the one proposed by Einstein in his letter to Stern (Schmidt-Böcking et al. 2019; pp. 128–129). It consisted of a current-carrying wire at right angles to the atomic beam but slightly displaced so that the beam would nearly miss it. The wire generated a spatially varying magnetic field that upon superposition with the field from the two sets of Stern-Gerlach magnets led to the field depicted in Fig. 12. The atomic beam traversing this field "felt" a rotation of the field direction by 360°.

A schematic of the apparatus constructed by Phipps and modified by Frisch and Segrè is shown in Fig. 13. With this apparatus, Frisch and Segrè were able to observe spin flips of the potassium atoms, Fig. 14. The curves show the beam intensity (ordinate) as measured by the hot-wire detector whose position could be vertically scanned (abscissa). Curve 1 shows the beam intensity distribution at the detector in the absence of the flipping field (the current through the wire D in the intermediate region was switched off, $i = 0$). Curves 2 and 3 were obtained with the intermediate field on ($i = 0.1$ A). The additional peaks to the right correspond to flipped atoms. Curve

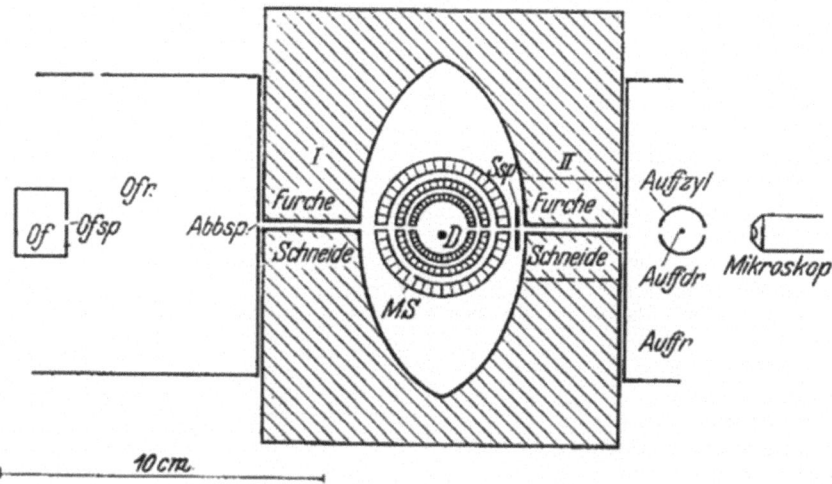

Fig. 13 The three-stage SGE of Frisch and Segrè: *Of* [Ofen] oven, *Ofsp* [Ofenspalt] oven aperture, *Ofr* [Ofenraum] source chamber, I and II Stern-Gerlach fields, *Abbsp* [Abbildungsspalt] entrance slit into the Stern-Gerlach field I, *Ssp* [Selektorspalt) selection slit, *MS* [Magnetischer Schutz – Panzerkugel] magnetic shield (later made out of high-permeability alloy obtained from *Heraeus*, *D* [Draht] current-carrying wire to produce the intermediate flipping field, *Auffzyl* [Auffangzylinder] detector chamber, *Auffdr* [Auffangdraht] wire detector. The angular deflection by either of the two Stern-Gerlach magnets was about 10 mrad. (Frisch and Segrè 1933)

3 was obtained for a different setting of the selection slit that picked out slower atoms. The separation between the two peaks of curves 2 and 3 corresponds to twice the deflection in a single Stern-Gerlach field and is larger for the slower atoms, as expected. However, the fraction of atoms whose magnetic dipole was flipped could not be reproduced quantitatively by theory. Ettore Majorana (1906–1938) developed a theory tailored to the Frisch and Segrè experimental setup, but his formula accounted only for about a half of the observed spin flips (Majorana 1932). Frisch and Segrè, Fig. 15, conjectured that this was likely because the flipping magnetic field was not properly accounted for in Majorana's model that only included effects arising from the vicinity of point *P*, see Fig. 12. However, as Isidor Rabi would point out in a 1934 letter to Stern, the discrepancy was in fact largely due to the neglect of the nuclear spin of the potassium atoms in Majorana's treatment (Schmidt-Böcking et al. 2019; p. 167).

In 1927, Isidor Rabi came to Europe as a Barnard Fellow (later a Rockefeller Fellow) and worked intermittently with Sommerfeld, Heisenberg, Bohr, and Pauli. As Norman Ramsey recounted (Ramsey 1993),

> The Stern-Gerlach experiment …had earlier sparked Rabi's keen interest in quantum mechanics and so, while working in Hamburg with Pauli, Rabi became a frequent visitor to Stern's molecular beam laboratory. During one of these visits Rabi suggested a new form of deflecting magnetic field; Stern in characteristic fashion invited Rabi to work on it in his laboratory, and Rabi in an equally characteristic fashion accepted. Rabi's work in Stern's laboratory was decisive in turning his interest toward molecular beam research.

Fig. 14 Intensity distribution of a potassium beam behind the second Stern-Gerlach magnetic field in the Frisch-Segrè experiment (Frisch and Segrè 1933). The smaller peaks to the right of the main maxima of curves 2 and 3 are due to reorientation of the magnetic dipole moments of the atoms (spin flips). Shown is also the current i through the wire D placed in the intermediate (flipping) region, cf. Figure 12. Curve 3 was obtained for a different setting of the selection slit whereby slower atoms were selected than those that gave rise to curves 1 and 2

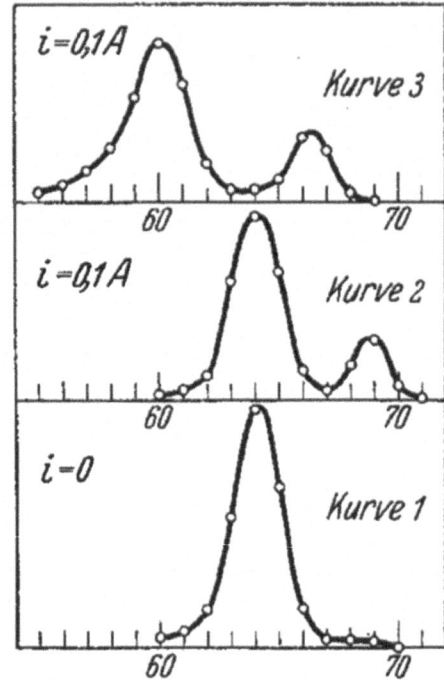

The new magnetic deflecting field alluded to above was based on Rabi's realization that magnetic dipoles can be deflected in a homogeneous magnetic field as well. Rabi's analysis was based on the analogy with Snell's law, i.e., on the change of the velocity of the atoms/molecules upon entering the conservative magnetic field due

Fig. 15 Otto Robert Frisch (left) and Emilio Segrè (right). Courtesy Fritz Thieme (University of Hamburg)

Fig. 16 Photograph of the splitting pattern of a potassium beam in a horizontal homogeneous magnetic field (Rabi 1929)

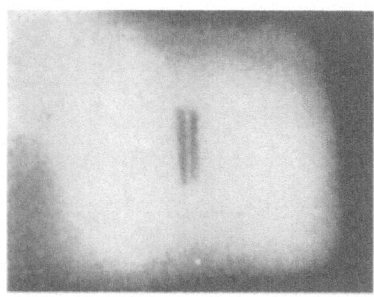

to a loss or gain of their Zeeman energy. Rabi showed that the deflection—which amounts to refraction—depends on the angle of incidence, initial kinetic energy, and the Zeeman energy. Rabi also carried out a proof-of-principle experiment in Stern's laboratory in which he measured the magnetic dipole moment of potassium (with a 5% accuracy) by splitting a beam of potassium atoms in the homogeneous field according to the different Zeeman energies of the spin-up and spin-down states (Rabi 1929).

The key advantage of using a homogeneous field was captured by Rabi in the following statement:

> [In the] new deflection method …only the energy difference of the molecules in the deflecting field matters, in consequence of which only the strength and not the inhomogeneity of the field is to be measured [controlled] …Homogeneous fields are not only easier to generate, but can be measured much more accurately.

Moreover, as shown in Fig. 16, the two traces corresponding to the $+1/2$ and $-1/2$ spin states of potassium are linear when the states are split by a homogeneous field.

Well-provided with ideas from Hamburg and elsewhere in Europe and flush with his own, Rabi departed for America in the summer of 1929 to assume a lecturership at Columbia University. Rabi's Molecular Beam Laboratory would become a major school of atomic, molecular, and optical physics and since about the mid-1930s play a pace-setting role in physics, see Chap. 7.

In December 1935, Rabi submitted a paper on spin reorientation (Rabi 1936), in which he discussed previous theoretical (Güttinger 1932; Majorana 1932) and experimental work (Phipps and Stern 1932; Frisch and Stern 1933). The next paper by Rabi on the spin reorientation problem, which appeared in the wake of related works (Motz and Rose 1936; Schwinger 1937), considered an applied field that changed its direction ("gyrated") at a fixed frequency (Rabi 1937). According to Norman Ramsey,

> A few months after the publication of that paper, following a visit by C. J. Gorter, Rabi directed the major efforts of his laboratory toward the development of the molecular beam magnetic resonance method with the magnetic fields oscillating in time.

The papers that introduced what became known as Rabi's magnetic resonance method followed in due course (Kellog, Rabi and Zacharias 1936; Rabi et al. 1939, 1938a,b).

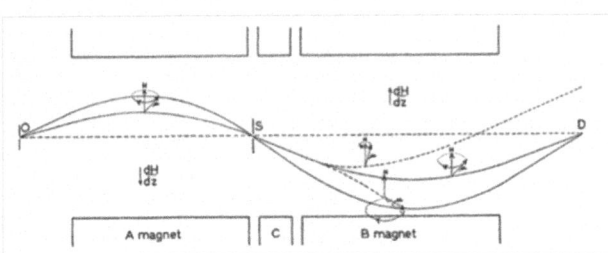

Fig. 17 The Rabi three-stage apparatus (Rabi et al. 1939)

In Rabi's method, see Fig. 17, a molecular beam is state-selected by passing through an inhomogeneous magnetic field (A field) and refocused by an identical but oppositely oriented inhomogeneous magnetic field (B field). Intermediate between the two fields A and B is a third field (C field), which is oscillatory. For an oscillation frequency of the C field that is resonant with an atomic/molecular transition, the atoms/molecules fail to refocus upon making the transition, which results in a dip in the signal. Thereby the energy differences between atomic/molecular levels, including hyperfine ones, could be accurately measured. One of the great virtues of Rabi's technique is that the refocusing is velocity-independent.

Rabi was awarded the 1944 Nobel Prize in Physics "for his resonance method for recording the magnetic properties of atomic nuclei."

Finally, we note that Heisenberg discussed a variant of the SGE in 1927 (Heisenberg 1927a) and remarked that Bohr had suggested earlier to make use of resonant photo-absorption in order to change the internal quantum state of the moving atom.

2.3   Experimental Evidence for de Broglie's Matter Waves

In his programmatic paper (Stern 1926), Stern envisioned "an experiment of the greatest fundamental significance" to demonstrate the existence of the de Broglie waves by examining whether "molecular beams, in analogy with light beams, exhibit diffraction and interference phenomena." Although he expected the de Broglie wavelengths of the molecular beams to be only on the order of an Ångström (0.1 nm), Stern was hopeful about the feasibility of the experiment. Stern's programmatic paper preceded the Davisson-Germer experiment on electron diffraction (Davisson and Germer 1927), whose serendipitous outcome was published on 1 December 1927.

When Stern—and his coworkers, Knauer, Estermann, and Frisch—succeeded, he would hardly contain his pride even thirty-five years hence: "I'm particularly fond of this experiment, which hasn't been properly appreciated" (Stern 1961).

The first attempt to find experimental evidence for the reality of matter waves was made in early 1927 in Stern's Hamburg laboratory. A preliminary report about its outcome was presented by Stern at the Lake Como conference in September 1927 and the first paper, written jointly with Friedrich Knauer (Knauer and Stern 1929a), published on 24 December 1928. This paper reflected the authors' struggle with a great number of daunting technical difficulties and reported only qualitative results—on the specular reflection and diffraction of molecular beams (mainly He and H_2) from optical gratings and crystal surfaces.

For the specular reflection off gratings, Stern and Knauer concluded that the reflected beam intensity increases with decreasing angle of incidence with respect to the surface (i.e., is at maximum at grazing incidence); the angle at which reflection becomes observable is on the order of mrad, in keeping with the calculated de Broglie wavelength of about 1 Å and a surface corrugation of 100–1000 Å; the reflected intensity sharply increases upon cooling the beam source/increasing the de Broglie wavelength, thereby conforming to the behavior expected for waves.

Of the crystal surfaces examined, the most intense reflection was obtained for a helium beam scattered from a rock salt (sodium chloride) crystal surface. For this system, it was found that at low angles of incidence (with respect to the crystal surface), the reflected intensity of the beam increases with the temperature of the beam source (lower de Broglie wavelength); at larger angles of incidence, such as 30°, it is the other way around. However, the most compelling evidence that the helium beam behaved in fact as a matter wave came from the observation of first-order diffraction maxima. For a cold helium beam (100 K), these could be observed at diffraction angles α fulfilling the condition

$$\cos\alpha - \cos\alpha_0 = n\frac{\lambda}{d} \qquad (1)$$

with $\lambda = 0.8$ Å the de Broglie wavelength, $d = 2$ Å the lattice constant, α_0 the angle of incidence, and n the diffraction order.

One of the great challenges of these experiments was dealing with the contamination of the surfaces by the adsorbed background gas in a vacuum chamber that could be evacuated to only about 10^{-5} torr. In order to keep the cleaved surfaces clean, the crystals—in fact much of the apparatus—were constantly heated to 100°C. Prior to an experiment, the crystals were baked out at 300°C.

The first, 1928 version of the Hamburg diffraction apparatus is shown in Fig. 18.

The incidence angle of the atomic beam on the crystal surface was fixed. The reflected/diffracted beam intensity was measured by a Pirani-type gauge (Knauer and Stern 1929b).

The first quantitative measurements of matter wave diffraction in Stern's laboratory were carried out using a more advanced apparatus built by Estermann and Stern that allowed to rotate the crystal surface (NaCl or LiF) with respect to the incident molecular beam (H_2 or He) as well as to scan the scattering angle for a fixed angle of incidence. Typical reflected/diffracted intensity distributions for a He beam incident on NaCl are shown in Fig. 19. The velocity distribution of the molecular beam was

Maxwell-Boltzmannian, controlled by the temperature of the beam source. The de Broglie wavelengths, obtained from the first-order diffraction maxima, cf. Equation (1), and the most probable velocities of the Maxwell-Boltzmann distribution, were found to be in the range 0.405 Å for a He beam produced at a source temperature 590 K to 1.37 Å for a H_2 beam produced at a source temperature of 100 K (Estermann and Stern 1930).

Direct verification of de Broglie's expression for the wavelength of matter waves was performed in two more machines, built by Estermann, Frisch, and Stern in 1932 (Estermann, Frisch, and Stern 1932). One apparatus allowed to velocity-select the molecular beam by reflection off a crystal, Fig. 20, the other by passing the incident beam through a pair of spatially offset cogwheels/slotted discs spinning about a common axis, Fig. 21. The latter method simultaneously allowed to accurately measure and control the beam velocity, v. Combined with the measured diffraction patterns, such as those in Fig. 22 which yielded the de Broglie wavelength, λ, Estermann, Frisch, and Stern were able to directly verify de Broglie's relationship $\lambda = h/(mv)$ for a beam of atoms or molecules of mass m—and thus the quantum-mechanical concept of matter-wave duality. In their landmark investigation, they used a helium beam impinging on a LiF crystal surface. The accuracy achieved in verifying de Broglie's relation was an admirable 1 %. As described in more detail in Chap. 23 by Peter Toennies, it would take decades before the next generation of matter wave diffraction experiments reached the accuracy of those by Stern and coworkers.

The series of papers written by Stern with Knauer, Estermann, and Frisch on the wave-particle duality are a paragon of thoroughness and ingenuity. They also illustrate Otto Stern's style of work in experimental physics. At the beginning there

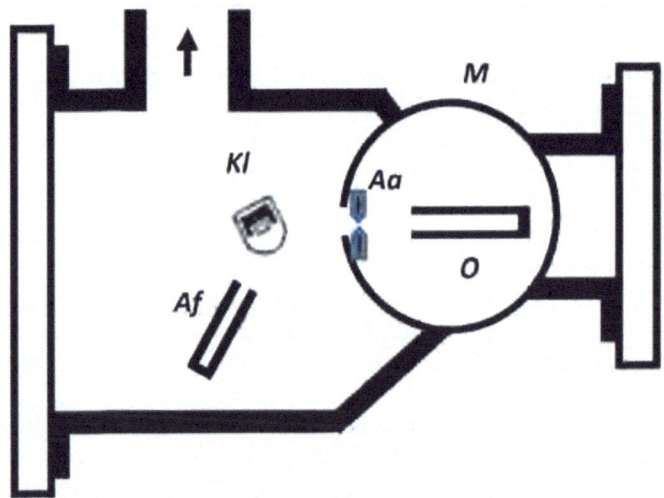

Fig. 18 Top view of the 1928 apparatus to measure the reflection of H_2 or He beams off crystals. O is the beam source orifice, Aa the collimating aperture, Kl the crystal and Af the detector (Knauer and Stern 1929a)

Fig. 19 Scattered intensity distributions as a function of the scattering angle for a He beam impinging on an NaCl crystal surface. The angle of incidence was fixed at 11.5° with respect to the surface (this angle defines the zero scattering angle). The He beam originated in a source of the indicated temperature. The first-order diffraction maxima (left and right from the reflection maximum in the center) are well resolved and allow for an accurate readout of their angular position (Estermann and Stern 1930)

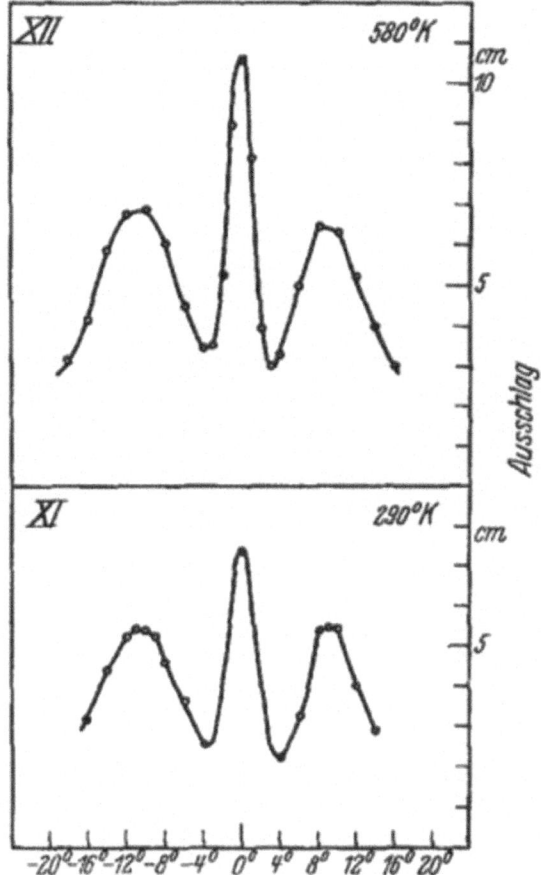

is a fundamental question and an idea how to answer it. After thorough feasibility considerations that include calculations of everything that can be calculated comes a series of experiments each of which teems with innovations and pushes the limits of the possible. No effort is spared in order to answer the question posed at the outset. Here's how Immanuel Estermann described Stern's work habits (Estermann 1962):

> [Stern] could sit in the laboratory, and when an experiment didn't want to go, he wouldn't give up. Well, he had no other interests in life practically. He would sit until 1:00 or 2:00, or 3:00 in the morning; it didn't matter to him at all; he wouldn't go out for dinner, he would bring an apple to the laboratory, and that was his dinner. And it was hard on the younger ones, especially those of us who were married. I think I was the only married one in the laboratory in those days.

The paper (Estermann, Frisch, and Stern 1932) provides an additional illustrative episode of the workings of Stern's research group. When evaluating the experimental results, the de Broglie wavelength was found to deviate by 3% from the one calculated from the molecular velocity as determined by the velocity selector. According to

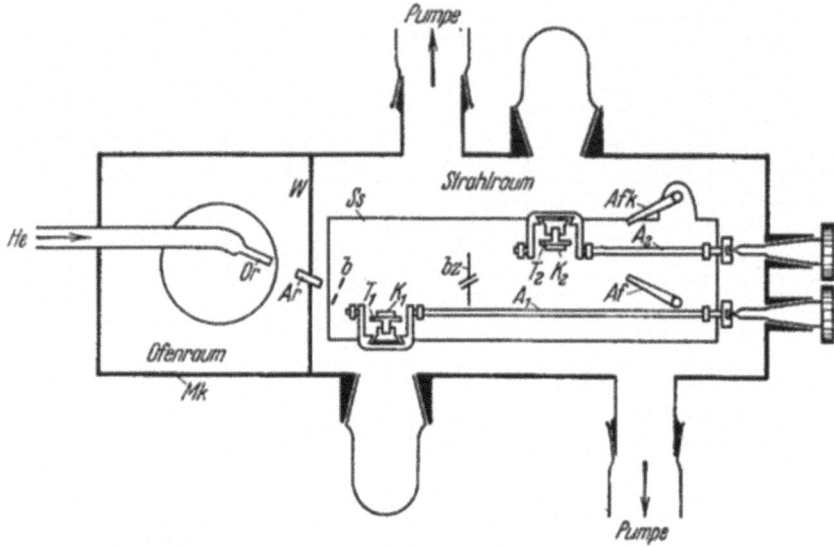

Fig. 20 Apparatus to verify the de Broglie wavelength formula for molecular beams with velocity selection by reflection. The He beam emanating from the source *Or* is collimated and velocity-selected by reflection off a crystal surface K_1. Upon collimation by slit bz, the velocity-selected beam is incident on the surface of crystal K_2. The twice-scattered He atoms are then detected in tube *Af* (Estermann, Frisch, and Stern 1932)

Stern's prior analysis, this lay outside the error bars of the measurements, which admitted a deviation of at most 1%. The problem was found upon inspecting the apparatus (Estermann, Frisch, and Stern 1932):

> The slotted discs had been made on a precision milling machine (*Auerbach-Dresden*), with the help of an indexing disc, which, according to the specifications, was supposed to divide the circumference of the wheel into 400 parts. Therefore, we took it for granted that the number of slits was 400. When we counted the slits, unfortunately only after completion of the experimental runs, we found that there were actually 408 of them (the indexing disk was indeed incorrectly labeled), which reduced the above mentioned deviation from 3 to 1%.

Thus Stern's masterful experiments on the diffraction of molecular beams provided definitive quantitative evidence for wave-particle duality.

More on matter waves can be found in Chaps. 23, 24 and 25.

2.4 Measurements of the Magnetic Dipole Moment of the Proton and the Deuteron

Measurements of nuclear magnetic moments were high on Stern's to-do list already in 1926 (Stern 1926). With the publication of Paul Dirac's "unified" quantum theory of the electron and the proton (Dirac 1930), the experimental determination of

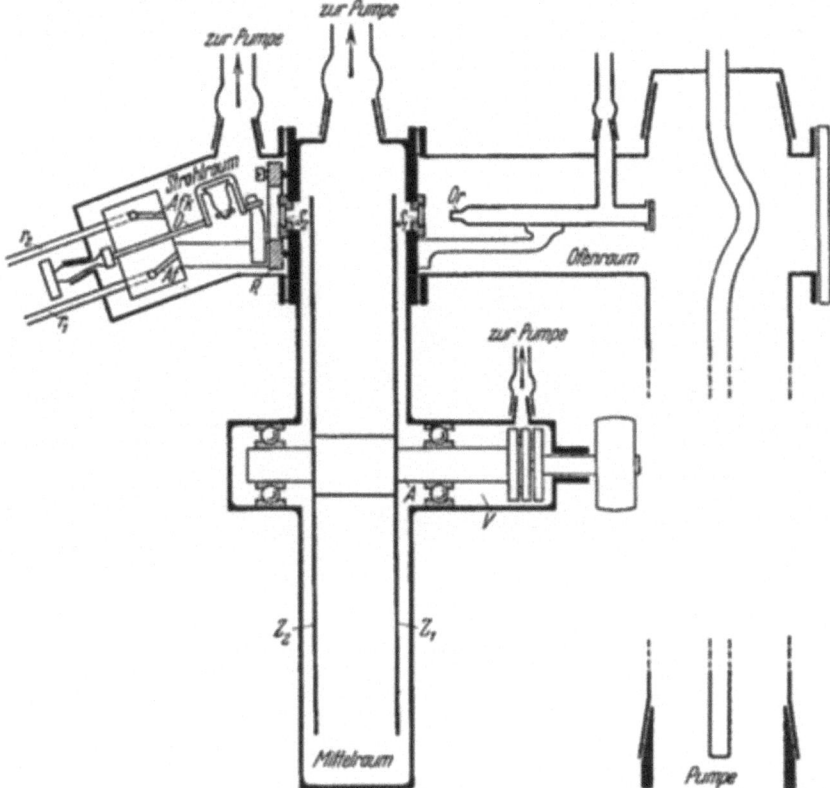

Fig. 21 Apparatus to verify the de Broglie wavelength formula for molecular beams with velocity selection by passage through spinning cogwheels/slotted discs–a *"Molekularstrahlspektrograph"*. The He beam produced in the *"Ofenraum"* is collimated and passed through spinning cogwheels Z_1 and Z_2 defining its velocity. The monochromatized beam is incident on the crystal surface attached to the rotatable axle D and its scattering detected at the catcher *Afk* (Estermann, Frisch, and Stern 1932)

proton's magnetic dipole moment became a priority for Stern and his coworkers. Dirac's theory posited that both the electron and the proton were point-like, carrying an elementary charge opposite in sign, and having magnetic dipole moments—the Bohr magneton and the nuclear magneton—whose magnitudes were mutually related by the ratio of their masses, i.e., $\mu_p/\mu_B = m_e/m_p$, with m_p and m_e the mass of the proton and of the electron, respectively, cf. Sect. 1. The feasibility of such an undertaking—the measurement of a dipole moment 1836-times smaller than the Bohr magneton—had only increased during the intervening time, thanks to both a refinement of the molecular beam detection methods (Knauer and Stern 1929b) and a better understanding of molecular hydrogen that became the species of choice to make the measurement on.

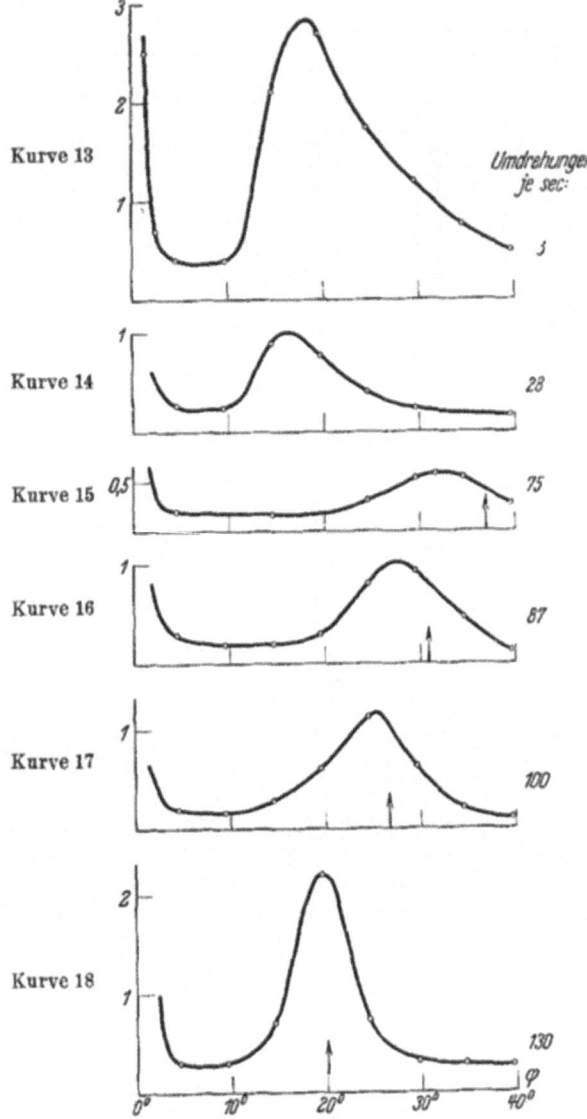

Fig. 22 Scattered intensity distributions as a function of the scattering angle for a He beam imping-ing on an LiF crystal surface. The individual curves, which correspond to the reflected and the right-hand diffracted peak of Fig. 19, were measured at different rotation rates (rpm shown on the right) of the spinning-cogwheel velocity selector. The diffraction peak of curve 13 (taken at a low rotation rate of 3 rpm) mirrors the Maxwell-Boltzmann velocity distribution. With increasing rotation rate (curves 14–18) of the spinning cogwheels, the distributions become narrower and eventually peak at a diffraction angle (shown by an arrow) corresponding to the de Broglie wavelength calculated via cf. Eq. (1) from the velocity defined by the velocity selector (Estermann, Frisch, and Stern 1932)

Prompted by the then mysterious line intensity alternations observed in the spectra of homonuclear diatomics (Slater 1926), Werner Heisenberg (Heisenberg 1927b) and Friedrich Hund (Hund 1927) postulated in 1927 the existence of two allotropic modifications of molecular hydrogen: ortho (parallel proton spins, odd-J rotational levels) and para (antiparallel proton spins, even-J rotational levels). In the same year, David Dennison (Dennison 1927) invoked these allotropic modifications to explain the anomalous behavior of molecular hydrogen's heat capacity at low temperatures, as observed by Arnold Eucken (Eucken 1912). Karl Friedrich Bonhoeffer (1899–1957) and Michael Polanyi (1891–1976) at Fritz Haber's Kaiser Wilhelm Institute for Physical Chemistry and Electrochemistry in Berlin-Dahlem (Friedrich et al. 2011; James et al. 2011; Friedrich 2016) took Heisenberg's and Hund's postulate literally and launched a search for molecular hydrogen in either of the two presumed allotropic forms. Their effort, joined by Paul Harteck (1902–1985), Adalbert (1906–1995) and Ladislaus Farkas (1904–1948) as well as Erika Cremer (1900–1996), provided in 1928-29 non-spectroscopic experimental evidence for the existence of molecular hydrogen's two allotropic modifications and led to the discovery of methods for their interconversion (Farkas and Sachsse 1933; Wigner 1933).

Stern and Frisch (Frisch and Stern 1933) recognized that the allotropic modifications of H_2 and the ability to vary their relative concentrations via interconversion were a godsend that would allow them to determine the contribution from molecular rotation to the overall magnetic dipole moment. The magnetic dipole moment of the hydrogen molecule arises namely from two sources: the nuclear spin dipole moments of the nuclei (protons) and from molecular rotation, i.e., from the spinning of the proton and electron charges. Whereas in ortho-hydrogen (parallel nuclear spins), both proton spin and molecular rotation contribute to the overall magnetic dipole moment, in para-hydrogen (antiparallel nuclear spins) the magnetic dipole moment is solely due to molecular rotation. Figure 23 shows schematically the two corresponding kinds of splittings. Hence by deflecting a beam of pure para-hydrogen, Stern and Frisch were able to determine the rotational contribution to the magnetic dipole moment. This came out as somewhat less than a nuclear magneton, $\mu_n (\mu_n = \mu_p)$. The rotational contribution could then be subtracted—in accordance with the schematic of Fig. 23—from the overall magnetic dipole moment found by deflecting a beam of ordinary hydrogen (25% para-H_2 and 75% ortho-H_2). This procedure yielded a magnetic dipole moment of the proton of 2.5 μ_n (with an error of about 20%)—and not 1 μ_n as predicted by the Dirac theory. The value of proton's magnetic moment would be refined in subsequent measurements by Stern and coworkers, see below. And so would the rotational magnetic moment. Its first theoretical estimate, by Hans Bethe (1906–2005), yielded a value of about 3 μ_n (Schmidt-Böcking et al. 2019; pp. 148–150); by including the effect of slippage of the electrons, recognized by Enrico Fermi (1901–1954), the theoretical value of the rotational magnetic dipole moment of H_2 in $J = 1$ dropped just below one nuclear magneton, in agreement with the measurements of Frisch and Stern.

That the magnetic dipole moment of the proton turned out to be quite different from one nuclear magneton brought the demise of Dirac's 1930 theory and a magnificent vindication of the imperative that guided Stern's work, namely to test the assumptions of theory—however plausible they may appear—by experiment. As Stern noted (Stern 1961):

> As the measurements of the magnetic moment of the proton were in progress, I was scolded by the theorists, who believed they knew what the outcome will be. Although our first runs had an error of 20%, the deviation [of our experimental results] from the expected theoretical value was [by] at least a factor of two.

The Frisch-Stern paper (Frisch and Stern 1933) with the revolutionary result was submitted on 27 May 1933. The technical details of the experiment described in it are astounding even today. A top view of the apparatus is shown in Fig. 24. The overall length of the molecular beam (from the source to the detector) was about 30 cm (nearly three times as much as in the SGE). The distance between the pole-pieces (edge and groove) of the Stern-Gerlach magnet was about 0.5 mm, producing a magnetic field gradient of about 2.2 T/cm. The deflection of a beam of H_2 molecules produced by a source at 90 K was about 40 μm per nuclear magneton. The molecular beam was collimated by a beak-like slit with platinum spacers 20 μm thick. The detector was a miniaturized Pirani gauge capable of registering pressure variation on the order of 10^{-8} torr within less than a minute. The entrance into the detector was defined by another 20 μm slit whose position along the direction of the deflection had to be scanned over a range of several tenths of a mm. Sample deflection data are shown in Fig. 25.

In a sequel, co-authored by Estermann and Stern (Estermann and Stern 1933a), and submitted on 12 July 1933, the error bars were reduced to just 10% for the magnetic dipole moment of the proton of 2.5 μ_n and the rotational moment per one rotational quantum of 0.85 μ_n. The main source of error were uncertainties in the inhomogeneity of the applied magnetic field, which were reduced by constructing the pole pieces of a new Stern-Gerlach magnet with greater accuracy. On 19 August 1933,

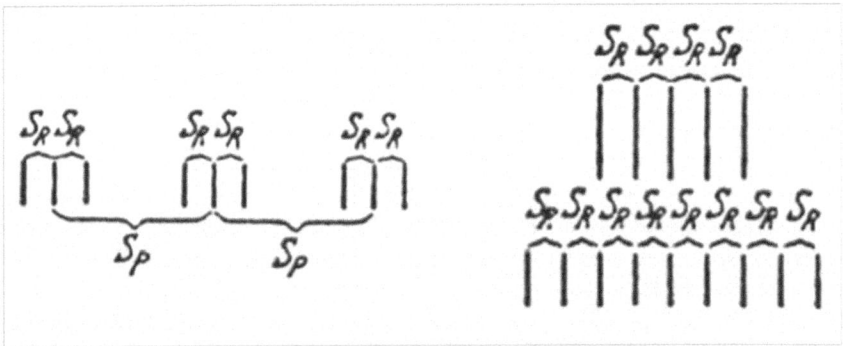

Fig. 23 Schematic diagrams of the splitting in a strong magnetic field of ortho-hydrogen (left) and of para-hydrogen (top right for $J = 2$ and bottom right for $J = 4$). Here S_P stands for the splitting due to the proton magnetic moment and S_R due to the rotational magnetic moment (Frisch and Stern 1933). The diagram on the left takes into account the Paschen-Back uncoupling of the rotational and proton moments in the magnetic field of 2 T used in the experiment

still from Hamburg, Estermann and Stern reported preliminary—and inconclusive—results (Estermann Stern 1933b) on the magnetic moment of the deuteron. It was Gilbert Newton Lewis (1875–1946) who is acknowledged for having provided 0.1 g of heavy water to his Hamburg colleagues for use in their experiment.

Upon their emigration—and settling with some of the Hamburg equipment at the Carnegie Institute of Technology in Pittsburgh— Estermann and Stern reported on 10 May 1934 their first conclusive result on the magnetic moment of the deuteron. This turned out to be only about 0.7 μ_n (Estermann and Stern 1934), which gave another jolt to the emerging nuclear physics community.

Given the paramount importance of the experimental values of the nuclear magnetic dipole moments of the proton and the deuteron, Stern and coworkers kept refining their measurements until 1937. Much of their effort went into reducing uncertainties in the inhomogeneity of the applied inhomogeneous magnetic field

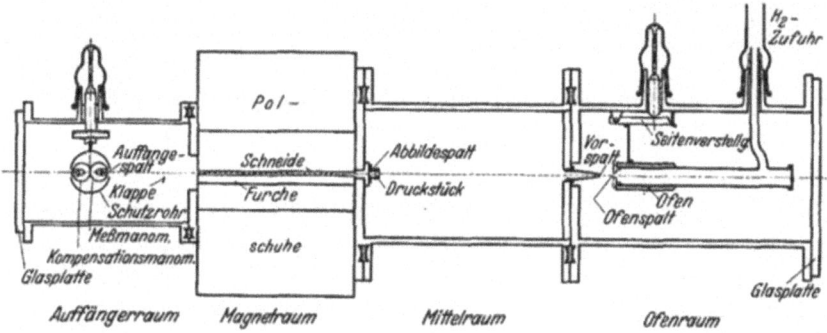

Fig. 24 Top view of the deflection apparatus designed for the measurement of nuclear and molecular magnetic moments (Frisch and Stern 1933)

Fig. 25 Deflection curve of a molecular beam of ordinary molecular hydrogen produced by a source held at 95 K (Frisch and Stern 1933). The wings that flank the central peak of undeflected molecules arise mainly from the magnetic deflection of ortho-H_2 in $J = 1$ as the population of $J = 2$ of para-H_2 is negligible at this beam temperature

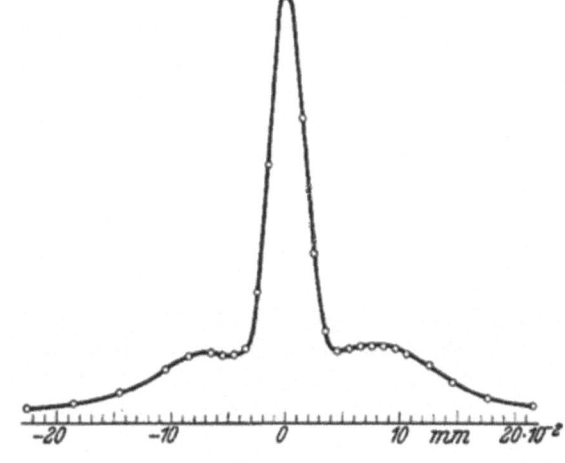

(Estermann et al. 1937). However, the molecular beams used in these experiments were not velocity-selected. This may have contributed to the deviation of the values obtained by Stern et al. for the magnetic moment of the proton and deuteron by about 10% from today's values of 2.793 μ_n and 0.855 μ_n, respectively. We note that a 1934 measurement by Isidor Rabi et al. on atomic hydrogen yielded 3.25 μ_n for the proton (Rabi et al. 1934).

Otto Stern and his Hamburg and Pittsburgh co-workers had thus provided unequivocal evidence that the proton has an internal structure and, unlike the electron, is not a point-like particle. Moreover, Stern's finding that the deuteron has a smaller magnetic dipole moment than the proton indicated that the neutron possessed a magnetic dipole moment as well, one oriented oppositely to that of the proton. Today we know that the magnetic dipole moment of the neutron is -1.913 μ_n, which implies that the neutron has an internal electric charge distribution that, however, perfectly "neutralizes itself" on the outside, as a neutron consists of one up quark (charge 2/3) and two down quarks (charge $-1/3$ each).

2.5   Experimental Demonstration of Momentum Transfer Upon Absorption or Emission of a Photon

The very last paper of the U.z.M. series, Number 30, was written by Otto Robert Frisch and submitted on 22 August 1933 (Frisch 1933a). Encouraged by Stern's programmatic paper (Stern 1926) as well as personal discussions, Frisch set out on a last-ditch effort to verify Einstein's 1917 premise (Einstein 1917) that atoms receive a tiny momentum kick upon absorption or emission of a photon.

Figure 26 shows the arrangement of Frisch's experiment: a beam of sodium atoms would be deflected by light from a sodium lamp (D-lines at 589.0 and 589.6 nm) propagating at right angles to the sodium beam either parallel (A) or perpendicular (B) to the collimation slit. The deflection would be detected by a hot-wire detector (tungsten, 10 μm diameter) whose position could be scanned perpendicular to the plane defined by the source and collimation slits. In the case of parallel illumination (A), only a broadening of the sodium beam was expected due to the photon recoil upon spontaneous emission whereas in the case of perpendicular illumination (B), the sodium beam was expected to be not only broadened but also shifted along the propagation direction of the photons from the sodium lamp due to the photon momentum transfer upon absorption.

The photon momentum involved was $h\nu/c$, with ν the frequency of the D-lines, which gave rise to a recoil velocity $h\nu/(m_{Na}c)$ of about 3 cm/s (m_{Na} is the mass of the sodium atom). Given that the mean velocity of the sodium atoms was about 9×10^4 cm/s, the angular deflection due to the absorption or emission of a photon was only about 29 μrad. For a length of the beam of about 30 cm (upon illumination behind the collimation slit), this corresponded to a perpendicular deflection of about 10 μm.

Fig. 26 Schematic of the photon-momentum transfer experiment (Frisch 1933a)

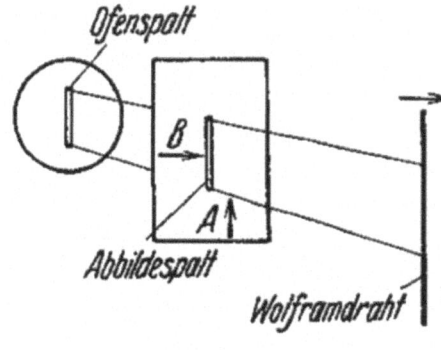

Fig. 27 Deflection of the sodium atom beam upon illumination perpendicular to the collimation slit (B), cf. Figure 26. Top dashed curve: illuminated sodium beam; full curve: 2/3 of the unilluminated sodium beam; bottom dashed curve: the difference of the two above curves corresponding to the distribution of the deflected sodium atoms (Frisch 1933a)

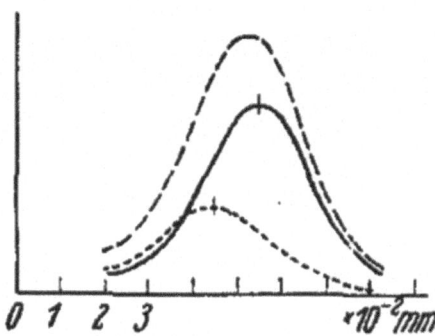

In order to estimate the fraction of the sodium atoms in the beam that were excited by the [*Osram*, double-filament] sodium lamp, Frisch determined from a photometric measurement on a sodium-vapor-filled resonance bulb that each atom was excited about 5×10^3 times a second, i.e., once in 2×10^{-4} s. Given that the atom would cover a distance of 20 cm during this time and that the illuminated stretch of the sodium beam by the *Osram* sodium lamp was 6 cm, Frisch concluded that about a third of the sodium atoms in the beam would be excited.

Figure 27 shows the results for an illumination perpendicular to the collimation slit, i.e., configuration B, see Fig. 26. The difference of the spatial distribution of the illuminated and unilluminated beam (after correction for the fraction of the atoms excited) gave the distribution of the deflected atoms. This distribution was found to peak at about 10 μm along the direction of the incident light from the sodium lamp, in agreement with the above theoretical expectation based on Einstein's theory. The deflection curve illustrates the key difference between (stimulated) absorption, which is directional, and spontaneous emission, which is not: Whereas the absorption momentum kick is imparted to the atom in the direction of the incident photon, the spontaneous emission (recoil) kick has a random direction and only results in a broadening of the spatial distribution.

The results presented by Frisch are convincing but only qualitative, as there was no time left for further work. The concluding sentence of the paper reads:

No doubt it would have been possible to achieve clearer and more impeccable results, for instance through more accurate measurements with narrower beams but, for external reasons, the experiments had to be interrupted.

Upon emigrating from Germany, Frisch would never return to this line of research. It would take more than four decades for the principles he demonstrated to surface in the work on laser cooling of atoms and ions by Theodor Hänsch and Arthur Schawlow (Hänsch and Schawlow 1975) and David Wineland and Hans Dehmelt (Wineland and Dehmelt 1975), who took up where Frisch left off. Chapters 20, 21 and 22 of this volume amply illustrate where the research on cold atoms and molecules has led so far.

2.6   The Experimental Verification of the Maxwell-Boltzmann Velocity Distribution via Deflection of a Molecular Beam by Gravity

The ability to measure tiny deflections of a molecular beam led Stern to revisit the topic that set him on his path to becoming a leading 20th century experimental physicist: the verification of the Maxwell-Boltzmann distribution of velocities. Unlike in his 1919 attempt (Stern 1920a,b), which was based on a deflection of a molecular beam by the Coriolis force (and that only provided a mean Maxwell-Boltzmann velocity), his 1937–1947 work relied on a deflection imparted by gravity. The idea for the experiment appeared in Stern's solo paper (Stern 1937) whose main concern, however, was the accurate determination of the fine-structure constant from a measurement of the Bohr magneton. Stern considered a horizontal atomic beam passing through a horizontal collimating slit placed half-way between the source and the horizontal wire of a Langmuir-Taylor detector, see Fig. 28. Assuming that the distance $AB = BC = \ell$, Stern obtained for the vertical distance S_v of free fall at the horizontal distance 2ℓ from the source A, $S_v = g\ell^2/v^2$. For cesium effusing from a source at a temperature 450 K and for $\ell = 100$ cm, this gives a free-fall distance for the most probable Maxwell-Boltzmann velocity $\alpha = \sqrt{2k_BT/m_{Cs}}$ of $S_\alpha = 0.177$ mm—by then an easily measurable deflection. Stern further considered compensating this free-fall deflection by an inhomogeneous magnetic field, H, whose gradient, $\partial H/\partial r$, would be oriented oppositely to the gravitational field and thus result in lifting up the atoms by interacting with their magnetic moment, μ. For a magnetic field gradient of a conductor (wire) running parallel to the atomic beam at a distance d and carrying an electric current I, the balance between the gravitational and magnetic force would be reached for $mg = \mu|\partial H/\partial r| = \mu(2I_0/d^2)$. In order to determine the compensating current I_0, Stern considered two options (Stern 1937): (a) lifting the atomic beam to the point C', see Fig. 28, by increasing the current I:

The instant I becomes larger than I_0, half of the atoms regardless of their velocity are deflected upwards and some atoms strike the wire. Since the amount of the deflection depends on the velocity, the slowest atoms strike the wire first, then with increasing $I - I_0$ the faster ones. No matter how far above the beam we set the detecting wire, we shall get an ion current as soon as I becomes larger than I_0.

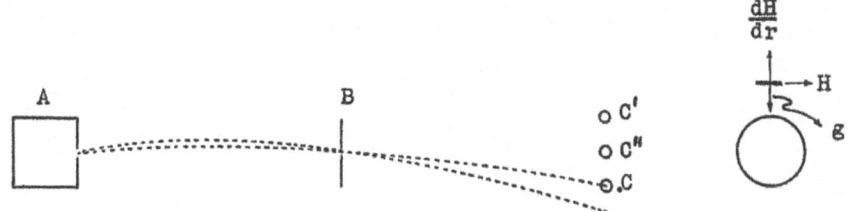

Fig. 28 Left: Schematic of the deflection of a horizontal atomic beam by the gravitational field. The beam originating in source A passes through a collimating slit B to a detector whose vertical position can be scanned through any of the points C. Right: View along the atomic beam. A current-carrying wire (circle) producing at a distance r a magnetic field H whose gradient is $\partial H/\partial r$. Note the opposite orientations of the magnetic gradient and the gravitational field whose acceleration is g (Stern 1937)

Option (b) would be to place the detector wire in the path of the beam, see point C" in Fig. 28, and

> measure [the ion current] i as a function of [the current through the conductor] I. Then i should have a maximum for $I = I_0$ because if I is larger or smaller than I_0 we [would] deflect atoms upward or downward and diminish the intensity.

Stern points out that the beam should be running in the north-south direction, as in this case the Coriolis force produced by Earth's rotation would have no vertical component that might reduce the accuracy of determining I_0.

A decade later, Estermann, Simpson, and Stern published a tour-de-force paper on the velocity distribution of cesium and potassium atoms based on gravitational deflection and its compensation by an inhomogeneous magnetic field (Estermann, Simpson, and Stern 1947a). The apparatus built for the purpose of the measurements was quite elaborate—and 2 m long. It entailed nothing less than two molecular beams that were run in parallel and whose deflections served to provide an average deflection intended to reduce the error due to mechanical distortions of the current-carrying "conductor tube" (of up to 50 A) producing the compensating magnetic field.

Representative data for a cesium beam obtained for a deflection by gravity are shown in Fig. 29. The beam intensity (ordinate) as determined by the hot-wire detector is plotted against the deflection S, i.e., the vertical position of the detector wire, in units of 10 μm (abscissa). Also shown on the abscissa are the multiples of the deflection S_α corresponding to the most probable velocity of Cs at a source temperature of 450 K. Note that slower atoms that suffer larger deflections are to the right. The authors conclude:

> The experiments serve as a demonstration that individual atoms follow the laws of free fall in the same way as other pieces of matter. Moreover, they permit a more accurate determination of the velocity distribution in molecular rays than those carried out earlier. The knowledge of this distribution is of great importance for many molecular beam experiments. It has usually been assumed that the Maxwell distribution law is valid as long as the mean free path of the molecules in the oven is several times as large as the width of the oven slit. These experiments show, however, that there is a considerable deficiency of slow molecules even at much lower

Fig. 29 Gravity deflection
of cesium atoms—both
calculated (full curve) and
measured (points). The
trapezoid on the left shows
the "shape" of the
undeflected beam, with b the
vertical size of the
collimation slit (Estermann,
Simpson, and Stern 1947a)

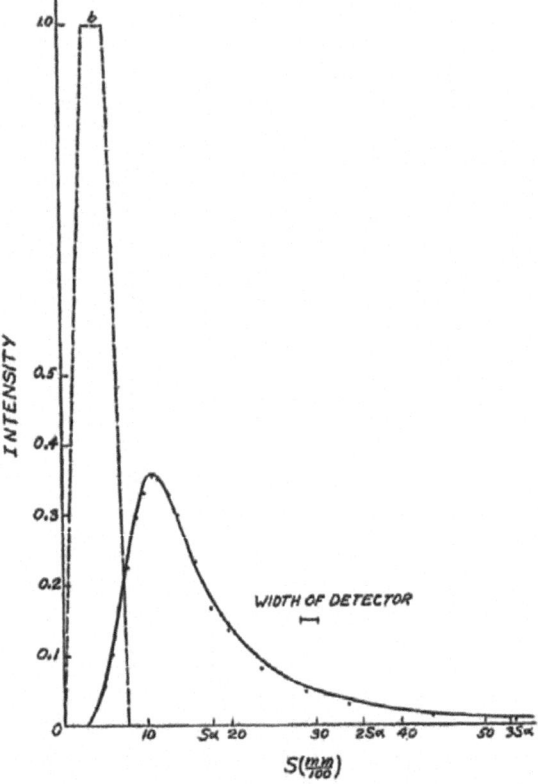

pressures. This deficiency is probably caused by collisions in the immediate vicinity of the
oven slit.

In his last molecular beam paper, submitted together with the above paper on 29
November 1946 and published back-to-back with it, Stern and coworkers reported
on gas-phase scattering of a cesium beam by helium, molecular nitrogen, and cesium
vapor and corroborated the above conclusion (Estermann et al. 1947b). The gravity
deflection curves served to infer the collision velocity.

3   Epilog

While at Hamburg, Otto Stern developed close friendships with a trio of colleagues
who are all captured, together with Stern, in the 1935 photo shown in Fig. 30: Niels
Bohr, Wolfgang Pauli, and Lise Meitner. Bohr was in fact the only (European) colleague with whom Stern was "*per Du*," i.e., on first-name terms. Stern's closeness
with the three can be inferred from Stern's correspondence. Stern had also a close,

Fig. 30 Copenhagen Conference 1935. From left: Niels Bohr, Wolfgang Pauli, Otto Stern, and Lise Meitner (OSC)

family-like relationship with Pauli's second wife Franca and with Bohr's wife Margarete. Judging, again, from his correspondence, Stern had a friendly rapport with all his colleagues and coworkers, although there may have been some clouds hanging over his relationship with Walther Gerlach. During the Nazi era, Gerlach would take up a leading role in the German nuclear program, see Chap. 8. Gerlach's brother was a high-ranking member of the SS, but apparently Gerlach himself would never join the NSDAP, see also Chap. 8. Unlike Stern, Gerlach enjoyed the limelight. In a note to Gerlach, Stern addressed him, apparently jocularly, as the "Grossbonze" [big shot]. In 1957, writing to Lise Meitner, Stern ostentatiously expressed a lack of interest to see Gerlach during his trip to Munich that year. In contrast, Gerlach expressed his admiration and fondness for Stern in his speech at the "Physikalischer Verein" in Frankfurt in 1960 (Gerlach 1960)—and did the same in his recollections of the Stern-Gerlach experiment—and of Stern—following Stern's death (Gerlach 1969).

Stern did his best to help those who needed help. This became especially manifest after Stern's emigration in 1933, when he would spend considerable amounts of time—and his own money—to help his colleagues to find a job or bridge times without one. He would be similarly helpful to his relatives (Schmidt-Böcking et al. 2018).

Stern enjoyed traveling, mostly by boat and train, although he would fly on occasions as well. He would be a frequent visitor in Copenhagen, attend the Solvay conferences in Brussels, meetings in Italy, Fig. 31, and later the meetings of the American Physical Society. Starting in 1946, Stern would spend several months each year in Europe, most notably in Zurich. Pension Tiefenau at Steinwiesenstrasse 8 became something of a second home to him, right after his house on 757 Cragmont Avenue in Berkeley where he lived since 1945, not far from his sister Berta's

Fig. 31 Stern in Rome at the Volta conference (OSC)

apartment. After 1950, he would come to Germany, at least eight times, to see his friends Max von Laue, Max Born, and Max Volmer (Schmidt-Böcking et al. 2018).

During his twelve-year tenure at the Carnegie Institute of Technology, Stern remained for the most part unnoticed by the Pittsburgh society. This changed abruptly following the arrival of a letter from Stockholm dated 14 November 1944, see Fig. 32. Here's what Stern said on 8 December 1944 at a gathering in Pittsburgh held in his honor (Schmidt-Böcking et al. 2019; p. 350):

I realize that this award is only in part a recognition of my personal work, but mainly of the work of all scientific physicists. Progress in pure science can only be achieved in a scientific atmosphere where everyone is allowed to choose his own problem and can discuss his work freely with other scientists. Both conditions for scientific work will be in danger in the future. First, the growing importance of the results of pure science for the industrial and military

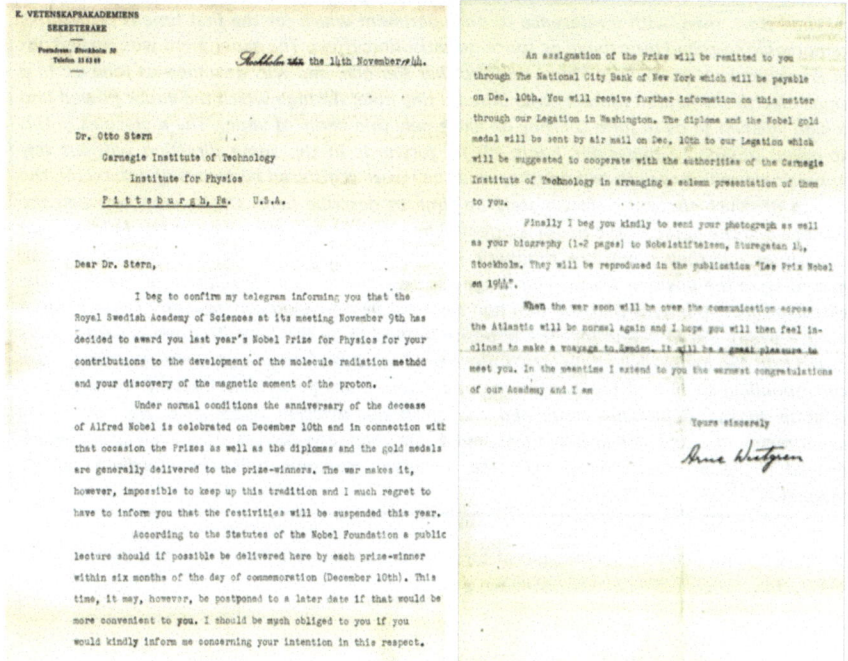

Fig. 32 Official letter from the Swedish Royal Academy to Stern (Bancroft Library, see (Schmidt-Böcking et al. 2018; p. 347)

development will make it necessary to maintain a certain degree of secrecy and will seriously impede the free interchange of ideas. Secondly, the basis and root of all scientific work is the absolute freedom of the scientist to choose his problems. Because of the fundamental importance of the results of scientific work for practical purposes the material resources for research will be concentrated on the solution of practical problems and the scientists themselves will hesitate to devote their work to problems without apparent significance for defense, social and industrial progress.

We must find the right balance between pure and applied science. We must maintain a high standard of pure science. We will have to do this even if we disregard the educational and cultural significance of science if only for the reason that without a vigorous pure science there will be no real progress in its applications.

For these reasons I am deeply grateful to the Royal Swedish Academy, not only for the great honor bestowed on me, but even more for the help given to pure science through the great prestige of Nobel and the Nobel foundation.

Although Stern's Nobel citation, see Sect. 1, mentions only the development of the molecular beam method and the measurement of the magnetic dipole moment of the proton, Eric Hulthén, a Nobel Committee member, extolled in his laudation— broadcast by the Swedish Radio—especially the Stern-Gerlach experiment, see Fig. 33.

> *I shall start, then, with a reference to an experiment which for the first time revealed this remarkable so-called directional or space-quantization effect. The experiment was carried out in Frankfurt in 1920 by Otto Stern and Walther Gerlach, and was arranged as follows: In a small electrically heated furnace, was bored a tiny hole, through which the vapor flowed into a high vacuum so as to form thereby an extremely thin beam of vapor. The molecules in this so-called atomic or molecular beam all fly forwards in the same direction without any appreciable collisions with one another, and they were registered by means of a detector, the design of which there is unfortunately no time to describe here. On its way between the furnace and the detector the beam is affected by a non-homogeneous magnetic field, so that the atoms - if they really are magnetic - become unlinked in one direction or another, according to the position which their magnetic axes may assume in relation to the field. The classical conception was that the thin and clear-cut beam would consequently expand into a diffuse beam, but in actual fact the opposite proved to be the case. The two experimenters found that the beam divided up into a number of relatively still sharply defined beams, each corresponding to one of the just mentioned discrete positional directions of the atoms in relation to the field. This confirmed the space-quantization hypothesis. Moreover, the experiment rendered it possible to estimate the magnetic factors of the electron, which proved to be in close accord with the universal magnetic unit, the so-called "Bohr's magneton".*

Fig. 33 Eric Hulthén's Nobel laudatio of Stern broadcast by the Swedish Radio on 10 December 1944 (CHS)

Felix Bloch together with his wife Lore penned the following poem to congratulate Stern on his Nobel Prize (Schmidt-Böcking et al. 2019; p. 381):

1. Twinkle, twinkle Otto Stern
how did Rabi so much learn?
He rose in the world so high
Like a diamond in the sky.
Twinkle, twinkle Otto Stern
how did Rabi so much learn?

2. The infant cried when he was born:
In Austria I feel forlorn.
And he said: The stupid stork
Should have brought me to New York.
Twinkle, twinkle Otto Stern
how did Rabi so much learn?

3. He crossed the sea a baby small
But that didn't hurt at all.
Great was his intelligence
In a certain narrow sense.
Twinkle, twinkle Otto Stern
how did Rabi so much learn?

4. Talmud and philosphie
Didn't really satisfy
So he thought as physicist
He perhaps would not be missed
Twinkle, twinkle Otto Stern
how did Rabi so much learn?

5. He together with his team
wiggled the atomic beam
Up and down through slits so fine
Saw the light of reason shine
Twinkle twinkle Otto Stern
How did Rabi so much learn.

6. Soon the moments made him
and he said: I'm awfully sorry.
Gentlemen, we have no chance
What we need is resonance.
Twinkle, twinkle Otto Stern
How did Rabi so much learn?

7. Well you know, he's always right,
This time he was even bright,
And a quadrupole he found.
Deuterons were no more round
Twinkle twinkle Otto Stern
How did Rabi so much learn.

8. At R.L. he said: Why not
Should I be a great big shot?
and again he was quite right
he almost made it, but not quite.
Twinkle twinkle Otto Stern
How did Rabi so much learn.

9. So he finally grew wise
Got himself the Nobel prize.
Back to physics now he is
With undreamt possibilities.
Twinkle twinkle Otto Stern
How did Rabi so much learn.

10. Twinkle, twinkle Otto Stern
How did Rabi so much learn?
He rose in the world so high
like a diamond in the sky.
Twinkle twinkle Otto Stern
How did Rabi so much learn.

Figure 34 shows the poem's dedication to Stern by Felix and Lore Bloch as well as its "endorsement" by the signatures of Isidor I. Rabi—the host of the celebratory gathering, George E. Uhlenbeck, Jerold Zacharias, Reg Turner, Wheeler Loomis, Walton N, J.H. Van Vleck, Luis Lederman, L.J. Haworth, Marshall, E.M. Purcell, James L. Lawson, Jane K. Lawson, Beth Purcell, Louis C. Turner, Edith Loomis, Anette Hugh, Goudsmit, Helen Rabi, John Slater, and others.

The molecular beam method made inroads into both physics and chemistry starting in the 1960s, especially in Europe and the U.S. Stern's former PhD student, Lester C. Lewis (1902-?),[5] drew a scientific family tree shown in Fig. 35. In his

---

[5]L. C. Lewis (born 1902) joined Stern in Hamburg in 1930 as a Charles A. Coffin Fellow and graduated there in 1931 with a thesis on chemical equilibria, "Die Bestimmung des Gleichgewichts

Fig. 34 Entstanden anlässlich einer Feier bei den Rabi's bei der wir alle an Sie dachten. Viele Herzliche Glückwünsche F. Bloch Auch von mir die herzlichsten Glückwünsche Lore Bloch

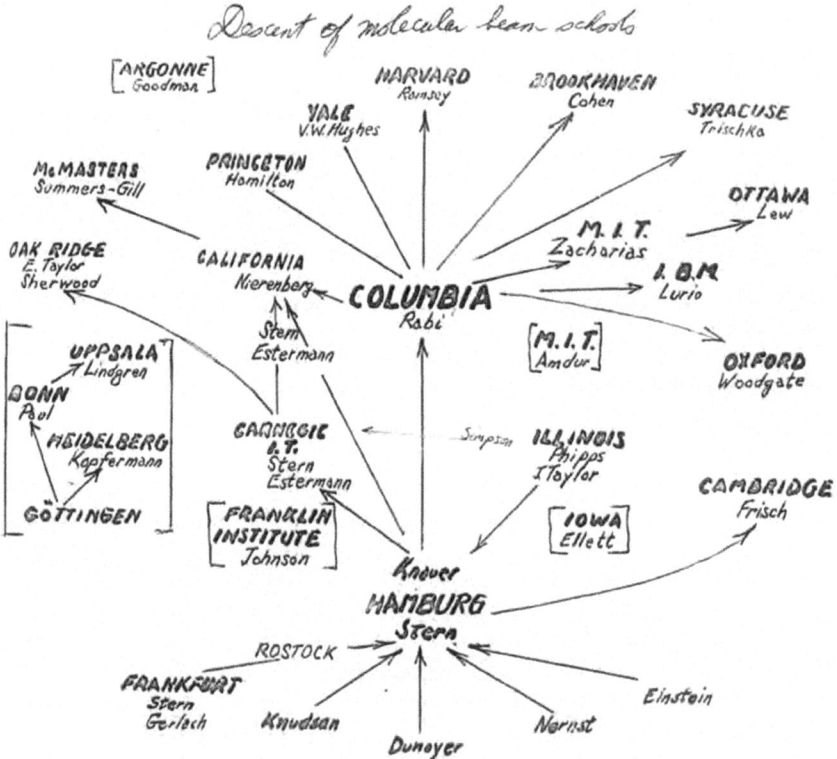

Fig. 35 Family tree of the molecular beam method in Europe and the United States by L.C. Lewis (Bancroft Library, see (Schmidt-Böcking et al. 2019; p. 257)

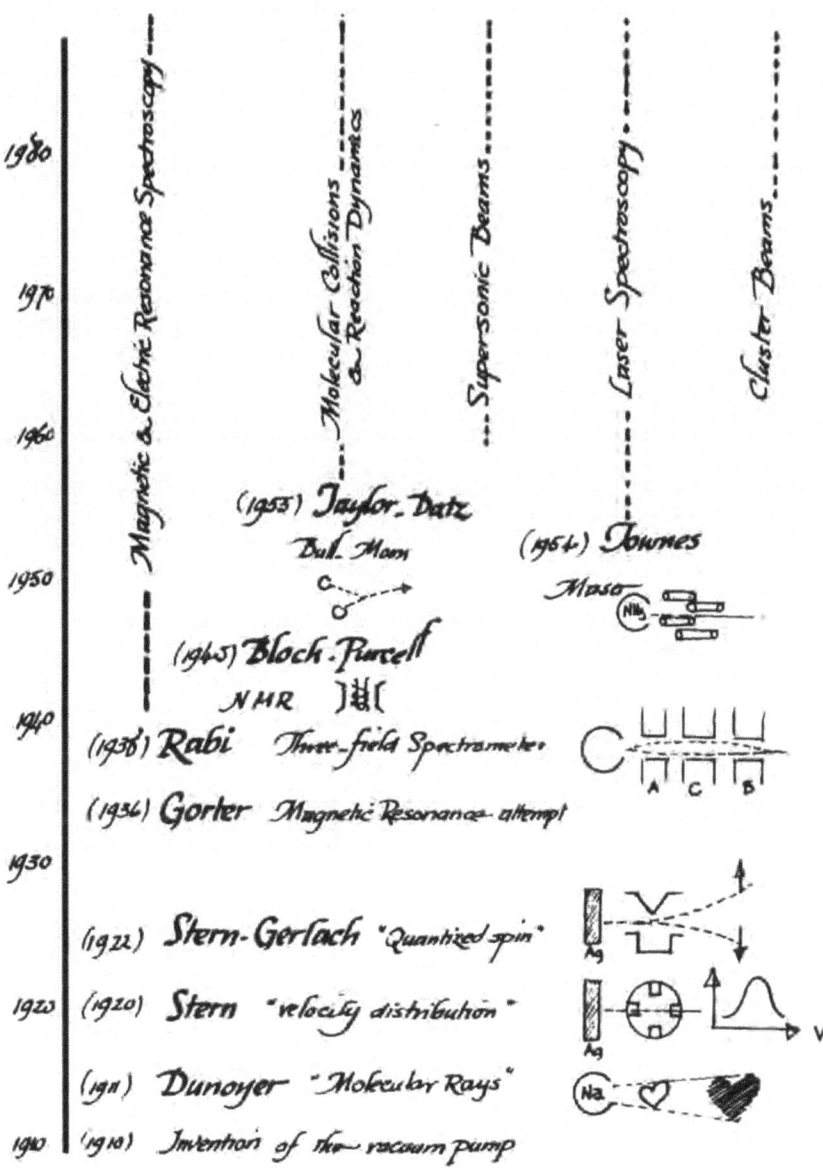

Fig. 36 Evolution of molecular beam and kindred methods according to Dudley Herschbach (Herschbach 1987)

1986 Nobel Lecture, Dudley Herschbach presented his depiction of the evolution and consequences of the molecular beam method, see Fig. 36.

In addition to the Nobel Prize, Stern received many additional honors, such as honorary degrees from Berkeley (1930) or Zurich (1960) or academy memberships in Europe and the U.S. After his death, the German Physical Society named its highest award in experimental physics the "Stern-Gerlach-Preis" (1988–1992), renamed in 1993 the "Stern-Gerlach-Medaille," see Fig. 37.

At the University of Hamburg, a lecture hall was named after Otto Stern and at the University of Frankfurt a new auditorium and library complex on the Riedberg campus was named the "Otto Stern Zentrum" (Fig. 38). Last but not least, the building in Frankfurt where Otto Stern launched his molecular beam experiments was declared in 2019 a "European Physical Society Historic Site." The plaque reads:

> This building housed Max Born's Institute for Theoretical Physics where key discoveries were made during the period 1919-1922 that contributed decisively to the development of quantum mechanics. The Institute launched experiments in 1919 via the molecular beam technique by Otto Stern, for which he was awarded the 1943 Nobel Prize in Physics. Experiments done in 1920 by Max Born and Elisabeth Bormann sent a beam of silver atoms measuring the mean free path in gases and probing various gases to estimate sizes of molecules. An iconic experiment in 1922 by Otto Stern and Walther Gerlach demonstrated space quantization of atomic magnetic moments and thereby also, for the first time, the quantization of atomic angular momenta. In 1921, Alfred Landé postulated here the coupling of angular momenta as the basis of the electron dynamics within atoms. This building is the seat of the Physical Society of Frankfurt (the oldest in Germany, founded in 1824).

Let us conclude with the words of one of Otto Stern's most prominent associates, Isidor Rabi, as told to John Rigden just a few days before Rabi's death (Rabi 1988):

> Stern had this quality of taste in physics and he had it to the highest degree. As far as I know, Stern never devoted himself to a minor question.

Archives

AEA   Albert Einstein Archives, The Hebrew University of Jerusalem, Jerusalem, Israel

CHS   Center for History of Science, The Royal Swedish Academy of Sciences, Box 50005, SE-104 05 Stockholm, Sweden

CMA   Archives Carnegie Mellon Institute, Pittsburgh, PA, USA

OSC   Otto Stern Picture Collection, Bancroft Library, Berkeley, CA, USA

Numbers (Sx) and (Mx) in the following list of references refer to the list of publications by Stern and his collaborators as given in *Otto Sterns Veröffentlichungen*, ed. H. Schmidt-Böcking *et al.*, Springer, 2016. Numbers (U.z.M. x) refer to the series of papers in Zeitschrift für Physik entitled Untersuchungen zur Molekularstrahlmethode.

---

zwischen den Atomen und den Molekülen eines Alkalidampfes mit einer Molekularstrahlmethode." He would become Curator and later Executive Director for Physical Sciences at the Smithsonian Museum in Washington, D.C.

Fig. 37 Stern-Gerlach
Medal of the German
Physical Society

Fig. 38 One of the authors (HS-B) in front of the "Otto Stern Zentrum" at the Riedberg Campus
of the University of Frankfurt

References

F. Aaserud, J.L. Heilbron (2013). *Love, Literature, and the Quantum Atom* (Oxford University Press, Oxford)

O. Amit, Y. Margalit, O. Dobkowski, Z. Zhou, Y. Japha, M. Zimmermann, M.A. Efremov, F.A. Narducci, E.M. Rasel, W.P. Schleich, R. Folman (2019). T^3 Stern-Gerlach matter-wave interferometer. Phys. Rev. Lett. **123**, 083601

F.W. Aston (1919). A positive ray spectrograph. Phil. Mag. **38**(228), 707–714

M. Badino, B. Friedrich (2013). Much polyphony but little harmony: Otto Sackur's groping for a quantum theory of gases. Phys. Perspect. **15**(3), 295–319

D. Buchwald, T. Sauer, Z. Rosenkranz, J. Illy, V.I. Holmes (2012). *The Collected Papers of Albert Einstein*, vol 10: *The Berlin Years: Correspondence: January May-December 1920* (Princeton University Press, Princeton)

D. Buchwald, J. Illy, Z. Rosenkranz, T. Sauer (2012). *The Collected Papers of Albert Einstein*, vol. 13: *The Berlin Years: Writings and correspondence January 1922-March 1923* (Princeton University Press, Princeton)

D. Buchwald, J. Illy, Z. Rosenkranz, T. Sauer, O. Moses (2015). *The Collected Papers of Albert Einstein*, vol. 14: *The Berlin Years: Writings and correspondence April 1923-May 1925* (Princeton University Press, Princeton)

C.G. Darwin (1928). Free Motion in the Wave Mechanics. Proc. R. Soc. Lond. A **117**, 193–258

C.J. Davisson, L.H. Germer (1927). Diffraction of electrons by a crystal of nickel. Phys. Rev. **30**, 705

P. Debye (1916). Quantenhypothese und Zeeman-Effekt. Physikalische Zeitschrift **17**(20), 507–512

D.M. Dennison (1927). A note on the specific heat of the hydrogen molecule. Proc. R. Soc. **115**, 483–486

M. Devereux (2015). Reduction of the atomic wave function in the Stern-Gerlach experiment. Can. J. Phys. **93**(11), 1382–1390

P.A.M. Dirac (1928). The quantum theory of the electron. Proc. R. Soc. **117**, 610–624, **118**, 351–361

P.A.M. Dirac (1930). A theory of electrons and protons. Proc. R. Soc. Lond. Ser. A **126**(801), 360–365

K.M. Downard (2007). Francis William Aston: The man behind the mass spectrograph. Eur. J. Mass Spectrom. **13**(3), 177–190

L. Dunoyer (1911). Sur la réalisation d'un rayonnement matériel d'origine purement thermique. Cinétique expérimentale. Le Radium **8**, 142–145

A. Einstein (1917). Zur Quantentheorie der Strahlung. Physikalische Zeitschrift **18**, 121. Reprinted in (Kox et al., 1996, Doc. 38)

A. Einstein, P. Ehrenfest (1922). Quantentheoretische Bemerkungen zum Experiment von Stern und Gerlach. Zeitschrift für Physik **11**, 31–34. Reprinted in (Buchwald et al., 2012, Doc. 315)

A. Einstein, O. Stern (1913). Some Arguments for the Assumption of Molecular Agitation at Absolute Zero. Annalen der Physik **40**, 551–560. Reprinted in (Klein et al., 1995, Doc. 11). (S5)

I. Estermann, O.R. Frisch, O. Stern (1931). Versuche mit monochromatischen de Broglie-Wellen von Molekularstrahlen. Physikalische Zeitschrift **32**, 670–674 (S43)

I. Estermann, O.R. Frisch, O. Stern (1932). Monochromasierung der de Broglie-Wellen von Molekularstrahlen. Physikalische Zeitschrift **32**, 348–365. (S42)

I. Estermann, O.R. Frisch, O. Stern (1933). Magnetic moment of the proton. Nature **132**, 169 (S51)

I. Estermann, O. Stern (1930). Beugung von Molekularstrahlen. Zeitschrift für Physik **61**, 95–125 (U.z.M. 15) (S40)

I. Estermann, O. Stern (1933a). Über die magnetische Ablenkung von Wasserstoffmolekülen und das magnetische Moment des Protons II. Zeitschrift für Physik **85**, 17–24 (U.z.M. 27) (S52)

I. Estermann, O. Stern (1933b). Über die magnetische Ablenkung von isotopen Wasserstoffmolekülen und das magnetische Moment des "Deutons. Zeitschrift für Physik **86**, 132–134 (U.z.M. 29) (S54)

I. Estermann, O. Stern (1934). Magnetic moment of the deuton [deuteron]. Nature **133**, 911 (S55)

I. Estermann, O.C. Simpson, O. Stern (1937). Magnetic deflection of HD molecules (Minutes of the Chicago Meeting, November 27–28, 1936). Phys. Rev. **51**, 64 (S58)

I. Estermann, O.C. Simpson, O. Stern (1947a). The free fall of atoms and the measurement of the velocity distribution in a molecular beam of cesium atoms. Phys. Rev. **71**, 238–249 (S65)

I. Estermann, S.N. Foner, O. Stern (1947b). The mean free paths of cesium atoms in helium, nitrogen and cesium vapor. Phys. Rev. **71**, 250–257 (S67)

I. Estermann (1962). Interviewed by John L. Heilbron. Oral History Interviews. American Institute of Physics https://www.aip.org/history-programs/niels-bohr-library/oral-histories/4593

I. Estermann (1975). History of molecular beam research: Personal reminiscences of the important evolutionary period 1919–1933. J. Am. Phys. **43**, 661

A. Eucken (1912). Die Molekularwärme des Wasserstoffs bei tiefen Temperaturen. Sitzungsberichte der Deutschen Akademie der Wissenschaften zu Berlin. **1912**(1), 141–151

L. Farkas, H. Sachsse (1933). Über die homogene Katalyse der Para-Orthowasserstoffumwandlung unter Einwirkung paramagnetischer Moleküle. I and II. Zeitschrift für Physikalische Chemie **B23**, 1–18; 19–27

B. Friedrich, D. Herschbach (1998). Space quantization: Otto Stern's lucky star. Daedalus **127**(1), 165–191

B. Friedrich, D. Herschbach (2003). Stern and Gerlach: How a bad cigar helped reorient atomic physics. Phys. Today **56**(12), 53–59

B. Friedrich, J. James Hoffmann (2011). One hundred years of the Fritz Haber Institute. Angew. Chem. Int. Ed. **43**, 10022–10049

J. James, T. Steinhauser, B. Friedrich Hoffmann (2011). *One Hundred Years at the Intersection of Chemistry and Physics* (De Gruyter, Berlin)

B. Friedrich (2016). Michael Polanyi (1891–1976): The life of the mind. Bunsen-Magazin **18**(5), 160–167

O.R. Frisch (1933). Experimenteller Nachweis des Einsteinschen Strahlungsrückstoßes. Zeitschrift für Physik **86**, 42–48 (U.z.M. 30) (M17)

O.R. Frisch, E. Segrè (1933). Über die Einstellung der Richtungsquantelung. II Zeitschrift für Physik **80**, 610–616 (U.z.M. 22) (M13)

O.R. Frisch, O. Stern (1933). Über die magnetische Ablenkung von Wasserstoffmolekülen und das magnetische Moment des Protons I. Zeitschrift für Physik **85**, 4–16 (U.z.M. 24) (S47)

W. Gerlach (1925). Über die Richtungsquantelung im Magnetfeld II. Annalen der Physik **76**, 163–197 (M0)

W. Gerlach (1960). Über die Entwicklungen atomistischer Vorstellungen, Physikalischer Verein Frankfurt, 2. März 1960; http://www.physikalischer-verein.de

W. Gerlach (1963a). Archive for the history of quantum physics, Interview on 15 July 1963. Also quoted in Mehra, J. and Rechenberg, H. (1982). The historical development of quantum theory, Vols. 1, Part 2 (Springer, Heidelberg)

W. Gerlach (1963b). Interview with Thomas Kuhn on February 1963 **18**

W. Gerlach (1969). Zur Entdeckung des""'Stern-Gerlach-Effektes. Physikalische Blätter **25**, 472

W. Gerlach, O. Stern (1922). Der experimentelle Nachweis des magnetischen Moments des Silberatoms. Zeitschrift für Physik **8**, 110–111 (S19)

W. Gerlach, O. Stern (1922b). Der experimentelle Nachweis der Richtungsquantelung im Magnetfeld. Zeitschrift für Physik **9**, 349–352 (S20)

W. Gerlach, O. Stern (1924). Annalen der Physik **74**, 673–699 (S26)

R.B. Griffiths (2015). Consistent quantum measurements. Stud. History Philos. Sci. Part B: Stud. History Philos. Mod. Phys. **52**, 188

P. Güttinger (1932). Das Verhalten von Atomen im magnetischen Drehfeld. Zeitschrift für Physik **73**, 169–184

T. Hänsch, A. Schawlow (1975). Cooling of gases by laser radiation. Opt. Commun. **13**, 68

W. Heisenberg (1927). Über den anschaulichen Inhalt der quantentheoretischen Kinematik und Mechanik. Zeitschrift für Physik **43**(3), 172–198

W. Heisenberg (1927). Mehrkörperprobleme und Resonanz in der Quantenmechanik II. Zeitschrift ür Physik **41**, 239–267

A. Hermann, K. von Meyenn, V. Weisskopf (1979). *Wolfgang Pauli. Wissenschaftlicher Briefwechsel mit Bohr, Einstein, Heisenberg a.o. Vol. 1: 1919-1929* (Springer, New York, Heidelberg, Berlin)

D. Herschbach (1987). Molecular dynamics of elementary chemical reactions (Nobel Lecture). Angew. Chem. **26**, 1221–1243

J.G. Huber (2014). Walther Gerlach (1888–1979) und sein Weg zum erfolgreichen Experimentalphysiker bis etwa 1925. PhD dissertation, LMU München. Dated 6 August 2014

F. Hund (1927). Zur Deutung der Molekelspektren II. Zeitschrift für Physik **42**, 93–120

H. Kallmann, F. Reiche (1921). Über den Durchgang bewegter Moleküle durch inhomogene Kraftfelder. Z. Phys. **6**, 352–375

J.M.B. Kellog, I.I. Rabi, J.R. Zacharias (1936). The Gyromagnetic Properties of the Hydrogens. Phys. Rev. **50**, 472

M.J. Klein, A.J. Kox, J. Renn, R. Schulmann (1995). *The Collected Papers of Albert Einstein*, vol 6: *The Swiss Years: Writings: 1912-1914* (Princeton University Press, Princeton)

F. Knauer, O. Stern (1929). Über die Reflexion von Molekularstrahlen. Zeitschrift für Physik **53**, 779–791 (U.z.M. 11) (S33)

F. Knauer, O. Stern (1929). Intensitätsmessungen an Molekularstrahlen von Gasen. Zeitschrift für Physik **53**, 766–778 (U.z.M. 10) (S36)

A.J. Kox, M.J. Klein, R. Schulmann (1996). *The Collected Papers of Albert Einstein*, vol 6: *The Berlin Years: Writings: 1914–1917* (Princeton University Press, Princeton)

A. Landé (1921). Über den anomalen Zeemaneffekt (Teil I). Zeitschrift für Physik **5**, 231–241

A. Landé (1921). Über den anomalen Zeemaneffekt (II. Teil). Zeitschrift für Physik **7**, 398–405

A. Landé (1923). Schwierigkeiten in der Quantentheorie des Atombaues, besonders magnetischer Art. Physikalische Zeitschrift **24**, 442

A. Landé (1929). Polarisation von Materiewellen. Die Naturwissenschaften **17**(32), 634–637

S. Machluf, Y. Japha, R. Folman (2013). Coherent Stern-Gerlach momentum splitting on an atom chip. Nat. Commun. **4**, 2424

E. Majorana (1932). Atomi orientati in campo magnetico variabile. Nuovo Cimento **9**, 43–50

Y. Margalit, Z. Zhou, S. Machluf, D. Rohrlich, Y. Japha, R. Folman (2015). A self-interfering clock as a "which path" witness. Science **349**, 1205

Y. Margalit, Z. Zhou, O. Dobkowski, Y. Japha, D. Rohrlich, S. Moukouri, R. Folman (2018). Realization of a complete Stern-Gerlach interferometer. arXiv:1801.02708

L. Motz, M.E. Rose (1936). On space quantization in time varying magnetic fields. Phys. Rev. **50**, 348

T.E. Phipps, O. Stern (1932). Über die Einstellung der Richtungsquantelung. Zeitschrift für Physik **73**, 185–191 (U.z.M. 17) (S41)

I.I. Rabi (1929). Zur Methode der Ablenkung von Molekularstrahlen. Zeitschrift für Physik **54**, 190–197 (U.z.M. 12) (M7)

I.I. Rabi (1937). Space quantization in a gyrating magnetic field **51**, 652

O.I. Rabi, On the process of spin quantization. Phys. Rev. **49**(4), 328–342 (1936)

I.I. Rabi et al. (1938a). A new method of measuring nuclear magnetic moment. Phys. Rev. **53**, 318

I.I. Rabi et al. (1938b). The magnetic moments of $_3L^6$, $_3L^7$ and $_9F^{19}$. Phys. Rev. **53**, 495

I.I. Rabi, S. Millman, P. Kusch, J.R. Zacharias (1939). The molecular beam resonance method for measuring nuclear magnetic moments. Phys. Rev. **55**, 526–535

I.I. Rabi, J. Kellogg, J. Zacharias (1934). The magnetic moment of the proton. Phys. Rev. **31**, 157–163

I.I. Rabi, as told to Rigden (1988), J.S., Otto Stern and the discovery of space quantization. Z. Phys. D - Atoms, Molecules and Clusters **10**, 119–120

N. Ramsey (1993). *I.I. Rabi 1898–1988* (National Academy of Sciences, A Biographical Memoir)

J.E.A. Ribeiro (2010). Was the Stern-Gerlach phenomenon classically described? Found. Phys. **40**, 1779–1782

O. Sackur (1911). Die Anwendung der kinetischen Theorie der Gase auf chemische Probleme. Ann. Phys. **36**, 958–980

O. Sackur (1913). Die universelle Bedeutung des sog. elementaren Wirkungsquantums. Ann. Phys. **40**, 67–86

T. Sauer (2016). Multiple perspectives on the Stern-Gerlach experiment, in *The Philosophy of Historical Case Studies*, ed. by T. Sauer, R. Scholl (Springer, Berlin), pp. 251–263

H. Schmidt-Böcking, K. Reich (2011). *Biographie Über Otto Stern* (Herausgegeben von der Goethe Universität Frankfurt, Societäts Verlag)

J. Schwinger (1937). On nonadiabatic processes in inhomogeneous fields. Phys. Rev. **51**, 648

M.O. Scully, W.E. Lamb, A. Barut (1987). On the theory of the Stern-Gerlach apparatus. Found. Phys. **17**(6), 575–583

H. Schmidt-Böcking et al. (2016). The Stern-Gerlach experiment revisited. Euro. Phys. J. H **41**, 327–364

H. Schmidt-Böcking, A. Templeton, W. Trageser (2018). *Otto Sterns gesammelte Briefe*. Band 1: *Hochschullaufbahn und die Zeit des Nationalsozialismus*. (Springer Spektrum)

H. Schmidt-Böcking, A. Templeton, W. Trageser (2019). *Otto Sterns gesammelte Briefe*. Band 2: *Sterns wissenschaftliche Arbeiten und zur Geschichte der Nobelpreisvergabe* (Springer Spektrum)

R. Schulmann, A.J. Kox, M. Janssen, J. Illy (1998). *The Collected Papers of Albert Einstein*. Vol. 8: *The Berlin Years: Correspondence: 1914-1918* (Princeton University Press, Princeton)

W. Schütz (1969). Persönliche Erinnerungen an die Entdeckung des Stern-Gerlach-Effektes. Physikalische Blätter **25**, 343–345

E. Segrè (1973). *Otto Stern (1888–1969* (National Academy of Sciences, A Biographical Memoir)

C. Slater (1926). Alternating intensities in band lines. Nature **117**, 555–556

A. Sommerfeld (1916). Zur Quantentheorie der Spektrallinien. Ann. Phys. **51**(1–94), 125–167

A. Sommerfeld (1924). *Atombau und Spektrallinien*, 4th revised edition (Braunschweig Vieweg)

O. Stern (1912). *Zur kinetischen Theorie des osmotischen Druckes konzentrierter Lösungen und über die Gültigkeit des Henryschen Gesetzes für konzentrierte Lösungen von Kohlendioxyd in organischen Lösungsmitteln bei tiefen Temperaturen* (Barth, Breslau, Grass), p. S1

O. Stern (1916a). Die Entropie fester Lösungen. Annalen der Physik **49**, 823–841 (S7)

O. Stern (1916b). Über eine Methode zur Berechnung der Entropie von Systemen elastisch gekoppelter Massenpunkte. Annalen der Physik **51**, 237–260 (S8)

O. Stern (1920a). Eine direkte Messung der thermischen Molekulargeschwindigkeit. Physik. Z. **21**, 582 (S14)

O. Stern (1920b). Eine direkte Messung der thermischen Molekulargeschwindigkeit. Z. Phys. **2**, 49–56 (S16)

O. Stern (1920c). Nachtrag zu meiner Arbeit: Eine direkte Messung der thermischen Molekulargeschwindigkeit. Zeitschrift für Physik **3**, 417–421 (S17)

O. Stern (1921). Ein Weg zur experimentellen Prüfung der Richtungsquantelung im Magnetfeld. Zeitschrift für Physik **7**, 249–253 (S18)

O. Stern (1924). Zur Theorie der elektrolytischen Doppelschicht. Zeitschrift für Elektrochemie **30**, 508–516 (S25)

O. Stern (1926). Zur Methode der Molekularstrahlen I. Zeitschrift für Physik **39**, 751–763 (U.z.M. 1) (S28)

O. Stern, F. Knauer (1926). Zur Methode der Molekularstrahlen II. Zeitschrift für Physik **39**, 764–779 (U.z.M. 2) (S29)

O. Stern (1929). Beugung von Molekularstrahlen am Gitter einer Krystallspaltfläche. Die Naturwissenschaften **17**, 391 (S37)

O. Stern (1937). A new method for the measurement of the Bohr Magneton. Phys. Rev. **51**, 852–854

O. Stern (1961). Erinnerungen. Interview with Res Jost. Tape recording, ETH-Bibliothek Zürich, Bildarchiv (Sign.: D 83:1-2); also available as transcript, ETH-Bibliothek Zürich, Archive (Sign.: Hs 1008:8) http://www.sr.ethbib.ethz.ch/

O. Stern (1962). On a proposal to base wave mechanics on Nernst's theorem. Helvetica Physica Acta **35**, 367–368 (S71)

H. Tetrode (1912). Die chemische Konstante der Gase und das elementare Wirkungsquantum. Ann. Phys. **38**, 434–442

J.P. Toennies, H. Schmidt-Böcking, B. Friedrich, J.C.A. Lower (2011). Otto Stern (1888–1969): The founding father of experimental atomic physics. Ann. Phys. **523**(12), 1045–1070

S.-I. Tomonaga (1997). *The Story of Spin* (Chicago University Press, Chicago)

G.E. Uhlenbeck, S.A. Goudsmit (1925). Ersetzung der Hypothese vom unmechanischen Zwang durch eine Forderung bezüglich des inneren Verhaltens jedes einzelnen Elektrons. Die Naturwissenschaften **13**, 953–954

I. Unna, T. Sauer (2013). Einstein, Ehrenfest, and the quantum measurement problem. Annalen der Physik (Berlin) **525**(1–2), A15–A19

M. Utz, M.H. Levitt, N. Cooper, H. Ulbricht (2015). Visualisation of quantum evolution in the Stern-Gerlach and Rabi experiments. Phys. Chem. Chem. Phys. **17**, 3867–3872

H. Wennerström, P. Westlund (2012). The Stern-Gerlach experiment and the effects of spin relaxation. Phys. Chem. Chem. Phys. **14**, 1677–1684

E. Wigner (1933). Über die paramagnetische Umwandlung von Para-Orthowasserstoff. III. Zeitschrift für Physikalische Chemie **B23**, 28–32

D. Wineland, H. Dehmelt (1975). Proposed $10^{14} \Delta v < v$ laser fluorescence spectroscopy on Tl^+ mono-ion oscillator III. Bull. Am. Phys. Soc. **20**, 637

Z. Zhou, Y. Margalit, D. Rohrlich, Y. Japha, R. Folman (2018). Quantum complementarity of clocks in the context of general relativity.Classical Quant. Gravity **35**, 185003

Z. Zhou, Y. Margalit, S. Moukouri, Y. Meir, R. Folman (2020). An experimental test of the geodesic rule proposition for the non-cyclic geometric phase. Sci. Adv. **6**, eaay8345

Chapter 6
Otto Stern—With Einstein in Prague and in Zürich

Hanoch Gutfreund

Abstract The two years that Otto Stern spent with Albert Einstein in Prague and Zürich, between the spring of 1912 and the spring of 1914, can be viewed as his apprenticeship in theoretical physics. This chapter describes that formative phase in Stern's scientific career, prior to his emergence as one of the greatest innovators in experimental physics.

1 One Semester in Prague

Otto Stern completed his studies in physical chemistry and worked under the supervision of Otto Sackur on his doctoral dissertation on the kinetic theory of the osmotic pressure of concentrated solutions. Stern chose the problem himself and Sackur suggested that the theoretical work be accompanied by measurements on solutions of dissolved carbon dioxide. In the spring of 1912 Stern submitted his thesis under the rather long title "On the Osmotic Pressure of Condensed Solutions and the Validity of the Henry Law for Condensed Solutions of Carbon di-Oxide in Organic Solvents at Low Temperatures."

In May 1912, Stern joined Einstein at the German part of Charles University in Prague. This was Einstein's first academic appointment as full professor and Stern was his first post-doctoral student. When asked, years later, why he preferred to go to Einstein, rather than, what seemed to be a more natural choice, to join Walther Nernst or Fritz Haber, Stern did not have a clear answer other than his wish to meet the great man that Einstein was. He acknowledged that this wish may have been an impudence, but Einstein agreed right away. The "matchmaker" between Stern and Einstein was Haber, who was friendly with both Sackur and Einstein.

The description of the relations between Stern and Einstein presented in this paper is based, to a large extent, on Stern's recollections expressed in two interviews from the early 1960s, with Res Jost in 1961 [1] and with Thomas S. Kuhn in 1962 [2].

H. Gutfreund (✉)
The Hebrew University of Jerusalem, Jerusalem, Israel
e-mail: mshanoch@mscc.huji.ac.il; hanoch.gutfreund@mail.huji.ac.il

© The Author(s) 2021
B. Friedrich and H. Schmidt-Böcking (eds.), *Molecular Beams in Physics and Chemistry*,
https://doi.org/10.1007/978-3-030-63963-1_6

Stern vividly remembered his first impression of Einstein:

> I expected to meet a very learned scholar with a large beard, but found nobody of that kind. Instead, sitting behind a desk was a guy without a tie who looked like an Italian road-mender. This was Einstein. He was terribly nice. In the afternoon he was wearing a suit and was shaven. I had hardly recognized him.

Einstein's time in Prague is best known as a formative chapter in his journey towards the General Theory of Relativity. In Prague he wrote 11 papers, 6 of which were devoted to relativity. Yet, he kept an open eye on the growing interest in the applications of quantum theory in solid state physics and in physical chemistry. He pioneered this development in 1907 with his paper on the specific heat of solids [3]. Although there were four local academic institutions with full physics professors, they were all doing classical work and Einstein had no one to talk to about the contemporary issues on the agenda of physics. Thus, there could not have been a better time for Stern to join Einstein. Despite the difference in background and experience they had a lot to talk about. They were both interested in molecular theory, both understood and appreciated the work of Boltzmann and both believed in the reality of atoms and molecules, which was not so common in those days.

Stern recalled that already then he realized that Einstein was one of the few contemporary physicists who insisted that thermodynamics was absolutely funda-mental, one of the few parts of physics that could never be changed. This observation is in accord with Einstein's own recollection about his fascination with thermody-namics in the early stage of his scientific career. In his *Autobiographical Notes*, we find a special reference to this domain of classical physics [4]:

> A theory is the more impressive the greater the simplicity of its premises, the more different kinds of things it relates, and the more extended its area of applicability. Hence the deep impressions it made upon me. It is the only physical theory of universal content which I am convinced will never be overthrown, within the framework of applicability of its basic concepts.

Stern's sojourn with Einstein in Prague lasted one semester. In August 1912, Einstein accepted an appointment at the Swiss Federal Institute of Technology (ETH) in Zurich. He invited Stern to join him there as his scientific assistant. About fifty years later, Stern summarized his first post-doctoral experience:

> With Einstein in Prague – this was a decisive element in my scientific development. I was then introduced to the real problems of those times.

2 Interacting with the Stars at ETH

In 1911, the mathematician Marcel Grossmann was appointed dean of the mathematics-physics department of ETH. One of the first initiatives as dean was to ask Einstein if he would be interested to return to Zürich, where he already held an academic position before going to Prague. At that time Einstein had already under-stood that in order to make progress in his effort to formulate a general theory of relativity, he needed mathematical methods which he did not know then. In Zürich, Grossmann became his mentor on tensor calculus and Riemannian geometry, and his partner and coauthor of a preliminary version of the general theory of relativity.

In spite of working almost exclusively on the general theory of relativity, Einstein was always willing to discuss issues and questions that Stern brought to his attention. The close relationship between them, which started in Prague, continued in Zürich. Stern would occasionally retreat into Einstein's office, because that was the only place in the institute where one was allowed to smoke and Einstein used to visit him in his laboratory. On such occasions they had lively discussions on the unsolved problems of quantum theory. They even wrote a paper together (see Chap. 5).

In Zurich, Stern also benefited greatly from interactions with other physicists, who were familiar with problems and debates at the forefront of research in physics. He enjoyed discussions with Paul Ehrenfest who spent an extensive visit at ETH with his wife Tatjana around the fall of 1913. Ehrenfest was also interested in the Nernst heat theorem, a topic that was on Stern's mind then and for years to come. It seems that they thought well of each other and Ehrenfest sent him a reprint of his volume on Statistical Mechanics with an inscription.

In Zürich, Stern met Max Laue (not yet von Laue) who had just made his monumental discovery of X-ray diffraction from crystals. Their acquaintance evolved into a lifelong friendship. Stern and Laue shared the same opinion on Bohr's model of the atom published in 1913. They were shocked by the departure from everything they learned in physics, implied by that model. To express their dismay, they vowed that "if this nonsense of Bohr should prove to be right in the end, we will quit physics." Stern recalled that Einstein had a better insight and foresight than they had: "Einstein mentioned to me that he had thought about something like Bohr's atom himself."

Stern's formal education in physics was a standard general physics course. This was greatly enriched by faithfully attending Einstein's lectures in Prague and in Zürich. Einstein's teaching in Zürich was mainly done in the colloquium, which was attended at various times also by Laue and Ehrenfest. Einstein did not prepare his lectures, but just talked at the board, sometimes getting lost in the argument or even making mistakes. The lectures were not good for ordinary students but were very stimulating for the better students who could see Einstein's mind at work. What Stern learned from those lectures was not to be ashamed of making mistakes and to be always ready to admit them. The other feature that he then learned from Einstein, which served him throughout his scientific life was "Querdenken" ("lateral thinking")—a kind of creative approach to solving problems via reasoning that is not immediately obvious. Today, one would refer to it as "thinking out of the box."

3 The "Zero-Point Energy" Paper

In 1911 Walter Nernst initiated a program of measuring the specific heat of solids. Results of measurements at sufficiently low temperatures showed a tendency of the specific heat to vanish as the absolute temperature approached zero. This was in accord with Einstein's specific heat formula, derived by applying Planck's quantum theory to atomic oscillations in a solid. This formula predicts that the specific heat of all solids vanishes at zero absolute temperature.

At sufficiently high temperatures, the specific heat of all elements in the solid state is the same—expressed by the classical Dulong-Petit law. Nernst wanted to extend these results to account for the contribution of rotational degrees of freedom to specific heat. To this end, Arnold Eucken, working in his laboratory, measured the specific heat of hydrogen [5].

Einstein was interested in Eucken's results, thinking that they might help to clarify another issue of basic importance. In 1911 Max Planck modified his black-body radiation theory [6], assuming a system of oscillators of frequency ν, which absorb energy continuously but emit energy in discrete energy units of $h\nu$. In Planck's original theory, the average energy of an oscillator, at temperature T, was:

$$E = \frac{h\nu}{\exp(h\nu/kT) - 1}.$$

In what is known as Planck's "second quantum theory," the radiation distribution is unchanged, but there is an additional energy term at all temperatures and particularly at zero temperature, which is referred to as "zero-point energy":

$$E = \frac{h\nu}{\exp(h\nu/kT) - 1} + \frac{h\nu}{2}$$

Einstein was looking for ways of detecting the existence of zero-point energy in physical phenomena. He believed that it may be reflected in Eucken's results on the specific heat of hydrogen. He engaged his assistant, Otto Stern, in this effort and shortly after arriving in Zürich, they published a joint paper [7].

The paper begins with the assumption that the mean kinetic energy of rotation acquired by the molecules under the influence of radiation must be equal to the mean kinetic energy acquired by collisions with other molecules. Hence, the question is for what mean value of rotational energy will a diatomic molecule be at equilibrium with radiation at a given temperature. Making the simplifying assumption that all molecules rotate at the same frequency ν, their rotational energy $E = J(2\pi\nu)^2/2$ has to be equal to Planck's expression for the radiation energy. As a result, the rotational frequency becomes temperature dependent and the specific heat will depend on the presence of the zero-point energy term. Einstein and Stern concluded that the specific heat calculated with the zero-point energy term agrees quite well with Eucken's measurements. When this term was omitted the result was very different from the measured curve.

The second part of the paper is a new derivation of Planck's radiation formula, including zero-point energy, based on a method used previously by Einstein and Hopf [8] to derive the classical Rayleigh-Jeans radiation law. This part of the paper was assigned to Stern. He recalled this episode as an unpleasant confrontation with Einstein. Stern concluded that the zero-point energy term was $h\nu$ and not $h\nu/2$.

Einstein thought that this was impossible and instructed him to redo the calculation. When he got the second time the same answer, Einstein was annoyed and decided to do the calculation himself and got the same result. In the paper there is a footnote stating that this discrepancy should be resolved in a more rigorous calculation.

The paper by Einstein and Stern generated broad interest and stimulated theoretical and experimental work. However, a short time after its publication Einstein announced that he and Stern encountered contradictions in their treatment of the specific heat of hydrogen and that their results are untenable. At the second Solvay Conference in October 1913 he even denounced the notion of zero point energy. In the interview with Res Jost, Stern refers to this paper as nonsense. I do not know how many "wrong" papers had Stern published in his scientific career. For Einstein, this was neither the first one, nor the last.

There was no way to treat correctly the specific heat of hydrogen before the distinct features of ortho-hydrogen and para-hydrogen were known.

It should be noted that the concept of zero-point energy remains a rigorous consequence of quantum mechanics.

In spite of its obvious discrepancies, it is appropriate to remember this paper in the present context because it played a role in the lively debates of those days and contributed to the experience of the young Stern as a novice in the international scientific arena.

4   The Habilitation Process

In German-speaking countries the habilitation process is required to be entitled to lecture at a university and to be eligible for appointment as a professor. The candidate for habilitation has to submit a summary of an original research. With his habilitation application, Stern submitted his paper "On the Kinetic Theory of Vapor Pressure of one-atomic solids", published under a broader title [9].

In this paper Stern calculated the change in entropy from zero temperature to temperatures where the classical molecular theory is valid. The expression of the entropy of a monoatomic gas contains a constant that affects the vapor pressure of the solid phase. This constant plays a fundamental role in the formulation of Nernst's theorem (the third law of thermodynamics). In his calculation, Stern used Nernst's theorem and Einstein's theory of the specific heat of solids. The change of entropy at vaporization was derived applying the classical statistical mechanics of gases, valid at the high temperature of vapor. He finally derived the entropy constant and obtained the same results as have been previously derived by Sackur [10] and Tetrode [11]. But, unlike their derivation, Stern's method was unquestionable. Stern remembered well, in the interview with R. Jost, this first physical-theoretical discovery—the derivation of vapor pressure using molecular theory. He was very proud of this result. Einstein liked this work and urged Stern to submit the habilitation application.

The Habilitation committee was composed of two physicists, Albert Einstein and Pierre Weiss, and a chemist, Emil Baur. In his recommendation to approve Stern's

Habilitation petition [12], Einstein refers to this paper as an entirely independent work. He is quite specific in his evaluation:

> The theoretical determination of the vapor pressure of solids is a problem that acquired great importance on account of Nernst's heat theorem and that many of today's ablest physicists have tackled, though these efforts have not achieved the desired goal. In the past year, Sackur finally found a formula that agreed with experience to within the margins of error, but Sackur's attempt to provide a theoretical foundation for this formula must be considered unsuccessful, because in order to carry out the derivation Sackur had to invoke hypotheses about molecular motion of gases that lacked any justification. Mr. Stern has now succeeded in deriving this formula using the methods of the kinetic theory of gases, without having to resort to any special hypotheses whatsoever.

Einstein concludes:

> This derivation is of lasting value. The method devised by Mr. Stern, which permitted him to achieve his goal in an astonishingly simple way, demonstrates unusual talent.

Emil Baur's opinion [13] is also worth quoting. He concurs with Einstein that the paper submitted by Stern, without any doubt, attests to great talent. He regrets that Stern summarized his work in a short paper rather than presenting his discovery in a monograph summarizing the whole phenomenon of vapor pressure. Baur believes that Stern did not do it because the physicists on the committee wanted to speed up the process so that Stern could join the teaching staff as soon as possible. With this goal he agrees. Baur describes the revival of interest in the classical problems of physical chemistry, basically due to Planck's radiation theory. Baur would like to see a lecture course on this new area of research at ETH and he thinks that Stern satisfies all the conditions to fulfill this task.

This is a remarkable accolade coming from two professionally mature scientists to their younger colleague, only about one year after completion of his doctoral thesis. Stern's application was approved on July 22nd, 1913, and he was awarded the title of *Privatdozent* in early August.

5   Concluding Remarks

In July 1913, Max Planck and Walter Nernst came to Zürich to present to Einstein a tempting proposal—election to the Prussian Academy with generous financial support, directorship of the Kaiser Wilhelm Institute of Physics and professorship at the University of Berlin. Einstein arrived in Berlin in March 1914 to complete his masterwork—The General Theory of Relativity. Stern remained for a short time at ETH and then embarked on his independent scientific odyssey, described in the other contributions to this volume. As a corollary to the Einstein and Stern collaboration, it is worth mentioning their correspondence in 1916, which is a direct extension of their discussions in Zürich. This correspondence ended in a disagreement on the validity of Nernst's heat theorem for solid solution of mixed crystals [14]. Einstein

Fig. 1   Albert Einstein (left) and Otto Stern during a meeting in the mid-1920s

thought that the theorem applies only to pure substances, but he changed his mind when he saw Stern's paper on this issue [15].

This may have been the last serious scientific exchange between Stern and Einstein, but their relation evolved into a lasting friendship. Their paths crossed occasionally in Berlin, see Fig. 1, and in the U.S. Both shared the common fate of many of their peers who became homeless in their homeland when the Nazis came to power and had to rebuild their personal lives and scientific career in a foreign land.

References[1]

1. ETH-Bibliothek Zürich, Archive, http://www.sr.ethbib.ethz.ch/, Otto Stern tape-recording Folder »ST-Misc.«, 1961 at E.T.H. Zürich by Res Jost; AEA, doc. 78-994
2. T. S. Kuhn (ed.), AIP oral history interviews. *Otto Stern Interviewed* (1962)
3. A. Einstein, Die Plancksche Theorie der Strahlung und die Theorie der spezifischen Theorie der Wärme, Annalen der Physik, **22**, 180 (1907); reprinted in CPAE, vol. 2, doc. 38
4. A. Einstein, in *Autobiographical Notes*, ed. by P.A. Schilpp. Albert Einstein: Philosopher-Scientist (Tudor Publishing Co., 1949), p. 33
5. A. Eucken, Die Molekulerwärme des Wasserstoffs bei tiefen Temperaturen (Berliner Berichte, 1912), p. 14
6. M. Planck, Eine Neue Strahlung Hypothese, Deutsche Physikalische Gesellschaft. Verhand-lungen **13**, p. 138, 1911; for a historical discussion of Planck's second theory see Thomas

---

[1] AEA—Albert Einstein Archives

CPAE—Collected papers of Albert Einstein

S. Kuhn, Black-Body Theory and the Quantum Discontinuity, 1894–1912, ch. 10 (Oxford University Press, 1978)

7. A. Einstein, O. Stern, Einige Argumente für die Annahme einer molekularen Agitation beim absoluten Nullpunkt, Ann. Phys. **40**, 551 (1913); this paper is reprinted with an editorial introduction in CPAE, vol. 4, doc. 11

8. A. Einstein, L. Hopf, Statistische Untersuchung der Bewegung eines Resonators in einem Strahlungsfeld. Annalen der Physik, **33**, 1105 (1910); reprinted in CPAE, vol. 3, doc. 8

9. O. Stern, Zur kinetischen Theorie des Dampfdrucks einatomiger festen Stoffe und über die Entropie-konstante einatomiger Gase, Physik. Zeitschr. **XIV**, 629 (1913)

10. O. Sackur, Ann. Phys. **40**, 67 (1913)

11. H. Tetrode, Ann. Phys. **38**, 434 (1912)

12. A. Einstein, *Expert Opinion on the Habilitation Petition of Otto Stern*, CPAE, vol. 5 doc 452

13. E. Baur, Zu dem Habilitationsgesuch der Herrn Dr. O. Stern, AEA, doc. 70-020

14. CPAE, vol. 8, doc. 191, 192, 198, 205

15. O. Stern, Die Entropie fester Lösungen. Ann. Phys. **49**, 823 (1916)

Chapter 7
Our Enduring Legacy from Otto Stern

Daniel Kleppner

Abstract Otto Stern's scientific legacy continues to animate discoveries on a rapidly advancing research frontier.

1 Introduction

Otto Stern's scientific legacy was commemorated and celebrated in 1988 at the centenary of his birth by a Festschrift [1]. In the three decades since then, scholarship has enriched our understanding of Stern's achievements. (See Chap. 5 of this volume). The goal of this essay is to show *how* Stern's legacy has grown through his links with new generations of scientists in Atomic, Molecular, and Optical (AMO) Physics. To keep the discussion tractable, it focuses on AMO's Nobel Laureates.

2 Preface: A View of Otto Stern's Legacy in 1988

For the centenary Festschrift, Norman F. Ramsey provided an overview of Stern's legacy. Ramsey was well-positioned to appreciate that legacy because he had worked with I. I. Rabi, since joining his group as a new graduate student in 1936. Rabi's career at Columbia had been launched by Stern in 1930. Stern was still active in Pittsburgh and Rabi spoke of him often. Ramsey became a leading figure in the world of physics, particularly molecular beam physics, and he was able to recognize Stern's achievements from first-hand knowledge.

In the decades since the Stern centenary, the field of Atomic, Molecular, and Optical Physics was transformed by advances that nobody could have predicted. Nevertheless, Ramsey's overview at the time of the Centennial provides a panoramic summary of Stern's impact on science at the dawn of this revolution in AMO physics. Here it is:

D. Kleppner (✉)
Massachusetts Institute of Technology, Cambridge, MA 02139, USA
e-mail: kleppner@mit.edu

© The Author(s) 2021
B. Friedrich and H. Schmidt-Böcking (eds.), *Molecular Beams in Physics and Chemistry*,
https://doi.org/10.1007/978-3-030-63963-1_7

97

MOLECULAR BEAMS, OUR LEGACY FROM OTTO STERN [2]

by Norman F. Ramsey

1. *Velocity distribution of molecules—Stern and others.*
2. *Space quantization—Stern and Gerlach.*
3. *Spin of electron = 1/2—Stem and Gerlach.*
4. *Anomalous magnetic moment of the proton—Frisch and Stern.*
5. *Nuclear spin measurements—Rabi and others.*
6. *Nuclear magnetic moments (stable and radioactive)—Rabi, Nierenberg and others.*
7. *Deuteron quadrupole moment and nucleon tensor force—Kellogg, Ramsey, Rabi and Zacharias.*
8. *Molecular beam electric and magnetic resonance methods—Rabi and Ramsey.*
9. *Anomalous magnetic moment of the neutron—Bloch and Alvarez.*
10. *Lamb shift in hydrogen hyperfine structure and quantum electrodynamics—Lamb and Retherford.*
11. *Anomalous H hyperfine structure and relativistic quantum electrodynamics—Nafe and Nelson.*
12. *Anomalous magnetic moment of the electron—Kusch.*
13. *Nuclear octupole moments—Zacharias and others.*
14. *First masers—Gordon, Zeiger and Townes.*
15. *Cs atomic clocks—Rabi, Zacharias, Essen, Ramsey and others.*
16. *Atomic hydrogen maser—Ramsey and Kleppner.*
17. *Accurate H, D and T hyperfine structure—Ramsey, Kleppner, Crampton and others.*
18. *Accurate atomic magnetic moments of H and D and reduced mass corrections—Ramsey, Valberg and Larson.*
19. *Rotational Magnetic moments of molecules—Stern, Ramsey and others.*
20. *Nuclear data—Magnetic moments, quadrupole moments and octupole moments.*
21. *Molecular data—Rotational moments, spin- rotational interaction, spin-spin interactions, quadrupole moment of molecules, orientation dependence of susceptibilities, etc.*
22. *Atomic scattering cross sections.*
23. *Reaction cross sections.*
24. *Van der Waals molecules.*
25. *Highly excited and Rydberg atoms.*
26. *Multiphoton atomic beam spectroscopy.*
27. *Jet sources and cluster beams.*
28. *Laser spectroscopy, excitation and detection of molecular beams.*
29. *Chemistry in detail—State to state reaction studies.*
30. *Measurement of parity non-conservation.*
31. *Laser cooling and trapping.*
32. *Tests of time reversal symmetry.*

3 Portraying Our Enduring Legacy Today

The "Our" in "Our Enduring Legacy" are members of the scientific community that is rooted in the work of Otto Stern. In the United States and abroad the community is known as Atomic, Molecular, and Optical Physics (AMO Physics). Stern's seminal research launched AMO Physics and his legacy continues to nourish it.

Portraying Stern's influence is a formidable challenge. In 1967 Rabi's influence was summarized by the Rabi Tree, an illustration of all the researchers influenced by Rabi [3]. At the roots of the Rabi Tree, prominently displayed, is the name *Otto Stern*. Considering the explosive growth of atomic physics in recent decades, a Stern Tree, with its additional trunks in chemistry and nuclear physics, would be intractable.

We shall summarize Stern's influence by showing connections from his work to the Nobel Prize laureates in AMO physics. The Nobel Prize is generally agreed to honor important advances, although focusing on it neglects important achievements that did not happen to have been awarded the Prize. Nevertheless, many of the laureates are linked to Stern by student-teacher or colleague-colleague experiences. It is difficult to think of stronger evidence for the value of Stern's legacy to science.

The achievements of the Nobel Laureates are well-documented elsewhere and will not be stressed here. The comments below focus on the laureates' personal links to Stern, or to Stern/Rabi. In large part, it is by such links that Stern's legacy thrives. A few cases in which no link is evident will be noted.

The origins of the legacy.

As described in Chap. 5 of this volume, Stern's scientific productivity was extraordinary. At the time that his work was interrupted by political interference in 1933, Stern's achievements included:

- The first measurements of molecular speeds
- The Stern-Gerlach experiment demonstrating spatial quantization
- The demonstration of recoil from absorption of a photon
- The first demonstration of atom diffraction
- Discovery of the anomalous magnetic moment of the proton
- Discovery of the magnetic moment of the neutron

This sequence of discoveries reveals Stern's uncanny ability to identify important problems. In addition, he obviously had great stamina and powers of concentration, what in German is called *Sitzfleisch*. According to his student and colleague Otto Frisch [4], Stern also had great talent for enjoying life. (See also the personal testimonials by his niece and great-nephew in Chaps. 3 and 4 of this volume.)

Stern's first position was as assistant to Einstein in Prague and Zurich from 1912 to 1914, see also Chap. 6. In his biographical memoir of Otto Stern, Emilio Segrè wrote [5]: "*It was from Einstein that he learned what were the really important problems of contemporary physics: the nature of the quantum of light with its double aspect of particle and wave, the nature of atoms, and relativity.*" Stern and Einstein remained scientific friends long after their careers parted.

Rabi was ever mindful of Stern's legacy and he preserved it for the coming generations. From time to time Rabi's friends—including this author—would hear him talk of it. In paying tribute to Wolfgang Pauli and Otto Stern, Rabi wrote [6]: *"From Stern and Pauli I learned what physics should be. For me it was not a matter of more knowledge. I learned a lot of physics as a graduate student. Rather it was the development of taste and insight; it was the development of standards to guide research, a feeling for what was good and what is not so good. Stern had this quality of taste in physics and he had it to the highest degree. As far as I knew, Stern never devoted himself to a minor question."*

In 1933 Stern's research was terminated by the Nazi regime. He was forced to abandon his University and his culture and flee his country. He was appointed as a professor at Carnegie Institute of Technology where he worked with Immanuel Estermann, one of his first Ph.D. students and a life-long collaborator. Stern's highly productive years had come to an end but his legacy continued to grow, nourished in part by his post-doctoral associate at Hamburg, I. I. Rabi.

Rabi met Stern when he visited the University of Hamburg for over a year in 1927–1929. He owed his research career to that time with Stern. He had planned to work with Wolfgang Pauli but happened to meet Stern and became interested in Stern's work. Rabi suggested an approach to magnetic deflection that avoided the magnetic gradients which bedeviled the Stern-Gerlach experiment. Stern invited the would-be theorist to try out his idea in the laboratory. It worked! Rabi was elated and his career abruptly headed in a new direction. He was appointed to a junior position at Columbia University and in 1930 he started a research program there.

From today's perspective, Stern and Rabi seem to be almost a single force. Rabi started building his research program at Columbia less than two years after his visit with Stern. Three years later Stern had to flee to his country. Stern's mainline research never recovered its momentum but as Stern's program was slowing, Rabi's was gathering speed. Rabi's research centered on the spins and magnetic moments of nuclei, atoms and molecules and on fundamental issues in the quantum properties of atoms—as had Stern's. Rabi's creative style, his experimental designs and his sense of scientific fitness were all evocative of Stern's.

Rabi made his major discovery—molecular beam magnetic resonance—in 1937 [7]. A second major discovery, the quadrupole moment of the deuteron, followed within two years [8]. Then, in 1940, war in Europe disrupted the research. In November, Rabi left Columbia to become the scientific director of the program to develop radar at the MIT Radiation Laboratory (the Rad Lab), bringing to a close a decade of innovative physics.

Stern and Rabi are viewed collectively here because their research created a continuous narrative. Because of Stern's influence on Rabi, and Rabi's deep appreciation for Stern's teachings and innovations, their legacies in AMO physics often meld.

4 The Nobel Prizes of Stern and Rabi

A natural place to launch this narrative is with the Nobel Prize awards to Stern and Rabi themselves.

1943 Nobel Prize: Otto Stern *"for his contribution to the development of the molecular ray method and his discovery of the magnetic moment of the proton."*

Otto Stern received the 1943 Nobel Prize and the following year the Nobel Prize was awarded to Rabi. The Prizes were awarded in New York City in 1944 at the same ceremony. As described in Chap. 5, Stern received an overwhelming number of Nobel Prize nominations—more than any previous winner. Considering his many significant scientific achievements, this is hardly surprising. Curiously, the Stern-Gerlach experiment is not mentioned in the citation. During Stern's career a revolution in physics was underway and one senses that the Nobel Committee was overwhelmed by the barrage of epochal discoveries and the confusions of the onslaught of perplexing new knowledge.

1944 Nobel Prize: Isidor Isaac Rabi *"for his resonance method for recording the magnetic properties of atomic nuclei."*

Isidor I. Rabi's Nobel Prize was awarded quickly following his discovery of molecular beam magnetic resonance in 1937. At that time one could point to two significant achievements: magnetic resonance and discovery of the deuteron quadrupole moment. The Prize Committee was prescient to realize the vast potential of magnetic resonance. Rabi's discovery of molecular beam magnetic resonance led to the creation of powerful new tools for atomic, molecular, and nuclear physics and his ideas diffused into adjacent fields. Rabi's enormous impact on science has been well documented [9]. He emerged from the war years as a statesman of science and his career as a statesman was extraordinary.

In the decade following the war Rabi conceived and led the creation of Brookhaven National Laboratory (with some help from Norman Ramsey [10]) and sparked the creation of CERN [11]. He also sparked the creation of the President's Science Advisory Committee (PSAC) which was influential for several presidencies, and he steered the U.S. national policy to keep nuclear technology under civilian control and he spent considerable effort trying to achieve international control.

Rabi's career as a statesman left him little time for basic research. After the war his single paper on fundamental physics was on the hyperfine structure of hydrogen (discussed below). Nevertheless, Rabi's impact on the Columbia Physics Department was enormous. The judgments about people and physics that he exercised while leading the Physics Department led to the discoveries of the Lamb Shift, the anomalous magnetic moment of the electron, the creation of the theory of relativistic quantum electrodynamics (QED) and the creation of the laser, and garnered seven Nobel prizes for work carried out at Columbia [12].

5   Links Connecting the AMO Nobel Laureates to Otto Stern

1952 Nobel Prize: Felix Bloch and Edward Mills Purcell *"for their development of new methods for nuclear magnetic precision measurements and discoveries in connection therewith."*

The invention of nuclear magnetic resonance was an early spin-off from Stern's measurement of the magnetic moment of the proton and Rabi's invention of the magnetic resonance technique.

Felix Bloch had a distinguished early career, having studied with Peter Debye at ETH in Zurich and Heisenberg in Leipzig. His career was interrupted when he had to flee from Germany in 1933. He was appointed to the faculty at Stanford University and became interested in Stern's discovery of the neutron magnetic moment. He designed a neutron beam magnetic resonance experiment which employed spin polarizers and analyzers that used magnetic scattering rather than Stern-Gerlach magnets. Rabi was visiting Stanford and the two worked together on preliminary version of the experiment. Its results were published in 1940 [13]. In the course of this he became interested in using magnetic resonance to measure magnetic fields. The proton magnetic moment was then known and Bloch realized that a sample of protons—for instance in water—would have magnetic susceptibility and that in a magnetic field there would be a significant difference in the spin-up and spin-down populations. A radio frequency field would cause transitions, creating a rotating polarization that would induce a current in a conducting loop. The frequency that induced the current would reveal the value of the field.

Edward Mills Purcell, then an assistant professor at Harvard, spent the war years at the MIT Radiation Laboratory (the Rad Lab) working on radar. Rabi was the scientific director of the Rad Lab and they interacted frequently. At the end of the war, Rabi asked Purcell and other experts to stay on to document their work. It was during this period that Purcell, working in the evenings with borrowed equipment, demonstrated Nuclear Magnetic Resonance. The magnetic moment of the proton was well known and Purcell realized that a mass of the protons would exhibit bulk magnetization that would absorb power at the resonance frequency. He detected absorption almost simultaneously with Bloch's discovery and they announced their discoveries in side-by-side abstracts at the 1946 spring meeting of the American Physical Society (APS), the first post-war meeting [14, 15].

At that meeting they realized that Bloch's Nuclear Induction and Purcell's Nuclear Magnetic Resonance were essentially identical. They had amiable personalities and in discussing their findings they agreed that it would be best to have a single name. They agreed on *Nuclear Magnetic Resonance* and the acronym NMR entered the lexicon of science.

1955 Nobel Prize: Polycarp Kusch *"for his precision determination of the magnetic moment of the electron"*, and **Willis Lamb** *"for his discoveries concerning the fine structure of the hydrogen spectrum."*

Suspicions that there were flaws in the Dirac theory of the electron inspired a series of AMO experiments at Columbia University in the late 1940s, where Rabi chaired the Physics Department. The results led to the creation of the theory of relativistic quantum electrodynamics (QED) shortly after the war—widely regarded as a triumph for Physics.

Three experiments were pursued in this quest.

1. The anomalous moment of the electron: Kusch and Foley [16]
   The magnetic moment of the electron was believed to be exactly one Bohr magneton. Detection of an anomaly—a departure from unity—would pose a fundamental problem in the theory. The Kusch-Foley experiment discovered such an anomaly. Polycarp Kusch had joined Rabi's group in 1937 and Rabi later appointed him to the Columbia faculty. With Henry M. Foley, Kusch carried out atomic beam magnetic resonance on three different atoms that had the same total angular momentum but different combinations of spin and orbital angular momentum. By studying radiofrequency resonances in a fixed magnetic field, they discovered a small anomaly and measured it to a precision of about 4%.

2. The Lamb Shift: Willis Lamb [17]
   According to the Dirac theory the energy levels in hydrogen with the same total angular momentum have the same energy. Willis E. Lamb showed that the energies of $2S_{1/2}$ and $2P_{1/2}$ states were not identical. The energy difference is known as the Lamb shift.
   In 1947 Rabi attracted Willis E. Lamb to the Columbia Physics Department from Berkeley where he had been working with Robert Oppenheimer. Somewhat to Rabi's surprise, Lamb capitalized on his experience developing microwave technology during the war and designed and executed an experiment. Using an atomic beam of metastable hydrogen atoms in the $2S_{1/2}$, and working with graduate student James Retherford, he observed the transition: $2S_{1/2} \rightarrow 2P_{1/2}$.

3. The Hyperfine Energy of Hydrogen: Rabi et al. [18]. The hyperfine energy of hydrogen depends on the product of the magnetic moments of the electron, proton, as well as other accurately known factors. A precision measurement of the hyperfine transition frequency would provide an independent value for the magnetic moment of the electron.
   Although the measurement did not add significant new knowledge, it had an important impact—it convinced Julian Schwinger to become engaged with the problem. Schwinger had been an undergraduate prodigy when Rabi brought him to Columbia before the war and mentored him. Schwinger received his undergraduate degree and also completed his Ph.D. thesis at the age of 19. He worked with Oppenheimer at Berkeley before the war and at the Rad Lab during the war. Schwinger then joined the faculty at Harvard where he returned to the problem of the anomalous magnetic moment of the electron. In 1965 he shared

the Nobel Prize in Physics with Richard P. Feynman and Sin-Itoro Tomanaga for creating Relativistic Quantum Electrodynamics to account for the values of the electron moment anomaly and the Lamb shift.

The 1964 Nobel Prize: Charles Hard Townes, Nikolay Basov and Alexander Prokhorov *"for fundamental work in the field of quantum electronics, which has led to the construction of oscillators and amplifiers based on the maser-laser principle."*

The maser preceded the laser and provided the foundation for its invention. The invention of the laser advanced essentially every branch of science and it transformed society. The maser was a new type of molecular beam resonance device.

In 1947 Rabi persuaded Charles Townes to join Columbia's physics faculty. Townes received his Ph.D. from California Institute of Technology in 1939 and joined the staff at Bell Laboratories. Townes appointed Arthur Schawlow to his postdoctoral staff and they co-authored the magisterial monograph *Microwave Spectroscopy* [28]. Townes had a particular interest in detecting the ammonia molecule by microwave spectroscopy of its inversion line, about 23 GHz. He employed a molecular beam with an electrostatic state separator, essentially the first half of a molecular beam resonance apparatus. He conceived the idea of observing a resonance transition by detecting the energy the molecules radiated as they passed through a resonator tuned to the molecular resonance. The operation of the maser was reported in 1955 [19].

N. G. Basov and A. M. Prokhorov also published a proposal for a similar device although few details are available [20].

As an amplifier, the maser found applications in radio astronomy and it inspired Ramsey's creation of the hydrogen maser, a device employed in frequency control laboratories and in GPS systems. The biggest impact of the maser is that it inspired Townes and Schawlow to propose a maser that could operate at optical frequency—the laser [21].

Nobel Prize 1966: Alfred *Kastler* *"for the discovery and development of optical methods for studying Hertzian resonances in atoms."*

The invention of optical pumping created a major new stream of AMO physics. The work of Stern and Rabi helped Kastler to develop optical pumping and its first application: optical double resonance. One of Kastler's early papers is entitled (in French) *Some suggestions concerning the production and detection by optical means of inequalities in the populations of levels of spatial quantization in atoms. Application to the Stern and Gerlach and magnetic resonance experiments* [22].

Alfred Kastler was born in Alsace in 1902 and studied at École Normale Superieur (ENS) from 1931 to 1936. His career started as a teacher in lycées in Alsace and Bordeaux. He became engaged in optics and spectroscopy and the transfer of angular momentum with circularly polarized light. This led him to conceive the idea of polarizing atomic nuclei by successive absorption of polarized photons [23]. Kastler became a professor at Bordeaux in 1938 and in 1941 he was invited to ENS to help establish the physics teaching program.

In 1945 Kastler was approached by a young ENS graduate, Jean Brossel, who asked to pursue research with him. Brossel had entered ENS in 1938 and spent two years in the Army before returning to finish his studies.

Kastler had had a correspondence with Francis Bitter, a professor at the Massachusetts Institute of Technology best known for his creation of high magnetic fields, and Kastler asked him if he could take Brossel into his laboratory for thesis research. Bitter agreed and Brossel started with him in 1948. During his time abroad, Brossel and Kastler kept in touch by frequent correspondence. (The correspondence is preserved in the MIT Bitter archives). Brossel demonstrated nuclear polarization by the successive absorption of circularly polarized photons, soon to be named *optical pumping*. It was first observed using a simple atomic beam of mercury. Later, at ENS, they discovered that the nuclear polarization is stable against gaseous and surface collisions, allowing the effect to be observed in a glass cell rather than a molecular beam. This enormously simplified its usage.

A technique for polarizing and analyzing atoms provides a natural platform for magnetic resonance. The technique is called *double resonance* and the possibility was recognized early in the Kastler-Brossel collaboration.

During his time at MIT Brossel developed the complete theory of double resonance [24]. He received his Ph.D. for this work shortly after returning to Paris in 1951.

The invention of optical pumping and double resonance opened a new branch of atomic physics. The Stern/Rabi methods center on interactions of atoms and molecules with magnetic fields while optical pumping centers on their interactions with light. This encompasses a much broader range of phenomena including light-induced energy level shifts, multiphoton processes and quantum optics.

There are no direct links between Kastler and Rabi although Bitter knew both of them. He corresponded extensively with Kastler, and he remained a close friend of Rabi after their graduate student days at Columbia. Bitter invited Kastler to visit MIT and arranged an invited talk at an APS meeting, which Kastler accepted, but the state Department denied him a visa. The U.S. was suffering a "red scare" and Kastler had been in a left-leaning organization. His visit to the U.S. never took place.

In viewing the scientific heritage of AMO physics, one sees Kastler standing alongside of Stern and Rabi.

Nobel Prize 1965: Sin-Itiro Tomonaga, Julian Schwinger and Richard P. Feynman *"for their fundamental work in quantum electrodynamics, with deep-ploughing consequences for the physics of elementary particles."*

None of these theorists would be identified as members of the AMO community although the overwhelming experimental evidence that led them to create their theories of relativistic quantum electrodynamics (QED) all came from Columbia under Rabi's reign. Furthermore, Schwinger's engagement with QED was directly due to a Rabi experiment.

Julian Schwinger had been an undergraduate prodigy when Rabi brought him to Columbia, see above. After the war Schwinger joined the faculty at Harvard.

The Rabi, Nafe, and Nelson experiment started just as the war ended and the first results, though not definitive, were strong enough to cause Schwinger to start work on his theory of QED [25].

Nobel Prize 1981: Nicolaas Bloembergen and Arthur Leonard Schawlow, *"for their contribution to the development of laser spectroscopy."*

Nicolaas Bloembergen studied at the University of Utrecht for two years before emigrating to the United States to work with Purcell at Harvard University. He arrived shortly after NMR had been discovered by Purcell, Torrey and Pound [14] and became interested in nuclear relaxation. The results of his work with Purcell and Pound led to the publication "Nuclear Relaxation," [26] which became a citation classic. When Townes reported operation of the ammonia beam maser [19], Bloembergen recognized that the essential element of maser operation was an inverted population and that many other systems should be capable of displaying this. He chose an ionic crystal system to illustrate his ideas, using microwave pumping to invert the populations. The solid-state maser he proposed was realized and became a useful tool for radio-astronomy, including the discovery of the cosmic background radiation by Penzias and Wilson.

When Townes and Schawlow published their analysis of an optical maser—the laser [21]—Bloembergen realized that operations must always involve the nonlinear response of a medium to the incident radiation. Nonlinear optics became the central theme of his research career and it revealed a cornucopia of new effects: optical doubling, three-and four-wave mixing, parametric generation, high-harmonic generation, line narrowing methods. His entire career was at Harvard in the Division of Engineering and Applied Physics, close to Ramsey and Purcell.

Arthur L. Schawlow received his graduate degree in molecular spectroscopy from the University of Toronto and joined Townes at Columbia in 1949. They worked together on microwave spectroscopy of molecules, work which was summarized in what became the classic monograph on the subject, *Microwave Spectroscopy* by Townes and Schawlow [27]. When Townes invented the maser [19], he was interested in extending its operation to shorter wavelengths and he and Townes together wrote a paper proposing how to do this [21]. The short wavelength maser was soon renamed the laser: this paper launched its creation.

In 1961 Schawlow joined Stanford University and started a program in laser spectroscopy with a young colleague, Theodor W. Hänsch. Previously, spectroscopy was carried out with incoherent light sources—thermal sources of gaseous discharges. Laser light is coherent and tunable, providing vastly improved resolution and a tool for investigating previously inaccessible states and, eventually manipulating the atoms themselves. They rapidly made the laser a practical research tool, inspiring new research, launching Hänsch in a lifetime career of ever-increasing precision and innovations in optics.

1989 Nobel Prize: Norman Ramsey *"for the invention of the separated oscillatory fields method and its use in the hydrogen maser and other atomic clocks."*

1989 Nobel Prize: Hans G. Dehmelt and Wolfgang Paul *"for the development of the ion trap technique."*

In September 1936, Norman Ramsey joined Rabi's group as a graduate student. (Rabi famously tried to discourage him on the grounds that the interesting things with molecular beams had essentially all been done. A few months later, Rabi discovered molecular beam magnetic resonance [28].) Among the group's most important discoveries was that the deuteron has a quadrupole moment, in which Ramsey played a major role [29]. After the war he helped Rabi found Brookhaven National Laboratory and he served as its first Head of the Physics Department. He started a group in molecular beam research whose summer schools eventually morphed into the International Conference on Atomic Physics (ICAP). This meeting continues today, providing an ongoing monument to the vitality of Otto Stern's heritage.

In 1947 Ramsey joined the faculty at Harvard, where he remained for the rest of his career. His class in molecular beams educated generations of graduate students and his monograph *Molecular Beams* [30] became the standard text on that topic. The book is noteworthy for its attention to Stern's work.

The separated oscillatory field method: [31] In 1950 Ramsey invented the separated oscillatory fields method, a technical advance that improved the accuracy of molecular beam magnetic resonance for his studies of magnetic interactions in molecules. This topic remained at the core of his research throughout his long career. The method also extended the Rabi method to high frequency, enabling the creation of the first atomic clock—the cesium beam clock—which remains in use until today [32] and has numerous metrological applications. Figure 1 shows Ramsey together with Rabi in 1959.

In recent years, a different aspect of the separated oscillatory field method has been recognized: In the region between the oscillating fields the atom can exist in an entangled state, thus providing a tool for research in quantum optics and quantum information theory.

The Hydrogen Maser: [33] Increasing the precision of a quantum measurement of energy or frequency, such as in an atomic clock, requires increasing the measurement time. Ramsey hit upon the idea of *storing* the atoms during the measurement process by confining them in some sort of container. The goal for creating the maser was to confirm the effect of gravity on the rate of a clock, which was eventually achieved.

Hans G. Dehmelt was a student of Hans Kopfermann at the University of Göttingen. Dehmelt initially studied NMR problems based on Bloch and Purcell's work as well as the magnetic resonance techniques of Rabi and Kastler. He moved to the University of Washington at Seattle and innovated techniques for trapping charged particles based on the radiofrequency trapping techniques developed by Paul as well as static magnetic-electric confinement. He refined his methods to the point where he could observe a single ion and trap a single electron, a "mono electron oscillator." [34] This initiated single particle spectroscopy and opened the way to a measurement of the electron magnetic moment to an accuracy of 0.28 parts per trillion [35], which remains the most precise measurement achieved in physics.

Fig. 1 Norman Ramsey (left) and Isidor Rabi at the Brookhaven Conference on Molecular Beams held in Heidelberg, Germany, in June 1959

Wolfgang Paul was also a graduate student of Kopfermann and moved with him to Göttingen where there was an active molecular beams group. Detecting atoms and molecules was a perpetual problem for molecular beam physics. Paul invented a mass spectrometer based on static and oscillating electric fields which provided high mass resolution and high efficiency. He went on to develop methods for trapping ions in oscillating fields—the "Paul Trap." The trap was useful for the spectroscopy of ions and was employed in the first observations of a single particle. He also developed the Penning trap which was used by Dehmelt and Gerald Gabrielse to probe the limits of QED through measurements of the magnetic moment of the electron [35].

1997 Nobel Prize: William D. Phillips, Claude Cohen-Tannoudji and Steven Chu *"for development of methods to cool and trap atoms with laser light."*

Observation of Bose-Einstein condensation (BEC) in an atomic gas was announced in the summer of 1995 [36, 37]. The achievement was immediately recognized as a major discovery and was awarded the Nobel Prize in 2001. In anticipation of that award, the 1997 Prize was awarded for the breakthrough that made the discovery possible—laser cooling, an optical technique for cooling atoms to unbelievably low temperature.

The history of laser-cooling constitutes a saga of experimental physics that is narrated in the Nobel Prize lectures of the laureates: Phillips [38], Cohen-Tannoudji [39], and Chu [40]. Principal events include:

- The demonstration of atom-slowing by laser light by William D. Phillips and Harold Metcalf and the discovery of excess cooling [41].
- The demonstration of three-dimensional cooling by Steven Chu [42].
- The theory for the unexpected cooling mechanism, "Sisyphus cooling", by Claude Cohen-Tannoudji [43].

Claude Cohen-Tannoudji was a student of Alfred Kastler and was deeply immersed in theory and experiment in optical pumping and optical double-resonance from the start of his research career. Previously, in magnetic resonance phenomena the oscillating field was treated classically, following Rabi's approach. Early in his career Cohen-Tannoudji developed, with assistance from Serge Haroche, a quantum theory for the atom and field, the "dressed atom" theory [44]. This provided a new language for describing magnetic resonance and the interactions of atoms with electromagnetic fields. The dressed atom theory ultimately explained and guided the development of laser cooling, including the surprising "Sisyphus effect."

Steven Chu was a graduate student of Eugene D. Commins at Berkeley: Commins did his Ph.D. research in Rabi's group at Columbia. Chu joined the staff at Bell Laboratories and became interested in Arthur Ashkin's research on manipulating small particles with light. (Ashkin received the Nobel Prize for this work in 2018, see below.) Chu extended the research to manipulating atoms with light. He joined the faculty at Stanford University and, with Schawlow, devised a method for reducing the speed of atoms by using laser light tuned slightly below the resonance frequency. The Doppler shift would retard the motion of atoms approaching the laser. In a standing wave, the motion would be opposed in either direction. In three perpendicular standing waves, *all* motion would be retarded [41].

Such a gas was called "optical molasses" because atoms behaved as if they were in a viscous medium. This technique was key to the cooling schemes that ultimately achieved BEC.

William D. Phillips did his graduate research at MIT with me: I was a student of Ramsey. Phillips, disregarding his advisor's advice, took a position at the National Institute of Science and Technology (NIST) rather than a university. At NIST he developed a research group that studied light forces and atom slowing. He carried out the first demonstration of atom slowing by laser light in an experiment with Harold Metcalf: An atomic beam of sodium was retarded by a laser beam tuned to the principal transition. As the atoms slowed, their resonance wavelength shifted due to the Doppler effect, but by applying a tailored longitudinal magnetic field the Zeeman energy shift effect compensated the Doppler shift for the length of the apparatus [45]. The atoms were slowed—their motion could even be reversed—though in one dimension only. Nevertheless, this set the stage for laser-cooling.

After optical molasses had been discovered, Phillips developed a method for measuring the temperature of the atom cloud. He turned off the confining radiation causing the cloud to drop and imaged the expansion. The temperature found was significantly *lower* than theory predicted. This discrepancy led Cohen-Tannoudji to develop the theory of Sisyphus cooling.

A postscript on Otto Stern and laser cooling: The scientific legacy of Otto Stern animates the history of laser cooling, even though the direct connection was not appreciated until after the discovery. The roots of atom cooling lie in Einstein's 1917 paper on radiation. The first part of the paper introduces the concepts of absorption, stimulated emission and spontaneous emission and the Einstein A and B coefficient. The second part, not as well known, is responsible for the discovery that photons ("light quanta") carry momentum. Einstein showed how a gas of atoms comes into equilibrium with a thermal radiation field by absorbing and emitting radiation, taking into account Doppler-shifts. He proved that equilibrium is possible only if the radiation field is described by the well-known black body thermal distribution, and only if photon carries momentum = energy/c.

Light momentum was exactly the type of phenomenon that attracted Stern because of its underlying fundamental nature, although its detection would be extremely difficult. Nevertheless, he searched for the deflection of an atomic beam of sodium that was transversely irradiated by light from a sodium discharge. The deflection was minute but was observed by Otto Robert Frisch and Stern in the final moments of Stern's Hamburg laboratory (see Chap. 5). Stern omitted his name from the publication likely to assist Frisch in his search for a new position.

2001 Nobel Prize in Physics: Eric A. Cornell, Wolfgang Ketterle and Carl E. Wieman *"for the achievement of Bose-Einstein condensation in dilute gases of alkali atoms, and for early fundamental studies of the properties of the condensates."*

The histories of the prize winners are of particular interest. Eric Cornell and Carl Wieman worked as a team at the Joint Institute for Laboratory Astrophysics (JILA) of the University of Colorado and the National Institute of Standards and Technology (NIST) in Boulder Colorado. The history of their discovery is described in a joint paper based on their Nobel Prize lectures [46]. Wolfgang Ketterle worked at the Massachusetts Institute of Technology Cambridge, Massachusetts. His Nobel lecture is also published [47].

With respect to the Stern/Rabi heritage, the laureates personal histories reveal some commonalities. Eric Cornell did his graduate research at MIT with David

E. Pritchard, working on high precision mass spectroscopy. At the time Cornell received his Ph.D., Pritchard had become interested in atom cooling and had made some valuable contributions. Cornell was intrigued and went to JILA as a postdoc, where he started collaborating with Carl Wieman. Wieman had worked with me as an undergraduate at MIT and then went to Stanford for graduate training where he worked with Ted Hänsch. Wolfgang Ketterle was a student of Herbert Walther in Garching but he had no experience in atom cooling when Pritchard appointed him to a postdoctoral position in his group. Ketterle quickly revealed talents that called for a faculty position. In a discipline such as AMO physics, the MIT Physics Department does not appoint a junior person to collaborate with a senior faculty member. To resolve the dilemma, Pritchard stepped aside, turning over the laboratory for atom-cooling to Ketterle. To complete the connections: David Pritchard was my graduate student when I was an Assistant Professor at Harvard. (We worked on molecular beam differential spin-exchange scattering.) When I moved to MIT, Pritchard came along to finish his research. We had all been members of Ramsey's group at Harvard, and Ramsey took over formal responsibility for Pritchard. There were few places where Stern's heritage burned as brightly as it did in Ramsey's group.

The search for Bose-Einstein condensation in an atomic gas (BEC) is one of the great scientific adventure stories of twentieth century physics. Laser-cooling was an essential development but that was only one part of the final success. New concepts needed to be created and new technologies needed to be developed. Many groups were involved and many postdocs launched their careers working in the search. Histories of the discoveries of BEC and developments since then have been presented in the Nobel lecturers that are referenced by Proukakis et al. [48].

2005 Nobel Prize in Physics: Roy J. Glauber *"for his contribution to the quantum theory of optical coherence."* (Stern/Rabi links not identified)

2005 Nobel Prize in Physics: John H. Hall and Theodor W. Hänsch *"for their contributions to the development of laser-based precision spectroscopy, including the optical frequency comb technique."*

John L. Hall did his undergraduate, graduate and postdoctoral research at the Carnegie Institute of Technology. In 1962 he went to the Joint Institute for Laboratory Astrophysics (JILA) and dedicated his career to the pursuit of high precision [49]. His influence on the AMO community is widespread through JILA's programs for students and visiting scientists which over the years brought many of today's AMO leaders to Boulder.

Theodor Hänsch graduated from Heidelberg University and pursued graduate research there in laser physics—then in its infancy—with Peter Toschek, a former student of Wolfgang Paul. In 1970 he joined Arthur Schawlow at Stanford University. The collaboration sparked a revolution in spectroscopy and metrology, culminating thirty years later in the creation of the optical frequency comb [50]. In 1986 Hänsch returned to Germany to become a professor at the Ludwig-Maximilians-Universität of Munich and to lead the Division of Laser Spectroscopy at Max-Planck-Institut für Quantenoptik in Garching.

The optical frequency comb: a revolutionary advance in metrology and control that extends radiofrequency and microwave techniques into the optical regime, is cited in both of these awards. It was developed independently and essentially simultaneously by the two laureates.

2012 Nobel Prize in Physics: Serge Haroche and David W. Wineland *"for groundbreaking methods that enable measuring and manipulation of individual quantum systems."*

Serge Haroche did his graduate research at Ecole Normale Supérieure when the laboratory was under the direction of Alfred Kastler and Jean Brossel. Haroche collaborated with Claude Cohen-Tannoudji in developing the dressed atom theory and then exploring its applications experimentally. In 1970 he went to Stanford University and worked with Arthur Schawlow. His scientific history—which could be summarized as the evolution from using photons to study and control atoms to using atoms to study and control photons—is described in his Nobel Lecture [51].

David J. Wineland was well linked to the Stern/Rabi tradition through his graduate research with Norman Ramsey. He did postdoctoral research with Hans Dehmelt at the University of Washington and went to the National Bureau of Standards (now NIST) in Boulder, Colorado. There he directed a program of research on trapped ions, precision measurements, quantum logic and other quantum phenomena including ion cooling, as recounted in his Nobel lecture [52].

2017 Nobel Prize in Physics: Rainer Weiss, Barry C. Barish and Kip S. Thorne *"for decisive contributions to the LIGO detector and the observation of gravitational waves"*

The concept of LIGO and the experimental search for gravitational waves originated when Rainer Weiss—then an MIT dropout—wandered into the laboratory of Jerrold Zacharias and volunteered to help out on electronics. He became fascinated with research on an atomic clock whose goal was to observe the effect of gravity on time. Zacharias mentored Weiss through graduate school and in his early career on the MIT faculty. The history of the birth of gravitational astronomy is narrated in Weiss's Nobel lecture. Zacharias was a postdoctoral fellow in Rabi's laboratory in the 1930s and worked on the first demonstration of molecular beam magnetic resonance. Following the war Zacharias started a molecular beams laboratory at MIT.

2018 Nobel Prize in Physics: Arthur Ashkin *"for the optical tweezers and their application to biological systems."*

Arthur Ashkin was an undergraduate in physics at Columbia University. He graduated in 1947 and went to Cornell University to study nuclear physics where he received the Ph.D. in 1952. He went to Bell Laboratories for the rest of his career. He initiated the use of laser light to control the motion of small particles and later collaborated with Steven Chu in the development of "optical tweezers" for manipulating molecules and atoms.

2018 Nobel Prize in Physics: Gérard Mourou and Donna Strickland *"for their method of generating high-intensity, ultra-short optical pulses."* Links to the scientists in the Stern/Rabi chain have not been identified.

6 Otto Stern's Heritage in Chemistry

This study has focused on Otto Stern's heritage in AMO Physics but his influence reaches well beyond that. His molecular beam method was a direct influence on those in pursuit of chemistry "under single-collision conditions."

The following is a summary of Nobel Prize winners who have benefited from the heritage of Otto Stern and passed it on.

1986 Nobel Prize in Chemistry: Dudley R. Herschbach, Yuan T. Lee and John C. Polanyi *"for their contributions concerning the dynamics of chemical elementary processes."*

Otto Stern's molecular beam method was a direct influence on those in pursuit of chemistry "under single-collision conditions." In Chap. 1 of this volume, Dudley Herschbach details his path to "doing chemistry" in crossed molecular beams. Herschbach set out on that path after taking Norman Ramsey's course in molecular beams at Harvard in 1955 (the author was a classmate). Ramsey was enthusiastic about Dudley's ideas and encouraged him to pursue them.

Dudley Herschbach had a lion's share in raising awareness about the legacy of Otto Stern—through the centennial Festschrift and numerous publications since as well as his many talks, including his Nobel Lecture. Herschbach also served as the honorary chair—together with Jan Peter Toennies—of the Otto Stern Fest in 2019 in Frankfurt.

Apart from Nobel laureates who were under direct influence—or spell—of Otto Stern and Isidor Rabi, there are a number of awardees whose connection to the founders of AMO Physics was more tangential or remote. Their work was nevertheless nourished by the AMO and Chemical Physics communities that produced the directly related laurates. Prominent among them are:

1991 Nobel Prize in Chemistry: Richard Ernst *"for his contributions to the development of the methodology of high resolution nuclear magnetic resonance (NMR) spectroscopy."*

1996 Nobel Prize in Chemistry: Robert Curl, Harold Kroto, and Richard Smalley *"for their discovery of fullerenes,"* using molecular beams and mass spectrometry. The carbon polyhedron C_{60} was named for a geodetic dome designed by the architect Richard Buckminster Fuller; also the C_{60} pattern exacts a soccer ball!

1999 Nobel Prize in Chemistry: Ahmed Zewail *"for his studies of the transition states of chemical reactions using femtosecond spectroscopy."*

2002 Nobel Prize in Chemistry: John Fenn and Koichi Tanaka *"for their development of soft desorption ionization methods for mass spectrometric analyses of biological macromolecules"* (a.k.a. electrospray).

2007 Nobel Prize in Chemistry: Gerhard Ertl *"for his studies of chemical processes on solid surfaces."*

7 Epigraph

The advances in AMO physics from the time of Otto Stern to the present follow a persistent theme; ever increasing control. The Stern-Gerlach experiment permitted control of the electronic spin state of a beam of atoms; Rabi discovered how to transfer atoms from a hyperfine state to one of the many hyperfine levels and Ramsey discovered how to transfer atoms into a coherent superposition state, which we would now describe as an entangled state. In inventing the laser, Townes made it possible to generate radiation in a single mode of the radiation field and create lasers that can transfer atoms to any desired electronic state. Kastler discovered how to transfer the nuclei in a gas of atoms into a single nuclear spin state. Dehmelt discovered how to capture and study a single electron; Paul discovered how to catch and hold ions in an ion trap. Stern's first beam measurements—of the speeds of atoms—initiated a history of increasingly precise control of atomic motion, culminating in laser cooling that gives total control of *all* the quantum states of atoms, external and internal. Beyond that lies the world of ultra-cold chemistry where molecules can be assembled one atom at time and the world of optical lattices where the spatial structure of a many-atom array can be controlled; atoms can be transferred to known vibrational states and their interactions with neighboring atoms can be controlled. The frontiers of atomic physics have been pushed into many-body physics where the many bodies are controlled with the full precision that quantum mechanics permits, and their dynamics can be observed as the systems are manipulated. The discovery of gravitational waves by the LIGO interferometer is the most recent advance in this ongoing process of ever-increasing control. By controlling space and time at the level of 1 part in 10^{21}, LIGO revealed a world of cosmic black-hole events never before seen. LIGO grew from the dream of Rainer Weiss when he was a postdoc in Jerrold Zacharias' molecular beams laboratory. Zacharias was the first postdoc in the laboratory of I. I. Rabi, a protégé of Otto Stern.

Acknowledgements I thank Theodor Ducas, Dudley Herschbach, Bretislav Friedrich, and Charles Holbrow for their assistance.

Appendix: A Summary of Links between the AMO Nobel Laureates and Stern/Rabi

Key   → was a student or post-doc of
      … indicates some other association
      Nobel laureates are in bold face

Bloembergen→**Purcell**→**Stern/Rabi**
Chu → Commins → **Stern/Rabi**
Cohen-Tannoudji → **Kastler**…**Stern/Rabi**
Cornell → Pritchard → (Kleppner/**Ramsey**) → **Stern/Rabi**
Haroche → **Cohen-Tannoudji** → **Kastler**… **Stern/Rabi**
Kastler…(indirect links)…**Stern/Rabi**
Ketterle → (Pritchard) → (Kleppner) → **Ramsey** → **Stern/Rabi**
Kusch → **Stern/Rabi**
Lamb…**Stern/Rabi**
Phillips → Kleppner → **Ramsey** → **Stern/Rabi**
Purcell… **Stern/Rabi**
Ramsey → **Stern/Rabi**
Schawlow → **Townes**…**Stern/Rabi**
Townes…**Stern/Rabi**
Weiss →Zacharias→ **Stern/Rabi**
Wieman → **Hänsch** → **Schawlow** → **Townes**…**Stern/Rabi**
Wineland → **Ramsey** → **Stern/Rabi**

The autobiographies of the Nobel Laureates are available at the NobelPrize.org website.

References

1. D. Herschbach (ed.), *Zeitschrift für Physik D Atoms, Molecules and Clusters*, vol. 10(2/3) (1988)
2. N.F. Ramsey, Our legacy from Otto Stern. Z. Phys. D **10**, 121 (1988)
3. J. Rigden, *I. I. Rabi: Statesman and Scientist* (Basic Books, New York, 1997), p. 11
4. O. Frisch, *What Little I Remember* (Cambridge University Press, Cambridge, 1980)
5. E. Segrè, *A Biographical Memoir, Otto Stern (1888–1969)* (National Academy of Sciences, Washington, D.C., 1973)
6. N.F. Ramsey, *A Biographical Memoir, I. I. Rabi (1898–1988)* (National Academy of Sciences, 1993)
7. I.I. Rabi, J.R. Zacharias, S. Millman, P. Kusch, Molecular beam magnetic resonance. Phys. Rev. **55**, 525–539 (1939)
8. J.M.B. Kellogg, I.I. Rabi, N.R. Ramsey Jr., J.R. Zacharias, An electrical quadrupole moment of the deuteron. Phys. Rev. **55**, 318 (1939)
9. J. Bernstein, *Rabi*, vol. 12 (New Yorker, 1975)
10. R. Crease, *Making Physics* (University of Chicago Press, Chicago, 1999)
11. J. Krige, I.I. Rabi and the Birth of CERN. Phys. Today **57**, 44 (2004)

12. Nobel Prize winners whose careers at Columbia were encouraged by Rabi: Polykarp Kusch, 1948; Willis E. Lamb, 1948; Tsung-Dao Lee, 1957; Charles H. Townes, 1964; James Rainwarer, 1975; Leon M. Lederman, 1988; Norman F. Ramsey, 1989

13. L.W. Alvarez, F. Bloch, A quantitative determination of the neutron moment in absolute nuclear magnetons. Phys. Rev. **55**, 111 (1940)

14. E.M. Purcell, H.C. Torrey, R.V. Pound, Resonance absorption by nuclear magnetic moments in a solid. Phys. Rev. **69**, 37 (1946)

15. F. Bloch, W.W. Hansen, M. Packard, Nuclear induction. Phys. Rev. **69**, 127 (1946)

16. P. Kusch, H.M. Foley, The magnetic moment of the electron. Phys. Rev. **74**, 250 (1948)

17. W.E. Lamb, J.R. Retherford, Fine structure of the hydrogen atom by a microwave method. Phys. Rev. **72**, 241 (1948)

18. I.I. Rabi, J.E. Nafe, E.B. Nelson, The hyperfine structure of atomic hydrogen and deuterium. Phys. Rev. **71**, 914 (1947)

19. J.P. Gordon, H. Zeiger, C.H. Townes, The maser a new type of microwave amplifier, frequency standard, and spectrometer. Phys. Rev. **95**, 282 (1954)

20. N.G. Basov, A.M. Prokhorov, Primenenie molekulyarnykh puchkov dlya radiospektroskopich-eskogo izucheniya vrashchatelnykh spektrov. J. Exptl. Theoret. Phys. (U.S.S.R.) **27**, 431 (1954)

21. A.L. Schawlow, C.H. Townes, Infrared and optical masers. Phys. Rev. **112**, 1940 (1958)

22. A. Kastler, Some suggestions concerning the production and detection by optical means of inequalities in the populations of levels of spatial quantization in atoms. Application to the Stern and Gerlach and magnetic resonance experiments. Le Journal de Physique et Radium **11**, 255 (1950)

23. A. Kastler, Thesis. Ann. Phys. (Paris) **6**, 663 (1936)

24. J. Brossel, F. Bitter, A new 'double resonance' method for investigating atomic energy levels. Application to Hg $^3P_1$. Phys. Rev. **86**, 308 (1952)

25. Norman Ramsey, Private Communication

26. N. Bloembergen, E.M. Purcell, R.V. Pound, Relaxation effects in nuclear magnetic resonance absorption. Phys. Rev. **73**, 679 (1948)

27. C.H. Townes, A.L. Schawlow, *Microwave Spectroscopy* (McGraw Hill, 1956)

28. D. Kleppner, *A biographical memoir, norman Foster Ramsey (1915–2011)* (National Academy of Sciences, Washington, D.C., 2015)

29. J.M.B. Kellogg, I.I. Rabi, N.F. Ramsey Jr., J.R. Zacharias, An electrical quadrupole moment of the deuteron: the radiofrequency spectra of HD and D_2 molecules in a magnetic field. Phys. Rev. **55**, 318 (1939)

30. N.F. Ramsey, *Molecular Beams* (Cambridge University Press, Cambridge, 1956)

31. N.F. Ramsey, A molecular beam resonance method with separated oscillatory fields. Phys. Rev. **78**, 695 (1950)

32. L. Essen, J.V.L. Parry, An atomic standard of frequency and time interval: a caesium resonator. Nature **176**, 285 (1955)

33. H.M. Goldenberg, D. Kleppner, N.F. Ramsey, Atomic hydrogen maser. Phys. Rev. Lett. **5**, 361 (1960)

34. D. Wineland, P. Ekstrom, H. Dehmelt, Monoelectron oscillator. Phys. Rev. Lett. **31**, 1279 (1973)

35. D. Janneke, S. Fogwell, G. Gabrielse, New measurement of the electron magnetic moment and the fine structure constant. Phys. Rev. Lett. **100**, 12081 (2008)

36. M.H. Anderson, J.R. Ensher, M.R. Matthews, C.E. Wieman, E.A. Cornell, Bose-Einstein condensation of a gas of sodium atoms. Science **269**, 198–201

37. K.B. Davis, M.-O. Mewes, M.R. Andrews, N.J. van Druten, D.S. Durfee, D.M. Kurn, W. Ketterle, Bose-Einstein condensation of a gas of sodium atoms. Phys. Rev. Lett. **75**, 3969 (1995)

38. W.D. Phillips, Laser cooling and trapping of neutral atoms. Rev. Mod. Phys. **70**, 721 (1998)

39. C. Cohen-Tannoudji, Manipulating atoms with photons. Rev. Mod. Phys. **70**, 707 (1998)

40. S. Chu, The manipulation of neutral particles. Rev. Mod. Phys. **70**, 685 (1998)

41. P.D. Lett, R.N. Watts, C.I. Westbrook, W.D. Phillips, P.L. Gould, H.J. Metcalf, Observation of atoms laser cooled below the Doppler limit. Phys. Rev. Lett. **61**, 169 (1988)
42. S. Chu, L. Hollberg, J.E. Bjorkholm, A. Cable, A. Ashkin, Three-dimensional viscous confinement and cooling of atoms by resonance radiation pressure. Phys. Rev. Lett. **55**, 48 (1985)
43. J. Dalibard, C. Cohen-Tannoudji, Laser cooling below the Doppler limit by polarization gradients: simple theoretical models. JOSA B **6**, 2023 (1989)
44. C. Cohen-Tannoudji, S. Haroche, Absorption et diffusion de photons optiques par un atome en interaction avec des photons de radiofrequence. J. Physique **30**, 153 (1969)
45. W. Phillips, H. Metcalf, Laser deceleration of an atomic beam. Phys. Rev. Lett. **48**, 596 (1982)
46. E.A. Cornell, C.E. Wieman, Bose-Einstein condensation in a dilute gas, the first 70 years and some recent experiments. Rev. Mod. Phys. **74**, 85 (2002)
47. W. Ketterle, When atoms behave as waves: Bose-Einstein condensation and the atom laser. Rev. Mod. Phys. **74**, 1131 (2002)
48. N.P. Proukakis, D.W. Snoke, P.B. Littlewood (eds.), *Universal themes of Bose-Einstein condensation* (Cambridge University Press, Cambridge, 2017)
49. J.L. Hall, Defining and measuring optical frequencies. Rev. Mod. Phys. **78**, 1279 (2006)
50. T. Hänsch, Passion for precision. Rev. Mod. Phys. **85**, 1083 (2013)
51. S. Haroche, Controlling photons in and exploring the quantum to classical boundary. Rev. Mod. Phys. **85**, 1083 (2013)
52. D.J. Wineland, Superposition, entanglement, and raising Schrödinger's cat. Rev. Mod. Phys. **85**, 1103 (2013)

Chapter 8
Walther Gerlach (1889–1979): Precision Physicist, Educator and Research Organizer, Historian of Science

Josef Georg Huber, Horst Schmidt-Böcking, and Bretislav Friedrich

Abstract Walther Gerlach's numerous contributions to physics include precision measurements related to the black-body radiation (1912–1916) as well as the first-ever quantitative measurement of the radiation pressure (1923), apart from his key role in the epochal Stern-Gerlach experiment (1921–1922). His wide-ranging research programs at the Universities of Tübingen, Frankfurt, and Munich entailed spectroscopy and spectral analysis, the study of the magnetic properties of matter, and radioactivity. An important player in the physics community already in his 20s and in the German academia in his later years, Gerlach was appointed, on Werner Heisenberg's recommendation, Plenipotentiary for nuclear research for the last sixteen months of the existence of the Third Reich. He supported the effort of the German physicists to achieve a controlled chain reaction in a uranium reactor until the last moments before the effort was halted by the Allied *Alsos Mission*. The reader can find additional discussion of Gerlach's role in the supplementary material provided with the online version of the chapter on SpringerLink. After returning from his detention at Farm Hall, he redirected his boundless elan and determination to the reconstruction of German academia. Among his high-ranking appointments in the Federal Republic were the presidency of the University of Munich (1948–1951) and of the Fraunhofer Society (1948–1951) as well as the vice-presidency of the German Science Foundation (1949–1961) and the German Physical Society (1956–1957). As a member of *Göttinger Achtzehn*, he signed the Göttingen Declaration (1957) against arming the *Bundeswehr* with nuclear weapons. Having made history *in* physics, Gerlach

Electronic supplementary material The online version of this chapter (https://doi.org/10.1007/978-3-030-63963-1_8) contains supplementary material, which is available to authorized users.

J. G. Huber
Am Schnepfenweg 35a, 80995 München, Germany

H. Schmidt-Böcking
Institut für Kernphysik, Universität Frankfurt, 60348 Frankfurt, Germany

B. Friedrich (✉)
Fritz-Haber-Institut der Max-Planck-Gesellschaft, Faradayweg 4-6, 14195 Berlin, Germany
e-mail: bretislav.friedrich@fhi-berlin.mpg.de

became a prolific writer *on* the history of physics. Johannes Kepler was his favorite subject and personal hero—as both a scientist and humanist.

1 Introduction

What Walther Gerlach said about his academic mentor, Friedrich Paschen (1865–1947), could also be said about Gerlach himself (Gerlach 1935):

> The physicists saw him as a master of experimental physical research who carried on the great tradition of precision physics …With his unusual manual dexterity, he built the finest [scientific instruments], tirelessly trying to get the last out of them, in the conviction that every instrumental advance in physical research opens up new possibilities—and will enable new insights. And the fact that he succeeded in this …made him love his [scientific instruments] almost tenderly.

By the time he earned his Ph.D. in Paschen's Tübingen laboratory in 1912 at age 23, Gerlach was a major player in the research area of black-body radiation. He would pursue a related topic, that of light pressure, after an interruption due to World War One and his crucial involvement in the epochal Stern-Gerlach experiment during 1921–1922. In 1925, Gerlach would assume the chair of his mentor and in 1929 move on to Munich as the successor of Wilhelm (Willy) Wien (1864–1928), thereby receiving the accolade due to a leading experimental physicist. Gerlach's tenure at Munich, which lasted until his retirement in 1957, would only be interrupted by his detention at Farm Hall (1945–1946) and a stint at the University of Bonn (1946–1948), then in the British Zone of Occupation.

In 1944, upon consulting Werner Heisenberg (1901–1976), Otto Hahn (1879–1968), and Paul Rosbaud (1896–1963), Gerlach became the head of the Physics Section at the Reich Research Council and *Reichsmarschall's* Plenipotentiary for nuclear physics responsible for the German *Uranprojekt*. Thereby, Gerlach entered higher echelons of Third Reich's establishment (Walker 1995). As available testimonials, including his own, suggest, in this capacity, Gerlach saved many young physicists from the service on the front—and, unbeknownst to him, likely kept the Allies abreast of the German nuclear research via Paul Rosbaud (1896–1963), a scientist and publisher who had become a British agent (Kramisch 1986). In his character testimonial about Gerlach, Rosbaud stated (Rosbaud 1945):

> Gerlach hated the Nazis, he had to suffer under their denunciations …he loved his country and wished the best to her and did not want her to perish …. During the last period of the war he only was interested in advancing pure research work and in saving the lives of scientists. He exceeded many times his competencies to save people …In contrast to many others, he was absolutely incorruptible and in consequence, despite [receiving] 2 or 3 Führerpakete,[1] sometimes half starved.

---

[1] A food allocation provided during WWII once a year to the military and other choice personnel on behalf of Adolf Hitler.

In the aftermath of World War Two and beyond, Gerlach directed his boundless elan and determination to the reconstruction of German academia. He built up anew the Institute of Physics at Munich's *Ludwig-Maximilans-Universität* and served as the university's Rector (1948–1951); during the same period he served as the founding President of the *Fraunhofer-Gesellschaft* for applied research; was Vice-President of the *Deutsche Forschungsgemeinschaft* (1949–1961) and of the *Deutsche Physikalische Gesellschaft* (1956–1957). "Making friends and cultivating friendships was one of his greatest talents" (Gentner 1980), which Gerlach amply deployed throughout these years.

Gerlach was also engaged in attempts to limit the spread of nuclear weapons and signed as a member of *Göttinger Achtzehn* the Göttingen Declaration opposing the move by the West-German government to arm the *Bundeswehr* with tactical nuclear weapons (12 April 1957).

Since the late 1940s, Walther Gerlach's interest turned increasingly to the history of science. He would write about 500 didactic, biographical, and memorial articles— apart from about 320 research papers and monographs (Nida-Rümelin 1982). His essay on Max Planck (Gerlach 1948) or book on Johannes Kepler (Gerlach 1980) belong to his most acclaimed history works.

Gerlach was co-nominated, with Otto Stern, thirty-one times for the Nobel Prize in Physics for the Stern-Gerlach experiment, Fig. 1. Gerlach's contributions to the fields of black body radiation, light pressure, magnetism, and spectroscopy were no less demanding but remain much less known. In this chapter, we revisit Gerlach's seminal works in an attempt to do justice to his scientific legacy. We conclude by showcasing his work in the history of science.

2 Walther Gerlach's Social Background, Upbringing, and Education

Walther Gerlach was born on 1 August 1889 in Wiesbaden-Biebrich (Huber 2015). His father, Valentin Gerlach (1858–1957), came from a family of craftsmen based in Frankfurt and became a doctor. However, he only practiced medicine for a short time and soon turned to experimental chemistry. His mother, Maria, neé Niederhaeuser (1868–1941), also came from a family of craftsmen, from the nearby Wiesbaden area. Figure 2 shows Walther Gerlach in the first year of his life. When he turned two, his twin brothers Werner and Wolfgang were born.

Formal upbringing in the family was primarily set by the father and took place within the framework of the conservative value system of the time. Figure 3 shows Gerlach as a school child. However, more strongly yet, it was shaped by the Enlightenment ideas of the Freemasons, of whose order the father was a member. Freedom, Equality, Brotherhood, Tolerance, and Humanity were at the foundation of their creed. The father, Figs. 4 and 5, was also an admirer and connoisseur of Johann Wolfgang Goethe, whose understanding of education played an important role in the

122 J. G. Huber et al.

Nomination	Year	Nominator	Country
1	1924	Albert Einstein	Germany
2	1925	Ernst Wagner	Germany
3	1927	Max Born	Germany
4	1927	James Franck	Germany
5	1927	Heinrich Rausch von Traubenberg	Czechoslovakia
6	1928	James Franck	Germany
7	1928	Max Reich	Germany
8	1928	Pierre Weiss	France
9	1928	Julius Wagner-Jauregg	Austria
10	1929	Eduard Haschek	Austria
11	1929	Gustav Jäger	Austria
12	1929	Stefan Meyer	Austria
13	1929	Karl Przibram	Austria
14	1929	Johannes Stark	Germany
15	1929	William Campbell	U.S.A.
16	1930	William Campbell	U.S.A.
17	1931	Max von Laue	Germany
18	1932	Friedrich Hund	Germany
19	1932	Erwin Meyer	Switzerland
20	1934	Gustav Jäger	Austria
21	1934	Stefan Meyer	Austria
22	1934	Egon von Schweidler	Austria
23	1934	Hans Thirring	Austria
24	1934	Dirk Coster	Netherlands
25	1936	Pierre Weiss	France
26	1937	Anton von Eiselsberg	Austria
27	1937	Stefan Meyer	Austria
28	1937	Egon von Schweidler	Austria
29	1937	Hans Thirring	Austria
30	1940	Dirk Coster	Netherlands
31	1944	Manne Kai Siegbahn	Sweden

Fig. 1 Walther Gerlach's nominations for a Nobel prize in Physics. The compilation is based on the information available at the nomination archive https://www.nobelprize.org/nomination/archive/. The 1924 nomination by Albert Einstein was not a valid one, as Einstein nominated additional candidates apart from Stern and Gerlach that year

Gerlach family as well. Not to forget Valentin Gerlach's membership in a student association *Corps Alemannia* to whose events he would often take his children along.

The upbringing in the Gerlach family was both highly demanding and encouraging, characterized by rigor and devotion. The father himself had learned that one can only achieve something in life through determined work and self-discipline and wanted to pass on this realization to his children. The parents set at first narrow boundaries but gradually expanded them as the children grew older and could increasingly take responsibility for their own actions. Walther Gerlach's first diary tells of extensive hikes, preoccupation with flora, fauna and minerals, visits to the theater, literary, artistic and musical activities as well as photography and much more. He played the piano and organ and tried his hand at drawing and poetry.

Fig. 2 Walther Gerlach in 1889 (Heinrich and Bachmann 1989)

Fig. 3 Walther Gerlach as a pupil (Heinrich and Bachmann 1989)

Fig. 4 Walther Gerlach with his father in 1909. Courtesy of Werner Kittel, Hamburg

Walther Gerlach later found the term "aimless determination" for his own understanding of how education works. What he meant was that, for instance, at high school, one should not pursue subjects with an eye on their utility for a future profession but rather give free rein to one's inclinations and interests "without a plan" but "with determination."

Walther received Protestant baptism shortly before starting school. In keeping with liberal attitudes, the family members were not practicing Christians, but rather sought the divine in natural phenomena.

Walther entered elementary school in 1896 and switched to the *Königliches Gymnasium zu Wiesbaden* (now *Diltheyschule-Wiesbaden*) in 1899, where he took the Abitur exam in 1908. Walther Gerlach's school performance was unspectacular. He was a good student, but not an outstanding one. In his *Abitur* certificate, Mathematics and Philosophy were noted as the desired courses of study. Upon his admission to the University of Tübingen at Easter 1908, Gerlach indeed began studying these two subjects. However, when he attended a lecture and laboratory course by the physicist Friedrich Paschen, Fig. 6, he was so impressed by Paschen's experiments that he gave up philosophy in favor of physics.

Fig. 5 Walther Gerlach
(left) with his brothers
Werner (2nd from left) and
Wolfgang (right) and their
father (seated). Courtesy of
Werner Kittel, Hamburg

Fig. 6 Friedrich Paschen.
Creative Commons

Fig. 7 Walther Gerlach in
Frankfurt, early 1920s.
Courtesy of the Archive of
the University of Frankfurt

At the outset of his studies in Tübingen, Gerlach joined the student association *Corps Borussia*, a fencing fraternity like his father's *Corps Alemannia*—and another formative influence. Figure 7 shows Gerlach in his early thirties with a fencing wound on his left cheek. Gerlach would leave the fraternity as late as 1954, likely to indicate his view that German universities should foster international spirit rather than parochial student associations.

Gerlach's physics studies progressed at a rapid pace: In the 5th semester he started work on his doctoral thesis, in the 6th semester he became Paschen's assistant, and at the end of the 8th semester, on 29 February 1912, he took his doctoral examination.

There was strict discipline at Paschen's institute but also an open international atmosphere. Gerlach's time at the institute proved formative for both his personality and his experimental abilities. Either became a key prerequisite for later success in performing the Stern-Gerlach experiment and other precision measurements where Gerlach pushed the limits of the possible. Paschen requested from his assistants to be almost permanently present at the institute and to work hard all the time, quipping "How's the crap going?" Paschen's manner earned him the epithet "Institute Tyrant" (Gerlach 1908–1950). Nevertheless, Gerlach remained grateful to and respectful of Paschen. Apparently, the mentoring by Paschen was for Gerlach just a continuation of his father's upbringing.

Gerlach stayed at Paschen's institute for two more years despite the hard time he was having. He greatly valued the stimulating discussions at the institute of all the exciting developments that were taking place in physics and remained highly productive throughout. In spite of his heavy workload, Gerlach maintained numerous contacts with researchers from a wide variety of disciplines, which rhymed well with his curiosity and fostered his versatility.

After the outbreak of World War One, Gerlach worked in the X-ray laboratory of the gynecological clinic at the University of Tübingen, whose director was a close friend. There he developed an astonishingly simple X-ray device for locating projectiles and metal splinters in soldiers' bodies that was, moreover, well suited for the rough field conditions.

On 24 August 1915, Gerlach was drafted into military service in Ulm as a *Landsturm* recruit, but released again in December because of rheumatoid arthritis.

In May 1916 he was called up again, this time to *Technische Abteilung der Funkertruppen*, abbreviated as *Tafunk*, with which he stayed until the end of the war. Its head was Max Wien, Willy Wien's cousin. The task of Gerlach's department was to develop and test radio equipment based on the new technology of tube amplifiers. His stay at *Tafunk* was interrupted twice by illness (appendicitis and the "Spanish flu"). While on sick leave in May 1916, he completed his *Habilitation*.

In the Fall of 1916, he took part in the fighting of the VIth Army in Flanders and Artois and directly experienced the horrors of war. After a dispute with Paschen, who wanted his assistant back at his institute in Tübingen, Gerlach did an *Umhabilitation*, in 1917, in Göttingen. He continued his scientific work and even managed to publish several papers based on his previous research. Most importantly, at *Tafunk*, Gerlach met other physicists, among them Max Born (1882–1970), James Franck (1882–1964), Wilhelm Westphal (1882–1978), but also Richard W. Pohl (1884–1976) and Peter Debye (1884–1976), who helped with his move to Göttingen. He also worked for an extended period with Gustav Hertz (1887–1975), Fig. 8, Heinrich Hertz's

Fig. 8 Walther Gerlach with Gustav Hertz (left) working at *Tafunk* in Jena, May 1917. The hand-written note by Gerlach reads: "Hertz und ich am Schreibemfänger [Hertz and I at the telegraph], Jena-May 1917" (Heinrich and Bachmann 1989)

Fig. 9 Richard Wachsmuth.
Courtesy of the Archive of
the University of Frankfurt

nephew and future Physics Nobel laureate, jointly with James Franck, for 1925. From September 1917 to March 1918 he was on an inspection tour in Belgium and northern France. Upon his return, Gerlach married Wilhelmine Mezger and in 1918 their daughter Ursula was born. On January 27, 1919, he was released from the military as chief engineer. In order to be able to provide for his family, Gerlach opted for an industrial rather than an academic job and landed a managerial position at the physical laboratory of the Elberfeld paint factory. However, he soon realized that industrial research was not his cup of tea and returned to academia once the University of Frankfurt offered him a position. As of 1 October 1920, Gerlach became the first assistant to the director of Frankfurt's Institute of Experimental Physics, Richard Wachsmuth (1868–1941), Fig. 9.

Frankfurt was the first station on Gerlach's academic path at which he had his own position. Three more would follow. A detailed timeline of Gerlach's life and career is given in Appendix A.

3   Precision Physics

In his first book, written in Frankfurt, Gerlach provided the following definition of "precision measurement" (Gerlach 1921):

> By 'precision measurement' we mean an investigation in which all sources of error are taken into account and all observed phenomena are clarified: It is also characteristic of [a precision] measurement that each individual step is theoretically and numerically justified, its influence on the course of the experiments thoroughly tested, spelled out, and presented in all detail; in short, the reader must be able to form a judgment from the description of the experiments about the evidential value and the [degree of] certainty of the results.

What Gerlach meant was best exemplified by his own work, which became a standard of precision physics.

3.1   Black-Body Radiation

There is a record of what Gerlach thought about the state of Physics in about 1910 when he entered the 5th semester at Tübingen and started working on his dissertation under Paschen (Gerlach 1978a), p. 200:

> [There] were special fields of general interest such as long-wave infrared, gas discharge, spectroscopy, radioactivity, canal rays, which had been worked on at various institutes; the theoretical foundations were thermodynamics, kinetic theory of gases, electromagnetism, electron theory of the electrical and optical properties of matter. But there was probably no such thing as central questions; these were certainly not relativity or quantum physics.

The dissertation topic that Paschen assigned to Gerlach had nothing to do with any of the above but rather entailed revisiting one of the major themes that Paschen had worked on a decade earlier, namely black body radiation. Paschen's interest was revived by a discrepancy between the "canonical" value of the constant σ in Stefan-Boltzmann's law as determined in 1898 by Ferdinand Kurlbaum (1857–1927) (Kurlbaum 1898) and a new value published in 1909 by the reputable Ch. Féry (Féry 1909). Strangely enough, Max Planck's 1900 law (Planck 1900) governing the spectral distribution of black-body radiation—and the first salvo of the quantum revolution—was neither mentioned nor cited in Gerlach's thesis completed in 1912 (Gerlach 1912). This in spite of the fact that Planck's law not only allowed to derive the Stefan-Boltzmann law but also to express the constant σ in terms of fundamental constants. Had Gerlach made this connection, it would have lent his effort a fundamental character as well, at least from a more recent perspective. At the time, however, only few—among them Albert Einstein (1879–1955)—regarded Planck's law (and Planck's constant) as fundamental (Frisch 1963), i.e., as more than a mathematical representation of empirical data.

The Stefan-Boltzmann law obtains by integrating Planck's spectral intensity, $I(\lambda, T)$, of black body radiation

$$I(\lambda, T) = \frac{2\pi hc^2}{\lambda^5} \left[\exp\left(\frac{hc}{\lambda kT}\right) - 1 \right]^{-1} \tag{1}$$

over the wavelenght λ at temperature T

$$I(T) \equiv \int_0^\infty I(\lambda, T)d\lambda = \sigma T^4, \tag{2}$$

yielding

$$\sigma = \frac{2\pi^5 k^4}{15c^2 h^3} \tag{3}$$

with k Boltzmann's constant, h Planck's constant, and c the speed of light. This derivation of the Stefan-Boltzmann law was carried out for the first time by Planck himself (Planck 1901).

In his dissertation, Gerlach set out to clarify the discrepancy between Kurlbaum's and Féry's values of σ—however without resorting to the ultimate arbiter, namely Eq. 3. This would not have been feasible at the time anyway, as Planck's constant was not known accurately enough at the time.

While Kurlbaum obtained a value of 5.32×10^{-12} W cm^2 K^{-4} (Kurlbaum 1898) using the bolometer method, Féry obtained a significantly larger value, of 6.30×10^{-12} W cm^2 K^{-4} (Féry 1909), using a thermocouple. Upon a thorough inspection of Féry's paper, Paschen concluded that Kurlbaum's method was likely the less accurate one and tasked Gerlach with recreating Kurlbaum's apparatus while avoiding possible sources of error, such as replacing a bolometer with a thermopile (i.e., an array of thermocouples) to measure the temperature.

Gerlach's apparatus is shown in Fig. 10. A *Hohlraum* realization of a black body (Valentiner 1910), produces black-body radiation at 0° or 100 °C, defined, respectively, by the freezing and boiling points of water at atmospheric pressure. Upon passage through a diaphragm, the radiation is absorbed by detection stripes made of manganin (an alloy of copper, manganese, and nickel with a low thermal expansion coefficient) electroplated with platinum black (in order to suppress selective absorption). The detection stripes were held at a distance of half a millimeter from a thermopile, with an insulating layer of ambient air in between. The thermopile was of the type developed earlier by Paschen for his spectroscopic investigations (Gerlach 1912). The current produced by the thermopile was measured by a sensitive galvanometer. The measurement procedure was as follows: (a) the black body at 100 °C irradiates the detection stripes for as long as the galvanometer reading increases, reaching a steady-state value of, say, i_0; (b) the black body at 100 °C is replaced with a black body at 0 °C and the detection stripes are electrically heated up until the galvanometer reading becomes equal to i_0; (c) The measured Joule heat (electric power) equals the difference of the radiant power carried by the black-body radiation at 100 and 0 °C. In order to achieve good statistics, the black bodies were swapped every minute or two and the galvanometer read every 15 s. The value that came out of Gerlach's measurements was $\sigma = (5.9 \pm 0.057) \times 10^{-12}$ W cm^2 K^{-4} (after a correction for reflected radiation). Gerlach's detection scheme is sometimes referred to as Ångström-type pyrheliometer (Coblentz 1913).

Paschen lavished the highest praise on Gerlach's achievement (Paschen 1912b):

> [Gerlach] was able to justify *ab ovo* every single aspect of the new method, which is one of the most difficult tasks of physics altogether.

However, when Gerlach's result, accompanied back-to-back by Paschen's endorsement, was published (Gerlach 1912), see also Fig. 11, the competitors, Ferdinand Kurlbaum and Siegfried Valentiner (1876–1971)—both from the *Physikalsch-Technische Reichsanstalt* (PTR) in Berlin—disagreed. A rather acrimonious public debate ensued that called for more work on Gerlach's and Paschen's part and led to two more investigations by Paschen and nine more by Gerlach, including Gerlach's *Habilitation* thesis.

Developing into a "war of attrition," the exchanges slowed down after the outbreak of World War One and ceased in 1916 (Gerlach 1916)—without resolving the issue. Throughout, Gerlach was troubled by the realization that a physics problem could not be brought to a closure, if possible in his favor. He would devise and implement new experimental schemes with a great persistence—but to no avail. In the end, the PTR made plans for resuming the measurements of σ—using Gerlach's method. Gerlach would demonstrate both his persistence and inventiveness in his later work

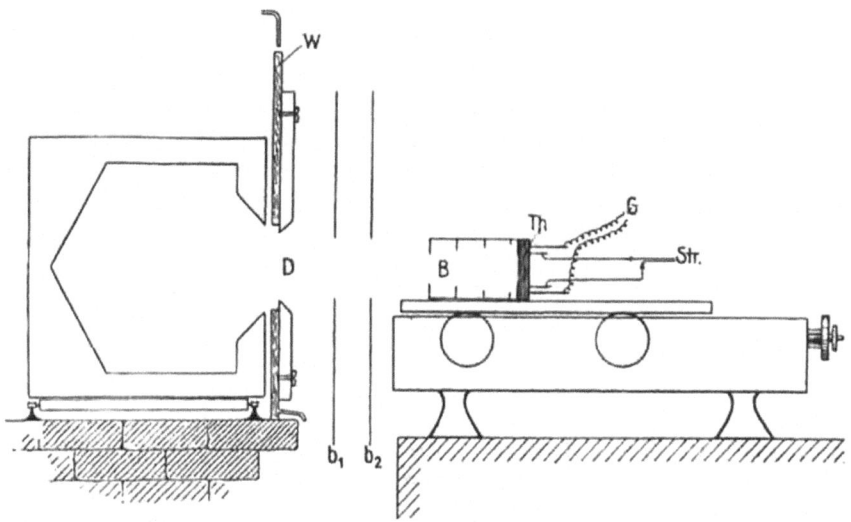

Fig. 10 Schematic of the apparatus Gerlach built in Paschen's laboratory in Tübingen to perform precision measurements of the proportionality constant σ in the Stefan-Boltzmann law (Gerlach 1912). Gerlach's realization of the black body together with a diaphragm (D) and slits (b_1 and b_2) is shown on the left. The right-hand side shows the detector with the detection strips and thermopile (Th), the galvanometer (G), and apertures (B). The detector assembly is mounted on a dividing engine whose position can be accurately controlled

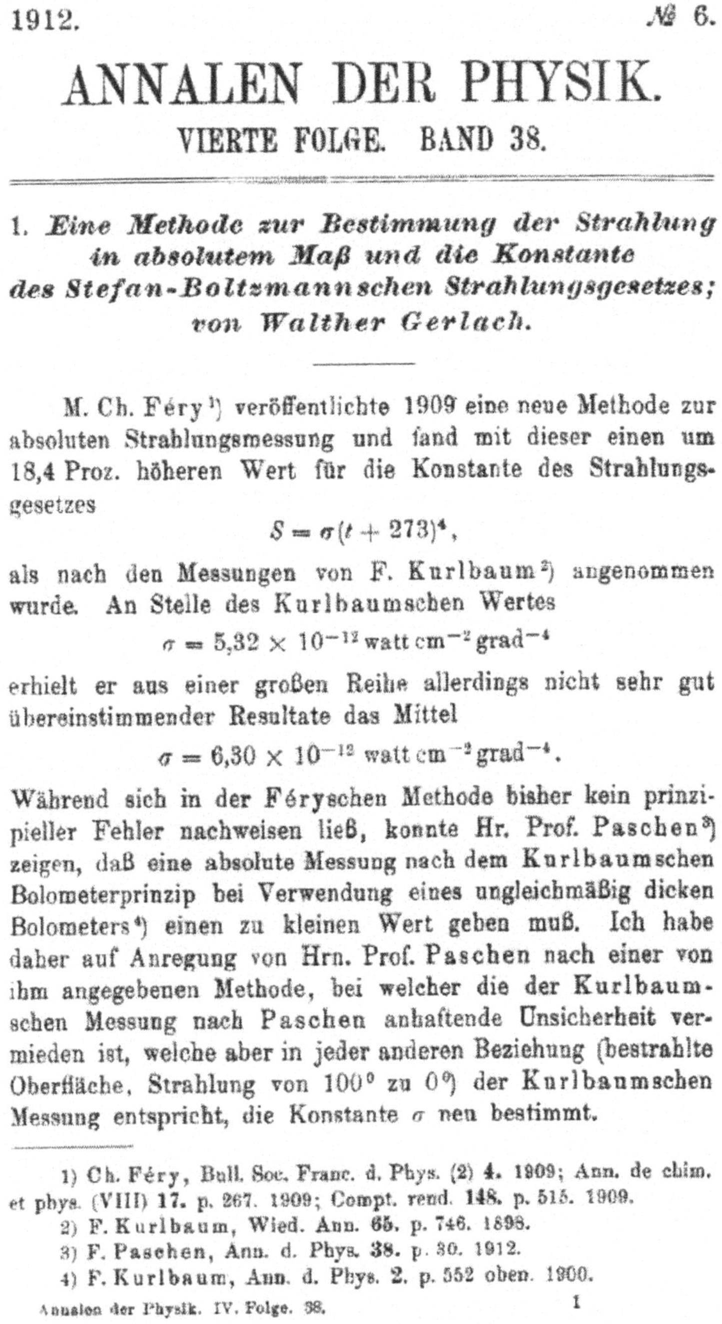

1912.   № 6.

ANNALEN DER PHYSIK.
VIERTE FOLGE. BAND 38.

1. *Eine Methode zur Bestimmung der Strahlung in absolutem Maß und die Konstante des Stefan-Boltzmannschen Strahlungsgesetzes; von Walther Gerlach.*

M. Ch. Féry[1]) veröffentlichte 1909 eine neue Methode zur absoluten Strahlungsmessung und fand mit dieser einen um 18,4 Proz. höheren Wert für die Konstante des Strahlungsgesetzes

$$S = \sigma(t + 273)^4,$$

als nach den Messungen von F. Kurlbaum[2]) angenommen wurde. An Stelle des Kurlbaumschen Wertes

$$\sigma = 5{,}32 \times 10^{-12}\,\text{watt}\,\text{cm}^{-2}\,\text{grad}^{-4}$$

erhielt er aus einer großen Reihe allerdings nicht sehr gut übereinstimmender Resultate das Mittel

$$\sigma = 6{,}30 \times 10^{-12}\,\text{watt}\,\text{cm}^{-2}\,\text{grad}^{-4}.$$

Während sich in der Féryschen Methode bisher kein prinzipieller Fehler nachweisen ließ, konnte Hr. Prof. Paschen[3]) zeigen, daß eine absolute Messung nach dem Kurlbaumschen Bolometerprinzip bei Verwendung eines ungleichmäßig dicken Bolometers[4]) einen zu kleinen Wert geben muß. Ich habe daher auf Anregung von Hrn. Prof. Paschen nach einer von ihm angegebenen Methode, bei welcher die der Kurlbaumschen Messung nach Paschen anhaftende Unsicherheit vermieden ist, welche aber in jeder anderen Beziehung (bestrahlte Oberfläche, Strahlung von 100° zu 0°) der Kurlbaumschen Messung entspricht, die Konstante σ neu bestimmt.

1) Ch. Féry, Bull. Soc. Franc. d. Phys. (2) 4. 1909; Ann. de chim. et phys. (VIII) 17. p. 267. 1909; Compt. rend. 148. p. 515. 1909.
2) F. Kurlbaum, Wied. Ann. 65. p. 746. 1898.
3) F. Paschen, Ann. d. Phys. 38. p. 30. 1912.
4) F. Kurlbaum, Ann. d. Phys. 2. p. 552 oben. 1900.

Fig. 11 Publication that came out of Walther Gerlach's Ph.D. thesis (Gerlach 1912)

as well, most conspicuously in the Stern-Gerlach experiment. Apparently, when he got something into his head, it was difficult to dissuade him from it.

Interestingly, as part of the debates between Gerlach and Paschen on the one side and the PTR scientists on the other, Paschen pointed out, (Paschen 1912a), that the new value of σ would be of consequence for the values of the fundamental constants it was made out of according to Planck's law, cf. Eq. 3. Let us note that the currently accepted value of the Stefan-Boltzmann constant is (CODATA 2020)

$$\sigma = 5.670374419 \times 10^{-12}\,\text{W}\,\text{cm}^{-2}\,\text{K}^{-4} \tag{4}$$

i.e., like Gerlach's value, between Kurlbaum's and Féry's values.

3.2 Walther Gerlach and the Stern-Gerlach Experiment

A detailed account of the purpose, outcome, and significance of the Stern-Gerlach experiment (SGE) can be found in Chap. 5 of this volume. Herein we emphasize Gerlach's contribution to the realization of the SGE and glean what the relationship between Stern and Gerlach was like from their mutual correspondence as well as from their correspondence with others.

As noted, in October 1920 Gerlach landed an assistantship at Wachsmuth's Institute for Experimental Physics at Frankfurt. The Frankfurt university recognized his *Habilitation* and, in addition, promoted him to the rank of *Extraordinarius* a month later. Max Born's adjacent Institute for Theoretical Physics was a more congenial environment for the curious and enterprising Gerlach than Wachsmuth's operation. All the more so that Born, with his assistants Otto Stern, Elisabeth Bormann, and Alfred Landé, was engaged in experiments as much as in theory and encouraged Gerlach to partake in their discussions as well as to give them a hand with their experiments. Born would even publish with Gerlach—on electron affinity (Gerlach and Born 1921a) and on light scattering (Gerlach and Born 1921b). However, Gerlach would also pursue his own agenda: it was at Frankfurt that he launched his investigations into the magnetic properties of materials that would bring him together with Stern and later take center stage in his research at Frankfurt and his subsequent stations. In particular, Gerlach was interested in the relationship between magnetization and structure (Bachmann and Rechenberg 1989), p. 10. In connection with his investigation of the magnetic properties of a bismuth alloy, the question arose as to whether atomic bismuth was para- or diamagnetic. Gerlach set out to answer this question in a molecular beam experiment, in which the deflection of a beam of bismuth atoms by an inhomogeneous magnetic field would be examined (Mehra and Rechenberg 1982), p. 436. Born tried to dissuade Gerlach from what seemed to be a hopelessly difficult undertaking. Whereupon Gerlach invoked a quip he heard from Edgar Meyer (1879–1960), his professor of theoretical physics at Tübingen: "No experiment is so dumb that it should not be tried" (Estermann 1975) and continued setting up his bismuth beam experiment and thus collecting experience in much of

Fig. 12 Otto Stern (2nd from left), Edgar Meyer (5th from left), Walther Gerlach (6th from left) in Tübingen in about 1926. Courtesy of the Otto Stern Collection, Berkeley

what was needed for the SGE. Let us add that Edgar Meyer, Fig. 12, with whom Gerlach had worked on the photo-effect, contested the separation of physics into theoretical and experimental. Max Born was apparently of the same persuasion in this respect. In February 1921, he reported to Einstein (Born 1969), p. 82:

> We have now Gerlach here with us, who is awesome: energetic, knowledgeable, skillful, ready to help.

In his 1977 talk, Gerlach told the story of his recruitment by Otto Stern for the SGE as follows (Gerlach 1977):

> One day Stern would come to me and say: 'Do you know what space quantization is?' I would say: 'No, I have no idea.' 'But you should actually know that. Recently Debye and Sommerfeld published [papers] suggesting that the [anomalous] Zeeman effect can be explained by a quantum effect, by the so-called space quantization. That is, [the magnetic dipole of] a silver or sodium atom can only have two settings [orientations] in a magnetic field, it cannot adjust itself at will or precess, but can only have two very specific settings [orientations], or actually even three, namely perpendicular to the magnetic field or in ... the direction or against the direction [of the magnetic field] ...

> Repeated discussions with Stern during our daily visits at *Café Rühl* finally led to a plan to make the experiment in such a way that there was hope of seeing space quantization.

Gerlach perhaps thought that he would just have to modify his current experiment on the magnetic properties of bismuth. Finally, he agreed: "Yes, I want to try it" (Gerlach 1977). But then

[Stern] would come back again: 'It isn't worth it, I've miscalculated, power of ten too little.' And then, it went back and forth a couple of times for a week or a fortnight and one day he would come back and say: 'Yes, now I've done [the calculations] properly and the thing only works if you get fields with an inhomogeneity of about ten or fifty thousand Oersted per centimeter—and that's not possible.' And then I said to him: 'Yes, I am almost there, I already have ten thousand [Oersted per centimeter], namely for my planned bismuth experiment'. 'So,' he said, 'let's try it.'

And they did. Stern first published the concept of what was to become the SGE, accompanied by feasibility calculations (Stern 1921), prompted to "patent" the idea by seeing the page proofs of a paper by Harmut Kallmann and Fritz Reiche on an electric analog of the SGE, see also Chaps. 5 and 20. The collaboration that ensued between Stern and Gerlach was in part so successful because of the complementarity of their skills and perhaps even working habits: while Stern had gained experience with molecular beams, Gerlach developed expertise in designing strong inhomogeneous magnetic fields. While Stern preferred to call it a day around 6 p.m. at that time, have dinner and go to the cinema, Gerlach liked to work at night, often doing with just three hours of sleep.

As described in Chap. 5, it took a tremendous effort to make the experiment work. Stern, who did not believe in the reality of space quantization to begin with, left on 1 October 1921 to assume a professorship at Rostock. Gerlach continued improving their apparatus and during the night of 4 November 1921 observed for the first time a broadening of a silver beam sent through an inhomogeneous magnetic field. This provided evidence that silver atoms carried a magnetic dipole moment—but did not suffice to demonstrate the existence of space quantization. During the Christmas recess, Gerlach and Stern reconfigured their apparatus again, but Gerlach's subsequent attempts to see space quantization had failed. At their meeting in Göttingen in early February 1922, Gerlach and Stern decided to try the experiment one more time. On the train back to Frankfurt, Gerlach remembered a modification he made earlier when examining crystals by X rays using the Debye-Scherrer method, namely to use a *slit* instead of a pinhole to boost both flux and spatial resolution. Gerlach had even reported on the improvement he thereby achieved at the German Physics Day in Jena in September 1921 (Huber 2015). Upon arrival in Frankfurt, Gerlach replaced the pinhole (of 0.05 mm diameter) defining the silver beam at the entrance into the inhomogeneous magnetic field by a rectangular 0.03×0.8 mm^2 slit with its longer side perpendicular to the field direction (Gerlach and Stern 1922)—and during the night of 7 February 1922 achieved the ultimate success.

Wilhelm Schütz (1900–1972), who was in 1922 Gerlach's Ph.D. student, described the difficulties of the SGE as well as the final triumph as follows (Schütz 1969):

The old apparatus had only yielded a broadening of the silver beam [deposit on the glass plate] of the expected magnitude …due to the inhomogeneous magnetic field. A major improvement of the apparatus with the aim to further increase its resolution was [therefore] necessary. During this rebuilding period, Stern moved to Rostock to assume a Professorship for Theoretical Physics there. He would show up in Frankfurt every now and then (during Christmas 1921 and Easter 1922) for discussions and to measure the inhomogeneity of the magnetic field …Soon came the time when I was able to enter the holy premises of the laboratory and take a look at the pumps, when [the technician Mr.] Schmidt was not on duty and Prof. Gerlach had to sleep once in a while …Anyone who has not been through it cannot at all imagine how great were the difficulties with an oven to heat the silver up to about 1300° K within an apparatus which could not be heated in its entirety [the seals would melt] and where a vacuum of 10^{-5} Torr had to be produced and maintained for several hours. The cooling was done with solid carbon dioxide and acetone or with liquid air. The pumping speed of the Gaede mercury backing pumps and the Volmer mercury diffusion pumps was ridiculously low compared with the performance of modern pumps. And then their fragility; the pumps were made of glass and quite often they broke, either from the thrust of boiling mercury …or from the dripping of condensed water vapor. In that case the effort of several days of pumping, required during the warming up and heating of the oven, was lost. Also, one could be by no means certain that the oven would not burn through during the four- to eight-hour exposure time. Then both the pumping and the heating of the oven had to be started from scratch. It was Sisyphus-like labor and the main load of responsibility lay on the broad shoulders of Prof. Gerlach. In particular, W. Gerlach would take over the night shifts. He would get in at about 9 p.m. equipped with a pile of reprints and books. During the night he then read the proofs and reviews, wrote papers, prepared lectures, drank plenty of cocoa or tea and smoked a lot. When I arrived the next day at the institute, heard the intimately familiar noise of the running pumps, and found Gerlach still in the lab, it was a good sign: nothing broke during the night.

Then I arrived at the institute one morning in February 1922; it was a wonderful morning: with cool air and fresh snow! W. Gerlach was once again at it, developing the deposit of an atomic beam that had been passing through an inhomogeneous magnetic field for eight hours. Full of expectation, we applied the development process, whereupon we experienced the success of several months of effort: The first splitting of a silver beam in an inhomogeneous magnetic field. After Master Schmidt and, if I remember correctly, E. Madelung had seen the splitting [the deposit was about 1.1 mm long and the splitting only about 0.06 to 0.1 mm], we went to Mr Nacken to the Mineralogical Institute to have the finding recorded on a microphotograph. Then I was tasked with sending a telegram to Professor Stern in Rostock, with the text: "Bohr is right after all!"

The consequences and impact of the stroke of luck for the emerging quantum physics that the collaboration between Otto Stern and Walther Gerlach at Frankfurt was are described in Chap. 5. We note that Albert Einstein and Paul Ehrenfest coined the term Stern-Gerlach experiment, in recognition of the fact that it was Stern who conceived it, although Gerlach largely carried it out (Einstein and Ehrenfest 1922). Moreover, Otto Stern was the pioneer of *quantitative* experiments with molecular beams.

In 1924, Einstein nominated, alongside with others, both Stern and Gerlach for the Nobel Prize in Physics (Schmidt-Böcking et al. 2019). By 1944, Gerlach and Stern had been nominated together thirty-one times for the Nobel Prize, cf. Fig. 1. Stern received fifty-two additional nominations for his other experiments with the molecular beam method and was awarded the Nobel Prize in Physics in 1944 for the year 1943. Gerlach ended up empty-handed, although Manne Siegbahn (1886–1978), then chairman of the Nobel Committee for Physics, proposed Stern, together with Gerlach, in 1944 as the sole candidates. And Eric Hulthén (1891–1972) in his broadcast on Swedish Radio on 10 December 1944 honoring the award of the Nobel prize to Stern extolled almost exclusively the SGE. In the documents and reports of the Nobel Archives there is no indication as to why Gerlach was left out. The reason may have been Gerlach's high-level involvement in the Nazi research establishment, especially in the management of the nuclear program, see Sects. 1 and 4.

The personalities of Stern and Gerlach were quite different: while Gerlach enjoyed being in the driver's seat, Stern preferred the back seat. Only a few letters exchanged between them have been preserved. The following one, from 16 January 1924, concerns the last (Gerlach and Stern 1924) of their four joint publications, all of which dealt with the SGE (Schmidt-Böcking et al. 2019), p. 125:

> Dear Gerlach, many thanks for your messages. I thought our paper had arrived at the *Annalen* [der Physik] a long time ago. In any case, I totally vote for the *Annalen*, and you do too, for such long claptrap is nothing for the [Zeitschrift für Physik]. I couldn't come in during the week, not to [Frankfurt], because I had to go to Breslau, and [going to] both [places] was a little too much for me. For [molecular beams] I invent ever more ingenious apparatus that only keeps working worse, z.[um] K.[otzen]! In contrast, the [electric molecular beams] are quite endurable. But it all goes so terribly slowly!
>
> I hear that Schaefer got a call from Freiburg. He has to go there! Cordial greetings to all friends, your family, and yourself. Yours Otto Stern

When Gerlach succeeded Paschen at Tübingen, Stern sent him, on 16 November 1925, the following telegram (Schmidt-Böcking et al. 2019), p. 125:

> = Cordial congratulations to the *Grossbonzen* [big shot] from Stern +

Whereupon Gerlach replied (Schmidt-Böcking et al. 2019), p. 126:

> Dear Stern, it is Sunday, 22 November, and I just got your telegram. As I started writing the above, the furniture trucks have arrived … So I begin my rant in the hope that my wife will leave me alone for a moment.
>
> … Mr. S. made statements about the evaluation of our magneton experiments which—as we noticed from his multiple inquiries—give rise to the impression that our calculation could be 100% wrong; and furthermore that the evaluation did not take into account possible sources of error and uncertainties, and that, in particular, we missed out on taking the width of the slit into account. Although Mr S.'s reasoning is correct, his note is indeed likely to lead to misunderstandings.
>
> Mr. S. namely always speaks about the distance between the locus of maximum intensity on the deflected strip and the locus of the … narrow undeflected strip, for which case the formula we use would indeed give an almost 100% error. However, our measurements always refer to the center of the deflected strip, which Mr. S. only discusses at the end of his note; for this case, Mr. S. himself calculates a deviation of at most 20%.
>
> Furthermore, Mr. S. seems to assume that we were not aware of the influence of the distance of the slit. [In our paper] we refer to the work of Stern where this influence was discussed and the corresponding formula … that takes into account the Coriolis force was derived. Mr S. could have easily figured that out from the literature. At the time we just remarked as much … and stated a possible error on the order of magnitude of ±10%. We insist that Mr. S.'s note doesn't bring forth any new thoughts and that its content pretty much coincides with our presentation. We only object to the manner of his attack.
>
> Dear Stern, how are you health-wise? It's a pity that you weren't in Göttingen. Here [in Tübingen], there's a terrible mess [due to Gerlach's move]. Hopefully, it will sort itself out soon. I will then write to you about the atomic beam experiments. Please do publish

something with [Immanuel] Estermann again! Cordial greetings, also from my wife, Yours W. Gerlach

Next in the chronology of the preserved letters that bear upon the relationship between Stern and Gerlach is a note written by Stern from Zurich to Lise Meitner (1878–1968) in 1957:

> Dear Lise Meitner, ... So let's meet in Munich. However, I can only come for 1-2 days, for two reasons: (1) [I cannot be away from Zurich for more than 1–2 days, because I expect a visitor]; (2) I don't care about seeing the Munich physicist Mr. Gerlach. Therefore, I leave it entirely up to you when you and I will meet. Please just let me know as soon as possible. It was very nice to see [Otto Robert] Frisch again and to get to know his wife; they seem to fit very well together.

> The two of us, the old ones, will have a lot to chat about and I'm hugely looking forward to seeing you again. Most cordially, Yours Otto Stern

Then there is a postcard to Stern penned jointly by Walther Gerlach, Otto Robert Frisch (1904–1979), Immanuel Estermann (1900–1973), William Nierenberg (1919–2020), Hans Kopfermann (1895–1963), and Peter Toschek (1933–2020) from the Brookhaven Molecular Beam Conference that was organized by Hans Kopfermann and held at Heidelberg in 1959 (Schmidt-Böcking et al. 2019), p. 245:

> Lichtstrahlen sind zum Brechen, Atomstrahlen z. K.! [zum Kotzen]. [This is a kind of affectionate "secret code" between Stern and Gerlach from their Frankfurt time—a pun expressing their occasional disgust with their difficult atomic/molecular beam experiments. "Brechen" means refraction as well as vomiting; "Kotzen" is a vulgar word for vomiting. A free translation, without the pun, would be: Light beams refract, atomic beams disgust.] Too bad that you aren't here, but we think of you warmly! Yours Walther Gerlach

> Remarkably, I got to know Mr. Gerlach only here. But molecular beams have become awfully complicated! With cordial greetings, yours OR Frisch

> Cordial greetings, Estermann

> Best regards will see you soon! Nierenberg

> We were very sorry not to have you here. Yours Hans Kopfermann

> Cordial Greetings from yours P. Toschek

It can be gleaned from many letters held at Otto Stern's Estate (Schmidt-Böcking et al. 2019) that he had quite a friendly relationship with all his correspondents. The above-quoted letter to Lise Meitner from 22 April 1957 suggests that Stern's feelings towards Gerlach were/became less than cordial, at least at the time. Conversely, Walther Gerlach wrote and spoke about Stern with the highest respect and much affection. This transpires in particular in the obituary of Stern that Gerlach wrote for the *Physikalische Blätter* (Gerlach 1969):

> Those who knew him appreciated his open-mindedness—he was a grand seigneur!—his unconditional reliability, the fruitful and—due to his fast thinking—difficult discussions, and—for those ho had a sense for it—his often nearly sarcastic but well-conceived assessments of things and people; bossing people or poor manners were anathema to him.

> Although a theoretician by nature, Stern was full of experimental ideas, never at a loss for a new proposal if the implementation of the previous one failed. At our farewell from Frankfurt, I gave him, in memory of the months of hopeless striving to see space quantization, an ashtray

with the inscription [Stern's and Gerlach's "secret code" in our translation] "Light beams refract, atomic beams disgust;" this ashtray endured all those years till Berkeley—but our experimental apparatus, lab books, and the originals of our results had burned during the Second World War.

A special tribute to the *"Stern-Stunden"* in Frankfurt and their importance for the development of quantum physics was given by Walther Gerlach in his lecture on 2 March 1960—still during Stern's life—at the *Physikalischer Verein Frankfurt* (Gerlach 1960):

> Around 1910, the French physicist Dunoyer developed the method of the so-called atomic or molecular beams. These are atoms that fly along straight lines from an oven through a small orifice into a highly evacuated chamber. Here at this institute, Max Born, Elisabeth Bormann, and, foremost, Otto Stern took up this idea in 1920 and experimentally developed the atomic beam method. That was a risky undertaking as at the time the means to produce high vacuum were still extremely limited ... Stern succeeded in measuring the mean velocity of the atoms, Born and Bormann measured their mean free path, and in later years Stern also succeeded in measuring the velocity distribution in an atomic beam. In the meantime, this method has been so refined by [Immanuel] Estermann, who is now at Chicago, that it affords the best temperature measurement of gases or vapors at 2000 degrees or more. Finally, Stern was able to demonstrate that free-flying atoms follow a free-fall parabola like a projectile. Moreover, at this institute, the reality of space quantization was successfully demonstrated in an experiment that provided direct access to an atomic state predicted by quantum theory.

Upon finishing the SGE, Gerlach would return to what he called his "hobby," namely his research on radiation pressure that he had started already in 1913 in Tübingen (Huber 2015). The pursuit of this "hobby" was deemed to be about as difficult as the SGE (Rollwagen 1980). Gerlach's interest was likely triggered by the inherent connection between radiation pressure and the Stefan-Boltzmann law.

3.3 Radiation Pressure

Ludwig Boltzmann (1844–1906) succeeded in 1884 to derive the law, $I(T) \propto T^4$, cf. Eq. 2, that his teacher, Josef Stefan (1835–1893), found in 1879 empirically (Boltzmann 1884). In his derivation, Boltzmann invoked Maxwell's theory of electromagnetism and the second law of thermodynamics, prompted by an earlier attempt by Adolfo Bartoli (1851–1896) to arrive at Stefan's law by the same route. Boltzmann was able to show that substitution of the pressure $p = I(T)/(3c)$ exerted by blackbody radiation of energy density $I(T)/c$ into the second law of thermodynamics in the form $T\,dp - p\,dT = [I(T)/c]dT$ yields

$$\frac{dI(T)}{4I(T)} = \frac{dT}{T} \tag{5}$$

which upon integration indeed gives Stefan's law—since then also known as the Stefan-Boltzmann law.

During his detention at Farm Hall (see below), Gerlach reminisced (Gerlach 1945) about his early attempts to come to terms with the effects he observed with a Crookes radiometer (light mill), a contraption invented by William Crookes (1832–1919) in 1873:

> In Tübingen in 1913/14, I tried to enhance the sensitivity of the radiometer [consisting of vanes mounted on a spindle in a partially evacuated bulb] by implementing alternative shapes of the vanes. This is when I observed a "negative" rotation of the vanes, i.e., in the direction opposite to that of the incident light.

Gerlach's original idea that he could measure radiation pressure with a Crookes radiometer turned out to be overly optimistic, as the processes involved in the radiometer physics are all but simple. It would take Gerlach and his coworkers two decades (1913–1932) to clarify the "positive" and "negative" radiometer effects and to carry out an absolute measurement of radiation pressure. Was it worth the effort? For sure it was, as those who were (and, in some quarters, still are) credited with first measurements of radiation pressure—Pyotr Lebedev (1866–1912), Ernst Nichols (1869–1924), and Gordon Hull (1870–1956)—did not and could not have measured anything else than spurious radiometer effects. As Gerlach and coworkers would show in their work, these only disappear at a vacuum better than 10^{-6} torr, which was not attainable during the period 1901–1903 when Lebedev, Nichols, and Hull published their radiation pressure studies.

Gerlach reentered the fray in 1919 when he published, jointly with Wilhelm Westphal, a theory of the radiometer (Gerlach and Westphal 1919) that, however, had to be quickly retracted (Westphal 1919):

> More detailed considerations have shown ... that the theory is untenable, despite a very good agreement with experiment. In particular, Mr [Albert] Einstein gave me a friendly hint that [our theory] contradicts momentum conservation.

At the 1920 meeting of the German Physical Society in Berlin, Westphal noted (Westphal 1920):

> The goal of the investigations [of the radiometer effects] is to collect a complete set of experimental data needed for a theory of the radiometer.

Gerlach answered the challenge implied by Westphal's talk with a series of four papers entitled *Untersuchungen an Radiometern I–IV* [Investigations of the Radiometer I–IV] published between 1923 and 1932. The first paper of the series opens with the bold statement (Gerlach and Albach 1923):

> As is well known, there is no complete theory of the radiometer available.

The paper then describes a compensation radiometer consisting of a single vane with thermally insulated sides enclosed in a bulb filled with gas of variable pressure (in the range of 10^{-1}–10^{-4} torr). One side of the vane is a receptor of radiation, the other is an electrically heatable bolometer. Like in his pyrheliometer, see Sect. 3.1, the carefully controlled electric heating of the bolometer side made it possible to compensate for the heating of the other side by the incident radiation. The compensation was carried

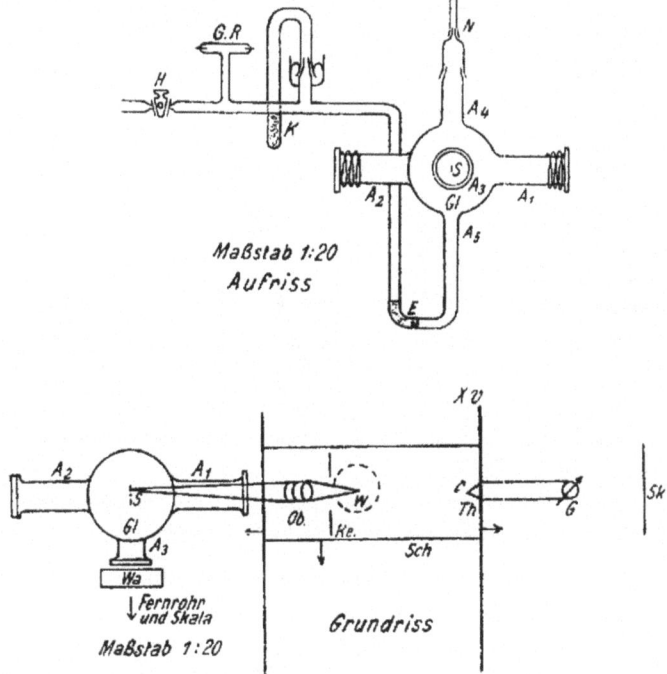

Fig. 13 The torsional radiometer of Gerlach and Golsen in side-view (top) and top-view (bottom) (Golsen 1924). The vane (not shown) used in the first quantitative measurement of radiation pressure was made of platinum foil (1.45 × 1.05 cm^2 and 7 μm thick). Its weight was balanced out by a platinum wire. The radiometer was housed in a glass ball (*Gl*) equipped with arms (A_1–A_5) for pumping and access and to allow to bring the radiation in and to take it out. It was evacuated by a Volmer diffusion pump combined with a cryo- and sorption pump (a *Volmeraggregat*) separated by a valve (*H*). The pressure was measured using a McLeod gauge and below 10^{-5} torr inferred from the damping of the torsional oscillations of the radiometer suspended on a 11 cm long quartz filament. A mirror (*S*) was attached to the filament to facilitate the read-out of the amplitude of the torsional oscillations. The radiation source was a tungsten arc lamp (*W*) whose output was focused on the vane by a camera lens (*Ob*). The power of the lamp was calibrated using a Hefner lamp and monitored during the measurements by a thermopile (*Th*) connected to a galvanometer (*G*). Except for the windows, the glass ball was shielded by a cotton-wool wrapping

out as a function of pressure for various absorption and thermal isolation materials. The instrument proved to be capable of sensitively measuring small changes of intense radiation.

However, Gerlach's goal was to directly measure light pressure rather than to investigate radiometer effects. To that end, he teamed up with Alice Golsen (Gerlach 1945):

> With Ms. Alice Golsen from Wiesbaden—who, as it turned out, was my classmate in 1896—I did the first measurement of radiation [pressure] as a precision measurement—with absolutely measured radiation energy. It was arduous but beautiful, clean work, a recuperation

Fig. 14 The dependence of the vane amplitude (ordinate) on the logarithm of gas pressure, $\log p$ (abscissa). The negative and positive amplitudes of the platinum vane refer, respectively, to deflections against and along the direction of the incident light beam. The various series of data points (•, ×, and ⊙) correspond to different irradiances and are all found to follow the same curve (Golsen 1924)

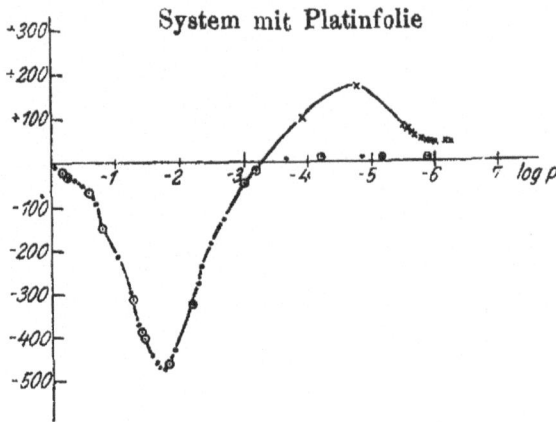

of sorts from the perpetual failures of the space-quantization experiments. In Ms. Golsen I found a wonderful collaborator, both scientifically and as a person.

Their collaboration resulted in the second paper (Gerlach and Golsen 1923) of the series as well as a detailed summary written by Alice Golsen (Golsen 1924). The aim of the experiment was to provide an unequivocal measurement of radiation pressure, free of radiometer effects and any disturbances. That meant that the radiometer measurements had to be done as a function of gas pressure all the way down to 10^{-6} or even 10^{-7} torr where a pressure dependence would vanish. A new apparatus was built, Fig. 13, that amounted to a torsion balance with a platinum vane attached to a quartz filament suspended in a glass ball. Its "rest-amplitude" observed at pressures below 10^{-6} torr was then attributed to radiation pressure. The measurements proceeded as follows: after several days of pumping, the dependence of the amplitude of the vane would be measured as a function of gas pressure at constant irradiation by a tungsten arc lamp, see caption to Fig. 13. The power of the lamp was monitored [normalized] by a thermopile. Achieving a steady-state amplitude lasted often for hours and was perturbed by outgassing as well as by the vibrations of the institute building. A typical dependence of the amplitude on gas pressure is shown in Fig. 14; it would take on the order of 100 h to acquire the data points shown. As one can see, at gas pressures between 1 torr and 10^{-4} torr, the amplitude is "negative," meaning that, upon irradiation, the vane moves against the incoming light beam. Only at pressures below 10^{-3} torr would the amplitude become "positive" (i.e., along the light beam direction), inching towards the pressure-independent "rest-amplitude" at pressures below 10^{-6} torr. In order to access the requisite pressure range, sorption pumping with charcoal and cryo-pumping with liquid oxygen (!) had to be applied—for days ... As stated by Gerlach and Golsen, cryo-pumping with dry ice had not sufficed to reach the "rest-amplitude" regime. The radiation pressure was then evaluated from the observed "rest-amplitude" and the measured properties of the torsion balance, such as its force constant. The measured light pressure (light force per illuminated surface area of the vane), p, and the calibrated irradiance, I^*, were then compared and found to obey the relationship

$$p = \frac{I^*}{c} \qquad (6)$$

with an accuracy of about 2%. This was the first-ever quantitative measurement of radiation pressure.

Gerlach and Golsen summarized their results thus (Gerlach and Golsen 1923):

1. In a vacuum from about 10^{-6} to 10^{-7} torr a constant amplitude ["rest-amplitude"] of the radiometer was found that is interpreted as purely due to radiation pressure.

2. This amplitude is proportional to the incident energy [power] and independent of the wavelength of the radiation.

3. The radiation pressure calculated from the constant amplitude agrees with the theoretical value.

In the third paper of the radiometer series (Gerlach and Madelung 1923), Gerlach and Erwin Madelung (1881–1972) debunk the radiometer theory published in 1922 by Edith Einstein. Finally, in 1932 Gerlach and Wilhelm Schütz publish the final, fourth sequel of the series (Gerlach and Schütz 1932) that deals with the radiometer effects at "high pressures" and corroborates the recent model put forward by Paul Epstein (Epstein 1929).

In 1975, Gerlach wrote a rebuttal (Gerlach 1975) to an article published in *Physik in unserer Zeit* whose author repeated the claim that radiation pressure was measured for the first time in the experiments of Lebedev, Nichols, and Hull. We note that Gerlach provided an impetus in 1970 for the founding of *Physik in unserer Zeit*.

It is mind-boggling that Gerlach's work on radiation pressure is still not widely known and that most textbooks keep attributing the first measurements of radiation pressure to experiments in which it could have not been observed.

After completion of the radiation pressure work at Frankfurt, Gerlach moved to his second academic station—his alma mater—as *Ordinarius*. His appointment at Tübingen received a strong push from Albert Einstein (Rechenberg 1979). Figure 15 shows Gerlach during the Tübingen period. Figure 16 shows his extended family during that time.

In addition to his time-consuming research projects at Frankfurt, Gerlach wrote two books: *Experimentelle Grundlagen der Quantentheorie* (Gerlach 1921) and the acclaimed *Materie, Elektrizität, Energie* (Gerlach 1923), a survey of the development of atomism over the previous decade.

We note that among Gerlach's students at Frankfurt was Hans Bethe (1906–2005), who began his physics studies in 1924. In his reminiscence (Bernstein 1979), Bethe acknowledged that Gerlach's stimulating lectures on atomic physics became a decisive influence on his further work in physics.

Fig. 15 Walther Gerlach as
director of the Physics
Institute in Tübingen.
Courtesy of Werner Kittel,
Hamburg

4   Gerlach's Involvement in the *Uranprojekt*

The German *Uranprojekt* was no precision physics. Launched in reaction to the
discovery of nuclear fission by Otto Hahn, Lise Meitner, Fritz Strassmann, and Otto
Robert Frisch and in the wake of subsequent theoretical work by Niels Bohr and John
Wheeler, the project started taking shape already several months before the outbreak
of World War Two. Paul Harteck, the successor at Hamburg of the exiled Otto Stern,
had written in April 1939 to the *Reichswehrministerium* [Ministry of Defence] about
the promise of both a nuclear reactor and a nuclear weapon, amply described in the
publications by the above. Harteck's letter ended up at the *Heereswaffenamt* [Army
Ordnance Bureau]. In September 1939, the Bureau's Kurt Diebner (1905–1964)
and former Heisenberg student Erich Bagge (1912–1996) enlisted leading German
physicists—Walther Bothe (1891–1957), Hans Geiger (1882–1945), Heisenberg,
Hahn, Harteck, and Carl Friedrich von Weizsäcker (1912–2007)—in a wide-ranging
war-time nuclear program. This received additional support through an initiative
by Göttingen's Wilhelm Hanle (1901–1993) and Georg Joos (1894–1959) from the
Ministry of Education. The members of the group, also known as the *Uranverein*, got
promptly down to work. Heisenberg produced a secret report in which he described
a uranium nuclear reactor (*Uranmaschine*) and urged the Bureau's leadership to
support isotope separation not only as the surest path to a functional reactor but also

Fig. 16 The Gerlach family in Weimar in about 1927. From left: Walther Gerlach, Wolfgang Gerlach (brother of Walther Gerlach), Ruth Gerlach, neé Probst (2nd wife of Walther Gerlach), Valentin Gerlach (Walther Gerlach's father), Ingeborg Gerlach (elder daughter of Werner Gerlach and his wife Henriette "Henny" Syffert, who in 1943 married Wolfgang Kittel; they had two sons: Werner Kittel, born in 1945, and Gerd Kittel, born in 1948), Marie Gerlach, neé Niederhaeuser (mother of Walther Gerlach), Henny Gerlach, neé Syffert (wife of Werner Gerlach), and Werner Gerlach (brother of Walther Gerlach). Courtesy of Werner Kittel, Hamburg

to a nuclear bomb, without specifying the critical mass of U-235 needed (Cassidy 2017), p. 49. Based on the flawed research by Bothe on neutron capture by carbon, the *Heereswaffenamt* introduced the fatal mistake into the German nuclear program by branding graphite as an unsuitable moderator and relying on heavy water instead (Walker 1995), p. 225. Enrico Fermi's reactor at Chicago went critical in December 1942 using highly-purified graphite as a moderator. The loss of the heavy-water plant Norsk Hydro in Nazi-occupied Norway in early 1943 would then in effect upend the German nuclear program that relied on heavy water as a moderator. The *Uranprojekt* would continue, however, until the seizure of the German nuclear equipment by the American-led *Alsos Mission* in April-May 1945.

In 1941, several centers of German nuclear research emerged, all at first coordinated by Diebner and Bagge and concerned with aspects of the nuclear reactor as outlined by Heisenberg in his report. The most significant among them were Heisenberg's own institute at Leipzig and the Kaiser Wilhelm Institute (KWI) for Physics in Berlin, which fell under military command with Diebner installed as its acting director. Further reorganization saw Heisenberg appointed director of the KWI and Diebner relegated to an army research station in Gottow near Berlin. In August, Fritz Houtermans (1903–1966) and, independently, von Weizsäcker, demonstrated theoretically that Pu (plutonium) 239, produced in a uranium reactor from U-238 by neutron capture and subsequent β-decay, was at least as fissionable as U-235. As a result, an atom bomb suddenly appeared feasible. A controversial trip of Heisenberg and von Weizsäcker to see Bohr in Copenhagen followed. With the *Wehrmacht* defeated at Moscow and stuck at Leningrad, and the consequent mobilization of the German economy, the Army Ordnance Bureau approved funding, in February 1942, essentially only for Diebner's operation in Gottow (Cassidy 2017), p. 54. Heisenberg's KWI, however, had a sponsor in Abraham Esau of the Reich Research Council of the Ministry of Education and eventually of the *Reichsminister* Bernhard Rust himself. After a tantalizing conference, in February 1942, chaired by Rust on the prospects of a nuclear reactor, including its ability to breed fissionable plutonium, Esau was appointed, in December 1942, *Reichsbevollmächtigter* [Reich Plenipotentiary] for nuclear physics. But then the new Minister of Armaments, Albert Speer, induced Hitler to appoint Hermann Göring as head of the Reich Research Council whereby Esau became Göring's representative for nuclear issues. Already in July 1942, Heisenberg received a dual appointment in Berlin—as director of the KWI for Physics and professor of physics at the Berlin University. Heisenberg would use his expanded influence to push for Esau's replacement by a kindred spirit—Walther Gerlach. And indeed, as of 1 January 1944, Gerlach would become *Reichsmarschall's* Plenipotentiary for nuclear physics and remain in this position for sixteen months until his capture by the *Alsos Mission*.

Gerlach moved to his third academic station, *Ludwig-Maximilians-Universität* in Munich, on 1 October 1929 as the successor of Willy Wien. In 1935, a battle with the proponents of the so-called *Deutsche Physik*—Johannes Stark, Philipp Lenard, and their followers (Walker 1995)—flared up for the succession of the recently retired Arnold Sommerfeld (1868–1951), who held Munich's chair in theoretical physics. Gerlach headed the university's hiring committee, which chose Sommerfeld's former pupil, Werner Heisenberg—then already a Nobel laureate—to fill the vacant chair. The battle, which went through several stages and included public Nazi denunciations of the "White Jew" Heisenberg as well as an intervention by Heinrich Himmler (1900–1945) on Heisenberg's behalf, raged until September 1939 when Heisenberg was finally exonerated after an extensive SS investigation. However, in the meantime, the Munich chair went to a Nazi, Wilhelm Müller (1880–1968), an applied physicist. Whereupon Gerlach declared physics "dead" in Munich … Heisenberg stayed put in Leipzig, until he received the call from Berlin.

Heisenberg and his *Uranverein* would hold additional presentations for both Speer and Göring and their staffs, carefully tailored to secure an autonomy of the physicists

Fig. 17 From left: Otto Hahn, Walther Gerlach, and Carl Friedrich von Weizsäcker in Göttingen, late 1950s. All three were members of Göttinger Achtzehn. Creative Commons

in setting the goals for the nuclear program and avoiding being "ordered to build the bomb; since failure to do so at the height of war would surely have meant execution" (Cassidy 2017), p. 55. We note that Heisenberg's understanding of the functioning of the bomb was inadequate all the way down to Farm Hall, as his recorded lecture to and conversations with his detained colleagues attest. As a result, his estimate of the critical mass of U-235 was orders of magnitude too high and so was the time needed to accumulate it by isotope separation (Bernstein 2001), pp. 129–131. Figure 17 shows Gerlach later on with two of his Farm Hall fellow detainees and interlocutors, Hahn and von Weizsäcker.

In June 1942, a heavy non-nuclear accident damaged the nuclear research laboratory at Leipzig. Afterwards, significant reactor research continued at two locations only—Heisenberg's KWI in Berlin and Diebner's facility in Gottow. Based on his calculations, Heisenberg concluded that about three tons of cast uranium and one and half tons of heavy water were needed in order to achieve a chain reaction in a cylindrical arrangement with rolled uranium plates interspersed with heavy water, a reactor design Heisenberg started building in a bunker at his KWI. Diebner, on the other hand, bet on using cast uranium in the form of cubes suspended on chains and immersed in frozen heavy water. When the ordered amounts of uranium finally arrived from the *Auergesellschaft*, Diebner's design produced a much higher neutron multiplication than Heisenberg's. Once Gerlach took over as Plenipotentiary for nuclear research, he diverted resources toward Diebner's facility, but enabled Heisenberg's operation to run in parallel, thereby thinning key resources, especially the wanting heavy water. By then, the Allied aerial bombing raids on Berlin became heavy enough for the city to start evacuating. On Speer's order, a large part of the personnel of Heisenberg's

KWI was moved to Hechingen, a rural place in Württemberg, not far from Tübingen. When Otto Hahn's KWI, a stone's throw from Heisenberg's, was destroyed in a targeted air raid, its personnel was moved to Tailfingen, not far from Hechingen. However, Heisenberg, his close associate Karl Wirtz (1910–1994) and their coworkers would stay on at the KWI for Physics in Dahlem and continue their attempts to get their reactor going. But at the end of 1944, with the Soviet Army reaching the left bank of the Oder river, Gerlach ordered both Heisenberg's and Diebner's groups to load their research equipment on trucks and move along with it to Hechingen. Once the convoy reached the experimental station of the *Reichsforschungsrat* in Stadtilm, about halfway, Gerlach pressed Diebner to stay there and make a final attempt to achieve chain reaction. Heisenberg's group, upon reaching Hechingen, set up a reactor in a cave—in fact a wine cellar—in a nearby village called Haigerloch. Their attempts, joined by von Weizsäcker, ended when the Haigerloch reactor was seized by the *Alsos Mission*. Gerlach's decision to enable Diebner his last-ditch effort is somewhat reminiscent of Gerlach's stubbornness in his own research that had so often paid off …

Apparently, Heisenberg and Gerlach—and most others involved—struggled until the last moment not only out of scientific interest but also to salvage their scientific reputation. As David Cassidy put it (Cassidy 2017), p. 58:

> For Heisenberg, success would have demonstrated the survival of decent German physics, and, perhaps equally [importantly], would have made German physicists influential figures in the postwar reconstruction of Germany.

In his conversation with Otto Hahn at Farm Hall secretly recorded after the atomic bombing of Hiroshima, Gerlach made a similar point but added yet another dimension to it (Hoffmann 1993), p. 157:

> When I took [the *Uranprojekt*] over, I talked it over with Heisenberg and Hahn, and I said to my wife: "The war is lost and the result will be that as soon as the enemy enters the country I will be arrested and taken away." I only did it because, I said to myself, that [fission] is a German affair and we must see [to it] that German physics be preserved. I never thought for a moment of a bomb but I said to myself: "If Hahn has made this discovery, let us at least be the first to make use of it." When we get back to Germany we will have a hard time. We will be looked upon as the ones who have sabotaged everything. We will not remain alive [for] long there. You can be certain that there [will be] many people in Germany who [will] say that it is our fault. Now please leave me alone.

Gerlach withdrew from Haigerloch to Munich, "where he quietly resumed his pre-war work in his university laboratory" and was captured there on 20 April 1945 (Cassidy 2017), p. 75. He was first interned with a group of high-ranking Nazis and only on 15 June reunited with a group of detained German nuclear physicists. From 3 July 1945 until 2 January 1946, he was "detained as guest of His Majesty" (Gerlach 1978b) at Farm Hall in Cambridgeshire (Operation Epsilon), together with Erich Bagge, Kurt Diebner, Otto Hahn, Paul Harteck, Werner Heisenberg, Horst Korsching, Max von Laue, Carl Friedrich von Weizsäcker, and Karl Wirtz. The daily life at Farm Hall was described by Gerlach as follows (Gerlach 1978b):

Five prisoners of war were taking care of cooking, house cleaning, and service. There were no interrogations or tasks so that we could use most of our time for work, for which the necessary literature was provided; radio, a good library, and a large park were all at our disposal; there were occasional trips to London or Middle-England. Hahn was the "doyen," who would smooth out occasional disagreements with the American and British officers. The rapport with the two British attending officers, who would also partake in common lunches and dinners, was amicable to the point of being personal. The good atmosphere would be only seldom disturbed by a visit by a high inspector of the secret service.

In the Farm Hall Protocols, Gerlach was characterized as "cheerful" and "cooperative" but, "based on the recorded conversations," under suspicion of "having had connections to the Gestapo" (Hoffmann 1993), p. 64. We have not found evidence in support of this suspicion, but Gerlach's involvement with the Nazi regime still remains an open question. However, as for instance Paul Rosbaud's testimonial suggests, see Sect. 1, Gerlach harboured a strong anti-Nazi sentiment. And he apparently never joined the NSDAP. But his brother Werner Gerlach (1891–1963), a professor of pathology, was an early NSDAP member and held a high honorary rank in the SS (Simon 2002). Werner would have a falling out over his NSDAP membership with his principled father, Valentin Gerlach. We hope that ongoing research will provide more clarity.

Ironically, the Farm Hall Protocols recorded the following conversation (Hoffmann 1993), p. 100:

Diebner: I wonder whether there are microphones installed here?

Heisenberg: Microphones installed? (laughing) Oh no, they are not that cunning. I don't think they know the real Gestapo methods; they're a little old fashioned in this respect.

Upon his release from Farm Hall, Gerlach, along with his fellow detainees, was confined to the British Zone of Occupation. Nevertheless, within the British Zone, he was free to accept a professorship at the University of Bonn. In April 1948 he would be free to return to Munich, in the American Zone of Occupation. In postwar Munich, Gerlach dedicated much of his time and effort to the restoration of the German academia in general and the *Ludwig-Maximilians-Univerität* in particular, including the resurrection of its Institute of Physics. Figure 18 shows Gerlach at the General Assembly of the Max-Planck-Gesellschaft, on whose Senate he served since 1951. His success in helping to raise the country from the ashes would earn him the highest honours in the Federal Republic, such as the Order *Pour le Mérite für Wissenschaften und Künste* awarded to him in 1970 by the President of Germany. In the context of this volume we note that, in 1988, the Stern-Gerlach Prize (since 1993 the Stern-Gerlach Medal) was established as the most prestigious German award for work in experimental physics, cf. Chap. 5. As a further example of Gerlach's stature we show a recently recovered silver plate, Fig. 19, that Gerlach received on his 70th birthday from the Senate of the Max-Planck-Gesellschaft in recognition of the services he provided as a member of the body over several decades.

Fig. 18 Walther Gerlach at
the general assembly of the
Max-Planck-Gesellschaft in
Stuttgart in 1956. Courtesy
of the Archiv der
Max-Planck-Gesellschaft

Fig. 19 Silver Plate
presented to Walther Gerlach
on the occasion of his 70th
birthday by the Senate of the
Max-Planck-Gesellschaft.
Long after Gerlach's death it
was passed on by his second
wife Ruth, see Fig. 16, to her
nephew, Werner Kittel. From
him it was acquired in 2020
by Horst Schmidt-Böcking
for the *Physikalischer Verein
Frankfurt*. Photo H.
Schmidt-Böcking, 2020

5 Gerlach's Work in the History of Science

From early on, Walther Gerlach cultivated a sense for the history of physics, perhaps in keeping with Goethe's maxim that "the history of a science is that science itself." Gerlach's first piece in the history *of* physics (Gerlach 1924) appeared at a time when he himself was making history *in* physics. As Gerlach's bibliography compiled by Margret Nida-Rümelin reveals (Nida-Rümelin 1982), this would be followed by about 500 additional publications on the history of physics/science, including about 60 scientific biographies, as well as outreach articles. During his distinguished career, Gerlach gave numerous talks on issues ranging from scientific funding to epistemological considerations, some of which would later be published. These are also included in the above number of 500.

Gerlach's sense for the history of science would also come to the fore in his capacity as educator. Like his academic mentor, Friedrich Paschen, Gerlach indulged his students in the spectacle of well-prepared experiments, some of which recapitulated chapters from the history of physics. The demonstration of Otto von Gericke's hemispheres, refuting the *horror vacui* theory, evacuated by Gerlach, a pioneer of high-vacuum technology, must surely have been a treat! Gerlach would also ask his students history questions during exams (Bachmann and Rechenberg 1989, p. 145). As Bachmann and Rechenberg report (Bachmann and Rechenberg 1989, p. 146):

> When [Gerlach] realized *how* Newton brought out certain optical phenomena or Goethe observed phenomena that seemingly disproved them, he would be perhaps more pleased than if he discovered an altogether new physical effect.

Gerlach's writings on the history of science are based on his detailed knowledge of the subject—and its literature. He would have likely concurred with Steven Weinberg when he remarked (Weinberg 1998): "By assuming that scientists of the past thought about things the way we do, we make mistakes; what is worse, we lose appreciation for the difficulties, for the intellectual challenges, that they faced."

One of Gerlach's personal heroes was Johannes Kepler (1571–1630), whom he extolled not only as the first physicist in history worthy of the name, but also as a forerunner of humanism—"a priest of the book of Nature" (Gerlach 1972):

> It was an unbearable thought … for Kepler that, on the one hand, human reason enables insight into the wonders of Nature (and "only science reveals wonders"), into the *harmonic* order of the world, but, on the other, that human life generally passes in *disharmony*, driven by quarrel, conflict, hate, and war.

Gerlach also details Kepler's relationship with Galileo (1564–1642), who kept snubbing Kepler, whether about celestial mechanics or optics. But it was Kepler, Gerlach points out, who provided, through his third law (published in 1619) relating quantitative properties of the orbits of different planets, the most irrefutable evidence for the heliocentric system. Galileo would, however, never use it in his defense during the 1633 trial by the Inquisition. The lack of appreciation for Kepler in some quarters may have aroused special sympathy in Gerlach, as he too had not always received due recognition, see Sects. 1 and 3.3. However, there's no trace of complaint

Fig. 20 Plaque at the entrance of the former *Physikalisches Institut* of the University of Frankfurt, Robert-Mayer Str. 2–4. Photo H. Schmidt-Böcking, 2002

about it in Gerlach's correspondence or any other source available to us. Secondly, Tycho de Brahe's measurements and their interpretation and analysis by Kepler of the eccentricity (0.0934) of Mars' orbit were revolutionary (in this case, also literally) precision measurements! And finally, Gerlach and Kepler were connected by the vicissitudes of their religious identity: they were both Protestants living in Catholic environments.

History of science was Gerlach's main preoccupation during the last twenty years of his life. His wide-ranging erudite historical writings deserve to be better known.

Fig. 21 Double-portrait of Otto Stern and Walther Gerlach by Jürgen Jaumann. The schematic of the Stern-Gerlach experiment and its outcome was drawn by Theodor Hänsch. Photo H. Schmidt-Böcking, 2020

6   In Conclusion

Walther Gerlach lives on through his enduring legacy in physics, higher learning, and history of science. His estate, held at the Deutsches Museum in Munich, is comprised of sixteen thousand items. Walther Gerlach also lives on in a number of public depictions, among them the memorial plaque, Fig. 20, designating *Die Alte Physik* building in Frankfurt as the site where the Stern-Gerlach experiment was carried out. The Physics Department at Frankfurt also features a double-portrait of Stern and Gerlach, Fig. 21.

We close with Gerlach's credo (Gerlach 1978):

Etwas Gutes kommt nie zu spät. [It's never too late for something good to happen.]

Appendix: Timeline of Walther Gerlach's Life and Career

The timeline below has been translated and adapted from the catalogue of the 1989 centennial exhibition *Walther Gerlach—Physiker—Lehrer—Organisator* at the *Deutsches Museum* in Munich curated by Rudolf Heinrich und Hans-Reinhard Bachmann (Heinrich and Bachmann 1989).

- **August 1889–March 1908 Childhood, Youth**
- 1 August 1889 Walther Gerlach was born in Biebrich am Rhein near Wiesbaden at 8:15; his mother was Maria Wilhelmine, neé Niederhaeuser; his father Dr. med. Valentin Gerlach, physician and chemist, Freemason and Goethe-expert
- 4 September 1891 Birth of twin brothers Werner and Wolfgang, joint Protestant baptism of all three brothers on 26 April 1896 in Bergkirche Wiesbaden
- 1895–1896 Volksschule [elementary school]
- April 1896–March 1899 City Middle School Wiesbaden
- April 1899–March 1908 Royal Humanities High School [Königliches Humanistisches Gymnasium] in Wiesbaden
- 9 March 1908 Abitur [finals] at the Royal Humanities High School in Wiesbaden
- **April 1908–Juli 1915 University studies in Tübingen**
- April 1908–February 1911 Studies at the Eberhard-Karls-Universität Tübingen: Since the 1st semester prepares to major in philosophy and mathematics; since the 5th semester in physics and chemistry. Gerlach attends lectures on philosophy by Ernst Adickes, mathematics by Alexander von Brill, experimental physics by Friedrich Paschen, theoretical physics by Richard Gans and Edgar Meyer
- April 1908 Joins Corps Borussia
- 15 November 1910 Student-Assistant of F. Paschen at the Institute of Physics, University of Tübingen (received an annual stipend of 1850 RM)
- March 1911 Exmatriculation
- 29 February 1912 Graduated "magna cum laude" with a thesis entitled "Eine Methode zur Bestimmung der Strahlung in absolutem Mass und die Konstante des Stefan-Boltzmannschen Strahlungsgesetzes." Adviser: Friedrich Paschen
- **August 1915–October 1920 First World War and First Employment**
- August 1915 Drafted to serve with the Infantry Regiment 247 in Ulm
- December 1915 Dismissed due to illness
- 29 April 1916 *Habilitationskolloquium* in Tübingen
- May 1916 Named *Privatdozent* at the University of Tübingen
- 6 May 1916 Drafted by the Pioneer Battalion Berlin-Schöneberg, subordinated to the *Prüfungskommission* [Examining Board]; Military rank: *Pioniergefreiter* [pioneer private]

- 2 June 1916 Assigned by the Tübingen Faculty to give a lecture "Über die Existenz eines Elektrizitätsatoms" [On the existence of an atom of electricity]
- 22 July 1916 Submitted habilitation thesis entitled "Experimentelle Untersuchungen über die Messung und Grösse der Konstanten des Stefan Boltzmannschen Strahlungsgesetzes" (Adviser F. Paschen)
- Fall 1916 Promoted to the rank of *Oberingenieur* [chief engineer] at the Inspectorate of the Radio Units. Assigned to the technical Department of the Radio Units (Tafunk), deployed to the test stations and factories in Würzburg, Stuttgart (at Bosch), and Jena
- Fall 1916 Drafted by the VIth Army in Flanders and Artois
- Dezember 1916–January 1917 Hospitalized at the surgical clinic of *Lazarett* Jena
- January–September 1917 With Tafunk in Berlin and Jena
- August 1917 Habilitation in Göttingen co-sponsored by Waldemar Voigt and Peter Debye; appointed as *Privatdozent*
- 12 September 1917 Relinquished the right to teach at the University of Tübingen
- 5 March 1918 Assigned to the back-up radio company Döberitz; takes part in the campaign in Champagne and Flanders
- 20 June 1918 Contracted the "Spanish flu;" at the *Lazarett* Mannheim
- Oktober 1918 Relocated to Tafunk in Berlin-Stahnsdorf
- December 1918 Carried out demobilization tasks for the Ministry of War
- 27 January 1919 Dismissed from Tafunk Berlin
- **February 1919–October 1920 Head of the Physics Laboratory of the *Farbenfabriken* Elberfeld**
- **October 1920–December 1924 *Privatdozent* and *Extraordinarius* Professor at the University of Frankfurt**
- 1 October 1920 First Assistant and *Privatdozent* at Richard Wachsmuth's Institute for Experimental Physics at the University of Frankfurt
- 1 November 1920 Senior Assistant and *Privatdozent* with the title *Extraordinarius* at the University of Fankfurt
- 8 Februar 1922 Evidence for space quantization of silver atoms in a magnetic field (Stern- Gerlach effect)
- 1 March 1923 Reported the first quantitative measurement of radiation pressure (with Alice Golsen)
- **January 1925–September 1929 Professor in Tübingen**
- 1 January 1925 *Ordinarius* Professor and Director of the Institute of Physics of the University of Tübingen as successor of his mentor Friedrich Paschen (Paschen left to become the President of the Physikalisch-Technische Reichsanstalt in Berlin)
- 2 December 1926 Public inaugural lecture in Tübingen: "Über das Wesen physikalischer Erkenntnis und Gesetzmässigkeiten"
- 3 June–5 July 1927 On leave at the University of Zurich working with Edgar Meyer
- 1928 Dean of the Faculty of Mathematics and Physics of the University of Tübingen
- **October 1929–May 1945 Professor in Munich (1st tenure)**
- 1 October 1929 *Ordinarius* Professor at the Ludwig-Maximilians-Universität Munich as successor to the deceased Willy Wien
- 22 February 1930 Elected Member of the Bavarian Academy of Science

- 15 June 1931 Member for life of the [governing] Committee of the Deutsches Museum
- Fall 1933 Banned from lecturing and administering exams for being allegedly unsuited to educate German Youth
- Beginning of 1934 Lifting of the lecturing ban
- 31 January 1935 Elected to a three-year membership in the administrative committee of the Deutsches Museum
- 20 March 1936 Participation at a conference on gravitation in London
- 1936 Lifting of the ban to administer examinations
- 1937 Elected Senator of the Kaiser-Wilhelm-Gesellschaft (forerunner of Max-Planck-Gesellschaft)
- 18 August 1938 Attended the symposium "Modern Methods of Chemical Analysis" in London, organized by the British Association, Cambridge
- Beginning of 1939 Founding of the international journal "Spectrochimica Acta" with Paul Rosbaud
- May 1939 Lecture tour in Poland
- November 1939 Prof. Dr.-Ing. Ernst August Cornelius from the *Technische Hochschule Charlottenburg* in Berlin entrusted by the Supreme Command of the Navy to establish a work group named after him—Arbeitsgruppe Cornelius (AGC); Gerlach together with about fifteen additional scientists from industry and universities called upon to join AGC, which cooperated, among others, with Askania-Werke in Berlin—a manufacturer of torpedos
- 27 November 1939 Gerlach tasked, within the AGC, with the development of methods for demagnetization of ships and torpedos, defusing magnetic mines and the development of magnetic fuses
- 1 October 1943 AGC was dissolved
- 1 January 1944 Hermann Göring named Gerlach head of the Physics Section in the *Reichsforschungsrat* and Plenipotentiary for nuclear physics, as successor to Abraham Esau
- April 1944 Gerlach founded the journal "Reichsberichte für Physik" [Reich Reports on Physics] which is slated explicitly for internal use only
- 31 January 1945 Relocation of part of the nuclear program (Diebner's group) to Stadtilm in Thuringia
- End of February 1945 Relocation of the rest of the nuclear program (Heisenberg's group) to Hechingen and Haigerloch in Württemberg
- **May 1945–March 1948 Detention, Professorship in Bonn**
- 3 May 1945 Relocation to Heidelberg by U.S. Army officers; meeting with Samuel Goudsmit
- 10 May–15 Juni 1945 Detention in France and Belgium (Le Vésniet, Le Grand Chesnay, Faqueval)
- 3 July 1945–2 January 1946 Detention at Farm Hall in England
- January 1946 In Alswede near Hannover
- 5 February 1946 Arrival in Bonn; ordered not to leave the British Zone of Occupation

- February 1946–31 March 1948 Assumed the duties of the chair and director of the Institute of Physics of the University of Bonn
- Spring 1946 President of the *Notgemeinschaft der Deutschen Wissenschaft* [German Science Foundation] in North Rhine-Westphalia
- 11 September 1946 Founding Member of the Max-Planck-Gesellschaft (in the British Zone)
- **April 1948–September 1957 Professor in Munich (2nd tenure)**
- 1 April 1948 Resumption of the professorship at Munich after the lifting of the ban on leaving the British Zone (Gerlach's substitute since May 1945 was Eduard Rüchardt)
- 7 May 1948 Elected to a three-year membership in the administrative committee of the Deutsches Museum
- 1948–1951 Rector of the *Ludwig-Maximilians-Universität* in Munich
- January 1949–June 1961 Vice-President of the *Notgemeinschaft der Deutschen Wissenschaft* and its successor organization, the *Deutsche Forschungsgemeinschaft* (DFG)
- 7 May 1949 Elected to a three-year term in the Governing Board of the Deutsches Museum; in 1963 Gerlach would be elected again for a three-year term and finally, in 1968, for life
- 1949 Founding President of the Fraunhofer-Gesellschaft
- 1951–1969 Member of the Senate of the Max-Planck-Gesellschaft
- 1956 –1957 President of the Association of the German Physical Societies
- 12 April 1957 Involvement in the preparation and signing of the Declaration of the Göttingen Seven
- 1957 Member of the Kepler Committee of the Bavarian Academy of Sciences
- October 1957–August 1979 Emeritus in Munich
- 1959 Founding Member of the Vereinigung Deutscher Wissenschaftler (VDW)
- **1965–1979 Research Fellow at the Forschungsinstitut für die Geschichte der Naturwissenschaften und der Technik at the Deutsches Museum**
- 1970 awarded Order *Pour le Mérite für Wissenschaften und Künste* by the President of Germany
- 26–28 August 1971 Attended the International Congress on the History of Science in Leningrad; talk on Johannes Kepler
- 16 May 1979 Received an honorary degree from the Faculty of Physics of the University of Tübingen
- 10 August 1979 Walther Gerlach died in Munich shortly after his 90th birthday

References

H.-R. Bachmann, H. Rechenberg (1989). *Walther Gerlach (1889–1979): Eine Auswahl aus seinen Schriften und Briefen* (Springer, Heidelberg)

J. Bernstein (1979). Profiles: master of the trade-I (Hans Albrecht Bethe). The New Yorker (December 3), pp. 50–107

J. Bernstein (2001). *Hitler's Uranium Club The Secret Recordings at Farm Hall* (Springer, Heidelberg)

L. Boltzmann (1884). Ableitung des Stefan'schen Gesetzes betreffend die Abhängigkeit der Wärmestrahlung von der Temperatur aus der elektromagnetischen Lichttheorie. Annalen der Physik **22**, 291–294

M. Born (1969). Max Born to Albert Einstein 12 February 1921, in *Albert Einstein, Hedwig und Max Born: Briefwechsel*, ed. by M. Born (Nymphenburger Verlagshandlung, München), pp. 1916–1955

D. Cassidy (2017). *Farm Hall and the German Atomic Project of World War II: A Dramatic History* (Springer, Heidelberg)

W.W. Coblentz (1913). Der gegenwärtige Stand der Bestimmung der Strahlungskonstanten eines schwarzen Körpers. Jahrbuch der Radioaktivität und Elektronik **10**, 340–367

CODATA (2020). The NIST *Reference on Constants, Units, and Uncertainty, Fundamental Physical Constants*. https://physics.nist.gov/cgi-bin/cuu/CCValue?sigmalShowFirst=Browse

A. Einstein, P. Ehrenfest (1922). Quantentheoretische Bemerkungen zum Experiment von Stern und Gerlach. Zeitschrift für Physik **11**, 31–34. Reprinted in (Buchwald et al., 2012, Doc. 315)

P.S. Epstein (1929). Zeitschrift für Physik **54**, 537

I. Estermann (1975). History of molecular beam research: personal reminiscences of the important evolutionary period 1919–1933. Am. J. Phys. **43**, 661

Ch. Féry (1909). Annales de Chimie et de Physique (VIII) **17**, 267; Comptes rendus de l'Académie des Sciences **148**, 515

O.R. Frisch (1963). Interview of Otto Frisch by Thomas S. Kuhn on 1963 May 8, Niels Bohr Library & Archives, American Institute of Physics, College Park, MD U.S.A. www.aip.org/history-programs/niels-bohr-library/oral-histories/4615

W. Gentner (1980). Gedenkworte für Walther Gerlach," in*Orden Pour le Mérite für Wissenschaften und Künste, Reden und Gedenkworte*, Sechzehnter Band (Verlag Lambert Schneider, Heidelberg), http://www.orden-pourlemerite.de/band/1980

W. Gerlach, Autobiographische Notizen 1908–1950. Nr. 27, Archiv des Deutschen Museums in München, NL 80/053 (1908–1950)

W. Gerlach (1912). Eine Methode zur Bestimmung der Strahlung in absolutem Mass und die Konstante des Stefan-Boltzmannschen Strahlungsgesetzes. Annalen der Physik **38**, 1–29 (Ph.D. Thesis)

W. Gerlach (1916). *Experimentelle Untersuchungen über die absolute Messung und Grösse der Konstante des Stefan-Boltzmannschen Strahlungsgesetzes.* (J.A. Barth, Leipzig) (Habilitation Thesis) (1916). See also W. Gerlach, Die Konstante des Stefan-Boltzmannschen Strahlungsgesetzes; neue absolute Messung zwischgen 20 und 450 °C. *Annalen der Physik* **50**, 259–269

W. Gerlach (1921). *Experimentelle Grundlagen der Quantentheorie* (Vieweg, Braunschweig)

W. Gerlach (1923). *Materie, Elektrizität, Energie. Die Entwicklung der Atomistik in den letzten 10 Jahren.* Dresden und Leipzig: Wissenschaftliche Forschungsberichte, Naturwissenschaftliche Reihe 7

W. Gerlach (1924). Alte und neue Alchimie. Frankfurter Zeitung (July 25)

W. Gerlach (1935). Friedrich Paschen zum siebzigsten Geburtstage. Forschungen und Fortschritte **11**, 50–51

W. Gerlach (1945). *Geschichtliche Notizen-geschrieben in Farm Hall 1945* (Archiv des Deutschen Museums in München, Privat Nachlass Gerlach)

W. Gerlach (1948). *Die Quantentheorie: Max Planck, sein Werk und seine Wirkung* (Universitätsverlag, Bonn)

W. Gerlach (1960). Über die Entwicklungen atomistischer Vorstellungen. Physikalischer Verein Frankfurt, 2 March

W. Gerlach (1969). Zur Entdeckung des 'Stern-Gerlach-Effektes. Physikalische Blätter **25**, 472

W. Gerlach (1972). Johannes Kepler zum 400. Geburtstag. Festrede, in *Bachmann and Rechenberg* 1989, pp. 181–202

W. Gerlach (1975). Lebedew misst nicht Strahlungsdruck. Zu dem Artikel K. Treml 'Der Strahlungsdruck des Lichtes'. Physik in unserer Zeit **5**, 100 (1974); Physik in unserer Zeit **6**, 32

W. Gerlach (1977). Umstände der Entdeckung der Richtungsquantelung, in *Jorrit de Boer Lecture in Experimemtal Physics* (Ludwig-Maximilains-Universität Munich, 26 January)

W. Gerlach (1978a). Erinnerungen an Albert Einstein, in *Albert Einstein, sein Einfluss auf Physik, Philosphie und Politik,* ed. by P.C. Aichelburg, R.U. Sexl (Vieweg, Braunschweig)

W. Gerlach (1978b). Otto Hahn, Lise Meitner, Fritz Strassmann. Die Spaltung des Atomkerns. Die Grossen der Weltgeschichte **11**, 51–71

W. Gerlach, A. Butenand (1978). Letter dated 24 March 1978, in *Bachmann and Rechenberg,* 1989, p. 253

W. Gerlach, M. List (1980). *Johannes Kepler. Leben und Werk* (Piper, München)

W. Gerlach, H. Albach (1923). Untersuchungen an Radiometern. I. Über ein Kompensationsradiometer. Zeitschrift für Physik **14**, 285–290

W. Gerlach, M. Born (1921a). Elektronenaffinität und Gittertheorie. Zeitschrift für Physik **5**, 433–441

W. Gerlach, M. Born (1921b). Über die Zerstreuung des Lichtes in Gasen. Zeitschrift für Physik **5**, 374–375

W. Gerlach, A. Golsen (1923). Untersuchungen an Radiometern. II. Eine neue Messung des Strahlungsdrucks. Zeitschrift für Physik **15**, 1–7

W. Gerlach, E. Madelung (1923). Untersuchungen an Radiometern. III. Notiz zur Radiometertheorie von E. Einstein. Annalen der Physik **74**, 674–699

W. Gerlach, W. Schütz (1932). Untersuchungen an Radiometern IV. Experimentelle Beiträge zur Prüfung der Theorien des gewöhnlichen Einplatten-Radiometers. Zeitschrift für Physik **78**, 43–58

W. Gerlach, O. Stern (1922). Der experimentelle Nachweis der Richtungsquantelung im Magnetfeld. Zeitschrift für Physik **9**, 349–352

W. Gerlach, O. Stern (1924). Über die Richtungsquantelung im Magnetfeld. Annalen der Physik **74**, 673–699

W. Gerlach, W. Westphal (1919). Über positive und negative Radiometerwirkungen. Verhandlungen der Deutschen Physikalischen Gesellschaft **21**, 218–226

A. Golsen (1924). Über eine neue Messung des Strahlungsdrucks. Annalen der Physik **73**, 624–642

R. Heinrich, H.-R. Bachmann (1989). *Walther Gerlach. Physiker—Lehrer—Organisator. Dokumente aus seinem Nachlass.* München: Deutsches Museum. Catalog to an eponymous exhibition at the Deutsches Museum, 26 July–29 1989

D. Hoffmann (1993). *Operation Epsilon. Die Farm-Hall-Protokolle oder die Angst der Alliierten vor der deutschen Atombombe* (Rowohlt, Berlin)

J.G. Huber (2015). *Walther Gerlach (1888–1979) und sein Weg zum erfolgreichen Experimentalphysiker bis etwa 1925* (Dr. Erwin Rauner Verlag, Augsburg)

A. Kramisch (1986). *The Griffin: The Greatest Untold Espionage Story of World War II* (Houghton Mifflin Company, Boston)

F. Kurlbaum (1898). Über eine Methode zur Bestimmung der Strahlung in absolutem Maass und die Strahlung des schwarzen Körpers zwischen 0 und 100 Grad. Wiedemanns Annalen **65**, 746

J. Mehra, H. Rechenberg (1982). *The Historical Development of Quantum Theory*, vol. 1, Part 2. (Springer, Heidelberg)

M. Nida-Rümelin (1982). *Bibliographie Walther Gerlach. Veröffentlichungen von 1912–1979.* München: Forschungsinstitut des Deutschen Museums für die Geschichte der Naturwissenschaften und der Technik

F. Paschen (1912a). Über die absolute Messung der Strahlung (Kritisches). Annalen der Physik **38**, 30–42

F. Paschen (1912b). Beurteilung der Dissertation GERLACHs durch PASCHEN von 20 January 1912 (Universitäts Archiv Tübingen, 136/34)

M. Planck (1900). Über eine Verbesserung der Wien'schen Spectralgleichung. Verhandlungen der Deutschen Physikalischen Gesellschaft **2**, 202–204

M. Planck (1901). Über das Gesetz der Energieverteilung im Normalspectrum. Annalen der Physik **4**, 553–563

H. Rechenberg (1979). Walther Gerlach zum Neunzigsten. Physikalische Blätter **35**, 370–374

W. Rollwagen (1980). Gedenkkolloquium zu Ehren von Walther Gerlach. München, 25 February. Tape recording DMA AV-T 0433

P. Rosbaud (1945). Statement by Dr. Paul Rosbaud, in R. Heinrich, H.-K. Bachmann, *Katalog der Ausstellung anlässlich des 100* (Geburtstages von Walther Gerlach, Deutsches Museum Munich, 1989)

H. Schmidt-Böcking, A. Templeton, W. Trageser (eds.) (2019). *Otto Sterns gesammelte Briefe. Band 2: Sterns wissenschaftliche Arbeiten und zur Geschichte der Nobelpreisvergabe* (Springer Spektrum)

W. Schütz (1969). Persönliche Erinnerungen an die Entdeckung des Stern-Gerlach-Effektes. Physikalische Blätter **25**, 343–345

G. Simon (2002). *Wissenschaftspolitik im Nationalsozialismus und die Universität Prag* (Gesellschaft für interdisziplinäre Forschung Tübingen, e. V, Tübingen)

O. Stern (1921). Ein Weg zur experimentellen Prüfung der Richtungsquantelung im Magnetfeld. Zeitschrift für Physik **7**, 249–253

S. Valentiner (1910). Vergleichung der Temperaturmessung nach dem Stefan-Boltzmannschen Gesetz mit der Skale des Stickstoffthermometers bis 1600°. Annalen der Physik **31**, 275

M. Walker (1995). *Nazi Science: Myth, truth, and the German atom bomb* (Plenum Press, New York)

S. Weinberg (1998). Physics and History. Daedalus **127**, 151–164

W. Westphal (1919). Zur Theorie des Radiometers. Verhandlungen der Deutschen Physikalischen Gesellschaft **21**, 669

W. Westphal (1920). Messungen am Radiometer. Verhandlungen der Deutschen Physikalischen Gesellschaft **22**, 10–11

Chapter 9
100 Years Molecular Beam Method Reproduction of Otto Stern's Atomic Beam Velocity Measurement

Axel Gruppe, Simon Cerny, Kurt Ernst Stiebing, Cedric George, Jakob Hoffmann, Maximilian Ilg, Nils Müller, Alienza Satar, Vincent Schobert, Leander Weimer, Markus Dworak, Stefan Engel, Gustav Rüschmann, Viorica Zimmer, Erich Zanger, and Horst Schmidt-Böcking

Abstract The history of Otto Stern's pioneering measurement of the Maxwell-Boltzmann velocity distribution of a Silver atomic beam performed 1919 in Frankfurt is described. It is shown how Albert Einstein influenced Stern in his research. This experimental apparatus is not any more existing; therefore it was reconstructed in the workshops of the Physics faculty of the Goethe University in Frankfurt. The experimental verification of Stern's results was finally achieved by a team of Frankfurt high school students (Gymnasium Riedberg) under the supervision of their teachers Axel Gruppe and Simon Cerny. By fighting against a number of difficulties, they succeeded to get the reconstructed apparatus started and were able to reproduce the results from the early experiments of Stern.

1 Otto Stern's Historic Atomic Beam Velocity Measurement

Otto Stern was originally educated as a theoretical physical chemist. That he finally turned into one of the most genius experimenters in modern quantum physics is indeed astonishing. In 1912 he completed his dissertation with the title "Zur kinetischen Theorie des osmotischen Druckes konzentrierter Lösungen und über die Gültigkeit des Henryschen Gesetzes für konzentrierte Lösungen von Kohlendioxyd

A. Gruppe
Institute for Didactic of Physics, University Frankfurt, 60438 Frankfurt, Germany

K. E. Stiebing · M. Dworak · S. Engel · G. Rüschmann · V. Zimmer · E. Zanger ·
H. Schmidt-Böcking (✉)
Institute for Nuclear Physics, University Frankfurt, 60438 Frankfurt, Germany
e-mail: hsb@atom.uni-frankfurt.de; schmidtb@atom.uni-frankfurt.de

A. Gruppe · S. Cerny · C. George · J. Hoffmann · M. Ilg · N. Müller · A. Satar · V. Schobert ·
L. Weimer
Gymnasium Riedberg, Friedrich-Dessauer-Str. 2, 60438 Frankfurt, Germany

© The Author(s) 2021
B. Friedrich and H. Schmidt-Böcking (eds.), *Molecular Beams in Physics and Chemistry*,
https://doi.org/10.1007/978-3-030-63963-1_9

Fig. 1 Otto Stern and his brother Kurt as soldiers [3]

in organischen Lösungsmitteln bei tiefen Temperaturen" [1], which was partly experimental and partly theoretical. Thereafter he began his career in theoretical physics working with A. Einstein in Prague.

In the same year he followed Einstein to Zürich. In 1914 Einstein was appointed professor in Berlin. Stern accepted the offer by Max von Laue to become Laue's "Privatdozent" in Theoretical Physics at the newly founded University in Frankfurt. From 1914 until the end of 1918 Stern was soldier in World War One serving as weather observer (Fig. 1). In the second half of the year 1918 Stern was delegated to the Institute of Walter Nernst in Berlin, where, together with Max Volmer (Fig. 2), he performed several experimental investigations [2] in which Stern already demonstrated his ingenious skill of designing sophisticated physical experiments.

It therefore was not a surprise, that, after his return to Frankfurt in February 1919, Stern, the initially theoretically trained physical chemist continued performing experiments in physics. The Frankfurt Institute of Theoretical Physics (see faculty members in Fig. 3), directed by Max Born, owned a workshop with the young Adolf Schmidt as the only precision mechanic.

The first experiment that Otto Stern performed in 1919 in Frankfurt was the measurement of the Maxwell-Boltzmann-velocity distribution [5] of Ag atoms evaporated from a solid at the temperature of the melting point ($T_m = 962$ °C) [6]. He explained that he was interested in this experiment because, due to the influence of the "zero-point energy", he expected deviations from Maxwell's law at very low beam velocities. Together with Einstein he had published a theoretical paper on this issue in 1913 [7].

Fig. 2 Max Volmer and his wife Liselotte nee Pusch [4]

Fig. 3 Members of the Frankfurt Physics faculty in 1920. From right: sitting Otto Stern, unknown, Max Born, Hedi Born and Richard Wachsmuth, standing: 3rd from right Alfred Landé, 4th Walther Gerlach, and next to Gerlach probably Elizabeth Bormann [3]

This pioneering experiment was the corner stone for Stern's famous molecular beam method MBM, which enabled the first ultra-high precision measurements of momenta of moving atoms or molecules in vacuum. With this experiment Stern established a method allowing the observation of inner atomic or even inner-nuclear ground state properties with unprecedented resolution, which, at least in 1919, was not achievable by energy spectroscopy (see Stern-Gerlach-Experiment SGE in 1922 [8]).

At a first glance, the measurement of the Maxwell-Boltzmann velocity distribution [6] looks simple. However, the authors of this paper, in their attempt to repeat this experiment, had to learn how difficult it really was, in particular, if one considers the very poor economic conditions in the year 1919 when the seminal experiment was performed.

The priming condition for the development of the MBM was the revolutionary progress in vacuum technology. Diffusion pumps became available creating vacuum in the low 10^{-5} torr regime. In such vacuum the free-path-length of moving atoms reaches several meters before they undergo a second collision. Stern benefited from his friendship with Max Volmer, who had developed a glass-made mercury diffusion pump, patented in 1918, Figs. 4 and 5. The Volmer mercury diffusion pump was fabricated in Berlin by Hanff and Buest. The rough vacuum was created by a rotating mercury pump invented by Wolfgang Max Paul Gaede (1878–1945) [9].

Otto Stern's experiment was inspired by the atomic beam experiment of L. Dunoyer in 1911 [10]. Dunoyer observed in his experiment that the beam particles move in vacuum like photons on straight lines. This is expected as long as the particles are not deflected by an external force or scattered by a gas molecules in the vacuum chamber. Vice versa one can use the transverse deflection in x and y direction (z is the direction of the velocity vector) by an external electro-magnetic force and thus determine electric or magnetic properties of the moving particle.

To perform deflection measurements and to obtain quantitative information on inner atomic properties one must measure the absolute value of the transverse momentum change. Therefore one has to know very precisely the direction of motion as well as the mass and the absolute velocity of the particle. In order to achieve this, one has to carefully prepare the atomic or molecular beam by a well aligned system of accurately manufactured slits. The principle of the transverse beam collimation is shown in Fig. 6.

The direction of motion is known from the geometry of the slit system. The velocity distribution of the atoms in the beam, generated by evaporating the atoms from a source at a defined temperature T was in 1919 only theoretically predicted but experimentally never measured. Thus, in order to later use the MBM for absolute momentum measurements, Stern had to verify Maxwell's theory [5] by measuring the Maxwell-Boltzmann velocity distribution of atoms, evaporated in a sufficiently good vacuum.

To perform these measurements Stern invented a kind of "streak camera" which is an ingeniously simple apparatus but which is very difficult to set into operation [6]. It is therefore astonishing and highly meritorious that the experiment had been accomplished, in particular when one considers the short period of one year from

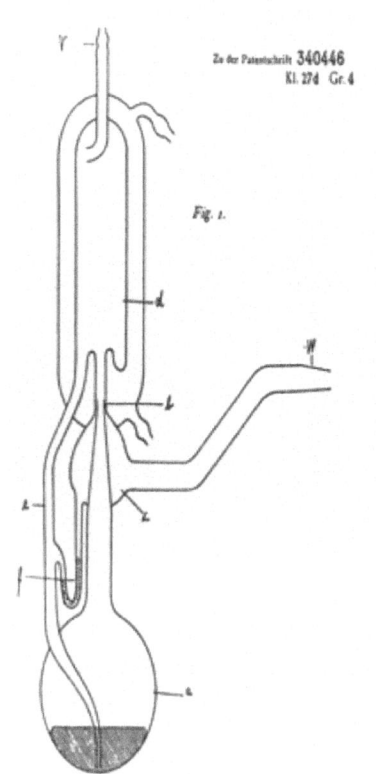

DEUTSCHES REICH

REICHSPATENTAMT

PATENTSCHRIFT

– № 340446 –

KLASSE 27d GRUPPE 4

Dr. Max Volmer in Berlin.

Quecksilberdampfstrahlpumpe.

Patentiert im Deutschen Reiche vom 9. Mai 1918 ab.

Fig. 4 The Volmer diffusion pump

beginning until getting a final result. One certainly has to anticipate that the help of the 26-year-old mechanic Adolf Schmidt was crucial for making the experiment a success.

In Fig. 7 Stern's "streak camera" is shown. The glass recipient had an inner diameter of 24 cm and was 30 cm high. With the help of the Gaede rotating mercury pump and the Volmer one-stage mercury diffusion pump the recipient could be evacuated to a pressure below 10^{-4} torr. The quality of the vacuum was measured by Geissler tubes. The pumping speed of both pumps was rather low (a few liters per second). The glass recipient was mounted vacuum-sealed on a $40 \cdot 40 \text{ cm}^2$ iron plate. The stationary frame (D) inside the recipient was fixed by screws on the iron plate. The streak camera (R) was adjusted inside D and could be rotated by a small motor (not seen on Fig. 7) with frequencies between 25 and 45 Hz. The axis of the motor was connected to the lower end of the main axis (A) of the apparatus by a short piece of vacuum hose. The other end of the motor axis was connected to a revolution counter by a flexible shaft. In the center of R a platinum wire (L) was mounted, the surface of

Fig. 5 Historic pumps used in the original SGE. Left the Volmer mercury diffusion pump, right the Gaede diffusion rough pump [photo HSB]

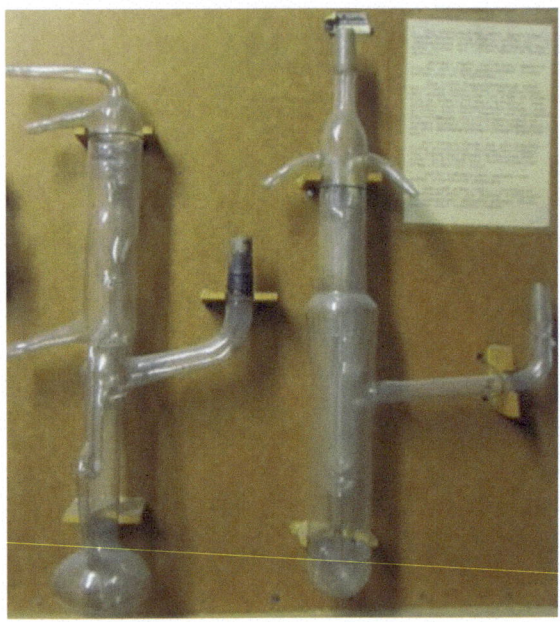

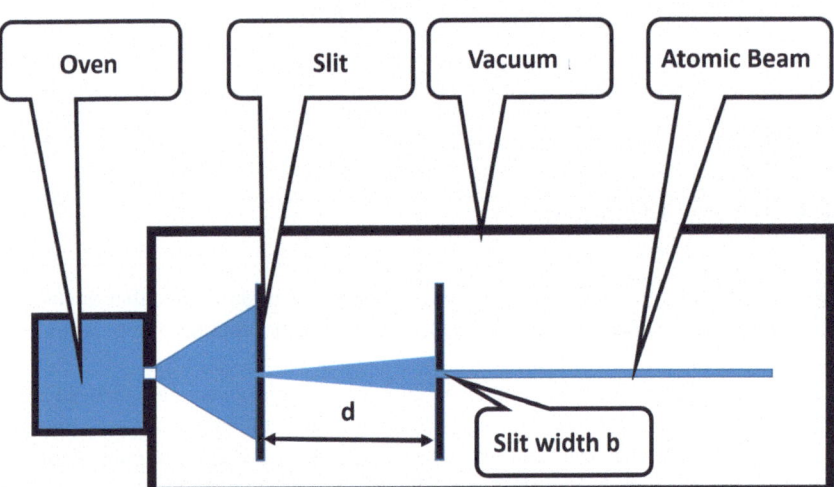

Fig. 6 Principle of the transverse beam collimation

which was covered by a thin layer (about 20 μm thick) of Ag. It could be heated by an electric current up to the melting temperature of Ag, emitting Ag vapor from the surface. It was very important that the wire remained stretched during heating. Two geometrically very thin beams were created by collimation by slits (S$_2$), mounted symmetrically on both sides of the wire. These beams condensed on two polished

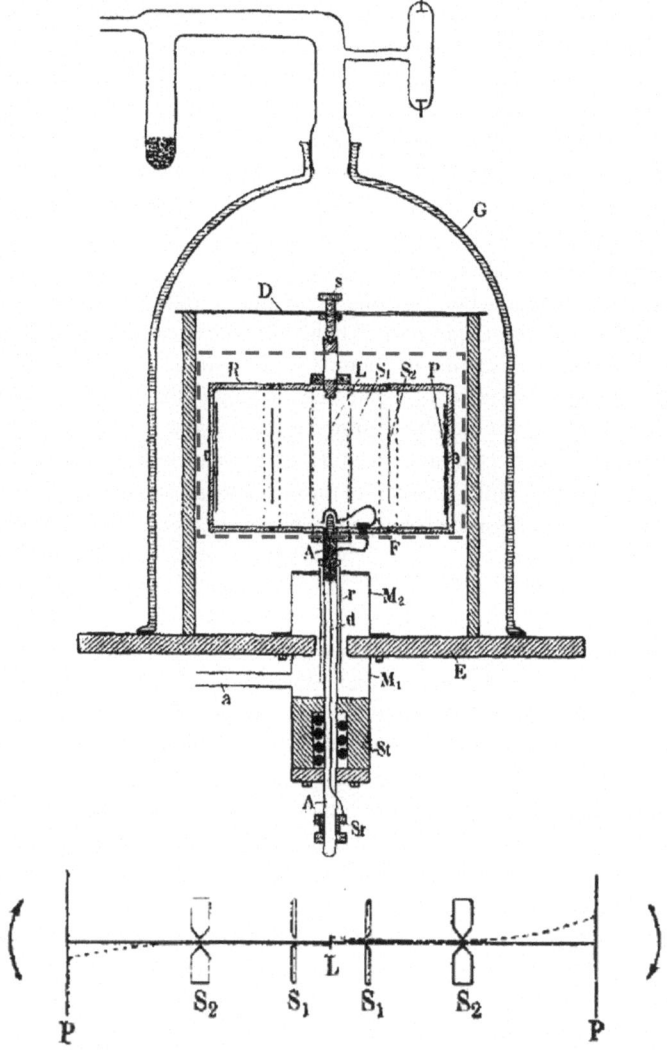

Fig. 7 Stern's "streak camera" apparatus. G = Glass recipient, R = "rotating streak camera" (within the red dashed line), D = stationary frame, in which the "camera" is rotating, L = Platinum wire, from where the silver atoms evaporate, S_1 and S_2 = slits, P = detection plate. The pumps connected to the glass tubes and the motor are not shown [6]

brass plates (P) in 6 cm distance from the wire. The slits (S_2) were halfway between L and P. A further set of two slits (S_1) (8 mm distance from the wire) was mounted to ensure that the well-defined position of the Ag beam source did not change during the time of the measurement.

In order to measure the atom velocity, the streak camera (R) had to be rotated around the axis A (see Fig. 7, lower part, where the clockwise rotation is indicated by the arrows). In case of no rotation the slits and the wire were aligned on one straight line leading to a small streak in the middle of P. In case of rotation the streak on the detector plate is shifted in opposite direction of the rotation by about 0.4 mm (at a rotation frequency of 25 Hz). The reason is that, depending on their velocity, the atoms need some time to fly from the slit to the detector P while the detector has rotated forward. To obtain a better separation the system was rotated in both directions yielding about 0.8 mm separation.

Although the working scheme of the apparatus is rather simple, it required a number of skills in different experimental fields to make it run successfully: precision engineering, pumping and sealing to obtain a good vacuum, frequency and temperature measurement etc. One may anticipate that the help of the young mechanic Adolf Schmidt was essential for Otto Stern to make the apparatus run. In order to get a good velocity resolution the whole segment R had to rotate with a constant frequency. According to Stern, the required balance of the rotating part and the necessary vacuum sealing at the feedthrough of the rotating axis were the most difficult problems. For sealing the axis they used oil-soaked asbestos rope (see (St) in Fig. 7). Since this sealing was too leaky, they additionally had to evacuate the space M1/M2 where the axis A rotates in a tight-fitting but not touching brass tube (see Fig. 7). Because of frequent heating and cooling the Platinum wire got stretched and had to be adjusted frequently, in order to avoid bending when glowing.

On both detector plates (P) Stern observed two clearly separated lines one for rotating the system clockwise and the other for rotating counterclockwise (see Fig. 8). From the measured separation, the geometry of the streak camera and the rotation frequency he deduced a mean velocity of the beam of about 600 m/sec. Maxwell's theory, however, predicted only 534 m/sec for a temperature of the Ag melting point at 962 °C. Stern assigned this difference to a possible deviation between the measured and the real temperature at the wire.

However, Einstein in Berlin recognized that Stern had made a mistake in his analysis. He had overseen that the transmission flux of the beam through a slit depends on the third power of the atom velocity but not on the square. Walter Grotrian, who reported on Stern's experiment in a seminar in Berlin, where Einstein, Planck, Laue and Nernst [12] were in the audience, wrote in a letter to Stern (on July 30, 1920) and informed him about the discussion in this seminar in Berlin [12]:

Dear Stern!

... Your experiment appeared to all, who listened, also Franck and Reiche astounding and convincingly. After long discussions we were convinced, that also in case of sublimation of a solid, e.g. coal, the mean kinetic energy of the emitted atoms or molecules is 3/2 kT. Thus the issue was settled.

Then followed the discussion, which I will present in detail. It began with Nernst. His remark was related to experimental details. First he mentioned the rotating electric contact into the vacuum ("Öhse") and named it a master piece. After some insignificant remarks Laue asked, whether the evaporating molecules do really have the mean energy 3/2 kT. I tried first following your letter to turn the concerns down which were related to the evaporation

Fig. 8 Detector image (see text). Left and right detector plate [11]

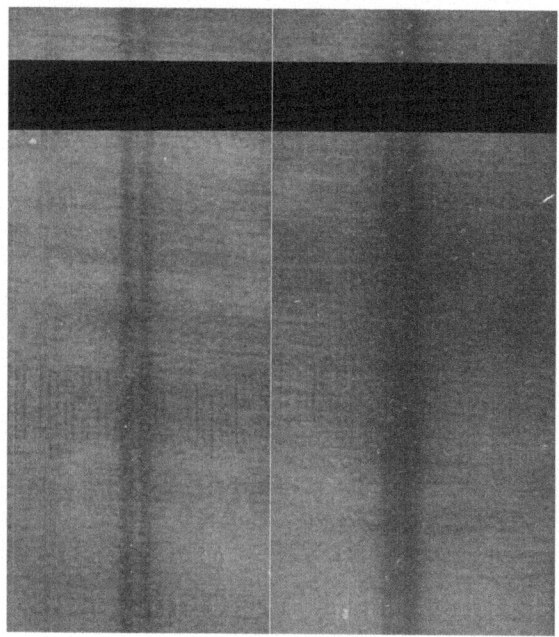

energy. Then Einstein stood up and went to the blackboard. He explained now, that one had to distinguish between velocity distribution per volume and the velocity distribution of these molecules which impact on or are emitted from a surface. The latter ones would be shifted to higher velocities.

From what he said and his later discussion with Planck it was not clear whether he was objecting your results or not.

We discussed yesterday again this issue in detail and came to the conclusion: … The question is: Is the mean square of the velocity distribution of a molecular beam penetrating in one direction per 1 cm^2 through a slit equal to the mean square of the velocity distribution per 1 cm^3 volume?…

We hope that you can inform us soon what is the answer to this question. It would be the best if you could visit us in Berlin.

Yours Walter Grotrian.

Einstein's concerns were proved to be true. On October 20 1920 Stern submitted an addendum to Z. Physik with a new analysis of his data based on Einstein's arguments [13]:

In the recent published communication [6] I have reported on experiments where the velocity of Ag atoms evaporated from a melting Ag surface into vacuum was measured with 600 m/sec. This value is within the error bars in agreement with the mean value calculated from the kinetic gas theory at the temperature of the melting point. This result seems to verify the assumption that the Ag atoms, which are emitted from the surface have the same velocity like the atoms of melting silver. But several people have now criticized this assumption, where the objection of Mr. Einstein is justified. The issue is the following: 1. We have a recipient filled with gas at a given temperature in equilibrium state. We now look at atoms that escape through a tiny hole into the vacuum. These atoms do not have the Maxwell velocity distribution of the equilibrium state inside the recipient in contrast to the analog case of black body radiation. The fast atoms have a higher probability to escape. Following known gas theoretical considerations the number dn'c of the molecules with velocity c escaping through the hole per time unit is equal to the number of atoms per volume unit dnc multiplied with the volume of a cylinder of length c. Therefore is dn'c not proportional to dnc but cdnc.

With this experiment Otto Stern for the first time confirmed that the Maxwell-Boltzmann theory on the velocity distribution of gases at temperature T agrees well with experimental results. By measuring the temperature T of the evaporating gas, Stern, with the help of Maxwell's theory, could deduce the velocity and thus the momentum of the moving atom or molecule. This experiment was the foundation of Stern's molecular beam method (MBM) enabling high-resolution momentum measurements. In the following decades even up to today numerous milestone experiments of quantum mechanics have been performed basing on this method.

In all his publications until his retirement in 1945, Stern never mentioned that his new method provided a high-resolution momentum spectrometer yielding a resolution never achieved before. He presented his data always as function of deflection angles. Therefor the extremely high momentum resolution was not obvious to the readers of his publications. It is important to note that the line width measured by Stern already in this first experiment corresponds to a transverse momentum width of sub-atomic size. This excellent momentum resolution can be estimated from Fig. 9 as follows: Let the velocity of the Ag atoms be 540 m/sec, corresponding to a momentum of about $p = 50$ au. The two lines in Fig. 9 are separated by 0.8 mm (on the detector plate) in 60 mm distance from the wire. Transformed into momentum space this distance corresponds to a momentum difference of $\Delta p = (0.8:60) \cdot 50$ au $= 0.67$ au (see Fig. 9). The width of the left line is then less than 0.2 au, demonstrating the excellent momentum resolution achieved in this experiment. Reducing the slit width, the momentum resolution could even be improved.

In the conclusion of his paper [6] Stern revealed the motivation for measuring the Maxwell velocity distribution:

Fig. 9 Line splitting as function of transverse momentum

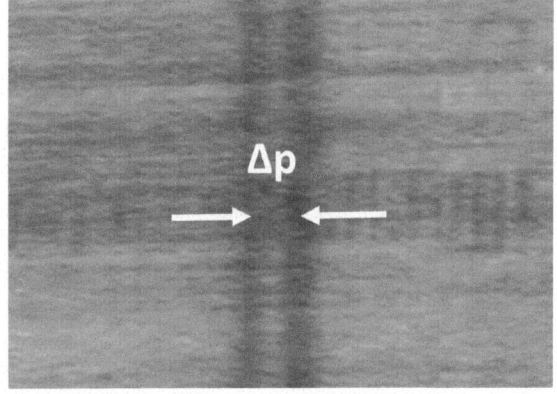

A very precise examination of Maxwell's velocity distribution law would be of particular interest. According to the Quantum theory, small deviations occur in gases with a small molecular weight at high pressures and low temperatures, which are estimated to be about 1 percent for hydrogen at the boiling point under atmospheric pressure. Unfortunately, it is not possible to give more precise information about the type and amount of these expected deviations - except, for example, that assuming zero-point energy, the low velocities will occur more rarely than according to Maxwell - because the Quantum theory of translational motion encounters previously insurmountable difficulties. The experimental investigation of these deviations would be thus even more important, and it was precisely this problem that gave me the reason for the present investigation. Unfortunately, the conditions here are also very unfavorable for the experiment, but perhaps gravity will provide sufficient dispersion for the analysis of the low speeds.

Finally, it should be noted that the above method allows for the first time to produce molecules of uniform speed, and e.g. to investigate whether condensation only takes place above or below a certain speed.

The first application of Stern's atomic beam method was performed by Max Born and Elisabeth Bormann [14]. They successfully used an Ag beam in 1920 in Frankfurt to determine the free path length λ of the Ag atoms in air. The Ag beam was collimated by a cascade of round copper screens, in each of which a centric hole was drilled for passing the beam through. When the air pressure in the recipient was gradually increased, more and more Ag atoms were scattered by the air molecule and were deposited on the copper diaphragms, which had been mounted in well-defined distances. The amounts of depositions were carefully measured and it was then possible to use a theoretical scattering model according to Jeans [15] to determine the "free path length" λ of the Ag atoms for the given pressure.

Stern himself used the atomic beam method for the first time in the famous Stern-Gerlach experiment SGE [8], which was carried out in Frankfurt from 1920 to 1922. The SGE demonstrated in an impressive manner what the MBM can achieve as a means for momentum measuring.

In 1928, when he worked as fellow in Stern's laboratory in Hamburg and later as professor at Columbia University in New York and at the MIT in Boston Isidor Rabi developed a new extremely powerful scheme for the application of MBM by using first a SGE approach to prepare atomic beams in well-defined quantum states. In a second interaction region the prepared states could be excited by resonant photon absorption into another quantum state with different magnetic quantum numbers. These states, excited by resonance absorption, moved into a third interaction region on a different trajectory and could be detected separately. He and his group very successfully applied this method to use the very narrow line width of photon absorption for high precision measurements of transition energies, like e.g. the Nobel Laureates [16] Willis Lamb and Polykarb Kusch for measuring the so-called Lamb-shift, Norman Ramsay to develop the Cs atomic clock (with 10^{-9} precision), Felix Bloch and Henry Purcell for developing the Nuclear Resonance technique etc.

Since 2002, when in Frankfurt the 80th anniversary of the SGE was commemorated, one of the authors (HSB) looked for remainders of the historical experimental set ups used by Stern at the various working places of Stern and Gerlach. The only parts he found were the microscope bought by Stern in 1919 from the company of

Seibert in Wetzlar (found in Berkeley) and the Volmer diffusion pump (found in Frankfurt). Obviously the historic apparatus did no longer exist. Therefore, the idea was born to reconstruct both experimental apparatus of Stern: the set-up to measure the Maxwell-Boltzmann velocity distribution and the famous Stern-Gerlach experiment. While the Stern-Gerlach experiment is still waiting for its reconstruction, the first one was now reconstructed and was put into operation. In the following the reconstruction and the successful commissioning of the first apparatus is described.

2 Reproduction of Otto Stern's Atomic Beam Velocity Measurement

2.1 Reconstruction of the Apparatus

On the occasion of the 100th anniversary of Stern's appointment to Frankfurt, the initiative was taken to reconstruct the historic set up, which did no longer exist, and to reproduce Stern's famous measurement of the velocity of Ag atoms in an atomic beam. Based on the drawings and the detailed description in Stern's publication [6], and sponsored by Roentdek Handels GmbH, a number of identical copies were fabricated in the workshops of the Institute for Nuclear Physics in Frankfurt.

One of the copies was given to the Gymnasium Riedberg at Frankfurt with the requirement to repeat Otto Stern's measurements and, if possible, to verify his results. This task was adopted by a team of high school students from the 10th grade, who founded an "Otto-Stern-Arbeitsgemeinschaft" (OSAG) and, under the supervision of their physics teacher, Axel Gruppe, started to work on this project in summer of 2015. Their work was presented at the VDI Student Forum 2015 [17].

First, the students read Otto Stern's biography to get an impression of his scientific achievements. Then they began to examine the optically very appealing replica (Figs. 10 and 11) in view of its suitability in a real experiment, which Otto Stern described in great detail in his work from 1920 [6]. The following points were studied in detail: the trajectories of the silver atoms, the measurement of the rotation frequency, the mean free path of the silver atoms and the measurement of the temperature of the platinum wire.

2.2 The Trajectories

Anticipating the rotating frame to be at rest, it appears beyond question that the silver atoms, emitted from the Platinum wire, will fly through the slit diaphragms S1 and S2, which are aligned on a straight line with the emitting wire, and will impact on the collecting plate P exactly at the point, where the straight line hits P. But how does the trajectory change when the frame rotates?

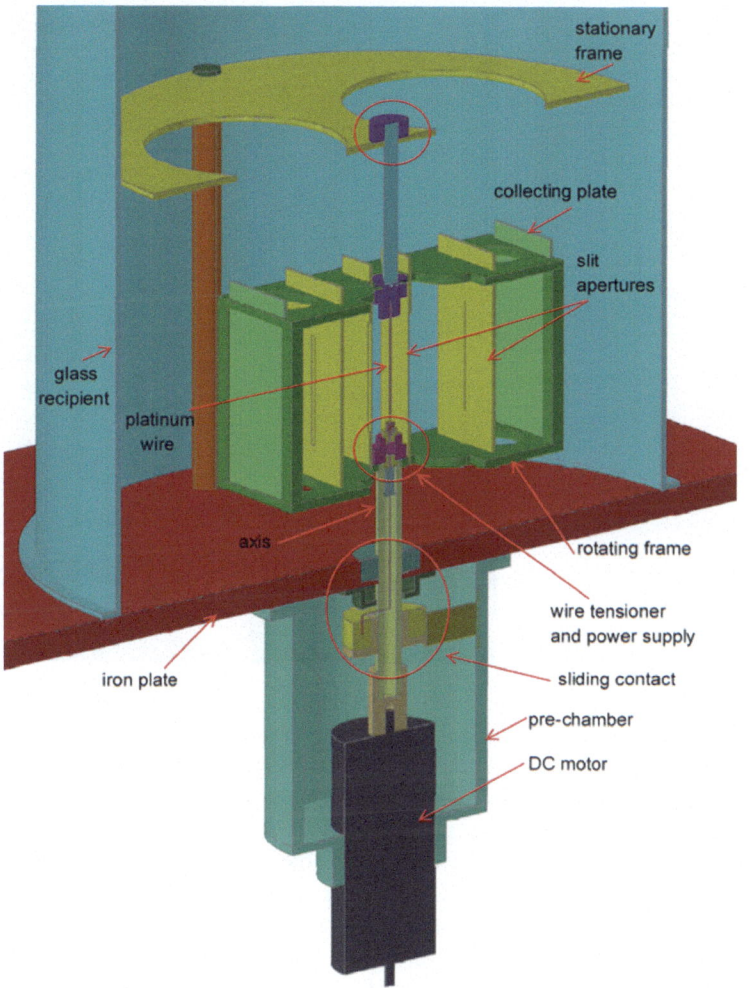

Fig. 10 Plan for the restoration of Stern's "streak camera" system

To understand this, one has to convert the path of the silver atoms, which in the laboratory system corresponds to a rectilinear, uniform motion, into the rotating coordinate system of the frame of the camera (R), with which the diaphragms and the collecting plate are firmly connected.

In [6] Stern describes the trajectory of the atoms in the co-rotating system as to represent a "horizontal throwing parabola" ("… by neglecting the centrifugal acceleration and the change in the Coriolis acceleration.").

In order to get an impression of the parameters of this parabola, the students developed an EXCEL worksheet for the coordinate transformation of the rectilinear,

uniform atomic motion into the rotating aperture system and thus studied the intri-
cacies of Stern's experiment. In the upper part of Fig. 12 one can see the wire L in
the coordinate origin of the rotating system X'/Y' as well as the slit diaphragms S1,
S2 and the collecting plate P.

Starting from L with a velocity of 500 m/s, the Silver atoms have to be emitted
at an angle $\alpha = -0.55°$ relative to the X'- axis, to pass through slit diaphragm S2.
It is evident that in this case the beam trajectory does not pass through S1 with the
result that the silver atoms will not hit the detection plate P. To make the trajectory
passing through S1 and S2, the location of the emission point must be shifted in the
Y' direction (Fig. 12, lower part).

This is the reason why Otto Stern had rolled down the initially round wire (0.4 mm
diameter) to a width of 0.6 mm in the Y' direction [6]. By this, he wanted to ensure
that the locations of emission of the Silver atoms enable trajectories that run through
both S1 and S2. With these calculations, the students also realized how small the
expected shift really is. At a rotation frequency of 25 Hz, the displacement of the
point of impact for silver atoms with v = 500 m/s is only about 0.5 mm in the Y'
direction!

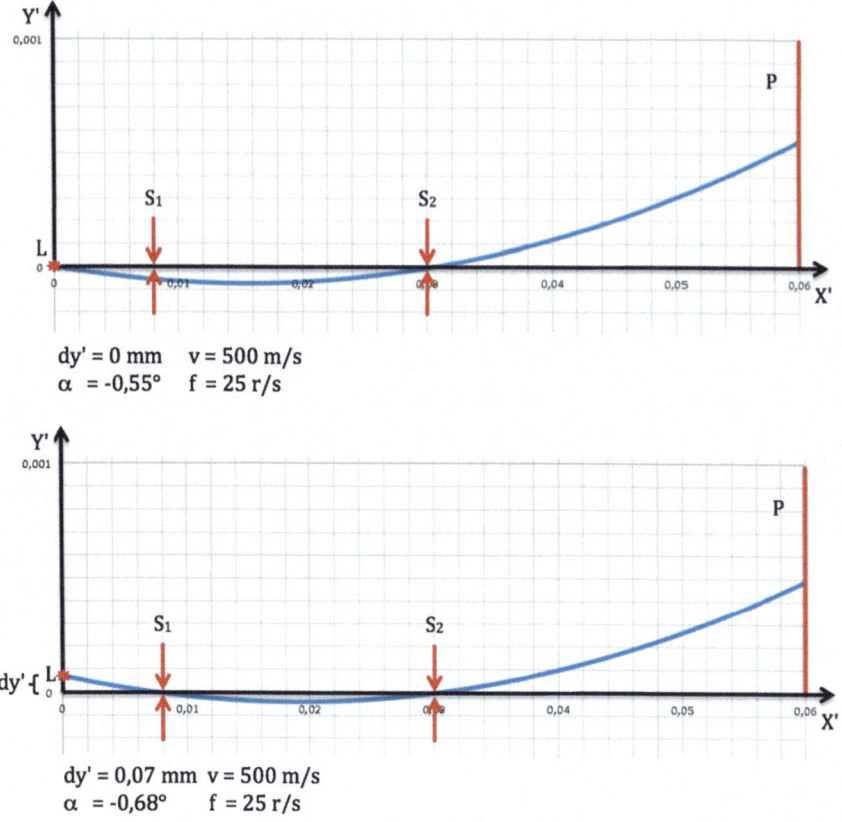

Fig. 12 Trajectories of the silver atoms in case of rotation (f = 25 Hz). Upper picture: the atoms start at L on the straight line passing through apertures S_1 and S_2. Since the trajectory has the shape of a parabola, no silver atom reaches the collecting plate P. Lower picture: The silver atoms must start from a point L shifted by dy′ in the Y′ direction from the center of rotation in order to reach the collecting plate P through the two apertures

2.3   Measurement of the Rotation Frequency

Since the slit diaphragms have a width of 0.2 mm, a 0.4 mm wide silver line is expected on the collecting plate. Under the most favorable conditions, the silver lines with the streak camera at rest and with the rotated camera would then be separated by just 0.15 mm. The students calculated that at a speed of 25 r/s a fluctuation of ±2 r/s would lead to a shift in the impact point of ±0.1 mm. Therefore the fluctuation of the rotation speed must not exceed this value if one wants to be able to separate the two silver lines and the rotation speed had to be controlled with sufficient accuracy. To achieve this a digital speed counter was designed and built by the students [18]. In this design, signals from chopping a light barrier by a slotted disc, which was attached to the axis of the built-in direct current motor, are read out and displayed as a frequency on a 4-digit 7-segment display by means of a tricky circuit.

2.4 Mean Free-Path and Quality of the Vacuum

Considering the dimensions of the original equipment, one may wonder, why it was not built larger, as the offset, and hence the effect would become larger, the longer the distance between source (L) and the screen (P) would be. At this point, the quality of the vacuum comes into play. The deposition lines of Silver will only sharply be imaged on the collecting plate (P) if the silver atoms do not undergo collisions with air molecules during their flight. Therefore the vacuum inside the apparatus must be good enough to avoid such collisions as far as possible. In order to estimate this influence, the students investigated the physics of the "mean free path" (MFP). MFP is the distance a molecule travels on average in a gas before it collides with another molecule. (According to the definition, the average number of molecules in an atomic beam, which have not yet collided with a residual gas atom, is only 1/3 after the beam has passed the length of MFP [19]). In addition to the temperature, MFP is essentially dependent on the gas pressure in the recipient.

For air at 20 °C the students found a value of MFP $= 6.8 \times 10^{-3}$ mbar cm/p [19]. Otto Stern mentions 1/10,000 mm (=1×10^{-4}, torr $= 1.33 \times 10^{-4}$ mbar) as the required vacuum. This corresponds to an MFP of approximately 50 cm in Stern's apparatus. According to the nomenclature of vacuum technology, such a pressure is already termed "high vacuum". This means that in Stern's experiment only about 8% of the silver atoms collided with an air molecule on their 6 cm path to the screen.

2.5 Measurement of the Temperature of the Filament

Another difficulty was the measurement of the temperature of the silver-plated platinum wire. Stern writes in [6]: "Now the temperature of the evaporating silver was certainly higher than 962° because the molten silver contracts to form droplets and the parts of the platinum wire that have been freed from silver assume a higher temperature due to their higher resistance, which increases due to conduction to the Silver droplets. According to the brightness, the temperature should have been around 1200° ...". The mention of the brightness of the wire by Stern suggests that he used a pyrometer to measure the temperature. Fortunately, a functional pyrometer from Hartmann and Braun ("Pyropto", manufactured in 1951 [21]) was found in the collection of the physical internship of the Institute of Applied Physics, and was kindly donated to OSAG.

2.6   The Improved Experimental Setup and the Decisive Measurement

After the essentially theoretical and preparatory work accomplished by this first OSAG at the Gymnasium Riedberg, the work was transferred back to the Goethe University. In discussions with physicists and mechanics of the Institute for Nuclear Physics Research (IKF), it became clear that the available replica of the Stern apparatus had to be changed in several details in order to successfully start with the next step, i.e. to reproduce the measurements from 1920:

- The sealing surface of the base plate for the glass bell was reworked.
- The cross section of the pump opening in the base plate was enlarged.
- The feed-through of the axis in the base plate was reconstructed. It was additionally encapsulated as in Stern's original equipment [6] so that this area could be evacuated separately by the backing pump.
- The power supply to the platinum wire was redesigned in order to be able to supply the comparatively high currents (up to 8A) at rotation frequencies of up to 45 Hz without interruption.
- The clamping device for the platinum wire was modified to withstand the high temperatures (up to about 1200 °C) and to maintain the mechanical tension of the platinum wire during the heating-related extension.
- During assembly, the unbalance of the rotating system was minimized with great mechanical effort.

In the school year of 2018/19, a new OSAG at the Gymnasium Riedberg started to perform measurements with the improved equipment. In the first test runs, it became clear that the necessary constancy of the rotation frequency could only be achieved with the help of a speed control of the DC motor. Using an Arduino® microcontroller, the students first built a simple linear control loop, the control oscillations of which, unfortunately, were still too large. However, reprogramming the Arduino to a more sophisticated proportional-integral-derivative (PID) controller led to the desired success (Fig. 13).

With the successful establishment of the PID control, all preparations for repeating the Stern experiment were completed in summer 2019. The final experimental setup is shown in Figs. 14, 15, 16, 17, 18 and 19.

The decisive measurement took place on August 30, 2019. Initially, the members of OSAG recorded a weak but clearly visible Ag line at a rotation speed of 45 r/s. During a second irradiation with the camera at rest, the lower part of the detector plate was covered with a plastic film to facilitate detection of the separation of the two lines. Figure 20 shows a scan of the brass plate which has been processed to enhance contrast. The evaluation shows a distance of the line centers of 0.62 mm with an estimated accuracy of 10% to 15%. This result must be compared with the value that Otto Stern presented in his publication [13]: "For 2700 tours (i.e. rotations/minute),

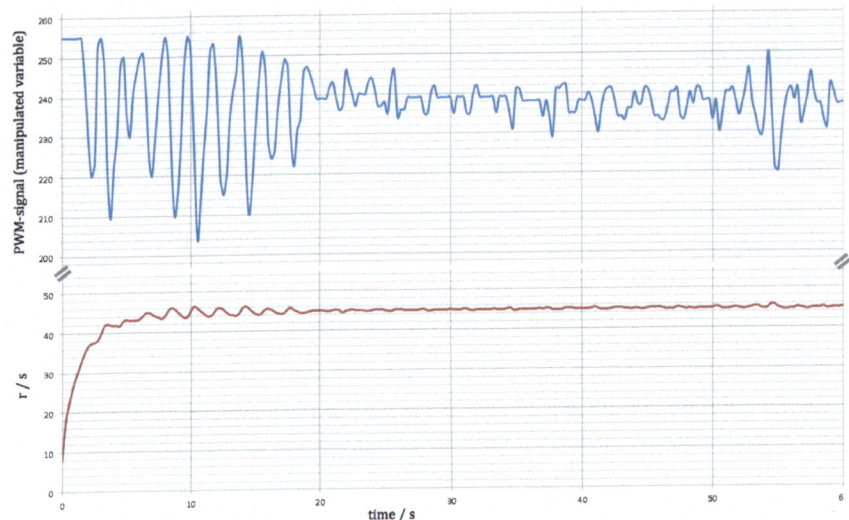

Fig. 13 The proportional-integral-derivative (PID) control of the rotational speed. Above: the time course of the manipulated variable (pulse wave modulation). Below: the time course of the regulated rotation frequency. In this case the PID controller has settled after 20 s and keeps the rotation speed constant at 45 ± 0.5 r/s

the distance between the centers of the two lines created with right and left rotation was 1.26 mm". Otto Stern's value divided by 2 yields 0.63 mm, which is in excellent agreement with the value measured by OSAG (0.62 mm at 45 r/s = 2700 tours with all other experimental parameters kept identical as far as possible). After a few measurements, the platinum wire had to be readjusted regularly (Fig. 21).

The OSAG team was very much impressed by this excellent agreement 100 years after the original measurement by Otto Stern. The participants were very proud that they had managed to successfully complete this experiment, which was easy to understand but technically difficult to carry out.

From the teacher's point of view, this project was an ideal example of how to offer students deep insights into the interplay of theoretical and experimental physics with the help of an ambitious topic and authentic material.

Fig. 14 The final experimental setup of the Otto Stern experiment in 2019: In the center: glass recipient with pressure gauge on an iron plate. Inside, the fixed frame with the rotating slit system is visible. On the right: parts of the vacuum system with backing pump (red, in front) and turbo-molecular pump (rear). In the foreground in front of the glass bell stands the pyrometer ("Pyropto", Hartmann & Braun®) for temperature measurement. Right front: PID controller for speed control on breadboard with an ARDUINO® controller. Rear right: controllable DC power supplies for the motor and the heating wire. Below is the vacuum measuring device with the vacuum display. The Pirani and Penning sensors (Balzers®) are positioned under the base plate and are not visible. On the left: Laptop for setting and logging the rotation speed during the measurement

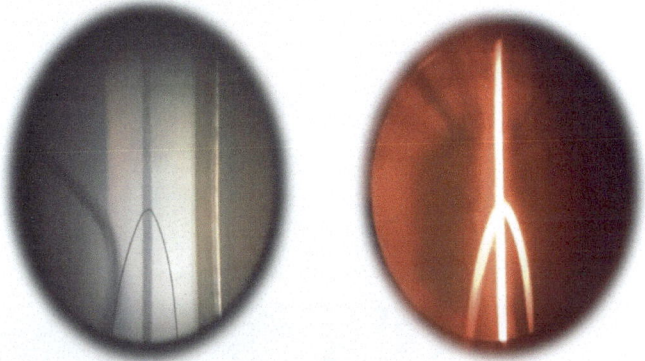

Fig. 15 The view through the pyrometer. The bent pyrometer wire is adjusted to the same brightness as the vertical platinum wire (left: at room temperature, right: at 1050 °C)

Fig. 16 The Members of the Otto-Stern-AG during a measurement (from left to right: Leander Weimer, Nils Müller, Simon Cerny, Jakob Hoffmann)

Fig. 17 The glowing platinum wire in the center of the rotating slit system at a rotation speed of 45 r/s

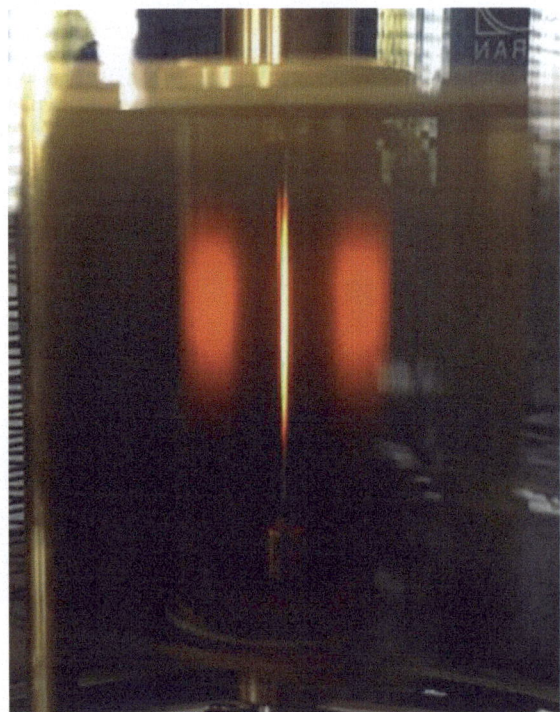

Fig. 18 The slit system with the detector plate after the experiment. On the right is the platinum wire; on the left, next to it, is the first slit aperture, which is completely covered with silver. To the left of it is the second slit aperture on which one can clearly see the "shadow" of the first slit aperture. On the far left, the detector plate is to be seen, on which a faint, brownish Ag line is just visible

Fig. 19 According to legend, Otto Stern "developed" the faint traces of Ag on the brass plates by cigar smoke, as Leander Weimer and Nils Müller also tried after the experiment

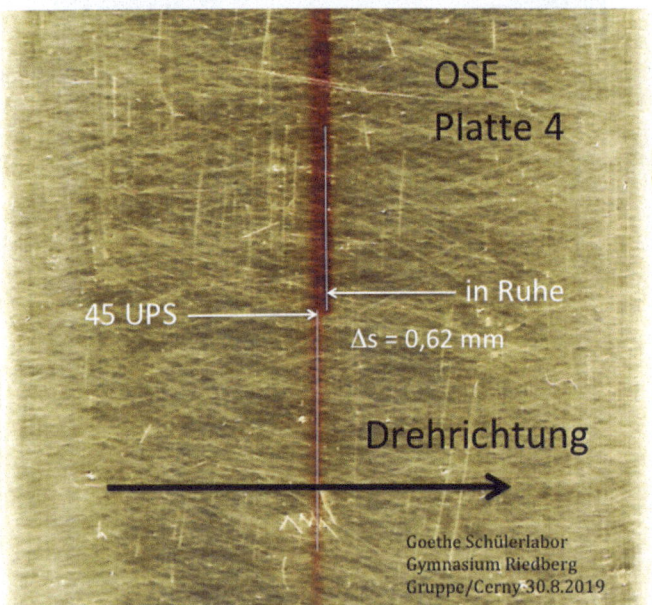

Fig. 20 Contrast-enhanced scan of the collecting plate. The longer thin line on the left was recorded at a speed of 45 r/s. The shorter line on the right was taken with the camera at rest. The distance between the line centers is 0.62 mm

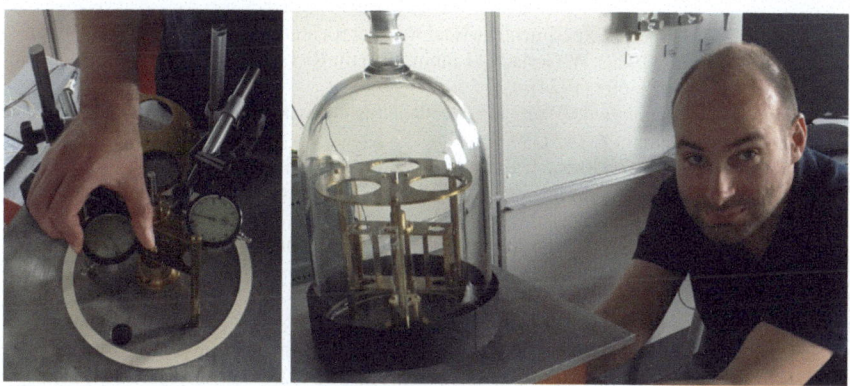

Fig. 21 The mechanic Stefan Engel balancing the equipment

Acknowledgements We would like to thank all those who have contributed to this project. In particular the commitment of the members of the IKF workshop and the accelerator group is gratefully acknowledged. We also express our gratitude to the Pfeiffer Vacuum Company [20] for providing us with a powerful HV pumping system consisting of a turbo-molecular pump in combination with a backing pump. The Pyrometer for measuring the temperature of the Platinum wire was donated by the Institute of Applied Physics. The experiments have been carried out within the framework and in the premises of the Goethe Lab-Schüler-Labor "Radioaktivität und Strahlung" at the Goethe University, Frankfurt. The project would not have been possible without the generous financial support by Roentdek Handels GmbH, Kelkheim, Germany.

References

1. O. Stern, Zur kinetischen Theorie des osmotischen Druckes konzentrierter Lösungen und über die Gültigkeit des Henryschen Gesetzes für konzentrierte Lösungen von Kohlendioxyd in organischen Lösungsmitteln bei tiefen Temperaturen. Z. Physik. Chem. **81**, 441–474 (1913)
2. O. Stern, M. Volmer, Über die Abklingungszeit der Fluoreszenz. Physik. Z., **20**, 183–188 (1919); O. Stern, M. Volmer, Sind die Abweichungen der Atomgewichte von der Ganzzahligkeit durch Isotopie erklärbar. Ann. Physik, **59**, 225–238 (1919); O. Stern, M. Volmer. Bemerkungen zum photochemischen Äquivalentgesetz vom Standpunkt der Bohr-Einsteinschen Auffassung der Lichtabsorption. Zeitschrift für wissenschaftliche Photographie, Photophysik und Photochemie, **19**, 275–287 (1920)
3. The Bancroft Library, University of California, Berkeley, Berkeley, Heritage Otto Stern, Picture collection
4. M. Volmer: Eine Biographie zum 100. Geburtstag 1985, ISBN, 37983 1053 x Univ. Bibliothek der TU-Berlin, Seite 20
5. https://de.wikipedia.org/wiki/Maxwell-Boltzmann-Verteilung
6. O. Stern, Eine direkte Messung der thermischen Molekulargeschwindigkeit. Physikal. Z., 2, 49–56 (1920); O. Stern, Eine direkte Messung der thermischen Molekulargeschwindigkeit, Physik. Z., **21**, 582 (1920)
7. A. Einstein, O. Stern, Einige Argumente für die Annahme einer Molekularen Agitation beim absoluten Nullpunkt. Ann. Physik **40**, 551–560 (1913)

8. W. Gerlach, O. Stern, Der experimentelle Nachweis der Richtungsquantelung im Magnetfeld. Z. Physik, **9**, 349–352 (1922); W. Gerlach, O. Stern, Über die Richtungsquantelung im Magnetfeld. Ann. Physik, **74**, 673–699 (1924)
9. https://de.wikipedia.org/wiki/Wolfgang_Gaede; https://de.wikipedia.org/wiki/Vakuumpumpe
10. L. Dunoyer, Le Radium **8**, 142 (1911)
11. Universitäts-Bibliothek, Goethe Universität Frankfurt, Otto Stern slide collection
12. H. Schmidt-Böcking, A. Templeton, W. Trageser; Otto Sterns gesammelte Briefe—Band 1: Hochschullaufbahn und die Zeit des Nationalsozialismus (2018) ISBN 978-3-662-55734-1 and Band 2: Sterns wissenschaftliche Arbeiten und zur Geschichte der Nobelpreisvergabe (2019)
13. O. Stern, Nachtrag zu meiner Arbeit: „Eine direkte Messung der thermischen Molekulargeschwindigkeit. Z. Physik **3**, 417–421 (1920)
14. M. Born, E. Bormann, Eine direkte Messung der freien Weglänge neutraler Atome. Phys. Zeitschr. **2**, 578–582 (1920). http://de.wikipedia.org/wiki/Mittlere_freie_Weglänge
15. J. H. Jeans, *The Dynamical Theory of Gases*, Ch. XI. § 284, 233 (Cambridge, 1904)
16. https://www.nobelprize.org/prizes/physics/1952/bloch/facts/; https://www.nobelprize.org/prizes/physics/1952/purcell/facts/; https://www.nobelprize.org/prizes/physics/1955/lamb/facts/; https://www.nobelprize.org/prizes/physics/1955/kusch/facts/; https://www.nobelprize.org/prizes/physics/1964/townes/facts/; https://www.nobelprize.org/prizes/physics/1989/ramsey/facts/
17. C. George, M. Ilg, A. Satar, V. Schobert, Das Otto-Stern-Experiment von 1920 - Ein Projekt am Gymnasium Riedberg in Zusammenarbeit mit der Universität Frankfurt. Tagungsband VDI-Schülerforum, 172–179 (2015)
18. G. Rüschmann, Drehzahlzähler für Stern-Versuchseinrichtung am Gymnasium Riedberg, PPP am Gymnasium Riedberg (2015)
19. https://physik.cosmos-indirekt.de/Physik-Schule/Mittlere_freie_Wegl%C3%A4nge
20. https://www.pfeiffer-vacuum.com/de/know-how/einfuehrung-in-die-vakuumtechnik/grundlagen/mittlere-freie-weglaenge/
21. https://www.alte-messgeraete.de/elektrotechnik/hartmann-braun-frankfurt-a-m/pyroptometer-bis-3500-c/

Chapter 10
Wilhelm Heinrich Heraeus—Doctoral Student at the University Frankfurt

S. Jorda and H. Schmidt-Böcking

The 702nd WE-Heraeus Seminar, conducted from September 1st–5th 2019 in the historic physics building in Frankfurt, commemorated the great discoveries in Quantum Physics made between 1919 and 1922 at the Frankfurt university in particular by Otto Stern and Walther Gerlach. These milestone discoveries were made in the theoretical institute of physics under the directorship of Max Born and Erwin Madelung by Otto Stern, Walther Gerlach, Max Born, Elisabeth Bormann and last not least by Alfred Landè. In this period in the early twenties Wilhelm Heinrich Heraeus was working on his doctoral thesis in Frankfurt and met Gerlach and probably Stern, Bormann, Landè and Born too (see Fig. 1). He was thus a contemporary witness to these great pioneering achievements, many decades before he and his wife Else would establish the Wilhelm and Else Heraeus Foundation.

Wilhelm Heinrich Heraeus was born on February 3rd 1900 in Hanau/Hessen, which is located about 25 km east of Frankfurt. He was the grandson of Wilhelm Carl Heraeus (* March 6th 1827; † September 14th 1904), the founder of the Heraeus company in Hanau. He described the first 23 years of his life in a short curriculum vita when he enlisted in 1923 at the University of Frankfurt for the Ph.D. examination. According to this, he had to leave grammar school, having just turned 17, with the emergency certificate, before he worked for a year, until Easter 1918, in the "patriotic emergency service". His subsequent studies of physics and natural sciences in Bonn were interrupted after only a few months by another military service, and the same was true of his subsequent studies in Göttingen, where Heraeus took part in lectures and practical exercises by Debye, Hilbert and Courant. From the fall of 1921 he finally studied in Munich, where he worked for "Geheimrat" (Privy Councilor) Wien and attended lectures by Sommerfeld, among others, before moving

S. Jorda
Wilhelm und Else Heraeus-Stiftung, Kurt-Blaum-Platz 1, 63450 Hanau, Germany

H. Schmidt-Böcking (✉)
Institut für Kernphysik, Goethe Universität, Max-von-Laue-Str.1, 60438 Frankfurt, Germany
e-mail: schmidtb@atom.uni-frankfurt.de

© The Author(s) 2021
B. Friedrich and H. Schmidt-Böcking (eds.), *Molecular Beams in Physics and Chemistry*,
https://doi.org/10.1007/978-3-030-63963-1_10

Fig. 1 Wilhelm Heraeus'
student card [Archiv,
Goethe-Universit"at
Frankfurt, Frankfurt,
Germany]

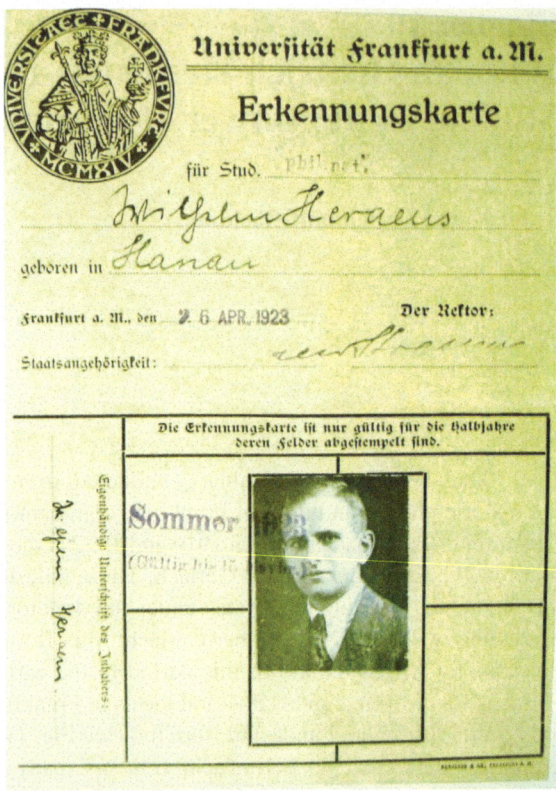

to Frankfurt in late 1922. There he took part in lectures and practical exercises by
Gerlach, Madelung, Lorentz, and carried out his doctoral thesis with "Geheimrat"
Wachsmuth and Gerlach.

Professor Richard Wachsmuth, the director of the experimental institute, together
with Professor Walther Gerlach (the co-author of the famous Stern-Gerlach experiment) supervised the dissertation of Wilhelm Heraeus. The experimental doctoral
research studies were performed in the laboratory of the Heraeus company in Hanau.
The results of Wilhelm Heraeus' work are described in the 26 pages of his dissertation with the title: "Die Abhängigkeit der thermoelektrischen Kraft des Eisens
von seiner Struktur" (The dependence of the thermo-electrical force of iron on its
structure). Wachsmuth had to apply to the dean Professor Fritz Drevermann, that
the submitted scientific work of Wilhelm Heraeus fulfilled the requirements of a
dissertation. Wachsmuth wrote in July 1923:

> A paper appeared in the Annals of Physics about 1 year ago, in which the author Borelius,
> by measuring thermoelectric forces between two samples of pure iron, one of which was
> pretreated by heating to temperatures up to 500°C and subsequent quenching, the other was
> untreated, has shown that in this interval a large number of internal iron conversion processes
> have now been found. Mr. Heraeus, to whom the resources of the father's company were
> available, seemed to me to be the suitable man for a revision. When the experiment was

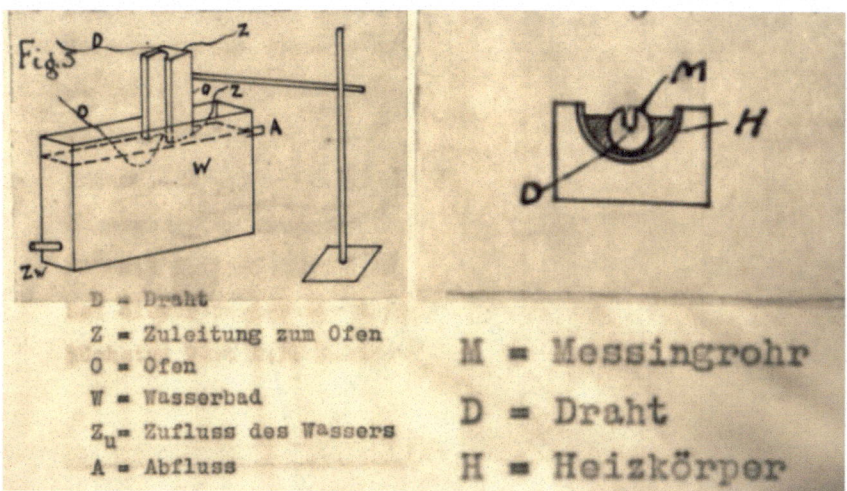

Fig. 2 The new experimental apparatus of Wilhelm Heraeus [Archiv, Goethe-Universit"at Frankfurt, Frankfurt, Germany]

extended up to 950 °C, the results were so interesting that this little extra work became a doctoral thesis.

Not only has the author of the present work discovered the errors in the Borelius's arrangement and the thermoelectric curves between "iron heated" and "iron unheated" up to the Borelius border without its discontinuities, but he has also been able to increase the temperature further. He found a decrease in the thermoelectric force between 500 and 790 °C and a new increase between 800 to 870 °C. He also associated this phenomenon with signs of recrystallization and checked it for clarification by means of appropriate tests and documented it with metallographic images. The work was carried out very carefully and conscientiously, but also completely independently. … I propose the work to the faculty for acceptance and apply for the rating "very good".

[Archiv, Goethe-Universit"at Frankfurt, Frankfurt, Germany]

In the introduction of his thesis Wilhelm Heraeus asserts: In order to develop a theory of the electrical conductivity of metals, detailed thermoelectric studies are necessary. So far, there is little data on this topic existing, as pure metals have so far hardly been available. Low impurities can have a decisive influence on the thermodynamic behavior of the metals. For this reason, for the present investigation, melted electrolytic iron was chosen, which was treated in the cold state with special care in order to avoid contamination from the carbon-containing iron of the rolls and from the material of the drawing dies. The investigation of the thermoelectric behavior of annealed versus un-annealed iron is also of particular interest insofar as there are already studies by Borelius and Gunnesson [1]. The results of this work have led to conclusions on structural changes in the temperature range from 60 to 500 °C, which are in sharp contradiction with other methods of investigation.

The new measuring device built by Wilhelm Heraeus allowed to heat an electrolytic iron wire in an electric oven with temperature steps of 10 °C (see Fig. 2).

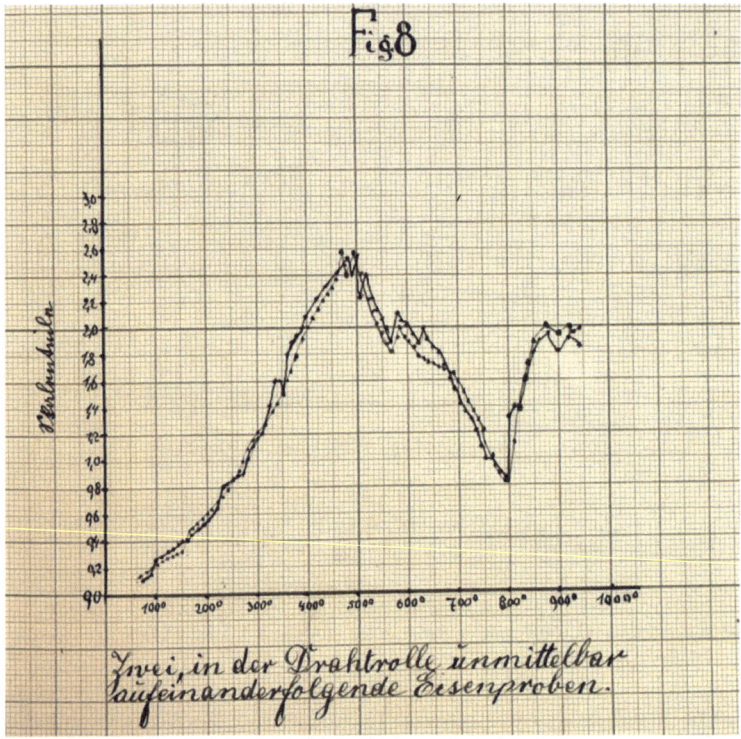

Fig. 3 Typical data sets [Archiv, Goethe-Universit¨at Frankfurt, Frankfurt, Germany]

After heating, the wire was quenched to room temperature in a water bath below. The two ends of the quenched iron wire were then connected to a galvanometer and the thermal force was measured. The measurement series were reproducible. Even if the galvanometer was connected in the opposite direction, the same values were obtained for both directions.

The Borelius' investigations were repeated by Wilhelm Heraeus. The samples were brought to a certain temperature T by means of an electric furnace and then quenched in water. Then the voltage difference was measured relative to another untreated sample. An experimental setup was chosen which was identical to that of Borelius. Even the samples were identical, since Borelius also obtained the iron wires from the Heraeus company. Wilhelm Heraeus then states: The curves (temperature on the abscissa and the measured voltage on the ordinate) of Borelius are characterized by their inconsistent, complicated course (see Fig. 3). The order of magnitude is the same for Borelius and Heraeus. However, what is completely missing is the reproducibility of each individual measurement. If the same wire was taken out of the arrangement several times and put back in again, the galvanometer deflections were completely different. It even turned out that the galvanometer showed a rash, even though the copper block was not warmed. The rashes could not therefore have resulted from a thermoelectric force; the investigation showed that there were contact forces between iron and copper, i.e. the pressing forces of the clamps. Even the smallest changes in temperature affect the measurement results (1 °C results in 0.05 to 0.15 microvolts). The fluctuations observed by Borelius can therefore all be explained by these influences.

Wilhelm Heraeus also took photomicrographs of the samples to identify structural changes. He was able to clearly identify recrystallization in his investigations, which also has a significant influence on the properties of iron. The type of recrystallization could also be influenced by the duration of the heating.

Wilhelm Heraeus concluded his written dissertation as follows.

The available studies therefore show:

1. That the experimental methods used by Borelius and Gunnesson when examining the thermocurve of un-annealed pure iron versus annealed pure iron led to incorrect results because (a) Contact forces occur, (b) the wires to be examined are mechanically stressed during the measurement, (c) the oxide formation is not taken into account. The curve of B and G could be reproduced with the Borelius arrangement, and the pure thermal curve was obtained further after the arrangement errors had been eliminated.
2. A new observation method for the observation of the thermal forces of annealed pure iron against un-annealed pure iron is described and the following observations were made: The temperature curve in the temperature range between 60 and 900 °C is a virgin curve in the sense that the curve is not reproducible at lower temperatures. The course of the thermal curve is thus determined during heating by the course of the recrystallization. The shorter the duration of the annealing, the sharper the kinks in the thermal curve between 60 and 900 °C.
3. The wires used for the investigation were examined microscopically, and the start of the recrystallization occurred approximately at 500 and up to 900 °C.

Finally in the acknowledgment Wilhelm Heraeus mentioned as his supervisors Wachsmuth and Gerlach as well as Professor Fränkel.

After passing the dissertation exam he acquired first experience in the precious metals company of his uncle in Newark, N.J., USA. In 1925 he joined the W.C. Heraeus company, and in 1927 he became a third generation member of the executive board. As head of technical development, he expanded the group's product range and, after 1945, also managed the reconstruction of the destroyed plant. In 1965 he changed from the management to the supervisory board of the family company.

Together with his wife Else († 1987) (they had no children of their own) he founded in 1963 what is now the Wilhelm and Else Heraeus Foundation, which supports scientific research and education with an emphasis on physics. Organizing WE-Heraeus Seminars is the most important and oldest funding activity of the foundation. In 1983, two years before he passed away, Wilhelm Heraeus became an honorary member of the German Physical Society (DPG).

Reference

1. G. Borelius, F. Gunnason, Annalen der Physik, 67 227 and 238 (1922); Annalen der Physik, 68 67 (1922)

Part II
Foundations of Quantum Physics and Precision Measurements

Chapter 11
Quantum or Classical Perception of Atomic Motion

John S. Briggs

Abstract An assessment is given as to the extent to which pure unitary evolution, as distinct from environmental decohering interaction, can provide the transition necessary for an observer to perceive quantum dynamics as classical. This has implications for the interpretation of quantum wavefunctions as a characteristic of ensembles or of single particles and the related question of wavefunction "collapse". A brief historical overview is presented as well as recent emphasis on the role of the semi-classical "imaging theorem" in describing quantum to classical unitary evolution.

1 Introduction

In describing the motion of the silver atom beam through their apparatus in 1922, Gerlach and Stern [1] naturally assumed classical mechanics; only the internal angular momentum was quantised. Later, following Schrödinger's 1926 introduction of wave mechanics, more attention was given to the quantum state of motion of the atoms. For example, Bohm in 1951 [2] described the beam translational motion by quantum mechanics and gave particular attention to the question of interference between the waves of the two beams leaving the magnet region. An extensive discussion of the classical and quantum aspects is given in Gottfried and Yan [3] in terms of particle wave packets.

Here, the general description of the motion of atomic particles over macroscopic distances is considered. It is shown how classical features appear autonomously and that the perception of classical or quantal behaviour depends upon the extent and accuracy of detection of such motion. This has consequences for the meaning assigned to a wave function and to its interpretation as to describing the motion of a single particle or the motions of an ensemble of identical particles.

J. S. Briggs (✉)
Institute of Physics, University of Freiburg, Freiburg, Germany
e-mail: briggs@physik.uni-freiburg.de

Department of Physics, Royal University of Phnom Penh, Phnom Penh, Cambodia

B. Friedrich and H. Schmidt-Böcking (eds.), *Molecular Beams in Physics and Chemistry*,
https://doi.org/10.1007/978-3-030-63963-1_11

1.1 Particle or Wave or Particle Ensemble?

In the scattering of electromagnetic waves around a sharp material object, the nature of the perceived outline depends upon the resolution and sensitivity of the instrument. For visible light, in the case of the human eye, usually a sharp outline of the object ascribable to a ray description of the light would be inferred. However, from measurement with an instrument able to resolve at sub-wavelength accuracy, a blurred outline corresponding to a diffraction pattern and ascribable to the wave nature of the light would be inferred. The instrument resolution is understood as the accuracy of position location. Important also is the sensitivity of the instrument, whose limit is taken here as the ability or not to register reception of a single quantum particle e.g. a photon, electron, atom or molecule.

If the detector sensitivity is sufficient one can monitor the arrival of individual particles. In the case of photons, their arrival at a screen appears in a seemingly arbitrary pattern until enough photons are counted. Then the statistical distribution gradually assumes the structured diffraction or interference form expected on the basis of the classical wave picture of electromagnetism. This is the wave-particle duality of light. With increasing resolution and sensitivity of the measurement, there are three levels of perception, classical ray trajectory, the wave picture and the ensemble of (quantum) particles picture.

A similar situation arises in the wave-particle duality of matter. For an ensemble of identical particles with a mass which is very large on an atomic scale, one assigns to their motion a unique classical trajectory so long as the resolution of, say position detection, is not itself on the atomic scale. When the mass of the particles *is* on the atomic scale it is necessary to calculate the average of their motion from the wave picture of quantum mechanics. Increasing the sensitivity to detect individual particles leads to a seemingly arbitrary pattern until the statistics are sufficient that a pattern predicted by the wave description emerges, for an experiment with electrons see Ref. [4]. Again there are three levels of perception of the ensemble; unique classical trajectory, many particles registered as a wave pattern or the statistical pattern from individual quantum particles. Which description is appropriate depends both upon the the resolution and the sensitivity of the measurement.

Indeed the analogy between the classical wave equations of electromagnetism (Helmholtz equations and paraxial approximation) and the wave equations of quantum mechanics (time-independent and time-dependent Schrödinger equations) is very close mathematically. This leads to the similarity of perception alluded to above. The semi-classical limit of quantum mechanics, used extensively below, corresponds to the eikonal approximation for electric wave propagation. Thus the quantum to classical limit for material particles corresponds to the wave to beam limit of electric field propagation. In optics, the large separation between source and observer is used to derive the Fraunhofer diffraction formula which is in close analogy to the asymptotic "Imaging Theorem" of quantum mechanics derived in section III.

One must be clear on this point. The quantum to classical transition in particle dynamics is the transition from wave to particle perception. Paradoxically, the quantum to classical transition in photon dynamics is the opposite, from particle to wave perception. It is the *wave to beam* transition of classical optical dynamics which is the analogue of the *quantum to classical* transition in particle dynamics.

A key element of quantum mechanics, not present for classical light, is the interpretation of the modulus squared of the wavefunction as a statistical probability. Here, two points of view have emerged. The first, to be called the ensemble picture, is that the probability describes the percentage of members of an ensemble of identical, and identically-prepared, particles having a particular value of a dynamical variable. The second, to be called the single-particle (SP) picture, considers that it is the probability with which an individual particle exhibits a given value out of the totality of possibilities. That is, on measurement the wavefunction "collapses" into one eigenstate of the observable with one definite value of the variable observed. The difference is that in the ensemble interpretation, only measurements on the whole ensemble are meaningful. In the SP picture meaning is assigned to a measurement on a single particle.

Here it is argued that only the ensemble interpretation of the wavefunction is tenable. However, it should be made clear from the outset that "ensemble" refers to an ensemble of *N measurements*, not necessarily *N* particles. The initial conditions have to be identical and the wavefunction gives statistical information on the outcomes. The measurements can be simultaneous or sequential. In the case of *N* particles the particles must be indistinguishable. This specification of ensembles of measurements is necessary since, unthinkable to the founders of quantum mechanics, experiments today can be made on trapped single electrons, atoms or molecules. Then the wavefunction gives only statistical information on a sequence of measurements in which the same particle initially is brought into the same state e.g. experiments on quantum jumps.

Furthermore, if the wavefunction refers to an entangled state of several particles, then that group of particles is to be regarded as a single member (single "particle") of the ensemble. Correspondingly, a feature that is very important but has been often neglected in the past, is the occurrence or otherwise of many-particle *good quantum numbers* describing eigenstates of some many-particle mechanical variable. This is because, for this special case, the measurement of the corresponding many-particle mechanical variable gives the same sharp value for all members of the ensemble. For the simple case of single-particle ensembles in an eigenstate, there is no difference between the SP and ensemble pictures, although a repeated sequence of measurements is still required to confirm the eigenstate unless defined uniquely by the preparation.

The object of this work is to re-appraise, in the light of the SP and ensemble pictures, the transition from quantum to classical mechanics by emphasising the role of the "Imaging Theorem" (IT) [5–10] in determining what an observer perceives as a consequence of the experimental resolution, sensitivity and information extraction. The IT was proved as long ago as 1937 by Kemble [5] whose aim was to show how a particle linear momentum vector could be measured and assigned in a collision experiment. Although largely forgotten until recently, here a more fundamental consequence of this theorem with regard to the quantum to classical transition is suggested, following Ref. [6].

The IT shows that any system of particles emanating from microscopic separations describable by quantum dynamics will acquire characteristics of classical trajectories simply through unitary propagation to the macroscopic separations at which measurements are made. Specifically, the IT equates the final position wavefunction $\Psi(r_f, t_f)$ at a detector at *macroscopic* position and time, to the initial momentum wavefunction $\tilde{\Psi}(p_i, t_i)$ at *microscopic* position and time but, importantly, where the variables r, p and t are related by a *classical trajectory*. This justifies the standard approach of experimentalists who use classical trajectories to trace the motion of particles from reaction zone to detector, even though the particle correlations indicate existence of a quantum wavefunction. In fact the relevance of great advances in multi-particle coincident detection [11] to the criteria for quantum or classical perception given here cannot be underestimated.

The IT connects the *momentum* wavefunction in the microscopic collision zone with the *position* wavefunction at macroscopic distance. The initial position in the collision zone cannot be defined precisely but, for free motion, the initial momentum is conserved out to the detector. This feature of the IT is completely in correspondence with the supposition of Schmidt-Böcking et.al. in Chap. 12 of this volume [12], who demonstrate that the momentum of particles emanating from a microscopic collision can be determined with sub-atomic precision but the initial position can never be measured with comparable precision.

As emphasised in the chapter by Schmidt-Böcking et al. the ensemble picture is essential for the correct interpretation of the uncertainty relation (UR) for position and momentum. According to the Robertson formulation of the UR [13] the state-dependent spread in the measurement of position and momentum refer to ensemble averages. Therefore, by definition, the individual measurements must have an *accuracy* much less than the spread. The fact that, in an individual single measurement, the product of accuracies of simultaneous momentum and position measurements can be less than $\hbar$ is demonstrated convincingly in Chap. 12.

The IT defines the asymptotic wavefunction of any quantum complex after interacting with particle or photon beams in a microscopic collision region. Such collisions *always* result in entangled many-particle wavefunctions. Indeed, although the recognition of entanglement is usually attributed to Schrödinger, initially he considered only effective one-particle problems, the hydrogen atom [14], the harmonic oscillator and the rotor [15].

It was Max Born [16] who first treated the two-particle problem of an electron colliding with an atom (to explain the experiment of Franck and Hertz) and wrote down an entangled collision wavefunction. In the same paper Born gave the statistical interpretation of the wave function which is of course the whole basis of the ensemble picture. The theory of entanglement in collisions, involving both the continuous variables of position and momentum and the discrete variables of binding energy, spin and angular momentum has been developed with great sophistication and in countless works in atomic, molecular and nuclear physics, since Born's pioneering paper, see for example [17–20].

The subject of many-particle entanglement has enjoyed enormously renewed interest in the last few years in the fields of quantum information and quantum computing, for example see Ref. [21]. However, this development, often with the limited purview of entangled photon states [22], has been made largely without reference to the study of entanglement in collision physics.

On the basis of the IT, it emerges that whether one ascribes (a) classical dynamics (a single trajectory analogous to a light ray), (b) a quantum wave description of the ensemble as a whole or (c) single particles registered separately whose statistical distribution corresponds to a wave, to the movement of material particles depends upon the precision and extent to which the dynamical variables of position and momentum are determined by the measurement. This is equally true for ensembles of many-particle systems involving entangled wavefunctions as it is for ensembles described by single particle wavefunctions.

The elimination of the overtly quantum effects of entanglement and coherence as a prerequisite for the transition to classical mechanics has been ascribed to interaction with the environment [23–26]. It goes under the broad name of "decoherence theory" (DT). This theory is part of the wider study of open quantum systems and these approaches usually involve propagation of the quantum density matrix in time.

The principal feature of such models is that the interaction with an environment leads to a suppression of the off-diagonal elements of the density matrix, which is considered a key element of the transition to classical behaviour.

Without doubt DT can explain many features of the quantum to classical transition but, according to DT, unitary evolution *in the system Hamiltonian alone* does not contribute to this transition. One main aim of the present work is to show that this is not the case.

Generically, from the result of the IT, a quantum system wavefunction or corresponding density matrix propagating in time without environmental interaction will develop such that the position and momentum coordinates change according to classical mechanics. In particular, the off-diagonal density matrix elements acquire oscillatory phase factors such that, except under a high-resolution measurement, they average to zero. In this sense, the IT does not negate any predictions of DT, rather it is complementary to it. However the unitary propagation transition occurs over time and position increments which are still of atomic dimensions and thus largely obviate any additional changes to the density matrix ascribable to the environment.

An exhaustive discussion of DT with an appraisal of its notable successes but also its limitations is given in the reviews of Schlosshauer [25, 26]. It is clear that this theory is anchored firmly in the SP interpretation of the wavefunction since wavefunction collapse, or apparent collapse, plays a prominent role.

In the present work only continuum quantum states are considered of relevance in the transition to classical mechanics. Bound states and quantised internal degrees of freedom (e.g. intrinsic spin) are viewed as wholly quantum features. They are not affected by the unitary propagation according to the IT of particles as a whole.

However, the quantised angular momentum may be used to affect the classical trajectories [27] exactly as in the original Stern-Gerlach experiment.

Here, there is no discussion of the measurement process itself. In the particle detectors considered in this work the quantum particle is intercepted by a macroscopic detector involving an enormous number of atomic degrees of freedom, giving a completely irreversible transformation. The particle energy is absorbed through ionisation or photon emission in the detector and amplified to give a recorded signal. Such measurements are often called "strong" measurements. Hence, here we do not consider so-called "weak" measurements, usually performed on light [28], which involve additional manipulation of the wavefunction e.g. change of polarisation state, during transit.

2 Interpretation of the Wavefunction

Here a simple but sufficient interpretation of the wavefunction is applied. This involves the minimum of supposition required to explain modern multi-hit coincident detection of particles emanating from complexes of atomic dimension. The following rules are adopted in connection with the detection of moving particles.

(1) The wavefunction always describes a statistical ensemble of identically-prepared particles. No meaning can be ascribed to the wavefunction of a single particle.
(2) The wavefunction $\Psi(r, t)$ contains information on the state of the ensemble. The wavefunction extent can be infinite or spatially confined.
(3) The quantity $|\Psi(r, t)|^2 dr$ gives the probability to detect a given particle from the ensemble at position r, at time t and with a resolution dr (Born's rule [16]). The quantity $|\tilde{\Psi}(p, t)|^2 dp$, where $\tilde{\Psi}(p, t)$ is the wavefunction in momentum space, gives the probability to detect a given particle from the ensemble with momentum p at time t.
(4) When information, either partial or total, is extracted by a measurement, the corresponding part of the quantum wavefunction has been utilised and no further information can be extracted from that part.

Consequent on this ensemble view, the popular expression that a particle can also behave as a wave is redundant. What is detected is always a particle. The wavefunction simply assigns a probability amplitude that a particle from an ensemble of identical particles will be detected to have particular values of the dynamical variables.

As will be shown in the following, the IT provides many of the features of wavefunction propagation ascribed to decoherence due to environmental interaction. However, since the propagation is unitary, classical features emerge autonomously.

Wavefunction "collapse" is a widely-accepted aspect of quantum mechanics. This concept is peculiar to the SP picture. In the ensemble picture the need to invoke collapse of the wavefunction does not arise.

3 The Imaging Theorem

The result known as the imaging theorem can be expressed in a few equations. Details of the original proof for free asymptotic motion can be found in the book of Kemble [5] and its generalisation for arbitrary motion, e.g. in external electromagnetic fields, in Ref. [6].

The propagation in time of a localised quantum state defined at time t' can be written

$$|\Psi(t)\rangle = U(t, t')|\Psi(t')\rangle, \tag{1}$$

where $U(t, t')$ is the time-development operator. Projecting this equation into position space gives

$$\Psi(r, t) = \int K(r, t; r', t')\, \Psi(r', t')\, dr', \tag{2}$$

where the function $K(r, t; r', t') \equiv \langle r| U(t, t')|r'\rangle$ is called the space-time propagator. The IT rests on the asymptotic large r, large t limit when the action becomes much greater than $\hbar$ and the propagator can be approximated by its semi-classical form [29]

$$K(r, t; r', 0) = \frac{1}{(2\pi i\hbar)^{3/2}}\left|\det\frac{\partial^2 S}{\partial r \partial r'}\right|^{1/2} e^{iS(r,t;r',0)/\hbar}, \tag{3}$$

where $S(r, t; r', 0)$ is the *classical* action function in coordinate space and the initial time t' is taken as the zero of time.

Now it is recognised that the r' integral is confined to a small volume, of atomic dimensions, around $r' \approx 0$, so that the action can be expanded around this point as

$$S(r, t; r', 0) \approx S(r, t; 0, 0) + \left.\frac{\partial S}{\partial r'}\right|_0 \cdot r'. \tag{4}$$

Then, using the classical relationship $\partial S/\partial r'|_0 \equiv -p$, substitution in the integral Eq. (2) gives a Fourier transform and the result

$$\Psi(r, t) \approx (i)^{-3/2}\left(\frac{dp}{dr}\right)^{1/2} e^{iS(r,t;0,0)/\hbar}\, \tilde{\Psi}(p, 0), \tag{5}$$

which is the IT of Kemble, here generalised to arbitrary classical motion.

One notes that the IT rests upon two approximations. The first is the semi-classical approximation of K in Eq. (2). However, in the integral over r', all possible values of r' contribute to the asymptotic wavefunction at r, t. It is the recognition that the quantum wavefunction at time zero is limited to a microscopic extent, Eq. (4), that associates a fixed classical momentum p to each final coordinate $r(t)$. That is, each initial $[(r' \approx 0), p]$ value is connected to a fixed final r, t by a classical trajectory. For free motion the connection is simply $r = p\, t/m$, where m is the particle mass.

The essence of the IT result is that the position and momentum coordinates evolve classically but within the shroud of the quantum wave functions. Indeed, one can show [8] that the asymptotic position wave function of Eq. (5) can be viewed as an eigenfunction of both the position and momentum operators. Hence, to the accuracy of the IT, these operators commute and, as emphasised in Ref. [12], there is no obstacle in measuring both momentum and position with arbitrary accuracy.

Then the probability density for detection of a particle of the ensemble is given by

$$|\Psi(\boldsymbol{r}(t))|^2 \approx \frac{d\boldsymbol{p}}{d\boldsymbol{r}} \, |\tilde{\Psi}(\boldsymbol{p}, 0)|^2. \tag{6}$$

Since the coordinates of the wavefunctions now conform to classical mechanics, this form has a wholly classical, statistical interpretation. An ensemble of particles with probability density $|\tilde{\Psi}(\boldsymbol{p}, 0)|^2$, defining the probability of occurrence of a certain initial momentum $\boldsymbol{p}$, move on classical trajectories and hence the ensemble members evolve to the position probability density $|\Psi(\boldsymbol{r}, t)|^2$.

The factor $d\boldsymbol{p}/d\boldsymbol{r}$ is the *classical* trajectory density of finding the system in the volume element $d\boldsymbol{r}$ given that it started with a momentum $\boldsymbol{p}$ in the volume element $d\boldsymbol{p}$ (see the books by Gutzwiller [29] or Heller [30]). Quantum mechanics provides the initial ensemble momentum distribution located at a microscopic distance $\boldsymbol{r}' \approx 0$. Each element of the initial momentum wave function is then imaged onto the spatial wave function at large distance $\boldsymbol{r}$, where the coordinates are related by classical mechanics.

That is, from Eq. (6) one has the asymptotic equality of probabilities in initial momentum space and final position space, i.e.

$$|\Psi(\boldsymbol{r}(t))|^2 \, d\boldsymbol{r} = |\tilde{\Psi}(\boldsymbol{p}, 0)|^2 \, d\boldsymbol{p}. \tag{7}$$

This shows that the loci of points of equal probability of particle detection are classical trajectories. Nevertheless, according to Eq. (5), the wavefunction remains intact.

Clearly, the IT can only be interpreted in the ensemble picture. The wavefunction spreading corresponds to the natural divergence of an ensemble of classical trajectories of differing initial momentum emanating from a microscopic volume and being detected after traversing a macroscopic distance. Nevertheless, estimates of the $\boldsymbol{r}$ and t values at which the semi-classical approximation becomes valid (Ref. [6]) show that this occurs for values which are still microscopic, typically only tens of atomic units, the precise value dependent upon particle masses and energies.

It is to be emphasised that the IT describes classical evolution of the wavefunction variables and the transition to this property arises from unitary quantum propagation i.e. the transition to classical behaviour is autonomous; external interactions are unnecessary. This justifies a routine assumption of experimentalists that one can use classical mechanics to trace a trajectory back from a point on the detector to the quantum reaction zone and is valid even for light particles such as electrons.

The consideration of the quantum to classical transition from a more mathematical viewpoint, so-called semi-classical quantum mechanics, began with the early WKB approximations and Van Vleck's work on time propagators [31]. It was formulated initially for scattering theory, for example by Mott and Massey [17], by Ford and Wheeler [32] and by Brink [33]. A completely general theory emerged later in the work of Berry and Mount [34] and of Miller [35] and Heller [30], for example. Major contributions made by Gutzwiller are to be found in Ref. [29].

In semi-classical scattering theory one examines the transition to a classical cross-section which occurs when the collision energy is much greater than the interaction energies of the collision complex, see for example, [36, 37]. This is to be contrasted with the IT in which quantum systems of atomic dimension are described fully by quantum mechanics but the transition to macroscopic distances by the semi-classical approximation. Then the semi-classical description is valid for all energies, after distances are traversed such that the classical action far exceeds $\hbar$. This is the autonomous aspect of the quantum to classical transition.

4   The Quantum to Classical Transition

4.1   Historical Context

The question of the transition from quantum to classical mechanics in the motion of particles is as old as wave mechanics itself. In the SP picture it is required that in the classical limit the wavefunction of a single particle describes a classical trajectory i.e. a narrow wavepacket. In the ensemble picture the limit is, as embodied in the IT, that the wavefunction describes an ensemble of particles following classical trajectories.

4.2   Schrödinger, Heisenberg and Kennard.

Schrödinger, immediately following his invention of wave mechanics in a sequence of papers in 1926, investigated the classical limit of wave mechanics. In a paper [38] entitled "On the continuous transition from micro- to macro-mechanics" he gave an example of how a packet of waves describing the harmonic oscillator can move in such a way that the displacement of the wavepacket as a whole follows the well-known classical dynamics of the one-dimensional harmonic oscillator. In this calculation Schrödinger repeatedly draws the analogy of superpositions of oscillator eigenfunctions to wavepackets formed from classical normal modes on an oscillating string.

The important point to note here is that Schrödinger was seeking, through the wave equation, to represent a *single particle* as a packet of quantum waves which is so localised in space that it can be perceived as a classical particle. Nevertheless he recognized the limitations of his model, pointing out, for example, that a nondispersive packet can only be built from *bound* eigenfunctions and any admixture of continuum states will result in an expanding wavepacket as in the optical case.

This latter point was taken up by Heisenberg [39] in a lengthy paper on the interpretation of the new quantum mechanics and its relation to classical mechanics. In a section also called "the transition from micro- to macro-mechanics", Heisenberg criticises the relevance of bound states in connection with classical mechanics. To illustrate the difficulty with continuum motion Heisenberg showed that an initial Gaussian wavepacket moving freely will spread in space as a function of time and so cannot represent a single material particle.

A more precise demonstration of the classical aspects of quantum motion can be traced back to 1927 in a paper by Kennard [40]. Kennard showed that the *centroid* of quantum "probability packets" moves according to classical mechanics. In retrospect, Kennard probably deserves recognition for the "Ehrenfest" theorem, but perhaps this is denied him since he couched his proof in the language of matrix mechanics, whereas Ehrenfest [41] used Schrödinger wave mechanics.

Kennard's paper is a very important landmark in the development of the meaning of the wavefunction. Interestingly, this is one of the last papers to utilise predominantly the Born, Heisenberg, Jordan [42] theory of matrix mechanics. Kennard defines a "probability amplitude" $M(q)$ for a variable q in matrix mechanics, which is later shown to be equivalent to the Schrödinger wavefunction $\psi(q)$.

He considers the motion of "probability packets" and shows that, for the cases of free motion or motion in constant electric or magnetic fields, the centroid obeys classical mechanics.

As perhaps the first to emphasise the ensemble picture, Kennard shows that Heisenberg's "proof" of the uncertainty principle is properly formulated as the statistical spread of momentum and position measured on an ensemble of identical systems. The spread, for the particular case of a free wavepacket, is calculated using the probability $MM^* dq$ which is identical to Born's probability interpretation of the Schrödinger wavefunction.

As mentioned above, the case of free motion had been solved already by Heisenberg [39] who showed that a Schrödinger free wavepacket spreads in time. Kennard, although he shows that his probability amplitude M is the same as a Schrödinger wavefunction ψ, uses this spreading as an argument against the superiority of Schrödinger wave mechanics with respect to matrix mechanics.

Kennard raises objections to the Schrödinger wave equation by pointing out that a spreading wavefunction of an electron must correspond to a spreading of charge density. Note that here, in contrast to his view of the M of matrix mechanics, in interpreting Schrödinger's ψ, Kennard is assuming that the SP picture applies to this wavefunction. Then he points out that a detection of the electron must localise its full charge at a point. Hence, because of the measurement, the original diffuse wavepacket "loses any further physical meaning" and must be replaced by "a new, smaller wavepacket". Kennard is using the necessity, in the particle picture, to invoke a "collapse of the wavefunction" as an argument against the use of a Schrödinger wavefunction.

Following this objection to the collapse scenario, Kennard then advances the ensemble interpretation of the probability amplitude of matrix mechanics. He writes "the wavepacket spreads, for example, like a charge of shot, in which each pellet describes a trajectory dependent upon its initial position and motion and the whole charge spreads in time as a consequence of differences in these initial conditions", precisely as described by the IT Eq. (6). In the ensemble picture, as distinct from the SP picture, there is no problem with the spreading of the wavepacket. Classical particles with different initial momenta will spread out as they move from micro- to macroscopic distances.

4.3 Ehrenfest and Einstein

Ehrenfest's paper was published a few months after Kennard's. Apparently, the clarification of the connection of quantum to classical mechanics received an enormous boost with this publication. Ehrenfest used the Schrödinger equation to prove the theorem that quantum position and momentum *expectation values* obey a law similar to Newton's law of classical mechanics. In one dimension, using Ehrenfest's notation, it is expressed as

$$m\frac{d^2\langle x\rangle}{dt^2} = \int dx\ \Psi\Psi^*\left(-\frac{\partial V}{\partial x}\right) = -\langle\frac{\partial V}{\partial x}\rangle \tag{8}$$

As often remarked, however, this is not Newton's Law which would require $-\partial\langle V\rangle/\partial x$ to appear on the r.h.s.. However, it turns out that for the cases $V = a$, $V = ax$ and $V = ax^2$, where a is a real constant, the theorem is the same as Newton's law. The spreading of wavepackets remains a problem since, if the wavepacket occupies a macroscopic volume of space, little meaning can be attributed to an average position. Also, for all other potentials with terms higher than quadratic one does not have motion according to Newton's law. Hence, for these two reasons and despite its appealing form, in general Ehrenfest's theorem cannot be considered as describing the transition to classical mechanics, as emphasised by Ballentine [43, 44].

Mindful of Heisenberg's proof of free wavepacket spreading, Ehrenfest is careful to stress that, within the particle picture, the motion of the mean value according to Newtonian mechanics is meaningful only "for a small wavepacket which remains small (mass of the order of 1gm.)". Clearly he was thinking of a single particle described by a small wave packet. The ideas that narrow wavepackets and Ehrenfest's theorem embody the nature of the quantum to classical transition for a single particle, pervade most elementary text books on quantum mechanics even today.

The SP and ensemble pictures were hotly discussed at the Fifth Solvay conference in October 1927. Einstein gave an example of electrons emerging from a small hole to impinge on a distant screen. He pointed out that in the ensemble picture the wave function simply gives the probability of electron detection at a given point. However, in the SP picture of the wave function, the wave which has spread to occupy a macroscopic space, must "collapse" to a point on the screen. Einstein objected to this and commented "one can only remove the objection in this manner, that one does not describe the process by the Schrödinger wave only but at the same time one localises the particle during propagation". Remarkably, after more than ninety years, we recognise that the IT wave function fulfills *exactly the property that Einstein was seeking*, a Schrödinger wave whose variables follow classical trajectories.

4.3.1   After 1927

It is interesting, although understandable in the first years of quantum and wave mechanics, that the SP and ensemble pictures are confused continually. This applies not only to Kennard, as outlined above, but also to Heisenberg and Schrödinger themselves. In discussing the uncertainty principle, Heisenberg describes exclusively measurements on a single particle, as is discussed in great detail in the accompanying paper by Schmidt-Böcking et.al. [12]. This is despite the Kennard paper quoted above and, most importantly, Robertson's proof [13] of the uncertainty principle. Both papers make clear that the spread of measured values of a variable refers to an ensemble statistical spread and not the uncertainty in measuring that property on a single particle. Similarly, Schrödinger, although a confirmed advocate of the SP picture, still admits the validity of Born's statistical interpretation and the necessity to consider a sequence of measurements, see the discussion of Mott's problem given below.

Although the Ehrenfest Theorem and narrow wavepackets are used as the classical limit in many elementary text books, reminders have been given continually since 1927 of the problems involved with this picture and the essential interpretation of a wavefunction as representing an ensemble and not a single particle.

Kemble in 1935 [49], comments that the interpretation of quantum mechanics "asserts that the wavefunctions of Schrödinger theory have meaning primarily as descriptions of the behaviour of (infinite) assemblages of identical systems similarly prepared".

Writing in 1970, Ballentine [43] advances several arguments "in favour of considering the quantum state description to apply only to an ensemble of similarly prepared systems, rather than supposing as is often done, that it exhaustively represents a single physical system". In addition, in 1972 [44] Ballentine, considers "Einstein's interpretation of quantum mechanics" and advances convincing evidence that Einstein was a firm proponent of the ensemble picture. Indeed, Ballentine must be considered as a prophet of the ensemble picture and many of his ideas are corroborated by the arguments advanced in the present paper.

In a scholarly essay in 1980, on the "Probability interpretation of quantum mechanics", Newton [45] emphasises that "the very meaning of probability implies the ensemble interpretation".

In 1994, Ballentine et al. [46] examined the Ehrenfest theorem from the point of view of the quantum/classical transition and concluded that "the conditions for the applicability of Ehrenfest's theorem are neither necessary nor sufficient to define the classical regime." Furthermore, in connection with the ensemble or SP pictures they pointed out that "the classical limit of a quantum state is an ensemble of classical orbits, not a single classical orbit." A comprehensive account of ensemble interpretations of quantum mechanics is given by Home and Whitaker [47].

5   Consequences of the IT and the Ensemble Picture

In this section three classic problems of quantum theory are analysed briefly within the IT and related ensemble picture. The problems are the subject of countless papers and the ensemble aspects have been discussed before. However, the consequences of the IT illuminate further the simplicity of the ensemble explanation. Then the reconciliation of the classical trajectory aspect of IT with the quantum interference effect is presented.

5.1   The Schrödinger Cat

The mere posing of this question by Schrödinger [50] attests to his adherence to the SP interpretation of the wavefunction. As has been observed earlier, in the ensemble picture the interpretation is trivial, as explained by Ballentine [48]. Since the wavefunction applies to many observations, one finds that half the cats are alive and half are dead. No meaning can be attached to the observation of a single cat, unless successive measurements are made over time and feline re-incarnation is allowed.

In the same paper, Schrödinger comments on the apparent problem that radioactive decay described by a spherically-symmetric wave does not lead to uniform illumination of a spherical screen but rather to individual points which slowly are seen to be uniformly distributed. However, although he states that "it is impossible to carry out the experiment with a single radioactive atom" he does not concede that this requires an ensemble interpretation of the wavefunction. This is precisely the problem of Mott which is considered next.

5.2   The "Mott Problem" of Track Structure

One of the oldest "problems" of the interpretation of a wavefunction for material particles is that posed by Schrödinger [50] and addressed in 1929 by Mott [51]. Certainly Mott's paper was at the instigation of his mentor Darwin, a confirmed adherent of the SP picture [52]. This is one of the most striking examples of erroneously assigning a wavefunction to a single particle. Mott remarked,

"In the theory of radioactive disintegration, as presented by Gamow, the α-particle is represented by a spherical wave which slowly leaks out of the nucleus. On the other hand, the α-particle, once emerged, has particle-like properties, the most striking being the ray tracks that it forms in a Wilson cloud chamber. It is a little difficult to picture how it is that an outgoing spherical wave can produce a straight track; we think intuitively that it should ionise atoms at random throughout space."

Mott presents a detailed argument based on scattering theory to argue that only atoms lying on the same straight line will be ionised successively by an α-particle emitted in a spherical wave. Although Mott repeatedly refers to the *probability* of ionisation he interprets the wavefunction as applying to a single α-particle.

However, according to the IT and the ensemble interpretation the proof of Mott is completely superfluous. There is absolutely no mystery attached to "how it is that an outgoing spherical wave can produce a straight track". This apparent dichotomy of wave mechanics is explained by the dual nature of the semi-classical wavefunction of Eq. (5); quantum wavefunction with classically-connected coordinates. Each coordinate of the initial momentum wavefunction corresponds to a specific momentum and therefore to a specific position $r(t)$ along the classical trajectory. The spherical S wavefunction applies to the ensemble as a whole and specifies equal probability of emission in all directions, i.e. uniform distribution of p on the unit sphere. Each α-particle is launched with a certain momentum p distributed according to the initial momentum wavefunction and, according to the IT, the position on a macroscopic cloud chamber scale follows a *straight line* classical trajectory. Hence it is obvious that only atoms lying along this trajectory can be ionised and the usual straight track in the cloud chamber is observed.

This is a prime example of the principle that what one perceives, in this case directed motion (a set of classical trajectories) or a spherically uniform distribution (a quantum S-wave probability) depends upon the nature and duration of the experiment.

5.3   Entanglement and Wavefunction Collapse

That wavefunction superposition applies to an ensemble is made clear also by the process of radioactive decay discussed above. Although usually thought of in the time domain, the stationary picture is simpler. The wave function of an ensemble of nuclei is described by a superposition of the state of a bound nucleus and the state of two separated product nuclei at the same total energy. The intrusion of a measuring device simply detects which state a given member of the ensemble occupies. The absence of a signal in a measuring device denotes undecayed state and a signal denotes a decay. The half-life is interpreted from a sequence of measurements on the ensemble. It is not a property of a single nucleus, although colloquially the half-life is often so ascribed. This aspect is emphasised particularly in the very clear exposition of Rau [53].

The paper of Einstein, Podolsky and Rosen [54], whose result often is referred to as the "EPR paradox", has been the subject of an enormous number of works on the subject of reality, action at a distance etc. Throughout the EPR paper appears the SP viewpoint of a partial wavefunction describing an independent particle.

Already in the first replies to the EPR paper, by Schrödinger [55] and Bohr [56], it was pointed out that it is essential to consider the *two-particle* commuting operators, ignored by EPR. Nevertheless the reply papers did not apply these considerations directly to the EPR entangled wavefunctions.

Here we infer the ensemble picture and show that the recognition of good two-particle quantum numbers is essential. Then, in the pure states considered in EPR, a good quantum number ensures that every pair of the ensemble will give the same value of the corresponding two-particle property upon measurement.

EPR consider a two-particle eigenstate written in the entangled form

$$\Psi(x_1, x_2) = \int \psi_p(x_2) u_p(x_1) \, dp \qquad (9)$$

where

$$u_p(x_1) = e^{\frac{i}{\hbar} p x_1} \quad \text{and} \quad \psi_p(x_2) = e^{-\frac{i}{\hbar} p(x_2 - x_0)} \qquad (10)$$

are eigenfunctions of one-particle operators p_1, p_2 with eigenvalues p and $-p$ respectively. The constant x_0 is arbitrary. Note that the single-particle momentum p can take any value.

The p integral in this equation can be carried out to give

$$\begin{aligned}
\Psi(x_1, x_2) &= 2\pi \delta(x_1 - x_2 + x_0) \\
&= 2\pi \int \delta(x_1 - x) \delta(x - x_2 + x_0) \, dx
\end{aligned} \qquad (11)$$

which is an entangled state in position space. However, again, all x values are possible.

Thus it has been shown that one and the same two-particle function can be expanded in terms of eigenfunctions of observables of particle 2, in this case p and x, which do not commute.

As shown by Schrödinger [55] and Bohr [56], the conserved quantities emerge from a transformation to relative and centre-of-mass (CM) coordinates for equal mass m particles. We define relative x_r and CM position X as

$$x_r = x_1 - x_2 \quad \text{and} \quad X = (x_1 + x_2)/2 \qquad (12)$$

and correspondingly relative and CM momenta

$$p_r = (p_1 - p_2)/2 \quad \text{and} \quad P_{CM} = p_1 + p_2 \qquad (13)$$

Immediately one sees from Eq. (11), that $\Psi(x_1, x_2)$ *is* an eigenfunction of the relative position coordinate $x_r = x_1 - x_2$ with eigenvalue $x_r = -x_0$. Similarly, from Eq. (9) with Eq. (10) one sees it is simultaneously an eigenfunction of CM momentum P_{CM} with eigenvalue zero. This is in order since these two operators commute. However it is readily checked, as must be, that $\Psi(x_1, x_2)$ *is not* an eigenfunction of X or p_r since these do not commute with P_{CM} and x_r respectively.

In summary, the two-particle wavefunction of EPR fixes the CM momentum at zero and the relative position of the particle pair is equal to $-x_0$. This is the only information in the two-particle wavefunction. One has, however, the clear requirement that the two-particle wavefunction should propagate intact to the detectors. In any measurement the two corresponding two-particle observables have the same precise value for all members of the ensemble of pairs.

Now one has two possible scenarios characterising entanglement.

(a) If one knows the two-particle good quantum numbers in advance e.g. by selection rules on state preparation, then the determination of the single-particle momentum to be $p = p_1$ fixes $p_2 = -p_1$. Similarly measurement of x_1 fixes $x_2 = x_1 + x_0$.

(b) If one does not know the quantum numbers in advance, one must perform measurements on many two-particle systems *in coincidence*. Then one can ascertain by experiment that, for all ensemble members, whatever the measured values of p_1 and x_1, one measures always $p_2 = -p_1$ and $x_2 = x_1 + x_0$.

The measured two-particle eigenvalues are sharp, $P_{CM} = p_1 + p_2 = 0$ and $x_r = x_1 - x_2 = x_0$ for all members of the ensemble with no statistical spread, in accordance with their commutation. Note that the specification of *two-particle* conserved observables allows one to assign precise values to both non-commuting *one-particle* observables. Hence there is absolutely no barrier to measuring both position and momentum of one or both of the particles with arbitrary accuracy. This is emphasised by the analysis of the Uncertainty Relation in the accompanying paper of Schmidt-Böcking et.al. [12]. Of course the *single-particle* p values have a distribution of probability predicted by projection of the one-particle probability amplitude out of the two-particle wavefunction.

Both scenarios require non-local information. The measurement in (b) requires communication between the two separated detectors to ensure coincidence. In case (a) only one detector is required but the non-local information is in the knowledge of the two-particle quantum numbers which are conserved for all particle separations.

The simultaneous fixing of position and momentum becomes apparent within the IT if, as is normal, detection is made at large distances from the volume from which the correlated pair is created. According to the IT there is a classical connection between position and initial momentum for detection of particles 1 and 2 at times t_1 and t_2 respectively. Then the space wavefunction can be written

$$\Psi(x_1, x_2) \propto \tilde{\Psi}(p_1, p_2 = -p_1). \tag{14}$$

In particular the IT gives the classical relation

$$x_1 = p_1 t_1 / m \text{ and } x_2 = -p_2 t_2 / m \tag{15}$$

so that from the second conservation law $x_2 = x_1 + x_0$ one has the restriction

$$x_0 = -(p_1 t_1 + p_2 t_2)/m. \tag{16}$$

Single-particle x and p can be measured simultaneously with sub-$\hbar$ accuracy, see Ref. [12].

A striking manifestation of such entanglement, which has been well-studied in experiments, is the full fragmentation of the helium atom by a single photon. This example is given since it comprises simultaneously both the momentum and position (continuous variable) entanglement of EPR *and the discrete variable spin entanglement*, as envisaged by Bohm [2], in a pure two-electron state. Furthermore, from the IT, the electrons can be assigned classical trajectories *within the two-electron quantum wavefunction*. This is not a "Gedankenexperiment" but a real measured system [57].

The two electrons emerging can be detected in coincidence and occupy a $^1P^o$ two-electron continuum state (this means their state is a spin singlet, has total orbital angular momentum one unit and odd parity). A selection rule [58] says that electrons of the same energy cannot be ejected back-to-back i.e at 180° such that $p_1 = -p_2$. That is, the *two-electron* state has a node for the EPR configuration as the coincidence experiments confirm.

If one of the electrons is left undetected a counter will register electrons of a given energy at a particular angle. However, if a detector diametrically opposed is switched on to detect electrons of the same energy in coincidence, the counts in both detectors will be zero. This coherent state can be made incoherent by switching off one of the detectors when electrons will be measured again. The essence is that this pure effect of wavefunction entanglement is evident, even though according to the IT, the electrons are moving on classical trajectories after they exit the reaction zone with well-defined momenta.

In interpreting the wavefunction, as in EPR, it is crucial that the ensemble is viewed as an ensemble of *two-electron* systems. This two-electron wavefunction is the single quantum entity and it must be transmitted to the macroscopic detection zone unchanged. Then there is no wavefunction interpretation problem within the ensemble picture. The wavefunction node says that there is zero probability that a given member of the ensemble (a coincident pair of electrons) will be emitted in the forbidden configuration.

The coincident detection of position and momentum extracts the information from the wavefunction of the ensemble of two-electron states. The non-coincident detection of electrons extracts information only on the ensemble of single electrons. The effect of entanglement is non-local simply because the two-electron wavefunction is non-local.

A comprehensive discussion of the implications of entanglement for the uncertainties in measured properties relevant to quantum information in the case of the continuous variable description of light rather than material particles, is given by Braunstein and van Loock [22].

5.4   Quantum Interference

The Davisson-Germer experiment of 1927 on electron-beam diffraction established
the validity of the description of particle ensembles by a wave function. The diffraction of heavy neutral particle beams was confirmed in the pioneering experiments
of Stern and co-workers as early as 1929 [59]. The demonstration of interference
even of large molecules has been achieved recently in the remarkable experiments
of Arndt and his group [60], reported in Chap. 24 of this volume.

The explanation of interference patterns in terms of semi-classical wavefunctions
and the underlying classical trajectories has been given in great detail by Kleber
and co-workers [61] and will not be repeated here. Based upon the IT (see eq.(1)
of Kleber [62]), their theory is used to interpret experiments such as those of Blondel et.al. [63]. Here the "photoionisation microscope" exhibits interference rings of
electrons ionised from a negative ion in the presence of an extracting electric field. In
the semi-classical explanation electrons can occupy two classical trajectories. Either
they proceed directly to the detector or, initially they are ejected moving away from
the detector but are turned around in the electric field. The imaging of the spatial
wavefunction squared is obtained by detection on a fixed flat screen i.e. only the
position of electrons is detected. Then an interference pattern from the two trajectories is observed.

However, were the vector position *and* vector momentum of the electrons to be
observed, that would correspond to a "which way" determination and the perception would be of two distinct non-interfering classical trajectories. Interestingly, as
distinct from entanglement, in this case it is a lack of information which gives rise
to wave perception. Blondel et al. [63] remark also that for ionisation from neutral
atoms the interference rings are there but are too small to be detected, again showing that perception depends upon resolution.

6   The Imaging Theorem and Decoherence Theory:
IT and DT

As stated in the Introduction, the suppression of state superposition, entanglement
and interference through environmental interaction can be seen as a requirement on
the way to a classical limit of quantum mechanics and has come to be known as
"decoherence theory" (DT). It is viewed as a universal phenomenon, extending even
to the classical limit of quantum gravity [64, 65] (for an interesting discussion see
Ref. [66]). In the following the transition from quantum to classical perception is
discussed.

6.1 Decoherence

There is an enormous literature on DT and associated theories describing "spontaneous localisation" due to stochastic interaction. Space does not permit a discussion of the many and varied aspects of these theories, so here consideration is given to those features relevant to the quantum to classical transition embodied in the IT and to the SP or ensemble interpretation of the wavefunction.

The essence of DT is given in the famous paper of Zurek [24] and in more detail in the reviews of Schlosshauer [25, 26]. A more exhaustive treatment with discussion of the $\hbar$ dependence of the environmental interaction terms is to be found in the stochastic Schrödinger equation approach [67]. Here the simpler original density matrix version of Zurek [24] is sufficient as illustration.

The basic mechanism of DT by which certain quantum aspects are eliminated is quite straightforward, accounting for the universality of this phenomenon. In the simplest case presented in Ref. [25], a one-dimensional two-state quantum system S, with states $|\psi_n\rangle$, is assumed to become entangled with an "environment" with corresponding states $|E_n\rangle$. Limiting to two-state quantum systems, the ensuing entangled state vector is

$$|\Psi\rangle = \alpha|\psi_1\rangle|E_1\rangle + \beta|\psi_2\rangle|E_2\rangle \tag{17}$$

and gives a total density matrix $\rho = |\Psi\rangle\langle\Psi|$. According to Ref. [25], "the statistics of all possible local measurements on S are exhaustively encoded in the reduced density matrix ρ_S", given by

$$\begin{aligned}\rho_S = Tr_E\rho &= |\alpha|^2|\psi_1\rangle\langle\psi_1| + |\beta|^2|\psi_2\rangle\langle\psi_2|\\ &+ \alpha\beta^*|\psi_1\rangle\langle\psi_2|\langle E_2|E_1\rangle + \alpha^*\beta|\psi_2\rangle\langle\psi_1|\langle E_1|E_2\rangle.\end{aligned} \tag{18}$$

Then a measurement of the particle's position is given by the diagonal element,

$$\begin{aligned}\rho_S(x,x) &= |\alpha|^2|\psi_1(x)|^2 + |\beta|^2|\psi_2(x)|^2\\ &+ 2\text{Re}[\alpha\beta^*\psi_1(x)\psi_2^*(x)\langle E_2|E_1\rangle]\end{aligned} \tag{19}$$

where "the last term represents the interference contribution". The assumption of DT is that in general the states of the environment are orthogonal and so the interference term disappears. More importantly, from Eq. (18) the off-diagonal terms disappear and one has a diagonal density matrix only. From Eq. (19) this has two "classical" terms interpreted as classical probabilities.

A slightly different model is adopted in Ref. [24] in that the two states comprising the system S are taken as two spatially-separated Gaussian wavefunctions. The corresponding system density matrix exhibits four peaks. This density matrix is propagated in time subject to a temperature-dependent environment interaction. The result is to give a density matrix of diagonal form with only two peaks along the diagonal.

In this case the decoherence reduces the off-diagonal elements to zero and the diagonal term does not contain the "interference" contribution since the Gaussians do not overlap. This removal of coherence between different spatial parts of the wavefunction is considered to correspond to the emergence of classicality. In connection with the classical transition Schlosshauer writes [26] "the interaction between a macroscopic system and its environment will typically lead to a rapid approximate diagonalisation of the reduced density matrix in position space and thus to spatially localised wavepackets that follow (approximately) Hamiltonian trajectories". This following of classical trajectories however, is not proven in detail.

Implicit is the SP picture in which the diagonal elements represent narrow wavepackets giving classical behaviour via Ehrenfest's theorem. The ultimate spreading of these wavepackets is not considered, although suitable environmental interaction can lead to the wavepackets remaining narrow. In short, the transition to classicality is viewed as an elimination of quantum coherence effects and the vital feature of the emergence of classical dynamics according to Newton is not shown.

6.2   Unitary Evolution

In appendix A, following the example of Ref. [24], the free *unitary* propagation of two, initially narrow, Gaussian wavepackets within the IT is calculated. It is shown that, under low detector resolution, the density matrix also assumes the diagonal form

$$\rho(x, x, t) = \frac{1}{\sqrt{\pi} \eta(t)} \, (e^{-(x-X_1)^2/\eta^2} + e^{-(x-X_2)^2/\eta^2}) \tag{20}$$

where X_1, X_2 are the centres of the wavepackets and the time-dependent width is $\eta = \tilde{\sigma} t/\mu$, for initial width $\tilde{\sigma}$ and particle mass μ. Hence, the intrinsic spreading of the wavepacket with time emerges as expected in the ensemble picture. In this picture there is no problem of interpretation of the two probabilities; 50% of the ensemble members will be detected near to X_1 and 50% near to X_2. Wavefunction collapse is unnecessary. Most important however, in the IT, the propagation of the co-ordinates of the diagonal density matrix is according to *classical* mechanics. Nevertheless, if the resolution is on the microscopic scale then interference and manifestations of quantum propagation resulting from finite off-diagonal elements can be detected. Just as in optics, the perception of particle trajectory (ray) or wave is decided by the sharpness of vision.

The study of collision complexes in nuclear, atomic and molecular physics has long been concerned with the questions of measurement of interference and entanglement effects [18–20]. Coincidence detection of several collision fragments in entangled states are performed with increasing sophistication (see, for example, Ref. [11]). In line with the IT, classical motion of the collision fragments outside the reaction zone is shown to be appropriate. Nevertheless quantum coherence is preserved showing that environmental decoherence does not occur in such experiments.

The degree of decoherence assigned to a many-body entangled state depends upon which particles are *not* observed or even which dynamical properties are observed and which are not. Coherence can be fully or partially removed according to the experiment. In the language of the experimentalist, either one registers the "coincidence" spectrum or the "singles" spectrum. Again this illustrates that perception of quantum effects depends upon the measurement. Non-detection of collision variables corresponds to a partial trace of the full density matrix, as in DT.

7 Conclusions

The imaging theorem corresponds only to the ensemble interpretation. According to the IT, an initial momentum wavefunction decides the spatial wavefunction at macroscopic distance. The modulus squared of the initial momentum wavefunction corresponds to an ensemble of classical particles with the same initial momentum distribution. Each particle appears to move along a classical trajectory to be registered at well-defined position at a distant screen. Nevertheless the quantum wavefunction is preserved so that the loci of points of equal probability are the classical trajectories but that probability is given by the quantum position wavefunction.

Indeed, all collision experiments support the ensemble picture. One counts many particles at different locations and times on a detector and so builds an image of the initial momentum distribution. Particularly striking in this respect is the observation of the gradual assembly of an interference pattern. Using electron diffraction through a pair of slits, it has been shown [4] that the wave interference pattern is built up slowly by registering many hundreds of hits of individual electrons on a detector screen. Even more strikingly, the experiment has been performed with very large organic molecules [69]. This shows convincingly that it is the ensemble of hits at the detector that builds up the wave interference pattern.

In contrast, the SP picture is that the wavefunction, extending over macroscopic distance, is carried by each molecule and the wavefunction collapses at different points on the screen. That is, the detector is required to instigate decoherence leading to instantaneous wavefunction collapse (from macroscopic to microscopic extent) and the electron being registered at a single localised point on the detector. Again, one is faced with the dilemma of Kennard and Einstein in accepting the plausibility of such a transition.

To summarise, it has been shown that;

(1) The IT preserves the quantum wavefunction out to macroscopic distances but the momentum and position *coordinates* change in time according to classical mechanics. With the IT asymptotic wave function, position and momentum can be measured with arbitrary accuracy [12].

(2) As a result of the IT, unitary evolution of quantum systems, even after propagation over relatively microscopic distances, leads to perception of an ensemble of particles as following classical trajectories.

(3) Standard measurement techniques, either on single or multiple particles, can lead to perception or otherwise of the quantum properties of interference and entanglement according to the information registered. The inference of classical or quantum behaviour depends ultimately upon the resolution and detail of the measurement performed.

Without environment influence, within the IT, unitary evolution of quantum systems results in effective decohering effects. This "decoherence" is of a different nature than in DT. It occurs due to cancellation of oscillating terms of different phase, which leads to non-resolution of oscillatory terms in the propagation of the density matrix to macroscopic times, as in Eq. (27) and Eq. (28). Hence, lack of sufficient resolution results in effective decoherence although paradoxically it arises from the very terms, oscillatory phase factors, which are the hallmark of quantum coherence in the wavefunction.

Already in 1951, long before the formulation of DT, in his discussion of the motion of atom beams in the Stern-Gerlach experiment [1], Bohm [2] points out the decoherence arising from interaction with a macroscopic detector. Interestingly, he attributes the lack of interference also to the impossibility of resolving oscillatory energy phase factors in the unitary time propagation (exactly as in the Appendix) but does not really emphasise the distinction between this and the DT interactions.

The preservation of the wavefunction can lead to interference. However, the perception of interference patterns, or not, again depends upon the nature of the measurement performed. The observation of interference patterns implies that, although resolution is high, incomplete information as to the different trajectories encoded in the wavefunction variables is extracted by the measurement. That is, a "which way" detection is not performed. Then, whether one perceives quantum or classical dynamics depends simply upon the precision of the measurement performed and the amount of information extracted from the wavefunction. This is all in close analogy, both physically and mathematically, to the optical case of perception of particle, wave or ray properties.

In the case of the detection of the effects of particle entanglement it is necessary to treat the ensemble entity as corresponding to the *many-particle* wavefunction and its quantum numbers. Incomplete extraction of the information encoded in the many-particle ensemble wavefunction, for example detection of only some of the particles or incomplete specification of vector variables, corresponds to an effective decoherence.

Acknowledgements The author is extremely grateful to Prof. James M. Feagin for several years of close cooperation on the derivation and meaning of the imaging theorem. The insight of Prof. Horst Schmidt-Böcking on experimental methods and their significance for the uncertainty relation, described in the accompanying paper (Chap. 12 of this volume), is acknowledged also.

8   Appendix

As in the discussion of decoherence by Zurek [24] the time development of a one-dimensional single-particle ensemble wavepacket is considered. The wavepacket is composed of two Gaussians centred at $x = X_1$ and $x = X_2$ with width such that there is essentially no overlap at $t = 0$. The initial state is then

$$\Psi(x, t = 0) = (\pi\sigma^2)^{-1/4} \sum_{i=1,2} e^{-x_i^2/(2\sigma^2)} \tag{21}$$

where $x_i \equiv x - X_i$. For $t > t_0$ this initial wavefunction propagates freely in time and has the exact form

$$\Psi(x, t) = (\sigma^2/\pi)^{1/4} \left(\sigma^2 + \frac{i\hbar t}{\mu}\right)^{-1/2}$$
$$\times \sum_{i=1,2} \exp\left[-\frac{x_i^2}{2\left(\sigma^2 + \frac{i\hbar t}{\mu}\right)}\right], \tag{22}$$

where μ is the particle mass. The IT condition emerges in the limit of large times and distances. Large times corresponds to $\hbar t/\mu \gg \sigma^2$. Then the spatial wavefunction assumes the IT form,

$$\Psi(x, t) \approx \left(\frac{\sigma^2}{\pi}\right)^{1/4} \left(\frac{\mu}{i\hbar t}\right)^{1/2}$$
$$\times \sum_{i=1,2} e^{-(\mu x_i \sigma/(\sqrt{2}\hbar t))^2} e^{i\mu x_i^2/(2\hbar t)} \tag{23}$$

The IT limit giving the classical trajectory is such that x_i and t both become large but the ratio is a constant classical velocity. To emphasise this we introduce the momenta $p_i = \mu x_i/t$. We also define, as the width of the Gaussian in momentum space, $\tilde{\sigma} \equiv \hbar/\sigma$. Then we can simplify the asymptotic spatial wavefunction using

$$(\mu x_i \sigma/(\sqrt{2}\hbar t))^2 \equiv p_i^2/(2\tilde{\sigma}^2) \tag{24}$$

and the energy phases

$$\mu x_i^2/(2\hbar t) = p_i^2 t/(2\mu\hbar). \tag{25}$$

The asymptotic spatial wavefunction is then,

$$\Psi(x, t) \approx \left(\frac{\mu}{i\sqrt{\pi}\tilde{\sigma}t}\right)^{1/2} \sum_{i=1,2} e^{-p_i^2/(2\tilde{\sigma}^2)+ip_i^2 t/(2\mu\hbar)} \tag{26}$$

which looks exactly like a pair of free momentum Gaussians propagating in time and corresponds to the 1D form of the IT of Eq. (5), with $dp_i/dx_i = \mu/t$ for free motion.

The diagonal element of the density matrix is defined as $\rho(x, x, t) = \Psi^*(x, t)\Psi(x, t)$ and is

$$
\begin{aligned}
\rho(x, x, t) =& \frac{\mu}{\sqrt{\pi}\tilde{\sigma}t} \sum_{i=1,2} e^{-p_i^2/\tilde{\sigma}^2} \\
&+ 2\cos\left[(p_1^2 - p_2^2)t/(2\mu\hbar)\right] e^{-(p_1^2+p_2^2)/(2\tilde{\sigma}^2)}
\end{aligned}
\tag{27}
$$

The off-diagonal density matrix is defined as $\rho(x, x', t) = \Psi^*(x, t)\Psi(x', t)$ and consists of four terms,

$$
\rho(x, x', t) = \frac{\mu}{\sqrt{\pi}\tilde{\sigma}t} \sum_{i,j=1,2} e^{-(p_i^2+p_j'^2)/(2\tilde{\sigma}^2)} e^{-i(p_i^2-p_j'^2)t/(2\mu\hbar)}
\tag{28}
$$

At $t = 0$ this gives rise to four Gaussian peaks, as in Ref.[24]. It reduces to the diagonal element when $p_i = p_i'$, i.e. $x = x'$ as it should.

One sees that the diagonal matrix element shows two peaks at $p_1 = 0$, $p_2 = 0$ or equivalently $x = X_1$, $x = X_2$. There is also an interference term. In the off-diagonal element there are four peaks, with the two additional peaks at $x' = X_1$ and $x' = X_2$. These also contain oscillatory phase factors giving interference.

Clearly, to observe interference effects the temporal resolution must typically be less than one oscillation, i.e. $t < 4\mu\pi/(p_1^2 - p_2^2)$. Consider that the particles are electrons with mass unity in atomic units (a.u.). If we take the two peaks to be separated by 1 a.u. of distance, then we have $t < 4\pi \approx 10^{-16}$ s. However, typical resolutions are nanoseconds, that is seven orders of magnitude larger than this. If the resolution is $\delta t \equiv \tau$ then the measurement must be integrated over this time period. Typically the oscillatory terms will then give, omitting constants

$$
\int_{-\tau/2}^{\tau/2} e^{i(p_1^2-p_2^2)t} \, dt \approx \delta(p_1^2 - p_2^2).
\tag{29}
$$

and similarly for the off-diagonal element when p_i is replaced by p_i'. In other words, the oscillations will average to zero under low resolution of measurement on an atomic time scale. From Eq. 27 this implies that the density matrix will exhibit only two diagonal gaussian peaks for such measurements,

$$
\rho(x, x, t) = \frac{\mu}{\sqrt{\pi}\tilde{\sigma}t} \left(e^{-p_1^2/\tilde{\sigma}^2} + e^{-p_1^2/\tilde{\sigma}^2}\right)
\tag{30}
$$

with $p_i = \mu(x - X_i)/t$. For the off-diagonal elements, from Eq. (28), all the terms will average to zero under normal time resolution to give zero off-diagonal elements. This is exactly the limit, elimination of off-diagonal density matrix elements, given by

Zurek [24] as the classical limit and resulting from time propagation in the presence of an interacting environment. However, we emphasise again that the wavepackets on the diagonal are spreading and only in the limit that particles are macroscopically massive can this be ignored to give localised single particles as envisaged in [25].

By contrast the IT proves that classicality emerges from unitary Hamiltonian propagation under low temporal resolution, in that the density matrix then has only two diagonal peaks. Quantum coherence is lost except where the temporal and spatial resolution are extremely high. The peaks represent an ensemble of classical particles spreading on classical trajectories and distributed according to the initial Gaussian momentum wavefunction.

References

1. W. Gerlach, O. Stern Zeits. f. Phys. **9**, 349 (1922)
2. D. Bohm, *Quantum Theory* (Prentice-Hall, New York, 1951)
3. K. Gottfried, T.-M. Yan, *Quantum Mechanics: Fundamentals*, 2nd ed. (Springer, New York, 2003)
4. R. Bach, D. Pope, S.-H. Liou, H. Batelaan, New. J. Phys. **15**, 033018 (2013)
5. E.C. Kemble, *Fundamental Principles of Quantum Mechanics with Elementary Applications* (McGraw Hill, 1937)
6. J.S. Briggs, J.M. Feagin, New J. Phys. **18**, 033028 (2016)
7. J.S. Briggs, J.M. Feagin, J. Phys. B: At. Mol. Opt. Phys. **46**, 025202 (2013)
8. J.M. Feagin, J.S. Briggs, J. Phys. B: At. Mol. Opt. Phys. **47**, 115202 (2014)
9. M.R.H. Rudge, M.J. Seaton, Proc. Roy. Soc. London, A **283**, 262 (1965); E.A. Solovev, Phys. Rev. A **42**, 1331 (1990); T.P. Grozdanov, E.A. Solovev, Eur. Phys. J. D **6**, 13 (1999); M. Kleber, Phys. Rep. **236**, 331 (1994); V. Allori, D. Dürr, S. Goldstein, N. Zanghí, J. Opt. B: Quantum Semiclass. Opt. **4**, S482 (2002); M. Daumer, D. Dürr, S. Goldstein, N. Zanghí, J. Stat. Phys. **88**, 967 (1997)
10. J.H. Macek in *Dynamical Processes in Atomic and Molecular Physics*, ed. by G. Ogurtsov and D. Dowek (Bentham Science Publishers, ebook.com, 2012)
11. J. Ullrich, R. Moshammer, A. Dorn, R. Doerner, L. P. H. Schmidt, and H. Schmidt-Boecking, Rep. Prog. Phys. 66, 1463 (2003), M. Gisselbrecht, A. Huetz, M. Lavolle, T. J. Reddish, and D. P. Seccombe, Rev. Sci. Instr. **76**, 013105 (2013), P. C. Fechner and H. Helm, Phys. Chem. Chem. Phys. **16**, 453 (2014)
12. H. Schmidt-Böcking, S. Eckart, H.J. Lüdde, G. Gruber, T. Jahnke Chapter 12 of this volume
13. H.P. Robertson, Phys. Rev. **34**, 163 (1929)
14. E. Schrödinger, Ann. der Phys. **79**, 361 (1926)
15. E. Schrödinger, Ann. der Phys. **79**, 489 (1926)
16. M. Born, Zeits. Phys. **38**, 803 (1926)
17. N.F. Mott, H.S.W. Massey, *Theory of Atomic Collisions* (OUP, Oxford, 1965)
18. U. Becker, A. Crowe (eds.), *Complete Scattering Experiments* (Kluwer Academic, New York, 2001)
19. U. Fano, J.H. Macek, Rev. Mod. Phys. **45**, 553 (1973)
20. K. Blum, *Density Matrix Theory and Applications* (Plenum, New York, 1981)
21. R. Horodecki, P. Horodecki, M. Horodecki, K. Horodecki, Revs. Mod. Phys. **81**, 865 (2009)
22. S.L. Braunstein, P. van Loock, Revs. Mod. Phys. **77**, 513 (2005)

23. From the very extensive literature, see for example E. Joos, H. D. Zeh, C. Kiefer, D. Guilini, J. Kupsch, I.-O. Stamatescu (eds.), *Decoherence and the Appearance of a Classical World in Quantum Theory*, 2nd Ed. (Springer, New York, 2003) and references therein, W.H. Zurek, Phys. Today **67**, 44 (2014) and Los Alamos Science, Number 27 (2002) and references therein, J.J. Halliwell, Phys. Rev. D **39**, 2912 (1989)
24. W.H. Zurek, Physics Today October (1991)
25. M. Schlosshauer, Rev. Mod. Phys. **76**, 1267–1305 (2004)
26. M. Schlosshauer in M. Aspelmeyer, T. Calarco, J. Eisert, F. Schmidt-Kaler (eds.), *Handbook of Quantum Information* (Springer, Berlin/Heidelberg, 2014)
27. M. Brouard, D.H. Parker, S.Y.T. van de Meerakker, Chem. Soc. Rev. **43**, 7279 (2014)
28. S. Kocsis et al., Science **332**, 1170 (2011)
29. M.C. Gutzwiller, *Chaos in Classical and Quantum Mechanics*, 2nd edn. (Springer, New York, 1990)
30. E.J. Heller, *The Semiclassical Way* (Princeton U.P., Princeton and Oxford, 2018)
31. J.H. Van Vleck, P.N.A.S. **14**, 178 (1928)
32. K.W. Ford, J.A. Wheeler, Ann. Phys. (NY) **7**, 259 (1959)
33. D.M. Brink *Semi-classical Methods for Nucleus-Nucleus Scattering* (CUP, Cambridge, New York, 1985)
34. M.V. Berry, K.E. Mount, Rep. Prog. Phys. **35**, 315 (1972)
35. W.H. Miller, Acct. Chem. Res. **4**, 161 (1971)
36. J.M. Rost, E.J. Heller, J. Phys. B **27**, 1387 (1994)
37. J.M. Rost, Phys. Rep. **297**, 271 (1998)
38. E. Schrödinger, Die Naturwissenschaften **14**, 664 (1926)
39. W. Heisenberg, Zeit. f. Phys. **43**, 172 (1927)
40. E.H. Kennard, Zeit. f. Phys. **44**, 326 (1927)
41. P. Ehrenfest, Zeit. f. Phys. **45**, 455 (1927)
42. M. Born, W. Heisenberg, P. Jordan, Zeit. f. Phys. **35**, 557 (1926)
43. L.E. Ballentine, Revs. Mod. Phys. **42**, 358 (1970)
44. L.E. Ballentine, Am. J. Phys. **40**, 1763 (1972)
45. R.G. Newton, Am. J. Phys. **48**, 1029 (1980)
46. L.E. Ballentine, Am. J. Phys. **55**, 785 (1987)
47. D. Home, M.A.B. Whitaker, Phys. Reports **210**, 223 (1992)
48. L.E. Ballentine, Ensembles in Quantum Mechanics in *Compendium of Quantum Physics*, ed. by D. Greenberger, K. Hentschel, F. Weinert (Springer, Berlin, 2009), p. 199
49. E.C. Kemble, Phys. Rev. **47**, 973 (1935)
50. E. Schrödinger, Die Naturwissenschaftern **48**, 807 (1935)
51. N.F. Mott, Proc. Roy. Soc. London, A **126**, 79 (1929)
52. C.G. Darwin, Proc. Roy. Soc. London, A **117**, 258 (1927)
53. A.R.P. Rau, Phys. Essays **30**, 60 (2017)
54. A. Einstein, B. Podolsky, N. Rosen, Phys. Rev. **47**, 777 (1935)
55. E. Schrödinger, Proc. Cam. Phil. Soc. **31**, 555 (1935)
56. N. Bohr, Phys. Rev. **48**, 555 (1935)
57. J.S Briggs, V. Schmidt, J. Phys. B **33**, R1 (2000)
58. F. Maulbetsch, J.S. Briggs, J. Phys. B **26**, 1679 (1994)
59. I. Estermann, O. Stern, Zeits. f. Physik **61**, 95 (1930)
60. F. Kialka et al., Phys. Scr. **94**, 034001 (2019)
61. T. Kramer, C. Bracher, M. Kleber, J. Phys. A **35**, 8361 (2002)
62. M. Kleber, Phys. Rep. **236**, 331 (1994)
63. C. Blondel, C. Delsart, F. Dulieu, Phys. Rev. Lett. **77**, 3755 (1996)
64. J.J. Halliwell, Phys. Rev. **D39**, 2912 (1989)
65. C. Kiefer in E. Joos, H.D. Zeh, C. Kiefer, D. Guilini, J. Kupsch, I.-O. Stamatescu (eds.) *Decoherence and the Appearance of a Classical World in Quantum Theory*, 2nd Ed. (Springer, New York, 2003), p. 181

66. S.E. Rugh, H. Zinkernagel in K. Chamcham, J. Silk, J. Barrow and S.Saunders (eds.) *The Philosophy of Cosmology* (CUP, Cambridge, 2016)
67. L. Diósi, W.T. Strunz, Phys. Lett. A **235**, 569 (1997)
68. D. Akoury et al., Science **318**, 949 (2007)
69. T. Juffmann et al., Nature Nano **7**, 297 (2012)

Chapter 12
The Precision Limits in a Single-Event Quantum Measurement of Electron Momentum and Position

H. Schmidt-Böcking, S. Eckart, H. J. Lüdde, G. Gruber, and T. Jahnke

Abstract A modern state-of-the-art *"quantum measurement"* [The term "quantum measurement" as used here implies that parameters of atomic particles are measured that emerge from a single scattering process of quantum particles.] of momentum and position of a **single** electron *at a given time* ["at a given time" means directly after the scattering process. (It should be noticed that the duration of the reaction process is typically extremely short => attoseconds).] and the precision limits for their experimental determination are discussed from an **experimentalists point of view**. We show—by giving examples of actually performed experiments—that in a single reaction between quantum particles *at a given time* only the momenta of the emitted particles but not their positions can be measured with sub-atomic resolution. This fundamental disparity between the conjugate variables of momentum and position is due to the fact that during a single-event measurement only the total momentum but not position is conserved as function of time. We highlight, that (other than prevalently perceived) Heisenberg's "Uncertainty Relation" UR [1] does not limit the achievable resolution of momentum in a **single-event measurement**. Thus, Heisenberg's statement that in a single-event measurement only either the position or the momentum (velocity) of a quantum particle can be measured with high precision contradicts a real experiment. The UR states only a correlation between the mean statistical fluctuations of a large number of repeated single-event measurements of two conjugate variables. A detailed discussion of the real measurement process and its precision with respect to momentum and position is presented.

H. Schmidt-Böcking (✉) · S. Eckart · G. Gruber · T. Jahnke
Institut für Kernphysik, Universität Frankfurt, 60438 Frankfurt, Germany
e-mail: hsb@atom.uni-frankfurt.de; schmidtb@atom.uni-frankfurt.de

H. J. Lüdde
Institut für Theoretische Physik, Universität Frankfurt, 60438 Frankfurt, Germany

© The Author(s) 2021
B. Friedrich and H. Schmidt-Böcking (eds.), *Molecular Beams in Physics and Chemistry*,
https://doi.org/10.1007/978-3-030-63963-1_12

1   Introduction

Otto Stern was the pioneer in high-resolution momentum spectroscopy of atoms and molecules moving in vacuum. Gerlach and Stern performed between 1920 to 1922 in Frankfurt their famous Stern-Gerlach experiment (SGE). They obtained for Ag atoms a sub-atomic momentum resolution in the transverse direction of 0.1 a.u. [2]. Today, modern state-of-the-art spectrometer devices such as the Scienta electron spectrometers [3] or the COLTRIMS Reaction Microscope C-REMI [4] can provide even a much better resolution. The imaging system C-REMI can even measure several particles in coincidence by detecting the momenta of all charged fragments emitted in a quantum process. Thus the complete entangled dynamics of such a single quantum process can be visualized. However, in such high resolution experiments the experimenter cannot obtain any direct information on the relative positions of particles. For a single event the absolute and also relative positions inside the quantum reaction are not measureable. The purpose of this paper is to illustrate for a single-event scattering measurement, as discussed by Heisenberg [1], the precision limits of electron momentum and position by presenting experimental examples.

The goal of a quantum measurement, e.g. scattering of a quantum projectile on a target atom in vacuum, is to obtain information on the quantum mechanical collision process. How can such a measurement be performed? The experimenter must prepare projectiles and target objects in a well-defined momentum and as far as possible also in a well-defined position state. This is typically achieved by classical methods. As shown below the momentum state of projectile and target object can be prepared with sub-atomic precision, but positions at a given time can never be controlled with atomic size accuracy. The reason is, no particle in vacuum can be brought completely at rest in the system of measurement and thus positions are not conserved with time. In other words the experimenter cannot predict with sub-atomic precision the impact parameter of the collision and the impact parameters are statistically distributed. Thus numerous single event measurements one after the other have to be summed up to obtain a statistical distribution.

The statistical distribution contains two sources of errors: First, the systematical error of each single-event measurement. This is de facto the horizontal error bar which is given by the quality of preparation and of the classical detection device only. This error bar depends very little on Heisenberg's uncertainty relation. Second, there is a statistical error in the ordinate values, which depends on the number of detected events and which does not depend on the precision of the single-event measurement. The sum of all single-event measurements, i.e. the statistical distribution, is relevant for comparison with theory.

What are the precision limits for parameters in the quantum experiment? The detection apparatus delivers only auxiliary values, from which then information on the quantum process can be deduced. Such auxiliary values are: The time, when the collision occurs. It can be determined by classical methods (see below) with about 50 pico-second precision. During the collision electrons and ionic fragments can be emitted each with a so-called final momentum. Immediately after emission they move in a spectrometer device in which the charged quantum particles exchange momentum due to the electro-magnetic force with the macroscopic detection device. Finally the particles impact on a detector, which can be placed at any distance from the collision zone. The auxiliary values, that are measured by the detector, are: detector impact position and time. Typically they are measured with a precision of 50 μm and 50 pico-seconds. It is to be noticed that these auxiliary quantities allow the experimenter to deduce the particle trajectories in the detection device and to determine the final momentum in the laboratory system from the trajectory of the particle (see below).

To obtain sub-atomic momentum precision (laboratory system) in a single event, the velocity vector (i.e. the total momentum) of the center-of-mass of the single-event collision system must be known from the method of preparation. Conservation laws are therefore of fundamental importance for the implementation of a single-event quantum measurement. An observable can only be measured with sub-atomic accuracy if time-dependent conservation properties are strictly fulfilled during the generally very short duration of the measurement. Total linear momentum, total angular momentum and total energy are conserved but not location. The measured momenta of all fragments can then be corrected for the center-of-mass motion because the total momentum is conserved. For position no conservation law exists, thus a large uncertainty in the location measurement cannot be avoided. Therefore Heisenberg's suggestions that a high resolution position measurement is possible and this position measurement would be even the basis of any quantum measurement completely contradicts real experiments.

Since 1927 numerous papers have been published discussing the consequences of the UR on a quantum measurement within the wave-picture. To the best of our knowledge there is no publication available, where the constraints and the purely classical experimental limits of a single-event quantum measurement are analyzed from the view of an experimenter. Although in the introduction of his paper [1], Heisenberg considered the kinematics and mechanics of a single particle and the measurement of the position and the velocity (momentum) of a single electron "at a given moment", Heisenberg's UR ($\Delta x \cdot \Delta p \geq \hbar$) applies, however, only for the mean statistical fluctuations of a large number of repeated single-event measurements of two conjugate variables and can be viewed to be a prediction of the future particle properties.

We deploy therefor the following two statements:

Statement 1. The UR applies for the statistical distribution of a large ensemble or for repeated measurements but not for the resolution of a single-event measurement.

This statement is in line with previous work, that revisited this discussion, as well. For example, Ballentine [5], Park and Margenau [6] as well as Briggs [7, 8] contradicted the single-event interpretation of Heisenberg and concluded that it applies only to a large ensemble of similarly prepared systems.

Statement 2. A single-event measurement can only provide information on the particle's properties back in the past but never allow a prediction of future properties, since the impact of a particle on the detector changes the particles momentum and position state.

Which parameters of a quantum reaction are measurable and what is the achievable precision? We discuss this question by illustrating the concept of a "real" quantum experiment. As paradigm example for a typical quantum measurement we have chosen the scattering of an electron or ion on a gaseous target atom followed by the coincident detection of all reaction fragments with modern "state-of-the-art" detection devices.

We will discuss the following three findings:

1. **One can measure the final momenta of all emitted charged fragments.** Since each single-event measurement takes some time (from preparation until detection), the conservation with time of the total momentum of the whole scattering system is a crucial property in order to obtain excellent resolution in real measurements. During the short period of measurement the momenta of all particles, are "correlated" due to the law of momentum conservation, i.e. they are even dynamically entangled, for the whole time until they finally impact on a classical detector (see Ref. [7] and comments therein connected to this paper).

2. The angular momentum of a single freely-moving electron emitted in a quantum reaction appears undetectable. However, the quantum states (whose quantum numbers) can be deduced, if the electron kinetic energy can be assigned to a well-defined transition. In an ion-atom collision process, however, a coincidence measurement can provide information on the **angular momentum** of a single particle. In the case of a complete multi-particle coincidence measurement, when the nuclear collision plane is determined, this additional information can be employed in some cases to deduce the angular momentum, as, for example, certain angular momentum states are emitted due to space quantization only into distinct regions like in the Stern-Gerlach experiment (see e.g. data in Fig. 4 of this paper).

3. One can also precisely determine the **amount of the electronic excitation energy** from the measured momenta of all particles in the preparation and final states, because the total energy is also conserved (assumption: projectile and target in the preparation state are in the ground-state). The excitation energy is then the difference between the kinetic energies in the initial and final states.

The UR imposes, in contradiction to Heisenberg's claim, no limit on the achievable momentum (velocity) resolution of a single quantum measurement. The UR affects the resolution of such a measurement only indirectly, as it has an impact on the quality of preparation of the pre-collision states of projectiles and target atoms. This has already been highlighted by Kennard in 1927 [9] who theoretically considered the passage of scattered electrons in a classical detection device and concluded:

„In den hier behandelten Fällen haben wir keinerlei quantentheoretische Abweichung gefunden von den klassischen Ergebnissen. Die einzige quantenhafte Eigentümlichkeit in solchen einfachen Fällen liegt in der durch das Heisenberg'sche Unbestimmtheitsgesetz festgesetzten prinzipiellen Unbestimmtheit der Anfangswerte" ("In the cases discussed here, we did not find any quantum theoretical deviation from the classically calculated values. The only quantum influence in such cases originates from the effect that the preparation state values are indeterminate in accordance to Heisenberg's Uncertainty Relation."). Today, the debate over the statistical versus single-event interpretation is still not converged (see [5–16] for proponents of the single-event interpretation and for papers opposing this interpretation).

In the following chapters we discuss the **purely experimental aspects** and the limits of experimental precision in a single-event measurement of momentum (velocity) and position and present examples:

In Sect. 2: The **scheme and time evolution of a single-event measurement** is discussed beginning with the preparation of the measurement followed by the quantum reaction process and concluding with the detection of the charged fragment in a classical measurement apparatus.

In Sect. 3: The **electron momentum (velocity) measurement by Time-of-Flight (TOF) trajectory imaging** is presented. We consider realistic experimental scenarios for electrons based on experimental results.

In Sect. 4: The determination of the **angular momentum state** of a single electron by a multi-fragment coincidence technique.

In Sect. 5: The **experimental limits for an electron position measurement** are discussed. We also show that Heisenberg's "Gedankenexperiment" on the γ-microscope is not feasible.

In Sect. 6: We consider the **product of precisions in momentum and position measurement of a freely moving single electron**. New experimental techniques for measuring momentum and position of a freely moving electron simultaneously in a one-step approach are provided for the moment of impact on a detector. Within this approach the product of the experimental error bars in electron momentum and detector impact position can be below $\hbar$ by several orders of magnitude.

2 Scheme of a Quantum Measurement

We consider an experiment where a projectile beam intersects in ultra-high vacuum with a gaseous target to ensure controlled single-event conditions i.e. that only one reaction process occurs during each measurement period. Because of the statistical nature of quantum measurements (to yield statistical distributions) one must prepare numerous projectiles in the "nearly identical" pre-collision state and numerous target objects in controlled "nearly identical" momentum and position states. In the preparation of the pre-collision state "nearly identical" means this preparation is still limited by Heisenberg's UR with respect to the large ensemble projectile and target momentum and position fluctuation widths. E.g. in an ion-atom collision the experimenter cannot precisely adjust the impact parameter to obtain the same deflection angle. The selection of impact parameters is of pure statistical nature. Thus, the experiment has to be repeatedly performed with numerous of such single projectiles and target objects. Finally summing over a huge number of single-events the experimenter obtains a statistical distribution that allows for the retrieval of the final-state fluctuation width (with the help of theory also quantum mechanical properties or properties of the wave function).

2.1 Time Evolution of a Quantum Measurement

In Fig. 1 the scheme of a single-event quantum experiment and the time evolution of such a complete quantum measurement process are shown. The measurement may be separated in three sequential steps: the time of preparation (pre-collision step, zone A), the time of reaction (zone B), and the time after the reaction (post-collision step, zone C) before the reaction products impact on the detector. In the view of the experimenter the momenta and trajectories of the particles in the macroscopic preparation stage A (pre-collision) as well as in the macroscopic spectrometer system C (post-collision) can be treated by the laws of classical physics. The very tiny reaction region B (typically of atomic to micrometer size) is a purely quantum mechanical region and must be treated accordingly. The dynamics in region B cannot be directly observed by the experimenter. The classical behavior in A and C is justified theoretically by the Imaging Theorem of the accompanying papers [7, 8]. This result shows

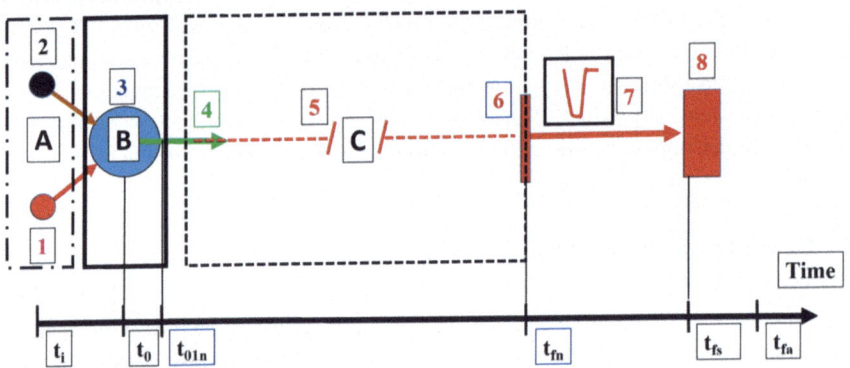

Fig. 1 Time evolution of a quantum measurement. A indicates the time interval before the interaction of projectile (1) and target (2), B is the very short time interval of the quantum scattering process (3) (occurring at the time t_0) and C the time interval in which the emitted reaction particles (4) are travelling inside the classical detection setup. The particle is finally detected on a detector (6). The detector yields an electronic signal (7) (typically a nanosecond long) providing time information on the quantum scattering event, which is stored electronically in a computer (8)

that, after propagation to or from macroscopic distances, the position and momentum variables of the quantum wave function obey classical relations.

The reaction products emitted in the quantum reaction are interacting with the macroscopic measurement apparatus in zone C. In the macroscopic apparatus they can be treated as classical particles with classically defined momenta (moving on classical trajectories) since they exchange in zone C de facto only momentum with the measurement device due to applied electric or magnetic spectrometer fields. Any interaction of the fragments with the rest gas in the spectrometer can be excluded because of the very low vacuum pressure (typically below 10^{-8} millibar). At the end of the macroscopic detection device position-sensitive detectors measure the impact position in the laboratory system of each fragment and also the time of impact for each fragment separately (if required all fragments can be measured in coincidence).

As Popper pointed out [17], after completion of a measurement the experimenter determines always the kinematical parameters of the "past" for each single event, whereas the UR makes predictions into the future for the outcome of statistical distributions of many repeated single-event measurements.

3   Electron Momentum (Velocity) Measurement by Time-of-Flight (TOF) Trajectory Imaging

3.1   The Experimental Scheme for Momentum (Velocity) Measurement

In the following we describe a quantum measurement of charged particles from an ionization process using a momentum-imaging approach. After leaving the reaction zone B (see Fig. 1 at time t_{01n}) the charged fragments begin to move in zone C on "quasi-classical" trajectories (see Refs. [7, 8]) with classically defined momenta, since in zone C they nearly exclusively exchange momentum with the spectrometer via classical forces. The distance d from the reaction point, from where one can neglect quantum mechanical post-collision interaction, can be crudely estimated by comparing the strengths of interacting forces, i.e. the magnitude of momentum exchange. In zone B the force between electron and ion dominates and in zone C the force imposed on the charged particles by the spectrometer fields is dominating. This is because the force between electron and ion depends on their distance d. Assuming the ion is singly charged then the electron-ion force is $F_{ion} = e^2/d^2$ (in a.u.). For d = 1000 a.u. one obtains $F_{ion} = 10^{-6}$ a.u., for d = 1 μm one obtains $F_{ion} = 2.8 \times 10^{-9}$ a.u. The strength of the classical force in the fields of the measurement device can be estimated from the electric field strength in the spectrometer. The field is typically larger than 10 V/cm, thus for an electron the acting spectrometer force is $F_{eS} > e \cdot 10$ V/cm $= 1.92 \times 10^{-9}$ a.u. Therefore, for distances d larger than a few tens of micrometers the electron-ion force strength can be neglected and the

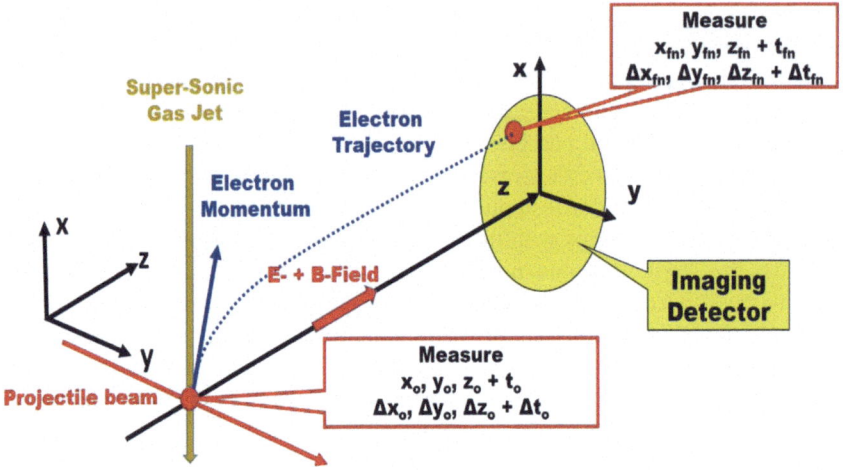

Fig. 2 Scheme of trajectory imaging technique for charged quantum particles in a classical spectrometer [4]. The electron momentum vector (blue arrow) is the so-called "final momentum", with which the electron is emitted from the collision process with respect to the center-of-mass system of the reaction process

emitted fragments are only interacting with the spectrometer field yielding a well-defined classical trajectory due to momentum conservation (charged fragment plus spectrometer are entangled). The momentum change and thus the classical trajectory of the fragment in the spectrometer depend on the electric-magnetic field design and on the final fragment momentum p_{fn}.

A static electric field accelerates electrons and positively charged ionic fragments into opposite directions. The fragments are finally detected by two position- and time-sensitive detectors placed in opposite directions (only one direction is shown in Fig. 2). Since the spectrometer provides for positively and negatively charged particles nearly a 4π-detection efficiency it can capture a complete image of the reaction process in momentum space.

The measurement of the final momentum of an emitted fragment can thus be achieved through a precise determination of the particle trajectory in part C in the classical detection device. To determine the complete classical trajectory of each particle one has to measure only the classical location parameters $r_0 = (x_0, y_0, z_0)$ and $r_{fn} = (x_{fn}, y_{fn}, z_{fn})$ as well as times t_0 and t_{fn} (see Fig. 2). Both time parameters can be determined with a precision of about 50 pico-seconds, t_0 can be measured by using a timed-bunched projectile beam and t_{fn} by using a "state-of-the-art" classical detection device [4]. Target location and position of impact on the detector can be measured with a precision of better than 50 μm (even 10 μm are achievable). Knowing (or calibrating) the electro-magnetic field configuration and measuring the above listed parameters, the final momentum vector of the fragment can easily be deduced by using simple classical equations [4]. **Although all auxiliary parameters**

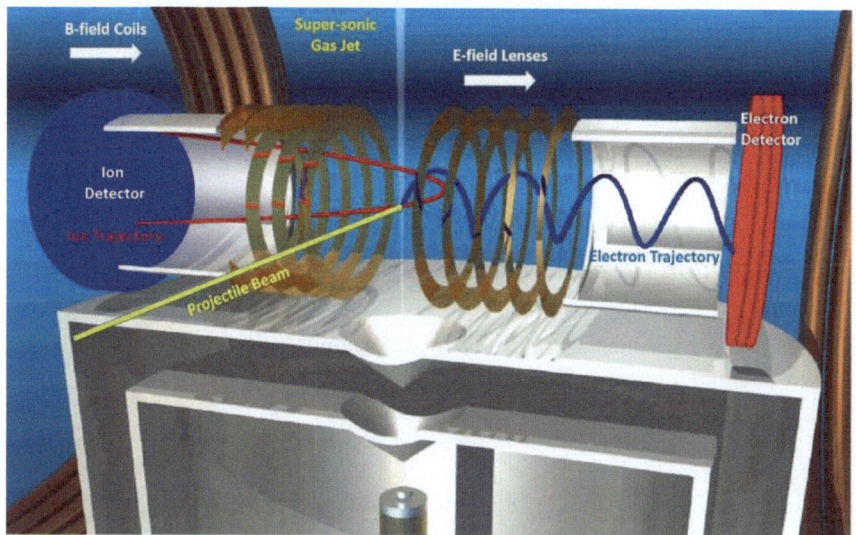

Fig. 3 Scheme of the C-REMI [4] which can image with 4π solid angle all emitted charged particles (ions: red trajectory, electrons: blue trajectory) in coincidence. A projectile beam intersects in the center of the C-REMI with a super-sonic gas jet (from below) inducing the quantum reaction process. The applied electric field super-imposed by a magnetic field (see the brown coils) projects all charged fragments/electrons on position- and time-sensitive detectors

are measured with macroscopic accuracy only, sub-atomic resolution for the electron and ion momenta (velocities) can be obtained.

The C-REMI [4] is such a "state-of-the-art" momentum-imaging device. In Fig. 3 the scheme of such a detection approach is presented. The reaction takes place within the tiny intersection region of projectile and target beams (e.g. internally very cold super-sonic gas jet). The blue and red curves in Fig. 3 indicate the classical trajectories of ionic fragments (red line) and electrons (blue line) in the spectrometer. With the help of electric and magnetic fields nearly all fragments are projected on position-sensitive detectors yielding a very high multi-coincidence detection efficiency.

Before we discuss a real experimental scenario, we first define "good" and "bad" resolution in a single-event quantum measurement with respect to the standard dimensions in an atomic system. The standard sizes of atomic parameters are defined by the classical features of an electron in a hydrogen atom. The classical K-shell radius is $r_K = 5.29 \times 10^{-9}$ cm, which is used to define the atomic unit of length (a.u.). The classical electron velocity of the electron in the hydrogen K-shell is $v_K = 2.18 \times 10^8$ cm/s, which defines 1 a.u. of velocity. The classical momentum of the electron in the hydrogen K-shell is $p = m_e v_K = 1$ a.u. An atomic unit of time is defined by the ratio of the hydrogen K-shell radius divided by the corresponding electron velocity, or 5.29×10^{-9} cm divided by 2.18×10^8 cm/s yielding 24 attoseconds. Furthermore, the electron charge e and mass are also set to 1 a.u. and hence $\hbar$ results to be 1 a.u., too.

Thus, it appears very reasonable when we define resolution of single-event quantum measurements with respect to these atomic units. "Good" sub-atomic resolution is on the order of a few percent of one a.u. and "very good" resolution is on the order of a per mill or even better. Bad resolution is larger than one a.u.

3.2 Momentum (Velocity) Measurement and Its Achievable Resolution for an Electron

The achievable experimental precisions for momentum (velocity) are discussed here for two quantum processes. First, the transfer ionization process which is

$$10\,\text{keV He}^{2+} + \text{He} => \text{He}^{1+} + \text{He}^{2+} + e$$

investigated by Schmidt et al. [18]. This experiment was performed to search for vortices in the electron current which should be visible in the velocity/momentum distribution of the emitted electrons. To visualize such effects in the electron momentum distribution, a high experimental momentum resolution ($\delta p = 0.01$ a.u.) in a single event is required. Additionally, a coincidence measurement with the ejected ions is necessary in order to determine the orientation of the quasi-molecule during the collision. This was achieved with the C-REMI approach. During such slow collisions quasi-molecular orbitals are formed and electrons are promoted to the continuum via a few selected angular momentum states.

In Fig. 4 the measured electron-momentum distributions are shown together with the achieved single-event resolution δp (black square) and with one example of a momentum fluctuation width $<\Delta p>$ (varies with electron energy). It is to be noticed that in this experiment of Schmidt et al. the electron-detector distance from the intersection region (gas jet-projectile beam) was only 3 cm due to other experimental requirements. This short distance limits the momentum resolution, because of the very short TOF. Nevertheless a resolution of 0.01 a.u. was obtained. The resolution can be improved by increasing this distance.

To demonstrate the high resolving power for electron momenta of the C-REMI a numerical example, a kind of "*Gedankenexperiment*", i.e. the process of electron impact ionization of He

$$e + \text{He} => \text{He}^{1+} + 2e$$

is discussed here. Today such an experiment would be feasible.

In Appendix A the preparation of the required electron beam quality is described which enables the high required momentum accuracy of the projectiles. To yield the required excellent "Time-of-Flight" TOF resolution the detectors should be located as far as possible from the zone B, i.e. the spectrometer should be as large as possible. For a trajectory length inside region C (from zone B to the electron detector surface)

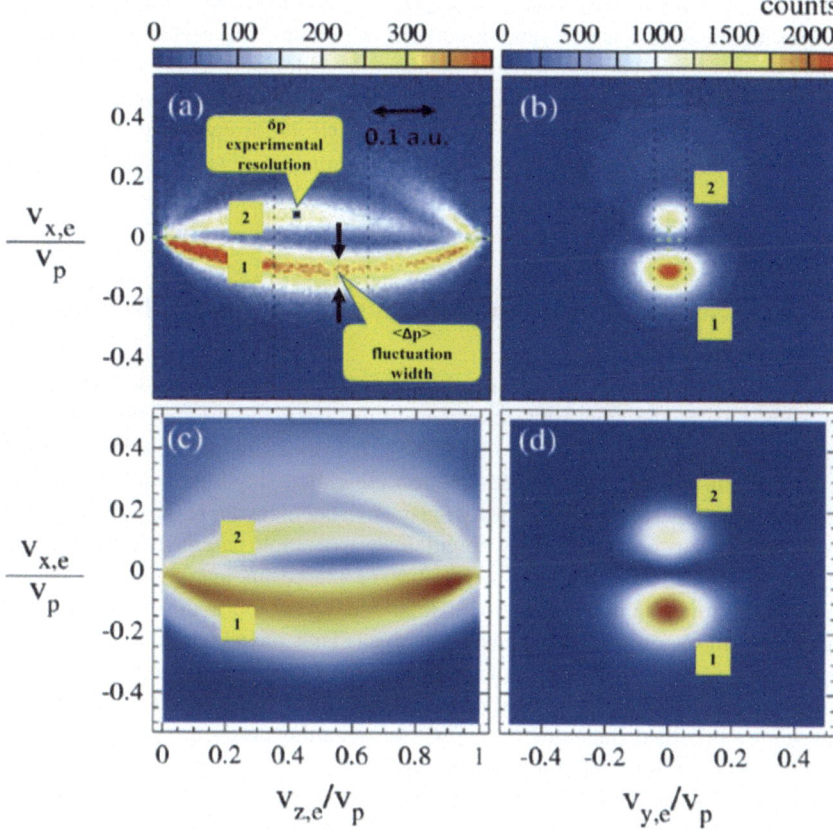

Fig. 4 a Measured electron velocity distribution in the nuclear collision plane in units of the projectile velocity $v_p = 0.63$ a.u. for small nuclear scattering angles <1.25 mrad. **b** Perpendicular to the nuclear collision plane. **c, d** corresponding theoretical predictions. An electron moving with the projectile velocity $v_p = 0.63$ a.u. has a momentum of 0.63 a.u. [18]. The experimental resolution in a single event of $\delta p = 0.01$ a.u. corresponds to an energy resolution of approximately 1 meV

of 2 m the angular resolution of the trajectory measurement is of the order of the sum of the intersection width of projectile beam and target beam and detector position resolution divided by the trajectory length. This ratio is about $2 \cdot 50 \ \mu m/200$ cm $\approx 0.5 \times 10^{-4}$. This geometrical ratio limits the transverse momentum resolution in x- and y-direction. The longitudinal momentum resolution (in z-direction) depends on the TOF resolution. An electron moving with 2 a.u. momentum has a velocity of $4.38 \times 10^{+8}$ cm/sec and its total TOF inside C is 200 cm/$4.38 \times 10^{+8}$ (cm/sec) $= 450$ ns. Thus the relative TOF resolution $\Delta TOF/TOF$ is about 10^{-4} yielding an overall momentum precision for an electron of 2 a.u. momentum of 2×10^{-4} a.u..

We would like to notice, that in C-REMI the velocities and masses of moving particles are measured, which yield directly the momenta. The velocities are macroscopically large and therefore directly measurable with macroscopic classical TOF devices. Heisenberg considered the measurement of the velocity (momentum) of an electron bound in an atom too. His approach will be discussed in Appendix B together with the possibility of momentum measurements of bound electrons via the process of Compton scattering.

4 Measurement of Angular Momentum of a Single Electron

Any bound electron usually has an orbital angular momentum in addition to its own spin. Due to the spin-orbit coupling, all electrons in an atom form one unit providing a quantized total angular momentum. If an experimenter can only measure the momentum of only one emitted electron (so-called single parameter measurement), then an experimenter can hardly make any statement about the quantum state in which the electron was originally bound. In case of single parameter measurement only from the electron momentum distribution of a large amount of identical ionization processes one can make a statement about the type of multipole distribution and thus on the angular momentum transfer involved. Thus the angular momentum of a single freely-moving electron emitted in a quantum reaction appears undetectable.

However, if the electron kinetic energy can be assigned to a single transition between well-defined quantum states, whose quantum numbers can be deduced. Furthermore in an ion-atom collision process and in the case of a complete multiparticle coincidence measurement, when the nuclear collision plane is determined too, this additional information can be employed in some cases to deduce the angular momentum states of a single ejected electron. In a slow ion-atom collision process, quasi-molecular electronic orbitals are formed during the collision, which are sharply angularly quantized with respect to the nucleus-nucleus scattering plane. Thus different angular momentum states are emitted due to space quantization only into distinct regions like in the Stern-Gerlach experiment (see e.g. data in Fig. 4 of this paper). If e.g. in a transfer-ionization process an electron passes over from these quasi-molecular states into the continuum [18] then the electrons in the x-y plane perpendicular to the nucleus-nucleus scattering plane are emitted with discrete transverse momenta (Fig. 4) and the different quasi-molecular orbitals e.g. 1 and 2 in Fig. 4 can clearly distinguished. Just as in the Stern-Gerlach experiment, these discrete transverse momenta correspond to certain angular momentum states which can be discerned in a coincidence measurement.

This clearly proves (Fig. 4: comparison of experiment and theory) that in a coincidence experiment the directional quantization of the quasi-molecular states becomes measureable and thus in selected collision systems the angular momentum states of single emitted electrons can be determined too.

5   Electron-Position Measurement and Achievable Resolution

Heisenberg described the position measurement of single electrons at a given moment as the foundation of any parameter measurement. He proposed to measure the velocity by detecting the electron positions at two succeeding moments. He explained his view on position measurements by thought experiments: *"If one wants to under-stand, what the definition of 'position of a particle', e.g. of the electron (relative to the reference system of measurement) means, one must describe well-defined experimental approaches, how the 'position of an electron' can be measured; otherwise the definition of position is meaningless.* He continued: **"There is no shortage of such experimental approaches, which can measure the 'position of an electron' with unlimited precision."** (page 174) [1]. Therefore he viewed a trajectory as a discontinuous path because of discontinuous observations. On page 185 he continued: *"I believe that the appearance of a classical trajectory is manifested by its observation"*.

Heisenberg proposed to use a so-called γ-microscope to measure the position of a quantum object, e.g. an electron at a given moment. He ascertained [1]: *"The resolution of the light microscope is only limited by the wave length of the light. Using short wave length x-rays the resolution should have no limitation."* The scheme of such a photon microscope measurement can only be explained in the wave-picture (thus many photons must be detected). But one has to make sure that the object is not changing its position during the exposure time of the measurement. With the help of such a microscope (combination of lenses) one can magnify tiny objects and project their image on a detector, e.g. photo plate. There is an one-to-one correspondence between position on the object and the position on the detector (only valid in the transverse plane). Thus with the help of lenses relative positions on very small quantum objects can be enlarged and thus become observable. It should be noted, that a "microscope" device for magnifying the geometrical size of an atom (about 10^{-8} cm diameter) and also magnifying the relative positions of atoms in a molecule to the macroscopic size of 1 mm must have a magnification factor of more than 10^6.

Heisenberg was convinced that the position of an electron at a given time could be measured even with "ultimate" precision using the technique of such a light microscope if the wavelength of the light would be small enough to resolve sub-atomic structure.[1] At a "given time" means always an exposure time period in which

---

[1]Several reasons prevent that Heisenberg's so-called γ-microscope can measure the position of one selected electron inside an atom at a given time with a required resolution of 10^{-10} cm or even better: First: Since the focus of the γ-pulse is of macroscopic size (larger than 1 μm^2) the scattered photons of the γ-pulse interact with different electrons in the atom or molecule and the measurement can on principal not identify which photons were scattered on the one special electron. Second: The Compton cross sections for scattering photons with a wave length of 10^{-10} cm (or $h\nu = 1.2$ MeV) on an electron are smaller than 10^{-25} cm^2. Therefore it requires per attosecond pulse more than 10^{+19} photons in a focus of 1 μm^2 to scatter about 100 photons on this electron. Such a photon pulse carries an energy power equivalent with 1% of the total energy emission of the sun. Third a technical reason: The γ-microscope needs a high precision lens system for 1 MeV photons to magnify the 1 μm^2 focus size to make the different electrons on a macroscopic detector distinguishable.

the location of a moving electron must be considered as "frozen". Such a time period for an electron detection must be shorter than one attosecond.

Therefore, in order to obtain an image of the position of an electron with sub-atomic resolution using a γ-microscope, one would have to scatter on the **same electron** numerous photons in a one attosecond "exposure" time period (since the electron is moving with a typical velocity of 1% of the speed of light). Because these γ-scattering cross sections (Compton scattering) are of nuclear size the photon pulse intensity in one attosecond must exceed 10^{19} photons per pulse in a focus of 1 μm diameter. A further problem in such a measurement is that the experimenter has no control on which electron in the target atom or molecule the photons are scattered. Both effects make such a γ-microscope measurement physically not feasible.

Furthermore, each Compton scattering process, as mentioned above, is destructive for the electronic state, thus the electronic state changes immediately. This disturbing effect of momentum transfer to the electron and thus changing the electron's position subsequently was already realized by Bohr [19]. These arguments show that Heisenberg's γ-microscope is not suited to measure the position of an electron at a given moment.

In one attosecond exposure time because of the tiny cross sections at most one photon might be scattered on the same electron. Thus the only information the experimenter obtains with Compton scattering is the detection of only one single photon providing one momentum vector. Even if this photon momentum vector is measured with sub-atomic precision the location of the reaction can never be deduced from this one vector with a precision better than the preparation of the target position before the scattering.

In contrast, position-measurements of heavy nuclei or atoms can be performed with a γ-microscope, since the velocities of atoms or nuclei are typically a factor of 10.000 smaller. Thus, the heavy particle position can be considered as "frozen" even for an exposure time of a few femtoseconds. Such relative position measurements of heavy atoms in molecules are now routinely performed with FEL X-ray pulses [20], where a lateral position resolution of about 5 Å is achieved. A slightly better resolution of about 3 Å is achieved with CRYO-electron microscopy [21].

One may expect that when performing a multi-coincidence measurement, i.e. measuring the momentum vectors of several fragments of the same reaction with excellent resolution, one could deduce the position, where the reaction took place,

---

Heisenberg proposed also to use energetic α-particles as scattering projectiles (because of their even shorter de Broglie wave-length) for a super high-resolution microscope and estimated even a position resolution of 10^{-12} cm as possible. He wrote on page 175: "*When two very fast particles succeeding each other scatter in a very short time distance Δt on the same electron, then the distance between the positions of both collisions is Δl. From the scattering laws, which has been observed for α-particles, we can conclude, that Δl can be made as small as 10^{-12} cm, if Δt can be made sufficiently small and the α-particles fast enough.*"

For a several MeV α-particle beam this requires a relative distance of the α-particles in the beam of about 10^{-11} cm. This relative distance of the α-particles is about one thousand times smaller than the normal inter-nuclear distance in a hydrogen molecule. α-particles with a relative distance of 10^{-11} cm would repel each other with a huge Coulomb repulsion force creating huge non-controllable transverse momenta of the α-beam.

by reconstructing the intersection point of all momentum vectors. Even if the impact positions of all fragments on the detector could be measured with atomic position resolution, the momentum vectors have still a finite angular uncertainty limited again by the target preparation. Because of the macroscopic dimensions of the detection device even a tiny angular uncertainty of these vectors would spoil any precise position measurement of the reaction region within the laboratory frame.

6   Product of Precisions in Momentum and Precision in Position in a Real Measurement of a Freely Moving Single Electron

The paradigmatic demonstration experiment for the UR given in textbooks for measuring simultaneously position and momentum of an electron (wave picture) is the scattering of this "wave" on a narrow slit (first step). The scattered wave yields an interference pattern of the electron wave on a screen (second step). According to most of the textbooks position and momentum of these electrons can only be measured in such a **two step-approach**, where in the first step the position is measured by the slit width and in the second-step by the interference pattern on the screen the momentum. The electron is theoretically described by wave functions which are different before and after the slit: before passing the slit the electron is described as plane wave with well-defined momentum eigenvalue but not localized in x-position; just after the slit the electron is described as a wave packet with some distribution in position and momentum. Thus on its way to the screen the electron is in a state which is not an eigenstate of the momentum operator. In both of the two time steps the UR is fulfilled. This is a result of the fact that the two operators do not commute. Thus the wave function is disturbed and a conceptually unavoidable uncertainty in the second-step measurement (momentum measurement) is generated. From the interference structure in the transverse momentum distribution of many single-event measurements the de Broglie wave length λ and thus electron momentum p can be determined.

We will now estimate how small the product of the two precision widths $\Delta x \cdot \Delta px \geq \hbar$ can be made in a **single-event process** by using a modern state-of-the-art detection device. We are in particular interested in whether the product of the experimental position resolution times the experimental momentum resolution can be made smaller than $\hbar$. With today's detection technique the two-step detection scheme can be replaced for single electron detection by a quasi **one-step detection approach**, where the narrow slit is "upgraded" to a very small pixel detector, which measures position and time of impact too. Thus we consider the momentum and position at the time when a single moving electron impacts on the position-sensitive detector. One can construct detectors which can measure the impact position of the electron on the detector with a few a.u. precision $\delta xy = 2$ a.u. (see Appendix C).

In a single event this detector provides, at the instant of electron impact, also a very fast electronic signal (time resolution <50 pico-seconds) which yields precise information on the electron velocity. If furthermore the location of the interaction region and the interaction time, from where this electron is emitted, are known with macroscopic precision, one can determine the electron Time-of-Flight TOF. Knowing precisely the distance d between emission point to detector (e.g. d = 2 m and Δd = 0.1 mm) one can precisely calculate the electron velocity: Assuming the measured TOF is e.g. 456.6 ± 0.1 ns and d is 200.00 ± 0.01 cm the electron velocity is then $v_e = 2 \cdot 200.00$ cm/456.6 nanosecond $= 4.3800 \times 10^{+8}$ cm/s with an error bar of $\pm 0.025\%$. Transforming the velocity in a.u. we obtain for $v_e = p_z = 2.0021 \pm 0.001$ a.u. (p_z is the electron momentum in flight direction). In perpendicular direction the errors in momentum are $\delta p_{x,y} = (0.1$ mm/2000 mm$) \cdot 2.0021$ a.u. $\approx 10^{-4}$ a.u. Thus, in case of a single event measurement the product of the experimental error bars in the momentum and position measurement can be made $\delta p_{xy} \cdot \delta xy \approx 10^{-4}$ a.u. $\cdot$ 2 a.u. $= 2 \times 10^{-4}$ a.u. which is much smaller than $\hbar$.

One could argue, however, that the detection plus preparation is still a two-step measurement. But nevertheless in a single event the product of precisions can be made much smaller than $\hbar$. Thus, once the particle has been detected, the trajectory, that the particle has travelled on in the past, can be defined such that the product of precisions of momentum and position measurement of this freely moving single electron is not limited by $\hbar$.

7 Conclusion

We have shown that in **a single-event quantum measurement** the momenta of emitted electrons or ions can be measured with high sub-atomic precision and the limits of precision for the momentum measurement are not restricted by Heisenberg's UR if one assigns trajectories to particles that have been detected. The precision in measuring positions in a single event can never approach or being better than 1 a.u. in a single-event measurement because the two conjugate parameters position and momentum do not have the apparent physical symmetry suggested by the UR, i.e., there exists a disparity in momentum compared to position measurement. The fundamental reason is: in a single event momenta are conserved with time (i.e. they are dynamically entangled), but positions are not conserved. This fundamental difference between momentum and position measurement as function of time in a quantum reaction is also apparent from the wave description (see Appendix D). The position wave functions broaden with increasing time even during a very short single-event measurement.

For a single freely moving particle in the moment of impact on the detector momentum and also position on the detector can be simultaneously detected in a single-step approach using position-sensitive detectors combined with a time-of-flight measurement. The product of the experimental momentum resolution δp times position resolution δx on the detector can be made much smaller than $\hbar$.

Acknowledgements We are indebted to John Briggs, Tilman Sauer, Bretislav Friedrich, Reinhard Dörner, Robert Griffiths and Lothar Schmidt for many very valuable discussions and corrections to the manuscript.

Appendix A

Electrons with a very well-defined momentum $p_z = 2.000$ a.u. can be created by photoionization. The primary photon energy h · ν is chosen to be 78.988 eV ± 0.007 eV (linearly polarized photons). These photons can singly ionize the He atom. In our example a single electron with a kinetic energy of 54.392 ± 0.007 eV and He-recoil ion with a kinetic energy of 7.4 meV are emitted back-to-back. In the He center-of-mass system both freely moving particles, electron and recoil ion, have the identical momentum with opposite direction ($p_e = -p_{rec} = 2.000$ a.u. with a precision of about 10^{-3} a.u.). The angular distribution of electrons emitted by photon ionization and recoil ions is a perfect dipole distribution.

Since the ionized He atom is not at rest in the laboratory system at the moment of ionization, one must correct the electron momentum vector for the motion of the He-atom in the Lab frame. The final electron velocity is about 5000 km/sec and the internal velocity spread of the cold super-sonic He jet is below 50 m/sec. Because the He jet is moving along the negative y direction, the correction for the electron in the z-direction can be performed with sufficient precision. Furthermore, the very small momentum kick of the incoming photon ($p_{photon} \approx 0.021$ a.u.) to the center-of-mass of the He atom changes the He velocity by only about 10 m/sec. Thus the absolute value of the electron momentum is known with about 10^{-3} a.u. precision. The direction of the final electron beam can now be defined by a collimation system (collimation in x and y direction). Using a double slit system at 2 m distance in the z-direction with slit widths of 10 microns each (in x- and y- direction) the so-collimated electron beam has an angular divergence of $\delta = 20/(2 \times 10^{+6})$ rad = 10^{-5} rad. The momentum exchange with electrons inside the slits due to image-charge formation is also insignificant. Thus the momenta of the electron beam have a width $\mathbf{\Delta p_{x, y, z}} < \mathbf{10^{-3}}$ **a.u.**, i.e. each electron in this electron beam has a momentum of 2.000 a.u. in the z-direction (with a precision of about 10^{-3} a.u. in all three dimensions x, y, and z). By changing the primary photon energy, the electron beam's momentum can be varied. Similar momentum resolution results can be also obtained for photon beams and slow ion beams, too.

Appendix B

Heisenberg described also a way to measure the velocity (momentum) of an electron bound in an atom. Heisenberg's approach to measure the velocity of a bound electron was copied from classical physics where the velocity of a particle is determined from the quotient of measured distance divided by measured time difference. Heisenberg's concept was to detect by γ-photon scattering at two instants in time the electron locations (for a bound electron separated by less than 10^{-11} cm). This would require a very precise simultaneous measurement of location and time within sub-attosecond time separation. As we discussed in Sect. 4 such a measurement of position and time with the required resolution is physically impossible to perform. Furthermore, Heisenberg worried that the electron velocity (momentum) just before and just after the photon (Compton) scattering process is not the same one due to the momentum kick by the photon and thus the velocity measurement before the kick seemed to be not accessible.

In 1927, Heisenberg was not aware that one could measure nevertheless both—the electron momenta just before and after the scattering by performing **a coincidence measurement of the scattered Compton-photon and the ejected electron**.

Both electron momenta just before and just after scattering can be determined with very high resolution ($\delta p = 10^{-2}$ to 10^{-3} a.u.) due to momentum conservation during the scattering process. This is possible at high photon energies and large photon momentum transfers where the "impulse approximation" is well justified [22]. In the "impulse approximation" for Compton ionization the ejected electron is treated as a quasi-unbound electron and thus the momentum change of the whole atom by Compton ionization is rather small and the remaining ion acts only as a spectator. The coincidence measurement allows therefore for a precise determination of $P_{e\,ionized}$ which is the momentum vector of the ejected electron after the Compton scattering and $P_{e\,bound}$, which is the unknown momentum vector of the bound electron just before scattering (see vector equation in Fig. 5). When Heisenberg wrote his paper in 1927, the experimental techniques to study quantum processes were in an "archaic" state compared with today. Precise timing measurements in the nanosecond regime required for coincidence measurements were beyond imagination. The first generation of coincidence technique was just invented in 1924 by Bothe and Geiger [23], but Bothe's and Geiger's time resolution was limited to a fraction of a millisecond. Instead of using high-energetic photon impact one can perform the same kind of momentum spectroscopy of bound electrons by using very fast ion [24] or electron impact [25].

Heisenberg also considered a velocity measurement of a bound electron by using the Doppler-effect (wave length-shift) of scattered red light. Heisenberg wrote on page 177 [1]: "*The velocity of a particle can easily be defined by a measurement, if the particle velocity is constant (no acting forces). One can scatter red light on the particle und measures by the Doppler shift its velocity. The measurement becomes more precise, as longer the wave length of the light is, since then the velocity change of the particle per photon becomes smaller. The determination of position becomes accordingly more unprecise predicted by equation (1) (UR). To measure the velocity at a given moment, the Coulomb forces of the nucleus and the of the other electrons must suddenly disappear, to ensure from this moment a constant velocity, which is necessary to perform the measurement.*" The Doppler-effect approach would, however, only allow the determination of the particle velocity component in the

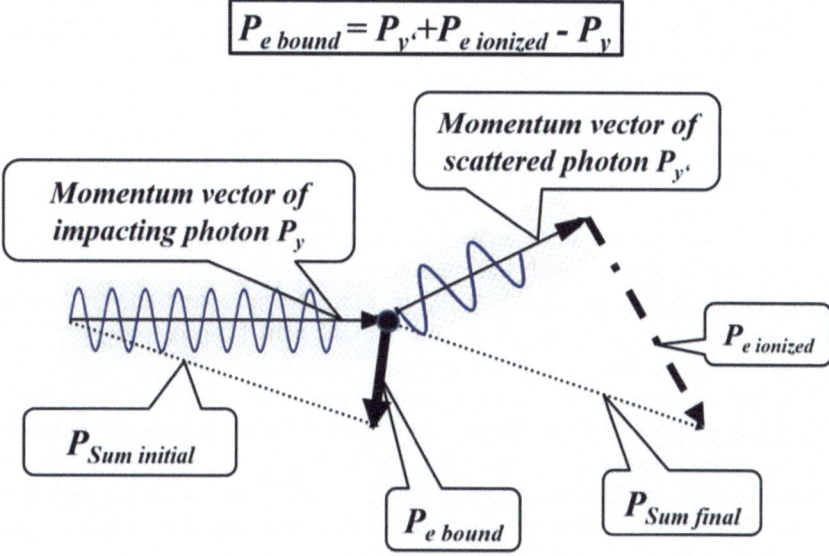

$$\boxed{P_{e\,bound} = P_{y'} + P_{e\,ionized} - P_y}$$

Fig. 5 Momentum vector diagram for Compton scattering. The dotted vectors $P_{Sum\,initial}$ and $P_{Sum\,final}$ represent the sum vectors in the initial state of impacting photon P_y (prepared and precisely known) and of bound electron $P_{e\,bound}$ (unknown) as well as in the final state of ejected electron $P_{e\,ionized}$ and of scattered photon $P_{y'}$ (both precisely measured)

direction of the incoming red light. Furthermore, the red light photon would be scattered on the whole atom and not on a single electron and would thus probe the atom velocity only.

Appendix C

Today, in principle one could build a macroscopic position-sensitive electron or ion detector with better than 10 a.u. position and 50 pico-seconds timing resolution. This detector can be a very small pixel detector or a large area position sensitive detector. Such a detector can be reassembled from two components: a commercially available position-sensitive channel-plate detector with a standard position resolution of 50 μm in x and y direction, respectively and a very thin (nanometer thickness) mask of regularly positioned holes of <10 a.u. diameter each, where electrons and ions can only be detected when they impact into such a hole (distance mask to detector surface a few nano-meter only). Then they induce in the detector an electronic signal. From this single electronic signal simultaneously position and timing information is obtained. The distance from hole to hole is 100 μm, thus the detection device is able to determine each particle impact on an absolute scale in the laboratory system for the position measurement with 10 a.u. precision. Therefore, one can detect the location of this particle impact in x and y direction with $\delta x \approx \delta y \approx 10$ a.u. resolution.

Appendix D

Time-dependent conservation laws are of fundamental importance for the implementation of a high-resolution single-event quantum measurement. This conservation is valid for the total linear momentum, for the total angular momentum and for the total energy, but not for the location. This is obvious for a particle picture, but also in the wave picture. In Fig. 6 one can see that the spatial wave function widens linearly over time, but the momentum wave function maintains its narrow width, which is determined by the preparation of the measurement. If several fragments are emitted in the reaction process, momentum conservation applies to each of the fragments on their flight to the detector. The momentum exchange with the classical electro-magnetic fields of the measuring apparatus can be determined from the classically measured trajectory of each fragment and thus corrected with high resolution.

The asymmetry as function of time for position and momentum space of the freely moving particle is also apparent in the wave approach and is due to the time propagation with $U = \exp(\Delta T \cdot E)$ with $E(p) = p^2/2$ that breaks the symmetry of position vs. momentum space.

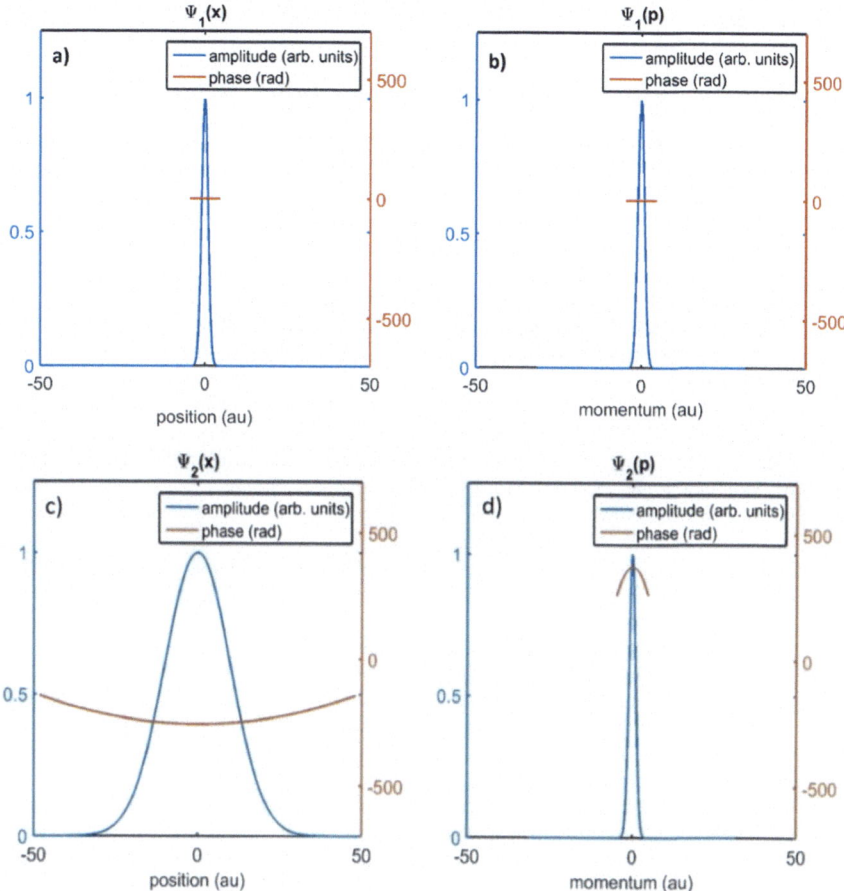

Fig. 6 Fourier limited wave function in position (**a**) and momentum (**b**) space. The phase is flat in position and momentum space and the amplitude is a Gaussian distribution. Wave function **c** from **a** after propagation for the time $\Delta T = 10$ a.u. The resulting wave function **d** in momentum space is $\Psi_2 = \Psi_1 \cdot \exp(\Delta T \cdot E)$ with $E(p) = p^2/2$. Ψ_2 has the same amplitude in momentum space as in **b** but a quadratic phase. In position space the amplitude distribution broadens compared to **a**

References

1. W. Heisenberg, Über den anschaulichen Inhalt der quantentheoretischen Kinematik und Mechanik. Zeit. f. Phys. **43**(3), 172–198 (1927)
2. W. Gerlach, O. Stern, Der experimentelle Nachweis der Richtungsquantelung im Magnetfeld. Z. Physik, **9**, 349–352 (1922); W. Gerlach, O. Stern, Über die Richtungsquantelung im Magnetfeld. Ann. Physik, **74**, 673–699 (1924)
3. https://www.scientaomicron.com/en/system-solutions/electron-spectroscopy
4. R. Dörner et al., Phys. Rep. **330**, 95 (2000) and J. Ullrich et al., Rep. Prog. Phys. **66**, 1463 (2006)
5. L.E. Ballentine, Rev. Mod. Phys. **42**(4), 358–381 (1970)
6. J.L. Park, H. Margenau, Simultaneous measurability in quantum theory. Int. J. Theor. Phys. **1**(3), 211–283 (1968); H. Margenau, Measurements in quantum mechanics. Ann. Phys. **23**, 469–485 (1963)
7. J.S. Briggs, Quantum or classical perception of Atomic Motion, Chapter 11 of these proceedings arXiv:1707.05006
8. J.S. Briggs, J.M. Feagin, New J. Phys. **18**, 033028 (2016); J.S. Briggs, J.M. Feagin, J. Phys. B At. Mol. Opt. Phys. **46**, 025202 (2013); J.M. Feagin, J.S. Briggs, J. Phys. B At. Mol. Opt. Phys. **47**, 1155202 (2014)
9. H. Kennard, Z. f. Phys. **44**, 326 (1927)
10. E.C. Kemble, Fundamental Principles of Quantum Mechanics with Elementary Applications (McGraw Hill, 1937)
11. D. Deutsch, Uncertainty in quantum measurements. Phys. Rev. Lett. **50**, 631–634 (1983)
12. M.R.H. Rudge, M.J. Seaton, Proc. Roy. Soc. London **A283**, 262 (1965)
13. W.E. Lamb, Nucl. Phys. B (Proc. Suppl,) **6**, 197–201 (1989)
14. M. Schlosshauer, Rev. Mod. Phys. **76**, 1267–1305 (2004)
15. P. Busch, P.J. Lathi, The Standard Model of Quantum Measurement Theory: History and Applications in Foundations of Physics, **26**, 7 (1996), pp. 875–893; P. Busch, T. Heinonen, P. Lathi, Heisenberg's uncertainty principle. Phys. Rep. **452**, 155–176 (2007)
16. R.B. Griffiths, What quantum measurements measure. Phys. Rev. A **96**, 032110 (2017) 14. D. Sen, The uncertainty relations in quanten mechanics. Curr. Sci. **107**(2) (2014)
17. K.R. Popper, Quantum theory and the schism in physics-from the "Postscript to the Logic of Scientific discovery page 22–23, Routledge (1989) and The Logic of Scientific Discovery page 225–226, Hutchinson
18. L.Ph.H. Schmidt, C. Goihl, D. Metz, H. Schmidt-Böcking, R. Dörner, S.Yu. Ovchinnikov, J.H. Macek, D.R. Schultz, Vortices associated with the wave function of a single electron emitted in slow ion-atom collisions. Phys. Rev. Lett. **112**, 083201 (2014); L.Ph.H. Schmidt, M. Schöffler, C. Goihl, T. Jahnke, H. Schmidt-Böcking, R. Dörner, Quasimolecular electron promotion beyond the 1 sσ and 2 pπ channels in slow collisions of He^{2+} and He. Phys. Rev. A, **94**, 052701 (2016)
19. N. Bohr, Nature **121**, 580 (1928)
20. H.N. Chapman et al., Nature **470**, 73 (2011)
21. W. Kühlbrandt, Microscopy: Cryo-EM enters a new era. eLife **3**:e03678, 4 p (2014)
22. B.K. Chatterjee, L.A. LaJohn, S.C. Roy, Investigations on compton scattering: new directions. Rad. Phys. Chem. **75**, 2165 (2006)
23. W. Bothe, H. Geiger, Ein Weg zur experimentellen Nachprüfung der Theorie von Bohr, Kramers, und Slater. Z. Phys. Band **26**, S. 44 (1924); W. Bothe, H. Geiger, Über das Wesen des Comptoneffekts; ein experimenteller Beitrag zur Theorie der Strahlung. Z. Phys. Band **32**, S. 639–663 (1925)
24. R. Moshammer, et al., Low-energy electrons and their dynamical correlation with the recoil-ions for single ionization of helium by fast, heavy-ion impact. Phys. Rev. Lett. **73** (1994) 3371; R. Moshammer, et al., The dynamics of target ionization by fast highly charged projectiles. Nucl. Instr. Meth. B **107**, 62 (1996); R. Moshammer, et al., The dynamics of target single

and double ionization induced by the virtual photon field of fast heavy ions x-ray and inner-shell processes. AIP Conf. Proc. **389**, 153 (1996); J. Ullrich, et al., Recoil ion momentum spectroscopy. J. Phys. B **30**, 2917 (1997) Topical Review

25. E. Weigold, I. McCarthy, Ion Electron Momentum Spectroscopy (Springer, 1999). ISBN 978-1-4615-477

Chapter 13
Precision Physics in Penning Traps Using the Continuous Stern-Gerlach Effect

Klaus Blaum and Günter Werth

1 Introduction

"A single atomic particle forever floating at rest in free space" (H. Dehmelt) would be the ideal object for precision measurements of atomic properties and for tests of fundamental theories. Such an ideal, of course, can ultimately never be achieved. A very close approximation to this ideal is made possible by ion traps, where electromagnetic forces are used to confine charged particles under well-controlled conditions for practically unlimited time. Concurrently, sensitive detection methods have been developed to allow observation of single stored ions. Various cooling methods can be employed to bring the trapped ion nearly to rest. Among different realisations of ion traps we consider in this chapter the so-called Penning traps which use static electric and magnetic fields for ion confinement. After a brief discussion of Penning-trap properties, we consider various experiments including the application of the "continuous Stern-Gerlach effect", which have led recently to precise determinations of the masses and magnetic moments of particles and antiparticles. These serve as input for testing fundamental theories and symmetries.

2 Penning-Trap Properties

The Penning traps used in the experiments described herein consist of a symmetric stack of cylindrical electrodes as shown in Fig. 1.

K. Blaum
Max-Planck-Institut für Kernphysik, Heidelberg, Germany

G. Werth (✉)
Johannes-Gutenberg-Universität, Mainz, Germany
e-mail: werth@uni-mainz.de

© The Author(s) 2021
B. Friedrich and H. Schmidt-Böcking (eds.), *Molecular Beams in Physics and Chemistry*,
https://doi.org/10.1007/978-3-030-63963-1_13

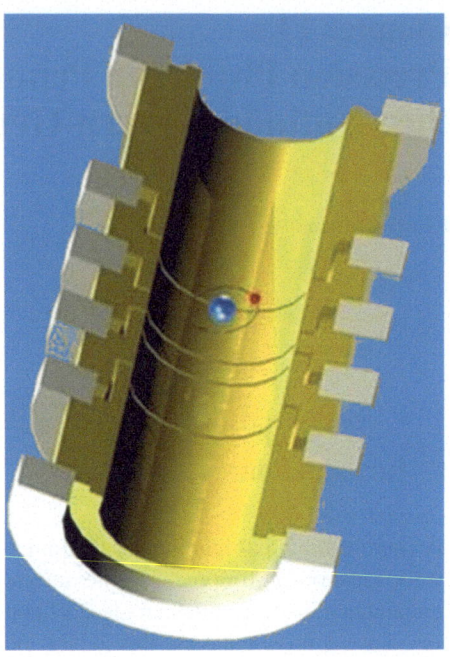

A static voltage is applied between the connected outer electrodes and the central ring. At a positive polarity at the endcaps a positively charged particle is confined in the axial direction. Escape in the radial direction is precluded by a homogeneous magnetic field directed along the trap's axis. Additional voltages applied to the so-called correction electrodes, placed between the ring and endcaps, serve to provide an electric potential which depends—at least for small amplitudes of the trapped particle—only on the square of the distance from the trap's centre. This is a prerequisite for achieving high resolution in the determination of the motional frequencies of the trapped ions. It is further required to avoid perturbation of the confined ion by collisions with background gas molecules. In our case, the trap electrodes and their container box are in thermal contact with a liquid helium bath. The cryopumping results in a residual pressure of less than 10^{-16} mbar. As a result, not even a single collision of the trapped ion with a neutral molecule was observed during typical trapping times of several months.

The ion's motion in the trap arrangement described above can be calculated analytically. As a result one obtains a superposition of three harmonic oscillations: An axial one at frequency ν_z, which depends on the ions mass M, the size of the trap, and the voltage, and two radial oscillations with frequencies ν_+ and ν_-. The frequency ν_+ is near the cyclotron frequency $\nu_c = qB/(2\pi M)$ of the free ion with charge q in the magnetic field B, slightly perturbed by the presence of the electric trapping field. The centre of this motion orbits around the trap centre at a low "magnetron frequency" ν_-. Figure 2 shows a sketch of the ion's motion. An important relation connects the motional frequencies to the free ion's cyclotron frequency [1]:

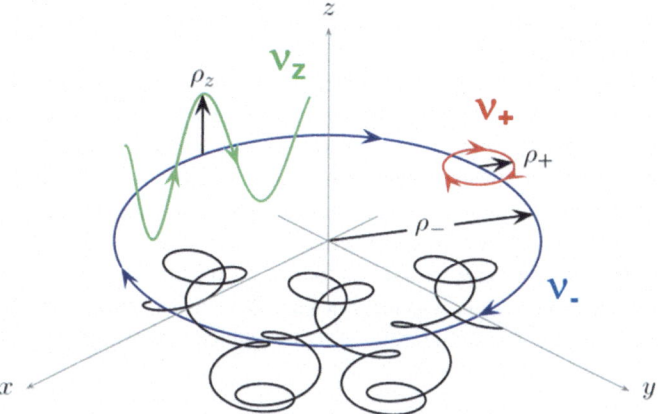

Fig. 2 Sketch of the ion's motion in a Penning trap. For details see text

$$\nu_+^2 + \nu_-^2 + \nu_z^2 = \nu_c^2.$$ (1)

Further details on Penning traps can be found in Refs. [2, 3].

3 Single Ion Detection by Induced Image Currents

Achieving high precision in trap experiments requires the use of single trapped particles to avoid perturbations by Coulomb interaction with other ions. The standard way to detect single trapped particles is by observation of laser induced fluorescence. This requires, however, optical transitions in the ions which are in reach of laser wavelengths. This is not the case for highly charged ions or elementary particles such as electrons or protons and their antiparticles. In these cases, detection can be performed by the image current that the oscillating ion induces in the trap's electrodes [4]. This current is on the order of a few fA and requires very sensitive detection methods. This can be realized by a superconducting high-Q tank circuit, kept at the temperature of the surrounding He-bath and tuned to the resonance frequency of the ion oscillation. Figure 3 shows a scheme of the detection of a radial frequency. The noise power of the circuit can be amplified and Fourier analysed. In case of

Fig. 3 Scheme for detection of the radial ion oscillation by a superconducting tank circuit attached to segments of the trap's split ring electrode

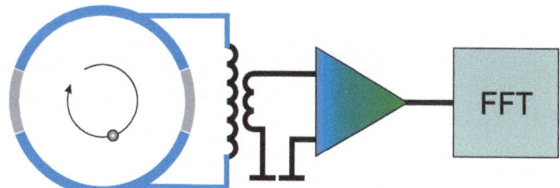

an ion present in the trap, the Fourier spectrum shows a maximum at the resonance frequency of the ion oscillation, as shown in Fig. 4.

The ion signal as shown in Fig. 4 indicates that the ion's kinetic energy is well above the thermal energy of the circuit. In order to reduce the ion's energy as required for high-precision measurements, the ion is kept in resonance with the circuit. The extra energy which the hot ion transfers into the circuit is then dissipated into the helium bath (resistive cooling) [5]. As consequence the ion adopts the temperature of the environment. The signal is then converted into a minimum in the noise spectrum as shown in Fig. 5. This can be understood based on the fact that the equivalent electronic circuit of the oscillating ion is a series resonance circuit which shortcuts the noise power of the detection circuit at its resonance frequency.

Fig. 4 Fourier analysis of the axial detection resonance circuit showing the induced image current of a single trapped ion on top of the Johnson noise of the detection circuit

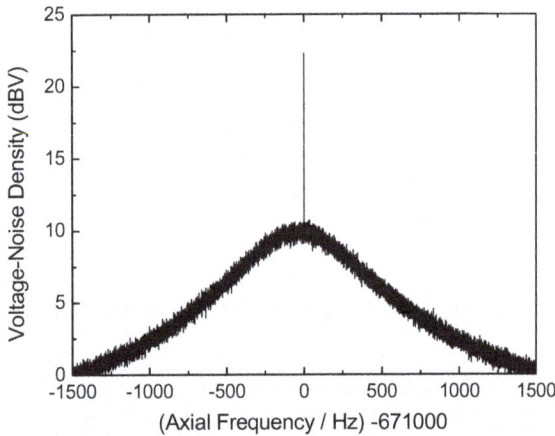

Fig. 5 Signal of a single ion's axial resonance in thermal equilibrium with the detection system immersed in a liquid He bath

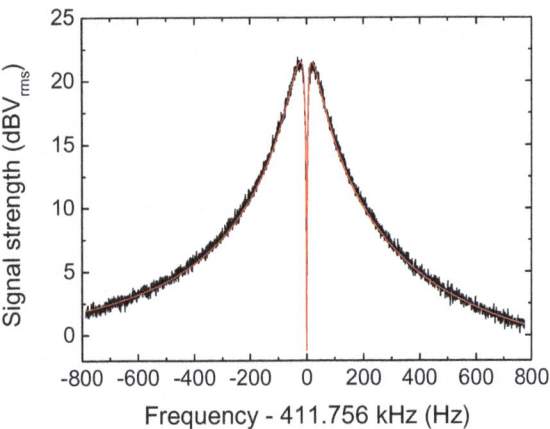

4 The Masses of the Proton and Antiproton

The ion signals such as those shown in Fig. 5 can be used for high-precision mass measurements. As recent examples, we consider the atomic masses of the proton and the antiproton. Their comparison serves as a test of the CPT invariance theorem. Quite general determination of atomic masses requires the comparison with the standard atomic mass, the carbon atom. The comparison is performed through measurements of the cyclotron frequencies of the particle under investigation $v_c(P) = eB/(2\pi M_P)$ (e.g. the proton P with mass M_P) and that of a carbon ion of charge state q: $v_c(C^{q+}) = qB/(2\pi M_C^{q+})$ in the same magnetic field B provided by a superconducting solenoid. In the case of the proton we used C^{6+} and correct the measured frequency by the masses of the missing electrons and their respective binding energies:

$$M_P = \frac{e}{q}\frac{v_C(C)}{v_C(P)} M_C \qquad (2)$$

By resolving the central part of the ion detection signal as shown in Fig. 5, we can determine the centre frequency with an uncertainty of a few mHz. In our experiments we measure exclusively the axial frequency. In order to determine the radial frequency as required for the determination of the cyclotron frequencies according to Eq. (1), we couple the radial frequencies to the axial one by an additional r.f. field applied to the trap electrodes at their difference frequency to the axial. This leads to a split of the axial detection signal which allows to determine the radial frequencies with the same uncertainty.

In order to perform the measurements of the respective cyclotron frequencies at the same position of the magnetic field, we extend our Penning trap by a number of additional electrodes. In different potential minima we can store simultaneously a single proton and a single carbon ion. By changing the potentials at the electrodes we can transport one ion into the central part of the trap structure where the homogeneity of the magnetic field is highest, while the other ion is stored in one of the remaining potential minima. Frequent exchange of protons and carbon ions eliminates to a large extent the influence of possible time variations of the magnetic field which is provided by a superconducting magnet. Figure 6 shows the complete setup.

Our result for the proton's atomic mass including the statistical and systematic uncertainties is [6]

$$M_P = 1.007\ 276\ 466\ 583(15)(29)\ \text{u}. \qquad (3)$$

i.e. a relative mass uncertainty of 3×10^{-11} has been achieved. It improves earlier results by a factor of 3 and determines the proton mass value in the most recent CODATA compilation of fundamental constants [7] (Fig. 7).

A similar experiment as described above using a nearly identical setup has been performed at CERN/Geneva, where single antiprotons have been confined and their cyclotron frequency measured [8]. The main difference was that, for comparison with a reference mass, not a carbon ion could be used since it would require a change

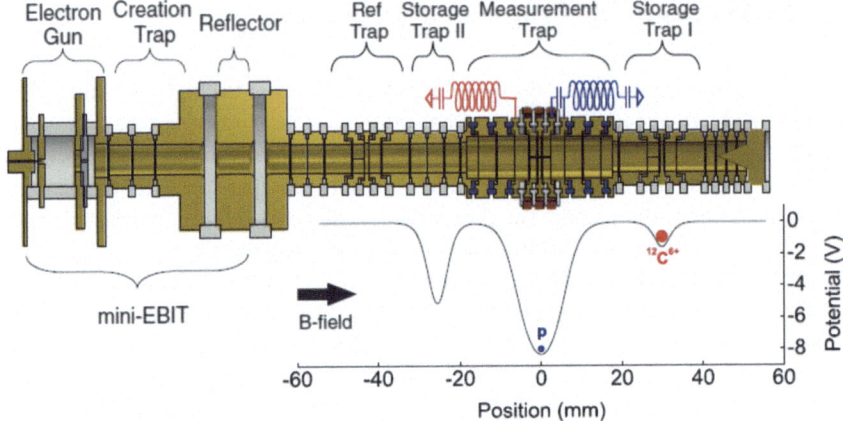

Fig. 6 Setup of the trap configuration for the determination of proton's atomic mass. The left part serves for the creation of protons and carbon ions by electron bombardment of a target. After ion creation and removal of unwanted species, single ions are transferred to one of the potential minima. Measurements are performed in the so-called measurement trap located at the most homogeneous part of the magnetic field. Shown are the resonance circuits attached to the measurement trap's electrode tuned to the different axial frequencies for the proton and C^{6+}. The trap configuration is placed in a hermetically sealed container in thermal contact with a liquid-He bath. The low temperature of the container walls provides a vacuum below 10^{-16} mbar by cryofreezing. Collisions of the stored particles with background molecules are absent for the period of several months which allows a nearly infinitely long perturbation-free storage time

Fig. 7 History of proton mass determinations (*courtesy* F. Heiße)

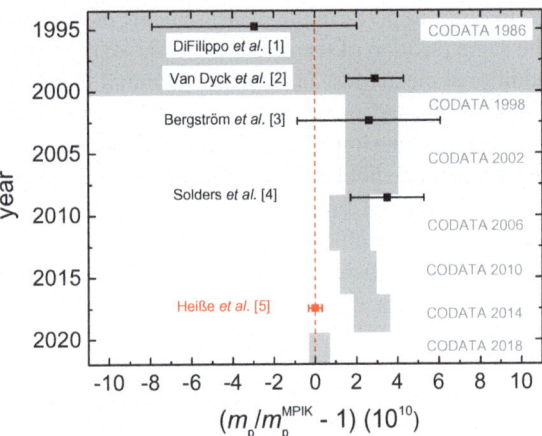

of the voltage sign at the trap because of the different charge signs of the particles. This, in turn, could lead to uncontrollable errors. Instead, the CERN experiment used the negative hydrogen ion H^-. Taking into account the masses of the 2 additional electrons, their binding energies E_a and E_b as well as the polarizability α_{pol,H^-} of H^-, the antiproton's m_{ap} mass can be compared to the protons mass through

$$m_{(H^-)} = m_{ap}\left(1 + 2\frac{m_e}{m_p} + \frac{\alpha_{pol,H^-}B_0^2}{m_p} - \frac{E_b}{m_p} - \frac{E_a}{m_p}\right) \tag{4}$$

The result of the proton/antiproton mass ratio is

$$(q/m)_{ap}/(q/m)_p = -1.000\,000\,000\,001\,(69) \tag{5}$$

at a relative uncertainty level of 7×10^{-11}. This result represents the most stringent test of the CPT invariance in the baryon sector.

5   The g-Factor of the Bound Electron

The g-factor of the electron is a dimensionless constant which relates the electron's magnetic moment μ_s to the spin S and the Bohr magnetron μ_B:

$$\mu_s = g\mu_B S \tag{6}$$

Dirac's relativistic treatment of the free electron predicts $g = 2$. Experimentally a deviation from 2 is found, which among others gave rise to the theory of Quantum Electrodynamics (QED), which describes the interaction of charged particles with electromagnetic fields by the exchange of virtual photons. Evaluation of Feynman diagrams to high orders allows calculating the g-factor of the free electron to extremely high precision [9]:

$(g - 2)/2 = 0.001\,159\,652\,181\,78\,(77)$. It agrees well with the experimental value [10]: $(g - 2)/2 = 0.001\,159\,652\,180\,73\,(28)$. The agreement represents the best test of QED in weak external fields.

In contrast to a free electron, an electron bound to an atomic nucleus experiences an extremely strong electric field. The strength of the field for a single electron bound in the 1S-state of a hydrogen-like ion of nuclear charge Z ranges from 10^9 V/cm in the helium ion ($Z = 2$) to $> 10^{15}$ V/cm in H-like uranium ($Z = 92$) (Fig. 8). This gives rise to a variety of new effects. The largest change of the bound electron's g-factor was analytically derived by Breit (1928) from the Dirac equation: $g_{Breit} = \frac{2}{3}\left(1 + 2\sqrt{1 - (Z\alpha)^2}\right) \approx 2\left(1 - \frac{1}{3}(Z\alpha)^2\right)$ [11], with $\alpha \approx 1/137$ the fine structure constant. The extremely high electric fields within atoms also require different methods to be used for calculating the QED contributions to the electron's magnetic moment. Feynman diagrams have to be calculated using the solution of the Dirac equation as an electron propagator. Contributions of high orders in $(Z\alpha)$ have been calculated by several authors [12]. In addition, nuclear structure and recoil effects must be considered [13]. Figure 9 summarises the different contributions to the electron's g-factor in the ground state of H-like ions as function of the nuclear charge Z.

Fig. 8 Electric field strength at the 1S- and 2S-states of hydrogen-like ions as function of the nuclear charge Z

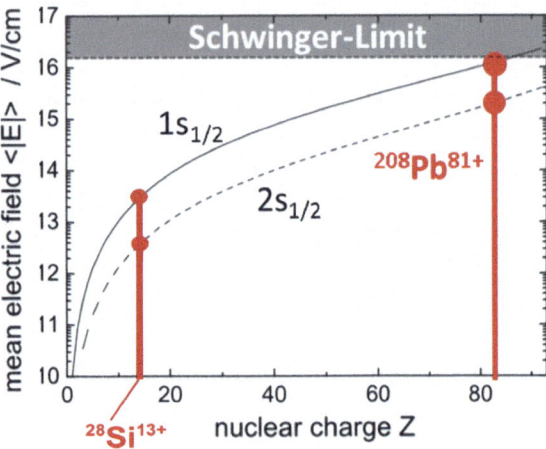

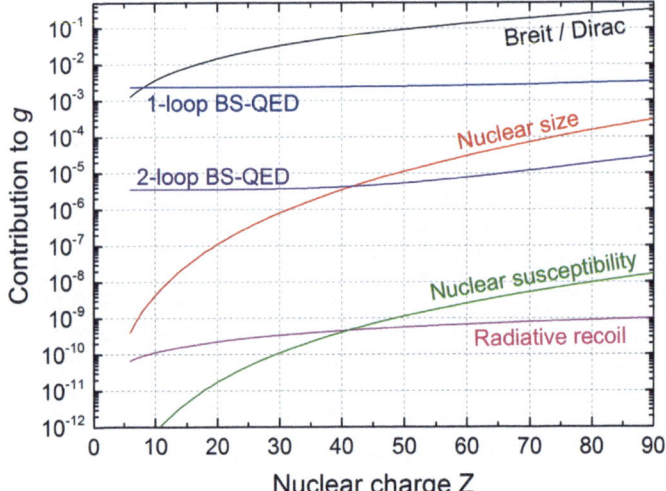

Fig. 9 Contributions from bound-state QED, nuclear size, structure, and recoil to the *g*-factor of the electron bound in hydrogen-like ions as function of the nuclear charge Z (*courtesy* Z. Harmann)

6 The Continuous Stern-Gerlach Effect

The determination of the electron's *g*-factor requires the measurement of the spin precession (Larmor) frequency ω_L:

$$\omega_L = \frac{g}{2} \frac{e}{m_e} B \tag{7}$$

The magnetic field strength B can be derived from the measurement of the three motional frequencies as described in Sect. 4. The spin precession, however, does not influence the ion's motion in a homogeneous magnetic field and consequently cannot be detected by observation of the axial ion resonance as only observable in our experimental set-up (see Sect. 3). The required coupling of the spin motion to the ion's oscillation can be provided by an inhomogeneity of the magnetic field. The force of $F = \mu_S grad B$ of a B-field inhomogeneity $grad B$ on the magnetic moment μ_S associated with the spin increases or reduces the electric trapping force of the Penning trap depending of the spin's direction. Consequently, a change in the spin direction leads to a change in the ion's axial frequency. This method has been first employed by Dehmelt in his experiment on the g-factor of the free electron and termed "continuous Stern-Gerlach effect" [14]. It was later adapted to experiments on highly charged ions [15].

The size of the change in the axial frequency upon a change in the spin direction is given by

$$\Delta\omega_z = \frac{g\mu_B B_2}{M\omega_z} \tag{8}$$

where M is the ions mass and B_2 is the quadratic part of the series expansion of the magnetic field $B_Z = B_0 + B_2 z^2 + ...$. The magnetic field inhomogeneity in our experiments is produced by a ferromagnetic central ring electrode (nickel) in the Penning trap assembly. It produces a bottle shaped B-field. Odd terms in the series expansion vanish. The measured value of B_2 in our trap is 10 mT/mm^2. The calculated change in the axial frequency, e.g., for H-like ^{28}Si^{13+} is 240 mHz in a total oscillation frequency of 412 kHz. The detection of such small frequency changes requires high stability of the electric trapping field. Figure 10 shows that changes in the spin direction, induced by a microwave field in the trap, can be unambiguously detected with nearly 100% probability.

Fig. 10 Change in the axial frequency of a single trapped ^{28}Si^{13+} when the spin direction is flipped by microwave-induced transitions. For details see text

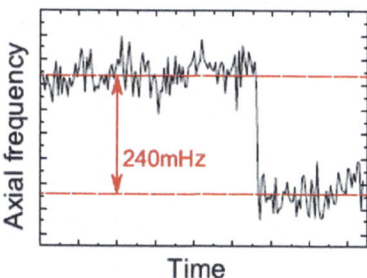

7 Measurement of *g*-Factors

Our first application of the continuous Stern-Gerlach effect was the determination of
the *g*-factor of the electron bound in hydrogen-like ions. We monitor induced spin flips
as shown in Fig. 10 while varying the microwave field's frequency. The maximum
spin flip probability occurs when the microwave frequency ω coincides with the
Larmor precession frequency ω_L. The *g*-factor can be derived from Eq. (5) when the
magnetic field B is known. B is obtained using the single ion detection signals at the
ion's oscillation frequencies as described in Sects. 3 and 4. The *g*-factor then follows
from the measurement of ω_L and ω_c:

$$g = 2\frac{\omega_L}{\omega_c}\frac{q_{ion}}{m_{ion}}\frac{m_e}{e} = 2\Gamma\frac{q_{ion}}{m_{ion}}\frac{m_e}{e} \tag{9}$$

Γ is the ratio of the applied microwave frequency ω to the measured cyclotron
frequency. However, in order to obtain high precision in the *g*-factor determination
we are faced with conflicting requirements: The continuous Stern-Gerlach effect
requires a strong inhomogeneity of the *B*-field at the ions position in order to detect
induced spin flips with high probability. High accuracy for *B*-field determination on
the other side requires a very homogeneous field at the ion's position. In order to
resolve this conflict we used a Penning-trap configuration similar to the one shown
in Fig. 6 but modified by introducing a nickel ring electrode in one of the storage
traps that provides the required *B*-field inhomogeneity. Step 1 of the measurement
procedure is the determination of the ion's spin direction in the inhomogeneous trap
by introduction of a spin flip and observation of the change in the axial oscillation
frequency as illustrated in Fig. 10. With a known spin direction, the ion is then transported into the measurement trap where the oscillation frequency ω_c is determined.
Simultaneously the ion is irradiated by a microwave field attempting to change the
spin direction. Then the ion is transported back into the inhomogeneous trap and,
as in step 1, its spin direction is determined again. A successfully induced spin
flip in the measurement trap is monitored in the inhomogeneous trap by the corresponding change in the axial frequency. Frequent repetition of this procedure with
varying microwave frequencies and monitoring the spin flip probability for different
frequencies results in a resonance curve with a maximum at the Larmor precession
frequency ω_L. The fact that ω_L and ω_c are measured at the same position as well
at the same time eliminates to a large extent the uncertainties due to time fluctuations of the *B*-field. Figure 11 shows an example of measurements on hydrogen-like
$^{28}Si^{13+}$, where the spin flip probability is plotted against the ratio Γ of ω and ω_C.
The maximum is at $\Gamma = \omega_L/\omega_C$.

The result of the experiment on $^{28}Si^{13+}$ for the *g*-factor of the single bound electron
is $g_{exp} = 1.995\ 348\ 958\ 7\ (5)(3)(8)$ [16]. The numbers in parenthesis correspond
respectively to the systematic and statistical uncertainties and the error of the electron
mass taken from the CODATA 2012 tables of fundamental constants [17]. The result
agrees well with the theoretical value $g_{theo} = 1.995\ 348\ 958\ 0\ (17)$ [18] and represents
the most stringent test of QED calculations in strong external fields.

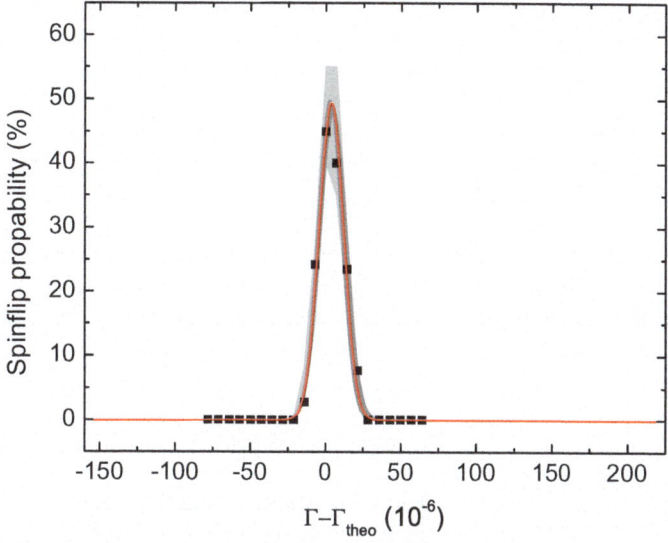

Fig. 11 Measured spin flip probability of ^{28}Si^{13+} as a function of the frequency ratio $\Gamma = \omega/\omega_C$. The maximum occurs at $\Gamma = \omega_L/\omega_C$. Γ_{theo} is the theoretical predicted value. Shifts caused by systematic effects are not yet corrected for

Similar results with comparable precision have been obtained on H-like ^{12}C^{5+} [19] and ^{16}O^{7+} [20]. Also experiments using lithium-like ions ^{40}Ca^{17+} and ^{48}Ca^{17+} [21], and ^{28}Si^{11+} [22] have been performed. Here the interaction of the additional 2 bound electrons modify slightly the value of the g-factor. The comparisons between theory and experiment test calculations of the inter-electronic interaction. Most recently the g-factor of the boron-like ion ^{40}Ar^{13+} has been determined with high precision [23]. The experimental result distinguishes between the conflicting predictions of the contribution of the electron-electron interaction.

Analogously to the electronic g-factors, the magnetic moments of the proton and antiproton were obtained. The particular challenge in these experiments had to do with the fact that the magnetic moment of these particles is about 500 times smaller than that of the electron. According to Eq. (6), the corresponding change in the axial frequency is significantly smaller than in the case of hydrogen-like ions. Its observation requires extremely high stability in the trap parameters. A measurable frequency change upon a spin flip was obtained by making the trap diameter about 3 times smaller than in the previous experiments and replacing the Ni-ring by a CoFe-ring and thus obtaining a larger magnetic field inhomogeneity. The result for the proton's and antiproton's magnetic moments are $\mu_p = 2.792\,847\,344\,62(82)\,\mu_N$ [24] and $\mu_{antip} = -2.792\,847\,344\,1(42)\,\mu_N$ [25], with μ_N the nuclear magnetron. The agreement of the two values within the error bars represents a test of the CPT invariance theorem.

8 The Electron Mass

As evident from the result of the g-factor determination in $^{28}\mathrm{Si}^{13+}$, the largest contribution in the error budget arises from the uncertainty in the electron mass taken from the CODATA 2012 compilation of fundamental constants. The QED contributions to the electronic g-factor of H-like ions scale approximately with the square of the nuclear charge Z. Since in the case of $^{28}\mathrm{Si}^{13+}$ with $Z = 14$ we find an agreement between theory and experiment on the level of 10^{-10} it can be reasonably assumed that in the case of H-like $^{12}\mathrm{C}^{5+}$ with $Z = 6$ the QED contributions to the g-factor have been calculated correctly. We can therefore rewrite Eq. (9) between the g-factor and the electron mass as

$$m_e = \frac{g}{2} \frac{e}{q} \frac{\nu_{\mathrm{cyc}}}{\nu_{\mathrm{L}}} m_{\mathrm{ion}} \equiv \frac{g}{2} \frac{e}{q} \frac{1}{\Gamma} m_{\mathrm{ion}} \tag{10}$$

We take now the g-factor from theory and determine the electron mass. $^{12}\mathrm{C}^{5+}$ is the natural choice as ion since there is virtually no uncertainty in its mass. In our experiment [26] we obtained as new value for the electron mass $m_e = 0.000\,548\,579\,909$ $067\,(14)(9)(2)$ a.u. The first two errors are the statistical and systematic uncertainties of the measurement, respectively, and the third one represents the uncertainties of the theoretical prediction of the g-factor and the electron binding energies in the carbon ion. The new value surpasses that of the CODATA 2012 literature value by a factor of 13 and represents the basis for the most recent adjustment of fundamental constants [7].

9 What Comes Next?

At the Max-Planck-Institut für Kernphysik in Heidelberg an electron beam ion trap (EBIT) [27] will allow production of hydrogen-like ions of high nuclear charge Z up to $^{208}\mathrm{Pb}^{81+}$. They can be extracted from the EBIT and injected into an improved Penning-trap arrangement (ALPHATRAP) placed at the center of a superconducting magnet [28]. Here Larmor-to-cyclotron frequency ratio measurements can be performed at the $dg/g \approx 10^{-12}$ level of accuracy. This will provide more stringent tests of bound state QED contributions to the electron's g-factor in an extremely strong electric field as evident from Figs. 8 and 9. In addition a new determination of the fine structure constant α by comparison of theoretical and experimental results seems possible. Figure 12 shows a sketch of the setup.

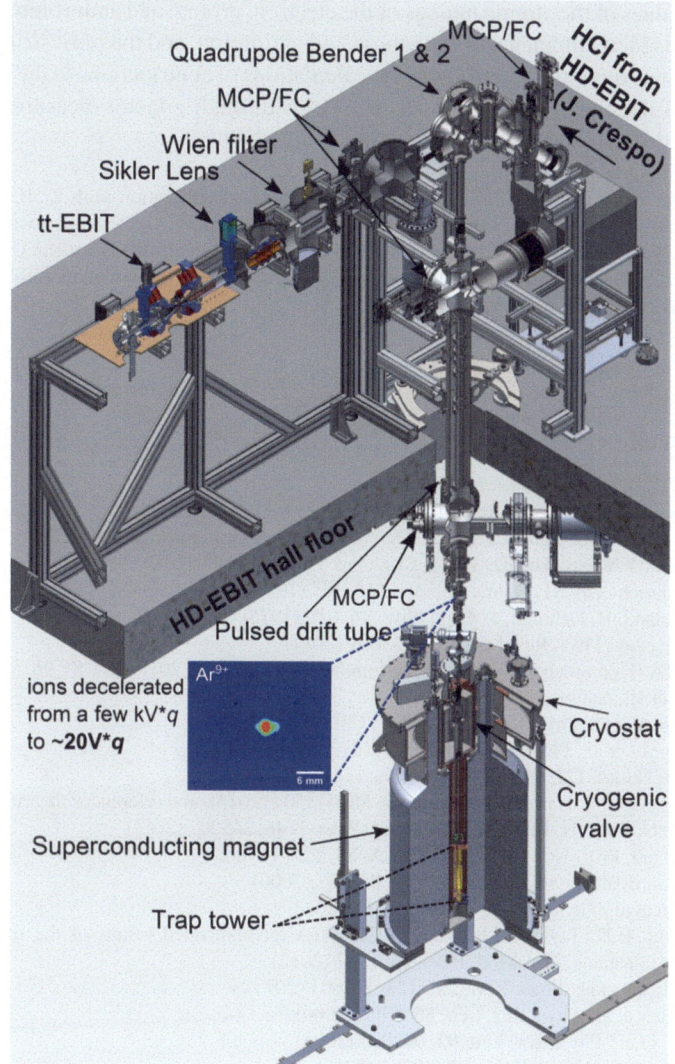

Fig. 12 Sketch of the ALPHATRAP set-up at MPIK Heidelberg to produce H-like ions of high nuclear charge from an EBIT to be injected into a Penning trap arrangement for g-factor determination [28]. A first experiment on ion $^{40}Ar^{13+}$ has been performed successfully [23]

10 Summary

Spectroscopy in Penning traps has reached an amazing level of precision even on exotic systems and has opened up many new fields of research. In this chapter we have summarised the results of recent experiments. They provide, to date, the most accurate values of the atomic masses of the electron, proton, and antiproton, the most accurate magnetic moments of the proton and antiproton, and the most stringent test of bound-state quantum electrodynamic calculations of contributions to the magnetic moment of the electron in strong electric fields through g-factor measurements on various hydrogen- and lithium-like highly charged ions.

Acknowledgements We gratefully acknowledge excellent collaboration with Z. Harman, Ch. Keitel, F. Köhler, W. Quint, V. Shabaev, D. Glazov, and S. Sturm. Financial support was provided by the Max-Planck-Gesellschaft, by IMPRS-PTFS and IMPRS-QD, by the European Union (ERC grant AdG 832848—FunI), and by the German Research Foundation (DFG) Collaborative Research Centre "SFB 1225 (ISOQUANT)."

References

1. L.S. Brown, G. Gabrielse, Rev. Mod. Phys. **58**, 233–311 (1986)
2. F.G. Major, V. Gheorghe, G. Werth, *Charged Particle Traps* (Springer, Heidelberg, 2002)
3. K. Blaum, Y.N. Novikov, G. Werth, Contemp. Phys. **51**, 149 (2010)
4. M. Diederich et al., Hyperfine Interact. **115**, 185 (1998)
5. D. Wineland, H. Dehmelt, J. Appl. Phys. **46**, 919 (1975)
6. F. Heiße et al., Phys. Rev. Lett. **119**, 033001 (2017)
7. CODATA Recommended Values of the Fundam. Phys. Const.: 2018 NIST SP 961 (2019)
8. S. Ulm et al., Nature **524**, 196 (2015)
9. T. Aoyama et al., Phys. Rev. Lett. **109**, 111807 (2012)
10. D. Hanneke et al., Phys. Rev. A **83**, 052122 (2011)
11. G. Breit, Nature **122**, 649 (1928)
12. V.A. Yerokhin, Z. Harman, Phys. Rev. A **88**, 042502 (2013). and references therein
13. V.A. Yerokhin, C.H. Keitel, Z. Harman, J. Phys. B **46**, 245002 (2013)
14. H. Dehmelt, Proc. Natl. Acad. Sci. U.S.A. **83**, 2291 (1986)
15. N. Hermanspahn et al., Phys. Rev. Lett. **84**, 427 (2000)
16. S. Sturm et al., Phys. Rev. A **87**, 030501(R) (2013)
17. P.J. Mohr, B.N. Taylor, D.B. Newell, CODATA recommended values of the fundamental physical constants. Rev. Mod. Phys. **84**, 1527 (2012)
18. D.A. Glazov et al., Phys. Rev. Lett. **123**, 173001 (2019)
19. H. Häffner et al., Phys. Rev. Lett. **85**, 5308 (2000)
20. J. Verdu et al., Phys. Rev. Lett. **92**, 093002 (2004)
21. F. Köhler et al., Nature Comm. **7**, 10264 (2016)
22. A. Wagner et al., Phys. Rev. Lett. **110**, 033003 (2013)
23. I. Arapoglou et al., Phys. Rev. Lett. **122**, 253001 (2019)
24. G. Schneider et al., Science **358**, 1081 (2017)
25. Ch. Smorra et al., Nature **550**, 371 (2017)
26. S. Sturm et al., Nature **506**, 467 (2014)
27. J.C. López-Urrutia et al., J. Phys. Conf. Ser. **2**, 42 (2004)
28. S. Sturm et al., Eur. Phys. J. **227**, 1425 (2019)

Chapter 14
Stern-Gerlach Interferometry with the Atom Chip

Mark Keil, Shimon Machluf, Yair Margalit, Zhifan Zhou, Omer Amit, Or Dobkowski, Yonathan Japha, Samuel Moukouri, Daniel Rohrlich, Zina Binstock, Yaniv Bar-Haim, Menachem Givon, David Groswasser, Yigal Meir, and Ron Folman

Abstract In this invited review in honor of 100 years since the Stern-Gerlach (SG) experiments, we describe a decade of SG interferometry on the atom chip. The SG effect has been a paradigm of quantum mechanics throughout the last century, but there has been surprisingly little evidence that the original scheme, with freely propagating atoms exposed to gradients from macroscopic magnets, is a fully coherent quantum process. Specifically, no full-loop SG interferometer (SGI) has been realized with the scheme as envisioned decades ago. Furthermore, several theoretical studies have explained why it is a formidable challenge. Here we provide a review of our SG experiments over the last decade. We describe several novel configurations such as that giving rise to the first SG spatial interference fringes, and the first full-loop SGI realization. These devices are based on highly accurate magnetic fields, originating from an atom chip, that ensure coherent operation within strict constraints described by previous theoretical analyses. Achieving this high level of control over magnetic gradients is expected to facilitate technological applications such as probing of surfaces and currents, as well as metrology. Fundamental applications include the probing of the foundations of quantum theory, gravity, and the interface of quantum mechanics and gravity. We end with an outlook describing possible future experiments.

M. Keil · S. Machluf · Y. Margalit · Z. Zhou · O. Amit · O. Dobkowski · Y. Japha · S. Moukouri · D. Rohrlich · Z. Binstock · Y. Bar-Haim · M. Givon · D. Groswasser · Y. Meir · R. Folman (✉)
Department of Physics, Ben-Gurion University of the Negev, Be'er Sheva 84105, Israel
e-mail: folman@bgu.ac.il

M. Keil
e-mail: mhkeil@gmail.com

S. Machluf
Analytics Lab, Amsterdam, The Netherlands

Y. Margalit
Research Laboratory of Electronics, MIT-Harvard Center for Ultracold Atoms, Department of Physics, Massachusetts Institute of Technology, Cambridge, MA 02139, USA

Z. Zhou
Joint Quantum Institute, National Institute of Standards and Technology and the University of Maryland, College Park, Maryland 20742, USA

© The Author(s) 2021
B. Friedrich and H. Schmidt-Böcking (eds.), *Molecular Beams in Physics and Chemistry*,
https://doi.org/10.1007/978-3-030-63963-1_14

1   Introduction

This review follows the centennial conference held in Frankfurt in the same building housing the original Stern-Gerlach (SG) experiments. Here we describe the SG interferometry performed in our laboratories at Ben-Gurion University of the Negev (BGU) over the last decade.

The trail-blazing experiments of Otto Stern and Walther Gerlach one hundred years ago [1–4] required a few basic ingredients: a source of isolated atoms with well-specified momentum components (provided by their atomic beam), an inhomogeneous magnetic field and, if we follow the historical account of events in [5], also a smoky cigar. In this review, we present our approach to these first two ingredients, with our sincere apologies that we will not be able to adequately address the third.

As Dudley Herschbach notes [4], the SG experiments formed the basis for a "symbiotic entwining of molecular beams with quantum theory" and, as shown in many of the papers at this centennial conference, this symbiotic relationship remains vigorous to the present day. In this review, our source of isolated atoms is instead provided by the new world of ultra-cold atomic physics, to which we couple inhomogeneous magnetic fields that are provided naturally by an atom chip [6]. Current-carrying wires on such chips were first realized as magnetic traps for ultra-cold atoms at the turn of the (twenty-first) century [7–9] and reviewed extensively since [6, 10–14]. We are using the atom chip as our basis for coherently manipulating atoms in a way that is complementary to the atomic and molecular beam techniques pioneered by Otto Stern and practiced so energetically and creatively by his scientific descendants.

The work presented here is performed with high-quality atom chips fabricated by our nano-fabrication facility [15]. The atom chip is advantageous for Stern-Gerlach interferometry (SGI) for 4 main reasons. First, the source (Bose-Einstein condensates, BEC) is a minimal-uncertainty wavepacket so it is very well defined in position and momentum. Second, the source of the magnetic gradients (current-carrying wires on the atom chip) is very well aligned relative to the atomic source. Third, due to the very small atom-chip distance, the gradients are very strong, and significant Stern-Gerlach splitting can be realized in very short times. Fourth, the gradients are very well defined in time since there are no coils and the inductance of the chip wires is negligible. We will describe how these advantages have overcome long-standing difficulties and have enabled different SG configurations to be realized at BGU (e.g., spatial interference patterns [16, 17] and a "full-loop" SGI [18, 19]) alongside several applications, such as spatially splitting a clock [20, 21]. Finally, let us mention that while the interferometers presented here are of a new type, it is worthwhile noting decades of progress in matter-wave interferometry [22].

The discovery of the Stern-Gerlach (SG) effect [1] was followed by ideas concerning a full-loop SGI that would consist of freely propagating atoms exposed to magnetic gradients from macroscopic magnets. However, starting with Heisenberg [23], Bohm [24] and Wigner [25] considered a coherent SGI impractical because it was thought that the macroscopic device could not be made accurate enough to ensure a reversible splitting process [26]. Bohm, for example, noted that the magnet would need to have "fantastic" accuracy [24]. Englert, Schwinger and Scully analyzed the problem in more detail and coined it the Humpty-Dumpty[1] (HD) effect [28–31]. They too concluded that for significant coherence to be observed, exceptional accuracy in controlling magnetic fields would be required. Indeed, while atom interferometers based on light beam-splitters enjoy the quantum accuracy of the photon momentum transfer, the SGI magnets not only have no such quantum discreteness, but they also suffer from inherent lack of flatness due to Maxwell's equations [32]. Later work added the effect of dissipation and suggested that only low-temperature magnetic field sources would enable an operational SGI [33]. Claims have even been made that no coherent splitting is possible at all [34].

Undeterred, we utilize the novel capabilities of the atom chip to address these significant hurdles. Let us briefly preview our most recent and most challenging realization, the full-loop SGI, in which magnetic field gradients act on the atom during its flight through the interferometer, first splitting, and then re-combining, the atomic wavepacket. We obtain a high full-loop SGI visibility of 95% with a spin interference signal [18, 19] by utilizing the highly accurate magnetic fields of an atom chip [6]. Notwithstanding the impressive endeavors of [35–45] this is, to the best of our knowledge, the first realization of a complete SG interferometer analogous to that originally envisioned a century ago.

Achieving this high level of control over magnetic gradients may facilitate fundamental research. Stern-Gerlach interferometry with mesoscopic objects has been suggested as a compact detector for space-time metric and curvature [46], possibly enabling detection of gravitational waves. It has also been suggested as a probe for the quantum nature of gravity [47]. Such SG capabilities may also enable searches for exotic effects like the fifth force or the hypothesized self-gravitation interaction [48]. We note that the realization presented here has already enabled the construction of a unique matter-wave interferometer whose phase scales with the cube of the time the atom spends in the interferometer [19], a configuration that has been suggested as an experimental test for Einstein's equivalence principle when extended to the quantum domain [49].

High magnetic stability and accuracy may also benefit technological applications such as large-momentum-transfer beam splitting for metrology with atom interferometry [50–52], sensitive probing of electron transport, e.g., squeezed currents [53], as well as nuclear magnetic resonance and compact accelerators [54]. We note that since the SGI makes no use of light, it may serve as a high-precision surface probe at short distances for which administering light is difficult.

For the purpose of this review, it is especially important to also realize that the atom chip allows our atoms to be completely isolated from their environment. This

---

[1]Can a fragile item be taken apart and be re-assembled perfectly? ... another tough problem, according to the popular English rhyme [27]
Humpty Dumpty sat on a wall,
Humpty Dumpty had a great fall.
All the king's horses
And all the king's men
Couldn't put Humpty together again.

is demonstrated, for example, by the relatively long-term maintenance of spatial coherence that can be achieved despite a temperature gradient from 300 K to 100 nK over a distance of just 5 μm [55]. Coherence of internal degrees of freedom close to the surface has also been measured to be very high [56].

This review is organized into the following sections:

Section 2. Particle Sources: a brief discussion of how the atom chip complements and extends the century-long use of atomic and molecular beams in Stern-Gerlach experiments;

Section 3. The Atom Chip Stern-Gerlach Beam Splitter: detailing relevant aspects of the atom chip and its basic operating characteristics as a platform for SGI;

Section 4. Half-Loop Stern-Gerlach Interferometer: first realization of SGI with spatial fringe patterns;

Section 5. Full-Loop Stern-Gerlach Interferometer: first realization of the four-field complete SGI with spin population fringes;

Section 6. Applications: clock interferometry and complementarity, the matter-wave geodesic rule and geometric phase, and a T^3 interferometer realizing the Kennard phase;

Section 7. Outlook: extending the atom-chip based SGI experiments to ion beams and to massive particles.

Finally, we note that the SG effect, in conjunction with the atom chip, may also be used for novel applications without the use of interferometry. For example, we have used the SG spin-momentum entanglement to realize a novel quantum work meter. In this work, done in conjunction with the group of Juan Pablo Paz, we were able to test non-equilibrium fluctuation theorems [57].

As we hope to show in this review, we believe that the atom chip provides a novel and powerful tool for SG interferometry, with much yet to learn as SG studies enter their second century. May we continue to find surprises, fundamental insights, and exciting applications.

2 Particle Sources

Molecular beam experiments exhibiting quantum interference, diffraction, and reflection have been brought very skillfully into the modern era in presentations at this Conference by Markus Arndt, Maksim Kunitski, and Wieland Schöllkopf, and as outlined in the keynote address by Peter Toennies. In particular, Stern's vision—and realization—of diffraction of atomic and molecular beams (see, for example [4]) have found their modern expression in the work of all these experts, and many others. Here we will concentrate on a complementary approach to precisely specify internal and external quantum states and how they can be used to study interference phenomena in particular.

Table 14.1 Parameters relating to diffraction experiments using He atomic beams [58, 59], Talbot-Lau interference experiments with macromolecules [60], and interference experiments using BEC's [17] as described in this review

Type	Source	Species	Temperature (K)	σ_z (μm)	σ_{v_z} (mm/s)	$k = \sigma_{p_z}/\hbar$ (μm^{-1})	$\sigma_z\sigma_{p_z}/\hbar$	Ref.
Diffraction	Beam	^{4}He	10^{-3}	20	14	0.9	18	[58]
Diffraction	Beam	^{4}He	Not given	50	43	2.6	130	[59]
T-L interference[a]	Beam	Macromolecules	Not given	0.266	0.04	16	4.3	[60]
Interference	BEC	^{87}Rb	40×10^{-9}	6	2.8	3.8	23	[17]
Particle-on-demand[b]	Ion trap	^{40}Ca$^+$	–	0.006	900	5×10^6	3×10^4	[61, 62]
First realization	Beam	Ag	1300	30	230	400	1×10^4	[1]

The temperature shown for the beam experiments corresponds to the velocity spread superimposed on the moving frame of the longitudinal most-probable velocity. The position spread σ_z for the beam experiments corresponds to the velocity spread superimposed on the moving frame of the longitudinal most-probable velocity. The position spread σ_z for the He beams is the beam collimator width, while the velocity spread σ_{v_z} is calculated from the beam angular divergence and its most-probable longitudinal velocity [63] ($v_x = v_{mp} \approx 288$ m/s for a He beam source temperature of 8 K). For the macromolecular beam, the parameters are taken from the grating period, interferometer length, and the stated longitudinal deBroglie wavelength. Corresponding parameters for the BEC are calculated using the Thomas-Fermi approximation and a temperature at which the BEC is about 90% pure. Parameters for the original Stern-Gerlach experiment are shown for comparison in the last line. All species are in their ground electronic state. The x- and z-co-ordinates refer to the horizontal and vertical directions respectively, where the beam experiments are horizontal (so z is the transverse direction) while the BEC experiments are vertical (so z is the longitudinal direction). We do not give parameters for the "beaded atom" experiments [36] since we believe that spatial interference fringes were not observed, as explained in [64]

[a]Talbot-Lau interference, as applied to matter-wave interference studies, is described in detail in [65]. The particle species in the quoted study are functionalized oligoporphyrin macromolecules with up to 2000 atoms and masses > 25000 Da [60].

[b]These parameters are for a proposal for SGI using ion beams that will be discussed in Sect. 7.1.

Let us begin by comparing experimental parameters used in the ultra-cold atomic environment in our laboratory, typically achieved with BECs of ^{87}Rb, with corresponding state-of-the-art parameters for atomic beams. Table 14.1 summarizes parameters that are most relevant for these experiments. Note that the beam experiments are conducted in a horizontal plane, transverse to the beam propagation direction, while our BEC interference experiments are conducted in an exclusively longitudinal direction with the atoms falling vertically due to gravity (and with all applied forces also acting in the longitudinal direction).

We see that ultra-cold atom localization and velocity spreads are on the same order as transverse localization from the exemplary atomic and molecular beam experiments quoted here but, of course, ultra-cold atoms are also localized in all three dimensions, whereas the beam techniques do not achieve localization along the beam propagation axis.

3   The Atom Chip Stern-Gerlach Beam Splitter

In order to apply Stern-Gerlach splitting, our ultra-cold atomic sample needs to have at least two spin states. However, our initial atomic sample is purely in the $|F, m_F\rangle = |2, 2\rangle$ state of ^{87}Rb. After preparing a BEC on the atom chip, our SG implementation therefore begins by first releasing the magnetic trap, and then applying a radio-frequency (RF) $\pi/2$ Rabi pulse to create an equal superposition of the two internal spin states $\frac{1}{\sqrt{2}}(|1\rangle + |2\rangle)$, where $|1\rangle$ and $|2\rangle$ represent the $m_F = 1$ and $m_F = 2$ Zeeman sub-levels of the $F = 2$ manifold in the ground electronic state [66]. Transitions to other m_F levels are avoided by retaining a modest homogeneous magnetic field even after trap release. A field of about 30 G is sufficient to create an effective two-level system by pushing the $m_F = 0$ sub-level about 200 kHz out of resonance with the $|2\rangle \rightarrow |1\rangle$ RF transition due to the non-linear Zeeman effect. The intensity of the RF Rabi pulses is calibrated such that a pulse duration of 20 μs corresponds to a complete population inversion between the two states, i.e., a π-pulse. This corresponds to a Rabi frequency of $\Omega_{RF} = 2\pi \cdot 25$ kHz.

We now consider the second factor crucial to the success of our SGI experiments: fast and precise magnetic fields, in both magnitude and direction, may be delivered by pulsed currents passed through micro-fabricated wires on the atom chip. Simple Biot-Savart considerations for atom chip wires, as used in our experiments, yield magnetic field gradients of about 200 G/mm at $\sim$100 μm from the chip, which is the starting distance for most of our experiments. Accurate control of this initial position, which is also crucial for the success of the experiments, is ensured by accurate control of chip wire currents and the homogeneous magnetic field referred to above. In addition, the straight atom chip wires have very low inductance, thereby enabling the generation of well-defined magnetic force pulses with currents that are typically tens of μs long. Such pulses are, in principle, able to induce momentum changes of hundreds of $\hbar k$.[2] Our earliest implementations of these experimental characteristics [67] were improved in subsequent apparatus upgrades [64].

Since the experiments proceed after turning off the magnetic trap, the observation time is limited by the time-of-flight (TOF) of the falling atoms and the field-of-view of our absorption imaging detection system. The latter is limited to about 4 mm, corresponding to a maximum TOF of about 28 ms. The optical detection system has a spatial resolution of about 5 μm, an important consideration for measuring spatial interference patterns (Sect. 4). Further experimental details may be found in several recent Ph.D. theses from our laboratory [64, 67, 68].

The Stern-Gerlach beam splitter (SGBS), first implemented in [16], begins with an equal superposition of $|1\rangle$ and $|2\rangle$ as described above and depicted schematically in Fig. 1. We then apply a magnetic field gradient $\nabla|\mathbf{B}|$ for duration T_1, which creates a state-dependent force $\mathbf{F}_{m_F} = m_F g_F \mu_B \nabla|\mathbf{B}|$ on the atomic ensemble, where μ_B, g_F, and m_F denote the Bohr magneton, the Landé factor, and the projection of the angular momentum on the quantization axis, respectively.

The magnetic potential created by the atom chip can be approximated as a sum of a linear part with characteristic force $\mathbf{F}$ and a quadratic part with characteristic frequency ω. After this magnetic gradient splitting pulse, the new state of the atoms is given by $\psi_f = \frac{1}{\sqrt{2}}(|1, p_1\rangle + |2, p_2\rangle)$, where $\mathbf{p}_i = \mathbf{F}_i T_1$ ($i = 1, 2$). This state represents a coherent superposition of two distinct momentum states, which are then allowed to separate spatially, thereby completing the operation of momentum and spatial splitting.

As we discuss further in the following sections, the SGBS can be extended as a tool for SGI. We describe two main configurations: a "half-loop" configuration in which the separated wavepackets are allowed to propagate freely, expand and eventually overlap, producing spatial interference patterns analogous to a double-slit experiment, and a "full-loop" configuration in which the wavepackets are actively re-combined, analogous to a Mach-Zehnder interferometer.

By applying additional pulses with different timing, these methods have been used to demonstrate, to the best of our knowledge, the first Stern-Gerlach spatial fringe interferometer (Sect. 4, [16, 17]), the first full-loop Stern-Gerlach interferometer (Sect. 5, [18, 19]), and several applications that we will describe in Sect. 6, including experiments to simulate the effect of proper time on quantum clock interference [20, 21].

4 Half-Loop Stern-Gerlach Interferometer

The two separated wavepackets generated by the SGBS initiate the pulse sequence shown in Fig. 2. Just after the SG splitting pulse, another RF $\pi/2$ pulse (10 μs dura-

---

[2]We express the momentum transfer in units of $\hbar k$, a reference momentum of a photon with 1 μm wavelength, in order to compare with atom interferometry based on optical beam splitters.

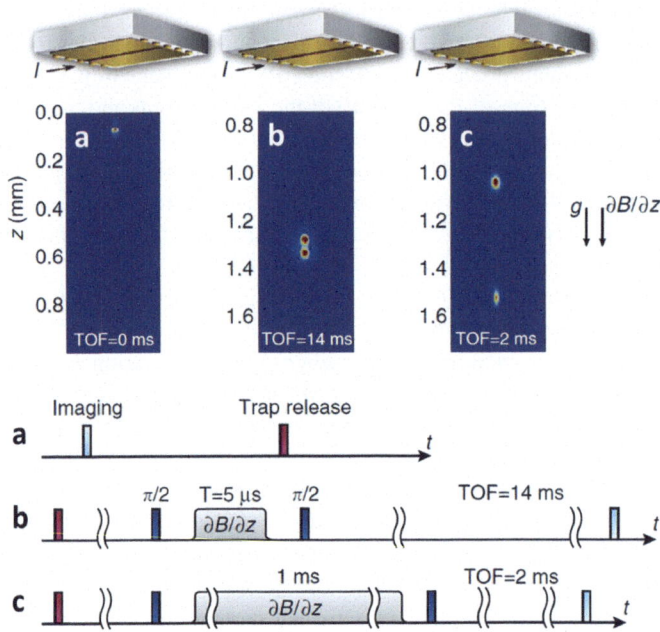

Fig. 1 The Stern-Gerlach beam-splitter (SGBS) at work [16, 67]. SGBS (a) input and (b, c) output images, and the corresponding schematic descriptions. The top row depicts our atom chip, with a pulsed current I being used to generate the magnetic gradient $\partial B / \partial z$ (we currently use three parallel wires with equal currents but opposing polarities). The chip faces downwards so that atoms can separate vertically during their free fall. (a) A magnetically trapped BEC in state $|2\rangle$ before release. (b) After a weak splitting of less than $\hbar k$ using a $5\,\mu s$ magnetic gradient pulse and allowing a TOF of 14 ms. (c) After a strong splitting of more than $40\,\hbar k$ using a 1 ms magnetic gradient pulse and allowing a TOF of 2 ms. Interferometric signals are formed either as spatial interference fringes by passively allowing overlap of the wavepackets (the "half-loop" SGI), or as spin-state population oscillations upon actively recombining them (the "full-loop" SGI), as described in Sects. 4 and 5 respectively. Adapted from [16].

tion) is applied, creating a wavefunction consisting of four wavepackets [67], of which we are concerned only with the two $|2\rangle$ wavepackets having momenta $\mathbf{p}_1$ and $\mathbf{p}_2$ (the $|1\rangle$ components can be disregarded since they appear at different final positions on completing the pulse sequence and a TOF).

The time interval between the two RF pulses (in which there are only two wavepackets, each having a different spin) is reduced to a minimum ($\sim 40\mu s$) to suppress the hindering effects of a noisy and uncontrolled magnetic environment, thereby removing the need for magnetic shielding. After a magnetic gradient pulse of duration T_2, designed to stop the relative motion of the two wavepackets, the atoms fall under gravity for a relatively long TOF, expanding freely until they overlap to create spatial interference fringes as shown schematically in Fig. 2 and experimentally in Fig. 3.

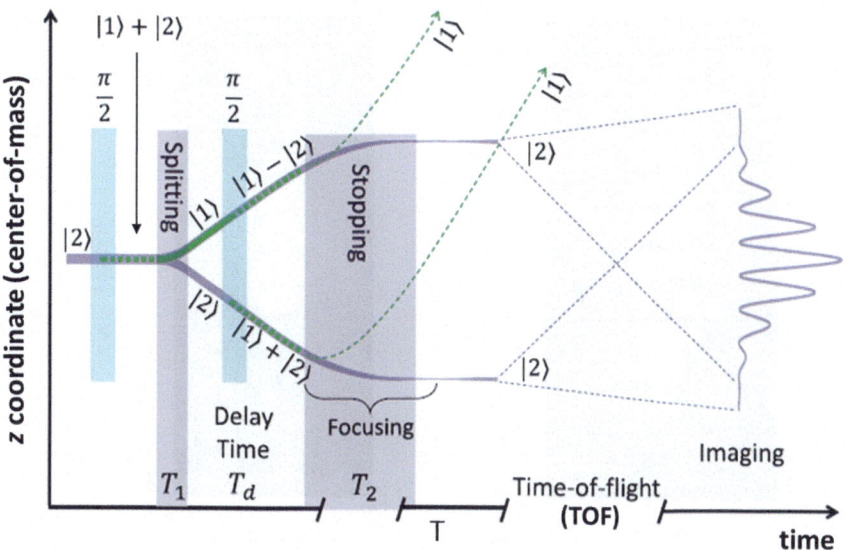

Fig. 2 Schematic depiction of the longitudinal half-loop SGI giving rise to spatial interference fringes (vertical position z in the center-of-mass frame vs. time). The initial wavepacket $|2\rangle$ (extreme left) is subjected to a $\pi/2$ pulse (blue column) that transfers the atoms into the superposition state $|1\rangle + |2\rangle$. The first magnetic gradient pulse of duration T_1 (purple column) induces a Stern-Gerlach splitting into $|1\rangle$ (green curve) and $|2\rangle$ (purple curve) having momenta $\mathbf{p}_1$ and $\mathbf{p}_2$, respectively. We then immediately apply a second $\pi/2$ pulse that places these diverging $|1\rangle$ and $|2\rangle$ states into equal superpositions $|1\rangle \mp |2\rangle$ as shown. The delay time T_d allows these wavepackets to spatially separate (in the z direction). The duration T_2 of a second gradient pulse is tuned to bring the momentum difference between the $|2\rangle$ components close to zero (see text), allowing their space-time trajectories to become parallel (solid purple curves) while expelling the $|1\rangle$ components (dotted green trajectories). The atoms then fall freely under gravity. Given sufficient time-of-flight, the two $|2\rangle$ wavepackets expand (dotted purple lines) and eventually overlap to generate spatial interference fringes, which are measured by taking an absorption image of the atoms. We note that due to the curvature of the magnetic field forming the magnetic gradient pulse, the long T_2 pulse also focuses the wavepackets, as depicted in the figure. In fact, this focusing accelerates the process of final expansion, thereby creating the two-wavepacket overlap in a shorter time. Adapted from [17] with permission © IOP Publishing & Deutsche Physikalische Gesellschaft. CC BY 3.0

The period of the interference fringes must be large enough to be observable with the spatial resolution of our imaging system (about $5\,\mu$m). This is accomplished if two conditions are fulfilled. First, the distance between the two wavepackets, d, should not be too large, since in principle the fringe periodicity varies as ht/md when the relative momentum is zero, where h, t, and m are the Planck constant, TOF duration, and the atomic mass, respectively. Second, the momentum difference between the two wavepackets should be smaller than their momentum width to avoid orthogonality. This is accomplished by tuning the duration T_2 of the second gradient pulse, which can stop the relative motion of the two $|2\rangle$ wavepackets; despite being in the same spin state, the slower wavepacket experiences a stronger impulse than

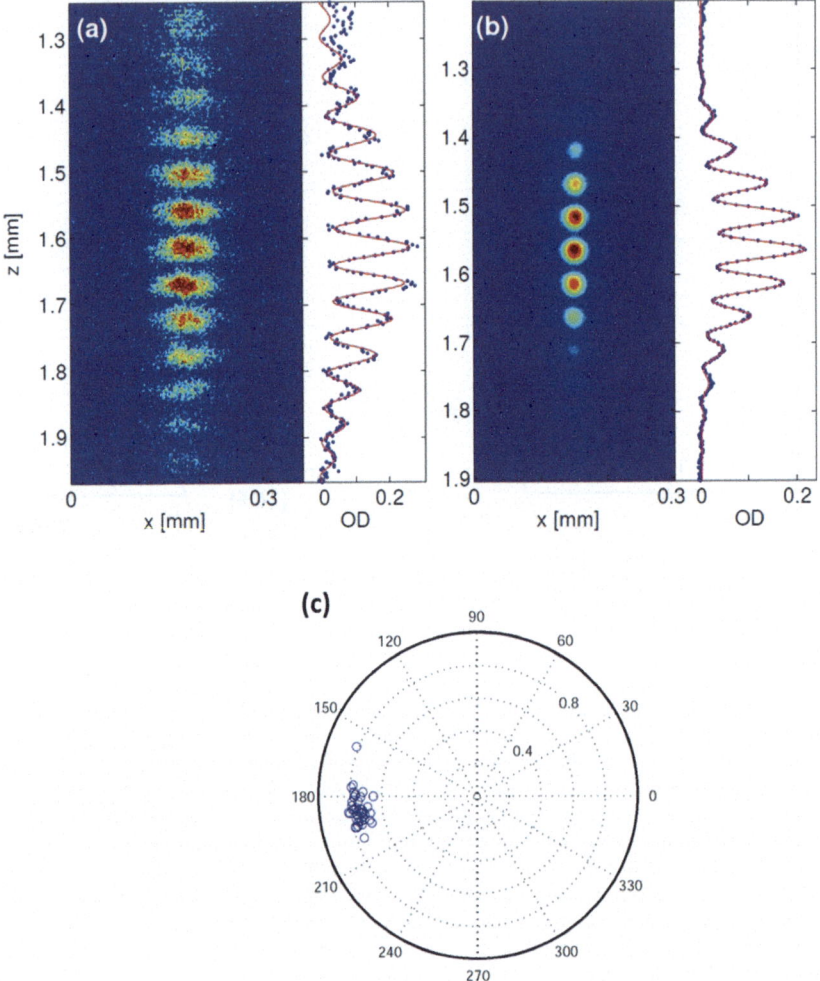

Fig. 3 Spatial interference patterns from the Stern-Gerlach interferometer. **a** A single-shot interference pattern of a thermal cloud with a negligible BEC fraction, fitted to Eq. (1) with a visibility of $V = 0.65$ (only slightly lower than single-shot visibilities typically measured for a BEC). **b** A multi-shot image made by averaging 40 consecutive interference images using a BEC (no correction or post-selection) with a normalized visibility of $V_N = 0.99$. **c** Polar plot of phase $0° \leq \phi \leq 360°$ versus visibility $0 \leq V \leq 1$ obtained from fitting each of the 40 consecutive images averaged in (**b**). The experimental parameters are $(T_1, T_d, T_2) = (4, 116, 200)\,\mu s$. Adapted from [64]

the faster one since it is considerably closer to the atom chip after the relatively long delay time T_d. We have found that zeroing the momentum difference between the two wavepackets is very robust [67].

Given that the final momentum difference between the two interfering wavepackets is smaller than their momentum spread, they overlap after a sufficiently long TOF and an interference pattern appears with the approximate form:

$$n(z, t) = A \exp\left[-\frac{(z - z_{CM})^2}{2\sigma_z(t)^2}\right]$$
$$\times \left[1 + V \cos\left(\frac{2\pi}{\lambda}(z - z_{ref}) + \phi\right)\right], \tag{1}$$

where A is the amplitude, z_{CM} is the center-of-mass (CM) position of the combined wavepacket at the time of imaging, $\sigma_z(t) \approx \hbar t / 2m\sigma_0$ is the final Gaussian width, $\lambda \approx 2\pi \hbar t / md$ is the fringe periodicity ($d = |z_1 - z_2|$ is the distance between the wavepacket centers), V is the interference fringe visibility, and $\phi = \phi_2 - \phi_1$ is the global phase difference. The vertical position z is relative to a fixed reference point z_{ref}. The phases ϕ_1 and ϕ_2 are determined by an integral over the trajectories of the two wavepacket centers. We emphasize that Eq. (1) is not a phenomenological equation, but rather an outcome of our analytical model [16].

In order to characterize the stability of the phase, which is the main figure of merit in interferometry, we average multiple experimental images with no post-selection or alignment (each single-shot image is a result of one experimental cycle). Large fluctuations in the phase and/or fringe periodicity in a set of single-shot images would result in a low multi-shot visibility, while small fluctuations correspond to high multi-shot visibility. The multi-shot visibility is therefore a measure of the stability of the phase and periodicity. Single-shot and multi-shot visibilities are all extracted by fitting to Eq. (1) after averaging the experimental images along the x direction (see Fig. 3) to reduce noise. We note that these procedures have been used over several years of half-loop SGI studies [16, 17], while the experimental results were simultaneously being greatly improved by significant modifications to the original apparatus [64, 67].

For a pure superposition state, as in our model, perfect fringe visibility V would be 1. A quantitative analysis of effects reducing V appears in [17, 64]. Some of these effects are purely technical, e.g., imperfect BEC purity and wavepacket overlap in 3D, as well as various imaging limitations etc. Such technical effects are irrelevant to the phase and periodicity stability shown by the multi-shot visibility, so we normalize the latter to the mean of the single-shot visibilities taken from the same sample: $V_N \equiv V_{avg}/\langle V_s\rangle$, where V_{avg} is the (un-normalized) visibility of the multi-shot average extracted from the fit and $\langle V_s\rangle$ is the mean visibility of the single-shot images which compose that multi-shot image. The normalized multi-shot visibility thus reflects shot-to-shot fluctuations of the global phase ϕ and the fringe periodicity λ. We note that some BEC intrinsic effects, such as phase diffusion, would not lead to a reduction of the single-shot visibility, but may cause the randomization of the shot-to-shot phase. However, such effects are expected to be quite weak, since atom-atom interactions rapidly become negligible as the BEC expands in free-fall, and the experiment may be described by single-atom physics.

Representative results from the above analysis are shown in Fig. 3. The very high (normalized) visibility shown in (b) demonstrates that the phase and periodicity are highly reproducible for each experimental cycle, the former being particularly emphasized in plot (c). High-visibility fringes ($V > 0.90$) were observed over a wide variety of experimental parameters, covering a range of maximum separations and velocities between the wavepackets. In particular, we conducted experiments at the apparatus-limited maximum value of $T_d = 600\,\mu s$ (which also required a long TOF = 21.45 ms) in order to maximize the spatial separation of the wavepackets during their time in the interferometer. These measurements achieved a separation $d = 3.93\,\mu m$, a factor of 20 larger than the atomic wavepacket size (after focusing, see Fig. 2), while maintaining a normalized visibility of $V_N = 0.90$ [17].

Given that our observed stable interference fringes arise from such well-separated paths, these experiments demonstrate what is, to the best of our knowledge, the first implementation of spatial SG interferometry. This achievement is due to three main differences compared with previous SG schemes. Firstly, we have used minimal-uncertainty wavepackets (a BEC) rather than thermal beams. Secondly, while the splitting is based on two spin states, the wavepackets in the two interferometer arms are in the same spin state for most of the interferometric cycle, thus reducing their sensitivity to disruptive external magnetic fields. Finally, chip-scale temporal and spatial control allows the cancellation of path difference fluctuations. It should also be noted that a longitudinal SGI, based on a particle beam source, cannot take images of spatial fringes due to the high velocity of the fringe pattern in the lab frame.

This, however, is not yet the four-field SGI originally envisioned shortly after the original Stern-Gerlach experiments (as recounted in [26]), since the separated wavepackets are not actively recombined in both position and momentum. The two remaining magnetic gradients required to complete such a "closed" SGI are discussed in the following section.

5   Full-Loop Stern-Gerlach Interferometer

Clearly, if a wavepacket can be coherently reconstructed after SG splitting and recombination in a four-field configuration [26], it should be possible to observe an interference pattern at the output of such an SGI. To the best of our knowledge however, no such interference pattern has heretofore been measured experimentally, and this is the task that we now describe, many details of which are taken from [64] and references therein.

The device envisioned consists of four successive regions of magnetic gradients giving rise to the operations of splitting, stopping, reversing and, finally, stopping the two wavepackets, as shown schematically in Fig. 4a. If executed perfectly, the two wavepackets would arrive at the output of such an interferometer with a minimal relative spatial displacement and momentum difference, so that an arbitrary initial spin state should be recoverable, using the spin state of the recombined wavepacket as the interference signal. However, the operation of such an interferometer was

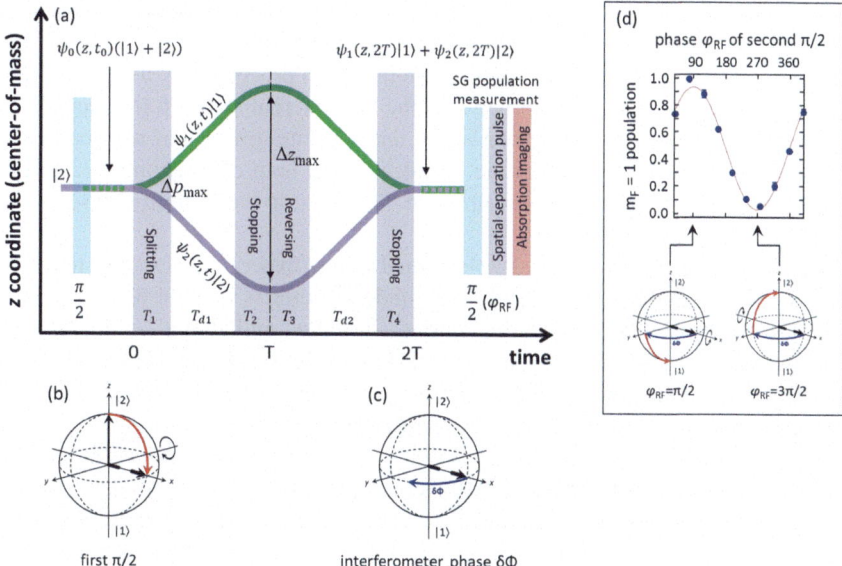

Fig. 4 The longitudinal full-loop SGI giving rise to spin population oscillations, plotted in the center-of-mass frame as in Fig. 2. **a** The sequence consists of RF pulses (blue) to manipulate the inner (spin) degrees of freedom and magnetic gradients (purple) to control the momentum and position of the wavepackets. The interferometer is prepared from the initial wavepacket $|2\rangle$ (extreme left) by applying a $\pi/2$ pulse that transfers the atoms into the superposition state $|1\rangle + |2\rangle$ [Bloch sphere shown in (**b**)]. The first magnetic gradient pulse at $t = 0$ induces a Stern-Gerlach splitting into $|1\rangle$ (green curve) and $|2\rangle$ (purple curve). Three additional magnetic gradient pulses are used to stop the relative motion of the wavepackets (at their maximum separation Δz_{max}), reverse their momenta, and finally stop them at the same position along z. The re-combined wavepacket at $t = 2T$ is therefore written as $\psi_1(z, 2T)|1\rangle + \psi_2(z, 2T)|2\rangle$, shown in (**c**) for an arbitrary interferometer phase $\delta\Phi$. After recombination, the population in $|1\rangle$ is measured by applying a second $\pi/2$ pulse with variable phase φ_{RF}, followed by a magnetic gradient to separate the populations and a subsequent pulse of the imaging laser. We expect to observe spin population fringes, i.e., oscillations in the $m_F = 1$ population, as we scan φ_{RF}, as indeed shown by the experimental results in (**d**), for which the measured visibility is 95%. The Bloch spheres in (**d**) show the particular case in which the initial vector (dashed black arrow) acquires an interferometer phase $\delta\Phi = \pi/2$ (blue arrow) followed by rotations about the $+x$ ($\varphi_{RF} = \pi/2$) or $-x$ ($\varphi_{RF} = 3\pi/2$) axes respectively (red arrows). The states $|F, m_F\rangle = |2, 2\rangle \equiv |2\rangle$ and $|2, 1\rangle \equiv |1\rangle$ are defined along the z axis in the Bloch spheres. Adapted from [64].

considered to be technically impractical, since coherent recombination of the two beam paths would require extremely precise control of the magnetic fields [24].

Our experiments begin, as before, with a $\pi/2$ pulse creating a superposition of the two spin states $|1\rangle$ and $|2\rangle$ of ^{87}Rb that is subsequently split into two momentum components by a magnetic gradient pulse (along the vertical axis z) as described in Sects. 3 and 4. Additional magnetic gradient pulses are needed to "close" the loop of such an interferometer, i.e., to overlap the wavepackets spatially and with zero relative momentum. To stop the relative motion of the two wavepackets after the first pulse, and to accelerate them backwards, we reverse the current on the atom chip, causing the force applied by the magnetic field gradient to be in the opposite direction. Alternatively, we can apply a spin inversion procedure by using a π Rabi pulse that inverts the population between the two internal states, following which a magnetic gradient pulse will then apply the opposite differential momentum to the two wavepackets. We obtain the signal with the help of a second $\pi/2$ pulse, followed by a spin population measurement. We measure the visibility by scanning the phase φ_{RF} of this $\pi/2$ pulse.

Our full-loop interferometer is implemented with an experimental system in which care is taken to reduce a wide range of hindering effects relative to our earliest work [16]. For example, a new atom chip was installed, utilizing a 3-wire configuration to produce a quadrupole magnetic field whose zero is at the precise height of the BEC. This reduces phase fluctuations by exposing the wavepackets to a weaker magnetic field while still generating strong magnetic gradients.

The practical difficulty encountered in re-assembling the original wavefunction was named the Humpty-Dumpty (HD) effect [28–30], implying that the initial wavepacket breaks under the SG field and cannot be reunited, as noted in the brief historical perspective given in Sect. 1. Quantitatively, the spin coherence, which is measurable as the visibility V of the observed spin fringes, is expressed as [29]

$$V = \exp\left\{-\frac{1}{2}\left[\left(\frac{\Delta z(2T)}{\sigma_z}\right)^2 + \left(\frac{\Delta p_z(2T)}{\sigma_p}\right)^2\right]\right\}, \qquad (2)$$

where $\Delta z(2T)$ and $\Delta p_z(2T)$ denote the mismatch between the wavepackets in their final position and momentum respectively, after the interferometer duration $2T$ (Fig. 4a), and σ_z and σ_p are the corresponding initial wavepacket widths. Equation (2) summarizes the main result of the HD papers in relation to our experimental observable. We emphasize that this reduction in visibility has nothing to do with effects of decoherence due to some coupling with the environment. We also note that the above HD calculation is done for a minimal-uncertainty wavepacket. For the general case, one can identify $l_z = \hbar/\sigma_p$ and $l_p = \hbar/\sigma_z$ as the relevant scales for coherence [26, 29], where l_z and l_p are the spatial coherence length and the momentum coherence width, respectively.

Let us discuss the meaning of this equation. The quantities σ_z and σ_p characterize the initial atomic wavefunction, and are thus microscopic quantities. The quantities Δz and Δp_z describe the experimental imprecision in the final recombination. In a "good" SG experiment (i.e., one which allows "unmistakable" splitting [29]) the maximum values of splitting in position and momentum should be much larger than their respective initial widths, meaning they should be macroscopic. On the other hand, according to Eq. (2), a nearly perfect maintenance of spin coherence ($V \simeq 1$) requires both $\Delta z \ll \sigma_z$ and $\Delta p_z \ll \sigma_p$. Consequently, Eq. (2) tells us that we need to recombine macroscopic quantities with a microscopic level of precision. This is the challenge facing SG interferometer experiments.

It is interesting to note that in the half-loop experiments, we found that Δp_z can be quite large (rendering the trajectories during the TOF period in Fig. 2 slightly non-parallel) without significantly reducing the measured spatial interference fringe visibility, so the stability of the half-loop experiments cannot be used to examine the HD equation. This robustness of the half-loop may also be understood by considering the fact that the expansion of the wavepackets creates an enhanced local coherence length, since for every region of space the k vector variance becomes smaller as TOF increases (see also [69, 70]).

A practical full-loop SG experiment must consider and address two effects. First, as noted above, the HD effect requires accurate recombination, namely, small Δz and Δp_z. These small values must be maintained for many experimental cycles, and thus a high level of stability in these values is also important. Achieving accurate recombination means that the overlap integral, calculated in Eq. (2), will have a significant non-zero value. Second, one must maintain a stable interferometer phase $\delta\Phi$, so that it has the same value shot-to-shot. This requires that the coupling to external magnetic noise is kept to a minimum, either by shielding the experiment and stabilizing the electronics (e.g., responsible for the homogeneous magnetic fields), or by conducting the experiment extremely quickly so that such environmental fluctuations do not have time to introduce significant phase noise.

Our full-loop SGI yields a visibility up to 95% (Fig. 4d), proving that we are able to use the SG effect to build a full-loop interferometer as originally envisioned almost a century ago. We note three differences between our realization and the scheme considered in the HD papers: (1) We use a BEC, which is a minimum-uncertainty wavepacket, whereas the HD papers considered atomic beam experiments with large uncertainties on the order of $\sigma_z\sigma_p \simeq 10^3$; (2) We implement fast magnetic gradient pulses generated by running currents on the atom chip, in contrast to using constant gradients from permanent magnets that were considered in the original proposals; (3) Our interferometer is a 1D longitudinal interferometer, while the originally envisioned SGI was 2D, i.e., it enclosed an area.

The full-loop experiments include a wide variety of optimizations and checks (see [64] for additional details). To make sure the spin superposition is not dephased due to some slowly varying gradients in our bias fields, we add π pulses giving rise to an echo sequence. To access a larger region of parameter space and to ensure the robustness of our results, we use several different configurations by, for example, implementing the reversing pulse (T_3) by inverting the sign of the atom chip currents vs. inverting the spins with the help of π pulses. We also utilize a variety of magnetic gradient magnitudes, and scan both the splitting gradient pulse duration T_1 and the delay time between the pulses T_d. All results are qualitatively the same. For weak splitting we observe high visibility ($\sim$95%), while for a momentum splitting equivalent to $\hbar k$ the visibility is still high ($\sim$75%), indicating that the magnet precision enabled coherent spin-state recombination to a high degree.

Finally, we briefly compare our experiments to previous work in an elaborate series of SGI experiments over a period of 15 years using metastable atomic beams [35–42, 44] and, more recently, thermal and ultra-cold alkali atoms [43, 45]. A detailed discussion is given in [64]. While these longitudinal beam experiments did observe

spin-population interference fringes, the experiments reviewed here are very different. Most importantly, an analogue of the full-loop configuration was never realized,
as only splitting and stopping operations were applied (i.e., there was no recombination) and wavepackets emerged from the interferometer with the same separation
as the maximal separation achieved within (see Fig. 2 of [40] and Footnote [10]
of [43]). We have not found anywhere in the many papers published by this group
(only some of which are referenced here) evidence of four operations being applied as
required for a full-loop configuration, whether the experiment was with longitudinal
or transverse gradients. In addition, no spatial interference fringes were observed,
as the spatial modulation they observed was a signature of multiple parallel longitudinal interferometers, each having its own individual relative phase between its two
wavepackets.

To conclude, we have shown that a full-loop has been realized [18, 121]. In addition, as previously shown in Heisenberg's argument, the momentum splitting is the
figure of merit in determining the phase dispersion. In our experiment, coherence is
observed up to a momentum splitting as high as $\Delta p_z(T_1)/\sigma_p = 60$. However, in contrast, the visibility is more sensitive to spatial splitting and we achieve $\Delta z(T)/\sigma_z = 4$,
much lower than for the half-loop, where we achieved $\Delta z/\sigma_z = 18$. The splitting
is coherent but its, limits in terms of the HD effect are yet to be explored quantitatively. Many mysteries remain to be solved, such as why is the observed reduction
not symmetric in momentum and spatial splitting, in contrast to Eq. 2. A simple
answer, which is yet to be examined in detail, is the existence of some sort of spatial
decoherence mechanism due to the environment.

Having now described the SG beam-splitter, the SG half-loop, and the SG full-
loop, we show in the next section how these techniques may be used for different
applications.

6 Applications

The pulse sequence in the half-loop experiments creates two spatially separated
wavepackets in the state $|2\rangle$ with zero relative momentum (left-most frame of Fig. 5a–
c). We now take advantage of the long free-fall period in the experiment (labelled TOF
in Fig. 2, i.e., after the "stopping pulse") to further manipulate these wavepackets
while they are allowed to expand and ultimately to overlap. The experiments are
based on imposing a differential time evolution between the two wavepackets, which
we measure as the interference patterns generated upon their recombination.

In particular, we create a "clock" state for each of the two wavepackets by first
applying an RF pulse that prepares the atoms in a superposition of two Zeeman
sublevels $|1\rangle$ and $|2\rangle$ whose coefficients depend on the Bloch sphere angles θ and ϕ.
This superposition state is a two-level system evolving with a known period, as in
the regular notion of an atomic clock. The RF pulse (duration T_R) controls the value
of $C = \sin\theta$, while a subsequent magnetic gradient pulse (duration T_G) controls
the value of $D_I = \sin(\phi/2)$ by changing the relative "tick" rate $\Delta\omega$ of the two

clock wavepackets, as illustrated in Fig. 5a–c. The quantities C and D_I describe the clock preparation quality and the ideal distinguishability between the two clock interferometer arms respectively, which we will find quantitatively useful in our discussion of clock complementarity [see Eqs. (4) and (5)]. We note that, although the magnetic gradient pulse applies a different SG force to each of the states within the clock, we have evaluated this effect for our experimental parameters and find that it is smaller than our experimental error bars ($\leq 2\%$, Supplementary Materials of [21]).

6.1 Clock Interferometery

Let us first discuss the motivation for clock interferometry [20]. Time in standard quantum mechanics (QM) is a global parameter, which cannot differ between paths. Hence, in standard interferometry [71], a height difference in a gravitational field between two paths would merely affect the relative phase of the clocks, shifting the interference pattern without degrading its visibility. In contrast, general relativity (GR) predicts that a clock must "tick" slower along the lower path; thus if the paths of a clock passing through an interferometer have different heights, a time differential between the paths will yield "which path" information and degrade the visibility of the interference pattern according to the quantum complementarity relation between the interferometric visibility and the distinguishability of the wavepackets [72]. Consequently, whereas standard interferometry may probe GR [73–75], clock interferometry probes the interplay of GR and QM. For example, loss of visibility because of a proper time lag would be evidence that gravitational effects contribute to decoherence and the emergence of a classical world [76].

Here we describe the use of this new tool—the clock interferometer—for its potential to investigate the role of time at the interface of QM and GR. Since the genuine GR proper time difference is too small to be measured with existing experimental technology, our experiments instead simulate the proper time difference between the clock wavepackets using magnetic gradients, thereby causing the clock wavepackets to "tick" at different rates. Our results in this proof-of-principle experiment show that the visibility does indeed oscillate as a function of the simulated proper time lag.

In the ultimate experiment, each part of the spatial superposition of a clock, located at different heights above Earth, would "tick" at different rates due to gravitational time dilation (so-called "red-shift"). We can easily calculate the proper time difference between two arms of the clock interferometer as a figure-of-merit for this effect. Using a first-order approximation of gravitational time dilation, and assuming a large separation between the arms of $\Delta h = 1\,\mathrm{m}$, an interferometer duration of $T = 1\,\mathrm{s}$ yields a proper time difference between the arms of only $\Delta\tau \simeq Tg\Delta h/c^2 \simeq 10^{-16}\,\mathrm{s}$. Such a small time difference means that a very accurate and fast-ticking clock must be sent through an interferometer with a large space-time area in order to observe the actual GR effect. Both requirements are beyond our current experimental capabilities. Our "synthetic" red-shift is created by applying an additional magnetic gradient

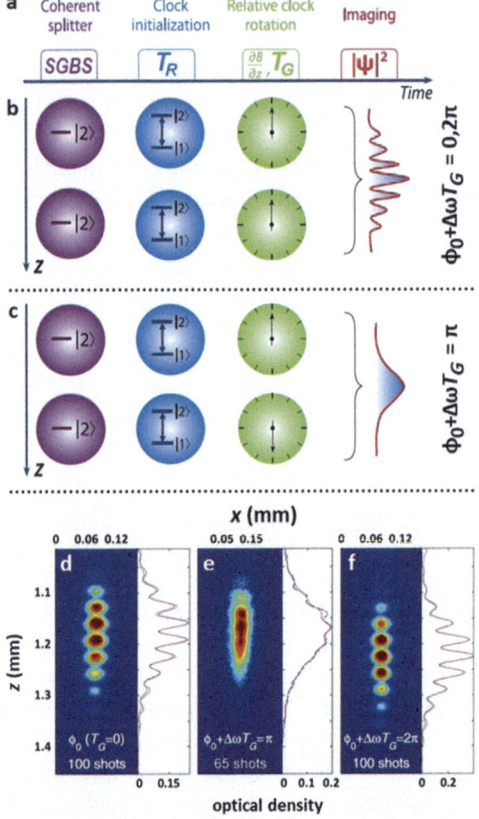

Fig. 5 Clock interferometry. **a** Timing sequence (not to scale): Following a coherent spatial splitting by the SGBS and a stopping pulse, the system consists of two wavepackets in the $|2\rangle$ state (separated along the z axis) with zero relative velocity, as in Sect. 4. The clock is then initialized with an RF pulse of duration T_R (usually a $\pi/2$ pulse, $T_R = 10\,\mu s$) after which the relative "tick" rate $\Delta\omega$ of the two clock wavepackets may be changed by applying a magnetic field gradient $\partial B/\partial z$ of duration T_G. Clock initialization occurs 1.5 ms after trap release, the first 0.9 ms of which is used for preparing the two wavepackets. The wavepackets are then allowed to expand and overlap and an image is taken. **b** Evolution in time, synchronized with (**a**). Each ball represents a clock wavepacket, where the hand represents its Bloch sphere phase ϕ_{BS}. When the clock reading (i.e., the position of the clock hand) in the two clock wavepackets is the same ($\phi_{BS} = \phi_0 + \Delta\omega T_G = 0, 2\pi$), fringe visibility is high. **c** When the clock reading is opposite (orthogonal, $\phi_{BS} = \phi_0 + \Delta\omega T_G = \pi$), there is no interference pattern. (d)-(f) Corresponding interference data of the two wavepackets, i.e., of the clock interfering with itself. All data samples are from consecutive measurements without any post-selection or post-correction. Single-shot patterns for $\phi_{BS} = \phi_0 + \Delta\omega T_G = \pi$ also show very low fringe visibility (see Fig. 2c of [20]). Adapted from [20] and reprinted with permission from AAAS; **e** is adapted from [64]

(of duration T_G) that causes the clock wavepackets to "tick" at different rates. We denote the "tick" rate difference by $\Delta\omega$.

Our results, some of which are presented in Fig. 5d–f, with more details in [20, 64], show that the relative rotation between the two clock wavepackets affects the interferometric visibility. In the most extreme case, when the two clock states are orthogonal, e.g., one in the state $\frac{1}{\sqrt{2}}(|1\rangle + |2\rangle)$ and the other in the state $\frac{1}{\sqrt{2}}(|1\rangle - |2\rangle)$, the visibility of the clock self-interference drops to near zero (Fig. 5e). By varying the duration of the magnetic gradient T_G and thereby scanning the differential rotation angle ϕ_{BS} between the two clock wavepackets, we show quantitatively that the visibility oscillates as a function of our "synthetic" red-shift with a period of $\Delta\omega T_G = 2\pi$ (Fig. 5d,f). As an additional test of the clock interferometer, we modulate its preparation by changing the duration of the clock initialization pulse T_R, which influences the relative populations of the two states composing the clock. This changes the state of the system from a no-clock state to a full-clock state in a continuous manner. The results show that the visibility behaves as expected in each case, further validating that it is the clock reading which is responsible for the oscillations in visibility that we observe as a function of T_R [20].

6.2 Clock Complementarity

These measurements of visibility may naturally be extended to study quantum complementarity for our self-interfering atomic clocks, which we again remark is at the interface of QM and GR. Our central consideration here is the inequality [77]

$$V^2 + D^2 \leq 1, \tag{3}$$

where V is the "visibility" of an interference pattern such as discussed throughout this review, and D is the "distinguishability" of the two paths of the interfering particle. The latter quantity can also be measured directly in the clock experiments by controlling the angle ϕ_{BS}, where $(\theta = \pi/2, \phi_{BS} = \Delta\omega T_G = \pi)$ prepares two perfectly distinguishable clocks such that $D = 1$ (Fig. 5e). A brief account of recent work theoretically and experimentally verifying this fundamental inequality is given by [21] and references therein.

It is important to investigate clock complementarity, particularly in view of recent theoretical work showing that spatial interferometers can be sensitive to a proper time lag between the paths [78] and speculation (see Table 1 in [72]) that the inequality of Eq. (3) may be broken such that $V^2 + D^2 > 1$ when the effect of gravity is dominant. Zhou et al. summarize the importance of this work as follows: "... on the one hand, if the 'ticking' rate of the clock depends on its path, then clock time provides which-path information and Eq. (3), developed in the framework of non-relativistic QM, must apply. Yet, on the other hand, gravitational time lags do not arise in non-relativistic QM, which is not covariant and therefore not consistent with the equivalence principle [79]. Hence our treatment of the clock superposition is a semiclassical extension of quantum mechanics to include gravitational red-shifts."

The experiments we conducted in [21] set out to test Eq. (3) quantitatively. Imperfect clock preparation (i.e., with $\theta \neq \pi/2$) reduces the measurable distinguishability D from its ideal value D_I as

$$D^2 = (C \cdot D_I)^2, \quad \text{where } C \equiv \sin\theta = 2\sqrt{P(1-P)} \tag{4}$$

and

$$D_I = |\sin(\Delta\phi/2)| \tag{5}$$

with P and $1 - P$ denoting the populations (occupation probabilities) of the two energy eigenstates of the clock and $\Delta\phi_{BS} \equiv \phi_{BS}^u - \phi_{BS}^d$, where u and d denote the upper and lower paths of the interferometer, respectively.

The experiment now has the task of measuring the three quantities V, D_I and C independently. We use the normalized visibility V_N as discussed in Sect. 4. We evaluate D_I independently by measuring the relative phases in two single-state interferometers, one for each of the two clock states, and we measure C, also independently, in a separate experiment by measuring P after the clock is initialized. Our results for these independently-measured quantities are shown in Fig. 6a, c, and e, where the results in (c) and (e) are based on analyzing the data in (b) and (d) respectively. We then combine these three quantities in the complementarity expression

$$(V_N)^2 + (C \cdot D_I)^2 \leq 1, \tag{6}$$

whereupon we see from Fig. 6f–g that the complementarity inequality [Eq. (3)] is indeed upheld for the clock wavepackets superposed on two paths through our SG interferometer.

While the relation in Eq. (6) is specific to clock complementarity, it is unusual in linking non-relativistic quantum mechanics with general relativity. A direct test of this complementarity relation will come when D_I reflects the gravitational red-shift between two paths which traverse different heights.

6.3 Geometric Phase

The geometric phase due to the evolution of the Hamiltonian is a central concept in quantum physics. In noncyclic evolutions, a proposition relates the geometric phase to the area bounded by the phase-space trajectory and the shortest geodesic connecting its end points [80–82]. The experimental demonstration of this geodesic rule proposition in different systems is of great interest, especially due to its potential use in quantum technology. Here, we report a novel experimental confirmation

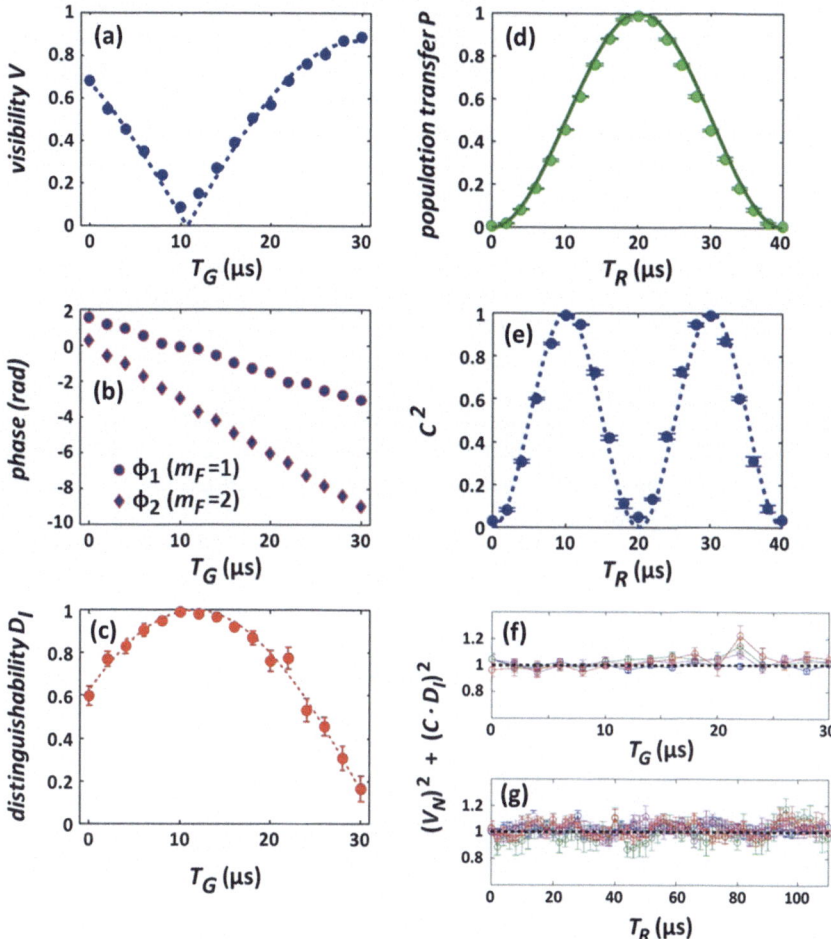

Fig. 6 Clock complementarity: **a–e** V, D_I, and C measured independently and **f–g** combined in the complementarity relations of Eqs. (3)-(6). **a** The visibility of an ideal clock ($C = 1$) interference pattern vs. T_G, fitted to $|\cos(\phi/2)|$; **b–c** the distinguishability is calculated from Eq. (5) using the difference in relative angles $\phi_2 - \phi_1$, each measured separately and shown in (**b**); and **d–e** the clock preparation quality C is calculated from Eq. (4) using the data in (**d**). Finally, **f** shows the combination of all three parameters $(V_N)^2 + (C \cdot D_I)^2$ for four values of C when D_I is scanned and **g** shows the same combination for four values of D_I when C is scanned. Only the data point in (**f**) for T_G near 22 μs differs from unity, due to a relatively large experimental error in measuring the interferometric phase. These data therefore verify clock complementarity. Adapted from [21] with permission © IOP Publishing & Deutsche Physikalische Gesellschaft, all rights reserved

of the geodesic rule for a noncyclic geometric phase by means of a spatial SU(2) matter-wave interferometer, demonstrating, with high precision, the predicted phase sign change and π jumps. We show the connection between our results and the Pancharatnam phase [83].

In the clock complementarity application just described, we scanned the third RF pulse (duration T_R) to vary the clock preparation parameter $C = \sin\theta$. In our case, a $\pi/2$ pulse typically corresponds to $T_R = 10\,\mu s$, so $T_R < 10\,\mu s$ places the Bloch vector in the northern hemisphere of the Bloch sphere with $P_1 < P_2$, while $10 < T_R < 30\,\mu s$ places the Bloch vector in the southern hemisphere ($P_1 > P_2$), i.e., the selected hemisphere is a periodic function of T_R such that an unequal superposition of $|1\rangle$ and $|2\rangle$ is created for each of the wavepackets unless θ lies on the equator. After applying this RF pulse (with some chosen duration T_R), we adjust the phase difference between the two superpositions by applying the third magnetic gradient pulse of duration T_G. This rotates the Bloch vectors along the latitude that was selected by the RF pulse to points A and B in the northern hemisphere (or A', B' in the southern hemisphere) as shown in Fig. 7a, thereby affecting the phase difference $\Delta\phi_{BS}$, which we simply call $\Delta\phi$ hereafter.

The two wavepackets are allowed to interfere as in our half-loop experiments, enabling a direct measurement of the geometric phase. As usual, we extract the "total" interference phase (labeled Φ) by fitting the fringe patterns using Eq. (1). For general values of θ and $\Delta\phi$ (i.e., after the application of both T_R and T_G), we write the total phase between the two wavepackets as [84]

$$\Phi = \arctan\left\{\frac{\sin^2(\theta/2)\,\sin(\Delta\phi)}{\cos^2(\theta/2) + \sin^2(\theta/2)\,\cos(\Delta\phi)}\right\}. \tag{7}$$

Measurements of Φ, combined with values of θ deduced independently from the relative populations of states $|1\rangle$ and $|2\rangle$, then allow us to fit $\Delta\phi$ to high precision as a function of T_G. These measurements verified that $\Delta\phi$ depends linearly on T_G, and we found that $\Delta\phi = \pi$ occurs at $T_G = 17\,\mu s$.

Figure 7b–c shows interference fringe images for this specific value of T_G, from which we extract the total phase as shown in Fig. 7e. We see immediately that this phase is independent of θ within each hemisphere, an observation we call "phase rigidity". Moreover, the (constant) phase in each hemisphere differs by π, which can also be deduced from the vanishing visibility shown in Fig. 7d in which we have combined the data from both hemispheres. Evidently, there is a sharp jump in the phase of the interference pattern as θ crosses the equator, as suggested by the singularities in Eq. (7) that arise when $\theta = \pi(n + 1/2)$ (integer n) and $\Delta\phi = \pi$.

To understand the non-cyclic geometric phase, we need to further examine the Bloch sphere. We see that the path from $A \to B$ along the latitude θ and returning along the geodesic (or "great-circle route") from $B \to A$ encloses an area [blue shading in Fig. 7(a)] in a counter-clockwise direction, whereas the corresponding path from $A' \to B'$ and back again in the southern hemisphere proceeds in a clockwise direction. One-half of this area is the "geometric phase" that we now wish to calculate.

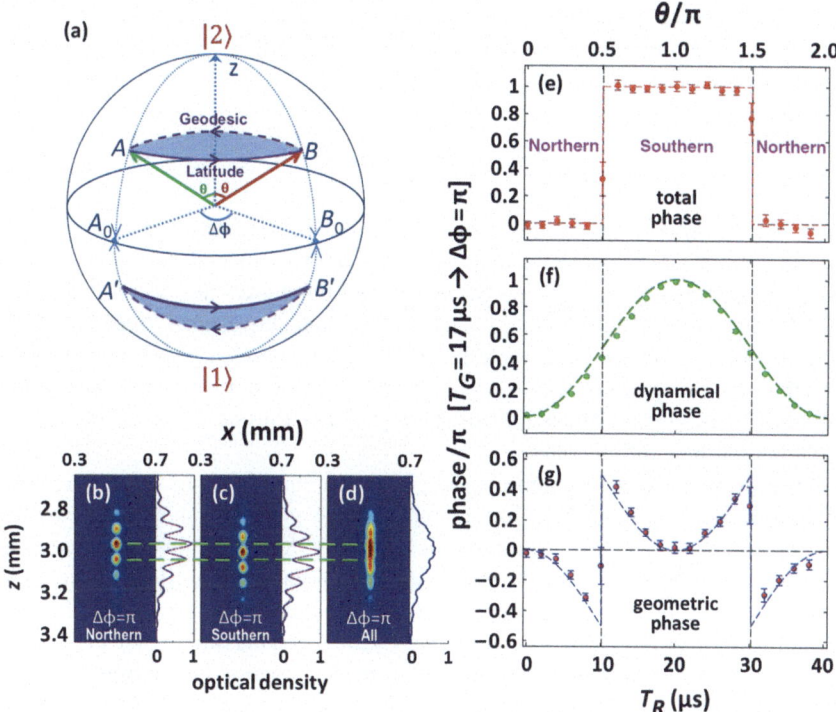

Fig. 7 Geometric phase. **a** Bloch sphere for the two wavepackets (green and red arrows labeled A and B, respectively) prepared by an RF pulse (duration T_R, rotation angle θ) and a subsequent magnetic gradient pulse (duration T_G) that induces a rotation angle difference of $\Delta\phi$. The rotation $A \to B$ lies along a constant latitude (solid purple line), while the returning geodesic $B \to A$ lies along the "great circle" curve (dashed purple line). Bloch vectors for corresponding wavepackets prepared in the southern hemisphere are shown as A' and B'. **b–c** Interference fringes generated by the half-loop SGI, averaged over a total of 330 experimental shots with varying $0 < T_R < 40\,\mu$s, while keeping a fixed value of $T_G = 17\,\mu$s (this value of T_G corresponds to $\Delta\phi = \pi$, see text). The dashed green lines show that the maxima in (**b**) lie exactly where the minima occur in (**c**), corresponding to Bloch vectors prepared in the northern and southern hemispheres, respectively. Adding all these interference patterns together in (**d**) shows near-zero visibility, i.e., they are completely out of phase. The fact that exactly the same pattern is observed while in the same hemisphere, independent of θ (duration of T_R), is called "phase rigidity". **e** Total phase extracted from the interference fringes measured as a function of the RF pulse duration (lower scale) and the corresponding latitude θ (upper scale). Phase rigidity is clearly visible. **f–g** Dynamical and geometric phases extracted from the data in (**e**) and independently measured values of θ and $\Delta\phi$ (see text). The range of T_R in (**e–g**) (T_G is fixed at $17\,\mu$s) corresponds to a full cycle from the northern hemisphere ($0 < T_R < 10\,\mu$s) through the southern hemisphere ($10 < T_R < 30\,\mu$s), and back to the north pole at $T_R = 40\,\mu$s. Adapted from [84] with permission © the authors, some rights reserved; exclusive licensee AAAS. CC BY 4.0

The total phase change Φ for closed paths like $A \rightarrow B \rightarrow A$ and $A' \rightarrow B' \rightarrow A'$ is a sum of two contributions, the dynamical phase Φ_D and the geometric phase Φ_G. The dynamical phase is given by [80]

$$\Phi_D = \frac{\Delta\phi}{2} \left(1 - \cos\theta\right), \tag{8}$$

which can be determined by measuring θ and $\Delta\phi$ independently. For the particular value of $\Delta\phi = \pi$ chosen as a sub-set of our experimental data, we are then able to present Φ_D in Fig. 7f. Finally, we subtract the phases Φ_D, as plotted in (f), from the total phases Φ plotted in (e) (which, as noted above, are extracted directly from the observed interference pattern) to obtain the phases Φ_G. Namely, we perform $\Phi - \Phi_D$ and get Φ_G, which is presented in Fig. 7g. Let us emphasize that the total phase Φ is also the Pancharatnam phase [83], and thus our experiment is also a direct measurement of this phase.

Our plot of Φ_G exactly confirms the prediction shown in Fig. 4d of [81], also reproduced as the dashed blue line in Fig. 7g. The predicted sign change as the latitude crosses the equator is clearly visible. The evident phase jump is due to the geodesic rule. When $\Delta\phi = \pi$, the geodesic must go through the Bloch sphere pole for any $\theta \neq \pi/2$. As the latitude approaches the equator (i.e., increasing θ), the blue area in Fig. 7a (twice Φ_G) grows continuously, reaching a maximum of π in the limit as $\theta \rightarrow \pi/2$. As the latitude crosses the equator, the geodesic jumps from one pole to the other pole, resulting in an instantaneous change of sign of this large area and a phase jump of π.

Finally, our approach for testing the geodesic rule is unique for the following reasons: (1) the use of a spatial interference pattern to determine the phase in a single experimental run (no need to scan any parameter to obtain the phase); (2) the use of a common phase reference for both hemispheres while scanning θ, enabling verification of the π phase jump and the sign change; and (3) obtaining the relative phase by allowing the two coherently-prepared wavepackets to expand in free flight and overlap, in contrast to previous atom interferometry studies that required additional manipulation of θ and $\Delta\phi$ to obtain interference.

6.4 T^3 Stern-Gerlach Interferometer

Here we consider an application of the full-loop SGI wherein we minimize the delay times between successive SG pulses as much as allowed by our electronics. In such an extreme scenario, it is expected that the phase accumulation will scale purely as T^3, thus representing the first pure interferometric measurement of the Kennard phase [19] predicted in 1927 [85, 86] (see also [87–89]). The theory for this experiment was done by the group of Wolfgang Schleich.

In order to describe the phase evolution of an atom moving in a time- and state-dependent linear potential, it is sufficient [90] to know the two time-dependent forces $\mathbf{F}_u \equiv F_u(t)\mathbf{e}_z$ and $\mathbf{F}_l \equiv F_l(t)\mathbf{e}_z$ acting on the atom along the upper and lower branches, respectively, of the interferometer shown in Fig. 8, where z is the axis of gravity, the axis of our longitudinal interferometer, and also the axis of our magnetic gradients.

In the present case, these forces comprise the gravitational force $F_g = mg$ and the state-dependent magnetic forces $\mathbf{F}_i = -\mu_B (g_F)_i (m_F)_i (\partial|\mathbf{B}|/\partial z)\,\mathbf{e}_z$, $(i = 1, 2)$:

$$F_{u,l}(t) = F_g + F_{2,1}\mathcal{F}(t), \tag{9}$$

where μ_B, g_F, and m_F are the Bohr magneton, the Landé factor, and the projection of the angular momentum on the quantization (y-)axis, respectively. The function $\mathcal{F}(t)$ provides the time-dependent modulation shown as the orange curve in Fig. 8(b):

$$\mathcal{F}(t) \equiv \Theta(t) - \Theta(t - T_1) - \Theta(t - T_1 - T_d) + \Theta(t - 3T_1 - T_d)$$
$$+ \Theta(t - 3T_1 - 2T_d) - \Theta(t - 4T_1 - 2T_d). \tag{10}$$

Here we are using the Heaviside step function $\Theta(t)$ and we are assuming that the duration of each gradient pulse is identical, i.e., $T_{2,3,4} = T_1$, as are the two delay times, $T_{d_1,d_2} = T_d$. We are also careful to ensure experimentally that the magnetic field is linear in the vicinity of the atoms and acts only along the vertical (z-)axis.[3]

As in the full-loop SGI experiments of Sect. 5, we measure the spin population in state $|1\rangle$ which, in this configuration, is a periodic function of the interferometer phase [91].

$$P_1 = \frac{1}{2}\left[1 - \cos(\delta\Phi + \varphi_0)\right], \tag{11}$$

where

$$\delta\Phi = \frac{1}{\hbar}\int_0^T dt\, \bar{F}(t)\delta z(t), \tag{12}$$

with the total time $T \equiv 4T_1 + 2T_d$. Note that the interferometer will be closed in both position and momentum provided that the differences

$$\delta p(t) = \int_0^t d\tau\, \delta F(\tau) \tag{13}$$

---

[3]Magnetic field linearity is ensured to a good approximation by the three-wire chip design and by carefully positioning the atoms very close to the center of the quadrupole field that they produce, as well as by the short distances that the atomic wavepackets travel ($\sim 1\,\mu$m) compared to their distance from the chip ($\sim 100\,\mu$m). We also adjust the duration of T_4 slightly, relative to T_1, to better optimize the visibility and account for any residual non-linearity. See [19, 68] for further details.

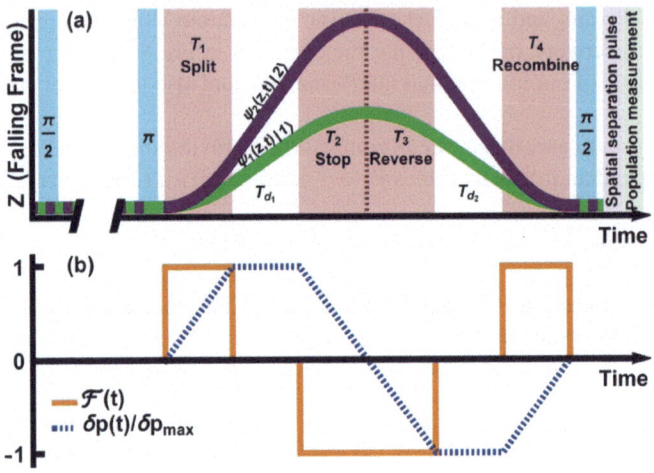

Fig. 8 Pulse sequence of our longitudinal T^3-SGI (not to scale). **a** Trajectories of the atomic wavepackets with internal states $|1\rangle$ (green curve) and $|2\rangle$ (purple curve). Here we are using the freely-falling reference frame (gravity upwards), distinct from the center-of-mass reference frame used for Figs. 2 and 4. Also shown are the RF (blue) and magnetic gradient (red) pulses. The magnetic field gradients result in a state-dependent force along the z-direction while the strong bias magnetic field along the y-direction defines the quantization axis and ensures a two-level system. **b** Time dependence of the relative force $\mathcal{F} = \mathcal{F}(t)$ [orange curve, Eq. (10)] and the corresponding relative momentum $\delta p(t)$ [blue dashed curve, Eq. (13)] between the wavepackets moving along the two interferometer paths. In the experiment, we achieved the maximal separation $\Delta z_{max} = 1.2\,\mu$m in position and $\Delta p_{max}/m_{Rb} = 17\,$mm/s in velocity. Reprinted from [19] with permission © (2019) by the American Physical Society

and

$$\delta z(t) = \frac{1}{m} \int_0^t d\tau\, \delta F(\tau)(t-\tau) \tag{14}$$

both vanish at $t = T$. Here φ_0 is a constant phase taking into account possible technical misalignment, while $\bar{F}(t) \equiv [F_u(t) + F_l(t)]/2 = F_g + \frac{1}{2}(F_1 + F_2)\mathcal{F}(t)$ and $\delta F(t) \equiv F_u(t) - F_l(t) = (F_2 - F_1)\mathcal{F}(t)$ are the mean and relative forces respectively. From Eq. (11) we finally obtain

$$\delta\Phi = \frac{mga_B}{\hbar}\left(\frac{\mu_1 - \mu_2}{\mu_B}\right)\left(2T_1^3 + 3T_1^2 T_d + T_1 T_d^2\right)$$
$$+ \frac{ma_B^2}{\hbar}\left(\frac{\mu_1^2 - \mu_2^2}{\mu_B^2}\right)\left(\frac{2}{3}T_1^3 + T_1^2 T_d\right), \tag{15}$$

with $a_B \equiv \mu_B \nabla B/m$ being the magnetic acceleration.

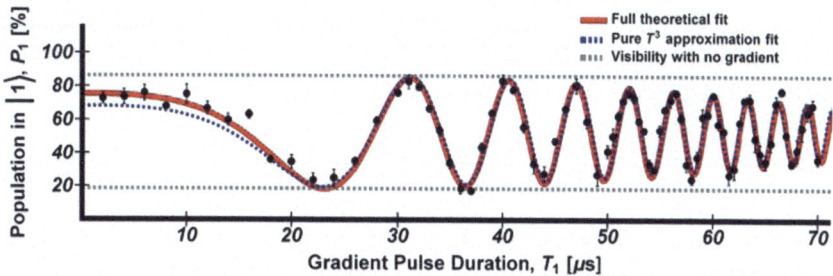

Fig. 9 Measurement of the cubic phase with the T^3-SGI presented in Fig. 8. The solid red line represents a fit based on Eq. (15), as described in the text. The dashed blue line is a fit with $T_d = 0$, showing that the interferometer phase scales purely as T_1^3 for $T_1 \gtrsim 20\,\mu s$. The visibility drops from 68% to 32% over 70 μs with a decay time of 75 μs. This reduction results from inaccuracies in recombining the two interferometer paths. The dashed gray horizontal lines depict the maximal and minimal values of the population P_1 measured independently without magnetic field gradients. Reprinted from [19] with permission © (2019) by the American Physical Society

As sketched in Fig. 8, the experiment begins with an on-resonance RF $\pi/2$-pulse that transfers the initially prepared internal atomic state $|2\rangle$ to an equal superposition, $\frac{1}{\sqrt{2}}(|1\rangle + |2\rangle)$. This $\pi/2$ pulse is applied 1 ms after the atoms are released from the trap in which they were prepared, in order to ensure that the trapping fields are fully quenched. Following a free-fall time of 400 μs (the first "dark time"), we apply an RF π-pulse that flips the atomic state to $\frac{1}{\sqrt{2}}(|1\rangle - |2\rangle)$. After a second dark time of another 400 μs, a second $\pi/2$ pulse completes the spin-echo sequence. The π-pulse inverts the population between the two states of the system thereby allowing any time-independent phase shift accumulated during the first dark time to be canceled in the second dark time. The experiment is completed by applying a magnetic gradient to separate the spin populations and a subsequent pulse of the detection laser to image both states simultaneously.

As with all our previous full-loop experiments, the four magnetic field gradient pulses are produced by current-carrying wires on the atom chip. This magnetic pulse sequence sends the spin states $|1\rangle$ and $|2\rangle$ along different trajectories in the SGI and ultimately closes the interferometer in both momentum and position. Careful calibration measurements verified that reversing the wire currents (the current flow is reversed during T_2 and T_3 relative to T_1 and T_4) provides magnetic accelerations that are equal in magnitude (but opposite in sign) to within our experimental uncertainty of $< 1\%$.

The experimental data shown in Fig. 9 are measured as a function of the time $2 < T_1 < 70\,\mu s$. From Eq. (15), it is apparent that the T^3 dependence will be most evident if $T_d \ll T_1$, which is satisfied for most of the experimental range by using a fixed experimental value of $T_d = 2.6\,\mu s$ (limited by the speed of our electronic circuits). Note that $T_1 \lesssim 100\,\mu s$ is limited by the duration of the second dark time.

The experimental data (dots) agree very well with the theory (solid red line) based on Eq. (15), where the fitting parameters are the magnetic acceleration a_B as well as the decay constant of the visibility and a constant phase φ_0. The dashed blue line is obtained by setting $T_d = 0$, leading to a pure T_1^3 scaling that is indistinguishable from the full theoretical fit for $T_1 \gtrsim 20\,\mu s$:

$$\delta\Phi^{(T^3)} \cong \frac{m a_B}{32\hbar}\left(\frac{\mu_1 - \mu_2}{\mu_B}\right)\left(g + \frac{\mu_1 + \mu_2}{3\mu_B}a_B\right)T^3. \tag{16}$$

The maximum visibility displayed by the gray lines is first measured by performing only the RF spin-echo sequence ($\pi/2 - \pi - \pi/2$) without the magnetic field gradients and changing the phase of the second $\pi/2$ pulse. The maximal visibility is limited by imperfections in the RF pulses. As discussed above, utilizing an echo sequence allows us to cancel out contributions to the interferometer phase from the bias magnetic field, and to increase the coherence time.

The excellent fit to these data allows a precise determination of the magnetic field acceleration, $a_B^{fit} = 246.97 \pm 0.09\,m/s^2$. Separate measurements were used to independently determine the magnetic field gradient using time-of-flight (TOF) techniques, which gave a value of $a_B^{TOF} = 249 \pm 2\,m/s^2$.[4] While these measurements agree with one another, the difference in measurement errors clearly shows that our T^3-SGI provides a much more precise measurement of the magnetic field gradient.

Let us now consider the case when $T_1 \ll T_d$, such that during T_d the relative momentum $\delta p_0 \equiv m a_B T_1 (\mu_1 - \mu_2)/\mu_B$ between the paths is kept constant, i.e., we take the magnetic field gradient pulses to be delta functions.

In this limit the interferometer phase from Eq. (15) becomes

$$\delta\Phi^{(T^2)} \cong \frac{\delta p_0}{4\hbar} g T^2, \tag{17}$$

scaling quadratically with the total time $T \cong 2T_d$, since we now maintain a piecewise constant momentum difference between the two arms. This is similar to the T^2-SGI [18] or the Kasevich-Chu interferometer [90], although the momentum transfer δp_0 is provided by the magnetic field gradient in the case of the T^2-SGI, rather than by the laser light pulse.

We conclude our discussion of this unique T^3 interferometer by comparing the scaling of the interferometer phases $\delta\Phi^{(T^3)}$ and $\delta\Phi^{(T^2)}$ with the total interferometer time T, as given by Eqs. (16) and (17) respectively. The data in Fig. 10 are taken from Fig. 9 and from our T^2-SGI (when experimentally realizing the condition $T_1 \ll T_d$), showing clearly that the T^3-SGI significantly outperforms the T^2-SGI with respect to total phase accumulation, even though the latter can currently operate for total times T up to three times longer than the former. Finally, let us briefly note that this T^3 realization has already been coined a proof-of-principle experiment for testing the quantum nature of gravity [49].

---

[4]These values for a_B^{fit} and a_B^{TOF} are different from those presented in [19] due to a different fitting procedure used there. A full analysis and fitting procedures are presented in the Appendices of [68].

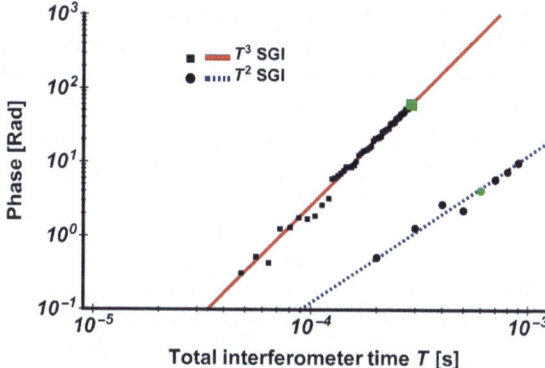

Fig. 10 Scaling of the interferometer phases $\delta\Phi^{(T^3)}$ [squares, Eq.(16)] and $\delta\Phi^{(T^2)}$ [circles, Eq. (17)], as functions of the total interferometer time T. The solid red line is fitted to our data for the T^3-SGI and the dashed blue line is fitted to our T^2-SGI data when experimentally realizing the condition $T_1 \ll T_d$. In its current configuration with $T_{max} = 285\,\mu s$, the phase of the T^3-SGI is almost six times larger than the phase of the best T^2-SGI, even though the magnetic field gradients and the maximal time $T_{max} = 924\,\mu s$ are larger than those of the T^3-SGI by factors of 2.3 and 3.2, respectively. For reference, the green square and green dot represent data for which the observed visibility is $\approx 30\%$ for both the T^3-SGI and T^2-SGI respectively. Adapted from [19] with permission © (2019) by the American Physical Society

Looking into the future, we may ask if one may extend the T^3 scaling to yet higher powers of time. In the Ramsey-Bordé interferometer [92], the phase shift that scales linearly with the interferometer time T originates from a constant position difference between two paths during most of this time. In the Kasevich-Chu interferometer [93, 94], the quadratic scaling of the phase with time is caused by a piecewise constant velocity difference, while a piecewise constant acceleration difference between the two paths results in the cubic phase scaling $\delta\Phi \propto T^3$, as presented above.

One can generalize this idea to achieve any arbitrary phase scaling by having a piecewise difference in the nth derivative of the position difference between the two paths. By designing an interferometer sequence consisting of pulses with a higher-order time-dependence of the forces, combined with careful choices of the relative signs and durations of the pulses, the total phase can be made to scale with the interferometer time as T^{n+1} for any chosen $n > 2$.

7   Outlook

7.1   SGI with Single Ions

The discovery of the Stern-Gerlach effect led to lively discussions early in the quantum era regarding the possibility of measuring an analogous effect for the electron itself (see e.g., [95, 96]). The Lorentz force adds the complicating factor of a purely classical deflection of the electron beam that would smear out any expected SG splitting. Here we summarize a generalized semiclassical discussion for any charged particle of mass m and charge e from [62] (though with the co-ordinate system in Table 14.1). Assuming a beam momentum p_x and a transverse beam spatial width Δz, we calculate the spread of the Lorentz force ΔF_L due to a transverse magnetic gradient B' as

$$\Delta F_L = \frac{e}{m} \, p_x B' \Delta z. \tag{18}$$

Since the beam would be well collimated, $\Delta p_z < p_x$, so

$$\Delta F_L > \frac{e}{m} \, B' \Delta p_z \Delta z \geq \frac{e\hbar}{2m} \, B' = \frac{m_e}{m} \times \left(\frac{e\hbar}{2m_e} \, B' \right) = \frac{m_e}{m} \, F_{SG}, \tag{19}$$

where the second inequality uses the uncertainty principle and we have introduced the electron mass m_e to relate F_L to the Stern-Gerlach force F_{SG}.

The spatially inhomogeneous Lorentz broadening is therefore larger than the SG splitting for electrons ($m = m_e$), at least in this semiclassical analysis [97], and this lively controversy has continued for decades though, as far as we know, without any conclusive experimental tests for electrons or for any other charged particles (see [98–100] for reviews of the early history of this issue and recent perspectives). In contrast, Eq. (19) shows no such fundamental problem if we take ions such that $m_e/m < 10^{-3}$, thereby motivating our proposals, including chip-based designs, for measurements using very high-resolution single ion-on-demand sources that have recently been developed using ultra-cold ion traps [61, 101]. As a practical matter, we note that a suitable ion chip could be fabricated and implemented either based on an array of current-carrying wires as analyzed in [62] or on a magnetized microstructure like those implemented in [102, 103].

Although we did not extend our analysis to include the coherence of the spin-dependent splitting, the suggested ion-SG beam splitter may form a basic building block of free-space interferometric devices for charged particles. Here we quote from our collaborative work with Henkel et al. [62]. In addition to measuring the coherence of spin splitting as in the "Humpty-Dumpty" effect (see Sect. 1), we anticipate that such a device could provide new insights concerning the fundamental question of whether and where in the SG device a spin measurement takes place. The ion interference would also be sensitive to Aharonov-Bohm phase shifts arising from the electromagnetic gauge field. The ion source would be a truly single-particle device [61] and eliminate certain problems arising from particle interactions in high-density sources of neutral bosons [104].

Such single-ion SG devices would open the door for a wide spectrum of fundamental experiments, probing for example weak measurements and Bohmian trajectories. The strong electric interactions may also be used, for example, to entangle the single ion with a solid-state quantum device (an electron in a quantum dot or on a Coulomb island, or a qubit flux gate). This type of interferometer may lead to new sensing capabilities [105]: one of the two ion wavepackets is expected to pass tens to hundreds of nanometers above a surface (in the chip configuration of our proposal [62]) and may probe van der Waals and Casimir-Polder forces, as well as patch potentials. The latter are very important as they are believed to give rise to the anomalous heating observed in miniaturized ion traps [106]. Due to the short distances between the ions and the surface, the device may also be able to sense the gravitational force on small scales [107]. Finally, such a single-ion interferometer may enable searches for exotic physics. These include spontaneous collapse models, the fifth force from a nearby surface, the self-charge interaction between the two ion wavepackets, and so on. Eventually, one may be able to realize a double SG-splitter with different orientations, as originally attempted by Stern, Segrè and co-workers [108, 109], in order to test ideas like the Bohm-Bub non-local hidden variable theory [110–112], or ideas on deterministic quantum mechanics (see, e.g., [113]). Since ions may form the basis of extremely accurate clocks, an ion-SG device would enable clock interferometry at a level sensitive to the Earth's gravitational red-shift (see the proof-of-principle experiments with neutral atoms in [20, 21]). This has important implications for studying the interface between quantum mechanics and general relativity.

7.2 SGI with Massive Objects

The main focus of our future efforts will be to realize an SGI with massive objects. The idea of using the SG interferometer, with a macroscopic object as a probe for gravity, has been detailed in several studies [46, 47, 114, 115] describing a wide range of experiments from the detection of gravitational waves to tests of the quantum nature of gravity. Here we envision using a macroscopic body in the full-loop SGI. We anticipate utilizing spin population oscillations as our interference observable rather than spatial fringes, i.e., density modulations. This observable, as demonstrated in the atomic SGI described above, is advantageous because there is no requirement for long evolution times in order to allow the spatial fringes to develop, nor is high-resolution imaging needed to resolve the spatial fringes. Let us note that there are other proposals to realize a spatial superposition of macroscopic objects [70, 116].

As a specific example, let us consider a solid object comprising $10^6 - 10^{10}$ atoms with a single spin embedded in the solid lattice, e.g., a nano-diamond with a single NV center. Let us first emphasize that even prior to any probing of gravity, a successful SGI will already achieve at least 3 orders of magnitude more atoms than the state-of-the-art in macroscopic-object interferometry [60], thus contributing novel insight to the foundations of quantum mechanics. Another contribution to the foundations of quantum mechanics would be the ability to test continuous spontaneous localization (CSL) models (e.g., [117] and references therein).

When probing gravity, the first contribution of such a massive-object SGI would simply be to measure little g. As the phase is accumulated linearly with the mass, a massive-object interferometer is expected to have much more sensitivity to g than atomic interferometers being used currently (assuming of course that all other features are comparable). This is also a method to verify that a massive-object superposition can be created [114, 118, 119]. A second contribution would measure gravity at short distances, since the massive object may be brought close to a surface while in one of the SGI paths, thus enabling probes of the fifth force. Once the SG technology allows the use of large masses, a third contribution will be the testing of hypotheses concerning gravity self-interaction [48, 116], and once large-area interferometry is also enabled, a fourth contribution would be to detect gravitational waves [46]. Finally, placing two such SGIs in parallel next to each other will enable probes of the quantum nature of gravity [47, 120]. Let us emphasize that, although high accelerations may be obtained with multiple spins, we intend to focus on the case of a macroscopic object with a single spin, since the observable of such a quantum-gravity experiment is entanglement, and averaging over many spins may wash out the signal.

To avoid the hindering consequences of the HD effect, one must ensure that the experimental accuracy of the recombination, as discussed in Sect. 5, will be better than the coherence length. Obviously it is very hard to achieve a large coherence length for a massive object, but recent experimental numbers and estimates seem to indicate that this is feasible. Another crucial problem is the coherence time. A massive object has a huge cross section for interacting with the environment (e.g., background gas), but the extremely short interferometer times, as discussed in this review, seem to serve as a protective shield suppressing decoherence. We are currently a detailed account of these considerations [121].

Acknowledgements We wish to warmly thank all the members—past and present—of the Atom Chip Group at Ben-Gurion University of the Negev, and the team of the BGU nano-fabrication facility for designing and fabricating innovative high-quality chips for our laboratory and for others around the world. The work at BGU described in this review was funded in part by the Israel Science Foundation (1381/13 and 1314/19), the EC "MatterWave" consortium (FP7-ICT-601180), and the German DFG through the DIP program (FO 703/2-1). We also acknowledge support from the PBC program for outstanding postdoctoral researchers of the Israeli Council for Higher Education and from the Ministry of Immigrant Absorption (Israel).
Disclosure Statement The authors declare that they have no competing financial interests.

References

1. W. Gerlach, O. Stern, Der experimentelle Nachweis der Richtungsquantelung im Magnetfeld. Z. Physik **9**, 349 (1922). https://doi.org/10.1007/BF01326983
2. B. Friedrich, D. Herschbach, H. Schmidt-Böcking, J.P. Toennies, An international symposium (Wilhelm and Else Heraeus Seminar #702) marked the centennial of Otto Stern's first molecular beam experiment and the thriving of atomic physics; a European Physical Society Historic Site Was Inaugurated. Front. Phys. **7**, 208 (2019). https://doi.org/10.3389/fphy.2019.00208
3. B. Friedrich, H. Schmidt-Böcking. Otto Stern's molecular beam method and its impact on quantum physics (2020)
4. D. Herschbach, Molecular beams entwined with quantum theory: a bouquet for Max Planck. Ann. Phys. (Leipzig) **10**, 163 (2001). https://doi.org/10.1002/1521-3889(200102)10:1/2<163::AID-ANDP163>3.0.CO;2-W

5. B. Friedrich, D. Herschbach, Stern and Gerlach: how a bad cigar helped reorient atomic physics. Phys. Today **56**, 53 (2003). https://doi.org/10.1063/1.1650229
6. M. Keil, O. Amit, S. Zhou, D. Groswasser, Y. Japha, R. Folman, Fifteen years of cold matter on the atom chip: promise, realizations, and prospects. J. Mod. Opt. **63**, 1840 (2016). https://doi.org/10.1080/09500340.2016.1178820
7. J. Reichel, W. Hänsel, T.W. Hänsch, Atomic micromanipulation with magnetic surface traps. Phys. Rev. Lett. **83**, 3398 (1999). https://doi.org/10.1103/PhysRevLett.83.3398
8. R. Folman, P. Krüger, D. Cassettari, B. Hessmo, T. Maier, J. Schmiedmayer, Controlling cold atoms using nanofabricated surfaces: atom chips. Phys. Rev. Lett. **84**, 4749 (2000). https://doi.org/10.1103/PhysRevLett.84.4749
9. N.H. Dekker, C.S. Lee, V. Lorent, J.H. Thywissen, S.P. Smith, M. Drndić, R.M. Westervelt, M. Prentiss, Guiding neutral atoms on a chip. Phys. Rev. Lett. **84**, 1124 (2000). https://doi.org/10.1103/PhysRevLett.84.1124
10. R. Folman, P. Krüger, J. Schmiedmayer, J. Denschlag, C. Henkel, Microscopic atom optics: from wires to an atom chip. Adv. At. Mol. Opt. Phys. **48**, 263 (2002). https://doi.org/10.1016/S1049-250X(02)80011-8
11. J. Reichel, Microchip traps and Bose-Einstein condensation. Appl. Phys. B **74**, 469 (2002). https://doi.org/10.1007/s003400200861
12. J. Fortágh, C. Zimmermann, Magnetic microtraps for ultracold atoms. Rev. Mod. Phys. **79**, 235 (2007). https://doi.org/10.1103/RevModPhys.79.235
13. *Atom Chips,* J. Reichel, V. Vuletić, eds. (Wiley-VCH, Hoboken, NJ, 2011). https://doi.org/10.1002/9783527633357
14. R. Folman, Material science for quantum computing with atom chips, in *Special Issue on Neutral Particles,*, ed. by R. Folman. Quantum Inf. Process. **10**, 995 (2011), https://doi.org/10.1007/s11128-011-0311-5
15. https://in.bgu.ac.il/en/nano-fab
16. S. Machluf, Y. Japha, R. Folman, Coherent Stern-Gerlach momentum splitting on an atom chip. Nature Commun. **4**, 2424 (2013). https://doi.org/10.1038/ncomms3424
17. Y. Margalit, Z. Zhou, S. Machluf, Y. Japha, S. Moukouri, R. Folman, Analysis of a high-stability Stern-Gerlach spatial fringe interferometer. New J. Phys. **21**, 073040 (2019). https://doi.org/10.1088/1367-2630/ab2fdc
18. Y. Margalit, Z. Zhou, O. Dobkowski, Y. Japha, D. Rohrlich, S. Moukouri, R. Folman, Realization of a complete Stern-Gerlach interferometer. arXiv:1801.02708v2 (2018)
19. O. Amit, Y. Margalit, O. Dobkowski, Z. Zhou, Y. Japha, M. Zimmermann, M.A. Efremov, F.A. Narducci, E.M. Rasel, W.P. Schleich, R. Folman, T^3 Stern-Gerlach matter-wave interferometer. Phys. Rev. Lett. **123**, 083601 (2019). https://doi.org/10.1103/PhysRevLett.123.083601
20. Y. Margalit, Z. Zhou, S. Machluf, D. Rohrlich, Y. Japha, R. Folman, A self-interfering clock as a which-path witness. Science **349**, 1205 (2015). https://doi.org/10.1126/science.aac6498
21. Z. Zhou, Y. Margalit, D. Rohrlich, Y. Japha, R. Folman, Quantum complementarity of clocks in the context of general relativity. Class. Quantum Grav. **35**, 185003 (2018). https://doi.org/10.1088/1361-6382/aad56b
22. A.D. Cronin, J. Schmiedmayer, D.E. Pritchard, Optics and interferometry with atoms and molecules. Rev. Mod. Phys. **81**, 1051 (2009). https://doi.org/10.1103/RevModPhys.81.1051
23. W. Heisenberg. *Die Physikalischen Prinzipien der Quantentheorie.* (S. Hirzel: Leipzig 1930); *The Physical Principles of the Quantum Theory* transl. by C. Eckart, F. C. Hoyt (Dover, Mineola, NY, 1950)
24. D. Bohm, *Quantum Theory* (Prentice-Hall, Englewood Cliffs, 1951), pp. 604–605
25. E.P. Wigner, The problem of measurement. Am. J. Phys. **31**, 6 (1963). https://doi.org/10.1119/1.1969254
26. H. J. Briegel, B.-G. Englert, M.O. Scully, H. Walther, Atom interferometry and the quantum theory of measurement, in *Atom Interferometry*, P.R. Berman, ed. (Academic Press, New York, 1997), p. 240

27. This popular version of the English nursery rhyme appears in https://en.wikipedia.org/wiki/ Humpty_Dumpty
28. B.-G. Englert, J. Schwinger, M.O. Scully, Is spin coherence like Humpty-Dumpty? I. Simplified treatment. Found. Phys. **18**, 1045 (1988). https://doi.org/10.1007/BF01909939
29. J. Schwinger, M.O. Scully, B.-G. Englert, Is spin coherence like Humpty-Dumpty? II. General theory. Z. Phys. D **10**, 135 (1988). https://doi.org/10.1007/BF01384847
30. M.O. Scully, B.-G. Englert, J. Schwinger, Spin coherence and Humpty-Dumpty. III. The effects of observation. Phys. Rev. A **40**, 1775 (1989). https://doi.org/10.1103/PhysRevA.40. 1775
31. B.-G. Englert, Time reversal symmetry and Humpty-Dumpty. Z. Naturforsch A **52**, 13 (1997). https://doi.org/10.1515/zna-1997-1-206
32. M. O. Scully, Jr. W.E. Lamb, A. Barut, On the theory of the Stern-Gerlach apparatus. Found. Phys. **17**, 575 (1987).https://doi.org/10.1007/BF01882788
33. T.R. de Oliveira, A.O. Caldeira, Dissipative Stern-Gerlach recombination experiment. Phys. Rev. A **73**, 042502 (2006). https://doi.org/10.1103/PhysRevA.73.042502
34. M. Devereux, Reduction of the atomic wavefunction in the Stern-Gerlach magnetic field. Can. J. Phys. **93**, 1382 (2015). https://doi.org/10.1139/cjp-2015-0031
35. J. Robert, C. Miniatura, S. Le Boiteux, J. Reinhardt, V. Bocvarski, J. Baudon, Atomic interferometry with metastable hydrogen atoms. Europhys. Lett. **16**, 29 (1991). https://doi.org/10. 1209/0295-5075/16/1/006
36. C. Miniatura, F. Perales, G. Vassilev, J. Reinhardt, J. Robert, J. Baudon, A longitudinal Stern-Gerlach interferometer: the "beaded" atom. J. Phys. II **1**, 425 (1991). https://doi.org/10.1051/ jp2:1991177
37. C. Miniatura, J. Robert, S. Le Boiteux, J. Reinhardt, J. Baudon, A longitudinal Stern-Gerlach atomic interferometer. App. Phys. B **54**, 347 (1992). https://doi.org/10.1007/BF00325378
38. J. Robert, C. Miniatura, O. Gorceix, S. Le Boiteux, V. Lorent, J. Reinhardt, J. Baudon, Atomic quantum phase studies with a longitudinal Stern-Gerlach interferometer. J. Phys. II **11**, 601 (1992). https://doi.org/10.1051/jp2:1992155
39. C. Miniatura, J. Robert, O. Gorceix, V. Lorent, S. Le Boiteux, J. Reinhardt, J. Baudon, Atomic interferences and the topological phase. Phys. Rev. Lett. **69**, 261 (1992). https://doi.org/10. 1103/PhysRevLett.69.261
40. S. Nic Chormaic, V. Wiedemann, C. Miniatura, J. Robert, S. Le Boiteux, V. Lorent, O. Gorceix, S. Feron, J. Reinhardt, J. Baudon, Longitudinal Stern-Gerlach atomic interferometry using velocity selected atomic beams. J. Phys. B **26**, 1271 (1993). https://doi.org/10.1088/0953-4075/26/7/011
41. J. Baudon, R. Mathevet, J. Robert, Atomic interferometry. J. Phys. B **32**, R173 (1999). https:// doi.org/10.1088/0953-4075/32/15/201
42. M. Boustimi, V. Bocvarski, K. Brodsky, F. Perales, J. Baudon, J. Robert, Atomic interference patterns in the transverse plane. Phys. Rev. A **61**, 033602 (2000). https://doi.org/10.1103/ PhysRevA.61.033602
43. E. Maréchal, R. Long, T. Miossec, J.-L. Bossennec, R. Barbé, J.-C. Keller, O. Gorceix, Atomic spatial coherence monitoring and engineering with magnetic fields. Phys. Rev. A **62**, 53603 (2000). https://doi.org/10.1103/PhysRevA.62.053603
44. B. Viaris de Lesegno, J.C. Karam, M. Boustimi, F. Perales, C. Mainos, J. Reinhardt, J. Baudon, V. Bocvarski, D. Grancharova, F. Pereira Dos Santos, T. Durt, H. Haberland J. Robert, Stern Gerlach interferometry with metastable argon atoms: an immaterial mask modulating the profile of a supersonic beam. Eur. Phys. J. D **23**,25 (2003). https://doi.org/10.1140/epjd/ e2003-00023-y
45. K. Rubin, M. Eminyan, F. Perales, R. Mathevet, K. Brodsky, B. Viaris de Lesegno, J. Reinhardt, M. Boustimi, J. Baudon, J.-C. Karam, J. Robert, Atom interferometer using two Stern-Gerlach magnets. Laser Phys. Lett. **1**, 184 (2004). https://doi.org/10.1002/lapl.200310047
46. R.J. Marshman, A. Mazumdar, G.W. Morley, P.F. Barker, S. Hoekstra, S. Bose, Mesoscopic interference for metric and curvature (MIMAC) & gravitational wave detection. New J. Phys. **22**, 083012 (2020). https://doi.org/10.1088/1367-2630/ab9f6c

47. S. Bose, A. Mazumdar, G.W. Morley, H. Ulbricht, M. Toroš, M. Paternostro, A.A. Geraci, P.F. Barker, M.S. Kim, G. Milburn, Spin entanglement witness for quantum gravity. Phys. Rev. Lett. **119**, 240401 (2017). https://doi.org/10.1103/PhysRevLett.119.240401

48. M. Hatifi, T. Durt, Revealing self-gravity in a Stern-Gerlach Humpty-Dumpty experiment. arXiv:2006.07420 (2020)

49. C. Marletto, V. Vedral, On the testability of the equivalence principle as a gauge principle detecting the gravitational t^3 phase. Front. Phys. **8**, 176 (2020). https://doi.org/10.3389/fphy.2020.00176

50. M. Gebbe, S. Abend, J.-N. Siemß, M. Gersemann, H. Ahlers, H.Müntinga, S. Herrmann, N. Gaaloul, C. Schubert, K. Hammerer, C. Lämmerzahl, W. Ertmer, E.M. Rasel, Twin-lattice atom interferometry. arXiv:1907.08416v1 (2019)

51. B. Canuel, S. Abend, P. Amaro-Seoane, F. Badaracco, Q. Beaufils, A. Bertoldi, K. Bongs, P. Bouyer, C. Braxmaier, W. Chaibi, N. Christensen, F. Fitzek, G. Flouris, N. Gaaloul, S. Gaffet, C.L. Garrido Alzar, R. Geiger, S. Guellati-Khelifa, K. Hammerer, J. Harms, J. Hinderer, M. Holynski, J. Junca, S. Katsanevas, C. Klempt, C. Kozanitis, M. Krutzik, A. Landragin, I. Làzaro Roche, B. Leykauf, Y.-H. Lien, S. Loriani, S. Merlet, M. Merzougui, M. Nofrarias, P. Papadakos, F. Pereira dos Santos, A. Peters, D. Plexousakis, M. Prevedelli, E.M. Rasel, Y. Rogister, S. Rosat, A. Roura, D. O. Sabulsky, V. Schkolnik, D. Schlippert, C. Schubert, L. Sidorenkov, J.-N. Siemß, C. F. Sopuerta, F. Sorrentino, C. Struckmann, G.M. Tino, G. Tsagkatakis, A. Viceré, W. von Klitzing, L. Woerner, X. Zou, Technologies for the ELGAR large scale atom interferometer array. arXiv:2007.04014v1 (2020); ELGAR–a European Laboratory for Gravitation and Atom-interferometric Research. Class. Quantum Grav. **37**, 225017 (2020). https://doi.org/10.1088/1361-6382/aba80e

52. J. Rudolph, T. Wilkason, M. Nantel, H. Swan, C.M. Holland, Y. Jiang, B.E. Garber, S.P. Carman, J.M. Hogan, Large momentum transfer clock atom interferometry on the 689 nm intercombination line of strontium. Phys. Rev. Lett. **124**, 083604 (2020). https://doi.org/10.1103/PhysRevLett.124.083604

53. D.V. Strekalov, N. Yu, K. Mansour, *Sub-shot Noise Power Source for Microelectronics.* NASA Tech Briefs (Pasadena, CA, 2011). http://ntrs.nasa.gov/archive/nasa/casi.ntrs.nasa.gov/20120006513.pdf

54. E. Danieli, J. Perlo, B. Blümich, F. Casanova, Highly stable and finely tuned magnetic fields generated by permanent magnet assemblies. Phys. Rev. Lett. **110**, 180801 (2013). https://doi.org/10.1103/PhysRevLett.110.180801

55. S. Zhou, D. Groswasser, M. Keil, Y. Japha, R. Folman, Robust spatial coherence 5 μm from a room-temperature atom chip. Phys. Rev. A **93**, 063615 (2016). https://doi.org/10.1103/PhysRevA.93.063615

56. P. Treutlein, P. Hommelhoff, T. Steinmetz, T.W. Hänsch, J. Reichel, Coherence in microchip traps. Phys. Rev. Lett. **92**, 203005 (2004). https://doi.org/10.1103/PhysRevLett.92.203005

57. F. Cerisola, Y. Margalit, S. Machluf, A.J. Roncaglia, J.P. Paz, R. Folman, Using a quantum work meter to test non-equilibrium fluctuation theorems. Nature Commun. **8**, 1241 (2017). https://doi.org/10.1038/s41467-017-01308-7

58. B.S. Zhao, W. Zhang, W. Schöllkopf, Non-destructive quantum reflection of helium dimers and trimers from a plane ruled grating. Mol. Phys. **111**, 1772 (2013). https://doi.org/10.1080/00268976.2013.787150

59. S. Zeller, M. Kunitski, J. Voigtsberger, A. Kalinin, A. Schottelius, C. Schober, M. Waitz, H. Sann, A. Hartung, T. Bauer, M. Pitzer, F. Trinter, C. Goihl, C. Janke, M. Richter, G. Kastirke, M. Weller, A. Czasch, M. Kitzler, M. Braune, R.E. Grisenti, W. Schöllkopf, L.P.H. Schmidt, M.S. Schöffler, J.B. Williams, T. Jahnke, R.Dörner, Imaging the He$_2$ quantum halo state using a free electron laser. Proc. Natl. Acad. Sci. USA **113**, 14651 (2016). https://doi.org/10.1073/pnas.1610688113

60. Y.Y. Fein, P. Geyer, P. Zwick, F. Kiałka, S. Pedalino, M. Mayor, S. Gerlich, M. Arndt, Quantum superposition of molecules beyond 25 kDa. Nature Phys. **15**, 1242 (2019). https://doi.org/10.1038/s41567-019-0663-9

61. G. Jacob, K. Groot-Berning, S. Wolf, S. Ulm, L. Couturier, S.T. Dawkins, U.G. Poschinger, F. Schmidt-Kaler, K. Singer, Transmission microscopy with nanometer resolution using a deterministic single ion source. Phys. Rev. Lett. **117**, 043001 (2016). https://doi.org/10.1103/PhysRevLett.117.043001

62. C. Henkel, G. Jacob, F. Stopp, F. Schmidt-Kaler, M. Keil, Y. Japha, R. Folman, Stern-Gerlach splitting of low-energy ion beams. New J. Phys. **21**, 083022 (2019). https://doi.org/10.1088/1367-2630/ab36c7

63. L.W. Bruch, W. Schöllkopf, J.P. Toennies, The formation of dimers and trimers in free jet ^{4}He cryogenic expansions. J. Chem. Phys. **117**, 1544 (2002). https://doi.org/10.1063/1.1486442

64. Y. Margalit, *Stern-Gerlach Interferometry with Ultracold Atoms*. (Ph.D. Thesis, Ben-Gurion University, 2018). http://www.bgu.ac.il/atomchip/Theses/Yair_Margalit_PhD_Thesis_2018.pdf

65. F. Kiałka, B.A. Stickler, K. Hornberger, Y.Y. Fein, P. Geyer, L. Mairhofer, S. Gerlich, M. Arndt, Concepts for long-baseline high-mass matter-wave interferometry. Phys. Scripta **94**, 034001 (2019). https://doi.org/10.1088/1402-4896/aaf243

66. D.A. Steck. Rubidium 87 D Line Data (2003). https://steck.us/alkalidata/rubidium87numbers.1.6.pdf

67. S. Machluf, *Coherent Splitting of Matter-Waves on an Atom Chip Using a State-Dependent Magnetic Potential* (Ph.D. Thesis, Ben-Gurion University, 2013). http://www.bgu.ac.il/atomchip/Theses/Shimon_Machluf_PhD_2013.pdf

68. O. Amit, *Matter-Wave Interferometry on an Atom Chip* (Ph.D. Thesis, Ben-Gurion University, 2020). http://www.bgu.ac.il/atomchip/Theses/PhD_Thesis_Omer_Amit_submitted.pdf

69. D.E. Miller, J.R. Anglin, J.R. Abo-Shaeer, K. Xu, J.K. Chin, W. Ketterle, High-contrast interference in a thermal cloud of atoms. Phys. Rev. A **71**, 043615 (2005). https://doi.org/10.1103/PhysRevA.71.043615

70. O. Romero-Isart, Coherent inflation for large quantum superpositions of levitated microspheres. New J. Phys. **19**, 719711 (2017). https://doi.org/10.1088/1367-2630/aa99bf

71. R. Colella, A.W. Overhauser, S.A. Werner, Observation of gravitationally induced quantum interference. Phys. Rev. Lett. **34**, 1472 (1975). https://doi.org/10.1103/PhysRevLett.34.1472

72. M. Zych, F. Costa, I. Pikovski, Č. Brukner, Quantum interferometric visibility as a witness of general relativistic proper time. Nature Commun. **2**, 505 (2011). https://doi.org/10.1038/ncomms1498

73. S. Dimopoulos, P.W. Graham, J.M. Hogan, M.A. Kasevich, Testing general relativity with atom interferometry. Phys. Rev. Lett. **98**, 111102 (2007). https://doi.org/10.1103/PhysRevLett.98.111102

74. H. Müntinga, H. Ahlers, M. Krutzik, A. Wenzlawski, S. Arnold, D. Becker, K. Bongs, H. Dittus, H. Duncker, N. Gaaloul, C. Gherasim, E. Giese, C. Grzeschik, T.W. Hänsch, O. Hellmig, W. Herr, S. Herrmann, E. Kajari, S. Kleinert, C. Lämmerzahl, W. Lewoczko-Adamczyk, J. Malcolm, N. Meyer, R. Nolte, A. Peters, M. Popp, J. Reichel, A. Roura, J. Rudolph, M. Schiemangk, M. Schneider, S.T. Seidel, K. Sengstock, V. Tamma, T. Valenzuela, A. Vogel, R. Walser, T. Wendrich, P. Windpassinger, W. Zeller, T. van Zoest, W. Ertmer, W.P. Schleich, E.M. Rasel, Interferometry with Bose-Einstein condensates in microgravity. Phys. Rev. Lett. **110**, 093602 (2013). https://doi.org/10.1103/PhysRevLett.110.093602

75. C.C.N. Kuhn, G.D. McDonald, K.S. Hardman, S. Bennetts, P.J. Everitt, P.A. Altin, J.E. Debs, J.D. Close, N.P. Robins, A. Bose-condensed, simultaneous dual-species Mach-Zehnder atom interferometer. New J. Phys. **16**, 073035 (2014). https://doi.org/10.1088/1367-2630/16/7/073035

76. I. Pikovski, M. Zych, F. Costa, Č. Brukner, Universal decoherence due to gravitational time dilation. Nat. Phys. **11**, 668 (2015). https://doi.org/10.1038/nphys3366

77. B.-G. Englert, Fringe visibility and which-way information: an inequality. Phys. Rev. Lett. **77**, 2154 (1996). https://doi.org/10.1103/PhysRevLett.77.2154

78. E. Giese, A. Friedrich, F. Di Pumpo, A. Roura, W.P. Schleich, D.M. Greenberger, E.M. Rasel, Proper time in atom interferometers: diffractive versus specular mirrors. Phys. Rev. Lett. **99**, 013627 (2019). https://doi.org/10.1103/PhysRevA.99.013627

79. M. Lugli, *Mass and Proper Time as Conjugated Observables* (Graduate Thesis, Università degli Studi di Pavia, Italy, 2017). http://arxiv.org/abs/1710.06504
80. J. Samuel, R. Bhandari, General setting for Berry's phase. Phys. Rev. Lett. **60**, 2339 (1988). https://doi.org/10.1103/PhysRevLett.60.2339
81. R. Bhandari, SU(2) phase jumps and geometric phases. Phys. Lett. A **157**, 221 (1991). https://doi.org/10.1016/0375-9601(91)90055-D
82. T. van Dijk, H.F. Schouten, T.D. Visser, Geometric interpretation of the Pancharatnam connection and non-cyclic polarization changes. J. Am. Opt. Soc. A **27**, 1972 (2010). https://doi.org/10.1364/JOSAA.27.001972
83. S. Pancharatnam, Generalized theory of interference, and its applications. Proc. Indian Acad. Sci. A **44**, 247 (1956). https://doi.org/10.1007/BF03046050
84. Z. Zhou, Y. Margalit, S. Moukouri, Y. Meir, R. Folman, An experimental test of the geodesic rule proposition for the noncyclic geometric phase. Sci. Adv. **6**, eaay8345 (2020). https://doi.org/10.1126/sciadv.aay8345
85. E.H. Kennard, Zur Quantenmechanik einfacher Bewegungstypen. Z. fur Physik **44**, 326 (1927). https://doi.org/10.1007/BF01391200
86. E.H. Kennard, The quantum mechanics of an electron or other particle. J. Franklin Inst. **207**, 47 (1929). https://doi.org/10.1016/S0016-0032(29)91274-6
87. G.G. Rozenman, M. Zimmermann, M.A. Efremov, W.P. Schleich, L. Shemer, A. Arie, Amplitude and phase of wave packets in a linear potential. Phys. Rev. Lett. **122**, 124302 (2019). https://doi.org/10.1103/PhysRevLett.122.124302
88. G.G. Rozenman, L. Shemer, A. Arie, Observation of accelerating solitary wavepackets. Phys. Rev. E **101**, 050201(R) (2020). https://doi.org/10.1103/PhysRevE.101.050201
89. G.G. Rozenman, M. Zimmermann, M.A. Efremov, W.P. Schleich, W.B. Case, D.M. Greenberger, L. Shemer, A. Arie, Projectile motion of surface gravity water wave packets: An analogy to quantum mechanics. Eur. Phys. J. Spec. Top. (2021). https://doi.org/10.1140/epjs/s11734-021-00096-y
90. M. Zimmermann, M. Efremov, W. Zeller, W. Schleich, J. Davis, F. Narducci, Representation-free description of atom interferometers in time-dependent linear potentials. New J. Phys. **21**, 073031 (2019). https://doi.org/10.1088/1367-2630/ab2e8c
91. M. Efremov, M. Zimmermann, O. Amit, Y. Margalit, R. Folman, W. Schleich, Atomic interferometer sensitive to time-dependent acceleration. In preparation (2021)
92. C.J. Bordé, Atomic interferometry with internal state labelling. Phys. Lett. A **140**, 10 (1989). https://doi.org/10.1016/0375-9601(89)90537-9
93. M. Kasevich, S. Chu, Atomic interferometry using stimulated Raman transitions. Phys. Rev. Lett. **67**, 181 (1991). https://doi.org/10.1103/PhysRevLett.67.181
94. A. Peters, K. Chung, S. Chu, High-precision gravity measurements using atom interferometry. Metrologia **38**, 25 (2001). https://doi.org/10.1088/0026-1394/38/1/4
95. L. Brillouin, Is it possible to test by a direct experiment the hypothesis of the spinning electron? Proc. Natl. Acad. Sci. **14**, 755 (1928). https://doi.org/10.1073/pnas.14.10.755
96. N. Bohr, Chemistry and the quantum theory of atomic constitution. J. Chem. Soc. **349**, (1932). https://doi.org/10.1039/JR9320000349
97. G.A. Gallup, H. Batelaan, T.J. Gay, Quantum-mechanical analysis of a longitudinal Stern-Gerlach effect. Phys. Rev. Lett. **86**, 4508 (2001). https://doi.org/10.1103/PhysRevLett.86.4508
98. H. Batelaan, T.J. Gay, J.J. Schwendiman, Stern-Gerlach effect for electron beams. Phys. Rev. Lett. **79**, 4517 (1997). https://doi.org/10.1103/PhysRevLett.79.4517
99. H. Batelaan, Electrons, Stern-Gerlach magnets, and quantum mechanical propagation. Am. J. Phys. **70**, 325 (2002). https://doi.org/10.1119/1.1450559
100. B.M. Garraway, S. Stenholm, Does a flying electron spin? Contemp. Phys. **43**, 147 (2002). https://doi.org/10.1080/00107510110102119
101. G. Jacob, *Ion Implantation and Transmission Microscopy with Nanometer Resolution Using a Deterministic Ion Source.* (Ph.D. Dissertation, Johannes-Gutenberg-Universität, Mainz, Germany, 2016). https://www.quantenbit.physik.uni-mainz.de/files/2019/10/DissSW.pdf

102. E.A. Hinds, I.G. Hughes, Magnetic atom optics: Mirrors, guides, traps, and chips for atoms. J. Phys. D **32**, R119–R146 (1999). http://iopscience.iop.org/article/10.1088/0022-3727/32/18/201/pdf

103. T.D. Tran, Y. Wang, A. Glaetzle, S. Whitlock, A. Sidorov, P. Hannaford, Magnetic lattices for ultracold atoms (2019). arXiv:1906.08918

104. G.M. Tino, M.A. Kasevich, eds., Atom iterferometry, in *Proceedings of International School of Physics "Enrico Fermi"*. http://ebooks.iospress.nl/volume/atom-interferometry, Vol 188 (IOS Press, Amsterdam, 2014)

105. F. Hasselbach, Progress in electron- and ion-interferometry. Rep. Prog. Phys. **73**, 016101 (2010). https://doi.org/10.1088/0034-4885/73/1/016101

106. M. Brownnutt, M. Kumph, P. Rabl, R. Blatt, Ion-trap measurements of electric-field noise near surfaces. Rev. Mod. Phys. **87**, 1419 (2015). https://doi.org/10.1103/RevModPhys.87.1419

107. R.O. Behunin, D.A.R. Dalvit, R.S. Decca, C.C. Speake, Limits on the accuracy of force sensing at short separations due to patch potentials. Phys. Rev. D **89**, 051301(R) (2014). https://doi.org/10.1103/PhysRevD.89.051301

108. T.E. Phipps, O. Stern, Über die Einstellung der Richtungsquantelung. Z. Phys. **73**, 185 (1932). https://doi.org/10.1007/BF01351212

109. R. Frisch, E. Segrè, Über die Einstellung der Richtungsquantelung II. Z. Phys. **80**, 610 (1933). https://doi.org/10.1007/BF01335699

110. D. Bohm, J. Bub, A proposed solution of the measurement problem in quantum mechanics by a hidden variable theory. Rev. Mod. Phys. **38**, 453 (1966). https://doi.org/10.1103/RevModPhys.38.453

111. R. Folman, A search for hidden variables in the domain of high energy physics. Found. Phys. Lett. **7**, 191 (1994). https://doi.org/10.1007/BF02415510

112. S. Das, M. Nöth, D. Dürr, Exotic Bohmian arrival times of spin-1/2 particles: an analytical treatment. Phys. Rev. A **99**, 052124 (2019). https://doi.org/10.1103/PhysRevA.99.052124

113. L.S. Schulman, Program for the special state theory of quantum measurement. Entropy **19**, 343 (2017). https://doi.org/10.3390/e19070343

114. C. Wan, M. Scala, G.W. Morley, A.A.T.M. Rahman, H. Ulbricht, J. Bateman, P.F. Barker, S. Bose, M.S. Kim, Free nano-object Ramsey interferometry for large quantum superpositions. Phys. Rev. Lett. **117**, 143003 (2016). https://doi.org/10.1103/PhysRevLett.117.143003

115. R.J. Marshman, A. Mazumdar, S. Bose, Locality and entanglement in table-top testing of the quantum nature of linearized gravity. Phys. Rev. A **101**, 052110 (2020). https://doi.org/10.1103/PhysRevA.101.052110

116. H. Pino, J. Prat-Camps, K. Sinha, B.P. Venkatesh, O. Romero-Isart, On-chip quantum interference of a superconducting microsphere. Quantum Sci. Technol. **3**, 25001 (2018). https://doi.org/10.1088/2058-9565/aa9d15

117. O. Romero-Isart, Quantum superposition of massive objects and collapse models. Phys. Rev. A **84**, 052121 (2011). https://doi.org/10.1103/PhysRevA.84.052121

118. M. Scala, M.S. Kim, G.W. Morley, P.F. Barker, S. Bose, Matter-wave interferometry of a levitated thermal nano-oscillator induced and probed by a spin. Phys. Rev. Lett. **111**, 180403 (2013). https://doi.org/10.1103/PhysRevLett.111.180403

119. M. Toroš, S. Bose, P.F. Barker, Atom-Nanoparticle Schrödinger Cats. arXiv:2005.12006 (2020)

120. C. Marletto, V. Vedral, Gravitationally induced entanglement between two massive particles is sufficient evidence of quantum effects in gravity. Phys. Rev. Lett. **119**, 240402 (2017). https://doi.org/10.1103/PhysRevLett.119.240402

121. Y. Margalit, O. Dobkowski, Z. Zhou, O. Amit, Y. Japha, S. Moukouri, D. Rohrlich, A. Mazumdar, S. Bose, C. Henkel, R. Folman, Realization of a complete Stern-Gerlach interferometer: towards a test of quantum gravity (2020). https://arxiv.org/pdf/2011.10928v1.pdf

Chapter 15
Testing Fundamental Physics by Using Levitated Mechanical Systems

Hendrik Ulbricht

Abstract We will describe recent progress of experiments towards realising large-mass single particle experiments to test fundamental physics theories such as quantum mechanics and gravity, but also specific candidates of Dark Matter and Dark Energy. We will highlight the connection to the work started by Otto Stern as levitated mechanics experiments are about controlling the centre of mass motion of massive particles and using the same to investigate physical effects. This chapter originated from the foundations of physics session of the Otto Stern Fest at Frankfurt am Main in 2019, so we will also share a view on the Stern Gerlach experiment and how it related to tests of the principle of quantum superposition.

1 Introductory Remarks

Experimentally, this research programme is about gas-phase experiments with large-mass particles, large compared to the mass of a single hydrogen atom, in order to test fundamental theories without the influence of the environment, which typically results in coherence-spoiling noise and decoherence effect. Tests of fundamental theories, such as quantum mechanics and gravity, are in the low-energy regime of non-relativistic velocities and therefore far away from a parameter regime of high-energy particle physics considerations. Fundamental theories will be tested in a new regime.

While Otto Stern's pioneering experiments [1], aligned with a fantastically bold and clear research programme, were about the study and control of freely propagating atoms and molecules in particle beams, we here make use of optical, magnetic and electric fields to trap and manipulate single particles, consisting of many atoms, in order to study the new physics and chemistry. The challenge here is to have a strong enough handle on the motion of the particle. For instance, the optical dipole force $F = \alpha \nabla E^2$ is strong and able to trap individual atoms and atomic ensembles making use of resonance effects. This is impossible for large molecules, again such

H. Ulbricht (✉)
School of Physics and Astronomy, University of Southampton, Highfield SO17 1BJ, UK
e-mail: h.ulbricht@soton.ac.uk

© The Author(s) 2021

B. Friedrich and H. Schmidt-Böcking (eds.), *Molecular Beams in Physics and Chemistry*,
https://doi.org/10.1007/978-3-030-63963-1_15

which consist of many atoms, as resonances are manifold and the oscillator strength is distributed across many different state transitions far away from the ideal two-level system situation which we luckily find in some atoms, which gave rise to a revolution in experimental physics. Cold atom experiments now allow for ultra-precise control of various degrees of freedoms—including the centre of mass motion of the atoms and to prepare non-classical sates, including collective ones such as atomic Bose Einstein Condensates (BEC). In a way our programme aims to achieve a similar level of control, but for particles of large mass and different cooling and manipulation techniques have to be developed and used for that. The off-resonant dipole force where α is a measure of the off-resonant detuning of all affected molecular states is however too weak to lead to a large enough effect to trap and manipulate individual molecules by coherent laser light [2, 3] . This situation changes dramatically if one increases the size (volume V) of the particle to trap and therefore its polarizability $\alpha \propto V$. Then dipole force becomes so strong to form a deep optical trap and optical fields can be used for controlling single particle motions again, which gave rise to the development of the new research field levitated optomechnaics [4], based on early pioneering work by Arthur Ashkin (Nobel Prize in Physics in 2018) [5] and already then in close relation to the then soon to be called cold atomic and optical physics. By now the field of levitated large-mass particle systems has seen the implementation of other than optical forces for trapping and manipulation, namely time-varying electrical fields in Paul traps [6] and magnetic traps [7], sometimes including superconductors [8]. All such technical developments give rise to the hope to soon perform experiments with truly macroscopic quantum systems, outperforming existing paradigms of large-mass matterwave interferometry [9]. Macroscopic here entails the involvement of a large-mass particle in a quantum superposition of large spatial separation [10].

There are two pillars of our research programme on testing fundamental physics are with a certain methodological approach. The *first* is the clearly distinctive predictions for the outcome of the same experiment originated from alternative theoretical descriptions. This is our approach for testing the universality of the quantum superposition principle in the context of collapse models [11]. Quantum mechanics and collapse models predict a different outcome of a matterwave interferometry experiment—if the experiment is performed in the right parameter regime. The *second* pillar of our research programme is to first perform a detailed analysis of the new physics to be tested and then to chose the best experiment to perform the test.

Outlook of this chapter. In the following, we will address new avenues to test quantum mechanics in Sect. 2 with the specific emphasis on experiments using levitated mechanical systems. Then we will address experimental tests of the interplay between quantum mechanics and gravity in Sect. 3 including the discussion of the semiclassical Schrödinger-Newton equation, gravitational deocherence of the wavefunction and the gravity of a quantum state. In the final Sect. 4 we will refer to using the Wigner function to simulate the original Stern Gerlach experiment.

2 Testing Quantum Mechanics with Collapse Models

There is an increasing interest in developing experiments aimed at testing collapse models, in particular the Continuous Localization Model (CSL), the natural evolution of the GRW model initially proposed by Ghirardi et al. [11–14]. Current experiments and related bounds on collapse parameters are partially discussed in other contributions in this review. Our aim here is to discuss some of the most promising directions towards future improvements. We will mostly focus on non-interferometric experiments. In Sect. 2.1 we will briefly outline proposals of matter-wave interference with massive nano/microparticles. Finally, in Sect. 2.2 we will discuss mechanical experiments, in particular ongoing experiments with ultracold cantilevers, ongoing and proposed experiments based on levitated nanoparticles and microparticles. We will not consider here two important classes of experiments which are separately discussed by other contributors in this review: matter-wave interference with molecules and space-based experiments. We will end in Sect. 2.3 with some ideas on how precision experiments can be used for testing collapse models. A summary of recent interferometric and non-interferomtric experiments which could set direct bounds on the CSL collapse model are summarized in Fig. 1.

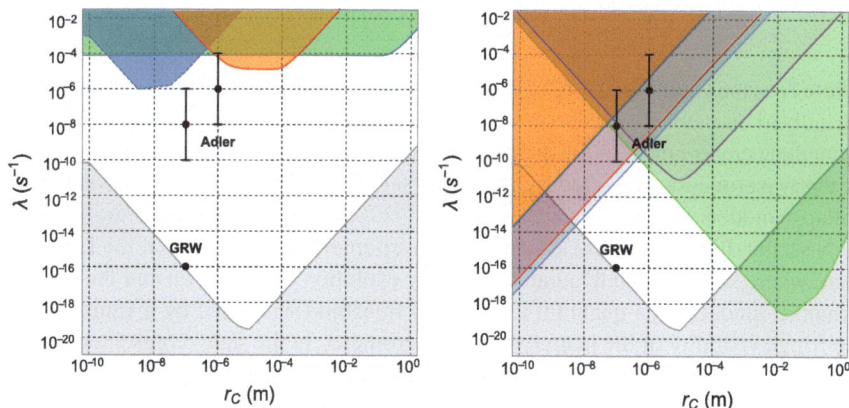

Fig. 1 Exclusion plots for the CSL parameters with respect to the GRW and Adler theoretically proposed values [12, 15]. Left panel—Excluded regions from *interferometric* experiments: molecular interferometry [16] (blue area), atom interferometry [17] (green area) and experiment with entangled diamonds [18] (orange area). Right panel—Excluded regions from *non-interferometric* experiments: LISA Pathfinder [19, 20] (green area), cold atoms [21] (orange area), phonon excitations in crystals [22] (red area), X-ray measurements [23] (blue area) and nanomechanical cantilever [24]. We report with the grey area the region excluded based on theoretical arguments [25]. Figure and caption taken from [26]

2.1 Tests of Quantum Mechanics by Matter-Wave Interferometry

Matterwave interferometry is directly testing the quantum superposition principle. Relevant for mass-scaling collapse models, such as CSL, are matterwave interferometers testing the maximal macroscopic extend in terms of mass, size and time of spatial superpositions of single large-mass particles. Such beautiful, but highly challenging experiments have been pushed by Markus Arndt's group in Vienna to impressive particle masses of 10^4 atomic mass units (amu), which still not significantly challenging CSL. Therefore the motivation remains to push matterwave interferometers to more macroscopic systems. Predicted bounds on collapse models set by large-mass matterwave interferometers are worked out in detail in [27].

As usual in open quantum system dynamics treatments, non-linear stochastic extensions of the Schrödinger equation on the level of the wavefunction [28] correspond to a non-uniquely defined master equation on the level of the density matrix ρ to describe the time evolution of the quantum system, say the spatial superposition across distance $|x - y|$, where the conserving von Neumann term $\partial \rho_t(x, y)/\partial t = -(i/\hbar)[H, \rho]$, is now extended by a Lindblad operator L term:

$$\frac{\partial \rho_t(x, y)}{\partial t} = -\frac{i}{\hbar}[H, \rho_t(x, y)] + L\rho_t(x, y), \tag{1}$$

where H is the Hamilton operator of the quantum system and different realisations of a Lindblad operator are used to describe both standard decoherence (triggered by the immediate environment of the quantum system) [29] as well as spontaneous collapse of the wavefunction triggered by the universal classical noise field as predicted by collapse models.

Now the dynamics of the system is very different with and without the Lindbladian, where with the Lindbladian the unitary evolution breaks down and the system dynamics undergoes a quantum-to-classical transition witnessed by a vanishing of the fringe visibility of the matterwave interferometer. In the state represented by the density matrix the off-diagonal terms vanish as the system evolves according to the open system dynamics, the coherence/superposition of that state is lost. The principal goal of interference experiments with massive particles is then to explore and quantify the relevance of the $(L\rho_t(x, y))$-term—as collapse models predict a break down of the quantum superposition principle for a sufficient macroscopic system. An intrinsic problem is the competition with known and unknown environmental decoherence mechanisms, if a visibility loss is observed. However solutions seem possible.

In order to further increase the macroscopic limits in interference some ambitious proposals have been made utilizing nano- and micro-particles, c.f. Fig. 2. The main challenge is to allow for a long enough free evolution time of the prepared quantum superposition state in order to be sensitive to the collapsing effects. The free evolution—the spatial spreading of the wavefunction $\Psi(r, t)$ with time—according to the time-dependent Schrödinger equation with the potential $V(r) = 0$,

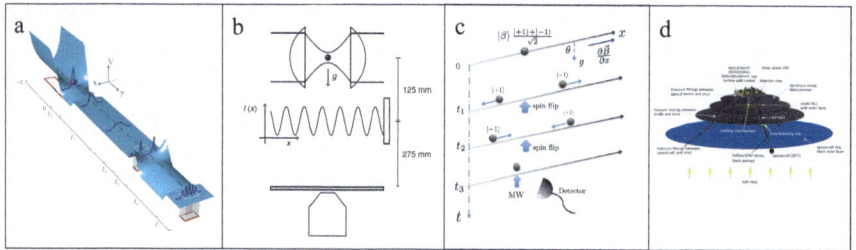

Fig. 2 Illustration of some of the proposed schemes for matterwave interferometry with nano- and micro-particles to test the quantum superposition principle directly, and therefore also collapse models. **a** The cryogenic skatepark for a single superconducting micro-particle (adapted from Ref. [31]); **b** The nanoparticle Talbot interferometer (adapted from Ref. [30]); **c** The Ramsey scheme addressing the electron Spin of a NV-centre diamond coupled to an external magnetic field gradient $(\partial B/\partial x)$ (adapted from Ref. [32]); **d** The adaptation of an interferometer at a free falling satellite platform in space to allow form longer free evolution times (adapted from Ref. [33])

$$\frac{\partial}{\partial t}\Psi(r,t) = -i\frac{\hbar}{2m}\nabla^2\Psi(r,t), \tag{2}$$

describes a diffusive process for probability amplitudes similar to a typical diffusion equation with the imaginary diffusion coefficient $(-i\hbar/2m)$. Therefore the spreading of $\Psi(r,t)$ scales inverse with particle mass m. For instance for a 10^7 amu particle it already takes so long to show the interference pattern in a matterwave experiment that the particle would significantly drop in Earth's gravitational field, in fact it would drop on the order of 100 m. This requires a dramatic change in the way large-mass matterwave interferometry experiments have to be performed beyond the mass of 10^6 amu [30].

Different solutions are thinkable. One could of course envisage building 100 m fountain, but that seems very unfeasible also given that no sufficient particle beam preparation techniques exist (and don't seem to be likely to be developed in the foreseeable future) to enable the launch and detection of particles in the mass range in question over a distance of 100 m. One can consider to levitate the particle by a force field to compensate for the drop in gravity, but here we face a high demand on the fluctuations of that levitating field, which have to be small compared to the amplitudes of the quantum evolution, which is not feasible with current technology. A maybe possible option os to coherently boost/accelerate the evolution of the wavefunction spread by a beam-splitter operation. The proposals in Refs. [31, 32] are such solutions, which are still awaiting their technical realisation for large masses. Alternatively and more realistic given technical capability is to allow for long enough free evolution by freely fall the whole interferometer apparatus in a co-moving reference frame with the particle. This is the idea of the MAQRO proposal, a dedicated satellite mission in space to perform large-mass matterwave interference experiments with micro- and nano-particles [33].

Another interesting approach is to consider the use of cold or ultra-cold ensembles of atoms such as cloud in a magneto optical trap (MOT) or an atomic Bose-Einstein Condensate (BEC) as also there we find up to 10^8 atoms of alkali species such as rubidium or caesium. On closer look it turns out that such weakly interacting atomic ensembles are not of immediate use for the purpose to test macroscopic quantum superpositions in the context of collapse model test. For instance testing the CSL model is build on a mass (number of particles N, more precisely the number of nucleons: protons and neutrons in the nuclei of the atoms) amplification which in principle can even go with N^2, if the condition for coherent scattering of the classical collapse noise treated as a wave with correlation length r_c scattered at the particle in the quantum superposition state. The central assumption of this amplification mechanism is that if the CSL noise is collapsing the wavefunction of only one of the constituent nucleons, then the total wavefunction of the whole composite object collapses. While in the case of a nanoparticle consisting of many atoms (and therefore nucleons), it is not the case for an weakly interacting atomic ensemble. If one atom is collapsing then the total atomic wavefunction remains intact and the one atom is lost from the ensemble.

This may change if the atoms in the cold or ultra-cold ensemble can be made stronger interacting, without running into the complications of chemistry which may forbid condensation of the atomic—then molecular—cloud at all. However there is hope that quantum optical state preparation techniques applied after a BEC has been formed such as collective NOON or squeezed states enable N and even N^2 scaling in the fashion fit for testing wavefunction collapse.

Interestingly, this might be different if the physical mechanism responsible for the collapse of the wavefunction, which remains highly speculative at present, is in any way related to gravity [34], then there might be hope that atomic ensembles even in the weakly interacting case can be used to test CSL-type models. The condition to fulfil is that the atomic ensemble is interacting gravitationally strong enough so that it acts collectively under collapse, even if just a single constituent atom (nucleon) is affected by the collapsing effect.

2.2   Non-interferometric Mechanical Tests of Quantum Mechanics

This class of experiments has emerged in recent years as one of the most powerful and effective ways to test collapse models. The underlying idea [35, 36] is that a mechanism which continuously localizes the wavefunction of a mechanical system, which can be either a free mass or a mechanical resonator, must be accompanied by a random force noise acting on its center-of-mass. This leads in turn to a random diffusion which can be possibly detected by ultrasensitive mechanical experiments.

In a real mechanical system such diffusion will be masked by standard thermal diffusion arising from the coupling to the environment, i.e. from the same effects which lead to decoherence in quantum interference experiments [37]. In practice there will be additional non-thermal effects, due to external non-equilibrium vibrational noise (seismic/acoustic/gravity gradient). Moreover, one has to ensure that the back-action from the measuring device is negligible.

Under the assumption that thermal noise is the only significant effect, the (one-sided) power spectral density of the force noise acting on the mechanical system is given by:

$$S_{ff} = \frac{4k_B T m \omega}{Q} + 2\hbar^2 \eta. \tag{3}$$

where k_B is the Boltzmann constant, T is the temperature, m is the mass, ω the angular frequency, Q is the mechanical quality factor.

η is a diffusion constant associated to spontaneous localization, and can be calculated explicitly for the most known models. For CSL, it is given by the following expression

$$\eta = \frac{2\lambda}{m_0^2} \int \int d^3 r \, d^3 r' \, \exp\left(-\frac{|\mathbf{r} - \mathbf{r}'|^2}{4r_C^2}\right) \frac{\partial \varrho(\mathbf{r})}{\partial z} \frac{\partial \varrho(\mathbf{r}')}{\partial z'} \tag{4}$$

$$= \frac{(4\pi)^{\frac{3}{2}} \lambda \, r_C^3}{m_0^2} \int \frac{d^3 k}{(2\pi)^3} \, k_z^2 \, e^{-\mathbf{k}^2 r_C^2} \, |\tilde{\varrho}(\mathbf{k})|^2 \tag{5}$$

with $\mathbf{k} = (k_x, k_y, k_z)$, $\tilde{\varrho}(\mathbf{k}) = \int d^3 x \, e^{i\mathbf{k} \cdot \mathbf{r}} \varrho(\mathbf{r})$ and $\varrho(\mathbf{r})$ the mass density distribution of the system. In the expressions above m_0 is the nucleon mass and r_C and λ are the free parameters of CSL. The typical values proposed in CSL literature are $r_C = 10^{-7}$ m and 10^{-6} m, while for λ a wide range of possible values has been proposed, which spans from the GRW value $\lambda \approx 10^{-16}$ Hz [12, 13] to the Adler value $\lambda \approx 10^{-8\pm2}$ Hz at $r_C = 10^{-7}$ m [15]. The possibility for such non-interferometric tests, which aim to directly test the non-thermal noise predicted by collapse models [24, 38–41].

An experiment looking for CSL-induced noise has to be designed in order to maximize the 'noise to noise' ratio between the CSL term and the thermal noise. In practice this means lowest possible temperature T, lowest possible damping time, or linewidth, $1/\tau = \omega/Q$, and highest possible η/m ratio. The first two conditions express the requirement of lowest possible power exchange with the thermal bath, the third condition is inherently related to the details of the specific model.

For CSL we can distinguish two relevant limits. When the characteristic size L of the system is small, $L \ll r_C$, then the CSL field cannot resolve the internal structure of the system, and one finds $\eta/m \propto m$. When the characteristic length of the system in the direction of motion L is large, $L \gg r_C$, then $\eta/m \propto \rho/L$, where ρ is the mass density [24, 38, 40]. The expressions in the two limits imply that, for a well defined characteristic length r_C, the optimal system is a plate or disk with thickness $L \sim r_C$ and the largest possible density ρ.

Among other models proposed in literature, we mention the gravitational Diosi-Penrose (DP) model, which leads to localization and diffusion similarly to CSL. The diffusion constant η_{DP} is given by [40]:

$$\eta_{DP} = \frac{G\rho m}{6\sqrt{\pi}\hbar} \left(\frac{a}{r_{DP}}\right)^3,$$ (6)

where a is the lattice constant and G is the gravitational constant, so that he ratio η_{DP}/m depends only on the mass density. Unlike CSL, there is no explicit dependence on the shape or size of the mechanical system.

2.2.1 Levitated Mechanical Systems

One of the most promising approaches towards a significant leap forward in the achievable sensitivity to spontaneous collapse effects is by levitation of nanoparticles or microparticles. The main benefits of levitation are the absence of clamping mechanical losses and wider tunability of mechanical parameters. In addition, several degrees of freedom can be exploited, either translational or rotational [41, 42]. This comes at the price of higher complexity, poor dynamic range and large nonlinearities, which usually require active feedback stabilization over multiple degrees of freedom. However, levitated systems hold the promise of much better isolation from the environment, therefore higher quality factor. One relevant example, in the macroscopic domain, is the space mission LISA Pathfinder, which is based on an electrostatically levitated test mass, currently setting the strongest bound on collapse models over a wide parameter range [43].

Several levitation methods for micro/nanoparticles are currently being investigated. The most developed is optical levitation using force gradients induced by laser fields, the so called optical tweezer approach [5]. While this is a very effective and flexible approach to trap nanoparticles, in this context it is inherently limited by two factors: the relatively high trap frequency, in the order of 100 kHz, and the high internal temperature of the particles, induced by laser power absorption, which leads ultimately to strong thermal decoherence. Alternative approaches have to be found, featuring lower trap frequency and low or possible null power dissipated in the levitated particle. The two possible classes of techniques are electrical levitation and magnetic levitation.

Electrical levitation has been deeply developed in the context of ion traps. The standard tool is the Paul trap, which allows to trap an ion, or equivalently a charged nanoparticle, using a combination of ac and dc bias electric fields applied through a set of electrodes [44]. The power dissipation is much lower than in the optical case, and the technology is relatively well-established. However, the detection of a nanoparticle in a Paul trap still poses some technological challenge (Fig. 3).

This issue has been extensively investigated in a recent paper [45], specifically considering a nanoparticle in a cryogenic Paul trap in the context of collapse model testing. Three detection schemes have been considered: an optical cavity, an optical tweezer, and a all-electric readout based on SQUID. It was found that to detect the nanoparticle motion with good sensitivity, optical detection has to be employed. Unfortunately, optical detection is not easily integrated in a cryogenic environment, and leads to a nonnegligible internal heating and excess force noise. On the other

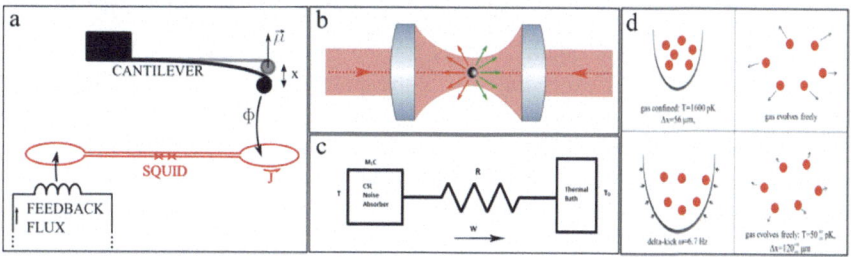

Fig. 3 Simplified sketch of some of the noninterferometric methods to test collapse models discussed in this contribution. **a** Measuring the mechanical noise induced by CSL using an ultracold cantilever detected by a SQUID (adapted from Ref. [24]); **b** Measuring the mechanical noise induced by CSL using a levitated nanoparticle detected optically (adapted from Ref. [45]); **c** Measuring the heating induced by CSL in a solid matter object cooled to very low temperature (adapted from Ref. [46]); **d** Measuring the increase of kinetic energy induced by CSL in a ultracold atoms cloud (adapted from Ref. [47])

hand, an all-electrical readout would potentially allow for a better ultimate test of collapse models, but at the price of a very poor detection sensitivity, which could make the experiment hardly feasible. The authors argue that a Paul-trapped nanoparticle, with an oscillating frequency of 1 kHz, cooled in a cryostat at 300 mK with an optical readout may be able to probe the CSL collapse rate down to 10^{-12} Hz at $r_C = 10^{-7}$ m. A SQUID-based readout , if viable, could theoretically allow to reach 10^{-14} Hz.

A recent experiment employing a nanoparticle in a Paul trap with very low secular frequencies at ~100 Hz and low pressure has demonstrated ultranarrow linewidth $\gamma/2\pi = 82\,\mu$Hz [48]. This result has been used to set new bounds on the dissipative extension of CSL. This experiment may be able to probe the current limits on the CSL model in the near future, once it will be performed at cryogenic temperature and the main sources of excess noise, in particular bias voltage noise, will be removed.

Magnetic levitation, while less developed, has the crucial advantage of being completely passive. Furthermore the trap frequencies can be quite low, in the Hz range. Three possible schemes can be devised: levitation of a diamagnetic insulating nanoparticle with strong external field gradients [49, 50], levitation of a superconducting particle using external currents [8, 51–53], and levitation of a ferromagnetic particle above a superconductor [54].

The first approach has been recently considered in the context of collapse models [50]. The experiment was based on a polyethylene glycol microparticle levitated in the static field generated by neodymium magnets and optical detection. The experiment has been able to set an upper bound on the CSL collapse rate $\lambda < 10^{-6.2}$ Hz at $r_C = 10^{-7}$ m, despite being performed at room temperature. A cryogenic version of this experiment should be able to approach the current experimental limits on CSL.

The second and third approach based on levitating superconducting particles are currently investigated by a handful of groups [8, 51–54], but no experiment has so far reached the experimental requirements needed to probe collapse models. However, a significant progress has been recently achieved: a ferromagnetic microparticle levitated above a type I superconductor (lead) and detected using a SQUID, has demonstrated mechanical quality factors for the rotational and translational rigid body mechanical modes exceeding 10^7, corresponding to a ringdown time larger than 10^4 seconds [53]. The noise is this experiment is still dominated by external vibrations. However, as the levitation is completely passive and therefore compatible with cryogenic temperatures, this appears as an excellent candidate towards near future improved tests of collapse models.

2.3   Concluding Remarks on Testing Quantum Mechanics in the Context of Collapse Models

We have discussed avenues for non-interferometric and interferometric tests of the linear superposition principle of quantum mechanics in direct comparison to predictions from collapse models which break the linear/unitary evolution of the wavefunction. As matters stand both non-interferometric and interferometric set already bounds on the CSL collapse model, while those from non-interferometric tests are stronger by orders of magnitude. The simple reason lyes in the immense difficulty to experimentally generate macroscopic superposition states, however a number of proposals have been made and experimentalists are set to approach the challenge.

We want to close by mentioning that there are possible other experimental platforms which could set experimental bunds on collapse models and it would be of interest to study those in detail. Collapse models predict a universal classical noise field to fill the Universe and in principle couple to any physical system. In the simplest approach the experimental test particle can be regarded as a two-level system, as typically described in quantum optics. Then the collapse noise perturbs the two-level system and emissive broadening and spectral shifts can be expected, unfortunately out of experimental reach at the moment [55]. The minuscule collapse effect on a single particle (nucleon) needs some sort of amplification mechanism which usually comes with an increase of the number of constituent particles. However, ultra-high precision experiments have improved a lot in recent years. For instance much improved ultra-stable Penning ion traps are used to measure the mass of single nuclear particles, such as the electron, proton, and neutron, with an ultra-high precision to test quantum electrodynamics predictions [56]. In principle also here the effect of collapse models should become apparent. Any theoretical predictions are difficult as relativistic versions of collapse models still represent a serious formal challenge. Other high potentials for testing collapse are ever more precise spectroscopies of simple atomic species with analytic solutions such as transitions in hydrogen [57] and needless to say atomic clocks [58].

As tests move on to set stronger and stronger bounds, we have to remain open to actually find something new. It is so easy to disregard tiny observed effects as unknown technical noise. In the case of direct testing collapse noise it is a formidable theoretical challenge to think about possible physics responsible for collapse, satisfying the constrains given by the structure of the collapse equation: the noise has to be classical and stochastic. Such concrete physics models will predict a clear frequency fingerprint, should we ever observe the collapse noise field.

3 Testing the Interplay Between Quantum Mechanics and Gravity

Here we will be concerned with table-top experiments in the non-relativistic regime as these experiments may provide a new access to shine light on the quantum—gravity interplay. Therefore the main emphasis is to explore possible routes to enter the new parameter regime, where both quantum mechanics and gravity are significant, see Fig. 4). This means the mass of the object has to be large enough to show gravity effects while also not being too large to still allow for the preparation of non-classical features of the behaviour of that massive object. That regime where both physical effects, the quantum and the gravity, could be expected to be relevant is at around the Planck mass, which is derived from the right mixture of fundamental constants ($\hbar$ Planck's constant, c speed of light, G gravitational constant) $m_{pl} = \sqrt{\hbar c / G} = 2.176470(51) \times 10^{-8}$ kg (the official CODATA, NIST) or below. No quantum experiment has been performed in that mass range.

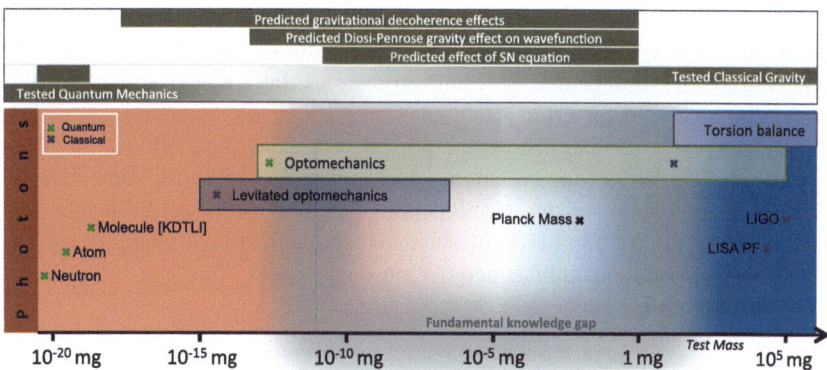

Fig. 4 Exploration map of mass: Mass range of the test mass as explored by experiments. Experiments to detect gravity have been done in the classical domain, *right hand side of picture*, with comparable large masses. Quantum experiments are routinely performed by using objects of much smaller masses so that gravity effects do not become visible or relevant. Neutron and atom matterwave interferometers are different as the test mass there is very small [the mass of a single neutron or atom], but in a spatial superposition state. The desired mass range for—at least some of—the experiments summarized in this review article is at the overlap between sufficiently large mass to see significant effects of gravity of the particle itself, while the particle can be maintained in a non-classical state. The domain where massive particles can be prepared in such non-classical states is *on the left hand side* of the picture

When we refer to quantum mechanical behaviour of massive systems, we mean the centre of mass motion of such a system, which may consist of many atoms. Surely, there are many other [we call those internal] degrees of freedom of the same system such as electronic states or vibrations and rotations which are described as relative motions of the atoms forming the large object, but here we are not concerned with those. When we talk about superpositions, we mean spatial superpositions, in the sense of a the centre of mass of a single particle, which can be elementary or composite, being *here* and *there* at a given time, the Schrödinger cat state. The most massive complex quantum system, which has been experimentally put in such a superposition state, are complex organic molecules of a mass on the order of $m_{max} = 10^{-22}$ kg [9].

Typically for gravity experiments there are two masses involved, the source mass which generates a gravitational field, potential or curvature of space-time (the source mass has usually a big mass) and the test mass which is probing the gravity effect generated by the source mass. Torsion balances are the classic device for typical gravity experiments. We think there are two regimes interesting for experimental investigation: (1) the regime where a quantum system is the test mass and interacts with a large external source mass. This is the regime where neutron and atom interferometry are already very successful and provide tools for precise measurements of gravity effects. (2) the regime where the quantum system itself carries sufficient mass to be the source mass and to allow for related quantum gravity effects to become experimentally accessible. So far there has been no convincing experiment in the second regime. Any experiment performed in that second regime will ultimately give insight into the interplay between gravity and quantum mechanics. Test of the Schrödinger-Newton equation and of quantum effects in gravity fall in the latter regime. It may very well be that there are surprises waiting for us if we become able to probe that regime by experiments.

In the following we shall discuss the prospects to experimentally test the semiclassical Schrödinger-Newton equation, which plays also a role for some ideas of gravity induced collapse of the wavefunction such as put forward by Roger Penrose [59], gravitational decoherence such as some ideas to investigate the gravity effects within a spatial quantum superposition state.

3.1 Proposals for Experimental Tests of the Schrödinger-Newton equation

What is the gravitational field of a quantum system in a spatial superposition state? The seemingly most obvious approach, the perturbative quantization of the gravitational field in analogy to electromagnetism, makes it alluring to reply that the spacetime of such a state must also be in a superposition. The non-renormalizability of said theory, however, has also inspired the hypothesis that a quantization of the gravitational field might not be necessary after all [60, 61]. Rosenfeld already expressed the thought that the question whether or not the gravitational field must be quantized can only be answered by experiment: *There is no denying that, considering the universality of the quantum of action, it is very tempting to regard any classical theory as a limiting case to some quantal theory. In the absence of empirical evidence, however, this temptation should be resisted. The case for quantizing gravitation, in particular, far from being straightforward, appears very dubious on closer examination.* [60]

Adopting this point of view, an alternative approach to couple quantum matter to a classical space-time is provided by a fundamentally semi-classical theory that is by replacing the source term in Einstein's field equations for the curvature of classical space-time, energy-momentum, by the *expectation value* of the corresponding quantum operator [60, 62]:

$$R_{\mu\nu} + \frac{1}{2} g_{\mu\nu} R = \frac{8\pi G}{c^4} \langle \Psi | \hat{T}_{\mu\nu} | \Psi \rangle. \tag{7}$$

Of course, such presumption is not without complications. For instance, in conjunction with a no-collapse interpretation of quantum mechanics it would be in blatant contradiction to everyday experience [63]. Moreover, the nonlinearity that the back-reaction of quantum matter with classical space-time unavoidably induces cannot straightforwardly be reconciled with quantum nonlocality in a causality preserving manner [64, 65]. Be that as it may, there is no consensus about the conclusiveness of these arguments [66–68]. The enduring quest for a theory uniting the principles of quantum mechanics and general relativity gives desirability to having access to hypotheses which could be put to an experimental test in the near future.

In the non-relativistic limit, the assumption of fundamentally semi-classical gravity yields a non-linear, nonlocal modification of the Schrödinger equation, commonly referred to as the Schrödinger–Newton equation [34, 69, 70]. After a suitable approximation [70], for the center of mass of a complex quantum system of mass M in an external potential V_{ext} it reads:

$$i\hbar \frac{\partial}{\partial t} \psi(t, \mathbf{r}) = \left(\frac{\hbar^2}{2M} \nabla^2 + V_{\text{ext}} + V_g[\psi] \right) \psi(t, \mathbf{r}) \tag{8a}$$

$$V_g[\psi](t, \mathbf{r}) = -G \int d^3 r' \, |\psi(t, \mathbf{r}')|^2 \, I_{\rho_c}(\mathbf{r} - \mathbf{r}'). \tag{8b}$$

The self-gravitational potential V_g depends on the wavefunction, and hence renders the equation nonlinear. The function I_{ρ_c}, which models the mass distribution of the considered system, will be defined below.

The Schrödinger–Newton equation has primarily been discussed in the context of gravitationally induced quantum state reduction [71, 72]. Its relevance for a possible experimental test of the necessity to quantize the gravitational field was pointed out by Carlip [61]. First ideas how to test such kind of nonlinear, self-gravitational effects focused on the spreading of a free wavefunction in matter-wave interferometry experiments [9, 37, 61, 69, 70]. Recently, other experimental test have been proposed

including one based on the internal dynamics of a squeezed coherent ground state of
a micron-sized silicon particle in a harmonic potential. We will now discuss further
ideas for testing the Schrödinger-Newton equation.

3.1.1 Proposed Direct Tests of Schrödinger-Newton Equation: Wavefunction Expansion

The direct test of the Schrödinger-Newton (SN) equation is by studying the free
expansion of the wavefunction of sufficiently massive objects. Then a contraction of
the wave function according to the SN self-gravity effect should have a consequence
on that expansion, competing with its natural Schrödinger's dynamics spread. Clearly,
because to the weakness of gravitation interaction, the mass has to be sufficiently
large while the object has to remain in a state which can be described by a centre of
mass quantum wavefunction, meaning the spatial extent of the wavefunction should
be detectable for the full duration of the evolution. See Fig. 5 for the mass-time
parameter space required to observe the predicted SN effect directly, which has been
studied extensively. While analytic solutions of the SN equation are difficult and
even numerical simulations are non-trivial.

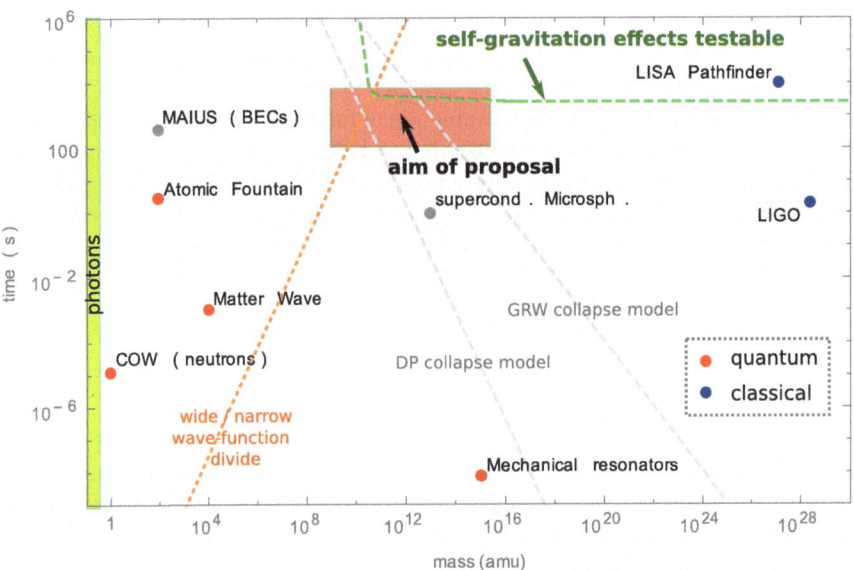

Fig. 5 Direct Test of Schrödinger-Newton (SN) wavefunction evolution: The mass-time plot
to illustrate the parameter range which needs to be reached for direct SN wavefunction evolution
experiments. This clearly needs to be done without external gravity and other forces/interactions
and therefore an experiment in space appears a likely option. The red area shows the parameter
range for a proposed space mission to test the SN effect

One possible experimental scenario would be a molecule interferometry experiment [9]. While such matterwave experiments probe spatial superposition states of large molecules—the SN contraction effect could also be observed for a free expansion of a singular wave function originated from a point in space. The key is that the mass of the evolving quantum object has to be comparable large, much larger than the mass achieved in present molecule interferometry experiments. Cold atoms and even BEC of atoms, which benefit from the multitude of coherent manipulation, control and cooling schemes do not seem to have large enough mass in order to show the SN expansion/contraction effect. Clearly one needs a high mass at the same time as access to the coherent quantum evolution of the objects wavefunction. The high mass and the long expansion times to be studied challenge the experimental realisation.

Therefore, should direct tests of the SN equation be done in space? Yes, at this point there seems to be no other way to allow the wavefunction expansion for long enough, typically some hundred seconds, see Fig. 5. Proposals to levitate massive particles (optically or magnetically) and therefore to compensate for the drop in Earth's gravity have not been realised and are more problematic for SN test. The levitated tests rely on proposed techniques to accelerate the wavefunction expansion artificially by optical or magnetic field gradients. That acceleration would have the potential to wash out completely the fragile SN effect.

3.1.2 Proposed Indirect Tests of SN Equation

Indirect SN effects have been predicted for optomechanics systems which are comparably massive and on the verge to be quantum, see Fig. 4. Such effects are very small, can be overwhelmed by noise effects in the experiments, but can be done on the table-top. Therefore these tests represent a serious experimental challenge, while proposed to be possible with available technology. Two optomechanics experimental cases and the study of the SN dynamics in non-linear optics analogs are mentioned:

A. *SN rotation of squeezed states* The mechanical motion of an optomechanical system, clamped or levitated, is squeezed. Quantum squeezing of clamped optomechanics has been realised experimentally already, while a classical analog has been demonstrated for a levitated system. An optical homodyne detection of both field quadratures of the mechanical state is plotted and shows the cigar-shaped state, see in Fig. 6 left. The SN equation predicts an extra rotation of the squeezed phase-space distribution [73].

B. *SN energy shifts of mechanical harmonic oscillator* A further theoretical study [74] predicts SN related shifts of the Eigenenergy levels of the quantum harmonic oscillator describing the optomechanical system, see for an illustration of the multiple energy shift effects the Fig. 6 right. There different effects for the so-called wide and narrow wavefunction regimes are predicted for the situations that the spatial extent of the centre of mass motion wavefunction is larger (wide wavefunction regime) or smaller (narrow wavefunction regime) than the physical size of the massive object. A detailed experimental scenario has been worked out and awaits its realisation in an actual laboratory.

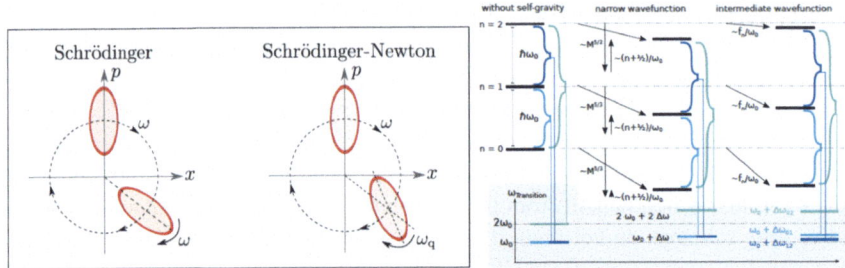

Fig. 6 Indirect Tests of the Schrödinger-Newton equation: *Left Panel*: Phase space plot of mechanical squeezed state with extra rotation of state distribution according to the SN effect. Left side: according to standard quantum mechanics, both the vector $(\langle x \rangle, \langle p \rangle)$ and the uncertainty ellipse of a Gaussian state for the centre of mass (CM) of a macroscopic object rotate clockwise in phase space, at the same frequency $\omega = \omega_{CM}$. Right side: according to the CM Schrödinger-Newton equation, $(\langle x \rangle, \langle p \rangle)$ still rotates at ω_{CM}, but the uncertainty ellipse rotates at $\omega_q = (\omega_{CM}^2 + \omega_{SN}^2)^2$. Picture taken from [73]. *Right Panel*: Schematic overview of the effect of the Schrödinger-Newton equation on the spectrum. The top part shows the first three energy eigenvalues and their shift due to the first order perturbative expansion of the Schrödinger-Newton potential. The bottom part shows the resulting spectrum of transition frequencies. In the narrow wavefunction regime (middle part), all energy levels are shifted down by an n-independent value minus an n-proportional contribution that scales with the inverse trap frequency. In the intermediate regime, where the wavefunction width becomes comparable to the localization length scale of the nuclei, this n-proportionality does no longer hold, leading to a removal of the degeneracy in the spectrum. Picture and caption taken from [74]

C. Non-linear optics simulation of the SN equation Specific delocalised non-linearities in optical systems, typically just a piece of glass with a large refractive index, show a very similar type of dynamics for the propagation of light though that system if compared to SN dynamics. The analog holds at least in (1+1) space-time dimensions. The analog provides an interesting option to study the dynamics of the SN equation in a parameter regime complementary to numeric simulations. Some experiments have been already performed [75, 76] to study cosmological settings of the SN equation such as exotic Boson stars. The main question remains, what can we ultimately learn from optics analog experiments. Do we really learn about gravity? No, but we learn about the formal analog dynamics which is hard to calculate or simulate otherwise.

3.2 Gravitational Decoherence Effects

Tests of gravitational decoherence are based on the the straight-forward approach to generate a spatial superposition state (or any other non-classical state) of a massive particle and test if such a state decoheres according to (classical or quantum) gravity. Clearly, the experimental challenge is the preparation of such a state of sufficient mass. Typical experiments involve matterwave interferometers and quantum optomechanics. While the larges mass is given again by molecule interferometry—some of the effects (such as time dilation) are more promising to be tested in smaller mass systems such as cold atom interferometry, as those can be prepared in larger size superposition states to pick up a larger dephasing or decoherence effect. While on first sight it appears that only massive systems can be used for the test, we know that GR effects also exist for photons [77].

A. Gravitational decoherence affecting superpositions One of the proposed effects is by GR time dilation [78, 79], which is picked up as a dephasing effect for a matterwave interferometer for the propagation of the wavefunction along the two different arms—ultimately resulting in a reduction of the visibility of the interference pattern. The effect has been predicted to scale with the number of all internal degrees of freedom, which are involved in the energy-momentum tensor on the right hand side of Einstein's equations and therefore to affect the spacetime curvature and therefore gravity.

Atom interferometry tests, profiting from the high control on the centre of mass motion of cold atoms, e.g. in 10 m fountain and with sensitively on the verge of 10^{-19}, of the time delation effect appear most promising at the moment, while the theoretical details of the effect are still debated. As a universal decoherence effect to explain the evident macroscopic quantum to classical transition, it is clear that that time dilation decoherence should it exist is weaker by many order of magnitude than know environmental effects such as decoherence due to collisions by an even very diluted background gas [80], which leaves the usefulness of the GR effect in question.

To be more precise, each (internal) degree of freedom of the particle is regarded as a clock running at a typical frequency, but depending via GR time dilation on the local gravitational environment. Then each single clock if separated between the two different paths of an interferometer will be sensitive to the relative duration of time and therefore dephase. This experiment has been realised as a proof of principle experiment with atomic chips [81], where the much larger spatial separation in other atomic interferometers [17] will help to improve the sensitivity to observe the predicted effect to test whether GR time dilation can be regarded as a universal source of decoherence to explain the macroscopic quantum to classical transition of physical systems, ultimately to explain the existence of the classical world.

B. Gravitational effect in dynamical reduction models Dynamical reduction or collapse models have been formulated to explain the quantum to classical transition on a fundamental level and in complement to decoherence models [14]. While the physics reason for the collapse to occur is explained by the existence of a universal classical and random noise field, the physics origin of that field is still debated. Gravity to be a candidate for the collapse field has to fulfil that two conditions of being classical and random. While the classicality is more straight forward, the implementation of a generic stochastic version of gravity represents a challenge. Some attempts have been undertaken and can also been seen as a stochastic modification of the Schrödinger-Newton equation, which was discussed in Sect. 3.1 [34, 82–84]. Tests of such gravity collapse models follow the same logic as tests of collapse models and in general a set of parameters has to be fulfilled. For more details related to

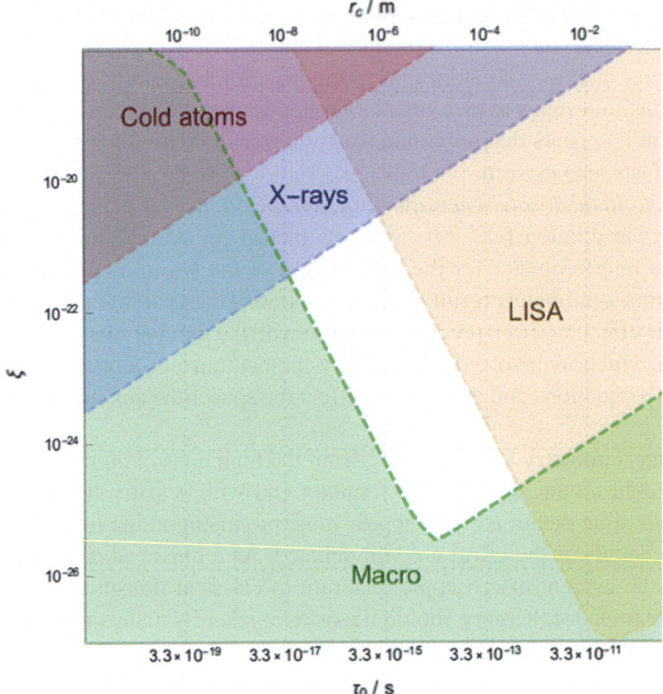

Fig. 7 Parameter map for gravity induced collapse models: (ξ, r_C) or equivalently (ξ, τ_0) parameter diagram of the gravity-induced collapse model. The white area is the allowed region. The blue shaded region (X-rays) is excluded by data analysis of X-rays measurements. The orange shaded region (LISA) is excluded from data analysis of LISA Pathfinder. The green shaded region (Macro) is an estimate of the region excluded by the requirement that the collapse is strong enough to localize macroscopic objects. Note that X-ray measurements sample the high frequency region of the spectrum (10^{18} Hz) and would disappear if the noise correlator has a cutoff below such frequencies, which is plausible. In such a case, the stronger upper bound on the left part of the plane is given by data analysis with cold atom experiments (Cold atoms) [35]. Picture and caption has been taken from [84]

experimental test we refer to [84], where the Fig. 7) has been taken from. While the bounds on gravity collapse models in Fig. 7 relate to experiments already done, future experiments proposed to close the raining gap in the parameter plot involve those to generate large and massive quantum superpositions [30–32, 85]. Such experiments are currently under development in the laboratories.

  C. Gravity induced collapse of the wavefunction Other ideas which are less related to the formalism of collapse models, but do explain the collapse of the wavefunction according to gravity are those independently by Diosí [86] and Penrose [87]. The best way to test those models is by large mass matterwave interferometry, where the mass has to be beyond the presently reached limit of molecule interferometry by many orders of magnitude. This means to test such models requires to preparation of large masses in non-classical states and optomechanical or magnetomechanical systems look most promising for the test [88–90]. Proposed experiments along those lines involve [31, 91].

D. Competing effects for matterwave interferometry In order to be able to see such gravity effects and how they collapse or decohere the wavefunction in matter-wave based experiments all competing environmental decoherence processes have to be suppressed, which is the major experimental challenge in order to perform the experiments. Dominating decoherence effects are due to collisions with background gas, collisional decoherence [92] and the effects because of exchange of thermal radiation between the quantum system and the environment [30, 85]. Magnetic levitation of superconducting microparticles by definition avoids all effects related to internal temperature radiation as the experiment is cryogenic and on top of that all noises related to lasers are removed as well [31] which represents a huge advantage compared to optomechanics test. Further vibrations set serious constraints to all mechanics based test of wavefunction collapse and gravity.

E. The case for space Ultimately a test of gravity decoherence and gravity induced collapse of the wavefunction would benefit from large masses of the particles in superposition states as well as long lifetimes of those superposition states in order to observe the extremely weak effects. The space proposal on macroscopic quantum resonators (MAQRO) [33] would be able to fulfil such all those conditions. A community has started to work towards such a test in space and to propose a related mission.

3.3   The Gravity of a Quantum State—Revisited

What gravitational field is generated by a massive quantum system in a spatial superposition? Despite decades of intensive theoretical and experimental research, we still do not know the answer. On the experimental side, the difficulty lies in the fact that gravity is weak and requires large masses to be detectable. However, it becomes increasingly difficult to generate spatial quantum superpositions for increasingly large masses, in light of the stronger environmental effects on such systems. Clearly, a delicate balance between the need for strong gravitational effects and weak decoherence should be found. We show that such a trade off could be achieved in an optomechanics scenario that allows to witness whether the gravitational field generated by a quantum system in a spatial superposition is in a coherent superposition or not. We estimate the magnitude of the effect and show that it offers perspectives for observability.

Quantum field theory is one of the most successful theories ever formulated. All matter fields, together with the electromagnetic and nuclear forces, have been successfully embedded in the quantum framework. They form the standard model of elementary particles, which not only has been confirmed in all advanced accelerator facilities, but has also become an essential ingredient for the description of the universe and its evolution.

In light of this, it is natural to seek a quantum formulation of gravity as well. Yet, the straightforward procedure for promoting the classical field as described by general relativity, into a quantum field, does not work. Several strategies have been put forward, which turned into very sophisticated theories of gravity, the most advanced being string theory and loop quantum gravity. Yet, none of them has reached the goal of providing a fully consistent quantum theory of gravity.

At this point, one might wonder whether the very idea of quantizing gravity is correct [59–64, 66, 66–68, 93, 94]. At the end of the day, according to general relativity, gravity is rather different from all other forces. Actually, it is not a force at all, but a manifestation of the curvature of spacetime, and there is no obvious reason why the standard approach to the quantization of fields should work for spacetime as well. A future unified theory of quantum and gravitational phenomena might require a radical revision not only of our notions of space and time, but also of (quantum) matter. This scenario is growing in likeliness [95–97].

From the experimental point of view, it has now been ascertained that quantum matter (i.e. matter in a genuine quantum state, such as a coherent superposition state) couples to the Earth's gravity in the most obvious way. This has been confirmed in neutron, atom interferometers and used for velocity selection in molecular interferometry. However, in all cases, the gravitational field is classical, i.e. it is generated by a distribution of matter (the Earth) in a fully classical state. Therefore, the plethora of successful experiments mentioned above does not provide hints, unfortunately, on whether gravity is quantum or not.

In a recent paper [98], we discuss an approach where a quantum system is forced in the superposition of two different positions in space, and its gravitational field is explored by a probe (Fig. 8). Using the exquisite potential for transduction offered by optomechanics, we can in principle witness whether the gravitational field is the

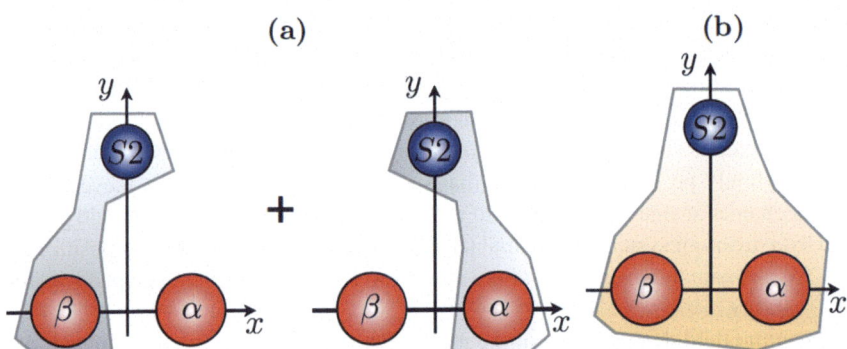

Fig. 8 Schematic representation of the two-body setup. S1 is prepared in a spatial superposition along the x direction (red balls). S2 is initially prepared in a localized wavepacket (blue ball), and it probes the gravitational field generated by S1. **a** The gravitational field acting on S2 is a linear combination of gravitational fields produced by S1 being in a superposed state. **b** The semi-classical treatment of gravity, where the gravitational field acting on S2 is that produced by a total mass m_1 with density $\frac{1}{2}\left(|\alpha(r)|^2 + |\beta(r)|^2\right)$

superposition of the two gravitational fields associated to the two different states of the system, or not. The first case amounts to a quantum behavior of gravity, the second to a classical-like one. We have illustrated the dynamics of an optomechanical system probing the gravitational field of a massive quantum system in a spatial superposition. Two different dynamics are found whether gravity is treated quantum mechanically or classically. Here, we propose two distinct methods to infer which of the two dynamics rules the motion of the quantum probe, thus discerning the intrinsic *nature* of the gravitational field. Such methods will be then eventually able to falsify one of the two treatments of gravity. A similar proposal has been made for angular superpositions [99].

The considered setup is formed of two systems interacting gravitationally. All non-gravitational interactions are considered, for all practical purposes, negligible. The first system (S1) has a mass m_1, and it is initially prepared in a spatial superposition along the x direction. Its wave-function is $\psi(r_1) = \frac{1}{\sqrt{2}}(\alpha(r_1) + \beta(r_1))$, where $\alpha(r_1)$ and $\beta(r_1)$ are sufficiently well localized states in position, far from each other in order to prevent any overlap. Thus, we can consider them as distinguishable (in a macroscopic sense), and we approximate $\langle \alpha | \beta \rangle \simeq 0$. The second system (S2) will serve as a point-like probe of the gravitational field generated by S1, it has mass m_2 and state $\phi(r_2)$. The state $\phi(r_2)$ is initially assumed to be localized in position and centered along the y direction [cf. Fig. 8]. The question we address is: which is the gravitational field, generated by the quantum superposition of S1, that S2 experiences? We probe the following two different scenarios.

Quantum Gravity Scenario. Although we do not have a quantum theory of gravity so far, one can safely claim that, regardless of how it is realized, it would manifest in S1 generating a superposition of gravitational fields. As discussed in the introduction, the assessment of this property precedes the quest to ascertain the existence of the graviton and the characterization of its properties, at least as far as the static, low-energy, non-relativistic regime we are considering is concerned. Linearity is the very characteristic trait of quantum theory, and one expects it to be preserved by any quantum theory of gravity.

The reaction of S2 is then to go in a superposition of being attracted towards the region where $|\alpha\rangle$ sits and where $|\beta\rangle$ does. The final two-body state will have the following entangled form

$$\Psi_{\text{QG}}^{\text{final}}(r_1, r_2) = \frac{\alpha(\mathbf{r}_1)\phi_\alpha(\mathbf{r}_2) + \beta(\mathbf{r}_1)\phi_\beta(\mathbf{r}_2)}{\sqrt{2}}, \tag{9}$$

where $\phi_\alpha(\mathbf{r}_2)$ ($\phi_\beta(\mathbf{r}_2)$) represents the state of S2 attracted towards the region where $|\alpha\rangle$ ($|\beta\rangle$) rests. The motion in each branch of the superposition is produced by the potential

$$\hat{V}_\gamma(\hat{r}_2) = -Gm_2 \int dr_1 \frac{\rho_\gamma(\mathbf{r}_1)}{|r_1 - \hat{r}_2|}, \qquad (\gamma = \alpha, \beta). \tag{10}$$

where $\rho_\gamma(\mathbf{r}_1)$ is the mass density of S1, centred in $\langle \hat{r}_1 \rangle_\gamma = \langle \gamma | \hat{r}_1 | \gamma \rangle$. We assume that S1 does not move appreciably during the time of the experiment (also quantum fluctuations can be neglected); clearly, such a situation can be assumed only as long as the S1 superposition lives. We further assume that its mass density is essentially spheric, so that the gravitational interaction can be approximated by

$$\hat{V}_\gamma(\hat{r}_2) \approx -\frac{Gm_1m_2}{|\langle \hat{\mathbf{r}}_1 \rangle_\gamma - \hat{\mathbf{r}}_2|}, \quad (\gamma = \alpha, \beta). \tag{11}$$

Semiclassical Gravity Scenario. The second scenario sees gravity as fundamentally classical. In this case, it is not clear which characteristics one should expect from the gravitational field generated by a superposition. However, in analogy with classical mechanics, one can assume that is the mass density $\rho(\mathbf{r}_1) = (\rho_\alpha(\mathbf{r}_1) + \rho_\beta(\mathbf{r}_1))/2$ of the system in superposition that produces the gravitational field. This is also what is predicted by the Schrödinger-Newton equation (see Sect. 3). The final two-body state will be of the form

$$\Psi_{\mathrm{CG}}^{\mathrm{final}}(r_1, r_2) = \frac{\alpha(\mathbf{r}_1) + \beta(\mathbf{r}_1)}{\sqrt{2}} \phi(r_2), \tag{12}$$

where the difference with Eq. (9) is clear. The gravitational potential becomes

$$\hat{V}_{\mathrm{cl}}(\hat{r}_2) \approx \frac{1}{2} \sum_{\gamma=\alpha,\beta} \hat{V}_\gamma(\hat{r}_2), \tag{13}$$

where $\hat{V}_\gamma(\hat{r}_2)$ can be eventually approximated as in Eq. (11).

Experimental progress with levitated mechanical systems makes is possible to reach a parameter regime to experimentally resolve the difference between the quantum and semiclassical scenarios as shown in our paper [98]. Other interferometric [100, 101] and non-interferometric [102] tests of the nature of gravity have been proposed. They are based on the detection of entanglement between two probes, respectively coupled to two different massive systems, which interact through gravity (NV center spins for [100] and cavity fields for [102]). Clearly, to have such entanglement, each of the three couples of interconnected systems (probe 1, system 1, system 2 and probe 2) there considered needs to be entangled on their own. Moreover, the entanglement between the two massive systems is inevitably small due to its gravitational nature. Conversely, our proposal benefits from having only a single massive system involved in the interconnection, which reduces correlation losses. In addition, we provide a second method for discerning the nature of gravity: the individuation of a second peak in the DNS. The latter does not rely on delicate measurements of quantum correlations but can be assessed through standard optomechanical detection schemes.

3.4 Concluding Remarks on Testing the Interplay of Quantum Mechanics and Gravity in the Low Energy Regime

While matterwave interferometer experiments have been performed in the low mass regime, see Fig. 4, the higher mass range, all the way up to milligram masses is unexplored by any experiment and especially not by any quantum experiment. Optomechanical devices and especially levitated particles are able to bridge this enormous mass gap; being in a quantum mechanical state and very massive at the same time. Levitated mechanical systems hold promise to test new physics in that new mass range. A variety of theoretical proposals and ideas for the interplay between quantum mechanics and gravity will become testable in this very mass range. The study of gravitational decoherence, the Schrödinger-Newton equation and the gravity of a quantum state provide concrete routes for experimental exploration.

4 Simulation of the Stern Gerlach Experiment Using Wigner Functions

The Stern-Gerlach (SG) experiment [103] is a seminal example of a quantum experiment involving coupling between internal and external degrees of freedom. In this experiment, an electron or nuclear spin interacts with a spatially inhomogeneous magnetic field through the magnetic Zeeman interaction. The outcome of the Stern-Gerlach experiment is, of course, "well-known": an incident molecular beam of particles with spin-1/2 is separated by the inhomogeneous magnetic field into two beams, each corresponding to particles with well-defined spin angular momenta along the field direction. But how does this separation happen *in detail*, on the level of the spatial quantum state?

In a recent article [104] we used an *extended Wigner function* (EWF) which includes the presence of internal degrees of freedom in the propagating particle, and the coupling of those internal degrees of freedom to inhomogeneous external fields (Fig. 9).

The Wigner function $W(x, p)$ is a joint quasi-probability density function defined over the combined domains of the spatial coordinate(s) x and its associated momentum (momenta) p. It is defined as a Weyl integral transform of the density operator $\hat{\rho} = |\psi\rangle\langle\psi|$, of the following form:

$$W(x, p) = \frac{1}{h} \int e^{-\frac{ips}{\hbar}} \langle x + \tfrac{s}{2}|\hat{\rho}|x - \tfrac{s}{2}\rangle \, ds. \tag{14}$$

Consider a particle with a finite number of internal quantum states. In the discussion below, we refer to these internal states as "spin states", although the same formalism applies to non-spin degrees of freedom, such as quantized rotational and

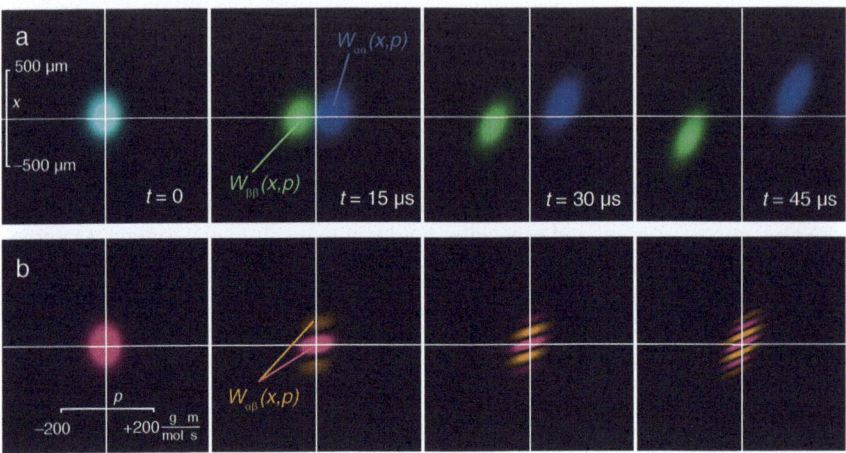

Fig. 9 **a** Evolution of $W_{\alpha\alpha}$ and $W_{\beta\beta}$ under the influence of a magnetic field gradient in a Stern-Gerlach experiment on Ag atoms in a field gradient of $10\,\mathrm{G}\,\mu\mathrm{m}^{-1}$, moving at a velocity of $550\,\mathrm{m/s}$ (rms velocity at an oven temperature of 1300 K). **b** Evolution of the real part of the off-diagonal element $W_{\alpha\beta}$, assuming a coherent state initially polarised along the x-axis. The strength of the magnetic field gradient has been reduced by a factor of 5×10^4 compared to A in order to make the spatial modulation visible. The shearing of the fine structure of the Wigner function represents decoherence. Figure and capture taken from Ref. [104]

vibrational states. We extend the Wigner function by combining it with the density operator formalism commonly used in the quantum description of magnetic resonance. The definition of the Wigner function is extended by projecting the density operator onto the spin-state specific position state $|x, \eta\rangle$, where $\eta = \alpha, \beta, \ldots$ denotes the spin state. This results in a Wigner probability density matrix $W_{\eta\xi}(x, p)$, whose elements depend parametrically on the positional variables and their associated momenta:

$$W_{\eta\xi}(x, p) = \frac{1}{h} \int e^{-\frac{ips}{\hbar}} \langle x + \tfrac{s}{2}, \eta | \hat{\rho} | x - \tfrac{s}{2}, \xi \rangle \, ds. \tag{15}$$

This means the extended Wigner function can be used to directly simulate the SG experiment. In the Stern-Gerlach experiment, a beam of spin-1/2 particles is exposed to a lateral magnetic field gradient. We define the axis of the molecular beam apparatus as z, and assume that the magnetic field varies in the transverse x-direction. The potential energy part of the Hamiltonian in the presence of an external magnetic field $\mathbf{B}$ is then given by

$$U(\hat{\mathbf{S}}, x) = -\hbar\gamma \mathbf{B}(\mathbf{x}) \cdot \hat{\mathbf{S}}. \tag{16}$$

The original magnet design used by Stern and Gerlach [103] produces divergent magnetic field lines at the location of the beam. This corresponds to a biaxial magnetic field gradient tensor, requiring two spatial dimensions to be included in the Wigner function. To avoid this complication, we use a different arrangement, in which the magnetic field gradient is uniaxial. In this case, the magnetic field lines are all parallel, but vary in density in the direction perpendicular to the magnetic field itself. Magnetic fields of this type occur in quadrupole polarisers.

We assume the magnetic field points along the y-axis, and varies linearly in magnitude along the x-axis, $\mathbf{B}(\mathbf{x}, \mathbf{y}, \mathbf{z}) = \left(\mathbf{B}_{\mathbf{y0}} + \mathbf{x}\,\mathbf{G}_{\mathbf{xy}}\right)\mathbf{e}_{\mathbf{y}}$, where B_{y0} is the magnetic field at $x = 0$, and $G_{xy} = \partial B_y/\partial x$. This field is fully consistent with Maxwell's equations, since it satisfies $\nabla \cdot \mathbf{B} = \mathbf{0}$. The field gradient has only a single non-zero cartesian component $\nabla\mathbf{B} = \mathbf{G}_{\mathbf{xy}}\,\mathbf{e}_{\mathbf{x}}\mathbf{e}_{\mathbf{y}}$. We choose the spin states $|\alpha\rangle$ and $|\beta\rangle$ as the eigenstates of $\hat{S}_y$, such that the matrix elements of the potential part of the Hamiltonian are

$$
\begin{aligned}
U_{\alpha\alpha}(x) &= -\tfrac{\gamma\hbar}{2} B_y(x) & U_{\alpha\beta}(x) &= 0 \\
U_{\beta\alpha}(x) &= 0 & U_{\beta\beta}(x) &= +\tfrac{\gamma\hbar}{2} B_y(x).
\end{aligned}
\tag{17}
$$

The resulting equations of motion for the EWF matrix elements are given in the SI.

In its original form, the Stern-Gerlach experiment was conducted on a beam of Ag atoms emanating from an oven at a temperature of about 1300 K. The magnetic field gradient was of the order of 10 G/cm over a length of 3.5 cm [105]. For simplicity, we ignore the nuclear spin of Ag, and treat the atoms as (electron) spin 1/2 particles. In the case of magnetic fields larger than the hyperfine splitting (about 610 G in the case of Ag), this is a good approximation, since the nuclear and the electron spin states are essentially decoupled. The root mean square velocity of Ag atoms 1300 K is approximately 550 m/s. After leaving the oven, the Ag atoms are collimated by a pair of collimation slits 30 μm wide and separated by 3 cm. The longitudinal momentum of the silver atoms is approximately 6×10^4 g mol^{-1} ms^{-1}. The collimation aspect ratio of 1:1000 therefore results in a transverse momentum uncertainty of $\Delta p = 60$ g mol^{-1} ms^{-1}, which corresponds to a 30 μm wide beam with a transverse coherence length of about $l_c = h/\Delta p \approx 7$ nm.

An unpolarised beam entering the magnetic field gradient is represented by a unity spin density matrix, such that $W_{\alpha\alpha}(t = 0) = W_{\beta\beta}(t = 0) = W_0(x, p)$, where the initial state $W_0(x, p)$ is a two-dimensional normalised Gaussian function centred at $(x, p) = (0, 0)$, with widths given by coherence length l_c and the beam width Δx (cf. SI). The off-diagonal Wigner functions vanish: $W_{\alpha\beta} = W_{\beta\alpha} \equiv 0$, and the diagonal ones can be obtained in closed form by integrating the equations of motion (cf. SI).

In conclusion, the SG magnet can be used as coherent beam splitter, but the original experiment did not do the recombination or any other protocol to demonstrate the quantum correlation. When a coherent superposition spin state is provided at the SG input then a coherent spatial superposition of the centre of mass motion of the particle can be achieved. This has been finally demonstrated by the group of Ron Folman [106] for the case of atom interferometry and is used as central ingredient for a recent proposal of the generation of macroscopic quantum superposition [100, 101].

Acknowledgements I would like to thank my research team members at Southampton and external collaborators for joint work on testing fundamental physics with mechanical systems in table-top experiments over the past ten years. We are getting there.

References

1. J.P. Toennies, H. Schmidt-Böcking, B. Friedrich, J.C. Lower, Otto Stern (1888–1969): The founding father of experimental atomic physics. Ann. Phys. **523**(12), 1045–1070 (2011)
2. S. Deachapunya, P.J. Fagan, A.G. Major, E. Reiger, H. Ritsch, A. Stefanov, H. Ulbricht, M. Arndt, Slow beams of massive molecules. Euro. Phys. J. D **46**(2), 307–313 (2008)
3. P. Asenbaum, S. Kuhn, S. Nimmrichter, U. Sezer, M. Arndt, Cavity cooling of free silicon nanoparticles in high vacuum. Nat. Commun. **4**(1), 1–7 (2013)
4. J. Millen, T.S. Monteiro, R. Pettit, A.N. Vamivakas, Optomechanics with levitated particles. Rep. Prog. Phys. **83**(2), 026401 (2020)
5. A. Ashkin, Optical trapping and manipulation of neutral particles using lasers: A reprint volume with commentaries (2006)
6. J. Millen, P.Z.G. Fonseca, T. Mavrogordatos, T.S. Monteiro, P.F. Barker, Cavity cooling a single charged levitated nanosphere. Phys. Rev. Lett. **114**(12), 123602 (2015)
7. D.C. Moore, A.D. Rider, G. Gratta, Search for millicharged particles using optically levitated microspheres. Phys. Rev. Lett. **113**(25), 251801 (2014)
8. C. Timberlake, G. Gasbarri, A. Vinante, A. Setter, H. Ulbricht, Acceleration sensing with magnetically levitated oscillators above a superconductor. Appl. Phys. Lett. **115**(22), 224101 (2019)
9. Y.Y. Fein, P. Geyer, P. Zwick, F. Kialka, S. Pedalino, M. Mayor, S. Gerlich, M. Arndt, Quantum superposition of molecules beyond 25 kDa. Nat. Phys. **15**(12), 1242–1245 (2019)
10. S. Nimmrichter, K. Hornberger, Macroscopicity of mechanical quantum superposition states. Phys. Rev. Lett. **110**(16), 160403 (2013)
11. A. Bassi, G. Ghirardi, Dynamical reduction models. Phys. Rep. **379**(5–6), 257–426 (2003)
12. G.C. Ghirardi, A. Rimini, T. Weber, Unified dynamics for microscopic and macroscopic systems. Phys. Rev. D **34**(2), 470 (1986)
13. G.C. Ghirardi, P. Pearle, A. Rimini, Markov processes in Hilbert space and continuous spontaneous localization of systems of identical particles. Phys. Rev. A **42**(1), 78 (1990)
14. A. Bassi, K. Lochan, S. Satin, T.P. Singh, H. Ulbricht, Models of wave-function collapse, underlying theories, and experimental tests. Rev. Mod. Phys. **85**(2), 471 (2013)
15. S.L. Adler, Lower and upper bounds on CSL parameters from latent image formation and IGM heating. J. Phys. A: Math. Theor. **40**(12), 2935 (2007)
16. S. Eibenberger, S. Gerlich, M. Arndt, M. Mayor, J. Tüxen, Matter–wave interference of particles selected from a molecular library with masses exceeding 10000 amu. Phys. Chem. Chem. Phys. **15**(35), 14696–14700 (2013)
17. T. Kovachy, P. Asenbaum, C. Overstreet, C.A. Donnelly, S.M. Dickerson, A. Sugarbaker, J.M. Hogan, M.A. Kasevich, Quantum superposition at the half-metre scale. Nature **528**(7583), 530–533 (2015)
18. K.C. Lee, M.R. Sprague, B.J. Sussman, J. Nunn, N.K. Langford, X.M. Jin, T. Champion, P. Michelberger, K.F. Reim, D. England, D. Jaksch, Entangling macroscopic diamonds at room temperature. Science **334**(6060), 1253–1256 (2011)
19. M. Armano, H. Audley, J. Baird, P. Binetruy, M. Born, D. Bortoluzzi, E. Castelli, A. Cavalleri, A. Cesarini, A.M. Cruise, K. Danzmann, Beyond the required LISA free-fall performance: new LISA Pathfinder results down to 20 Hz. Phys. Rev. Lett. **120**(6), 061101 (2018)
20. M. Armano, H. Audley, G. Auger, J.T. Baird, M. Bassan, P. Binetruy, M. Born, D. Bortoluzzi, N. Brandt, M. Caleno, L. Carbone, Sub-femto-g free fall for space-based gravitational wave observatories: LISA pathfinder results. Phys. Rev. Lett. **116**(23), 231101 (2016)

21. T. Kovachy, J.M. Hogan, A. Sugarbaker, S.M. Dickerson, C.A. Donnelly, C. Overstreet, M.A. Kasevich, Matter wave lensing to picokelvin temperatures. Phys. Rev. Lett. **114**(14), 143004 (2015)

22. S.L. Adler, A. Vinante, Bulk heating effects as tests for collapse models. Phys. Rev. A **97**(5), 052119 (2018)

23. K. Piscicchia, A. Bassi, C. Curceanu, R.D. Grande, S. Donadi, B.C. Hiesmayr, A. Pichler, CSL collapse model mapped with the spontaneous radiation. Entropy **19**(7), 319 (2017)

24. A. Vinante, M. Bahrami, A. Bassi, O. Usenko, G.H.C.J. Wijts, T.H. Oosterkamp, Upper bounds on spontaneous wave-function collapse models using millikelvin-cooled nanocantilevers. Phys. Rev. Lett. **116**(9), 090402 (2016)

25. M. Torol, G. Gasbarri, A. Bassi, Colored and dissipative continuous spontaneous localization model and bounds from matter-wave interferometry. Phys. Lett. A **381**(47), 3921–3927 (2017)

26. M. Carlesso, A. Bassi, Current tests of collapse models: How far can we push the limits of quantum mechanics? in *Quantum Information and Measurement* (Optical Society of America, 2019, April), pp. S1C-3

27. M. Toroš, A. Bassi, Bounds on quantum collapse models from matter-wave interferometry: Calculational details. J. Phys. A: Math. Theoretical **51**(11), 115302 (2018)

28. C. Gardiner, P. Zoller, *Quantum Noise: A Handbook of Markovian and Non-Markovian Quantum Stochastic Methods with Applications to Quantum Optics* Vol. 56 (Springer Science and Business Media, 2004)

29. H.P. Breuer, F. Petruccione, *The Theory of Open Quantum Systems* (Oxford University Press, Oxford, 2002)

30. J. Bateman, S. Nimmrichter, K. Hornberger, H. Ulbricht, Near-field interferometry of a free-falling nanoparticle from a point-like source. Nat. Commun. **5**(1), 1–5 (2014)

31. H. Pino, J. Prat-Camps, K. Sinha, B.P. Venkatesh, O. Romero-Isart, On-chip quantum interference of a superconducting microsphere. Quantum Sci. Technol. **3**(2), 025001 (2018)

32. C. Wan, M. Scala, G.W. Morley, A.A. Rahman, H. Ulbricht, J. Bateman, P.F. Barker, S. Bose, M.S. Kim, Free nano-object Ramsey interferometry for large quantum superpositions. Phys. Rev. Lett. **117**(14), 143003 (2016)

33. R. Kaltenbaek, M. Aspelmeyer, P.F. Barker, A. Bassi, J. Bateman, K. Bongs, S. Bose, C. Braxmaier, Brukner, Christophe, B., Chwalla, M., Macroscopic quantum resonators (MAQRO): 2015 update. EPJ Quantum Technol. **3**(1), 5 (2016)

34. M. Bahrami, A. Smirne, A. Bassi, Role of gravity in the collapse of a wave function: A probe into the diósi-penrose model. Phys. Rev. A **90**(6), 062105 (2014)

35. B. Collett, P. Pearle, Wavefunction collapse and random walk. Found. Phys. **33**(10), 1495–1541 (2003)

36. S.L. Adler, Stochastic collapse and decoherence of a non-dissipative forced harmonic oscillator. J. Phys. A: Math. Gen. **38**(12), 2729 (2005)

37. K. Hornberger, S. Gerlich, P. Haslinger, S. Nimmrichter, M. Arndt, Colloquium: Quantum interference of clusters and molecules. Rev. Mod. Phys. **84**(1), 157 (2012)

38. S. Nimmrichter, K. Hornberger, K. Hammerer, Optomechanical sensing of spontaneous wave-function collapse. Phys. Rev. Lett. **113**(2), 020405 (2014)

39. S. Bera, B. Motwani, T.P. Singh, H. Ulbricht, A proposal for the experimental detection of CSL induced random walk. Sci. Rep. **5**(1), 1–10 (2015)

40. L. Diósi, Testing spontaneous wave-function collapse models on classical mechanical oscillators. Phys. Rev. Lett. **114**(5), 050403 (2015)

41. D. Goldwater, M. Paternostro, P.F. Barker, Testing wave-function-collapse models using parametric heating of a trapped nanosphere. Phys. Rev. A **94**(1), 010104 (2016)

42. M. Carlesso, M. Paternostro, H. Ulbricht, A. Vinante, A. Bassi, Non-interferometric test of the continuous spontaneous localization model based on rotational optomechanics. New J. Phys. **20**(8), 083022 (2018)

43. M. Carlesso, A. Bassi, P. Falferi, A. Vinante, Experimental bounds on collapse models from gravitational wave detectors. Phys. Rev. D **94**(12), 124036 (2016)

44. W. Paul, Electromagnetic traps for charged and neutral particles. Rev. Mod. Phys. **62**(3), 531 (1990)
45. A. Vinante, A. Pontin, M. Rashid, M. Torol, P.F. Barker, H. Ulbricht, Testing collapse models with levitated nanoparticles: Detection challenge. Phys. Rev. A **100**(1), 012119 (2019)
46. R. Mishra, A. Vinante, T.P. Singh, Testing spontaneous collapse through bulk heating experiments: An estimate of the background noise. Phys. Rev. A **98**(5), 052121 (2018)
47. M. Bilardello, S. Donadi, A. Vinante, A. Bassi, Bounds on collapse models from cold-atom experiments. Phys. A **462**, 764–782 (2016)
48. A. Pontin, N.P. Bullier, M. Torol, P.F. Barker, Ultranarrow-linewidth levitated nano-oscillator for testing dissipative wave-function collapse. Phys. Rev. Res. **2**(2), 023349 (2020)
49. B.R. Slezak, C.W. Lewandowski, J.F. Hsu, D'Urso, B., Cooling the motion of a silica microsphere in a magneto-gravitational trap in ultra-high vacuum. New J. Phys. **20**(6), 063028 (2018)
50. D. Zheng, Y. Leng, X. Kong, R. Li, Z. Wang, X. Luo, J. Zhao, C.K. Duan, P. Huang , J. Du, Room temperature test of wave-function collapse using a levitated micro-oscillator (2019). arXiv preprint arXiv:1907.06896
51. O. Romero-Isart, L. Clemente, C. Navau, A. Sanchez, J.I. Cirac, Quantum magnetomechanics with levitating superconducting microspheres. Phys. Rev. Lett. **109**(14), 147205 (2012)
52. B. van Waarde, *The lead zeppelin: a force sensor without a handle*, Ph.D. Thesis, Leiden University (2016)
53. A. Vinante, P. Falferi, G. Gasbarri, A. Setter, C. Timberlake, H. Ulbricht, Ultralow mechanical damping with Meissner-levitated ferromagnetic microparticles. Phys. Rev. Appl. **13**(6), 064027 (2020)
54. J. Prat-Camps, C. Teo, C.C. Rusconi, W. Wieczorek, O. Romero-Isart, Ultrasensitive inertial and force sensors with diamagnetically levitated magnets. Phys. Rev. Appl. **8**(3), 034002 (2017)
55. M. Bahrami, A. Bassi, H. Ulbricht, Testing the quantum superposition principle in the frequency domain. Phys. Rev. A **89**(3), 032127 (2014)
56. S. Sturm, F. Köhler, J. Zatorski, A. Wagner, Z. Harman, G. Werth, W. Quint, C.H. Keitel, K. Blaum, High-precision measurement of the atomic mass of the electron. Nature **506**(7489), 467–470 (2014)
57. M. Weitz, A. Huber, F. Schmidt-Kaler, D. Leibfried, T.W. Hänsch, Precision measurement of the hydrogen and deuterium 1 S ground state Lamb shift. Phys. Rev. Lett. **72**(3), 328 (1994)
58. E. Oelker, R.B. Hutson, C.J. Kennedy, L. Sonderhouse, T. Bothwell, A. Goban, D. Kedar, C. Sanner, J.M. Robinson, G.E. Marti , D.G. Matei, Demonstration of 4.8×10^{-17} stability at 1 s for two independent optical clocks. Nat. Photon. 13(10), pp.714-719 (2019)
59. R. Penrose, On the gravitization of quantum mechanics 1: Quantum state reduction. Found. Phys. **44**(5), 557–575 (2014)
60. L. Rosenfeld, On quantization of fields. Nuclear Phys. **40**, 353–356 (1963)
61. S. Carlip, Is quantum gravity necessary? Class. Quantum Gravity **25**(15), 154010 (2008)
62. C. Møller, Les théories relativistes de la gravitation. Colloques Internationaux CNRS **91**(1) (1962)
63. D.N. Page, C.D. Geilker, Indirect evidence for quantum gravity. Phys. Rev. Lett. **47**(14), 979 (1981)
64. K. Eppley, E. Hannah, The necessity of quantizing the gravitational field. Found. Phys. **7**(1–2), 51–68 (1977)
65. N. Gisin, Stochastic quantum dynamics and relativity. Helv. Phys. Acta **62**(4), 363–371 (1989)
66. J. Mattingly, Why Eppley and Hannah's thought experiment fails. Phys. Rev. D **73**(6), 064025 (2006)
67. C. Kiefer, Why quantum gravity? in *Approaches to Fundamental Physics* (Springer, Berlin, Heidelberg, 2007), pp. 123–130
68. M. Albers, C. Kiefer, M. Reginatto, Measurement analysis and quantum gravity. Phys. Rev. D **78**(6), 064051 (2008)

69. D. Giulini, A. Großardt, Gravitationally induced inhibitions of dispersion according to the Schrödinger–Newton equation. Class. Quantum Gravity **28**(19), 195026 (2011)
70. D. Giulini, A. Großardt, Gravitationally induced inhibitions of dispersion according to a modified Schrödinger–Newton equation for a homogeneous-sphere potential. Class. Quantum Gravity **30**(15), 155018 (2013)
71. L. Diósi, Gravitation and quantum-mechanical localization of macro-objects. Phys. Lett. A **105**(4–5), 199–202 (1984)
72. R. Penrose, Quantum computation, entanglement and state reduction. Philos. Trans. R. Soc. London. Ser. A: Math. Phys. Eng. Sci. **356**(1743), 1927–1939
73. H. Yang, H. Miao, D.S. Lee, B. Helou, Y. Chen, Macroscopic quantum mechanics in a classical spacetime. Phys. Rev. Lett. **110**(17), 170401 (2013)
74. A. Großardt, J. Bateman, H. Ulbricht, A. Bassi, Optomechanical test of the Schrödinger-Newton equation. Phys. Rev. D **93**(9), 096003 (2016)
75. R. Bekenstein, R. Schley, M. Mutzafi, C. Rotschild, M. Segev, Optical simulations of gravitational effects in the Newton–Schrödinger system. Nat. Phys. **11**(10), 872–878 (2015)
76. T. Roger, C. Maitland, K. Wilson, N. Westerberg, D. Vocke, E.M. Wright, D. Faccio, Optical analogues of the Newton–Schrödinger equation and boson star evolution. Nat. Commun. **7**(1), 1–8 (2016)
77. M. Zych, F. Costa, I. Pikovski, T.C. Ralph, Č. Brukner, General relativistic effects in quantum interference of photons. Class. Quantum Gravity **29**(22), 224010 (2012)
78. M. Zych, F. Costa, I. Pikovski, Č. Brukner, Quantum interferometric visibility as a witness of general relativistic proper time. Nat. Commun. **2**, 505 (2011)
79. I. Pikovski, M. Zych, F. Costa, Č. Brukner, Universal decoherence due to gravitational time dilation. Nat. Phys. **11**(8), 668–672 (2015)
80. M. Torol, A. Bassi, Bounds on quantum collapse models from matter-wave interferometry: Calculational details. J. Phys. A: Math. Theor. **51**(11), 115302 (2018)
81. Y. Margalit, Z. Zhou, S. Machluf, D. Rohrlich, Y. Japha, R. Folman, A self-interfering clock as a "which path" witness. Science **349**(6253), 1205–1208 (2015)
82. S. Nimmrichter, K. Hornberger, Stochastic extensions of the regularized Schrödinger-Newton equation. Phys. Rev. D **91**(2), 024016 (2015)
83. S. Bera, S. Donadi, K. Lochan, T.P. Singh, A comparison between models of gravity induced decoherence. Found. Phys. **45**(12), 1537–1560 (2015)
84. G. Gasbarri, M. Torol, S. Donadi, A. Bassi, Gravity induced wave function collapse. Phys. Rev. D **96**(10), 104013 (2017)
85. O. Romero-Isart, A.C. Pflanzer, F. Blaser, R. Kaltenbaek, N. Kiesel, M. Aspelmeyer, J.I. Cirac, Large quantum superpositions and interference of massive nanometer-sized objects. Phys. Rev. Lett. **107**(2), 020405 (2011)
86. L. Diosi, A universal master equation for the gravitational violation of quantum mechanics. Phys. Lett. A **120**(8), 377–381 (1987)
87. R. Penrose, On gravity's role in quantum state reduction. Gen. Relativ. Gravit. **28**(5), 581–600 (1996)
88. S. Bose, K. Jacobs, P.L. Knight, Preparation of nonclassical states in cavities with a moving mirror. Phys. Rev. A **56**(5), 4175 (1997)
89. S. Bose, K. Jacobs, P.L. Knight, Scheme to probe the decoherence of a macroscopic object. Phys. Rev. A **59**(5), 3204 (1999)
90. M.R. Vanner, I. Pikovski, G.D. Cole, M.S. Kim, Č. Brukner, K. Hammerer, G.J. Milburn, M. Aspelmeyer, Pulsed quantum optomechanics. Proc. National Acad. Sci. **108**(39), 16182–16187 (2011)
91. W. Marshall, C. Simon, R. Penrose, D. Bouwmeester, Towards quantum superpositions of a mirror. Phys. Rev. Lett. **91**(13), 130401 (2003)
92. K. Hornberger, S. Uttenthaler, B. Brezger, L. Hackermüller, M. Arndt, A. Zeilinger, Collisional decoherence observed in matter wave interferometry. Phys. Rev. Lett. **90**(16), 160401 (2003)
93. S.L. Adler, Gravitation and the noise needed in objective reduction models Quantum Nonlocality and Reality: 50 Years of Bell's Theorem ed M Bell and S Gao (2016)

94. M. Bronstein, Republication of: Quantum theory of weak gravitational fields. Gen. Relativ. Gravit. **44**(1), 267–283 (2012)
95. B.S. DeWitt, D. Bryce Seligman, *The Global Approach to Quantum Field Theory* (Vol. 114) (Oxford University Press, USA, 2003)
96. A. Peres, D.R. Terno, Hybrid classical-quantum dynamics. Phys. Rev. A **63**(2), 022101 (2001)
97. C. Marletto, V. Vedral, Why we need to quantise everything, including gravity. npj Quantum Inf. 3(1), 1–5 (2017)
98. M. Carlesso, A. Bassi, M. Paternostro, H. Ulbricht, Testing the gravitational field generated by a quantum superposition. New J. Phys. **21**(9), 093052 (2019)
99. M. Carlesso, M. Paternostro, H. Ulbricht, A. Bassi, When Cavendish meets Feynman: A quantum torsion balance for testing the quantumness of gravity (2017). arXiv preprint arXiv:1710.08695
100. S. Bose, A. Mazumdar, G.W. Morley, H. Ulbricht, M. Torol, M. Paternostro, A.A. Geraci, P.F. Barker, M.S. Kim, G. Milburn, Spin entanglement witness for quantum gravity. Phys. Rev. Lett. **119**(24), 240401 (2017)
101. C. Marletto, V. Vedral, Gravitationally induced entanglement between two massive particles is sufficient evidence of quantum effects in gravity. Phys. Rev. Lett. **119**(24), 240402 (2017)
102. H. Miao, D. Martynov, H. Yang, A. Datta, Quantum correlations of light mediated by gravity. Phys. Rev. A **101**(6), 063804 (2020)
103. W. Gerlach, O. Stern, Der experimentelle Nachweis des magnetischen Moments des Silberatoms. ZPhy **8**(1), 110–111 (1922)
104. M. Utz, M.H. Levitt, N. Cooper, H. Ulbricht, Visualisation of quantum evolution in the Stern–Gerlach and Rabi experiments. Phys. Chem. Chem. Phys. **17**(5), 3867–3872 (2015)
105. B. Friedrich, D. Herschbach, Stern and Gerlach: How a bad cigar helped reorient atomic physics. Phys. Today **56**(12), 53–59 (2003)
106. S. Machluf, Y. Japha, R. Folman, Coherent Stern–Gerlach momentum splitting on an atom chip. Nat. Commun. **4**(1), 1–9 (2013)

Part III
Femto- and Atto-Science

Chapter 16
Inducing Enantiosensitive Permanent Multipoles in Isotropic Samples with Two-Color Fields

Andres F. Ordonez and Olga Smirnova

Abstract We find that two-color fields can induce field-free permanent dipoles in initially isotropic samples of chiral molecules via resonant electronic excitation in a one-3ω-photon versus three-ω-photons scheme. These permanent dipoles are enantiosensitive and can be controlled via the relative phase between the two colors. When the two colors are linearly polarized perpendicular to each other, the interference between the two pathways induces excitation sensitive to the molecular handedness and orientation, leading to uniaxial orientation of the excited molecules and to an enantio-sensitive permanent dipole perpendicular to the polarization plane. We also find that although a corresponding one-2ω-photon versus two-ω-photons scheme cannot produce enantiosensitive permanent dipoles, it can produce enantiosensitive permanent quadrupoles that are also controllable through the two-color relative phase.

1 Introduction

Chirality (handedness) is the geometrical property that allows us to distinguish a left hand from a right hand. Like hands, many molecules have two possible versions which are non-superimposable mirror images of each other (opposite enantiomers). This "extra degree of freedom" stemming from the reduced symmetry (lack of improper symmetry axes) of chiral molecules leads to interesting behavior absent in achiral molecules [1–6] with profound implications for biology [7, 8]. Furthermore, since

A. F. Ordonez
Max-Born-Institut, Max-Born-Str. 2A, 12489 Berlin, Germany
e-mail: ordonez@mbi-berlin.de

O. Smirnova (✉)
Max-Born-Institut, Max-Born-Str. 2A, 12489 Berlin, Germany

Technische Universität Berlin, Straße des 17. Juni 135, 10623 Berlin, Germany
e-mail: smirnova@mbi-berlin.de

© The Author(s) 2021
B. Friedrich and H. Schmidt-Böcking (eds.), *Molecular Beams in Physics and Chemistry*,
https://doi.org/10.1007/978-3-030-63963-1_16

335

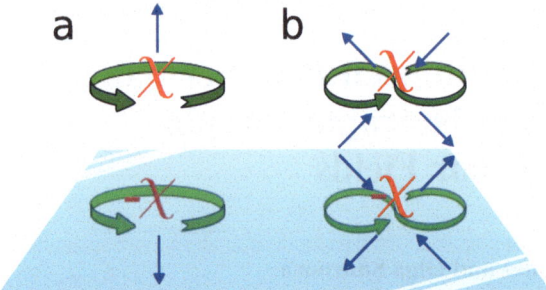

Fig. 1 Symmetry in ω and ω-2ω setups. **a** A circularly polarized field (circular arrow) interacts with an isotropic sample of chiral molecules (represented by χ) and produces a net photoelectron current (in general a vectorial signal) perpendicular to the polarization plane (arrow pointing up). The mirror reflection shows that the interaction of the same field with the opposite enantiomer (represented by $-\chi$) yields the opposite current. **b** A field with its fundamental and second harmonic linearly polarized perpendicular to each other (∞-like arrow) interacting with an isotropic chiral sample produces a quadrupolar photoelectron current (in general a quadrupolar signal). The mirror reflection shows that the interaction of the same field with the opposite enantiomer yields the opposite current. In both (**a**) and (**b**), the reversal of the signal when the polarization is changed follows from considering a rotation of 180° (not shown) of the full system, which changes the polarization but not the isotropic sample (see Figs. 2–4 in Ref. [14])

opposite enantiomers share fundamental properties like their mass and their energy spectrum, one must often rely precisely on this chiral behavior to tell opposite enantiomers apart—a task of immense practical importance in chemistry [9, 10].

An example of this chiral behavior is the phenomenon known as photoelectron circular dichroism (PECD) [4, 11–13], which consists in the generation of a net photoelectron current from an isotropic sample of chiral molecules irradiated by circularly polarized light [14–16]. This photoelectron current, which results from different amounts of photoelectrons being emitted in opposite directions, is directed along the normal to the polarization plane (because of the overall cylindrical symmetry) and changes sign when either the enantiomer or the circular polarization is reversed (see Fig. 1a). Importantly, PECD occurs within the electric-dipole approximation, which makes typical PECD signals orders of magnitude stronger than traditional enantiosensitive signals, such as circular dichroism (CD), which rely on interactions beyond the electric-dipole approximation [10, 17]. Furthermore, the electric-dipole approximation also rules out any influence of the wave vector of the incident light and hence of the momentum of the photons.

Given that: the molecules are randomly oriented in space, the electric field is circularly polarized, and the momentum of the photon does not play any role in PECD; it is only natural to wonder *why does a net current of photoelectrons perpendicular to the polarization plane occur?* From the point of view of symmetry, the question would be instead *what symmetry prevents this current from taking place in the case of achiral molecules?* The answer is simple: in the electric-dipole approximation[1] the

---

[1] Beyond the electric-dipole approximation the wave vector of the light breaks reflection symmetry.

system consisting of isotropic achiral molecules together with the circularly polarized electric field is symmetric with respect to reflection in the polarization plane[2] and therefore the current normal to the polarization plane must vanish. When achiral molecules are replaced by chiral molecules, this mirror symmetry is broken and the PECD current emerges [14].

While this symmetry analysis does not provide an answer in terms of the specific mechanism, the insight it provides applies to several other closely related effects occurring within the electric dipole approximation, which rely on electric field polarizations confined to a plane and yield enantiosensitive vectorial responses perpendicular to that plane [3, 14, 18–23]. For example, if the photon energy of the circularly polarized light is not enough to ionize the molecule, the lack of reflection symmetry due to the chiral molecules leads to oscillating bound currents normal to the polarization plane [14, 23]. In this case, the current results from the excitation of bound states and the associated oscillation of the expected value of the electric dipole operator. The enantiosensitivity is reflected in the phase of the oscillations, which are out of phase in opposite enantiomers.

Analogously, one may also expect that it should be possible to induce permanent electric dipoles (i.e. non-vanishing zero-frequency components of the expected value of the electric dipole operator) normal to the polarization plane and with opposite directions for opposite enantiomers. Indeed, such static electric dipoles have been investigated in the context of optical rectification [24–28], where two excited states close in energy are resonantly excited with monochromatic circularly polarized light. Very recently enantiosensitive static dipoles have also been studied in the context of molecular orientation induced by intense off-resonant light pulses [29–32]. Such light pulses excite rotational dynamics and cause orientation of one of the molecular axes that persists after the pulse is over. Here we show that field-free enantiosensitive permanent electric dipoles and the associated orientation can also be induced in the context of purely electronic excitation on ultrafast time-scales, without relying on rotational dynamics. We achieve this via interference of one- and three-photon excitation pathways.

Quite recently an extension of single-color PECD to two-color ω-2ω fields with orthogonal linear polarizations has been observed [33–35] (see Fig. 1b). As we discuss in Refs. [36, 37], this is an example of how molecular chirality can be reflected not only in scalar (e.g. CD) and vectorial observables (e.g. PECD), but also in higher-rank tensor observables. Here we show that two-color ω-2ω fields with linear polarizations perpendicular to each other can induce enantiosensitive permanent quadrupoles in samples of isotropic chiral molecules.

---

[2]Note that circularly polarized light is not chiral within the electric-dipole approximation, and therefore the chirality of the light itself does not play a role in PECD [14].

2 Exciting an Enantiosensitive Permanent Dipole

Consider the excitation scheme depicted in Fig. 2, where the interference of contributions from a one-3ω-photon pathway and a three-ω-photon pathway control the population of the state $|3\rangle$ of a chiral molecule. For simplicity, we first consider excitation via intermediate resonances in states $|1\rangle$ and $|2\rangle$. The presence of resonances in these states is not essential, as discussed later, but simplifies the analysis. The field is assumed to have the form

$$E^{\mathrm{L}}(t) = F(t)\left(E^{\mathrm{L}}_{\omega}e^{-i\omega t} + E^{\mathrm{L}}_{3\omega}e^{-3i\omega t}\right) + \text{c.c.}, \tag{1}$$

where $F(t)$ is a smooth envelope, E^{L}_{ω} and $E^{\mathrm{L}}_{3\omega}$ specify the polarizations and phases of each frequency, and the L and M superscripts indicate vectors and functions in the laboratory frame and in the molecular frame, respectively. For a given molecular orientation $\varrho \equiv \alpha\beta\gamma$, where $\alpha\beta\gamma$ are the Euler angles, the wave function after the interaction is

$$\Psi^{\mathrm{M}}(r^{\mathrm{M}}, \varrho) = \sum_{i=0}^{3} a_i(\varrho)\, e^{-i\omega_i t}\psi^{\mathrm{M}}_i(r^{\mathrm{M}}), \tag{2}$$

where $\psi^{\mathrm{M}}_i(r^{\mathrm{M}})$ is the coordinate representation of state $|i\rangle$ in the molecular frame. In the perturbative regime we have

$$a_3(\varrho) = A^{(1)}_3[d^{\mathrm{L}}_{3,0}(\varrho)\cdot E^{\mathrm{L}}_{3\omega}] + A^{(3)}_3[d^{\mathrm{L}}_{3,2}(\varrho)\cdot E^{\mathrm{L}}_{\omega}][d^{\mathrm{L}}_{2,1}(\varrho)\cdot E^{\mathrm{L}}_{\omega}][d^{\mathrm{L}}_{1,0}(\varrho)\cdot E^{\mathrm{L}}_{\omega}], \tag{3}$$

where $A^{(1)}_3$ and $A^{(3)}_3$ are first- and third-order coupling constants that depend on the detunings and the envelope (see Appendix). Analogous expressions apply for the other amplitudes a_i. The transition dipoles $d^{\mathrm{M}}_{i,j} \equiv \langle \psi^{\mathrm{M}}_i(r^{\mathrm{M}})|d^{\mathrm{M}}|\psi^{\mathrm{M}}_j(r^{\mathrm{M}})\rangle$ are fixed in the molecular frame and have been expressed in the laboratory frame using the rotation matrix $R(\varrho)$ according to $d^{\mathrm{L}}_{i,j}(\varrho) = R(\varrho)d^{\mathrm{M}}_{i,j}$.

Fig. 2 Excitation scheme used to produce an enantiosensitive permanent dipole in an isotropic sample of chiral molecules

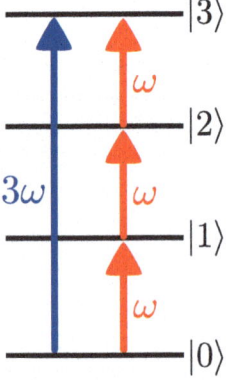

The expected value of the electric dipole operator in the molecular frame $\langle d^{\mathrm{M}}(\varrho)\rangle$ $\equiv \langle \Psi^{\mathrm{M}}(r^{\mathrm{M}}, \varrho)| d^{\mathrm{M}} |\Psi^{\mathrm{M}}(r^{\mathrm{M}}, \varrho)\rangle$ has a zero-frequency component of the form

$$\langle d^{\mathrm{M}}(\varrho)\rangle_{\omega=0} = \sum_{i=0}^{3} |a_i(\varrho)|^2 d_{i,i}^{\mathrm{M}}, \tag{4}$$

i.e. the permanent dipole for a given molecular orientation is the sum of the permanent dipoles of each state weighted by their orientation-dependent populations at the end of the pulse.

Transforming $\langle d^{\mathrm{M}}(\varrho)\rangle_{\omega=0}$ to the laboratory frame and averaging over all molecular orientations yields the permanent dipole

$$\langle d^{\mathrm{L}}\rangle_{\omega=0} \equiv \int d\varrho \langle d^{\mathrm{L}}(\varrho)\rangle_{\omega=0}. \tag{5}$$

The contribution of state $|3\rangle$ to this expression reads as[3]

$$\langle d_3^{\mathrm{L}}\rangle_{\omega=0} \equiv \int d\varrho \, |a_3(\varrho)|^2 d_{3,3}^{\mathrm{L}}(\varrho) = A_3^{(1)*} A_3^{(3)} \chi_3 Z^{\mathrm{L}} + \text{c.c.}, \tag{6}$$

where $\int d\varrho \equiv \int_0^{2\pi} d\alpha \int_0^{\pi} d\beta \int_0^{2\pi} d\gamma /8\pi^2$ is the integral over all molecular orientations and we defined

$$\chi_i \equiv \frac{1}{30} \left[(d_{2,1}^{\mathrm{M}} \cdot d_{1,0}^{\mathrm{M}}) d_{3,2}^{\mathrm{M}} + (d_{3,2}^{\mathrm{M}} \cdot d_{1,0}^{\mathrm{M}}) d_{2,1}^{\mathrm{M}} + (d_{3,2}^{\mathrm{M}} \cdot d_{2,1}^{\mathrm{M}}) d_{1,0}^{\mathrm{M}} \right] \cdot (d_{3,0}^{\mathrm{M}} \times d_{i,i}^{\mathrm{M}}), \tag{7}$$

$$Z^{\mathrm{L}} \equiv \left(E_\omega^{\mathrm{L}} \cdot E_\omega^{\mathrm{L}} \right) \left(E_\omega^{\mathrm{L}} \times E_{3\omega}^{\mathrm{L}*} \right). \tag{8}$$

χ_3 is a rotationally invariant molecular pseudoscalar, i.e. a molecular quantity independent of the molecular orientation. It has opposite signs for opposite enantiomers and vanishes for achiral molecules; χ_3 encodes the enantiosensitivity of $\langle d_3^{\mathrm{L}}\rangle_{\omega=0}$. Selection rules for χ_3 can be directly read off from Eq. (7). In particular, it vanishes if $d_{3,0}^{\mathrm{M}}$ and $d_{3,3}^{\mathrm{M}}$ are collinear. Z^{L} is a light pseudovector—it is a vector that depends only on the light's polarization and is invariant under the inversion operation; Z^{L} determines the direction of $\langle d_3^{\mathrm{L}}\rangle_{\omega=0}$. Selection rules for Z^{L} can be read off directly from Eq. (8). In particular, it vanishes if ω is circularly polarized ($E_\omega^{\mathrm{L}} \cdot E_\omega^{\mathrm{L}} = 0$) or if ω and 3ω are linearly polarized parallel to each other ($E_\omega^{\mathrm{L}} \times E_{3\omega}^{\mathrm{L}*} = 0$).

For example, if we choose ω and 3ω linearly polarized perpendicular to each other, say $E_\omega^{\mathrm{L}} = x^{\mathrm{L}}$ and $E_{3\omega}^{\mathrm{L}} = e^{-i\phi} y^{\mathrm{L}}$, with x^{L} and y^{L} the unitary vectors along each axis then

---

[3] We use Eq. (A16) in Ref. [14] for the interference term. The direct terms vanish because the possible non-zero field pseudovectors are purely imaginary, e.g. $(E_{3\omega}^* \times E_{3\omega})$, while the accompanying molecular pseudoscalars are real and the coupling coefficients appear within absolute values, see Ref. [38].

$$E^{\mathrm{L}}(t) = 2F(t)\left[\cos(\omega t)\,x^{\mathrm{L}} + \cos(3\omega t + \phi)\,y^{\mathrm{L}}\right] \tag{9}$$

and we obtain

$$\langle d_3^{\mathrm{L}}\rangle_{\omega=0} = 2\chi_3 \Re\left\{A_3^{(1)*}A_3^{(3)}e^{i\phi}\right\}z^{\mathrm{L}}, \tag{10}$$

i.e., $\langle d_3^{\mathrm{L}}\rangle_{\omega=0}$ is perpendicular to the polarization plane and its magnitude and sign can be controlled through the relative phase ϕ. Note that the relative phase of the coupling coefficients $A_3^{(1)}$ and $A_3^{(3)}$, which can be modified for example by changing the detunings, must also be taken into account.

The contributions from states $|1\rangle$, and $|2\rangle$ to the permanent dipole (5) have the same structure as Eq. (10), albeit with different coupling constants and molecular pseudoscalars χ_1 and χ_2, respectively [see Eq. (7)]. Since $|a_0|^2 = 1 - |a_1|^2 - |a_2|^2 - |a_3|^2$, the contribution from the ground state involves the coupling constants associated to $|1\rangle$, $|2\rangle$, and $|3\rangle$, and a molecular pseudoscalar χ_0 [see Eq. (7)]. Together, these contributions yield

$$\langle d^{\mathrm{L}}\rangle_{\omega=0} = 2\left[(\chi_1 - \chi_0)\Re\left\{A_1^{(1)}A_1^{(3)*}e^{i\phi}\right\} + (\chi_2 - \chi_0)\Re\left\{A_2^{(2)'*}A_2^{(2)}e^{i\phi}\right\}\right.$$
$$\left. + (\chi_3 - \chi_0)\Re\left\{A_3^{(1)*}A_3^{(3)}e^{i\phi}\right\}\right]z^{\mathrm{L}} \tag{11}$$

where $A_1^{(1)}$ and $A_1^{(3)}$ are the coupling coefficients for the transitions $|0\rangle \xrightarrow{\omega} |1\rangle$ and $|0\rangle \xrightarrow{3\omega} |3\rangle \xrightarrow{-\omega} |2\rangle \xrightarrow{-\omega} |1\rangle$, respectively; $A_2^{(2)}$ and $A_2^{(2)'}$ are the coupling coefficients for the transitions $|0\rangle \xrightarrow{\omega} |1\rangle \xrightarrow{\omega} |2\rangle$ and $|0\rangle \xrightarrow{3\omega} |3\rangle \xrightarrow{-\omega} |2\rangle$, respectively.

In the absence of the intermediate resonances through the states $|1\rangle$ and $|2\rangle$ the contribution from the third-order term in Eq. (3) turns into a sum over all intermediate states $|j\rangle$ and $|k\rangle$ weighted by a coefficient $A_{3;jk}^{(3)}$. The intermediate states retain no population at the end of the pulse and the permanent dipole takes the form

$$\langle d^{\mathrm{L}}\rangle_{\omega=0} = 2\sum_{j,k}(\chi_{3;jk} - \chi_{0;jk})\Re\left\{A_3^{(1)*}A_{3;jk}^{(3)}Z^{\mathrm{L}}\right\}$$
$$= 2F_0^4\left[2\pi\delta_\sigma(\Delta)\right]^2\sum_{j,k}\frac{\chi_{3;jk} - \chi_{0;jk}}{(\omega_{k,0} - 2\omega_L)(\omega_{j,0} - \omega_L)}\Re\left\{Z^{\mathrm{L}}\right\} \tag{12}$$

which is valid for arbitrary polarizations [see Eq. (8)]. Here $\chi_{i;jk}$ is given by Eq. (7) with the replacements $1 \to j$ and $2 \to k$. In the second equality we wrote the coupling constants explicitly, $\omega_{i,j} \equiv \omega_i - \omega_j$, $\Delta \equiv \omega_{3,0} - 3\omega_L$, and we took $\int_{-\infty}^{\infty} dt\, F(t)e^{i\omega t} \equiv 2\pi F_0\delta_\sigma(\omega)$ with $\delta_\sigma(\omega)$ equal to the Dirac delta in the limit of infinitesimal σ.

2.1 A Simple Picture of the Mechanism Leading to the Enantiosensitive Permanent Dipole

The orientation averaging procedure we applied [38], although very powerful, is also rather formal. Below we demonstrate that the mechanism leading to the generation of the permanent dipole $\langle d^L \rangle_{\omega=0}$ stems from the sensitivity of the excitation to the molecular orientation and handedness, which induces uniaxial and enantiosensitive orientation of the initially isotropic sample. We remark that the excitation induces *orientation* ($\uparrow$) as opposed to just *alignment* ($\updownarrow$) and that this orientation is further-more enantiosensitive.

Consider the interaction of the field (9) with a dummy molecule with $d^M_{1,0}$ and $d^M_{3,0}$ perpendicular to each other and $d^M_{3,2} = d^M_{2,1} = d^M_{1,0}$. For simplicity we again assume that intermediate states are resonantly excited and that only state $|3\rangle$ has a non-zero permanent dipole. The population $P_3(\varrho) \equiv |a_3(\varrho)|^2$ of the excited state $|3\rangle$ reads [see Eq. (3)]

$$P_3(\varrho) = |A_3^{(1)}|^2 \mathcal{P}_{3\omega}(\varrho) + |A_3^{(3)}|^2 \mathcal{P}_\omega(\varrho) + 2\Re\left\{A_3^{(1)*} A_3^{(3)} e^{i\phi}\right\} \mathcal{P}_{\omega,3\omega}(\varrho) \qquad (13)$$

where

$$\mathcal{P}_\omega(\varrho) \equiv \left[d^L_{1,0}(\varrho) \cdot x^L\right]^6, \quad \mathcal{P}_{3\omega}(\varrho) \equiv \left[d^L_{3,0}(\varrho) \cdot y^L\right]^2, \qquad (14)$$

$$\mathcal{P}_{\omega,3\omega}(\varrho) \equiv \left[d^L_{3,0}(\varrho) \cdot y^L\right]\left[d^L_{1,0}(\varrho) \cdot x^L\right]^3. \qquad (15)$$

$\mathcal{P}_\omega$ will select molecular orientations where $d^L_{1,0}$ is aligned along the x^L axis. $\mathcal{P}_{3\omega}$ will select molecular orientations where $d^L_{3,0}$ is aligned along the y^L axis. $\mathcal{P}_{\omega,3\omega}$ will select molecular orientations where $d^L_{1,0}$ is aligned along the x^L axis *and* $d^L_{3,0}$ is aligned along the y^L axis. These orientations are shown in Fig. 3b. While the direct terms $\mathcal{P}_\omega$ and $\mathcal{P}_{3\omega}$ do not distinguish between this subset of orientations $\{\varrho_i\}_{i=1}^4$, the interference term $\mathcal{P}_{\omega,3\omega}$ will be positive for orientations ϱ_1 and ϱ_3 and negative for orientations ϱ_2 and ϱ_4. This produces an imbalance between the number of excited molecules with orientations ϱ_1 and ϱ_3 and those with orientations ϱ_2 and ϱ_4. As can be seen in Fig. 3, this imbalance amounts to the molecular axis $d^M_{1,0} \times d^M_{3,0}$ being oriented. That is, the field (9) induces field-free uniaxial orientation of the molecular sample in the state $|3\rangle$. The emergence of a permanent dipole follows trivially, provided that $d^M_{3,3}$ has a non-zero component along the oriented axis, i.e. as long as $d^M_{3,3} \cdot (d^M_{1,0} \times d^M_{3,0}) \neq 0$. Note that, according to Eq. (7), this is in agreement with having $\chi_3 \neq 0$. If we consider the situation depicted in Fig. 3 now mirror reflected across the polarization plane, which is equivalent to swapping the enantiomer while leaving the field as it is, we immediately see that $d^M_{3,3}$ and therefore also $\langle d^L_3 \rangle_{\omega=0}$ point in the opposite direction, which explains the enantiosensitivity of $\langle d^L_3 \rangle_{\omega=0}$.

Since the emergence of a permanent dipole $\langle d^L \rangle_{\omega=0}$ relies on the molecules in the excited state $|3\rangle$ being oriented, we expect $\langle d^L \rangle_{\omega=0}$ to survive for at least a few picoseconds before decaying due to molecular rotation. A decay of the dipole

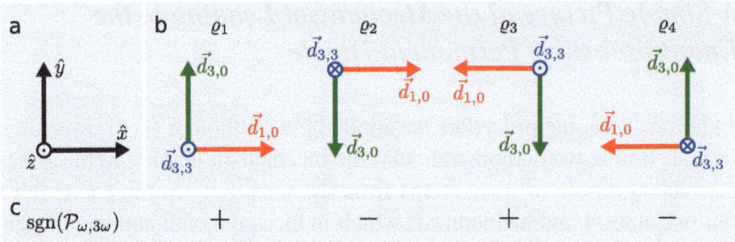

Fig. 3 Simplified analysis of the mechanism leading to an enantiosensitive permanent dipole for a field (9) and a dummy molecule with $d_{1,0}$ and $d_{3,0}$, perpendicular to each other and $d_{3,2} = d_{2,1} = d_{1,0}$. Only the component of $d_{3,3}$ perpendicular to the plane defined by $d_{1,0}$ and $d_{3,0}$ is shown. **a.** Laboratory frame. **b.** Molecular orientations with $d_{1,0}$ aligned along x and $d_{3,0}$ aligned along y. **c.** Sign of the interference term (15) for each molecular orientation. The interference distinguishes orientations ϱ_1 and ϱ_3 from orientations ϱ_2 and ϱ_4 and therefore causes the molecular axis $d_{3,3}$ to become oriented. This leads to a non-vanishing permanent dipole

on the picosecond time-scale should lead to broadband THz emission [39] with an enantiosensitive phase. Furthermore, a quantum treatment of the rotational dynamics might reveal revivals of the molecular orientation (see e.g. Ref. [31]) .

3   Exciting an Enantiosensitive Permanent Quadrupole

Let us now consider the control scheme depicted in Fig. 4, where the interference of contributions from a one-2ω-photon pathway and a two-ω-photon pathway control the population of the state $|2\rangle$ of a chiral molecule. In this case the field reads as

$$E^{L}(t) = F(t)\left(E_{\omega}^{L}e^{-i\omega t} + E_{2\omega}^{L}e^{-2i\omega t}\right) + \text{c.c.} \tag{16}$$

As in the previous section we begin assuming an intermediate resonance and then consider the case where the intermediate state is not resonant. The wave function reads as in Eq. (2) but with a sum up to $i = 2$,

$$a_2(\varrho) = A_2^{(1)}[d_{2,0}^{L}(\varrho) \cdot E_{2\omega}^{L}] + A_2^{(2)}[d_{2,1}^{L}(\varrho) \cdot E_{\omega}^{L}][d_{1,0}^{L}(\varrho) \cdot E_{\omega}^{L}]., \tag{17}$$

Fig. 4 Excitation scheme used to produce an enantiosensitive permanent quadrupole in an isotropic sample of chiral molecules

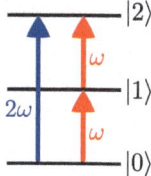

and an analogous expression for $a_1(\varrho)$. The expected value of the permanent electric quadrupole operator in the molecular frame $\langle Q_{p,q}^{\mathrm{M}}(\varrho)\rangle \equiv \langle \Psi^{\mathrm{M}}(\boldsymbol{r}^{\mathrm{M}},\varrho)|\,Q_{p,q}^{\mathrm{M}}\,|\Psi^{\mathrm{M}}(\boldsymbol{r}^{\mathrm{M}},\varrho)\rangle$, where $p,q = x,y,z$, will have a zero-frequency component of the form

$$\langle Q_{p,q}^{\mathrm{M}}(\varrho)\rangle_{\omega=0} = \sum_{i=0}^{2} |a_i(\varrho)|^2\, \langle Q_{p,q}^{\mathrm{M}}\rangle_{i,i} \tag{18}$$

where $\langle Q_{p,q}^{\mathrm{M}}\rangle_{i,i} \equiv \langle \psi_i^{\mathrm{M}}|Q_{q,p}^{\mathrm{M}}|\psi_i^{\mathrm{M}}\rangle$. Transforming $\langle Q_{p,q}^{\mathrm{M}}(\varrho)\rangle_{\omega=0}$ to the laboratory frame and averaging over all molecular orientations yields the permanent quadrupole

$$\langle Q_{p,q}^{\mathrm{L}}\rangle_{\omega=0} \equiv \int \mathrm{d}\varrho \langle Q_{p,q}^{\mathrm{L}}(\varrho)\rangle_{\omega=0}. \tag{19}$$

The contribution of state $|2\rangle$ to this expression reads as (see Appendix)

$$\langle (Q_2^{\mathrm{L}})_{p,q}\rangle_{\omega=0} \equiv \int \mathrm{d}\varrho\ |a_2(\varrho)|^2\, \langle Q_{p,q}^{\mathrm{L}}(\varrho)\rangle_{2,2} \tag{20}$$

$$= \langle (Q_2^{\mathrm{L}})_{p,q}\rangle_{\omega=0}^{(\mathrm{achiral})} + \left[A_2^{(1)*} A_2^{(2)} \chi_2' Z_{p,q}'^{\mathrm{L}} + \mathrm{c.c.} \right], \tag{21}$$

where $\langle (Q_2^{\mathrm{L}})_{p,q}\rangle_{\omega=0}^{(\mathrm{achiral})}$ results from the diagonal terms in $|a_2(\varrho)|^2$ and is not enantiosensitive. χ_2' is a rotationally invariant molecular pseudoscalar (zero for achiral molecules) encoding the enantiosensitivity of $\langle (Q_2^{\mathrm{L}})_{p,q}\rangle_{\omega=0}$ and defined according to

$$\chi_i' \equiv \frac{1}{30} \left\{ \left[(\boldsymbol{d}_{1,0}^{\mathrm{M}} \times \boldsymbol{d}_{2,0}^{\mathrm{M}}) \cdot (\langle Q^{\mathrm{M}}\rangle_{i,i}\boldsymbol{d}_{2,1}^{\mathrm{M}}) \right] + \left[(\boldsymbol{d}_{2,1}^{\mathrm{M}} \times \boldsymbol{d}_{2,0}^{\mathrm{M}}) \cdot (\langle Q^{\mathrm{M}}\rangle_{i,i}\boldsymbol{d}_{1,0}^{\mathrm{M}}) \right] \right\}, \tag{22}$$

with $\langle Q^{\mathrm{M}}\rangle_{i,i}$ a quadrupole matrix, i.e. $\langle Q^{\mathrm{M}}\rangle_{i,i}\boldsymbol{d}_{2,1}^{\mathrm{M}}$ and $\langle Q^{\mathrm{M}}\rangle_{i,i}\boldsymbol{d}_{1,0}^{\mathrm{M}}$ denote multiplications of a matrix and a vector. $Z_{p,q}'^{\mathrm{L}}$ is a symmetric field pseudotensor of rank 2. It encodes the dependence of $\langle (Q_2^{\mathrm{L}})_{p,q}\rangle_{\omega=0}$ on the field polarization according to

$$Z_{p,q}'^{\mathrm{L}} \equiv \left(\boldsymbol{E}_\omega^{\mathrm{L}} \times \boldsymbol{E}_{2\omega}^{\mathrm{L}*} \right)_p \left(\boldsymbol{E}_\omega^{\mathrm{L}} \right)_q + \left(\boldsymbol{E}_\omega^{\mathrm{L}} \times \boldsymbol{E}_{2\omega}^{\mathrm{L}*} \right)_q \left(\boldsymbol{E}_\omega^{\mathrm{L}} \right)_p. \tag{23}$$

This expression shows that all components of $Z_{p,q}'^{\mathrm{L}}$ vanish if ω and 2ω are linearly polarized parallel to each other, or if ω and 2ω are circularly polarized and counter-rotating.

For example, if we take ω and 2ω linearly polarized perpendicular to each other, say $\boldsymbol{E}_\omega^{\mathrm{L}} = \boldsymbol{x}^{\mathrm{L}}$ and $\boldsymbol{E}_{2\omega}^{\mathrm{L}} = e^{-i\phi}\boldsymbol{y}^{\mathrm{L}}$, then

$$\boldsymbol{E}^{\mathrm{L}}(t) = 2F(t)\left[\cos(\omega t)\,\boldsymbol{x}^{\mathrm{L}} + \cos(2\omega t + \phi)\,\boldsymbol{y}^{\mathrm{L}}\right], \tag{24}$$

and we obtain

$$\langle (Q_2^{\mathrm{L}})_{p,q}\rangle_{\omega=0} = \langle (Q_2^{\mathrm{L}})_{p,q}\rangle_{\omega=0}^{(\mathrm{achiral})} + 2\chi_2'\Re\left\{A_2^{(1)*}A_2^{(2)}e^{i\phi}\right\}\left(\delta_{p,z}\delta_{q,x} + \delta_{q,z}\delta_{p,x}\right). \tag{25}$$

Furthermore, one can show that for the field (24) the achiral terms vanish for $p \neq q$ (see Appendix) and therefore the enantiosensitive xz component reads as

$$\langle (Q_2^{\mathrm{L}})_{x,z}\rangle_{\omega=0} = 2\chi_2'\Re\left\{A_2^{(1)*}A_2^{(2)}e^{i\phi}\right\}, \tag{26}$$

i.e., it doesn't have an achiral background and can be controlled through the relative phase ϕ. The other non-diagonal components xy and yz vanish.

The contribution from state $|1\rangle$ to the permanent quadrupole (19) has the same structure as Eq. (26), although with different coupling constants and molecular pseudoscalar χ_1' [see Eq. (22)]. Since $|a_0|^2 = 1 - |a_1|^2 - |a_2|^2$, the contribution from the ground state involves the coupling constants associated to $|1\rangle$ and $|2\rangle$, and a molecular pseudoscalar χ_0' [see Eq. (22)]. Together, these contributions yield

$$\langle Q_{x,z}^{\mathrm{L}}\rangle_{\omega=0} = 2\left[(\chi_1' - \chi_0')\Re\left\{A_1^{(1)}A_1^{(2)*}e^{i\phi}\right\} + (\chi_2' - \chi_0')\Re\left\{A_2^{(1)*}A_2^{(2)}e^{i\phi}\right\}\right], \tag{27}$$

where $A_1^{(1)}$ and $A_1^{(2)}$ are the coupling coefficients for the transitions $|0\rangle \overset{\omega}{\to} |1\rangle$ and $|0\rangle \overset{2\omega}{\to} |2\rangle \overset{-\omega}{\to} |1\rangle$, respectively. The other non-diagonal elements of the permanent quadrupole vanish and the diagonal terms are not enantiosensitive.

As in the previous section, in the absence of an intermediate resonance through the state $|1\rangle$, the contribution from the second-order term in Eq. (17) turns into a sum over all intermediate states $|j\rangle$. The intermediate states retain no population at the end of the pulse and the permanent quadrupole takes the form

$$\begin{aligned}
\langle Q_{x,z}^{\mathrm{L}}\rangle_{\omega=0} &= 2\sum_j(\chi_{2;j}' - \chi_{0;j}')\Re\left\{A_2^{(1)*}A_{2;j}^{(2)}e^{i\phi}\right\} \\
&= 2F_0^3\,[2\pi\delta_\sigma(\Delta)]^2\sum_j \frac{\chi_{2;j}' - \chi_{0;j}'}{\omega_{j,0} - \omega_L}\cos\phi,
\end{aligned} \tag{28}$$

where $\Delta \equiv \omega_{2,0} - 2\omega_L$, $\chi_{i;j}$ is given by Eq. (22) with the replacement $1 \to j$ and the other symbols were introduced as in Eq. (12).

A simplified analysis analogous to that presented in Fig. 3 is shown in Fig. 5 for the case of the field (24) interacting with a dummy molecule with $\boldsymbol{d}_{1,0}^{\mathrm{M}}$, $\boldsymbol{d}_{2,0}^{\mathrm{M}}$, and $\langle Q^{\mathrm{M}}\rangle_{2,2}$ oriented as shown and with $\boldsymbol{d}_{2,1}^{\mathrm{M}} = \boldsymbol{d}_{1,0}^{\mathrm{M}}$. The population of state $|2\rangle$ is determined by

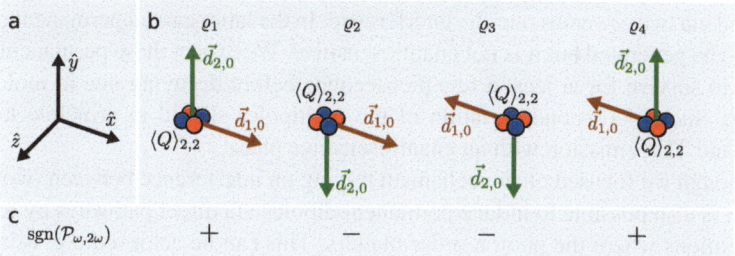

Fig. 5 Simplified analysis of the mechanism leading to an enantiosensitive permanent quadrupole for a field (24) and a dummy molecule with $d_{1,0}, d_{2,0}$, and $\langle Q \rangle_{2,2}$ oriented as shown with respect to each other and $d_{2,1} = d_{1,0}$. Blue and red balls stand for negative and positive charges. **a.** Laboratory frame. **b.** Molecular orientations with $d_{1,0}$ aligned along x and $d_{2,0}$ aligned along y. **c.** Sign of the interference term (29) for each molecular orientation. The interference causes the molecular axis $d_{2,0}$ to become oriented, which together with the alignment of $d_{1,0}$ along x explains the non-vanishing permanent quadrupole

the interference term

$$\mathcal{P}_{\omega,2\omega}\left(\varrho\right) \equiv [d_{2,0}\left(\varrho\right) \cdot y][d_{1,0}\left(\varrho\right) \cdot x]^{2}, \tag{29}$$

which is positive for orientations ϱ_1 and ϱ_4 and negative for orientations ϱ_2 and ϱ_3. This causes the molecular axis $d_{2,0}^{M}$ to become oriented. If $d_{2,2}^{M}$ has a non-zero component along $d_{2,0}^{M}$, then a permanent dipole emerges. However, this permanent dipole is contained in the polarization plane and will therefore not change upon reflection of the system across the polarization plane. Since this reflection is equivalent to a change of the enantiomer, the permanent dipole is not enantiosensitive. In contrast and as can be seen from Fig. 5, the imbalance between orientations ϱ_1 and ϱ_4 in comparison to orientations ϱ_2 and ϱ_3 is enough to produce an enantiosensitive permanent quadrupole that does change upon reflection in the polarization plane.

4   Conclusions

We have shown that permanent dipoles and quadrupoles can be induced in initially isotropic samples of chiral molecules using perturbative two-color fields that resonantly excite electronic transitions. These permanent multipoles are enantiosensitive and their sign can be controlled through the relative phase between the two colors. The mechanism leading to these permanent dipoles (or quadrupoles) stems from uniaxial orientation of the molecule, which occurs due to the selectivity of the excitation to the orientation of the molecule. Such orienting excitation can be accomplished using fields where the fundamental and its second (or third) harmonic are linearly polarized perpendicular to each other. The enantiosensitive permanent dipole is obtained via three-ω- versus one-3ω-photon interference. The enantiosensitive quadrupole is obtained via two-ω versus one-2ω interference. In the latter case, a permanent dipole can also be generated but it is not enantiosensitive. We expect these permanent multipoles to survive for at least a few picoseconds before decaying due to molecular rotation. Such picosecond variation of the multipoles should in principle lead to broadband THz emission with an enantiosensitive phase.

Although we focused on a mechanism relying on interference between two pathways, it is also possible to induce permanent dipoles via direct pathways by relying on transitions where the photon order matters. This can be achieved e.g. using the pulse sequence in Ref. [29].

Efficient generation of enantio-sensitive permanent dipoles and quadrupoles via orientation-sensitive excitations is possible in strong laser fields using efficient excitation of Rydberg states via the so-called Freeman resonances [40] in the regime when the pronderomotive potential is comparable to the laser frequency. Since Rydberg states have large polarizability, we expect significant contrast in the orientation of left and right enantiomers. Opposite orientation of left and right enantiomers and their respective induced permanent dipoles create opportunities for enantio separation using static electric fields.

Acknowledgements We gratefully acknowledge support from the DFG SPP 1840 "Quantum Dynamics in Tailored Intense Fields" within the project SM 292/5-1;

Appendix

Coupling coefficients $A_f^{(n)}$

Consider a Hamiltonian $H = H_0 + H'(t)$, where H_0 is the time-independent field-free Hamiltonian and $H'(t)$ can be treated as a perturbation. If at time $t = 0$ the system is in the state $|0\rangle$, the probability amplitude of finding the system in the state $|f\rangle$ at the time $t = T$ can be written as $a_f = a_f^{(1)} + a_f^{(2)} + \cdots$, where

$$a_f^{(N)} = \left(\frac{1}{i}\right)^N \int_0^T dt_N \ldots \int_0^{t_3} dt_2 \int_0^{t_2} dt_1 \langle f | H_I'(t_N) \ldots H_I'(t_2) H_I'(t_1) | 0 \rangle \qquad (30)$$

and $H_I'(t) = e^{iH_0 t} H'(t) e^{-iH_0 t}$. In the electric dipole approximation we have $H' = -\mathbf{d} \cdot \mathbf{E}(t)$. For a field $\mathbf{E}(t) = F(t)\mathbf{E}_\omega e^{-i\omega t} +$ c.c., the contributions to $a_f^{(N)}$ from absorption of N photons yield

$$a_f^{(N)} = \sum_{j_1, j_2, \ldots, j_{N-1}} a_{f:j_1, j_2, \ldots, j_{N-1}}^{(N)}, \qquad (31)$$

where the sum is over the different quantum pathways through the intermediate states $|j_1\rangle, |j_2\rangle, ..., |j_{N-1}\rangle$. The amplitude of each pathway can be written as

$$a^{(N)}_{f;j_1,j_2...,j_{N-1}} = A^{(N)}_{f;j_1,j_2,...,j_{N-1}}(\omega)\,(d_{f,j_{N-1}} \cdot E_\omega)\dots(d_{j_2,j_1} \cdot E_\omega)(d_{j_1,0} \cdot E_\omega), \qquad (32)$$

The coupling coefficient $A^{(N)}_{f;j_1,j_2,...,j_{N-1}}(\omega)$ carries the information about the frequency of the light, its envelope, and the detunings according to

$$A^{(N)}_{f;j_1,j_2,...,j_{N-1}}(\omega) = i^N \int_0^T dt_N F(t_N)\, e^{i(\omega_{f,j_{N-1}}-\omega)t_N} \dots$$

$$\times \int_0^{t_3} dt_2 F(t_2)\, e^{i(\omega_{j_2,j_1}-\omega)t_2} \int_0^{t_2} dt_1 F(t_1)\, e^{i(\omega_{j_1,0}-\omega)t_1}, \qquad (33)$$

where $\omega_{ij} \equiv \omega_i - \omega_j$. Contributions to $a^{(N)}_f$ from pathways involving photon emissions require exchanging E_ω by E^*_ω in Eq. (32)[4] and ω by $-\omega$ in Eq. (33) in the corresponding transitions.

In the case of a resonant pathway the sum in Eq. (31) reduces to a single term, which is the assumption in several parts of the main text. There we write $A^{(N)}_f$ as a shorthand for $A^{(N)}_{f;j_1,j_2,...,j_{N-1}}$.

Orientation integrals required in Sect. 3

Replacing Eq. (17) in Eq. (20) we obtain

$$\langle Q^{\mathrm{L}}_{p,q}\rangle_{\omega=0} = |A^{(1)}_2|^2 I^{(2\omega)}_{p,q} + |A^{(2)}_2|^2 I^{(\omega)}_{p,q} + \left[A^{(1)*}_2 A^{(2)}_2 I^{(\omega,2\omega)}_{p,q} + \mathrm{c.c.}\right], \qquad (34)$$

where the integrals $I^{(2\omega)}_{p,q}$, $I^{(\omega)}_{p,q}$, and $I^{(\omega,2\omega)}_{p,q}$ are defined by

$$I^{(2\omega)}_{p,q} \equiv \int d\varrho \left|d^{\mathrm{L}}_{2,0}(\varrho) \cdot E^{\mathrm{L}}_{2\omega}\right|^2 \langle Q^{\mathrm{L}}_{p,q}\rangle_{2,2}, \qquad (35)$$

$$I^{(\omega)}_{p,q} \equiv \int d\varrho \left|[d^{\mathrm{L}}_{2,1}(\varrho) \cdot E^{\mathrm{L}}_\omega][d^{\mathrm{L}}_{1,0}(\varrho) \cdot E^{\mathrm{L}}_\omega]\right|^2 \langle Q^{\mathrm{L}}_{p,q}\rangle_{2,2}, \qquad (36)$$

$$I^{(\omega,2\omega)}_{p,q} \equiv \int d\varrho\, [d^{\mathrm{L}}_{2,1}(\varrho) \cdot E^{\mathrm{L}}_\omega][d^{\mathrm{L}}_{1,0}(\varrho) \cdot E^{\mathrm{L}}_\omega][d^{\mathrm{L}}_{2,0}(\varrho) \cdot E^{\mathrm{L}*}_{2\omega}]\langle Q^{\mathrm{L}}_{p,q}\rangle_{2,2}. \qquad (37)$$

---

[4]If the transition dipoles are complex then one must also complex conjugate them. Here we assume they are real.

These integrals can be solved following the procedure in Ref. [38]. We will first solve $I_{p,q}^{(\omega,2\omega)}$ for arbitrary polarizations and then show that $I_{p,q}^{(2\omega)}$ and $I_{p,q}^{(\omega)}$ vanish when $p \neq q$, $E_\omega^L = x$, and $E_{2\omega}^L = e^{-i\phi}y^L$.

$I_{p,q}^{(\omega,2\omega)}$

We are dealing with an integral of the form

$$
\begin{aligned}
I_{i_4 i_5} &= \int d\varrho \left(a^L \cdot B^L\right)\left(b^L \cdot B^L\right)\left(c^L \cdot C^L\right) Q_{i_4,i_5}^L \\
&= I_{i_1 i_2 i_3 i_4 i_5; \lambda_1 \lambda_2 \lambda_3 \lambda_4 \lambda_5}^{(5)} a_{\lambda_1}^M b_{\lambda_2}^M c_{\lambda_3}^M Q_{\lambda_4,\lambda_5}^M B_{i_1}^L B_{i_2}^L C_{i_3}^L,
\end{aligned} \tag{38}
$$

where

$$
I_{i_1 i_2 i_3 i_4 i_5; \lambda_1 \lambda_2 \lambda_3 \lambda_4 \lambda_5}^{(5)} \equiv \int d\varrho\, l_{i_1 \lambda_1} l_{i_2 \lambda_2} l_{i_3 \lambda_3} l_{i_4 \lambda_4} l_{i_5 \lambda_5}, \tag{39}
$$

a^M, b^M, and c^M are arbitrary vectors fixed in the molecular frame, and $Q_{i_4 i_5}^M$ is an arbitrary symmetric second-rank tensor fixed in the molecular frame. The transformation to the laboratory frame is given by $v_i^L(\varrho) = l_{i\lambda}(\varrho) v_\lambda^M$ for vectors and $Q_{i_1,i_2}^L(\varrho) = l_{i_1 \lambda_1}(\varrho) l_{i_2 \lambda_2}(\varrho) Q_{\lambda_1,\lambda_2}^M$ for the second-rank tensor, where $l_{i\lambda}(\varrho)$ is the matrix of direction cosines, we sum over repeated indices and use latin indices for components in the laboratory frame and greek indices for components in the molecular frame. B^L and C^L are arbitrary vectors fixed in the laboratory frame. Using Eq. (31) in Ref. [38] we obtain

$$
\begin{aligned}
I_{i_4 i_5} = \frac{1}{30}\Bigg[&\epsilon_{\lambda_1 \lambda_3 \lambda_4} \delta_{\lambda_2 \lambda_5} \epsilon_{i_1 i_3 i_4} \delta_{i_2 i_5} + \epsilon_{\lambda_1 \lambda_3 \lambda_5} \delta_{\lambda_2 \lambda_4} \epsilon_{i_1 i_3 i_5} \delta_{i_2 i_4} + \epsilon_{\lambda_2 \lambda_3 \lambda_4} \delta_{\lambda_1 \lambda_5} \epsilon_{i_2 i_3 i_4} \delta_{i_1 i_5} \\
&+ \epsilon_{\lambda_2 \lambda_3 \lambda_5} \delta_{\lambda_1 \lambda_4} \epsilon_{i_2 i_3 i_5} \delta_{i_1 i_4} \Bigg] a_{\lambda_1}^M b_{\lambda_2}^M c_{\lambda_3}^M Q_{\lambda_4,\lambda_5}^M B_{i_1}^L B_{i_2}^L C_{i_3}^L
\end{aligned} \tag{40}
$$

where we used $\epsilon_{i_1 i_2 i_3} B_{i_2} B_{i_3} = \epsilon_{\lambda_1 \lambda_2 \lambda_3} Q_{\lambda_2 \lambda_3} = 0$. The first term can be rewritten as

$$
\begin{aligned}
\epsilon_{\lambda_1 \lambda_3 \lambda_4} \delta_{\lambda_2 \lambda_5} \epsilon_{i_1 i_3 i_4} \delta_{i_2 i_5} &a_{\lambda_1}^M b_{\lambda_2}^M c_{\lambda_3}^M Q_{\lambda_4,\lambda_5}^M B_{i_1}^L B_{i_2}^L C_{i_3}^L \\
&= \left(a^M \times c^M\right)_{\lambda_4} Q_{\lambda_4,\lambda_5}^M b_{\lambda_5}^M \left(B^L \times C^L\right)_{i_4} B_{i_5}^L \\
&= \left[\left(a^M \times c^M\right) \cdot \left(Q^M b^M\right)\right] \left(B^L \times C^L\right)_{i_4} B_{i_5}^L.
\end{aligned} \tag{41}
$$

Analogous operations for the rest of the terms yield

$$I_{i_4 i_5} = \frac{1}{30} \left\{ \left[(\boldsymbol{a}^{\mathrm{M}} \times \boldsymbol{c}^{\mathrm{M}}) \cdot (\boldsymbol{Q}^{\mathrm{M}} \boldsymbol{b}^{\mathrm{M}}) \right] + \left[(\boldsymbol{b}^{\mathrm{M}} \times \boldsymbol{c}^{\mathrm{M}}) \cdot (\boldsymbol{Q}^{\mathrm{M}} \boldsymbol{a}^{\mathrm{M}}) \right] \right\}$$
$$\times \left\{ (\boldsymbol{B}^{\mathrm{L}} \times \boldsymbol{C}^{\mathrm{L}})_{i_4} B_{i_5}^{\mathrm{L}} + (\boldsymbol{B}^{\mathrm{L}} \times \boldsymbol{C}^{\mathrm{L}})_{i_5} B_{i_4}^{\mathrm{L}} \right\} \tag{42}$$

Performing the substitutions $\{a, b, c, Q\} \to \{d_{2,1}, d_{1,0}, d_{2,0}, \langle Q \rangle_{2,2}\}$, $\{B, C\} \to \{E_\omega, E_{2\omega}^*\}$, and $\{i_4, i_5\} \to \{p, q\}$ and using Eqs. (34) and (37) yields Eqs. (21)-(23).

$I_{p,q}^{(2\omega)}$

Assuming a linearly polarized $E_{2\omega}$ we must deal with an integral of the form

$$I_{i_3 i_4} = \int \mathrm{d}\varrho \, [\boldsymbol{a}^{\mathrm{L}} \cdot \boldsymbol{B}^{\mathrm{L}}][\boldsymbol{a}^{\mathrm{L}} \cdot \boldsymbol{B}^{\mathrm{L}}] Q_{i_3 i_4}^{\mathrm{L}}$$
$$= I_{i_1 i_2 i_3 i_4; \lambda_1 \lambda_2 \lambda_3 \lambda_4}^{(4)} a_{\lambda_1}^{\mathrm{M}} a_{\lambda_2}^{\mathrm{M}} Q_{\lambda_3, \lambda_4}^{\mathrm{M}} B_{i_1}^{\mathrm{L}} B_{i_2}^{\mathrm{L}}, \tag{43}$$

where

$$I_{i_1 i_2 i_3 i_4; \lambda_1 \lambda_2 \lambda_3 \lambda_4}^{(4)} \equiv \int \mathrm{d}\varrho \, l_{i_1 \lambda_1} l_{i_2 \lambda_2} l_{i_3 \lambda_3} l_{i_4 \lambda_4}, \tag{44}$$

and we use the same notation as in the previous subsection. Using Eq. (19) in Ref. [38] we get

$$I_{i_3 i_4} = \boldsymbol{F}_{i_3 i_4}^{(4)} \cdot M^{(4)} \boldsymbol{G}_{i_3 i_4}^{(4)}, \tag{45}$$

where $\boldsymbol{F}_{i_3 i_4}^{(4)}$ is given by

$$\boldsymbol{F}_{i_3 i_4}^{(4)} = \begin{pmatrix} \delta_{i_1 i_2} \delta_{i_3 i_4} \\ \delta_{i_1 i_3} \delta_{i_2 i_4} \\ \delta_{i_1 i_4} \delta_{i_2 i_3} \end{pmatrix} B_{i_1}^{\mathrm{L}} B_{i_2}^{\mathrm{L}} = \begin{pmatrix} |\boldsymbol{B}^{\mathrm{L}}|^2 \delta_{i_3 i_4} \\ B_{i_3}^{\mathrm{L}} B_{i_4}^{\mathrm{L}} \\ B_{i_3}^{\mathrm{L}} B_{i_4}^{\mathrm{L}} \end{pmatrix} \tag{46}$$

For $\boldsymbol{B}^{\mathrm{L}} = \boldsymbol{y}^{\mathrm{L}}$ we have $B_i^{\mathrm{L}} = \delta_{iy}$ and therefore $B_{i_3}^{\mathrm{L}} B_{i_4}^{\mathrm{L}} = \delta_{i_3 y} \delta_{i_4 y} = \delta_{i_3 i_4} \delta_{i_3 y}$, which yields $\boldsymbol{F}_{i_3 i_4}^{(4)} \propto \delta_{i_3 i_4}$ and $I_{i_3 i_4} \propto \delta_{i_3 i_4}$. The substitutions $\{a, Q, B, i_3, i_4\} \to \{d_{2,0}, \langle Q \rangle_{2,2}, \boldsymbol{y}, p, q\}$ then yield $I_{p,q}^{(2\omega)} \propto \delta_{p,q}$.

$I_{p,q}^{(\omega)}$

Assuming a linearly polarized E_ω we must deal with an integral of the form

$$I_{i_5,i_6} = \int d\varrho [\boldsymbol{a}^{\mathrm{L}} \cdot \boldsymbol{B}^{\mathrm{L}}][\boldsymbol{b}^{\mathrm{L}} \cdot \boldsymbol{B}^{\mathrm{L}}][\boldsymbol{a}^{\mathrm{L}} \cdot \boldsymbol{B}^{\mathrm{L}}][\boldsymbol{b}^{\mathrm{L}} \cdot \boldsymbol{B}^{\mathrm{L}}]Q^{\mathrm{L}}_{i_5i_6}$$

$$= I^{(6)}_{i_1i_2i_3i_4i_5i_6;\lambda_1\lambda_2\lambda_3\lambda_4\lambda_5\lambda_6} a^{\mathrm{M}}_{\lambda_1} b^{\mathrm{M}}_{\lambda_2} a^{\mathrm{M}}_{\lambda_3} b^{\mathrm{M}}_{\lambda_4} Q^{\mathrm{M}}_{\lambda_5,\lambda_6} B^{\mathrm{L}}_{i_1} B^{\mathrm{L}}_{i_2} B^{\mathrm{L}}_{i_3} B^{\mathrm{L}}_{i_4}, \qquad (47)$$

where

$$I^{(6)}_{i_1i_2i_3i_4i_5i_6;\lambda_1\lambda_2\lambda_3\lambda_4\lambda_5\lambda_6} = \int d\varrho\, l_{i_1\lambda_1} l_{i_2\lambda_2} l_{i_3\lambda_3} l_{i_4\lambda_4} l_{i_5\lambda_5} l_{i_6\lambda_6}, \qquad (48)$$

and we use the same notation as in the previous subsections. Using Table II in [38] we have that

$$I_{i_5,i_6} = \boldsymbol{F}^{(6)}_{i_5i_6} \cdot M^{(6)} \boldsymbol{G}^{(6)}_{i_3i_4}, \qquad (49)$$

where $\left(F^{(6)}_{i_5i_6}\right)_r \equiv f^{(6)}_r B^{\mathrm{L}}_{i_1} B^{\mathrm{L}}_{i_2} B^{\mathrm{L}}_{i_3} B^{\mathrm{L}}_{i_4}$ and $f^{(6)}_r$ ($r = 1, 2, \ldots, 15$) is given in Table II in [38]. For $\boldsymbol{B}^{\mathrm{L}} = \boldsymbol{x}^{\mathrm{L}}$ we have $B^{\mathrm{L}}_i = \delta_{ix}$ and therefore

$$\left(F^{(6)}_{i_5i_6}\right)_r = \begin{cases} \delta_{i_5i_6}, & r = 1, 4, 7 \\ \delta_{i_5x}\delta_{i_6x}, & \text{otherwise} \end{cases} \qquad (50)$$

Since $\delta_{i_5x}\delta_{i_6x} = \delta_{i_5i_6}\delta_{i_5x}$, then $\boldsymbol{F}^{(6)}_{i_5i_6} \propto \delta_{i_5i_6}$ and $I_{i_5,i_6} \propto \delta_{i_5i_6}$. The substitutions $\{\boldsymbol{a}, \boldsymbol{b}, Q, \boldsymbol{B}, i_5, i_6\} \to \{\boldsymbol{d}_{1,0}, \boldsymbol{d}_{2,1}, \langle Q \rangle_{2,2}, \boldsymbol{x}, p, q\}$ then yield $I^{(\omega)}_{p,q} \propto \delta_{p,q}$.

References

1. E.U. Condon, Rev. Mod. Phys. **9**(4), 432 (1937). https://doi.org/10.1103/RevModPhys.9.432. URL https://link.aps.org/doi/10.1103/RevModPhys.9.432

2. K. Soai, T. Shibata, H. Morioka, K. Choji, Nature **378**(6559), 767 (1995). https://doi.org/10.1038/378767a0. URL https://www.nature.com/articles/378767a0. Number: 6559 Publisher: Nature Publishing Group

3. P. Fischer, F. Hache, Chirality **17**(8), 421 (2005). https://doi.org/10.1002/chir.20179. URL https://onlinelibrary.wiley.com/doi/abs/10.1002/chir.20179

4. S. Beaulieu, A. Ferré, R. Géneaux, R. Canonge, D. Descamps, B. Fabre, N. Fedorov, F. Légaré, S. Petit, T. Ruchon, V. Blanchet, Y. Mairesse, B. Pons, New J. Phys. **18**(10), 102002 (2016). https://doi.org/10.1088/1367-2630/18/10/102002. URL http://stacks.iop.org/1367-2630/18/i=10/a=102002

5. K. Banerjee-Ghosh, O.B. Dor, F. Tassinari, E. Capua, S. Yochelis, A. Capua, S.H. Yang, S.S.P. Parkin, S. Sarkar, L. Kronik, L.T. Baczewski, R. Naaman, Y. Paltiel, Science **360**(6395), 1331 (2018). https://doi.org/10.1126/science.aar4265. URL https://science.sciencemag.org/content/360/6395/1331. Publisher: American Association for the Advancement of Science Section: Report

6. D.S. Sanchez, I. Belopolski, T.A. Cochran, X. Xu, J.X. Yin, G. Chang, W. Xie, K. Manna, V. Süß, C.Y. Huang, N. Alidoust, D. Multer, S.S. Zhang, N. Shumiya, X. Wang, G.Q. Wang, T.R. Chang, C. Felser, S.Y. Xu, S. Jia, H. Lin, M.Z. Hasan, Nature **567**(7749), 500 (2019). https://doi.org/10.1038/s41586-019-1037-2. URL https://www.nature.com/articles/s41586-019-1037-2

7. S. Mason, Chemical Society Reviews **17**, 347 (1988). https://doi.org/10.1039/CS9881700347. URL https://pubs.rsc.org/en/content/articlelanding/1988/cs/cs9881700347

8. D.G. Blackmond, Proc. Nat. Acad. Sci. **101**(16), 5732 (2004). https://doi.org/10.1073/pnas.
    0308363101. URL https://www.pnas.org/content/101/16/5732
9. G.Q. Lin, Q.D. You, J.F. Cheng, *Chiral Drugs: Chemistry and Biological Action* (Wiley, Hoboken, NJ, 2011)
10. N. Berova, P.L. Polavarapu, K. Nakanishi, R.W. Woody, *Comprehensive Chiroptical Spectroscopy*, vol. 1 (Wiley, Hoboken, NJ, 2012)
11. B. Ritchie, Phys. Rev. A **13**(4), 1411 (1976). URL http://journals.aps.org/pra/abstract/10.1103/
    PhysRevA.13.1411
12. N. Böwering, T. Lischke, B. Schmidtke, N. Müller, T. Khalil, U. Heinzmann, Phys. Rev.
    Lett. **86**(7), 1187 (2001). https://doi.org/10.1103/PhysRevLett.86.1187. URL http://link.aps.
    org/doi/10.1103/PhysRevLett.86.1187
13. I. Powis, in *Advances in Chemical Physics* (Wiley, New York, 2008), pp. 267–329. URL http://
    onlinelibrary.wiley.com/doi/10.1002/9780470259474.ch5/summary
14. A.F. Ordonez, O. Smirnova, Phys. Rev. A **98**(6), 063428 (2018). https://doi.org/10.1103/
    PhysRevA.98.063428. URL https://link.aps.org/doi/10.1103/PhysRevA.98.063428
15. A.F. Ordonez, O. Smirnova, Phys. Rev. A **99**(4), 043416 (2019). https://doi.org/10.1103/
    PhysRevA.99.043416. URL https://link.aps.org/doi/10.1103/PhysRevA.99.043416
16. A.F. Ordonez, O. Smirnova, Phys. Rev. A **99**(4), 043417 (2019). https://doi.org/10.1103/
    PhysRevA.99.043417. URL https://link.aps.org/doi/10.1103/PhysRevA.99.043417
17. L.D. Barron, J. Am. Chem. Soc. **101**(1), 269 (1979). https://doi.org/10.1021/ja00495a071.
    URL http://dx.doi.org/10.1021/ja00495a071
18. J.A. Giordmaine, Phys. Rev. **138**(6A), A1599 (1965). https://doi.org/10.1103/PhysRev.138.
    A1599. URL https://link.aps.org/doi/10.1103/PhysRev.138.A1599
19. P.M. Rentzepis, J.A. Giordmaine, K.W. Wecht, Phys. Rev. Lett. **16**(18), 792 (1966). https://
    doi.org/10.1103/PhysRevLett.16.792. URL https://link.aps.org/doi/10.1103/PhysRevLett.16.
    792
20. D. Patterson, M. Schnell, J.M. Doyle, Nature **497**(7450), 475 (2013). https://doi.org/10.1038/
    nature12150. URL http://www.nature.com/nature/journal/v497/n7450/full/nature12150.html
21. D. Patterson, J.M. Doyle, Phys. Rev. Lett. **111**(2), 023008 (2013). https://doi.org/10.1103/
    PhysRevLett.111.023008. URL http://link.aps.org/doi/10.1103/PhysRevLett.111.023008
22. A. Yachmenev, S.N. Yurchenko, Phys. Rev. Lett. **117**(3), 033001 (2016). https://doi.org/
    10.1103/PhysRevLett.117.033001. URL https://link.aps.org/doi/10.1103/PhysRevLett.117.
    033001
23. S. Beaulieu, A. Comby, D. Descamps, B. Fabre, G.A. Garcia, R. Géneaux, A.G. Harvey,
    F. Légaré, Z. Mašín, L. Nahon, A.F. Ordonez, S. Petit, B. Pons, Y. Mairesse, O. Smirnova,
    V. Blanchet, Nat. Phys. **14**(5), 484 (2018). https://doi.org/10.1038/s41567-017-0038-z. URL
    https://www.nature.com/articles/s41567-017-0038-z
24. N.I. Koroteev, J. Experim. Theor. Phys. **79**(5), 681 (1994). URL http://jetp.ac.ru/cgi-bin/e/
    index/e/79/5/p681?a=list
25. S. Woźniak, G. Wagnière, Opt. Commun. **114**(1), 131 (1995). https://doi.org/10.1016/0030-
    4018(94)00498-J. URL http://www.sciencedirect.com/science/article/pii/003040189400498J
26. R. Zawodny, S. Woźniak, G. Wagnière, Opt. Commun. **130**(1), 163 (1996). https://doi.
    org/10.1016/0030-4018(96)00224-6. URL http://www.sciencedirect.com/science/article/pii/
    0030401896002246
27. B.S. Wozniak, Mol. Phys. **90**(6), 917 (1997). https://doi.org/10.1080/002689797171913. URL
    https://www.tandfonline.com/doi/abs/10.1080/002689797171913. Publisher: Taylor & Francis
28. P. Fischer, A.C. Albrecht, Laser Phys. **12**(8), 1177 (2002)
29. E. Gershnabel, I.S. Averbukh, Phys. Rev. Lett. **120**(8), 083204 (2018). https://doi.org/10.1103/
    PhysRevLett.120.083204. URL https://link.aps.org/doi/10.1103/PhysRevLett.120.083204
30. I. Tutunnikov, E. Gershnabel, S. Gold, I.S. Averbukh, J. Phys. Chem. Lett. **9**(5), 1105 (2018).
    https://doi.org/10.1021/acs.jpclett.7b03416. URL https://doi.org/10.1021/acs.jpclett.7b03416
31. I. Tutunnikov, J. Floß, E. Gershnabel, P. Brumer, I.S. Averbukh, Phys. Rev. A **100**(4),
    043406 (2019). https://doi.org/10.1103/PhysRevA.100.043406. URL https://link.aps.org/doi/
    10.1103/PhysRevA.100.043406

32. A.A. Milner, J.A.M. Fordyce, I. MacPhail-Bartley, W. Wasserman, V. Milner, I. Tutunnikov, I.S. Averbukh, Phys. Rev. Lett. **122**(22), 223201 (2019). https://doi.org/10.1103/PhysRevLett.122.223201. URL https://link.aps.org/doi/10.1103/PhysRevLett.122.223201

33. P.V. Demekhin, A.N. Artemyev, A. Kastner, T. Baumert, Phys. Rev. Lett. **121**(25), 253201 (2018). https://doi.org/10.1103/PhysRevLett.121.253201. URL https://link.aps.org/doi/10.1103/PhysRevLett.121.253201

34. P.V. Demekhin, Phys. Rev. A **99**(6), 063406 (2019). https://doi.org/10.1103/PhysRevA.99.063406. URL https://link.aps.org/doi/10.1103/PhysRevA.99.063406

35. S. Rozen, A. Comby, E. Bloch, S. Beauvarlet, D. Descamps, B. Fabre, S. Petit, V. Blanchet, B. Pons, N. Dudovich, Y. Mairesse, Phys. Rev. X **9**(3), 031004 (2019). https://doi.org/10.1103/PhysRevX.9.031004. URL https://link.aps.org/doi/10.1103/PhysRevX.9.031004

36. A.F. Ordonez, O. Smirnova, [physics] (2020). URL http://arxiv.org/abs/2009.03660

37. A.F. Ordonez, O. Smirnova, [physics] (2020). URL http://arxiv.org/abs/2009.03655

38. D.L. Andrews, T. Thirunamachandran, J. Chem. Phys. **67**(11), 5026 (1977). https://doi.org/10.1063/1.434725. URL http://aip.scitation.org/doi/citedby/10.1063/1.434725

39. D.J. Cook, R.M. Hochstrasser, Opt. Lett. **25**(16), 1210 (2000). https://doi.org/10.1364/OL.25.001210. URL https://www.osapublishing.org/ol/abstract.cfm?uri=ol-25-16-1210

40. R.R. Freeman, P.H. Bucksbaum, H. Milchberg, S. Darack, D. Schumacher, M.E. Geusic, Phys. Rev. Lett. **59**(10), 1092 (1987). https://doi.org/10.1103/PhysRevLett.59.1092. URL https://link.aps.org/doi/10.1103/PhysRevLett.59.1092

Chapter 17
Ultra-fast Dynamics in Quantum Systems Revealed by Particle Motion as Clock

M. S. Schöffler, L. Ph. H. Schmidt, S. Eckart, R. Dörner, A. Czasch,
O. Jagutzki, T. Jahnke, J. Ullrich, R. Moshammer, R. Schuch,
and H. Schmidt-Böcking

Abstract To explore ultra-fast dynamics in quantum systems one needs detection schemes which allow time measurements in the attosecond regime. During the recent decades, the pump & probe two-pulse laser technique has provided milestone results on ultra-fast dynamics with femto- and attosecond time resolution. Today this technique is applied in many laboratories around the globe, since complete pump & probe systems are commercially available. It is, however, less known or even forgotten that ultra-fast dynamics has been investigated several decades earlier even with zeptosecond resolution in ion-atom collision processes. A few of such historic experiments, are presented here, where the particle motion (due to its very fast velocity) was used as chronometer to determine ultra-short time delays in quantum reaction processes. Finally, an outlook is given when in near future relativistic heavy ion beams are available which allow a novel kind of "pump & probe" experiments on molecular systems with a few zeptosecond resolution. However, such experiments are only feasible if the complete many-particle fragmentation process can be imaged with high momentum resolution by state-of-the-art multi-particle coincidence technique.

M. S. Schöffler · L. Ph. H. Schmidt · S. Eckart · R. Dörner · A. Czasch · O. Jagutzki · T. Jahnke · H. Schmidt-Böcking (✉)
Institut für Kernphysik, Universität Frankfurt, 60348 Frankfurt, Germany
e-mail: hsb@atom.uni-frankfurt.de; schmidtb@atom.uni-frankfurt.de

A. Czasch · O. Jagutzki · H. Schmidt-Böcking
Roentdek GmbH, 65779 Kelkheim, Germany

J. Ullrich
PTB, Brunswick, Germany

R. Moshammer
MPI für Kernphysik, Heidelberg, Germany

R. Schuch
Physics Department, Stockholm University, 107 67 Alba Nova, Stockholm, Sweden

© The Author(s) 2021
B. Friedrich and H. Schmidt-Böcking (eds.), *Molecular Beams in Physics and Chemistry*,
https://doi.org/10.1007/978-3-030-63963-1_17

353

1 Introduction

To explore the nature of atomic matter scientists have developed during the last century sophisticated approaches to reveal the microscopic structure of matter and also the dynamics between atoms or even inside atoms and molecules. The resolving power for static structural features of molecular systems, e.g. measured by Cryo-electron microscopy [1] or X-ray spectroscopy [2], is presently in a range of a few 10^{-10} m, which is about a few times the diameter of a single atom. In these measurement approaches the momenta of electrons or photons scattered on a molecular object are detected. The measured momentum distributions, are converted by Fourier transformation into coordinate space, yielding a spatial image of the molecular structure.

To explore the dynamics of a reaction between quantum objects or to reveal the electron dynamics inside a quantum object the experimenter in general interacts with a fast projectile (photon, electron, ion etc.) in a first step (the excitation or ionization step) with the quantum object and observes after very short time delays (typically attoseconds) electron and ionic fragment emission. Thus, the experimenter obtains information on the dynamically changed final states or even intermediate states of the object.

In order to reveal the "entangled" electron dynamics inside the same molecule, it is typically not sufficient to perform single parameter measurements on the same molecular object at two shortly successive instants in time. To reveal entangled dynamics, the simultaneous detection of the momenta of all fragments emitted from the same single molecule is required. Thus, a multi-fragment coincidence measurement imaging the complete momentum space with high resolution is necessary. Such high-resolution multi-coincidence detection systems are available since about two decades: The COLTRIMS-reaction microscope C-REMI possesses all necessary properties to perform such high-resolution multi-coincidence investigations [3].

What are the time delays Δt of interest? The duration of a chemical reaction is typically in the order of a picosecond, and, accordingly, a nucleus can be considered as locally frozen during a time interval of a femtosecond [4]. Therefore, to explore such nuclear processes a time resolution of about one femtosecond (10^{-15} s) is required. This is the standard time range where modern femtosecond Laser pump-and probe-schemes can visualize chemical reactions, geometrical changes and their dynamics [5]. Intra-atomic or intra-molecular processes do proceed faster. The duration of a charge transfer process in ion-atom or ion/molecule collisions depends on the projectile velocity and on the quasi-molecular promotion path-way [6]. It can be as short as an attosecond (10^{-18} s). Hole migration in photon-excited molecules can proceed also in the attosecond range, as well as intra-atomic or intra-molecular Coulombic vacancy or energy transfer. Interatomic (or intermolecular) electronic decay processes occur in a wide range of durations from few femtoseconds to several picoseconds [7]. Energy transfer processes as in photosynthesis of chlorophyll, for example, proceed on the upper femtosecond level.

In fast ion-atom collisions intra-atomic and intra-molecular dynamics can take place even on the lower zeptosecond level (10^{-19} s), which is about 4 orders of magnitude shorter than the typical femtosecond Laser pulse can resolve [8–12]. In such a short time interval light travels only a distance of 0.3 Å. To visualize these ultra-fast dynamics, one needs detection methods which visualize its time dependence, e.g. in interference structures like in quantum-beats. As will be shown below when dynamical processes proceed via two different pathways they accumulate different phases yielding characteristic interference structures. From these structures phase differences can be determined and, as outlined before, by knowing the velocity of the fast ion, time delays even in the zeptosecond regime can be deduced [9].

It may be a more theoretical and philosophical issue whether ultra-short time scales below one attosecond may be of any relevance in atomic physics. But these time scales are doubtlessly of high interest in quantum physics, in general. For example, fundamental questions arise, as whether the so-called "collapse" of a wave function is a local process and starts in one location inside a molecule and proceeds then with speed of light through the whole molecular system. In this case the "collapse" would last about 300 zeptoseconds to stretch across a simple molecule. Or is the collapse a non-local process instantaneously present everywhere across the molecule? Measurements with 10 zeptosecond time resolution would allow to explore such a fundamental question e.g. in a triatomic molecule with a non-linear geometry (see Sect. 3.3).

2   Ultra-fast Chronometer Mechanisms Using Fast Moving Particles as Clock

Burgdörfer et al. [8] have recently presented a review on the historic development and the present status of attosecond physics performed in the field of ion-atom collisions and short-pulse Laser physics. By discussing the theoretical aspects, they have shown the similarity of ion- and Laser-induced processes. Since the chronometer scheme of the multi-photon pump & probe technique is discussed widely in [5, 8, 12], this paper will concentrate on ionization processes induced by ion impact, where the motion of particles provides the ultra-fast chronometers.

The method of "pumping" a quantum object to an excited or ionized state and "probing" this excited state, i.e. by observing the delayed fragment or photon emission, is in general the principle of any measurement in reactions between quantum particles*. So-called "Pump & Probe" measurements are today commonly identified with Two-pulse Laser Pump & Probe methods where the very short time delay Δt between the two Laser pulses can be well adjusted by two different geometrical path ways yielding a time resolution in the femto- or even attosecond range. In this Laser Pump & Probe approach the probing is processed via a delayed second photon pulse where the delay time can be chosen by the experimenter.

In ion-atom collisions the experimenter can never prepare two projectiles ions such that they interact with an atom or molecule at the same impact parameter with a well-controlled time delay of attosecond precision to undergo like in Laser physics a Pump & Probe process. The "pump & probe" process in ion-atom collisions must be induced by the same ion at two different locations in a molecule. Since the relative locations of atoms in a molecule are known with a precision of about 0.1 a.u., the delay-time between the ion reaction at two different locations can then be varied by changing the ion velocity—and typically achievable ion velocities correspond to 10 zeptosecond-pump-probe-resolution.

2.1 Historic Life-Time Measurements with Nano- and Picosecond Precision

Measurements with time-delay determination have been performed already 100 years ago, e.g. by Stern and Volmer [13], when they measured the mean decay time of photon-excited I_2 molecules (Fig. 1). Since many of the articles in these proceedings of the Otto-Stern conference are related to Otto Stern's scientific work we will shortly discuss here Stern's and Volmer's pioneering work of measuring life times of excited molecules, too. Stern and Volmer used the thermal motion of vapor molecules as chronometer. The excited molecules expanded from a tiny interaction spot of a few micrometer diameter, where they have been excited, according to their thermal motion (i.e. the motion of the molecules created streaking). Stern and Volmer observed the excitation and decay positions of the molecules using a light microscope. From the outreaching tails of the excitation spot Stern and Volmer derived the

Fig. 1 Inside the glass tube I_2 molecules were evaporated from solid Iodine. A very narrow collimated photon beam (1 μm diameter) excited the molecules (small quadratic box). The fluorescence light emission was observed in a greater halo region due to the thermal motion of the excited molecules. This halo distribution was measured using a lens system. From the halo distribution and the thermal properties, the mean decay lifetime was determined [13]

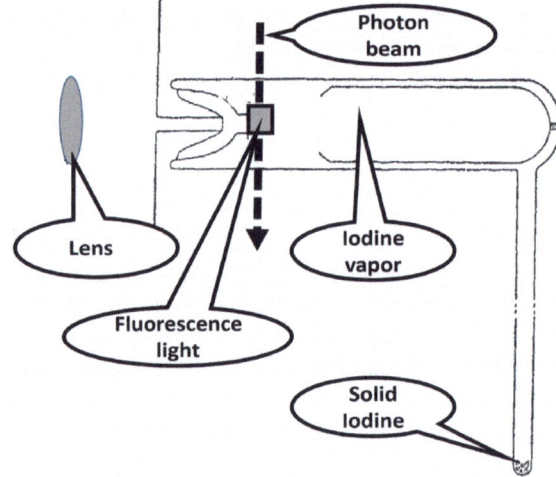

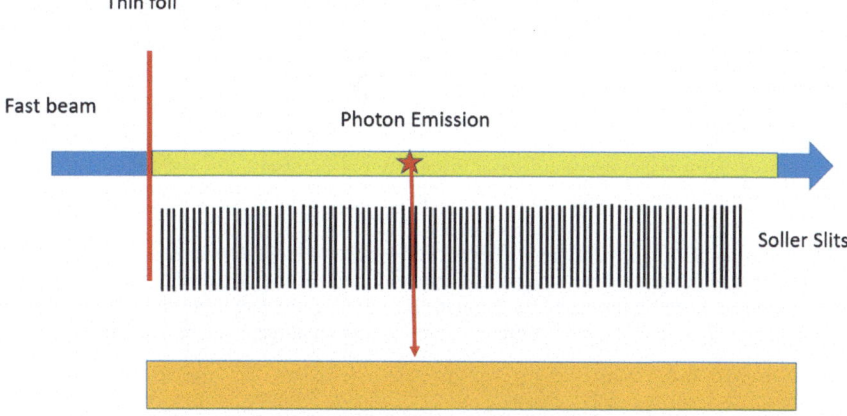

Fig. 2 Beam foil spectroscopy with fast ions [14]

mean life time for the decay process of the excited states with a time resolution of about 2 ns.

In the past, numerous methods have been developed and applied to measure decay times and explore dynamics in atomic and molecular systems. For historic reasons we describe here also the so-called "beam-foil" techniques [14]. A fast ion beam (kinetic energy typically 0.1 to 10 MeV/u) penetrates a very thin foil. Inside the foil ions get excited and decay downstream the moving beam. The emitted fluorescence photons are detected with a position-sensitive photon detector. A Soller-slit system allows only photons emitted transversely to reach the detector.

Thus, the exponential decay distribution as function of distance from the foil (i.e. decay time) is measured. From the exponential slope one can calculate the delay time with picosecond precision ("beam-foil" techniques see Fig. 2) [14]. This is a kind of streaking technique where from the observable positions (foil and decay) and from the ion velocity the decay time is deduced by macroscopic methods.

If inside the foil two nearby ionic levels can be excited simultaneously, the time-resolved fluorescence light emitted from these coherently excited levels can show a characteristic quantum beat structure.

2.2   *Quantum Beat Structures as Ultra-fast Chronometers*

Ultra-short time interval measurements can be performed also with fast moving particles. As "clock" the fast motion of ions is used whose trajectory can be considered as classical. Measuring transition probabilities resulting from two spatially localized interaction areas the two transitions amplitudes interfere yielding a characteristic interference pattern. Since the delay-time can be calculated from the classical motion

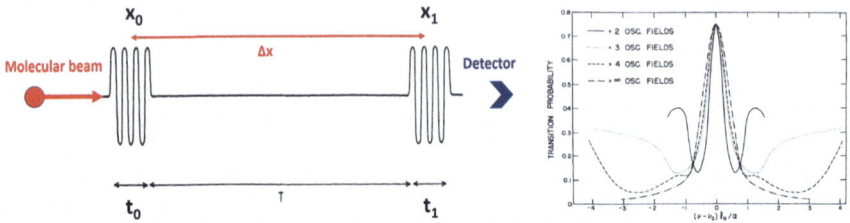

Fig. 3 The principle scheme of Ramsey's "Separated oscillating field" device. A fast-moving object (here indicated as molecular beam) passes two coherent cavities at time t_0 and t_1 respectively, and the object can be excited at time t_0 and t_1. From the interference structure measured in the excitation probability (see right side) the phase difference $\Delta\Phi$ of both amplitudes can be determined [15]

of the ion and from the locations where transitions occur, the phases can be determined from the measured interference pattern and thus information on the dynamics of the reaction process can be derived.

This superposition scheme of two wave amplitudes for moving atoms emitted at time t_0 and t_1 (see Fig. 3) was already applied in Ramsey's "Separated oscillating field method", which is the basis of the atomic clock [15]. In Ramsey's pioneering experiment a moving object (in Fig. 3 indicated as molecular beam) passes through two cavities and the moving object can be excited either at t_0 or t_1. Since the experimenter does not know in which cavity the excitation took place both excitation amplitudes at t_0 and t_1 add coherently. From the interference structure measured in the excitation probability (see right side of Fig. 3) the phase difference $\Delta\Phi$ of both amplitudes can be determined. From $\Delta\Phi$ and the known time delay T the transition energy can be deduced with high precision.

3 Experimental Examples of Quantum-Beat Measurements in Ion-Atom/Molecule Collisions

3.1 Quantum Beats in Quasi-molecular X-Ray Emission

In specially prepared ion-atom collision processes one can use the fast classical motion of an ion as a very fast clock to visualize even electronic dynamics with a time resolution in the lower zeptosecond regime. By measuring the quantum beat structure in the spectra of quasi-molecular X-ray emission Schuch et al. [9] have obtained in fast ion-atom-collisions even a time resolution of nearly 10 zeptoseconds. The X-rays emitted in a reaction visualize the streaking of the quasi-molecular orbitals by the two-center ion-atom nuclear potential, which provides a very fast, with time varying streaking force. The X-ray photon energies encode, thus, the fingerprints of the strength of the streaking force at the moment of emission and yield information on Δt.

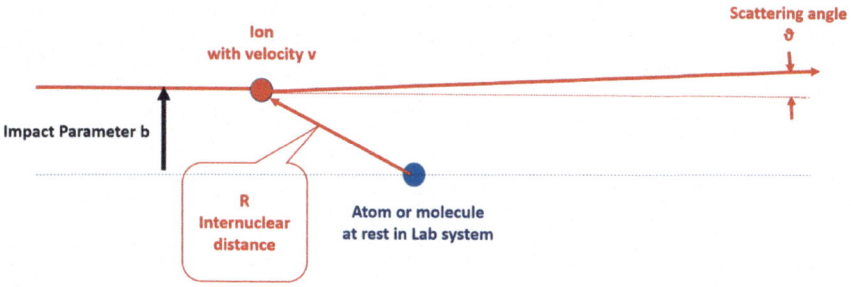

Fig. 4 Scheme of an ion-atom scattering process

The purpose of the experiment by Schuch et al. [9] was to measure the binding energy of the very short living quasi-molecular $1s\sigma$-state which was formed in fast ion-atom collisions for a time duration of only about 1 attosecond. More than 30 years ago several groups (quasi-molecular radiation) [10] and (K-K vacancy sharing) [11] have measured quantum beat structures in ion-atom collisions with oscillations in the atto- and zeptosecond range. From the interference structures of these quantum beats, phase differences were determined yielding energy or time domain information. In Fig. 4 the scheme of such an ion-atom scattering process is presented. An X-ray/scattered-projectile coincidence measurement is required to reveal such quantum-beat structures for a given impact parameter. To probe such a short living quasi-molecular state with the detection techniques of the eighties was extremely difficult since the achieved quasi-molecular X-ray/scattered-projectile coincidence rate was a few true counts per hour. In the laboratory system the projectile ion bypasses an atom on a quasi-straight line (very small deflection angles of about 1°, which are determined by detecting the scattered projectile deflection angle). From the deflection angle the impact parameter b can be deduced. Since the ion velocity is known the internuclear distance R (vector) can be calculated as function of the relative collision time t ($t_0 = 0$ is the time moment at distance of closest approach).

If R is much smaller than the projectile ion or target atom K-shell radii even the most-inner electronic orbitals steadily approach during this extremely short (sub-attosecond) collision time the united-atom electronic states due to the combined projectile and target nuclear Coulomb potentials. Thus, the combined nuclear potentials "streak" as function of $R = R(t)$ the energy values of the bound quasi-molecular states (see Fig. 5).

To reveal the streaked quasi-molecular energy values, one has to prepare the projectile in a very special ionic configuration to create observable quantum beat structures. If one bombards a hydrogen-like Cl^{16+} projectile ion on an Ar atom, thus, a $1s\sigma$ vacancy is already present on the incoming part of the collision and identical X-rays can be emitted on the way into the collision ($-t$ values) and on the way out of the collision ($+t$ values). Thus the transition amplitudes on the first half (way into the collision) interfere with those of the second half of the collision. Like in a double slit experiment (see Ramsey's atomic-clock [15] "separated-oscillating-field method")

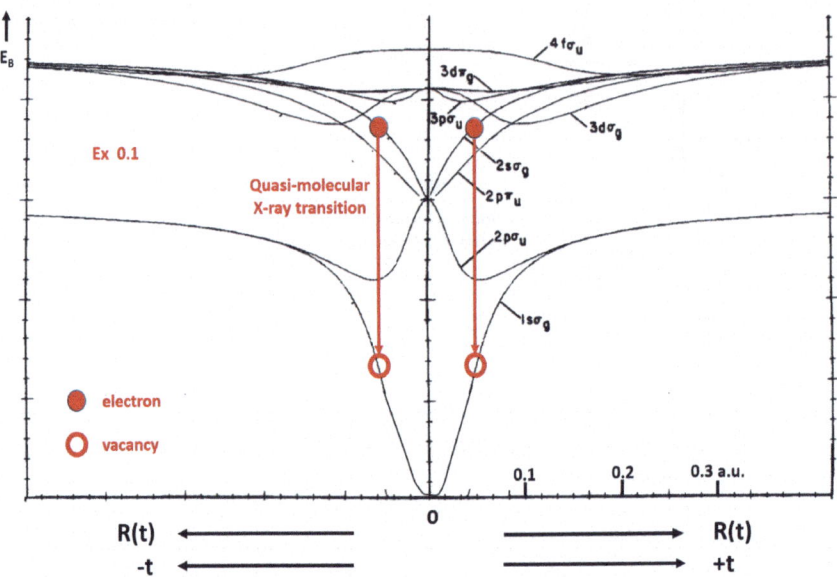

Fig. 5 Quasi-molecular correlation diagram for Cl^{16+} on Ar as function of the internuclear distance R. During the collision an electron from the $2p\pi$ orbital can pass over into the $1s\sigma$ orbital and an X-ray is emitted. The X-ray transition energy is the energy difference between the $2p\pi$ and the $1s\sigma$ orbital [9] at the particular R value

one does observe characteristic quantum beat structures in the spectra of quasi-molecular X-rays emitted during the collision (see Fig. 6). Since the quantum beat structures vary with impact parameter the X-rays must be detected in coincidence with scattered projectiles to select one given scattering angle (i.e. a fixed impact parameter-range). The X-ray transition process observed here is the quasi-molecular K_α-transition (electron transition from the $2p\pi$ into the $1s\sigma$ quasi-molecular state). The K_α-transition energy in the united-atom limit at very small internuclear distances ($Z_{UA} = Z_{Bromium} = Z_{Cl} + Z_{Ar} = 35$) is about 12 keV (see Fig. 7). These X-rays are emitted per definition at the time moment t_0. X-rays of lower energy are emitted at larger internuclear distances R, i.e. larger -t or +t values (see Fig. 5). Thus, the collision time parameter t = R(t) is zero at E_x of the united atom (E_x = 12 keV) and increases to larger R values, i.e. lower X-ray energies.

From Fig. 7 we derive that each X-ray transition energy corresponds to a well-defined internuclear distance and thus to a well-defined collision moment. Thus, we can visualize the variation of the quasi molecular Coulomb potential with nearly 10 zeptosecond resolution. Furthermore, the observed quantum beat structure yields a phase information with $\varphi(t) =_{t=0} \int^t E_x(R(t))/\hbar \cdot dt$. The highest X-ray energies correspond to R_{min} or t = 0. According to [9] one can now—vice versa—determine the transition energies $E_x(R(t))$ as function of the collision time, i.e. the internuclear distance R. The final result is shown in Fig. 7, where from all measured spectra

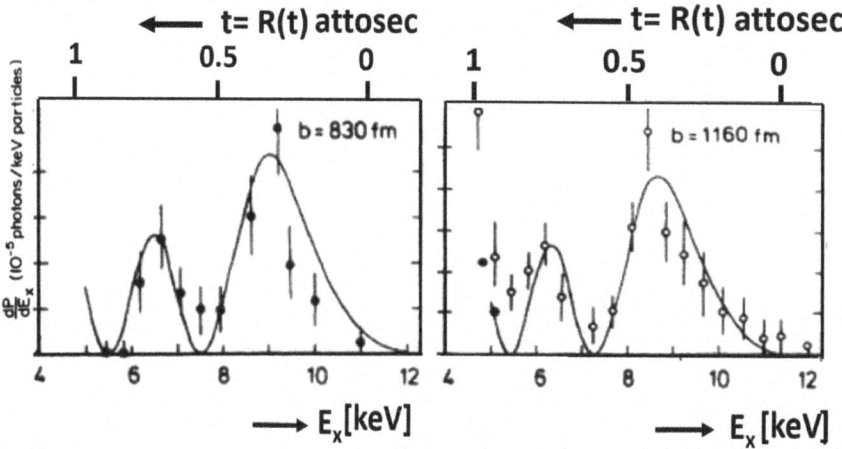

Fig. 6 Measured X-ray spectra for the 2pπ − 1sσ quasi-molecular transition in Cl^{16+}-Ar collisions [9] at fixed impact parameters. The X-ray energy is directly measured with a Si(Li)-detector and can be transferred into an internuclear distance R via the correlation diagram (Fig. 5). From the internuclear distance R and the ion velocity the time scale can be calibrated

(different impact parameters and different collision energies) the quasi-molecular energy values as function of the internuclear distance are displayed.

The analysis of the quantum beats shows that for E_x as function of R an universal curve is obtained independent of the ion-atom collision energy, i.e. independent of the streaking time. For 20 MeV collision energy the R scale (from R = 0 up to 0.1 a.u.) in Fig. 7 corresponds to 500 zeptoseconds, for 5 MeV to 1 attosecond. Such measurements [9–11] show that using ion-atom collisions the dynamics of quasi-molecular states could be explored with 10 zeptosecond resolution.

3.2   Young-Type Interference Structures in Slow H_2^+ +He Collisions

Schmidt et al. have [6] investigated "Young-type interference structures" in 10 keV H_2^+ + He => H_2^* + He^+ => H + H + He^+ collisions (relative velocity v = 0.45 a.u.). They measured the momentum vectors of all reaction fragments in the final state in coincidence using the C-REMI approach. Thus, the orientation of the H_2^+ molecule with respect to the projectile momentum vector was determined for each event yielding the He^+ scattering distribution in dependence of the angle θ (Fig. 8). In the moving projectile system, the He atom can be scattered by both H_2^+ projectile nuclei (double slit) (see Fig. 8). The scattered He^+ momentum wave function is then the coherent sum of the two amplitudes emerging from the two H nuclei scattering centers. From the resulting interference structures in the He^+ scattering distribution

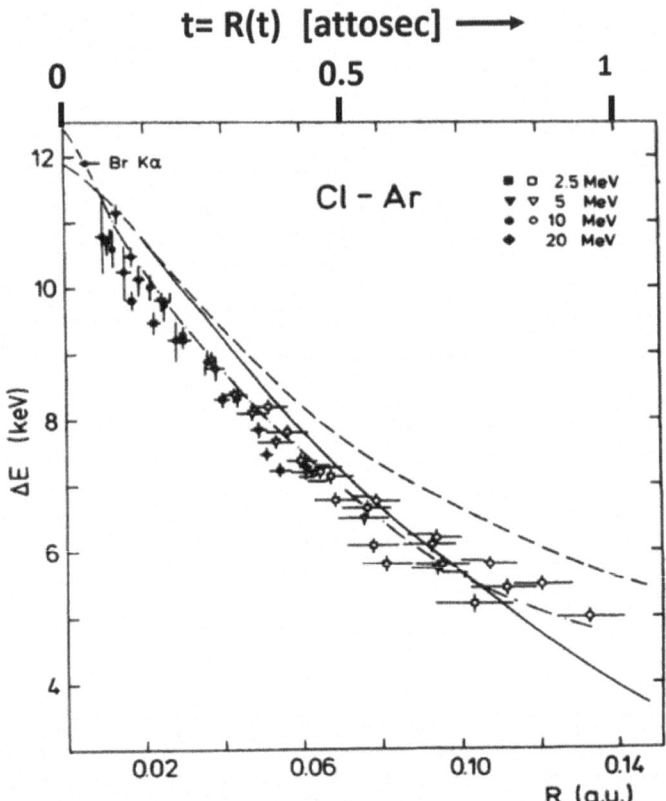

Fig. 7 Transition energies between $1s\sigma$ and $2p\pi$ quasi-molecular states as function of the internuclear distance R. The dashed-dotted line results from DFS calculations for the $1s\sigma - 2p\pi$ transition; solid and dashed lines are scaled from $H^+ + H$ for the $2p\sigma - 1s\sigma$ and $2p\pi - 1s\sigma$ transition, respectively [9]. The time scale corresponds to 5 MeV collision energy

the phase shifts between the two amplitudes can be deduced as function of Θ visualizing the tiny time delay between the interaction of the He atom with the first (t_1) and second (t_2) H atom (see Fig. 8).

The phase difference between both amplitudes due to the molecular orientation θ is proportional to the measurable delay time $(\Delta t = t_2 - t_1)$. In a multi-particle coincidence measurement one can also calculate it directly from the measured angle θ and from the ion velocity v.

Figure 9 clearly shows that the interference pattern varies with θ and even with the KER value (electronic excitation), too. Schmidt et al. presented a model calculation (red dashed line in Fig. 10) for the superposition of the scattering amplitudes at the two H atoms. The differential scattering cross sections can be expressed as $d\sigma/d\varphi \sim \cos^2(\beta/2)$ where the phase shift ß is $\beta = \pi + R \cdot \Delta p_{He}/\hbar + \Delta E \cdot \Delta t/\hbar$. The phase jump π accounts for the inversion of the molecular symmetry in the electronic transition

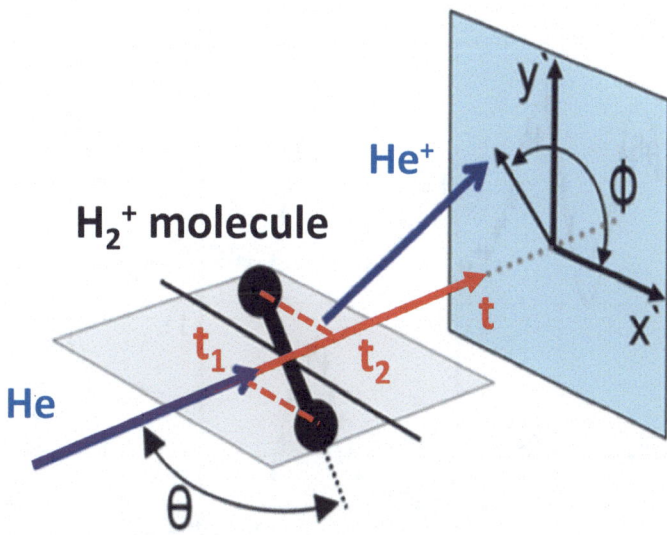

Fig. 8 Nuclear scattering scheme. In inverse kinematics the He target atom collides with the H_2^+ molecular ion and transfers one electron to the H_2 molecule which breaks up into neutral H atoms. The two H atoms are detected at small scattering angles in forward direction. The He$^+$ ion is detected with a C-REMI under 90°. The He is scattered into the azimuthal angle Φ. The relative orientation of the molecule to the He impact direction is defined by the angle Θ. The time difference $t_2 - t_1$ can be determined from the measured angle Θ and the ion velocity

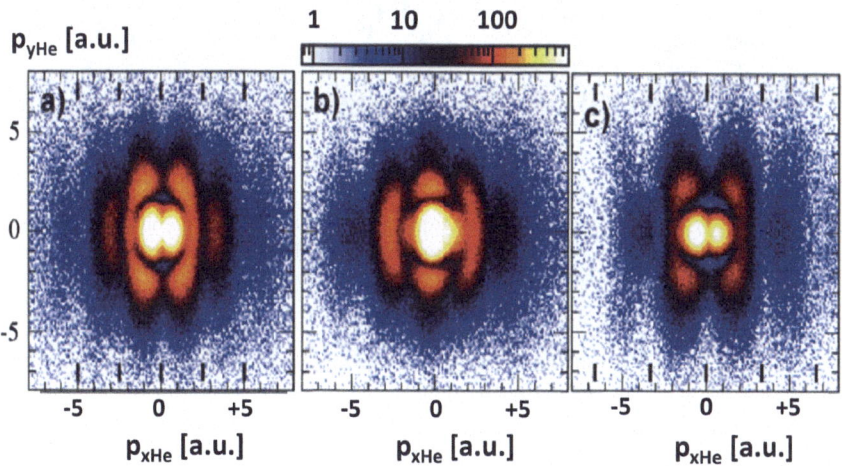

Fig. 9 Two-slit interference pattern in the plane perpendicular to the projectile momentum vector for three different θ angles. **a** Events for molecular orientation angles (with respect to the beam direction) from 80° to 90° and KER* values 1 to 2 eV. This KER corresponds to R values from 2.3 to 2.9 a.u. **b** Events for molecular orientation angles from 50° to 60° and KER 3 to 4 eV. This KER corresponds to R values from 2.3 to 2.9 a.u. **c** Events for molecular orientation angles from 80° to 90° and KER 3 to 4 eV. This KER corresponds to R values from 1.7 to 2.0 a.u. [6]

counts

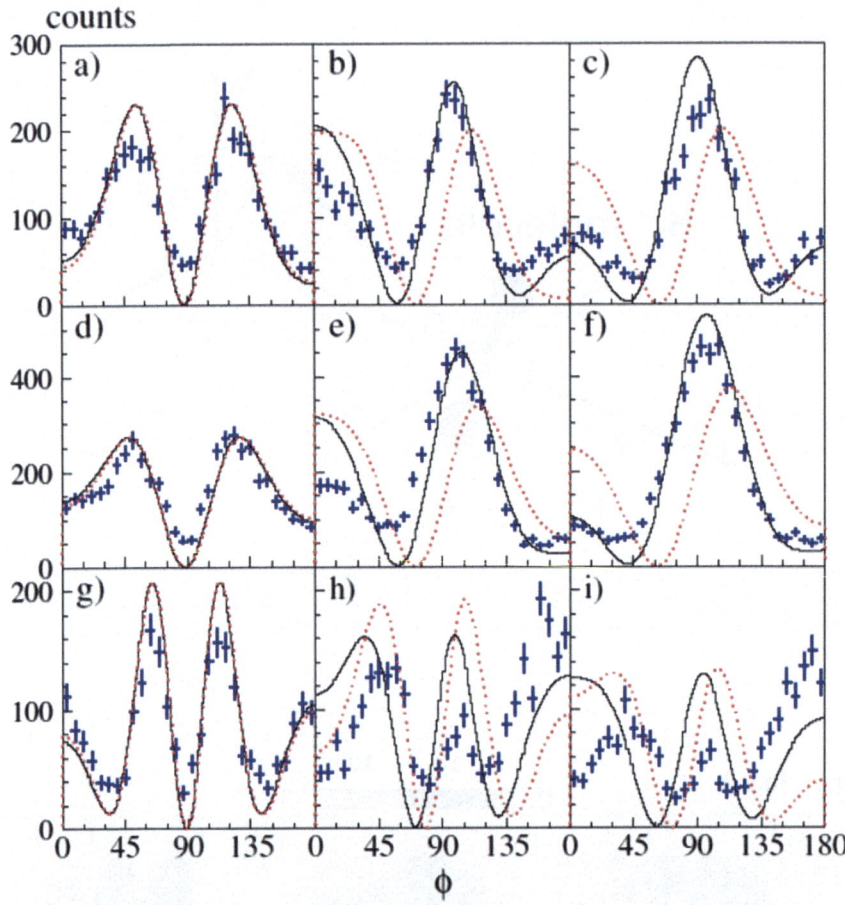

Fig. 10 Relative cross sections as function of the angle φ in comparison with model calculations (red dashed lines) and full theory (solid lines) [6]. The left column represents collisions where θ is 85–90°, the middle column where θ is 55–60° and the right column where θ is 45–50°. The rows show data for different transverse momenta and slightly different KER values

and the second term for the change of the He momentum Δp_{He} due to the scattering. which leads to a change of the de Broglie wave length of the scattered He and for the term $\Delta E \cdot \Delta t/\hbar$. It accounts for the correction of the so-called translation factor with $\Delta t = t_2 - t_1$ (see Fig. 8) as time difference of the interaction of the He projectile with the two H atoms

In Fig. 10, the measured interference pattern are compared with these model calculations (red dashed lines) and the full theory (black lines) [6]. The data are presented for a given polar scattering angle (i.e. impact parameter) as function of the azimuthal angle Φ. In the left column (a + d + g) of Fig. 10 the data are shown for $\theta = 90°$ and $\Delta t = 0$. In the middle column (b + c + h) $\theta = 60°$ and the right column (c + f + i) data are shown for $\theta = 45°$ where the influence of the delay Δt on the scattering as well as electronic excitation becomes strongly visible. The data of Schmidt et al. [6] prove in a convincing manner that the electronic excitation processes as function of the delay Δt does vary.

The full theory (only for the phase variation, not including electronic dynamics for transition probabilities) describes rather well (besides g and h) the measured phase variations. The model calculations (red dashed lines) do not include the nuclear Δt phase effect. They are shifted by about 20° to higher angles and indicate the large effect of Δt on the phases. The good agreement in phase of the full theory proves that the time delay calculated from the geometry and ion velocity are properly taken into account. The time delay Δt (derived from the collision geometry, see Fig. 8) is for θ angles between 45° to 50° $\Delta t = R \cdot \sin(45°)/v_{He} = 2.3$ a.u. $\cdot \sin(45°)/0.45$ a.u. $= 3.6$ a.u. $= 87$ attosecond. The disagreement in the absolute height shows, that the electron dynamics varies with Δt too. The measured data contain also information on the electon dynamics in such reactions.

The limits of the resolution for Δt determined from the experimentally observed phase shifts can be estimated from the data of Fig. 10 and from the comparison with the theoretical calculations. The resolution in determing phases is about 3°. This corresponds for the 10 keV He on H_2 collision system to 10 attosec time resolution in such collisions. If the ion velocity increases the ion-motion based clock would gain resolution. The He on H_2 collision system investigated by Schmidt et al. [6] clearly demonstrates that the effect of time delay between both scattering amplitudes is nicely visible in the interference structure. The absolute scattering intensities in the different final excitation states of the two H atoms are, however, only in modest agreement with the data. It is to notice, that only such channels were measured by Schmidt et al. where the final H fragments remained in the ground state. To observe more significant differences in the excitation of both atoms of the dipolar molecule one should investigate molecular species with higher Z values.

Since the overall momentum resolution is so excellent, the different channels of electronic excitation in these scattering processes can be resolved event by event, as well, and one can identify different electronic promotion channels during the collision. From Fig. 11 one can deduce that for each event the different electron promotion channels (different molecular orbitals (see Fig. 12 and 13)) with KER and Q value are fully separated. The different electronic promotion pathways are marked by the letters "a" to "f" and "A" plus "B". In the three-body (He and two proton nuclei) scattering process the He^+ ion is mostly scattered out of plane into the angle Φ.

In the 10 keV $H_2^+ + He => H_2^* + He^+ => H + H + He^+$ electron transfer process one He electron is captured to metastable H_2^* vibrational states (H(1s) + H(2l)) state with an energy minimum at about R = 2 a.u. [6]. Since the internuclear distance R is a function of time, the H_2^* fragmentation process provides a fast clock, too. Thus, one can estimate that the time period for the capture (in the He-H_2 system) lasts only about a few hundred attoseconds. In this reaction channel a fraction of these states $He^+ =$ $> H + H + He^+$ decays after the collision during the ongoing Coulomb explosion (duration about some tenth of femtoseconds) into the H(1s) + H(1s) ground state by

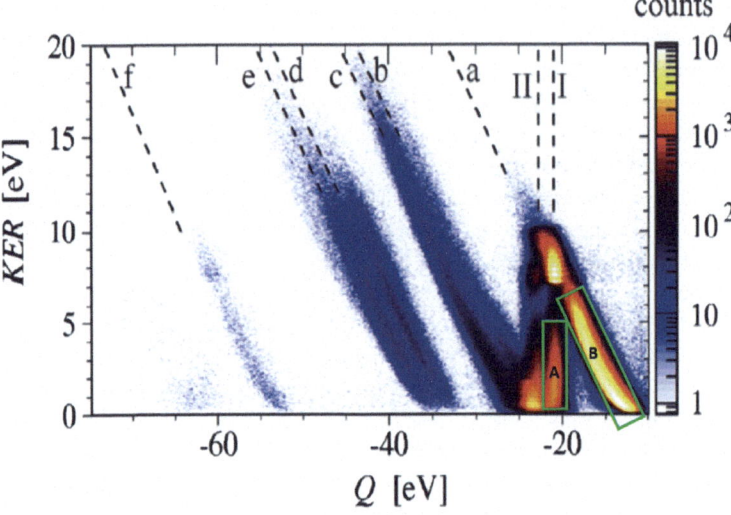

Fig. 11 Final state KER-Q-value correlation diagram of the different electronic excitation channels. The lines a to f mark different exited states. The channels A and B are discussed in more detail in the next Figs. 12 and 13 [6]

Fig. 12 Electron promotion and fragmentation scheme of region B in Fig. 11 [6]

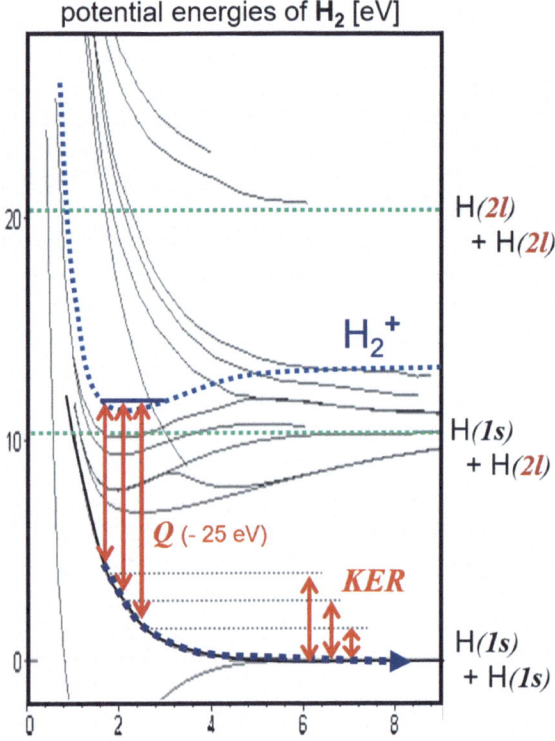

Fig. 13 Electron promotion and fragmentation scheme of region A in Fig. 11 [6]

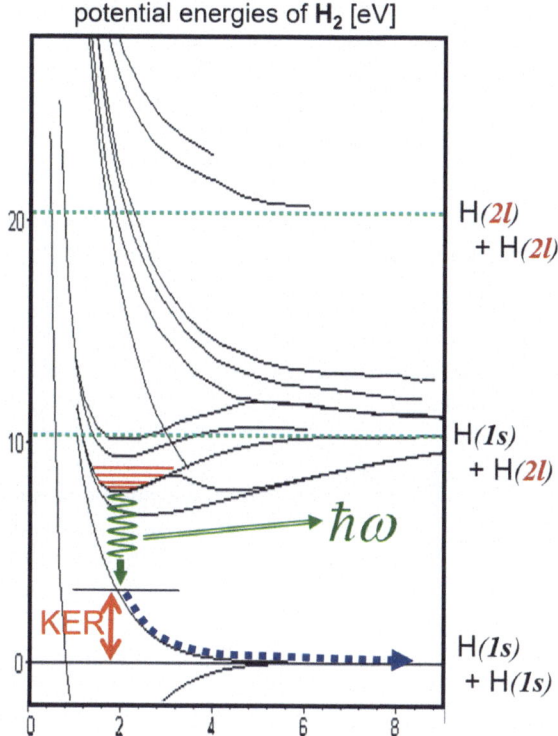

potential energies of H_2 [eV]

emitting photons (green oscillation $\hbar\omega$ in Fig. 13) with a variable KER energy (see channel A in Fig. 11). The remaining small amount of energy [relative to the ground state $H(1s) + H(1s)$] is detectable as final KER value.

In the collision 10 keV H_2^+ on He, one He electron can also be captured to the $H(1s) + H(1s)$ ground state (red arrows in Fig. 12) (region B in Fig. 11) [6]. Like for region A, this electron transfer occurs during the ongoing Coulomb explosion into the $H(1s) + H(1s)$ ground state. The remaining amount of energy (relative to the ground state $H(1s) + H(1s)$) converts to KER in the final state.

Similar experiments for molecules with higher Z-values have been performed recently by Iskandar et al. [16]. Iskandar et al. investigated collisions of low energy Ar^{9+} ions on Ar_2 dimer targets and measured all ionic fragments in coincidence. From the measured recoil-ion momenta the dimer orientation with respect to the projectile direction, the nuclear transverse momentum transfer (impact parameter) and the KER values could be measured for each event. They found clear evidence that the capture from the first hit atom in the dimer is favored. Because of the large distance of both atoms in the dimer the highly charged ion interacts preferable only with one atom with a high probability of multiple capture. The subsequent intra-molecular vacancy sharing probability between the two Ar atoms in the dimer is low because of the large distance of both atoms in the dimer. They analyzed their data in the "Over-barrier Model" and found reasonable agreement between theory and experiment describing the capture processes. Since their momentum resolution was not good enough to distinguish quasi-molecular promotion path-ways they were not able to explore the fast electronic dynamics.

3.3   A Proposal: Scheme of an Ion-Atom/Molecule Pump & Probe Technique Approaching 10 Zeptoseconds Time Resolution

As already pointed out above the ion-atom/molecule pump & probe scheme has little in common with the Laser pump & probe scheme, in which two Laser pulses can be created as an interlocked pair. No experimenter can produce an ion beam where always two of these ions move as group with a time delay adjusted to 1 attosecond precision and passing through a molecule on an identical trajectory. Even if one could prepare such an ion pair, these two ions would immediately repel each other by their nuclear Coulomb force. Thus, a pump & probe process with ions can only be performed by a single (solely) moving ion, interacting at two different locations in the same molecule. These locations (e.g. two different atoms in the same molecule) must be detectable with subatomic precision. Thus, the experimenter must be able to measure the position (impact parameters => transferred momenta) of the projectile-molecule reaction and the orientation of the ion trajectory with respect to the structure of the molecule. Measuring all ionic fragment momenta in the final state by a multi-coincidence-approach with high momentum resolution allows for a deduction of both—the orientation and the impact parameters. Thus, the experimenter knows in which time sequence and time delays the ion interacted with the different atoms in the molecule. E.g. using the C-REMI approach these requirements can indeed be satisfied.

Since the relative distances between atoms in a molecule can be calculated with about 0.1 a.u. precision, and as a fast ion (particularly a relativistic moving heavy ion) follows a perfect classical straight-line trajectory, the relative time delays between the impact at the two different atoms in a molecule can be determined with a few zeptoseconds resolution. But how can one utilize this kind of timing to investigate electronic dynamics in ion-molecule collisions?

An ion moving with a relativistic velocity (v_{ion} => c speed of light) interacts with atoms or molecules via a very sharp retarded dipole-like electric field where the opening angle scales with $1/\gamma$, where $\gamma = 1/\mathrm{sqrt}(1 - (v/c)^2)$. For $\gamma = 20$ the opening angle is about $20°$ yielding a short interaction time in the lower zeptosecond regime. As shown in [17] the virtual photon field of relativistic heavy ions as projectiles is very strong. It interacts simultaneously with nearly all electrons in the molecule resulting in multiple ionization of the ionic fragments. To control the electron dynamics in such a collision, the momenta of all ejected electrons must be measured in coincidence, too. Thus, the scheme of an ion induced pump & probe measurement presented here

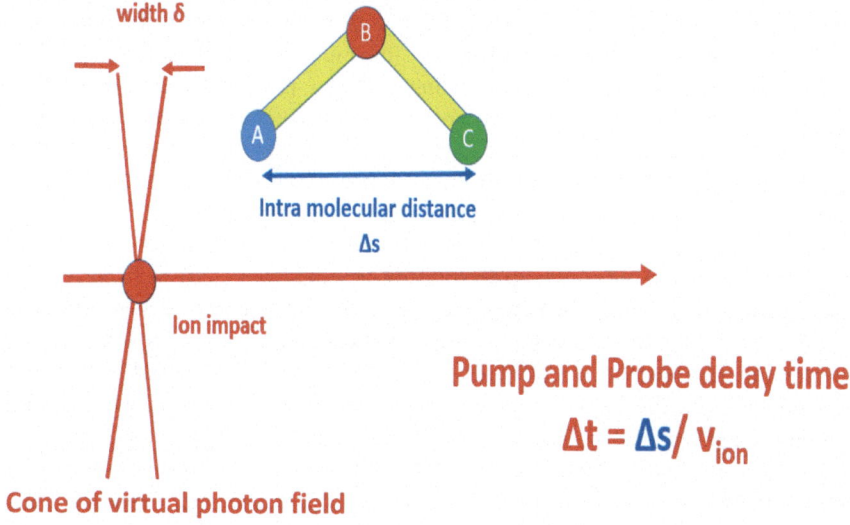

width δ

B

A

C

Intra molecular distance

Δs

Ion impact

Pump and Probe delay time

Δt = Δs/ v_ion

Cone of virtual photon field

Fig. 14 Ion-molecule pump & probe scheme

requires a multi-coincidence detection approach which must have a high detection efficiency and excellent momentum resolution for ions and electrons.

In Fig. 14 that scheme of an ion-molecule pump & probe measurement is shown. The direction of the impacting ion beam is precisely prepared, however, the molecules (occurring in the gas phase) are randomly oriented.

As an example a triatomic molecule is considered here, where the different atoms A, B, and C are bound in a non-linear formation. The distances between the atoms in the molecule are typically in the order of 2 to 3 a.u. In the example shown here, an ion travelling with the speed of light, interacts first with atom A (pump process) and induces the "collapse" of the molecular ground state wave function. After a time delay of about $\Delta t = \Delta s/c \approx 3$ a.u./137 a.u. ≈ 0.022 a.u. ≈ 480 zeptoseconds the same ion approaches atom C. One interesting question is, when does atom C "know" that atom A suffered the collapse of its wave function? If the collapse is instantly present everywhere in the molecule then atom C is very likely in an excited state, when the ion interacts after about 460 zeptosecond with the molecule at position C. If, however, the collapse emerges from atom A with the speed of light, then the transport of collapse information via atom B to atom C will arrive at best at the same time, but most likely a little later than the relativistic projectile ion. Thus, the observed final state of fragment C may depend on the collapse expansion time.

To explore the nature of the collapse expansion one has to measure the momenta of all ionic fragments and all emitted electrons in the final state in coincidence for all orientations of the molecule with respect to the projectile ion flight direction. A C-REMI provides for each fragment detection a nearly 4π solid angle with about 50% detection efficiency for each fragment. Thus, the total multi-coincidence detection efficiency is rather high (for three ionic fragments and 3 electrons is the total coincidence efficiency still about 2%). However, this multi-particle detection efficiency can be increased dramatically, by optimizing the single particle detection efficiency: enhancing the transmission and open area of the micro channel plates up to 90% [18]. The overall final state of the reaction process may strongly depend in which sequence the ion interacted with the different atoms in the molecule. The overall final state may differ e.g. by mirroring the projectile velocity vector.

A first, "simple" experiment is proposed here, where the final ionic states A* and C* are compared, when the projectile is impinging from "left" or from "right". Reducing the ion velocity far below the speed of light, the time delay range can be extended. This kind of measurement could already now provide new inside into the range of zeptosecond electronic dynamics. The method proposed here to use ions moving with the speed of light as ultra-fast clocks allows the investigation of fundamental features of "Locality" or "Non-Locality" in quantum systems, i.e., whether the information exchange occurs in such systems instantaneous or only by speed of light.

Theorists may be convinced that the questions raised here are already answered. But nevertheless one should experimentally verify any fundamental theoretical prediction. When Otto Stern decided in 1920 to perform the "Stern-Gerlach-Experiment" [19] and later in 1933 to measure the magnetic moment of the proton [20] theorists tried to convince him, such difficult experiments should not be done since theory had already answered these questions. Nevertheless, he performed his milestone measurements and could disprove theory.

4 Conclusion

This paper does show that since more than 100 years the motion of atoms, molecules or ions was successfully used to measure dynamical features (like lifetimes, phase shifts etc.) in quantum systems. Already more than 30 years ago quantum beat structures with 10 zeptoseconds resolution could be measured about factor 100 to 1000 shorter than present pump & probe two-pulse laser techniques can achieve. The future heavy ion facility FAIR [21] will provide relativistic heavy ion beams with which the here proposed ion-atom/molecule pump & probe technique can be performed to explore dynamics in the zeptosecond regime.

Acknowledgements We thank Siegbert Hagmann, Hans Jürgen Lüdde, John Briggs, Mike Prior, and C. L. Cocke for many helpful discussions. Furthermore, we are indebted to the Deutsche Forschungsgemeinschaft and the BMFT for financial support as well as Roentdek company for technical support in performing the experiments.

References

1. W. Kühlbrandt, The resolution revolution. Science, **343**, 1443 (2014) and other references therein
2. F. Calegari, G. Sansone, M. Nisoli, *Attosecond Pulses for Atomic and Molecular Physics* (2014) Book Lasers in Materials Science, pp. 125–141, Springer International Publishing; A. L. Cavaleri et al., Nature (London) 449, 1029 (2007)
3. R. Dörner, V. Mergel, O. Jagutzki, L. Spielberger, J. Ullrich, R. Moshammer, H. Schmidt-Böcking, Cold target recoil ion momentum spectroscopy: a 'momentum microscope' to view atomic collision dynamics. Phys. Rep. **330**, 95–192 (2000); R. Dörner, T. Weber, M. Achler, V. Mergel, L. Spielberger, O. Jagutzki, F. Afaneh, M.H. Prior, C.L. Cocke, H. Schmidt-Böcking, 3-D coincident imaging spectroscopy for ions and electrons, imaging in chemical dynamics, in *ACS Symposium Series*, vol. 770, ed. by A. Suits, R. E. Continetti, (Oxford Univ. Press, 2001), pp. 339–349; R. Dörner, H. Schmidt-Böcking, V. Mergel, Th. Weber, L. Spielberger, O. Jagutzki, A. Knapp, H.P. Bräuning, From atoms to molecules. *Many Part. Quantitative Dynamic Atomic Molecular Fragm.*, ed. by V.P. Shevelko, J. Ullrich (Springer Verlag, 2002); J. Ullrich, R. Moshammer, A. Dorn, R. Dörner, L.Ph.H. Schmidt, H. Schmidt-Böcking Recoil-ion and electron momentum spectroscopy: reaction-microscopes. Rep. Prog. Phys. **66** 1463–1545 (2003)
4. M. Born, R. Oppenheimer, Zur Quantentheorie der Molekeln. Ann. Phys. **389**(20), 457–484 (1927)
5. A.H. Zewail, Femtochemistry: recent progress in studies of dynamics and control of reactions and their transition states. J. Phys. Chem. (Centennial Issue) **100**, 12701 (1996); A.H. Zewail, Femtochemistry: atomic-scale dynamics of the chemical bond†. J. Phys. Chem. A. **104**(24), 5660–5694 (2000), adapted from the Nobel Lecture; F. Krausz and M. Ivanov, 2009 Rev. Mod. Phys. **81**, 163; M. Drescher, M. Hentschel, R. Kienberger, G. C. Tempea, C. Spielmann, G.A. Reider, P.B. Corkum, F. Krausz, Science **291**, 1923 (2001)
6. L.Ph.H. Schmidt, S. Schössler, F. Afaneh, M. Schöffler, K. Stiebing, H. Schmidt-Böcking, R. Dörner, Young-type interference in collisions between hydrogen molecular ions and helium. Phys. Rev. Lett. **101**, 173202 (2008)
7. T. Jahnke, L. Cederbaum T. Jahnke, A. Czasch, M. S. Schöffler, S. Schössler, A. Knapp, M. Käsz, J. Titze, C. Wimmer, K. Kreidi, R. E. Grisenti, A. Staudte, O. Jagutzki, U. Hergenhahn, H. Schmidt-Böcking, R. Dörner, Experimental observation of interatomic coulombic decay in neon dimers. Phys. Rev. Lett. **93**, 163401 (2004); T. Jahnke, A. Czasch, M. Schöffler, S. Schössler, M. Käsz, J. Titze, K. Kreidi, R.E. Grisenti, A. Staudte, O. Jagutzki, L.Ph.H. Schmidt, Th. Weber, H. Schmidt-Böcking, K. Ueda, R. Dörner, Experimental separation of virtual photon exchange and electron transfer in interatomic coulombic decay of neon dimers. Phys. Rev. Lett. **99**, 153401 (2007); L.S. Cederbaum, J. Zobeley, and F. Tarantelli; Giant intermolecular decay and fragmentation of clusters. Phys. Rev. Lett. **79**, 4778 (1997)
8. J. Burgdörfer, C. Lemmel, X. Tong, Invited Lecture at ICPEAC 2019, arXiv:2001.02900v1 [quant-ph] 9 Jan 2020
9. R. Schuch, M. Meron, B.M. Johnson, K.W. Jones, R. Hoffmann, H. Schmidt-Böcking, I. Tserruya, Quasimolecular X-ray spectroscopy for slow Cl^{16+}-Ar collisions. Phys. Rev. A **37**, 3313 (1988)
10. I. Tserruya, R. Schuch, H. Schmidt-Böcking, J. Barrette, Wang Da-Hai, B.M. Johnson, M. Meron, K.W. Jones, Interference effects in the quasimolecular K X-ray production probability for 10 MeV Cl^{16+}- Ar collisions. Phys. Rev. Lett. **50**, 30 (1983); R. Schuch, H. Schmidt-Böcking, I. Tserruya, B.M. Johnson, K.W. Jones, M. Meron, X-ray spectroscopy of Cl-Ar molecular orbitals from $1s\sigma$-$2p\pi$ transitions. Z. Phys. A **320**, 185 (1985)
11. R. Schuch, H. Ingwersen, E. Justiniano, H. Schmidt-Böcking, M. Schulz, F. Ziegler, Interference effects in K-vacancy transfer of hydrogen like S Ions with Ar. J. Phys. B **17**, 2319 (1984); R. Schuch, H. Ingwersen, E. Justiniano, H. Schmidt-Böcking, M. Schulz, F. Ziegler, Experiments with decelerated S^{15+}-beams: interferences in K-shell to K-shell charge transfer, atomic and

nuclear heavy ion interactions. Centr. Inst. of Phys. (1986); S. Hagmann, J. Ullrich, S. Kelbch, H. Schmidt-Böcking, C.L. Cocke, P. Richard, R. Schuch, A. Skutlartz, B. Johnson, M. Meron, K. Jones, D. Trautmann, F. Rösel, K-K-charge transfer and electron emission for 0.13 MeV/u F^{8+} + Ne collisions. Phys. Rev. A **36**, 2603 (1987); R. Schuch, M. Schulz, Y.S. Kozhedub, V.M. Shabaev, I.I. Tupitsyn, G. Plunien, P. H. Mokler, H. Schmidt-Böcking, Quantum Interference of K Capture in Energetic Ge^{31+}(1s)-Kr Collisions

12. J.M. Dahlström, A.I. Huillier, A. Maquet Introduction to attosecond delays in photoionization. J. Phys. B. At. Mol. Opt. Phys. **45**, 183001 (2012); J.M. Dahlström and E. Lindroth, Study of attosecond delays using perturbation diagrams and exterior complex scaling, J. Phys. B: Atomic Mol. Opt. Phys. **47**, 124012 (2014)

13. O. Stern, M. Volmer, Über die Abklingungszeit der Fluoreszenz. Physik. Z. **20**, 183–188 (1919)

14. S. Bashkin, Nucl. Instrum. Methods **28**: 88 (1964); S. Bashkin, ed., *Beam-Foil Spectroscopy* (Gordon and Breach, New York 1968); S. Bashkin and I. Martinson, J. Opt. Soc. Am. **61**: 1686 (1971); S. Bashkin (ed.), Beam-Foil Spectroscopy Springer, Berlin (1976); H G Berry 1977 Rep. Prog. Phys. **40**, 155. View the article online for updates and enhancements. Related content

15. N.F. Ramsey, A molecular beam resonance method with separated oscillating fields. Phys. Rev. **78**, 695 (1950); N. Ramsey, Rev. of Mod. Phys. **62**(3) 541–552 (1990)

16. W. Iskandar, J. Matsumoto, A. Leredde, X. Fléchard, B. Gervais, S. Guillous, Atomic site-sensitive processes in low energy ion-dimer collisions. Phys Rev Lett. **113**, 14, 143201

17. R. Moshammer, J. Ullrich, H. Kollmus, W. Schmitt, M. Unverzagt, H. Schmidt-Böcking, R.E. Olson, The dynamics of target single and double ionization induced by the virtual photon field of fast heavy ions, X-ray and inner-shell processes. AIP Conf. Proc. **389**, 153 (1996); R. Moshammer, W. Schmitt, J. Ullrich, H. Kollmus, A. Cassimi, R. Dörner, O. Jagutzki, R. Mann, R.E. Olson, H.T. Prinz, H. Schmidt- Böcking, L. Spielberger, Ionization of helium in the attosecond equivalent light pulse of 1 GeV/nucleon U^{92+} projectiles. Phys. Rev. Lett. **79**, 3621 (1997); R. Moshammer, J. Ullrich, W. Schmitt, H. Kollmus, A. Cassimi, R. Dörner, R. Dreizler, O. Jagutzki, S. Keller, H.J. Lüdde, R. Mann, V. Mergel, R.E. Olson, Th. Prinz, H. Schmidt-Böcking, L. Spielberger, Photodisintegration of atoms in the attosecond equivalent light pulse of highly charged relativistic ions. Phys. Rev. Lett. **79**, 3621 (1997); C.F. von Weizsäcker, Z. Phys. **88**, 612 (1934); E. J. Williams, Kgl. Danske Videnskab. Selskab Mat.-fys. Medd. **13**(4) (1935)

18. K. Fehre, D. Trojanowskaja, J. Gatzke, M. Kunitski, F. Trinter, S. Zeller, LPhH Schmidt, J. Stohner, R. Berger, A. Czasch, O. Jagutzki, T. Jahnke, R. Dörner, M.S. Schöffler, Absolute ion detection efficiencies of microchannel plates and funnel microchannel plates for multi-coincidence detection. Rev. Sci. Instrum. **89**, 045112 (2018)

19. O. Stern, Ein Weg zur experimentellen Prüfung der Richtungsquantelung im Magnetfeld. Z. Physik **7**, 249–253 (1921)

20. ETH-Bibliothek Zürich, Archive, http://www.sr.ethbib.ethz.ch/, Otto Stern tape-recording Folder» ST-Misc.«, 1961 at E.T.H. Zürich by Res Jost

21. https://www.gsi.de/forschungbeschleuniger/fair.htm

Chapter 18
High-Resolution Momentum Imaging—From Stern's Molecular Beam Method to the COLTRIMS Reaction Microscope

T. Jahnke, V. Mergel, O. Jagutzki, A. Czasch, K. Ullmann, R. Ali, V. Frohne,
T. Weber, L. P. Schmidt, S. Eckart, M. Schöffler, S. Schößler, S. Voss,
A. Landers, D. Fischer, M. Schulz, A. Dorn, L. Spielberger, R. Moshammer,
R. Olson, M. Prior, R. Dörner, J. Ullrich, C. L. Cocke,
and H. Schmidt-Böcking

Abstract Multi-particle momentum imaging experiments are now capable of providing detailed information on the properties and the dynamics of quantum systems in Atomic, Molecular and Photon (AMO) physics. Historically, Otto Stern can be considered the pioneer of high-resolution momentum measurements of particles moving in a vacuum and he was the first to obtain sub-atomic unit (a.u.)

T. Jahnke · O. Jagutzki · A. Czasch · K. Ullmann · L. P. Schmidt · S. Eckart · M. Schöffler ·
S. Schößler · S. Voss · R. Dörner · H. Schmidt-Böcking (✉)
Institut für Kernphysik, Universität Frankfurt, 60348 Frankfurt, Germany
e-mail: hsb@atom.uni-frankfurt.de; schmidtb@atom.uni-frankfurt.de

V. Mergel
Patentconsult, 65052 Wiesbaden, Germany

O. Jagutzki · A. Czasch · K. Ullmann · S. Schößler · S. Voss · H. Schmidt-Böcking
Roentdek GmbH, 65779 Kelkheim, Germany

R. Ali
Department of Physics, The University of Jordan, Amman 11942, Jordan

V. Frohne
Department of Physics, Holy Cross College, Notre Dame, IN 46556, USA

T. Weber · M. Prior
Chemical Sciences, LBNL, Berkeley, CA 94720, USA

A. Landers
Department of Physics, Auburn University, Auburn, AL 36849, USA

D. Fischer · M. Schulz · R. Olson
Department of Physics, Missouri S&T, Rolla, MO 65409, USA

A. Dorn · R. Moshammer
MPI für Kernphysik, 69117 Heidelberg, Germany

L. Spielberger
GTZ, 65760 Eschborn, Germany

© The Author(s) 2021
B. Friedrich and H. Schmidt-Böcking (eds.), *Molecular Beams in Physics and Chemistry*,
https://doi.org/10.1007/978-3-030-63963-1_18

375

momentum resolution (Schmidt-Böcking et al. in The precision limits in a single-event quantum measurement of electron momentum and position, these proceedings [1]). A major contribution to modern experimental atomic and molecular physics was his so-called molecular beam method [2], which Stern developed and employed in his experiments. With this method he discovered several fundamental properties of atoms, molecules and nuclei [2, 3]. As corresponding particle detection techniques were lacking during his time, he was only able to observe the averaged footprints of large particle ensembles. Today it is routinely possible to measure the momenta of single particles, because of the tremendous progress in single particle detection and data acquisition electronics. A "state-of-the-art" COLTRIMS reaction micro-scope [4–11] can measure, for example, the momenta of several particles ejected in the same quantum process in coincidence with sub-a.u. momentum resolution. Such setups can be used to visualize the dynamics of quantum reactions and image the entangled motion of electrons inside atoms and molecules. This review will briefly summarize Stern's work and then present in longer detail the historic steps of the development of the COLTRIMS reaction microscope. Furthermore, some bench-mark results are shown which initially paved the way for a broad acceptance of the COLTRIMS approach. Finally, a small selection of milestone work is presented which has been performed during the last two decades.

1 Introduction

What have Stern's Molecular Beam Method (MBM) [2] and the COLTRIMS reaction microscope (C-REMI)[1] [4–11] in common? Both methods yield a very high, sub-atomic unit (a.u.) momentum resolution for low energy particles moving in vacuum. In both approaches the high resolution is obtained because the initial momentum state of the involved quantum particles is very precisely prepared. Conceptually, there is no theoretical limitation for the achievable precision of a momentum measurement of a single particle—the precision is only limited by the design of the macroscopic appa-ratus [1]. Developing novel experimental detection techniques and achieving higher experimental resolution are often required for advancements in science. Already Stern's second MBM experiment, the famous Stern-Gerlach experiment, performed from 1920 to 1922 in Frankfurt, yielded for silver atoms moving in a vacuum a sub-a.u. momentum resolution in the transverse direction of about 0.1 atomic units (a.u.). Stern and Gerlach achieved this excellent momentum resolution due to a very close

J. Ullrich
PTB, 38116 Brunswick, Germany

C. L. Cocke
Department of Physics, Kansas State University, Manhattan, KS 66506, USA

[1]COLTRIMS is the abbreviation for "Cold Target Recoil Ion Momentum Spectroscopy". Another, widely employed name for this technique is "reaction microscope" (REMI). Throughout this article we will use a combination of both acronyms, i.e., C-REMI.

collimation of the atomic Ag beam [12]. That way, Stern was able to show that the Ag atoms evaporated from solid silver obeyed the Boltzmann-Maxwell velocity distribution causing a momentum broadening along the beam direction. Later in Hamburg Stern used a double gear system to chop the atomic beam, which yielded also in beam direction a quite mono-energetic beam, further improving the momentum resolution of his apparatus.

At the time Stern performed his experiments (1919–1945), detectors for the detection of individual particles did not exist. Therefore, he was only able to analyze distributions of a large ensemble of individual particles. Today, because of revolutionary developments in the recent decades, as for example, in the electronic detection techniques for low energy particles, in the target cooling, and the advances in multi-parameter data storage, the AMO experimenter can detect and obtain information on single particles and even perform so-called "complete" high-resolution measurements on atomic and molecular many-particle systems. The C-REMI approach [4–11] uses detectors that can detect the position of impact of single particles with very good position resolution (50 μm or even less) and measure the arrival time of the particles with a precision of <100 ps. From these quantities the flight times of the particles and thus their velocities are determined with—conceptually—unlimited resolution. A C-REMI setup can reach a single particle-momentum resolution of below 0.01 a.u. and it can detect all fragments emitted from an individual atomic or molecular fragmentation process in coincidence. With such properties, it has been shown in the past, that the entangled dynamics occurring during such processes can be visualized and, in special cases, relative timing resolution of 1 attosecond or better can be inferred [13]. A further important aspect of the C-REMI concept lies in the multi-parameter data handling technique employed. It provides the ability to store the raw data of each detected particle in list-mode on a computer. Thus, the experiment can be replayed during the analysis of the data applying different constraints to the data and investigating different physical aspects of the process under investigation. This advantage is common in nuclear and particle physics, but has become prominent in AMO research with the C-REMI methodologies.

2 History of Stern's Molecular Beam Method: The Technological Milestones

In 1919, when Otto Stern came back to Frankfurt he began to build his first atomic beam apparatus [3, 12] stimulated by Dunoyer's experiment [14]. Already in 1911, Louis Dunoyer had published his famous work on the generation of a so-called atomic beam in the journal Le Radium 8. He had observed that the molecules of a gas that flow from a higher pressure volume through a small aperture into a vacuum (pressure < 10^{-3} Torr) move on a straight line. The development of the molecular beam method MBM became technically possible due to the rapid improvement of vacuum techniques during World War I. Diffusion pumps were invented which enabled a vacuum of below 10^{-5} Torr. Thus for a vacuum of about 10^{-5} Torr the mean free path-length of particles moving with a velocity of about 500 to 1000 m/sec is in the order of 10 m. In such a high vacuum the experimenter can perform controlled deflection and scattering measurements with very high momentum resolution. By deflection of the particle due to an interaction with a known external force (e.g. from electric, magnetic or gravitational fields) Stern could determine atomic properties as, e.g., magnetic or electric dipole moments. The MBM allowed, furthermore, to study the ground-state properties of atoms, which were not accessible by means of photon- or electron spectroscopic methods. The deflection observed in a MBM experiment corresponds to a transverse momentum transfer. This transverse momentum transfer can be determined on an absolute scale when particle velocity and mass are known. However, in all experiments performed by Stern or his group members, beginning 1922 in Frankfurt with the famous "Stern-Gerlach-Experiment" [3, 15, 16] continuing 1923 until 1933 in Hamburg and from 1933 to 1945 in Pittsburgh only deflection angles were measured using different particle detection techniques [2, 15].

Although the Stern-Gerlach-Experiment had already demonstrated in an impressive manner what is achievable by the MBM, Stern and his colleagues continued to introduce improvements, especially during Stern's time in Hamburg. They tried to increase the sensitivity of the method and, more crucially, to further improve the momentum resolution and beam intensity.

In Frankfurt Stern used in his first experiment a heated platinum wire coated with Ag paste. Then in the Stern-Gerlach-Experiment the wire was replaced by an oven, which significantly increased the vapor pressure and thus the intensity of the atomic beam. A further increase in the beam intensity was achieved by using a slit diaphragm (see Fig. 1) instead of a small hole aperture. Since the MBM only required a high resolution in one transverse direction, the beam aperture could be made very narrow in the horizontal direction (see slit width "b" in Fig. 1) which improved the apparatus' resolution, but it could be enlarged in the other transverse direction (slit length "h" in Fig. 1) by a factor of nearly 100. Stern invented the so-called "Multiplikator" [2], where many parallel beams were created in the vertical direction, thus, de facto allowing for many measurements to be performed in parallel, without affecting the transverse momentum resolution. Stern described in [2] further efforts for improvements of the transverse momentum resolution. The path lengths r and l (see Fig. 1) were increased by about a factor of 10 compared to the setup employed for the Stern-Gerlach-Experiment and by introducing rotating gears Stern obtained also a quite well-defined longitudinal beam velocity. To be able to measure the tiny magnetic moment of the proton the magnetic deflection force had to be increased (yielding larger deflection angles) and the beams (particular H_2 and He beam) had to emitted from sources operating at the lowest possible temperature. In the last experiment performed in Hamburg before Stern's emigration in September 1933, Frisch [17] tried to observe the atom recoil momentum which is transferred when a photon is emitted or absorbed from/by an atom, which had been predicted by Albert Einstein. Frisch illuminated a sodium beam at right angle with sodium D_2 light, which caused a deflection of the atoms upon absorption of the light. Stern succeeded in his Hamburg time to improve strongly the momentum resolution thus

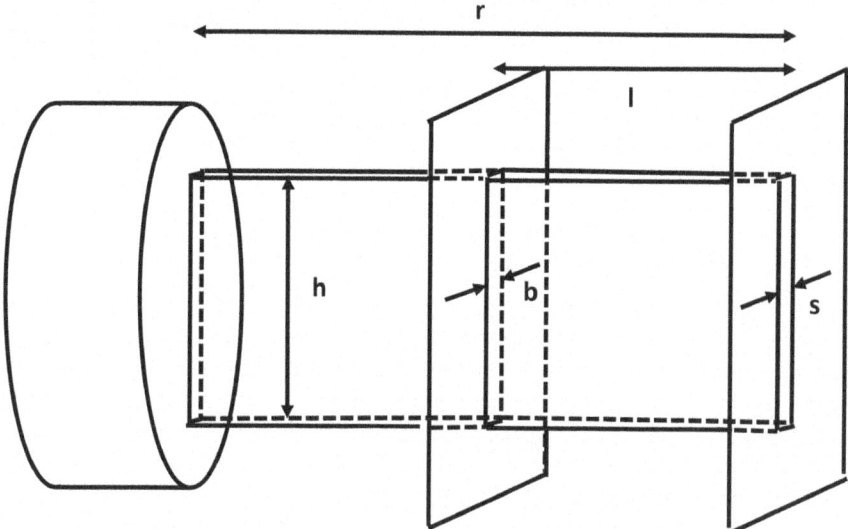

Fig. 1 Stern's method of beam intensity amplification with simultaneous improvement of the momentum resolution [2]

Frisch was able to detect this very small transverse momentum transfer of about 0.001 a.u. in this experiment, which is considered as the pioneering experiment for the Laser cooling approach. The momentum resolution obtained by Frisch is even nowadays a "state-of-the-art" benchmark achievement.

In the years 1919 to 1922 Stern employed detection techniques where a large number of the silver atoms were deposited on polished brass or glass plates in order to observe a beam spot. Later, by chemical treatments Stern was able to observe even a one-atom layer of beam deposition by silver sulfide formation observable as a black spot. In addition, the microscopic beam spot analysis (e.g. by photography) provided an excellent optical resolution in the low micron range, yet not allowing single atom counting. For beams consisting of lighter atoms or molecules, e.g., He and H_2, Stern used a different detection approach. He employed closed gas tubes with a tiny hole for beam entrance. Using very sensitive gas pressure meters he was able to obtain angle-resolved beam scattering distributions. When using H_2 beams, a further method he applied was to measure the heat increase on a metallic surface by a sensitive thermal element. Lastly, he used the Langmuir approach, as well, where the impacting atoms were ionized on a heated wire. The electric current in the wire was proportional the scattered beam intensity. The angular resolution of this method corresponded to the wire thickness [2, 16]. Stern was never able to detect single atoms or molecules.

Stern's followers, like Rabi and his scholars, used the MBM mostly for preparation of beams into selected atomic or molecular states. E.g. Townes used a Stern-Gerlach device to produce population inversion to create the first MASER device. Ramsey used two cavities with two separated oscillating fields to excite Cs atoms. In this case one could not decide in which of the cavities the atom was excited. One had to add the excitation amplitudes coherently creating sharp interference structures from which the transition frequency could be determined with excellent resolution (10^{-9}). Both Rabi's scholars were awarded the Nobel Prize in Physics (Townes in 1964 for Maser development and Ramsey in 1989 for the invention of the atomic clock) [18].

3 The C-REMI Approach

The multi-coincidence C-REMI approach [4–11] is a many-particle detection device imaging momentum space with high-resolution. The imaging is performed by measuring (in a high vacuum environment) the times-of-flight (TOF) and the positions of impact of low energy charged particles which started in a narrowly confined region in space. From these measurements the particles' trajectories inside the spectrometer volume are inferred yielding the particles' properties. This is similar to studies using the historic bubble chamber in high-energy particle physics.

In the late seventies many atomic physics groups worldwide working at accelerator laboratories investigated ionization processes in noble gas atoms induced by swift ion impact. Many research projects were dealing with the measurement of total and differential cross-sections for single and multiple ionization [19–26] (see in particular review article [26] and references therein). The resulting low-energy ions (referred to as "recoil ions" in the following) attracted interest mainly for two reasons: One research direction tried to measure the probability of ionization as function of the scattering angle by means of a projectile-recoil-ion coincidence. When measuring in thin gas targets at very small deflection angles (milli- and micro-rad) almost exclusively scattering of the projectiles from interaction with the collimation slits was observed. Therefore, to eliminate this slit-scattering problem, the projectile's deflection angle (i.e. it's very small transverse momentum) had to be measured in inverse kinematics, by measuring the transverse momentum of the recoil ion. The measurement in inverse kinematics would provide, furthermore, a tremendously improved momentum and energy loss resolution if one could bring the target atom before the collision to a nearly complete rest in the laboratory system (which was achieved later by using a super-sonic jet target [9, 11, 27] or an optical trap [28]). As an example, if in a collision of a 1 GeV/amu Uranium ion on He the projectile energy loss shall be determined, one can measure the momentum change either of the projectile or of the recoiling target atom. When detecting the projectile, the achievable resolution is limited by the properties of the preparation of the incoming beam. Even at the best existing accelerators or storage rings a relative resolution of 10^{-5} is the limit. In case of detecting the recoil ion an energy resolution of below 1 meV can be achieved, yielding a relative resolution in the energy loss of the projectile of far below 10^{-10}.

The other area of interest in the research on very low energy recoil ions, was triggered in the late seventies by the Auger-spectroscopy work of Rido Mann and coworkers [29]. He observed in high energy heavy ion-atom collisions, that (in contradiction to expectations) inner-shell Auger transitions had very narrow line widths. This indicated that the Auger electron emitting recoil ions created in these collisions stayed nearly at rest.

As first step towards developing the C-REMI approach, Charles Lewis Cocke (Kansas State University) and Horst Schmidt-Böcking (Goethe University, Frankfurt) performed together in 1979 at KSU a first test experiment to measure recoil momenta in collisions of MeV heavy ions on He atoms. In this test experiment (using a diffusive room temperature gas target, non-position-sensitive recoil detector and non-focusing recoil-ion extraction field) they measured the TOF difference between the scattered projectile and the recoil ion by performing a scattered projectile-recoil ion coincidence. The measured TOF spectrum could not be converted into absolute values of recoil ion energies, since the spectrometer could not determine the recoil ion emission angle. The results of this test experiment were therefor not published.

In the period 1982–1987 Joachim Ullrich started (as part of his Ph.D. thesis) to develop a new spectrometer approach to determine the absolute value of the transverse momentum of the recoiling target ion by measuring the TOF of the slow recoil ion emitted at 90° to the projectile beam. This development was, for the Frankfurt group, quite risky since the funding request at BMBF/GSI was officially not approved and thus one had to rely at the beginning on "self-made" equipment (e.g. detectors, spectrometers and electronic devices) [9]. Nevertheless, the project was started.[2] It was essential, that one could benefit from the experimental experience from the fields of nuclear and particle physics. C. L. Cocke and H. Schmidt-Böcking had both performed their Ph.D. research in nuclear physics and were trained in using coincidence techniques.

To accomplish the envisioned approach, novel experimental equipment, not commercially available, had to be developed. New self-made position-sensitive detectors [Micro-Channel Plate electron multipliers, MCP, with Backgammon or Wedge & Strip anodes (WSA)] for measuring recoil ions with kinetic energies between zero and several keV were developed and successfully tested. Since 1973 Schmidt-Böcking and his group had developed position-sensitive gas filled Parallel-Plate-Avalanche-Detectors PPAD for performing x-ray/electron heavy-ion coincidence measurements [30]. The work with such detectors required also experience with fast timing electronics and multi-parameter data handling and storing. With a self-made gas filled PPAD the impact time and deflection angle of the high-energy projectiles could be measured in coincidence with the recoil-ion impact time and position. This experience gave confidence that the envisioned C-REMI project was feasible. However, it took until about 1993–1995 before the first full functioning C-REMI was operating. There were moments in these years before 1990 where parts

---

[2]In the mid eighties at a small workshop on the physics at the planned TESR storage ring at MPI in Heidelberg HSB presented the perspectives on the physics with very "cold" recoiling ions. The GSI director of that time Paul Kienle heavily objected this kind of physics. Saying: we will not build a GeV accelerator to perform micro eV physics. The Frankfurt application to the BMBF to get financial support for this kind of physics and the technical developments was thus not approved, but surprisingly also not declined. Thus the Frankfurt group received support without official approval.

of the project seemed unsolvable. But Joachim Ullrich never gave up! Without his efforts and ideas C-REMI would probably not exist. Besides that, the history of C-REMI is not only a chain of recoil-ion milestone experiments performed by different groups, it is in particular the history of technological developments.

In order to finally obtain sub-a.u. momentum resolution, the target had to be prepared in a state of very small momentum spread which led to using a super-sonic jet source. A further crucial piece was specifically designed electro-magnetic spectrometer fields, that provided optimal momentum focusing with maximum detection efficiency. When, in the early nineties, the detection power of C-REMI became apparent to the atomic physics community, the Frankfurt group was ready to help other groups to build up their own C-REMI systems. Schmidt-Böcking founded in 1990 the company "Roentdek" [31] to produce the C-REMI equipment components or later even deliver complete C-REMI systems to other laboratories. Equipment was delivered to research groups worldwide by selling or in a few cases by loan. The commercial availability of C-REMI systems was essential for the propagation of the C-REMI to several new fields in Physics (single photon research, strong-field and ultrafast sciences etc.). This provided in the last two decades for many groups in AMO physics, as well as in chemistry and biology the support to perform many milestone experiments and pioneering breakthroughs. The C-REMI has enabled insight into many-particle quantum dynamics at the few attosecond scale.

3.1   The Development of C-REMI Components

In the late seventies and early eighties one of the main research activities in atomic collision physics was to measure total ionization cross sections as function of the recoil ion charge state in high energy heavy-ion rare-gas collisions using the TOF coincidence method [19–26], and to determine such cross sections differentially as function of the projectile scattering angle. These total cross sections had (for single ionization) sometimes macroscopic values $[10^{+6}$ Mbarn $= 10^{-12}$ cm$^2]$ and even the creation of completely ionized Ar^{18+} was possible (with a cross section about 1 Mbarn in 15.5 MeV/u U on Ar collisions [19]). Both types of experiments required a coincidence measurement between scattered projectiles and recoiling ions. Detecting the recoil ion yielded the start signal, detecting the scattered projectile with a PPAD provided the stop signal.

3.1.1   Detectors

For the detection of the high energy projectile since 1973 self-made gas filled PPADs were available, which could monitor rates up to one GHz very stably in gas flow mode (see Fig. 2) [30]. Adapted to the experimental task they measured only scattering angles by annular shaped anode structures. At that time electronics were made that enabled a simultaneous measurement of 16 scattering angles.

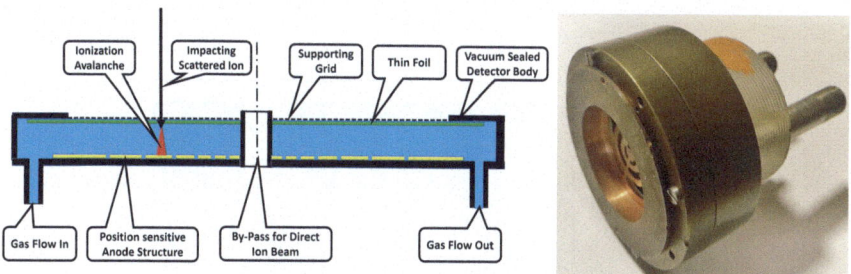

Fig. 2 Left: scheme of a PPAD. The impacting ion ejects from the entrance foil several electrons into the gas filled detector volume. In the high electric field (between entrance foil and anode structure) the electrons are accelerated and create a secondary electron avalanche which is detected as function of the anode position. This detector can have a central hole to allow the un-scattered beam to pass through. Right: A photograph of the first PPAD built in 1973 [30]. This detector had three annular anode rings and could handle rates up to 1 GHz

A position-sensitive recoil-ion detector for such low energy ions was not commercially available in the early eighties of the last century. As an initial part of a recoil-ion detector so-called micro-channel plates MCP were used [31]. The slow recoils were post-accelerated close to the MCP surface and released, upon impact on the MCP, secondary electrons that induced an avalanche inside the very narrow MCP channels. The single particle detection efficiency of standard MCP is limited by the open area ratio (e.g. how much "hole-area" is present in the total surface). Typical values are 60%. New developments of MCPs with surfaces, that look like a funnel, increase the efficiency up to 90% [33]. The position readout of the MCP was performed using a "Wedge and Strip" anode structure. Located behind the MCP, this anode structure yielded information on the position of impact of the primary particle by means of a charge partition method (see Fig. 3 [31, 32]). During the Ph.D. work of Ullrich the anode structures were fabricated as printed circuits. Prior to use, they all needed a careful restoring work by using optical microscopes. In later years such anodes were carefully printed on ceramics and did not need any initial reconditioning. A breakthrough in ion detection was achieved by Ottmar Jagutzki [33] using the delay-line approach for determining the impact positions of the particles on the detector. For such detectors the signal read-out proceeds via a delay-line structure (see Fig. 4: double-wire structure). From the arrival-time difference at both ends of the delay-line system, the position of particle impact can be determined with a resolution of better than 100 µm. The delay-line approach yields several important advantages as compared to the charge-partition method. It can handle much higher detection rates, since it does not rely on slow charge collection processes. It can detect more than one particle at (almost) the same time (i.e., being multiple-hit capable) because the induced timing signals are very short (in the range of 5–10 ns) and, lastly, the use of a "timing approach" fits perfectly to the digitized world of computers and is easy to adjust and much cheaper to build.

The first generation delay-line detectors consisted of two separate delay lines mounted at right angle. Later, Jagutzki developed a three-layer delay-line structure

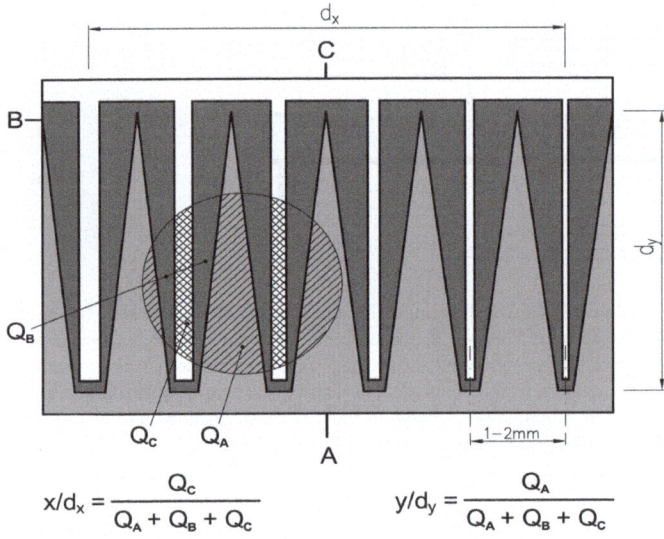

$$x/d_x = \frac{Q_C}{Q_A + Q_B + Q_C} \qquad\qquad y/d_y = \frac{Q_A}{Q_A + Q_B + Q_C}$$

Fig. 3 Scheme of the Wedge & Strip anode (WSA). If the charge cloud covers several (at least 2) pitches of three electrodes (B "wedge", C "strip" and A "meander"), measuring the relative charge portions Q_i allows to determine the centroid of the charge cloud [31, 32]

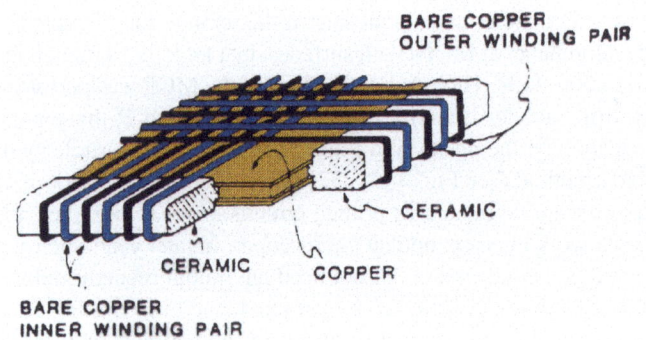

Fig. 4 The principle set-up of the delay line anode and other references therein [33, 34]

(a so-called hexanode) (see Fig. 5). The hexanode detector yields a better linearity and an improved multi-hit resolution with smaller dead-time blockade. The working diameter of these circular delay-line detectors can be as large as 120 mm diameter and recent developments target 150 mm. By using its three delay-lines the hexanode registers redundant information on each particle's position and impact time. Thus, it is possible to recover position and time information for several particles beyond the electronic dead-time limit: Even simultaneously arriving particle pairs can be detected as long as they have a minimum spatial separation of 10 mm.

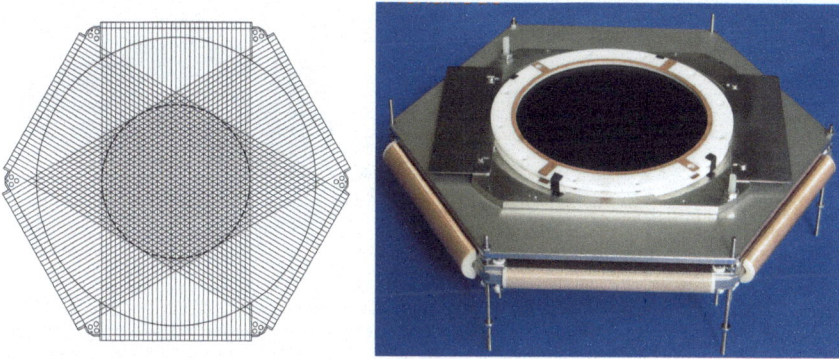

Fig. 5 Left: the Hexanode structure and right: working detector system (active area 80 mm diameter) [33–35]

3.1.2 Multi-parameter Data Handling

In the late seventies coincidence measurements, which were standard in nuclear physics, were very rarely performed in atomic physics. Thus, there was no need for fast electronics and many-parameter data handling. The electronic hardware for such measurements was quite expensive at that time and thus the comparably small groups of the atomic physics community could not afford to perform coincidence measurements where multi-parameter data had to be registered and stored. Only in nuclear and high-energy particle physics were multi-particle coincidence measurements commonly used. In order to have access to such measurement infrastructure, the Frankfurt atomic physics group, for example, performed all coincidence experiments in nuclear physics laboratories either at GSI-Darmstadt or at the MPI for Nuclear Physics in Heidelberg, where the needed electronics and data storing systems were available. The support by Ulrich Lynen [36] and Reinhold Schuch is highly acknowledged and was absolutely essential for the ongoing development of the C-REMI. Since about 1985 Ullmann [37] developed a PC based multi-parameter data storing system which was cheap and powerful enough to satisfy the needs of a two-particle or even 7 parameter coincidence measurement (implemented on an Atari ST mainstream personal computer). This development yielded a breakthrough enabling small groups to perform coincidence experiments. The inclusion of all these improvement steps took about one decade from about 1984 to 1994. The steady progress of this project was published in the annual reports of the IKF-University Frankfurt in the eighties and nineties, i.e., 1984 to 1995.

In the early years the charge signals of the WSA detectors were registered by charge-sensitive preamplifiers and subsequently amplified by standard modules. The timing-signal was created by the "Constant Fraction Discriminator" scheme [38]. These preamplifiers, constant fraction discriminator units etc. were built in the electronics workshops of the Physics Institute in Heidelberg and GSI-Darmstadt. After the foundation of Roentdek GmbH [35] several members of the Frankfurt group

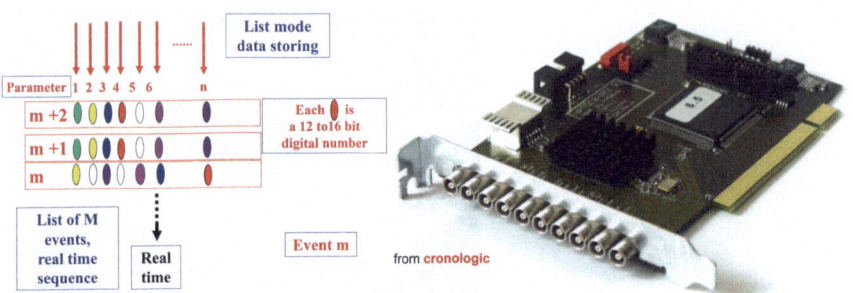

Fig. 6 Left: scheme of multi-parameter "list mode" data recording. Right: an 8-fold fast time-to-digital converter made by cronologic [39]

were employed at Roentdek and they developed their own electronic circuits. These circuits (based on modern "state-of-art" digital chips) could handle nearly unlimited numbers of parameters per event and allowed high repetition rates. Furthermore, they were inexpensive.

Simultaneously the interface between the electronic modules and the data storing PC changed from slow and expensive CAMAC to fast self-made TDC units (Time-to-digital Converters, with 25 ps timing resolution (see Fig. 6)) or even ADC units (Analog-to-Digital Converter) [39]. When using fast ADCs, the analog signal is sampled in e.g. 250 ps time-slices and its amplitude is digitized. A fast analysis program can determine several properties, as the "center" of each peak, its height or even disentangle double-peak structures (often referred to as "Camel peaks"). This development was crucial, since in case of the hexanode seven detector signals (2 for each of the three delay-line layers and one signal from the MCP to obtain the time of impact) needed to be detected for each impacting particle. Thus, a fast multi-hit recovery with very good timing resolution was needed (see Fig. 7). The present state-of-the-art C-REMI electronics including data list-mode storing can monitor coincidence rates up to several MHz.

3.1.3 Spectrometer Design

The first generation of recoil-ion spectrometers (Fig. 8) was designed to measure only total cross sections for recoil-ion production in energetic heavy-ion collisions as function of the recoil charge state and as function of the final projectile charge state [19–24]. The collimated (1) projectile beam (2) intersected with a diffusive gas jet (3). In the collision with the projectile the target atoms were multiply ionized by, e.g., pure ionization or electron capture. The projectiles (9) were deflected after the interaction with the target by a magnet (8) and detected in a position-sensitive PPAD (10). The final projectile charge states were distinguished by their bending angles behind the magnet. The low energy recoil ions were extracted by an electric field applied between plate (4) and a grid (5), which was on zero-potential. The extracted

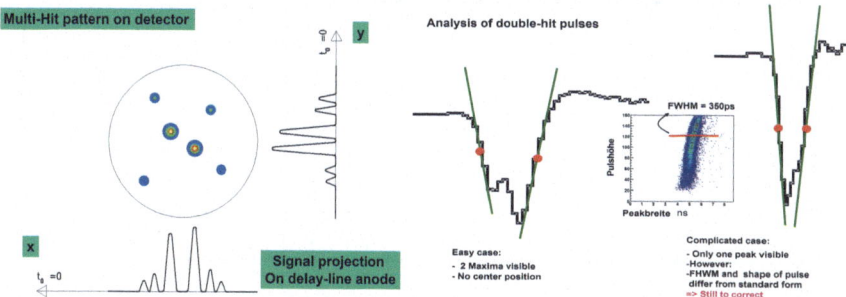

Fig. 7 Left: the circle represents the active area of the DL detector. In the event shown 6 particles impacted within about 100 ns on the detector creating electron avalanches that differ in height, which induce in the delay-line structure localized charge clouds. Each of the six ends of delay lines is connected to a fast sampling ADC. Right: the sampled multi-hit signal of one channel is shown. It is analysed later (i.e. after the actual measurement) in high detail using a PC, which allows to resolve the multi-hit pattern

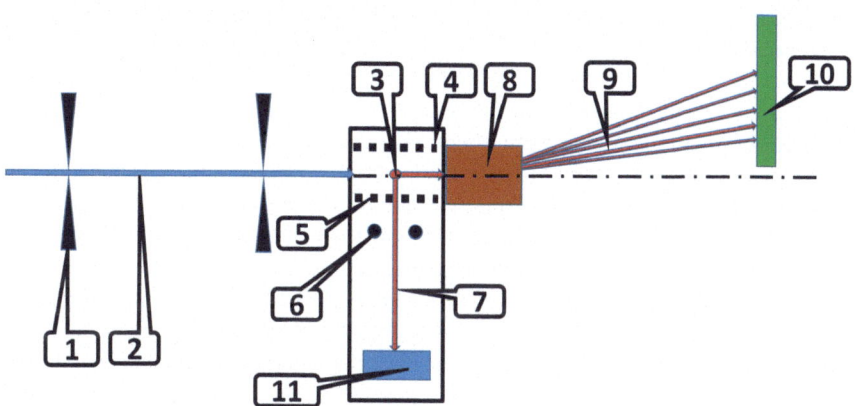

Fig. 8 Scheme of the recoil-ion deflected projectile ions (see text)

recoil ions (7) were focused on the recoil detector (11) with the help of an einzel lens (6). The recoil-ion charge state was determined by the TOF of the ions (see Fig. 9).

The second generation recoil-ion spectrometer was aimed at measuring the transverse recoil-ion momenta. This was the first working recoil-ion momentum spectrometer of the Frankfurt group. In Fig. 10 the spectrometer used by Ullrich et al. [40] is shown. The projectiles intersected with the diffusive gas target inside a field-free cylinder and were detected downstream with a PPAD. Between the inner and outer cylinder 700 V were applied to post-accelerate the recoil-ions transverse into the recoil-ion spectrometer. These accelerated ions were focused by an einzel lens. In a small magnet the recoil-ion charge states were separated and monitored by a one-dimensional position-sensitive channel-plate detector (the anode structure was only "backgammon"-like). Only such recoil ions were post-accelerated which passed

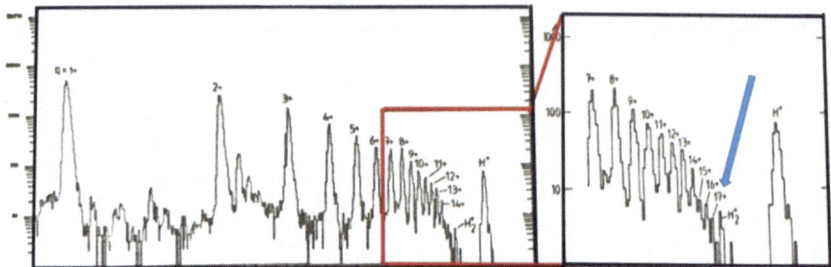

Fig. 9 Ar recoil-ion TOF spectrum after the collision with 15.5 MeV/amu U^{75+} projectiles [19]. Even the Ar^{17+} fraction is clearly visible (blue arrow). The Ar^{18+} contribution is covered by the H_2^+ molecular ion charge state. The Ar^{18+} production cross section is approximately 1 Mbarn

through a tiny hole in the inner cylinder. The recoil velocity (i.e., its momentum) was determined by performing a recoil-ion projectile coincidence yielding the recoil-ion TOF inside the inner (field free) cylinder. The first successful experiments investigating recoil-ion production could be performed in the mid-eighties at the Heavy-ion accelerators (UNILAC) at GSI. The first publication on the new recoil-ion momentum spectrometer with reliable small angle data appeared in Phys. Lett. A [40, 41].

For collisions of Uranium ions on Ne the transverse absolute recoil ion and the scattered projectile momenta were obtained in coincidence (see Fig. 11). The data showed that at the very small scattering angle of only μrad the sum of each recoil-ion and corresponding projectile transverse momentum did not add up to zero as expected for a two-body collision. By comparing the data to the CTMC theory of Olson et al. [42, 43], it became clear that the observed deviations were due to the influence of the emitted electron in the ionization process and due to the target temperature (the target was at room temperature), too.

From Ron Olson's calculation it became clear that internal motion of the gas target (due to its temperature) had to be strongly reduced to obtain quantitative information on the electron momenta in such measurements. Using ultra-cold targets, the method could be improved that much in resolution that electron momenta could be obtained solely by deducing them from the measured momenta of the involved ions. In his Ph.D. work Dörner et al. [44] started to build a cooled gas target. He achieved, using a static-pressure gas target, a temperature reduction down to approximately 15 K. This cooling improved the resolution, but by far not enough. In Fig. 12 the Dörner-spectrometer is shown. Inside a cooled gold-plated brass housing (connected to the head of a cryogenic pump) the cold He target is intersected by a fast, well-collimated proton beam. The He recoil ions can exit through a slit aperture towards the recoil-ion detector. Behind the slit the recoil ions are post-accelerated, focused by an einzel lens and magnetically deflected. The impact position of the recoil ions is measured by a position-sensitive MCP detector (back-gammon anode) and the TOF of the recoil ions by a coincidence with the scattered projectiles. The measurements of Reinhard Dörner showed that the expected two-body correspondence between projectile and recoil-ion transvers momentum was broken below angles of about 0.6 mrad. The first

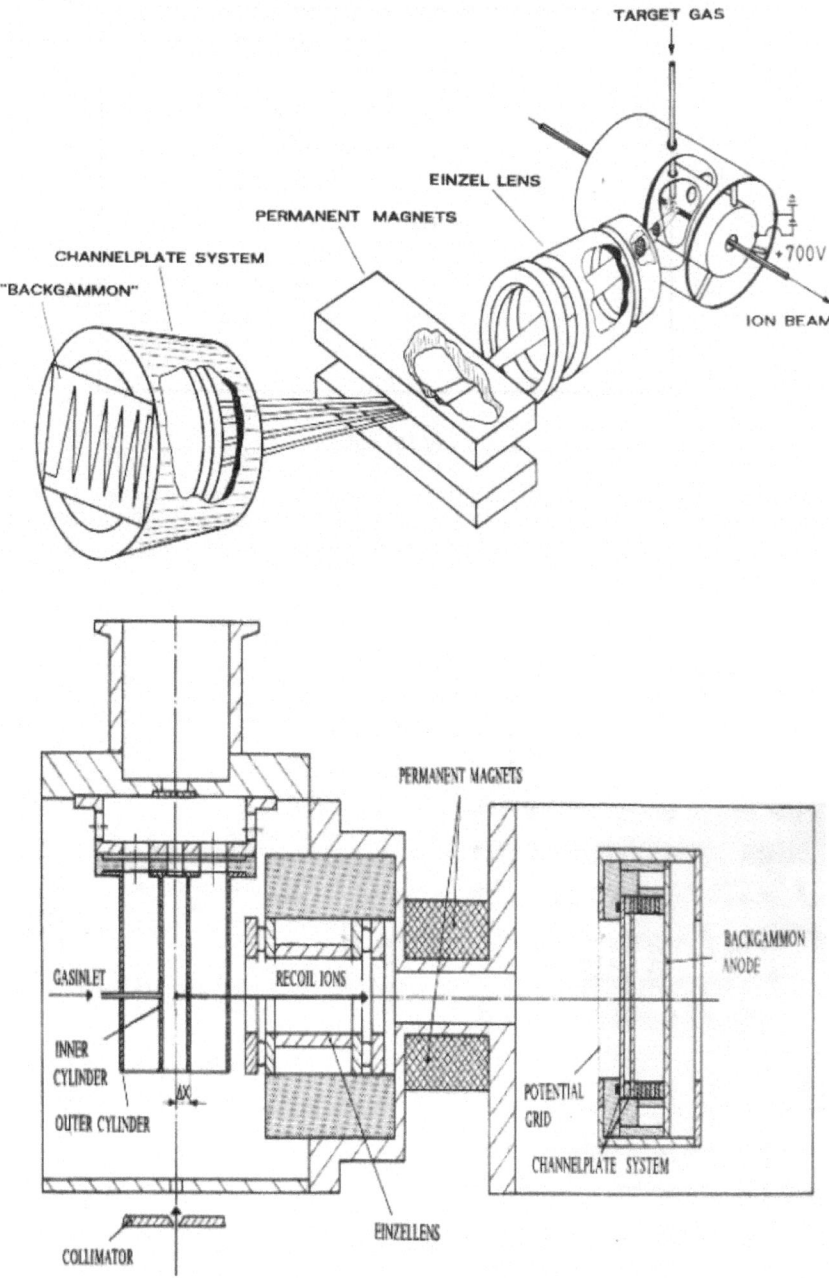

Fig. 10   Recoil-ion momentum spectrometer design [40, 41] Upper part: 3-dimensional view, lower part: cross section seen from above

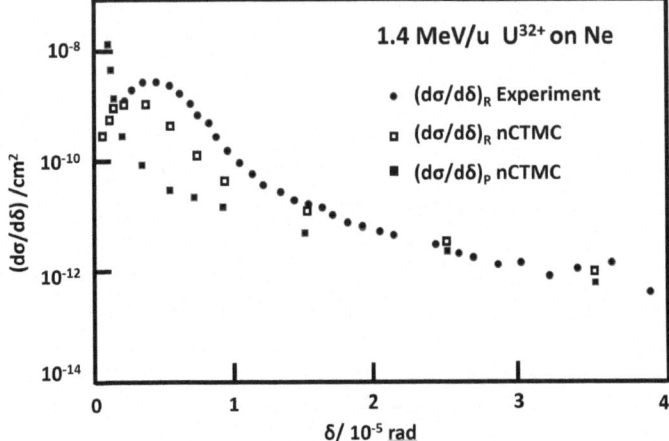

Fig. 11 These experiments were the first where—in high-energy heavy-ion rare-gas collisions ionization—probabilities at very small scattering angles <10 μrad were successfully measured. In parallel, the group of Ivan Sellin at Oak Ridge [45–47] measured mean energies of low energy recoil ions, too

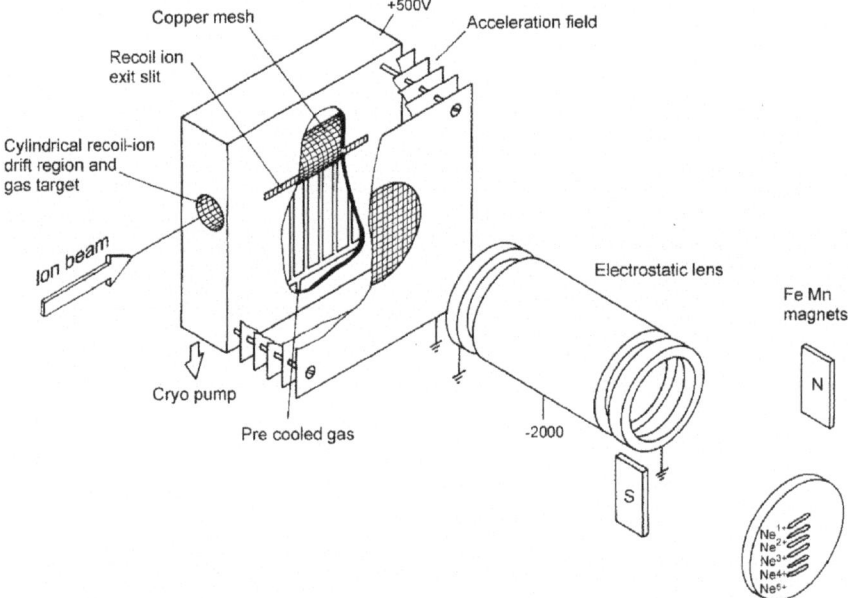

Fig. 12 Recoil-ion spectrometer of Dörner et al. [44] (see text)

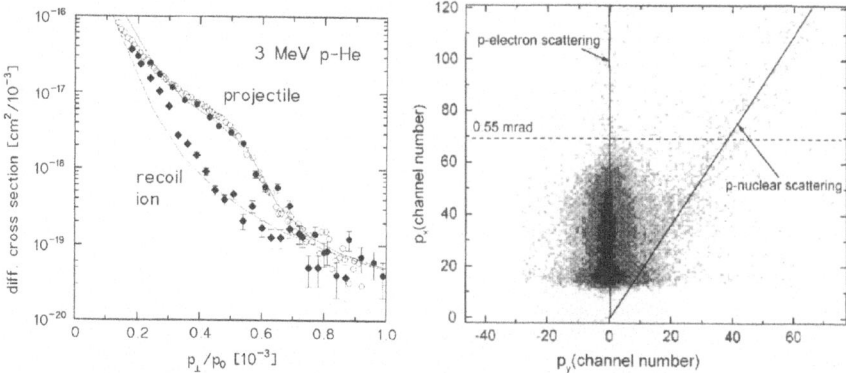

Fig. 13 Left: differential single-ionization cross sections of He in 3 MeV Proton collisions [44]. The circles show the identical data plotted versus the measured projectile transverse momentum, the diamonds versus the recoil-ion momenta. As seen in the right plot the projectile can be scattered by the He nucleus (diagonal line p-nuclear scattering) or by the He electrons (vertical line p-electron scattering). Above 0.55 mrad the projectile cannot be deflected by an electron at rest. The p-electron scattering above 0.55 mrad and the broadening below 0.55 mrad is due to the initial electron momentum (plus target temperature)

perception was: this is the principle limitation of recoil-ion momentum spectroscopy. However, with the help of CTMC calculations of Ron Olson and numerous discussions with him (Olson was a Humboldt Award fellow in Frankfurt from 1986 to 1987) it became clear, that the method (recoil-ion momentum spectroscopy) was not limited to projectile scattering angles above 10^{-5} rad, but that the method was at lower angles even sensitive to the momenta of the involved electrons if the target temperature could be decreased by several orders of magnitude. The experiment performed by Reinhard Dörner demonstrated that further target cooling would improve the momentum resolution and that it should be possible to measure the momentum exchange between nuclei and electrons with high resolution. This observation (Fig. 13) was a milestone perception towards the realization of C-REMI.

Rami Ali and Charles Lewis Cocke at KSU used the first recoil-ion extraction system [48, 49], which was time-focusing [50]. Thus the KSU group was the first to determine the Q-value (inelastic energy loss or gain) in an ion-atom collision process by measuring the longitudinal momentum component of the recoil ion [48] (i.e. parallel to the incident projectile momentum). They investigated the multiple electron capture process in 50 keV Ar^{15+} on Ar collisions and obtained a Q-value resolution of about 30 eV. Relative to the projectile kinetic energy this corresponds to a resolution just below the 10^{-3} level (see Fig. 14). The method of Q-value determination by the longitudinal recoil momentum component has been discussed before in an invited talk by Dörner et al. at the ICPEAC in Brisbane 1991 [51]. In this invited lecture, it was shown that in high-energy heavy ion collisions a relative Q-value resolution far below 10^{-6} can be obtained.

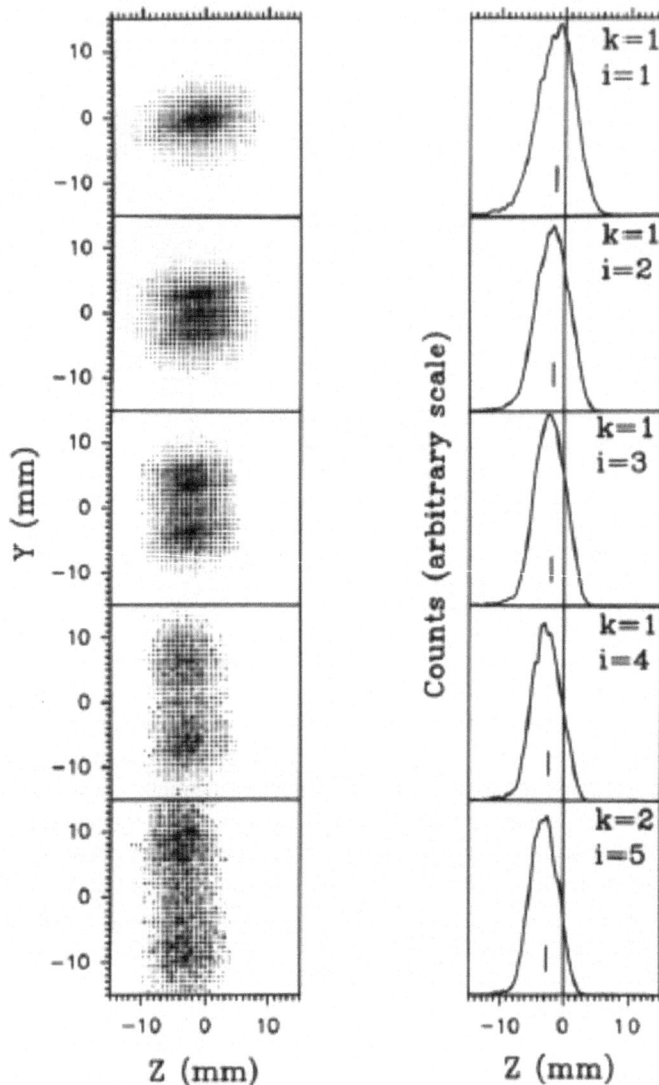

Fig. 14 Left: two-dimensional recoil-ion momentum distributions (z-abscissa: longitudinal component, y-axis: transverse component) for different projectile charge change k and recoil-ion charge state i. Right: their projections on the z-abscissa. The vertical bars indicate the center of the projections [48, 49]

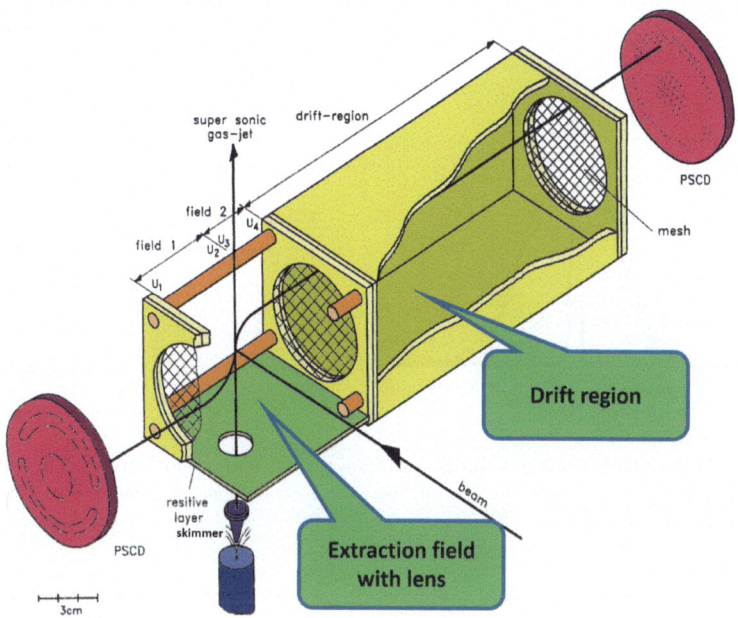

Fig. 15 Recoil-ion spectrometer used by Mergel et al. [52] with transverse extraction. The extraction field and drift zone are designed in length and field strength to obtain time focusing conditions. The detector PSCD on the right side monitors recoil ions and the detector on the left side electrons

The first breakthrough into the high-resolution domain was achieved in the "Diplomarbeit" of Mergel et al. [52] by performing an experiment with a He super-sonic jet as target and using a spectrometer design with strongly improved time-focusing properties. A further significant improvement was then achieved in the Ph.D.-thesis work of Mergel et al. [52], which used the first three-dimensionally focusing spectrometer. This spectrometer included the so called time focusing conditions and the focusing of the extension of the gas jet in the direction of the electrical field for the recoil-ion extraction. For the first time this spectrometer combined the time focusing with a two-dimensional focusing lens in the extraction field focusing the gas jet projectile reaction volume with respect to the two dimensions perpendicular to the extraction field (Fig. 15). Using this three-dimensional focusing in combination with a pre-cooled (17 K) super-sonic gas jet, a momentum resolution of 0.05 a.u. was achieved in all three dimensions, which was a breakthrough in momentum resolution of C-REMI and was the best resolution achieved at that time [52].[3]

---

[3] As Volker Mergel remembers: The three-dimensional focusing was invented in the early nineties during a night-shift performing an experiment at the tandem accelerator at KSU. In that night-shift, an experimental resolution was observed which was better than expected. These surprisingly good experimental results triggered a discussion between Charles Lewis Cocke and Volker Mergel searching for the reasons. Performing in that night-shift some calculations on the possible electric field configuration Volker Mergel could show that the reason for the improved resolution must be an inhomogeneity of the electrical field, which accidentally caused a focusing of the

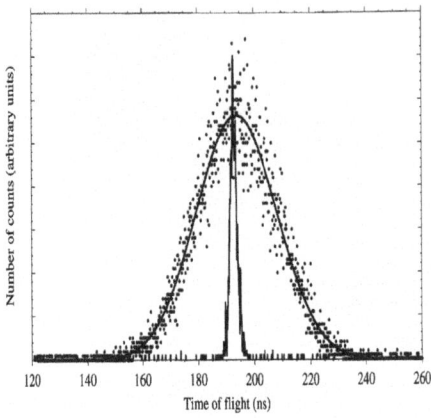 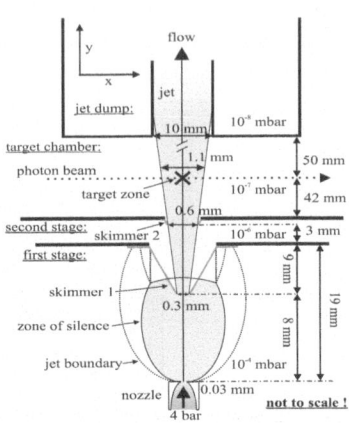

Fig. 16 TOF distribution of recoil ions emitted from a room temperature target in comparison to ions emitted from a super-sonic jet. The full line shows the thermal momentum distribution [9]. Right: Two-stage super-sonic jet [57]

In the early nineties, the Göttingen group of Udo Buck and Jan Peter Toennies [27] provided important support in constructing a super-sonic jet as target. Using this recoil technique, in the late nineties, Daniel Fischer in the group of Joachim Ullrich obtained a resolution of 10^{-6} in the Q-value measurement [11].

In the early nineties also the group of Amine Cassimi at Caen [53–56] started to use a super-sonic jet-target for recoil-ion production. Their original motivation was to build a source for intense, cold and highly-charged ions based on recoil-ion production. The effect of target cooling is visible in Fig. 16, where the recoil-momentum distribution for a room temperature and a super-sonic jet target are compared. The small transverse spread in the momentum distribution of a super-sonic jet can still be improved by collimating the super-sonic beam with skimmers (Fig. 16), thus nearly unlimited sub-atomic resolution in the transverse momentum space can be obtained.

The next very important milestone step towards today's "state-of-the-art" C-REMI system as a multi-particle momentum-imaging device was the incorporation of the magnetic field confinement of high-energy electrons (Figs. 17, 18 and 19). With this improvement, one could finally achieve a nearly 4π geometrical solid angle for the detection of even high energy electrons. Joachim Ullrich and Robert Moshammer conceived this benchmark development in the early nineties [5, 11]. It increased the multi-coincidence detection efficiency by orders of magnitude. As shown in [5, 11] the guiding magnetic field pointing in parallel to the axis of the electric extraction-field provides an unambiguous determination of the initial electron momentum, as long as the electron is not detected at flight times where its detection radius is re-approaching zero.

---

extension of the reaction volume. As a result of this discussion Volker Mergel built a new three-dimensionally focusing spectrometer as shown in Fig. 15. This technology was then also patented [DE 196 04 472 C1].

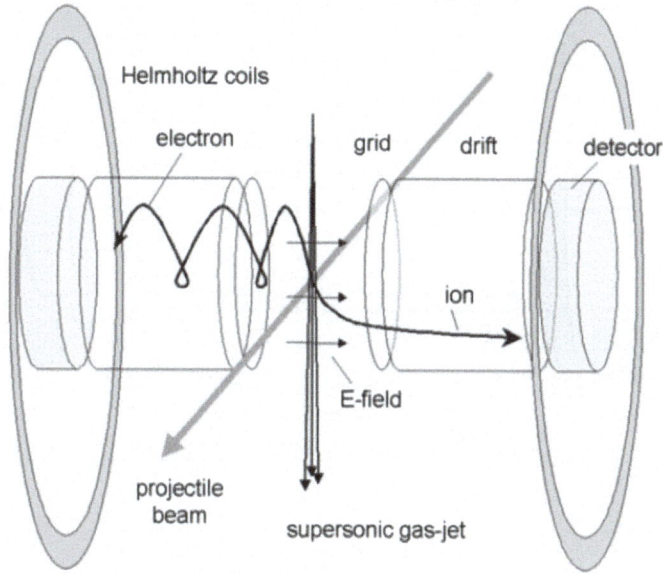

Fig. 17 The COLTRIMS reaction microscope with guiding magnetic field [11]

Fig. 18 Plotted is the TOF versus radius of electron trajectories inside the spectrometer in the presence of the magnetic field B [57]. The time between minima in the radius is the cyclotron motion period, which is used to measure the magnetic field

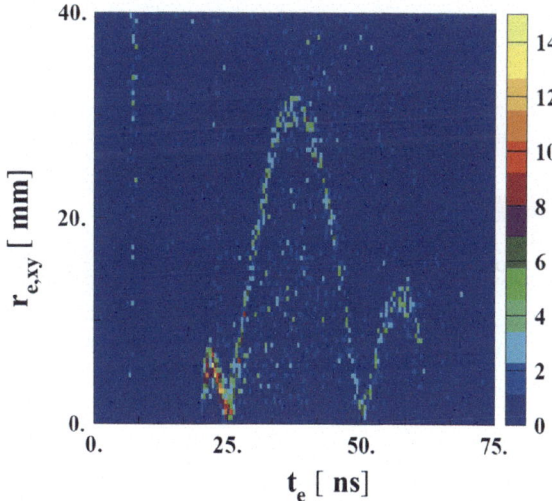

Joachim Ullrich, Robert Moshammer et al. performed several of their measurements on the physics of the recoil-ion momentum at the GSI storage ring ESR. The operation of the ESR required a wide-open spectrometer for the circulating ion beam. Therefore, they designed a spectrometer where a particle-extraction can be performed in principle in any direction (see Fig. 19), but with preferential extraction

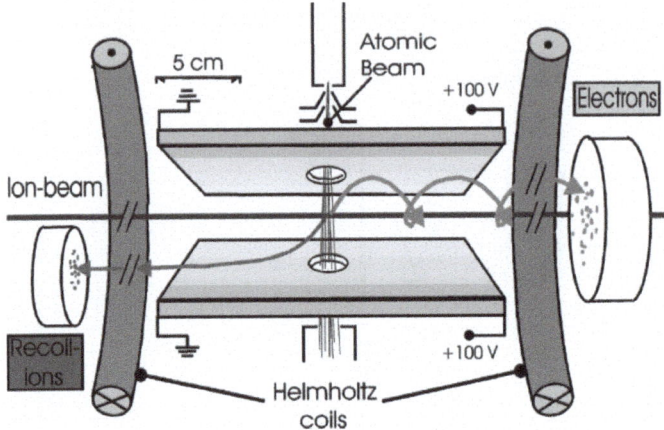

Fig. 19 The C-REMI system of Ullrich and Moshammer installed at the GSI storage ring (ESR) [5, 11]

in the longitudinal direction along the ion beam. The recoil-ion as well as the three electron detectors were positioned in time-focusing geometry [5, 11], which allowed a high Q-value resolution. Figures 20 and 21 show schematic views of modern, "state-of-the-art" C-REMI systems with transverse extraction and time-focusing. In Fig. 22 the out-side view of one of the Frankfurt C-REMI systems is shown.

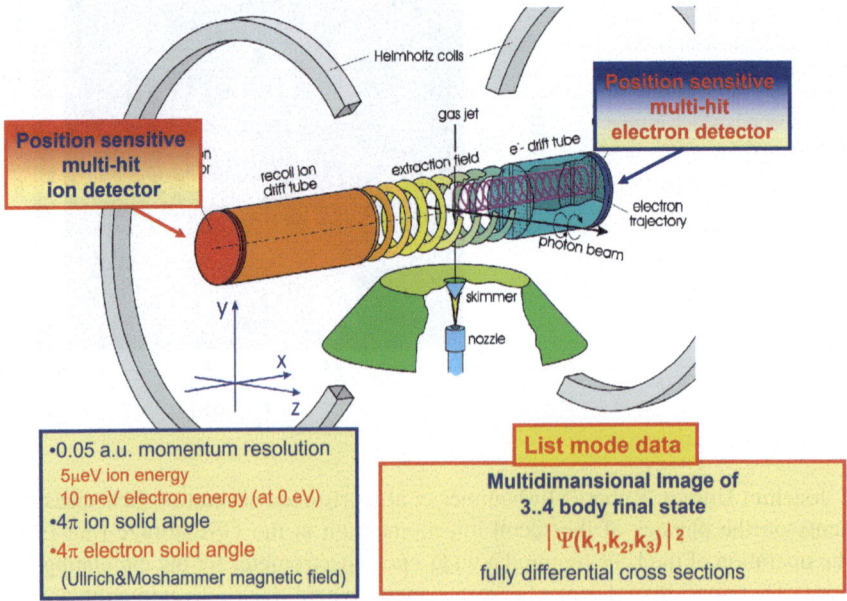

Fig. 20 Schematic view of a C-REMI system

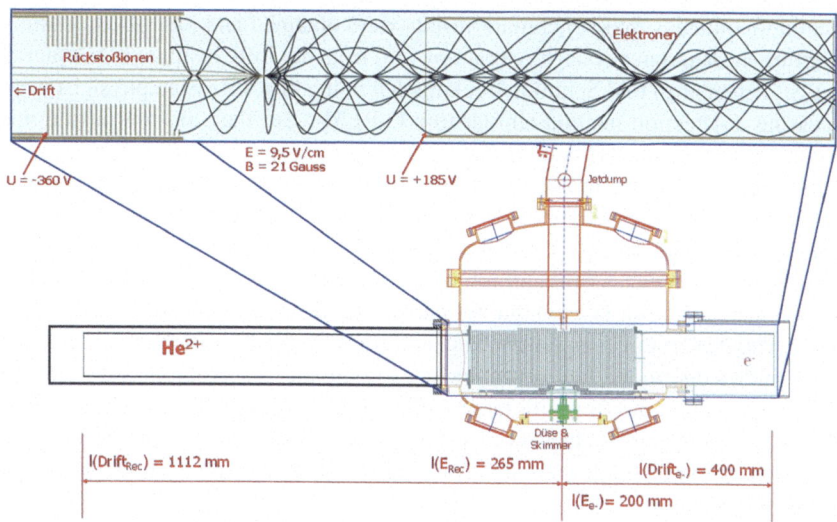

Fig. 21 Electrostatic lens-system of a C-REMI with transverse extraction showing ion and electron trajectories and time marker [58]

Fig. 22 One of the Frankfurt C-REMI systems used for MeV p on He collisions. Left: View from outside. Right: Inside the vacuum system with view on the C-REMI spectrometer [35]

4 The Early Benchmark Results

In the eighties many experts in atomic physics were skeptical that a C-REMI-like approach could actually work. The high resolution and detection efficiency of the C-REMI method was slowly recognized and acknowledged by the physics community, when in the nineties first benchmark results were obtained and published. Visiting Frankfurt, the Russian physicist Afrosimov [59] from the Joffe Institute in Leningrad (now St. Petersburg) told Schmidt-Böcking that in the fifties Russian physicist were discussing a detection method similar to a C-REMI. But they did not pursue this concept, since they did not believe that it could work because of the thermal motion of the target atoms and molecules. Using room temperature targets first angular resolved recoil-ion measurements were performed in the late seventies [60, 61]. In lectures on "Inelastic Energy-Loss Measurements in Single Collisions" Fastrup [62] discussed the advantages of recoil ion momentum spectroscopy, i.e. the method of inverse kinematics. But he also did not pursue recoil-ion momentum spectroscopy, since a target at room temperature did not allow a good momentum resolution. To our knowledge in the late seventies or early eighties no group was developing recoil-ion detection devices with larger solid angle imaging features. The required equipment like detectors, electronics, coincidence and vacuum equipment and cold target preparation methods were not available at that time to give a C-REMI a real chance of success.

4.1 Q-Value Measurements

At ICPEAC XVII in 1991 in Brisbane, the Frankfurt group presented theoretical estimates and first experimental results on the high-resolution obtainable by the cold-target recoil-ion method [51]. The prediction was that in MeV/u heavy-ion atom collisions a Q-value resolution relative to the projectile energy of below 10^{-8} and deflection angles below 10^{-8} rad could be measured. The relation between $\mathbf{Q}$-value and recoil longitudinal momentum $\mathbf{p}_{r\parallel}$ is: $\mathbf{Q} = -(\mathbf{p}_{r\parallel} + q/2) \cdot v_p$ [9, 11], where q is the number of electrons transferred from the target to the projectile and v_p is the projectile velocity (all values are in a.u.). The summand $(q/2 \cdot v_p)$ is due to the mass transfer of electrons from the target atom at rest into the fast moving projectile system. The experimental verification of the high-resolution power of the C-REMI approach was demonstrated in the period of 1992 to 1994 by Volker Mergel [52] when he assembled the first fully working **COLTRIMS** (**CO**ld **T**arget **R**ecoil **I**on **M**omentum **S**pectrometer) system (see Fig. 15). It included a super-sonic He jet as target. The He gas was pre-cooled down to about 15 K and expanded under high pressure (>10 bar) through a nozzle of 20 μm diameter into vacuum. By the expansion process the inner temperature of the super-sonic beam decreased to a few mK.

   The beam was collimated by a skimmer (about 1 mm circular opening) to reduce the transverse momentum spread of the gas jet. Furthermore, static electric extraction fields of the C-REMI were designed to provide perfect time focusing [9, 11, 52]. Using a predecessor of the spectrometer as shown in Fig. 15 in 250 keV He^{2+} on He collisions, Mergel et al. [52] obtained an energy loss/gain resolution of 0.26 a.u. (i.e. 7 eV) by measuring the longitudinal recoil-ion momentum (Fig. 23). Relative to the kinetic energy of the impacting He^{2+} projectiles this is an energy loss resolution

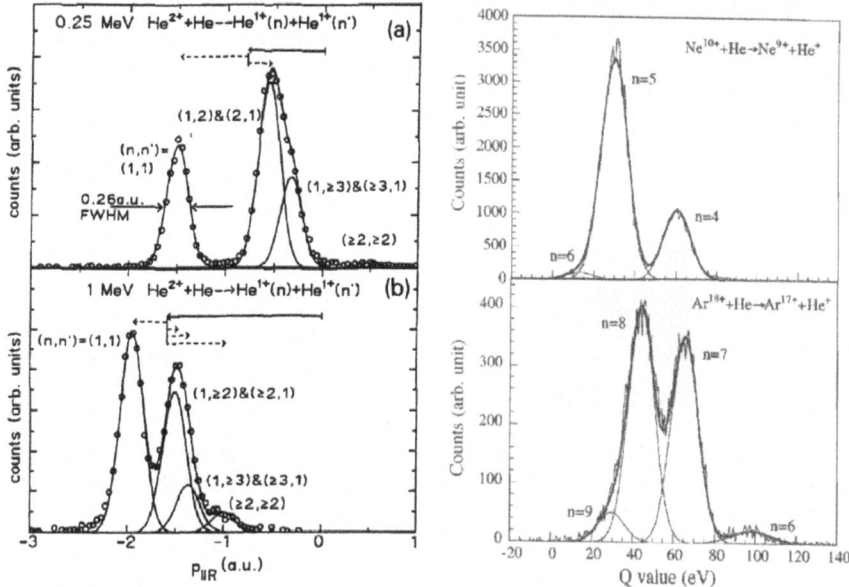

Fig. 23 Left: $p_{r\parallel}$ measurement of Mergel et al. [52]. The numbers in the brackets indicate the mean shell of the transferred electron in the initial and final state, respectively. 1 a.u. corresponds to a Q-value of 27.2 eV. Right: Q-value measurement of the GANIL group [53]. The kinetic energy of Ne^{10+} was 6.82 keV/u and 6.75 keV/u for Ar^{18+}

in the order of 10^{-5}. It is to notice that one can determine the energy loss/gain of projectiles extremely precisely without accurate knowledge of the projectile beam energy. By using the three-dimensional focusing technology of the spectrometer as shown in Fig. 15, the resolution could be improved by a factor of about 5 yielding a resolution of 0.05 a.u. in all three dimensions. The method allowed one to visualize details of electron transitions in a collision and to determine the involved electronic energy transfer with high resolution (see also [53–56, 63–65]).

4.2 Electron–Electron Contributions in the Ionization Process of Ion-Atom Collisions

In the nineties, several groups tried to separate the contributions of target-nuclei-electron (n_t-e) and electron–electron (e–e) interaction in ion-atom collisions. The electron-electron interaction can only knock out the bound electron if the mean relative velocity (projectile velocity) exceeds a certain barrier. Thus measuring the projectile ionization cross section as function of the projectile velocity the (e–e) contribution would contribute only above a certain velocity. Using the C-REMI

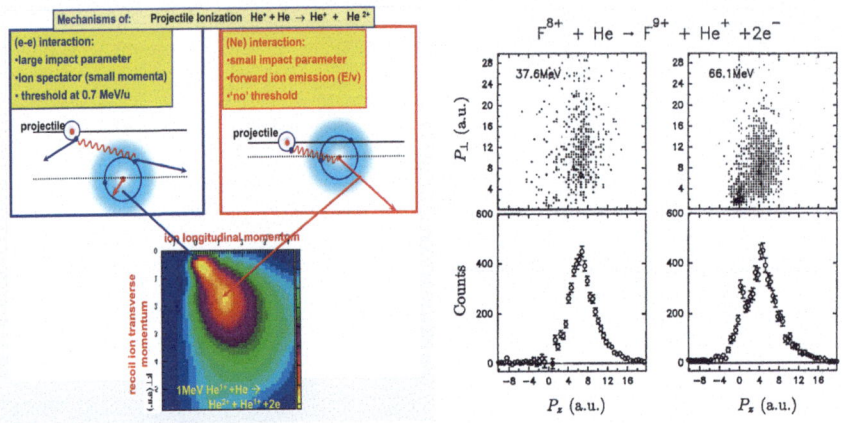

Fig. 24 Recoil-ion momentum plots for projectile ionization. Left: 1 MeV He^{1+} on He [66]. Right: 37.6 and 66.1 MeV F^{8+} on He. Density Plots and corresponding projections, the z-component is the longitudinal momentum axis [67]

approach, however, both contributions should become distinguishable in the recoil-ion momentum distribution. In the (e–e) process both involved projectile and target electrons are knocked-out and the recoil ion would act only as an observer, thus its final momentum remains at target temperature (close to zero momentum). In the (n$_t$-e) process the recoil ion must compensate the momentum of the electron knocked-out. In Fig. 24 (left side) the two mechanisms are explained by diagrams and the measured recoil ion momentum data for He$^+$ on He collisions are shown [66]. At this impact energy the two peaks in the distribution are clearly separated. On the right side of Fig. 24 the data are shown for F^{8+} on He [67] at two impact energies (left below the barrier, right above the barrier).

4.3 Momentum Spectroscopy in High-Energy Heavy Ion Atom Collisions

A further benchmark experiment by Moshammer et al. [68] demonstrated the high resolution power in measuring Q-values and deflection angles. In 3.6 MeV/u Ni^{24+} on He-collisions the full kinematics of the ionization process was measured by a recoil-ion electron coincidence. In Fig. 25 the sum-momentum of the electron and the He^{1+} recoil-ion is presented as function of the longitudinal momentum. More than 90% of all electrons are ejected in forward direction and their momentum is mainly balanced by the backward recoiling He^{1+} ion showing that binary projectile-electron collisions are of minor importance. The full width half maximum (FWHM) is 0.22 a.u., which corresponds to a relative projectile energy loss of $\Delta E/E_p = 3.4 \times 10^{-7}$. The obtained resolution in the transverse momentum corresponds to a resolution in

Fig. 25 Sum of electron and recoil-ion longitudinal momentum (in atomic units). Upper scale: $\Delta p_{r\parallel}$ relative to the incoming projectile momentum p_p. The histogram shows the experimental data and as a full line results from CTMC theory (normalized) [68, 69]

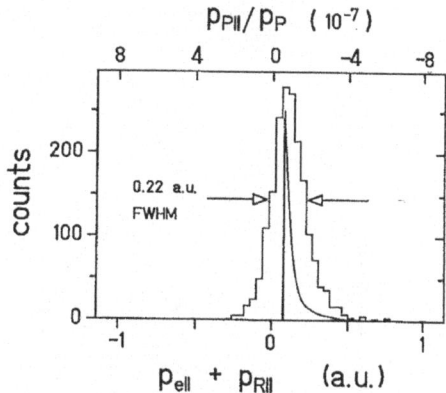

the deflection angle below 10^{-7} rad. In this publication for the first time the acronym COLTRIMS was defined.

In the nineties, the GSI group of Ullrich and Moshammer in cooperation with the Frankfurt group and partly with the CAEN group of Amin Cassimi explored in several research projects the mechanisms of multiple ionization of rare gas atoms in high-energy heavy ion impact. The C-REMI method allowed visualization even for GeV projectiles the energy loss at the few eV level (below 10^{-8} precision) as function of the projectile deflection angle. In Fig. 26 examples of such data are shown for He (left side) and Ne (right side) as a target [68–74]. At higher projectile velocities the momentum distribution of the ejected electrons and of the recoil ions [72] becomes more photon-like. The projectile provides by virtual photon interaction the energy for the ionization process. With higher projectile charge and slower the projectile velocity the electrons are increasingly ejected in the forward direction. In another kinematically complete benchmark experiment on single ionization of He in collisions with 100 MeV/amu C^{6+} the collaboration found small but significant discrepancies between experiment and theory which were interpreted to be the result of higher-order effects in ionization [75]. Today, the puzzling contribution of such presumed higher-order contributions remains a matter of discussion.

4.4   Single-Photon Ionization

Since 1993 the C-REMI technique also contributed strongly to the field of single-photon induced ionization processes. At HASYLAB/DESY-Hamburg and the ALS/LBNL-Berkeley first experiments with C-REMIs were performed. The C-REMI apparatus installed at Berkeley in the group of Michael Prior was mainly funded by the Max Planck Forschungspreis (200.000 DM) awarded together to Cocke and Schmidt-Böcking in 1991. Additionally, Kansas State University provided

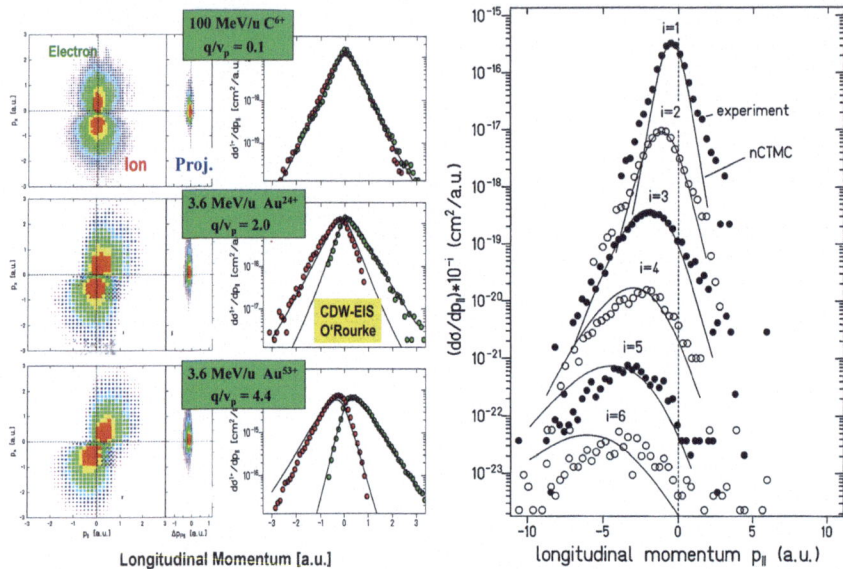

Longitudinal Momentum [a.u.]   longitudinal momentum $p_{\parallel}$ (a.u.)

Fig. 26 Left side: electron-recoil ion momentum plots for heavy-ion impact on He measured in single event coincidence (projected on the plane of incoming projectile—recoil momentum vectors). The plotted projectile momentum vector is the sum of recoil ion and the electron momentum vectors. In the middle the projection of the projectile momentum change is plotted in a.u. [69]. Right: projections of the recoil-ion momentum in multiply ionizing 5.9 MeV/m U^{65+} on Ne is plotted (in a.u.) [69]

fellowships for Ph.D. students and Postdocs. LBNL supported the Berkeley-KSU-Frankfurt collaboration with electronic and computer equipment.

The first achievement was the measurement of the ratio R of He^{2+} to He^{1+} by single photon ionization [76]. The absolute value of this ratio was debated since the methods did not allow a reliable calibration of detection efficiencies. The C-REMI approach had the advantage that the He^{2+} and He^{1+} ions were simultaneously recorded and hence the product of photon beam intensity times target thickness and geometrical solid angle was identical, only the detection efficiencies of the channel-plate detector for He^{2+} and He^{1+} could differ.

Since the height of the detector signal was recorded for every event, too (see Fig. 27 left side), it was evident that the efficiencies for both He charge states were identical, as well. When the ratios were analyzed, however, they did not agree on absolute scale with the standard data available in the literature at that time. The data by Dörner et al. were about 30% lower than the "official" numbers published as reference values. Thus, Dörner et al. began a long search for possible unknown systematic errors in their data analysis. On a meeting at RIKEN/Tokyo in 1995 the Dörner et al. data were compared to new theoretical calculations of Tang and Shimamura. These experimental and theoretical data agreed nicely on an absolute scale within their error bars. Consequently, both were immediately published. The

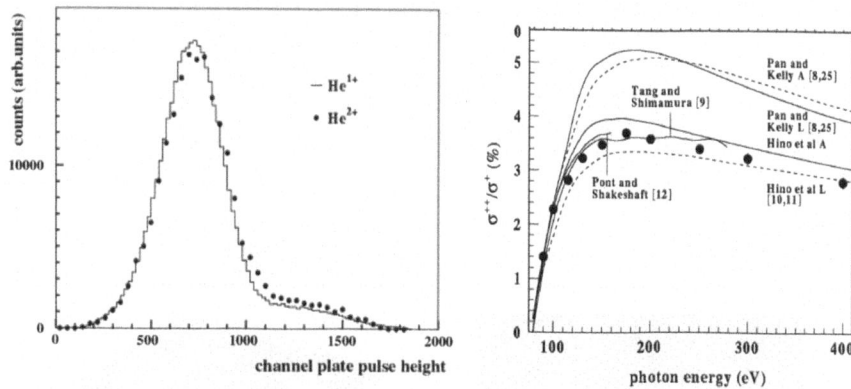

Fig. 27 Left: pulse height distribution from the channel-plate detector for He^{1+} and He^{2+}. Right: the ratio R (full circles) from Dörner et al. [76] as function of the photon energy

"photon ionization community" reacted friendly and acknowledged immediately that the standard data used so far were, for an unknown reason, increased by 40% and could be wrong. The new published data were then accepted as reliable reference.

The ratio of the total ionization cross sections of Helium occurring due to the photo-ionization and the Compton effect was another fundamentally important problem in photon physics at these times. The traditional methods of ion counting could not distinguish by which mechanism the atom was ionized. Both processes, however, differ in their recoil-ion momentum. In case of the photo effect the momentum p_e of the ejected electron is fully balanced by the recoil-ion momentum with $p_{recoil} = -p_e$ (see Fig. 28, left side) [77]. In case of the Compton effect the recoil ion acts only as a spectator and its final momentum peaks at zero (see Fig. 28, right side) [77]. Using the C-REMI approach, these different momentum distributions could quite easily be measured and separated. For photoionization one obtained, furthermore, information on two-electron correlations were the second He electron is simultaneously excited to higher n states (see Fig. 28, left side, rings of smaller electron momenta).

Correlated two-electron processes, like the double ionization of He by a single photon, were, in the nineties very hot topics in the field of photon physics performed at synchrotron machines. Pioneering, fully differential data on the subject were measured by Volker Schmidt's [79] and Alan Huetz's groups [80] by performing electron-electron coincidences. They used traditional electron spectrometers which had compared to C-REMI very small solid angles (resulting in a coincidence efficiency below 10^{-6}). The C-REMI approach has a coincidence efficiency of almost 50% and could image in quasi "one shot" the complete differential distribution (see Fig. 28, left side). Thus C-REMI revolutionized the field of double ionization processes by photon impact. Even the multi-TOF electron spectrometers of Becker and Shirley [81] did not reach the C-REMI coincidence efficiency.

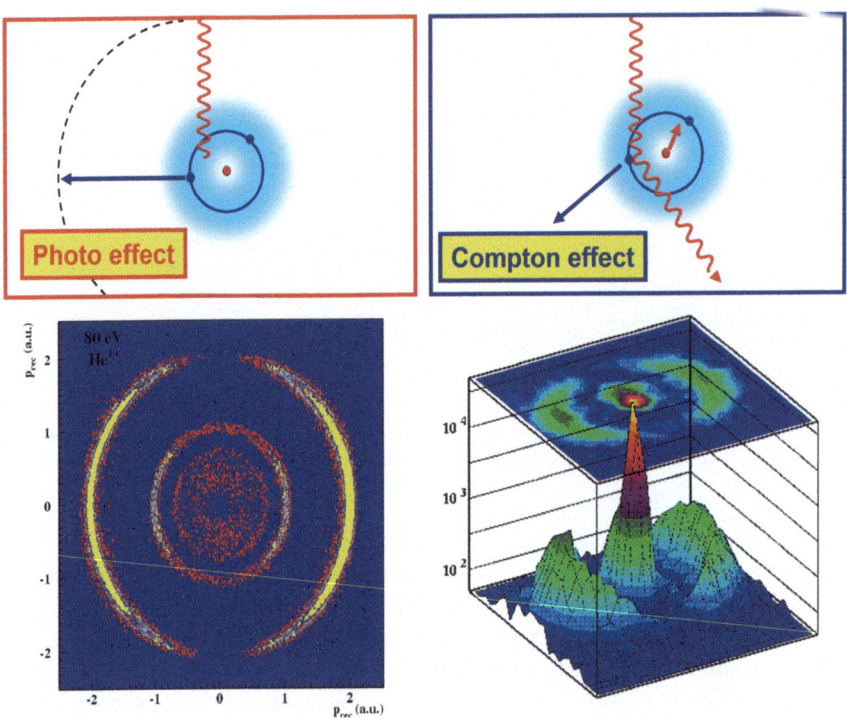

Fig. 28 C-REMI measurement of He photon ionization [77]. Left: Photo effect, right: Compton effect [78]

The first fully differential He double ionization data for circularly polarized photons were measured by Volker Mergel, Hiroshi Azuma and Matthias Achler [84] at the synchrotron machine at Tsukuba. Figure 29 shows the momentum distributions of one electron with respect to the momentum vector of the other electron for linearly polarized photons. In Fig. 30 the same plot is shown for circularly polarized photons. The asymmetric, chiral electron emission patterns are clearly visible in the distributions.

4.5 Saddle Point Ionization Mechanism in Slow Ion-Atom Collisions

In slow ion-atom collisions the mechanism of so-called saddle-point emission played an important role in the ionization process. Even when the projectile velocity was so slow that in a binary projectile nucleus-target electron collision the electron cannot be knocked out, theory predicted that the electron can be promoted to the continuum via quasi-molecular orbitals. Riding finally in the middle of the two nuclei like on a

Fig. 29 Fully differential He double ionization data for linearly polarized photons of 79 eV. The momentum distribution of one electrons is plotted with respect to the momentum vector of the other electron [82, 83]

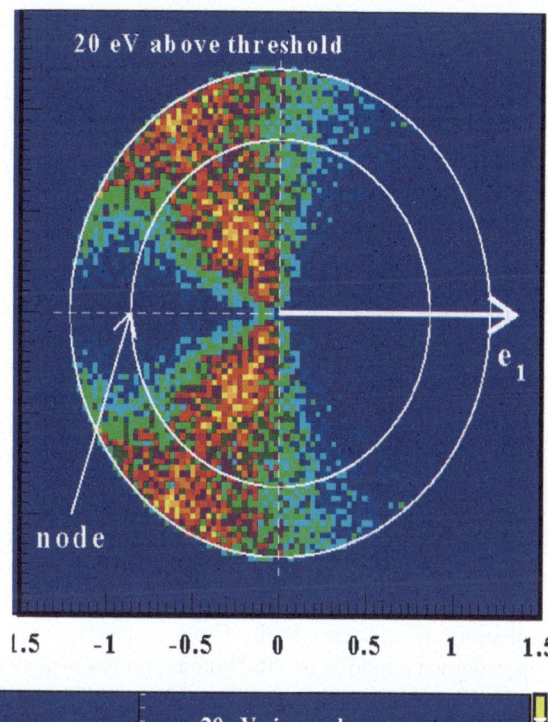

Fig. 30 Fully differential He double ionization data for circularly polarized photons of 99 eV. The momentum distribution of one electron is plotted with respect to the momentum vector of the other electron [84]

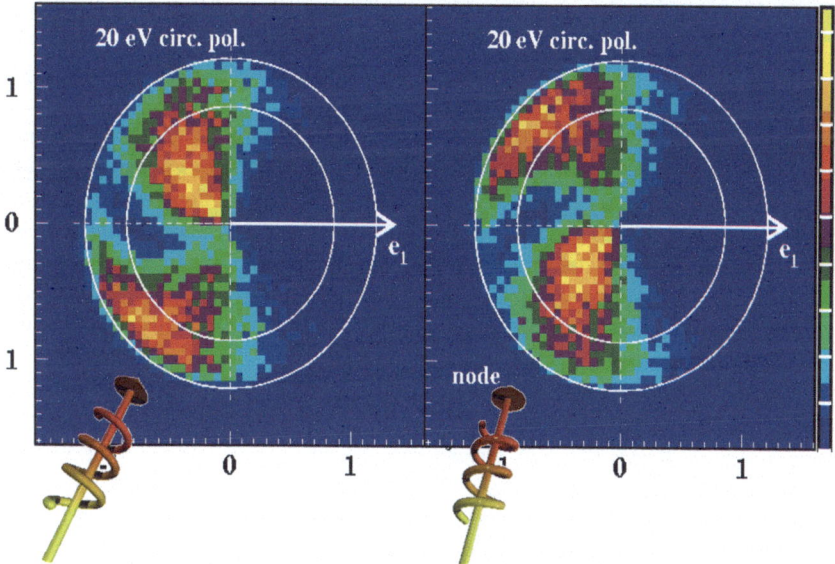

saddle, the electrons end up in the continuum in forward direction with about half the projectile velocity. Using the C-REMI approach Dörner et al. [85] investigated, at the Berkeley ECR ion source, the ionization process in slow p on He collisions, measuring the recoil-ion momentum vector in coincidence with two momentum vector components of the ejected electron. Because of the conservation of total momentum and total energy, the collision dynamics is kinematically fully defined. The surprising result was that the electrons did not ride on a saddle but their emission was kinematically steered by angular momentum conservation. The maxima of the "banana"-like electron distributions (see Fig. 31) vary in emission angle as function of projectile velocity. These shapes are centered in the nuclear collision plane.

4.6 Visualization of Virtual Contributions to the He Ground State

In 1983 Eric Horsdal Pedersen and Charles Lewis Cocke at KSU [86] and in 1986 Reinhold Schuch in Heidelberg [87] could verify, by examining the scattering angle dependence of the transfer ionization process in 7.4 MeV p + He => H° + He^{2+} + e collisions, the existence of the Thomas ionization mechanism [88]. These findings triggered great attention on the Thomas process in the whole atomic physics community. In this process the projectile nucleus can kick the bound He target electron 1 in a binary collision under 45°. On its way to the continuum electron 1 collides in a subsequent binary process with the second electron 2, thus one electron is ejected under 90° in the laboratory system and the other electron under 0°. This forward going electron is then captured by the parallel moving proton projectile resulting in He-double ionization. These billiard like two-step processes require that the projectile is deflected under the angle of $\delta_p = 0.55°$. Thus the He double ionization as function of δ_p should show a peak structure at 0.55°. Varying the projectile velocity Horsdal-Pedersen found that this maximum gets even more pronounced when the projectile velocity increases [89]. Theory, however, predicted a v_p^{-11} law [90]. Therefore, the question arose, is the peak structure at about $\delta_p = 0.55°$ really related to the Thomas process?

In the Ph.D. work of Volker Mergel the complete kinematics of the transfer ionization process in fast proton He collisions was measured by an H° and He^{2+} coincidence using a C-REMI [91, 92]. Determining the He^{2+} momentum vector and the H° transverse momentum components, the kinematics is fully controlled. In Fig. 32 the measured He^{2+} recoil-ion momentum distribution is shown for protons of 1 MeV scattered under 0.55°. Surprisingly two strong maxima appear. One at $p_{rec\parallel} = +$ 0.8 a.u. and $p_{rec\perp} = -0.5$ a.u. which coincides with the expected Thomas peak position, but the second unexpected maximum (named cKTI-p^2) at $p_{rec\parallel} = -2.8$ a.u. and $p_{rec\perp} = -1.8$ a.u. indicates there must be another, so far, unconsidered mechanism enabling transfer ionization at $\delta_p = 0.55°$. The analysis of the kinematics showed that one electron is captured at large impact parameters into the H° ground state

Fig. 31 Electron momentum distributions projected on the nuclear scattering plane for so-called "saddle point electron emission" in slow p-He collisions [85]. The momenta are plotted in relative units of the electron velocity (v_p 'is the projectile velocity)

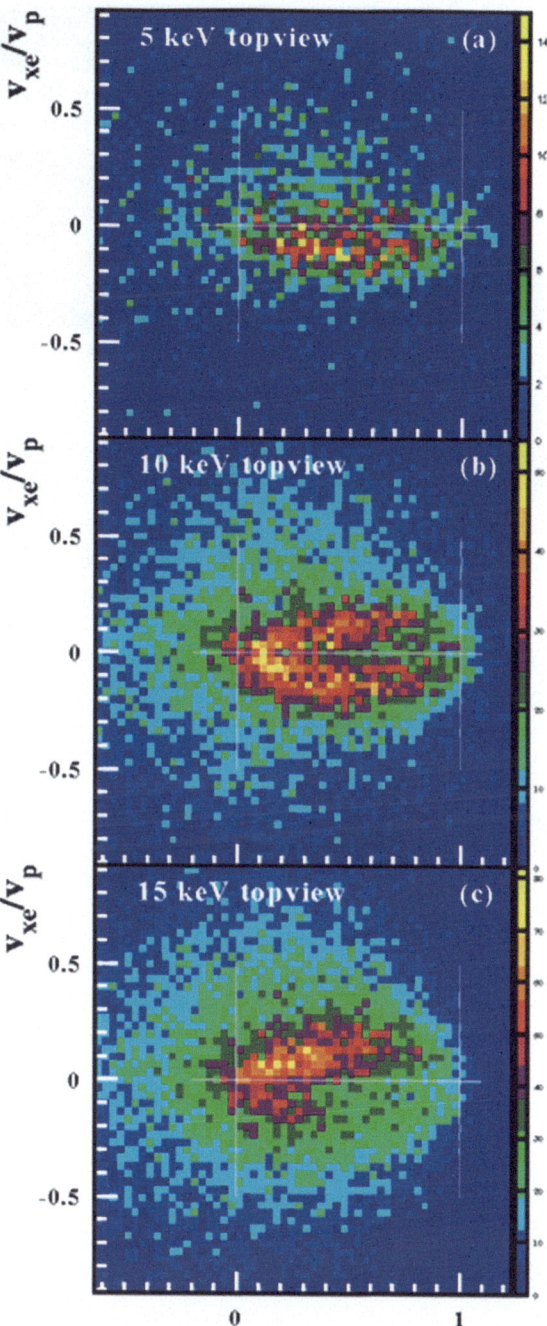

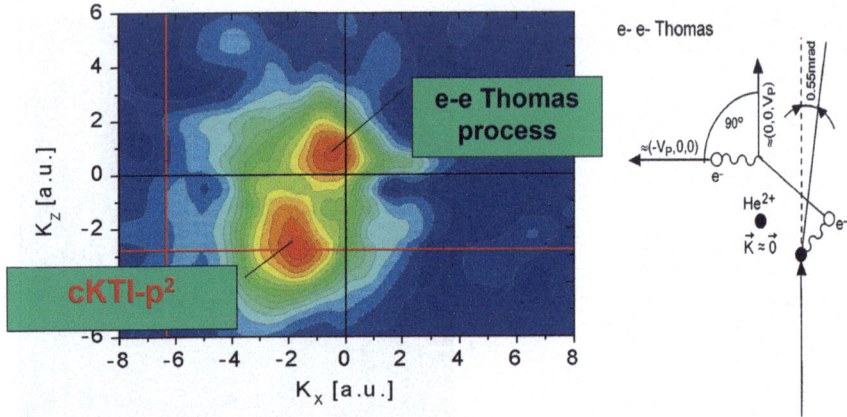

Fig. 32 He^{2+} recoil-ion momentum distribution in the nuclear scattering plane for 1 MeV on He transfer ionization process [91]. Right: The kinematics of the Thomas process

(Brinkmann Kramer mechanism) and the second He electron is emitted backward under about 135° with a momentum of approx. 3 to 4 atomic units. Mergel found the total cross section for this maximum follows a $v_p^{-7.4}$ law, thus, compared to the Thomas peak, it is the dominant transfer ionization channel at higher projectile velocities.

According to multi-configuration theory the He ground state contains a small contributions of 1–2% of the so-called pseudo-states like p^2, d^2 etc. In the p^2 pseudo-state the two electrons have opposite angular momenta and in a He atom at complete rest the target nucleus balances at any moment in a fully entangled motion the sum electron momentum to zero. If one electron in the p^2 state is captured at large impact parameters by the proton the electron 1 velocity and its direction are identical with that of the moving proton. In this moment the other He electron in the p^2 state and the nucleus must move backward. Because electron 2 is in a p pseudo-state it must enter in this moment a real continuum state. In Fig. 33 the angular distribution of the emitted electron 2 is shown in comparison to theoretical predictions. The agreement between experimental data and theory is rather good giving confidence that the presented explanation is valid. It is really surprising that such "virtual states" can be visualized in the real experimental environment. Even the kinematics at a given virtual excitation energy is visible. These tiny contributions represent an extremely small part of states contributing e.g. to the Lamb shift. However, the C-REMI approach is sensitive enough to probe the kinematics of such very small fractions of virtual states.

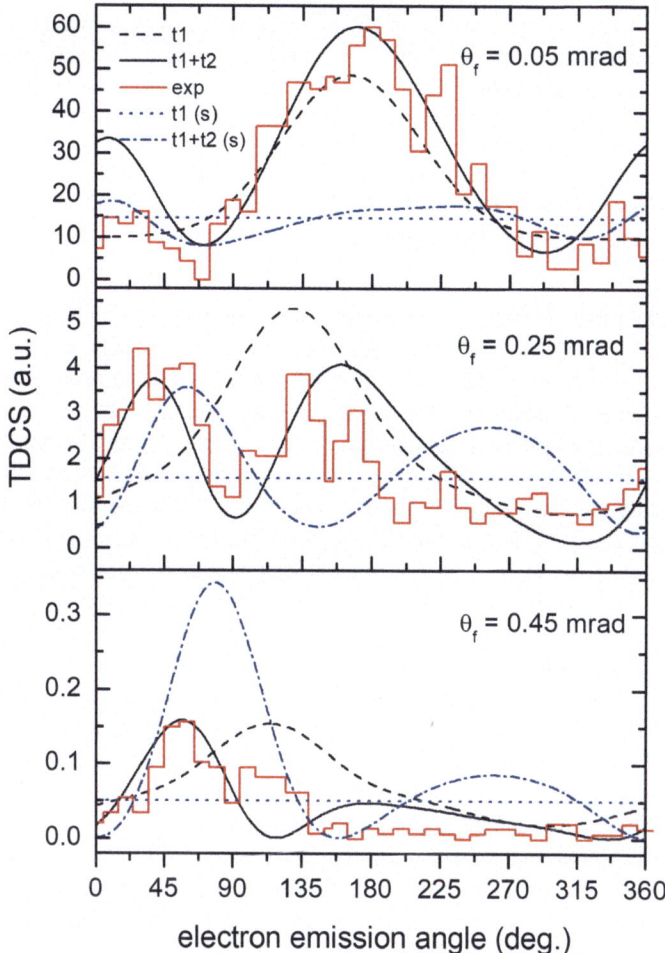

Fig. 33 Triple differential cross sections of Transfer Ionization in 630 keV p on He collisions and 20 eV kinetic energy of the electron corresponding to maximum two (Fig. 32) for three different projectile scattering angles. The black solid line is the theoretical prediction for the non-s^2 contributions. Theory and experiment are relatively normalized [93–95]

5   Milestone Discoveries

The C-REMI had grown into being an established experimental approach to study dynamics in quantum systems in Physics, Chemistry and other fields in the mid-nineties. In several hundred laboratories worldwide C-REMI systems are operating, partially commercially purchased or self-made. By using the C-REMI imaging technique many groups have produced numerous milestone discoveries. However, reference to all of these in this review paper would exceed the purpose of this article. To

present all milestone results produced by the authors of this paper would also over-
shoot the capacity of this review. Thus only a few of those achieved by the authors
of this paper are presented here.

5.1 Multi-photon Processes—Experimental Verification of Re-Scattering Mechanism

To explain the processes underlying multiple ionization and the high double ionization probability of He and other rare gases in intense Laser pulses, Paul Corkum
(1993) and Kenneth Kulander (1995) proposed the so-called re-scattering model
[96]. There, emitted electrons are oscillating in the strong Laser field and are re-
scattered at their parent atom. At that time period, the strong-field community did
not have the proper detection device to verify experimentally this hypothesis. To
visualize the dynamics of this re-scattering process one had to measure the momenta
of two or more ejected electrons (and if possible of the recoil ion, too) in coincidence
with high resolution. Thus, two independent collaborations, which stayed in very
close contact, performed, in parallel, such coincidence experiments.

The collaborations consisted of, first, the Heidelberg group of Joachim Ullrich
and Robert Moshammer, who supplied a C-REMI and joined the Laser group of
Wolfgang Sandner and Horst Rottke in Berlin [97] and, second, the Frankfurt group
of Reinhard Dörner and Thorsten Weber supplying the C-REMI and joined Harald
Giessen in Marburg providing the Laser [98]. Presented here, in Fig. 34, are only the
data of the experiment performed in Marburg [98]. For 220 femtosecond long Laser
pulses of 800 nm wave length at intensities of 2.9 till 6.6×10^{14} W/cm^2 the He^{1+}
and He^{2+} recoil ion momenta were simultaneously measured.

The surprising result was: the He^{1+} recoil momenta are strongly directed parallel to
the Laser electric field with much smaller momenta in the transverse direction- even
much less than in case of single photon ionization. The He^{2+} recoil ion momenta are in
transverse direction of the Laser field similar to the He^{1+} momenta, but parallel to the
field 5 to 10 times larger. In addition, they show two maxima separated by a minimum
at zero. In case of single photon ionization, the recoil-ion momentum distribution
reflects mainly the momentum distribution of the electron in its initial bound state,
in case of double ionization by a single photon it reflects possible electron-electron
correlations in the initial state. But the He^{2+} recoil momenta never exceed the He^{1+}
recoil momenta by more than a factor two. Thus, in case of Laser induced double
ionization only Corkum's re-scattering mechanism can explain the observation of
such large He^{2+} recoil-ion momenta parallel to the Laser field. In his model, the
electron can gain in the Laser field a high ponderomotive energy yielding finally a
large recoil momentum. Similar work can be found in [99]. This work provided the
experimental proof that the re-scattering process does explain the dynamics of the
double ionization in intense Laser field and that both electrons act coherently.

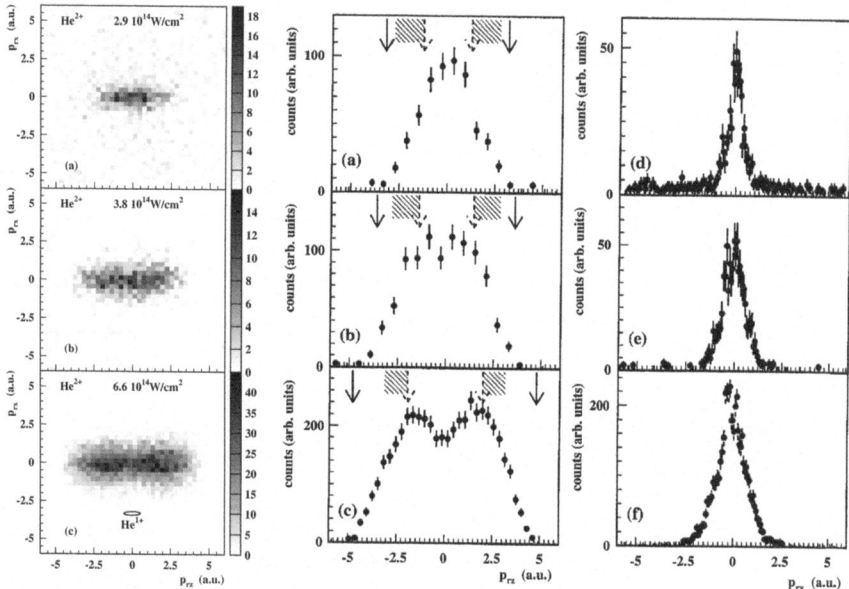

Fig. 34 Left column: He^{2+} recoil ion momentum plots for three different Laser intensities. The horizontal axis is parallel to the direction of the electric field of the Laser. The vertical axis is given by the Laser propagation. The small ellipse in (**c**) shows the half-width of the He^{1+} distribution. Middle and right column: projections of the plots of left column onto the horizontal and vertical axis. The arrows indicate the maximal momentum due to the ponderomotive energy. The dashed areas represent the in the rescattering model calculated values [98]

The two-Laser pulse pump-probe technique is well established to measure timing (delays) in the femtosecond-regime. However, using this technique in combination with the C-REMI coincidence imaging one can—despite of employing pulses of many femto-second duration—obtain timing information in the atto-second regime [100]. In 2005 the collaboration between Paul Corkums group in Ottawa and Reinhard Dörners group in Frankfurt performed such timing measurements in the attosecond regime [101] by using the Laser pump-probe scheme for ionization and for detection of the emitted low-energy electrons the C-REMI approach. The first Laser pulse aligned a nitrogen molecule and the subsequent strong probe pulse ionized the molecule. Recoil ions and electrons were detected in coincidence and from the measured recoil-ion momentum vectors the spatial alignment of the molecule in the lobaratory frame was determined. They found that both electrons did exit the molecule more likely in the same direction when the polarization of the probe pulse was parallel to the direction of the alignment. Double ionization was less probable and takes longer for the perpendicular alignment (a few hundred atto-seconds longer).

In a coincidence experiment where several momentum vectors of the emitted fragments resulting from the same reaction are detected, one can deduce from the relative angular vector directions phase differences and thus determine relative time delays. Thus a multi-coincidence momentum-imaging approach like the C-REMI method is the key to explore atto- and even zeptosecond dynamics by measuring streaking effects of Laser fields on the momenta of emitted particles. This technique allows to measure, as outlined before, time differences shorter than present Laser pump-probe technique can resolve. Ursula Kellers group at the ETH Zürich in cooperation with the Frankfurt group performed such measurements on the tunneling times in He [102]. From the observed phase shifts in the recoil ion and electron-momentum distributions it was claimed that the tunneling process takes a finite time of about 20 as, triggering strong debate on the topic in the following years. Many more important experiments have been performed in recent years in the field of ultrafast processes. References [103–112] are a few selected papers on this topic.

5.2 Single Photon Ionization of Molecules

Since Max von Laues X-ray diffraction experiment in 1912 in Munich the scattering of X-rays and electrons has been used to explore the structure of molecules. In all these studies the molecules had to be in an ordered structure (e.g. crystal) to know the molecular orientation. The C-REMI allows the study of freely moving non-oriented molecules in a gas phase. By performing multi-hit electron-ion coincidence measurements the orientation of the molecule with respect to the detection device is determined from the ionic momenta. The first successful experiments employing the idea of inferring molecular orientation from fragment emission directions were performed by Eiji Shigemasa et al. in 1995 [113] and Heiser et al. [114]. They used traditional electron spectrometers with small solid angles and had to scan the electron energy. Thus these measurements were very time consuming and gave results only for discrete angles.

The first such experiments on single photon ionization of simple molecules using a C-REMI were performed in Berkeley and in parallel in Paris. When the Advanced Light Source (ALS) started operation in 1993 Michael Prior of the LBNL in Berkeley, Charles Lewis Cocke and his group at KSU together Reinhard Dörner and Horst Schmidt-Böcking from the University Frankfurt installed a C-REMI system at the LBNL, which could be used either at the ECR source in the 88″ cyclotron building or the ALS. At the Oji-Workshop (Atomic Mol. Photoionization, September 1995) in Tsukuba, Paul Guyon and Horst Schmidt-Böcking arranged to use the C-REMI coincidence system with position-sensitive detectors to perform collaborative experiments on single photon ionization of molecules at the Paris synchrotron. Paul Guyon's group had used so far the ZEKE technique [115] to study such processes. This method had extremely small coincidence efficiency because of tiny solid angles accepted in the direction transverse of the photon beam. The Paris group provided the photon beam and gas target, the Frankfurt group the detection and data acquisition system.

First experiments on single photon ionization of simple molecules and their fragmentation by photo ionization started at Berkeley in the late nineties with Alan Landers (at that time at KSU) and Thorsten Weber (at that time in Frankfurt) being the responsible investigators. Landers et al. [116] measured the two fragment ions, their charge state and the photo electron upon C-K-shell ionization of CO in coincidence. Following the inner-shell photoionization, Auger electrons are emitted after a short delay leading to a Coulomb explosion of the molecule. Therefore, the ions' emission directions correspond to the molecular orientation at the instant of the photoionization and, from the ions' relative momenta, the kinetic energy released in the fragmentation was also obtained. For this concept to work, it is important that the delay between the fragmentation of the molecule and the initial photoionization is short compared to possible molecular rotation periods. As the photoelectron was measured in coincidence, its angular emission distribution with respect to the molecular axis was obtained.

Figure 35 shows the angular distributions of the C-K-shell photoelectrons in a polar representation, where the distance of a data point to the center of the plot represents the intensity. The double arrow with the two balls in each plot indicates the direction of the photon polarization and the molecular orientation. With the help of theory [116] details of the three-dimensional molecular potential could be deduced from such measurements. Parallel in time to the measurements by Alan Landers and Thorsten Weber et al., also the group of Anne Lafosse et al. located in Paris in cooperation with the Frankfurt group performed such measurements using C-REMI approach [117].

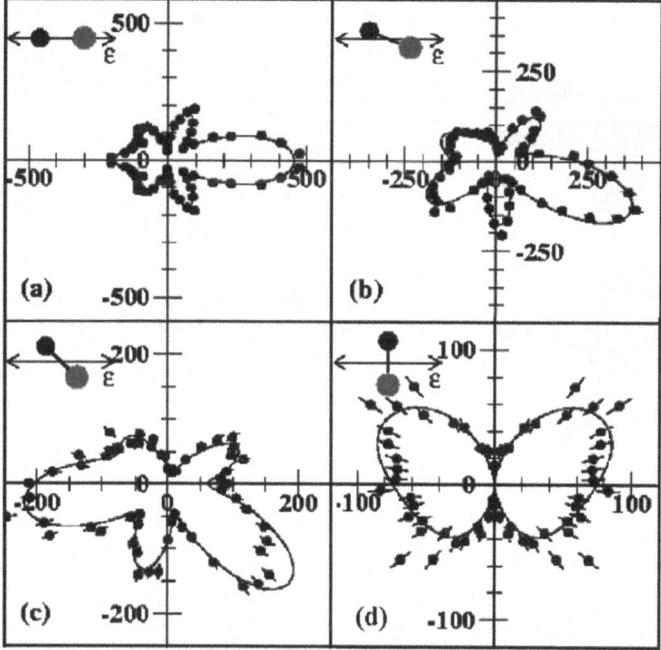

Fig. 35 Polar distribution of 10.2 eV photo-electrons in the frame of the CO molecule (small ball carbon, large ball oxygen). The solid line represents a fit to the data [118]

More photo-ionization measurements of molecules have been performed in the last two decades using the C-REMI approach (see, e.g., [118–122]). Jahnke et al. [118] performed corresponding measurements using circularly polarized photons providing first full 3-dimensional molecular frame photoelectron angular distributions, as shown in Fig. 36, left. Furthermore, they found a strong circular dichroism (CD) in the photoelectron emission (see Fig. 36, right part).

At the ALS in Berkeley Thorsten Weber and Michael Prior, in collaboration with the KSU and Frankfurt groups performed several further studies [120, 121] including the measurement of the complete photon induced fragmentation of D_2. With the support of the theory groups of Bill McCurdy in Berkeley and Fernando Martin in Madrid, fundamental information on symmetry breaking in the D_2 fragmentation processes was deduced.

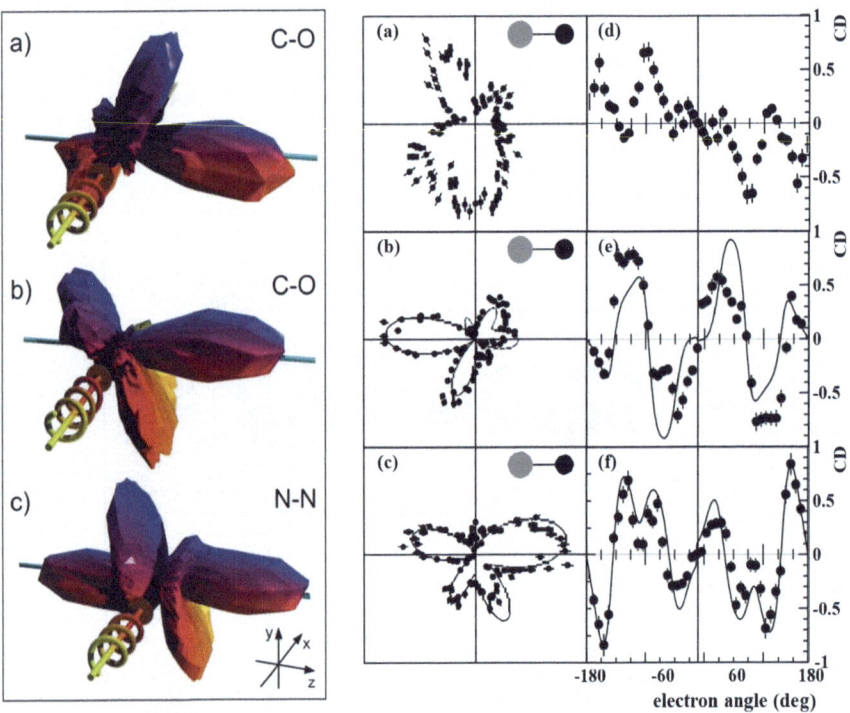

Fig. 36 Left: three-dimensional molecular-frame angular distribution of C and N 1s photoelectrons emitted from CO and N_2 molecules. The molecular orientation is indicated by the green line and the label. The handedness of the photons and their impact direction are indicated by the spirals. Right (**a, b, c**): projections of the data (left) on the plane perpendicular to the photon propagation, right (**d, e, f**): extracted circular dichroism [118]. The corresponding distributions for different molecular orientations can be found at: www.atom.uni-frankfurt.de/research/20_synchrotron/30_photon_mol ecule/20_K-shell_CO_N2/

5.3 Multi-fragment Vector Correlations in Inner Shell Single-Photon Ionization Processes of Atoms and Molecules—Dynamics of Entangled Systems

This kind of measurement approach delivers insight into two or more new fundamental aspects of atomic physics research:

1. the study of oriented very short living excited atomic and molecular ionic configurations, which can never be produced by any other preparation technique (e.g. like Laser orientation and excitation).
2. the study of dynamical entanglement in sequential cascading decay processes, exploring memory effects and dynamically induced symmetry breaking.

The multi-coincident fragment detection from a Coulomb-exploding molecule can provide insight into fundamental aspects of entangled many-particle Coulomb dynamics. In Fig. 37 the scheme of such a multistep process is indicated. From left a circular polarized photon with a well-defined energy (momentum) and angular momentum is absorbed by a two-atom molecule and creates a K vacancy thus a low energy photo electron is emitted (step 1). The electron-momentum vector (three dimensions) is measured. After a short delay, in step 2 a K-Auger electron (probably from the atom where the K vacancy was created) is ejected. The momentum vector of the Auger electron is measured too. Vacancies and excitation energy can be shared by the two atoms. In the following steps 3 to 5 more Auger electrons are emitted. From the measured Auger electron momenta, the experimenter knows the electron energy and thus the time sequence of the different steps. The delay times, however, remain unknown. Finally, with increasing degree of ionization the molecule undergoes Coulomb explosion. Measuring the ionic momenta and its final charge state the experimenter has a full dynamical control on the orientation of the molecule and on the dynamics of the fragmentation (i.e. dynamic entanglement). Finally, in this

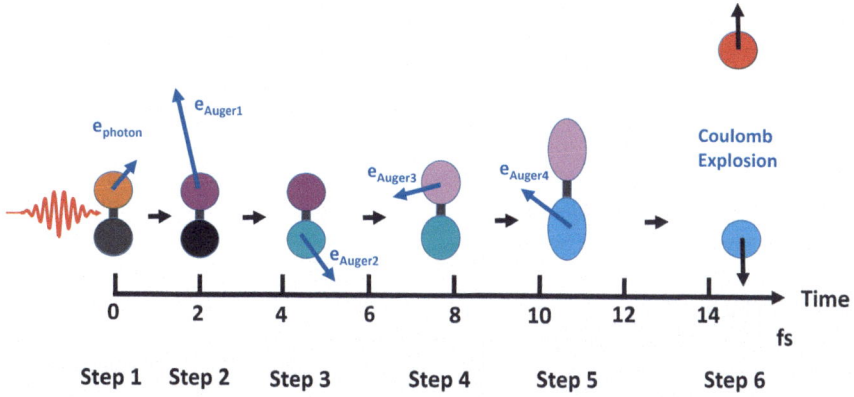

Fig. 37 Scheme of a fragmentation chain with intermediate steps 1 to 6

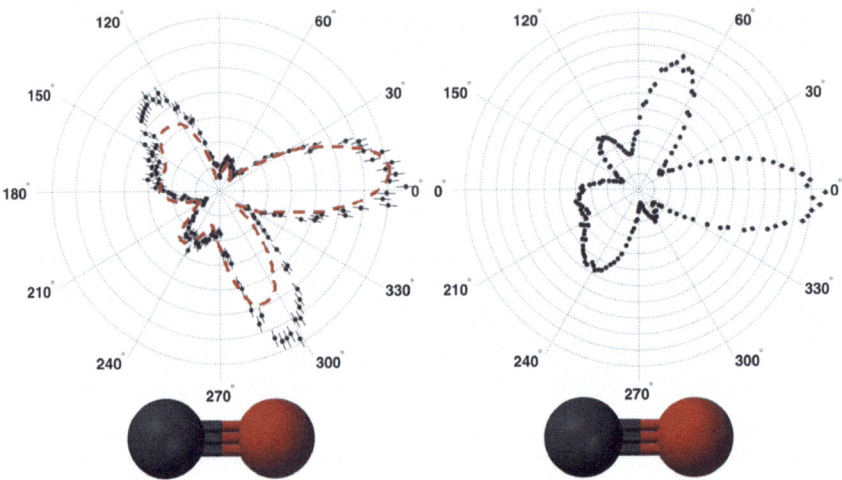

Fig. 38 Polar plot of angular photoelectron distributions in the plane perpendicular to the propagation of the photon. Left: The red dashed line represents the distribution of the right side (right handed photon) but mirrored in time. The molecular orientation is indicated by the bar-bell with Carbon on the left. Only events with a KER value >11 eV are selected, which ensures the axial recoil approximation [121]

example of fragmentation using C-REMI the experimenter has measured all together 24 momentum components and two charge states (the angular momentum vector of the photon, 5 electron momentum vectors and two momentum vectors of the ionic fragments).

Comparing this approach with the two-pulse Laser Pump & Probe technique one can "pump" (ionize) a molecule by a single high energy circular-polarized photon with subsequent photo-electron and multiple Auger-electron emission (i.e., multiple probe technique MPT). The angular momentum of the system recoil-ion and photoelectron is identical with the one of the photons (assumption: the initial state of the molecule has no angular momentum) and is therefore known by the experimenter. In this way the experimenter "pumps" the molecule by single photon absorption without destroying the dynamic entanglement of the system. Thus this MPT establishes a new field in atomic and molecular physics allowing the investigation of extremely short-lived excited molecular states.

The MPT allows one to ask whether the delayed emitted electrons have a "memory" of the earlier fragmentation steps and whether any dynamically induced symmetry breaking (in time or parity) may occur. From the measured vectors L_y and p_{fn} (L_y angular momentum vector of the photon and p_{fn} the momentum vector of the n-th emitted electron) one can define new dynamical coordinate systems, e.g. $L_y \times p_{f1} = Z_{y1}$ and $p_{f1} \times p_{f2} = Z_{12}$ and plot the delayed electron-emission probabilities emitted in step 2,3,.. with respect to these new coordinates [121]. This new **pump & multiple-probe MPT** approach enables the investigation of fundamental dynamical processes in many-particle systems, like time or parity symmetry breaking. If time

Table 1 Vector products with respect to time and parity symmetries

Vector product	$t \to -t$	$r \to -r$
$Z = A_\gamma \times p_{\text{ephoto}}$	$Z(t) = +Z(-t)$	$Z(r) = -Z(-r)$
$Z' = (A_\gamma \times p_{\text{ephoto}}) \times p_{\text{K-Auger}}$	$Z'(t) = -Z'(-t)$	$Z'(r) = +Z'(-r)$
$S = (A_\gamma \times p_{\text{ephoto}}) \cdot n$	$S(t) = +S(-t)$	$S(r) = +S(-r)$

symmetry is broken then the distribution of the Auger electron of step 2 with respect to vector $Z_{Ly1}(+t)$ (p_{e1} is the momentum vector of the photo-electron) and to vector $Z_{Ly1}(-t)$ should be asymmetrical. This vector equation shows

$$Z_{Ly1}(-t) = L_\gamma(-t)x\,p_{e1}(-t) = (-)L_\gamma(+t)x(-)p_{e1}(+t) = +Z_{Ly1}(t)$$

that in case of time inversion the vector does not change its sign.

In [121] for 306 eV right and left handed photons on Carbon Monoxide CO the vector correlations between the 10 eV photo electron, the K-shell Auger electron (Carbon) and the singly charged ionic fragments were measured (Fig. 38). Florian Trinter et al. [121] analyzed the coincidence data with respect to possible dynamically induced symmetry breaking. In Table 1 some "dynamical" vector products are shown with respect to time and parity symmetries.

Trinter et al. [121] have analyzed also the K-Auger electron distributions for different conditions on the momenta of the emitted photo electrons for both left and right handed photons. In Fig. 39 the K-Auger electron distributions are shown for left and right handed polarized photons with the identical conditions on the photo-electron momentum vector (in the same planes as in Fig. 38). In case of complete symmetry with respect to dynamics both distributions should have the same shape. I.e. mirroring the time the corresponding distributions did not agree within the statistical error bars. However, these preliminary measurements do not allow within their error bars any reliable conclusion on time reversal we only assert that such fundamental aspects of quantum dynamics can be explored with the C-REMI approach in these kinds of measurements.

5.4 Single Photon Induced Interatomic Coulombic Decay

Electronically excited atoms or molecules decay by photon or electron emission. More than twenty years ago Cederbaum et al. [123] predicted another very fast decay channel in loosely bound matter, where the excitation energy can be exchanged by means of a virtual photon between an excited atom and its neighboring atom. This decay channel was named "Interatomic Coulombic decay" (ICD). This process occurs in very weekly bound molecules. For example, the Ne dimer, which is a prototype system for ICD, is bound by the van der Waals forces with a binding energy of 2 meV at an inter-nuclear distance of 3.4 Å. First experimental evidence

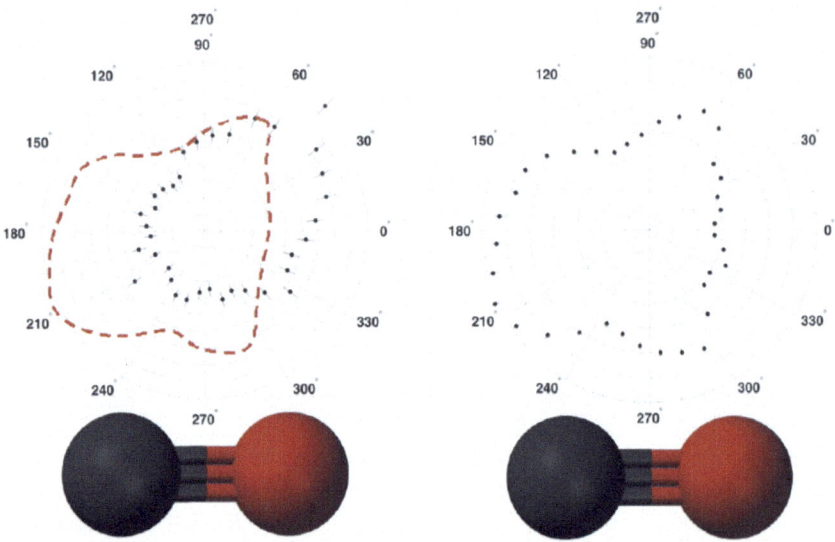

Fig. 39 Polar plot of angular K-Auger-electron distributions in the plane perpendicular to the propagation of the left and right handed circular polarized photons. Left: The red dashed line represents the distribution of the right side (right handed) [121]

for the existence of ICD was reported by observation of slow electrons emitted in large photon excited Ne clusters by Marburger et al. [124]. ICD in Ne dimers can, however, be unambiguously identified by coincident detection of two Ne^{1+} fragments and the low-energy ICD electron. To yield a unique fingerprint of this ICD process, Till Jahnke and Achim Czasch have performed at BESSY II in Berlin a corresponding multi-fragment coincidence experiment using the C-REMI approach [125]. The photon energy was chosen such that only a 2s electron in one Ne atom could be ejected, but a subsequent Auger transition in the same ionized atom was energetically not possible. As ICD occurs, the excitation energy is transferred to the other atom of the dimer causing the ionization of its outer shell. The energy released in the process is shared by the ICD electron and fragment ions. Therefore, the total sum of the kinetic energies is fixed and can be used for an unambiguous identification of the ICD process. A scheme of the ICD process is shown in Fig. 40. The quantitative values of the shared energies are plotted in a two-dimensional plot, KER energy versus electron energy in Fig. 41. The ICD feature forms a diagonal line depicting the constant energy sum, as predicted for the ICD process. In [126–131] more recent work on the ICD process is presented.

The experimental proof and verification of the existence of the ICD process was only possible by the coincident detection of all charged fragments occurring in the process. Most hydrogen or van der Waals bound systems, most prominently liquid water, will often release or transfer energy via the ICD channel. A recent, comprehensive review on ICD can be found in [132].

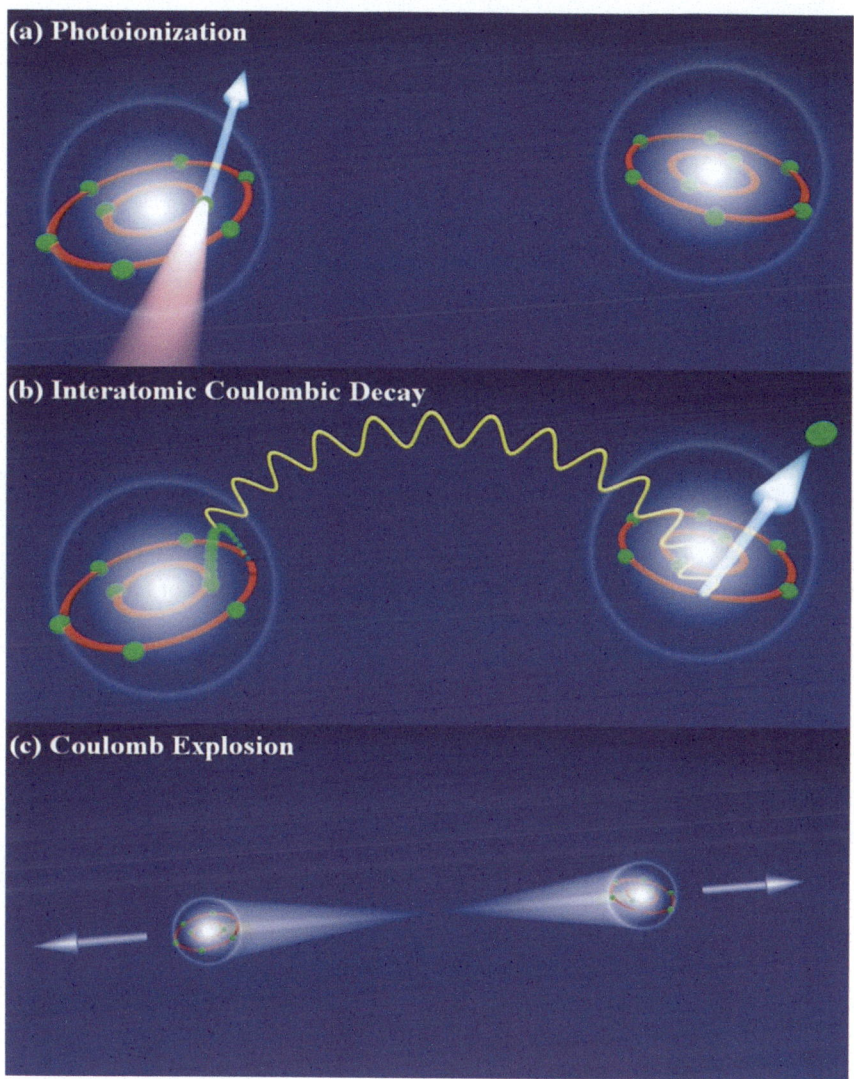

(a) Photoionization

(b) Interatomic Coulombic Decay

(c) Coulomb Explosion

Fig. 40 Scheme of the ICD process. **a** Photoionization with ejection of a 2s photoelectron; **b** Virtual photon transfer from the ionized atom to its neutral partner atom yielding the emission of 2p electron from the partner atom; **c** Coulomb explosion of the doubly charged dimer [125]

In 2013 Trinter et al. [131] have investigated the ICD process in van der Waals-bound HeNe molecules. Najjari et al. [133] predicted that in such molecules one of the atoms can act as a very efficient antenna to absorb photons. In case of HeNe, the ionization cross section is strongly enhanced (by a factor of 60) if the photons can first interact with the He atom. It absorbs the photon and in an ICD-process the energy is transferred to the neighboring Ne atom which is then ejecting an electron.

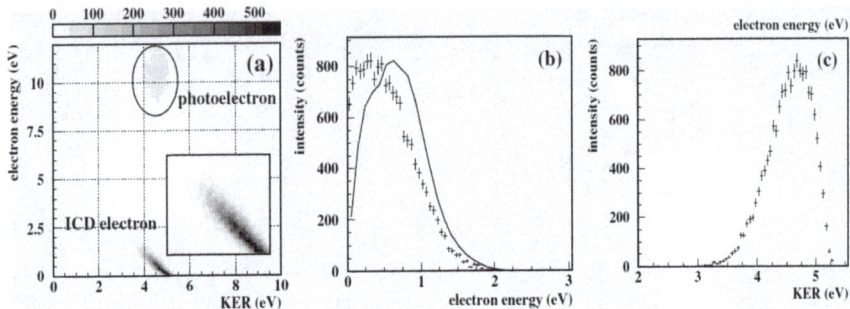

Fig. 41 Left: Kinetic energy release KER of the Ne ions versus the energy of photo electron and ICD electron. Right: Projections of electron and KER value distributions [127]

In Fig. 42 (left side) the different steps of this process are shown. The measurement was performed with a C-REMI system detecting the emitted electron and ion in coincidence. Florian Trinter et al. have experimentally verified that a single atom can act as a highly efficient antenna to absorb energy from a photon field and transfer the energy to a neighboring receiver atom within a few hundreds of femtoseconds. The resolved vibrational states of the resonance provided a benchmark for future calculations of the underlying energy transfer mechanism of ICD.

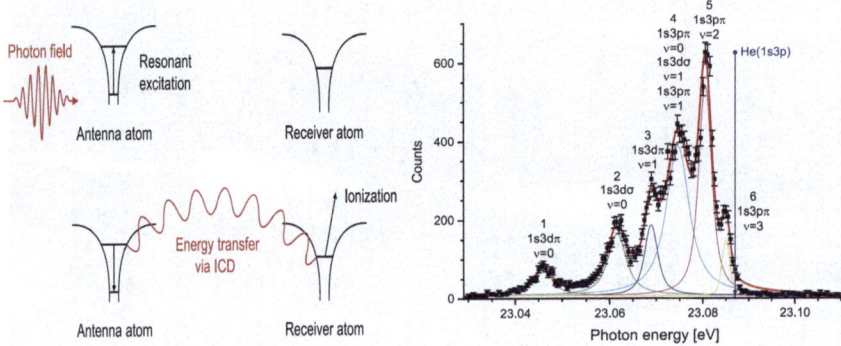

Fig. 42 Left: Scheme of absorption and decay steps. The photon coming from left is absorbed by the He atom, which is resonantly excited into the 1s, 3p state. Before it can decay by photon emission the excitation energy is transferred via resonant ICD to the neutral Ne atom, leading to its ionization. Right: the photon energy was scanned over the range of the He resonance below the actual ionization threshold. The vibrational states of the molecule can be nicely resolved (see theory [135])

5.5   Core-Hole Localization

Each atom or molecule represents one unified dynamical quantum state with a well-defined total energy, where all electrons together with the nuclei form by spin-orbit coupling one state with well-defined angular momentum and exactly ZERO total momentum in its own center-of-mass system—strictly conserved over varying time. Each atom or molecule is not a sum of single particle states, the experimenter cannot number and distinguish each electron e.g. as a specific K-shell or L-shell electron, which can be knocked-off to the continuum thus allowing an initial state localization of the ejected electron. One can only create an ionized atom/molecule in an excited new energy state with a K-shell or L-shell vacancy. In case of an inner-shell hole this vacancy may be localized for an extremely short time near the nucleus of one atom. In 2008 Schöffler et al. [134] have been able to explore this open problem by investigating the symmetry in the angular emission distributions of photo- and Auger electrons emitted from molecular N_2. In their experiment, they measured the photo- and the Auger electron, as well as the emitted ionic fragments in coincidence. The emitted electrons yielded de facto an ultra-fast probe of the shortly existing possible asymmetry of the electronic potential near both nuclei. Early theoretical calculations [135] suggested that even fully symmetric molecules consisted of asymmetric contributions in their ground-state in case of core-hole localization. This work resolved a decade of debate on possible core-hole localization with several experiments proving its existence and others concluding that core-holes are fully delocalized. The C-REMI work by Markus Schöffler et al. demonstrated, that the question of core-hole localization or delocalization remains not fully answered.

It is not only the core-hole (or the corresponding photoelectron) that needs to be considered, but the whole molecule as such. The emitted photoelectron (and thus the core-hole) forms an entangled state with the Auger electron and the fragment ions of the molecule. Depending on the properties of the entangled partners, the properties of the core-hole changes, as well. It was shown in [134] that fingerprints of a localized core-hole can be observed, if the Auger electron resides in a superposition of *gerade* and *ungerade* states, and inversely, the core-hole is delocalized if the corresponding Auger electron can be attributed to a distinct *gerade* or *ungerade* configuration. Figure 43 depicts the photoelectron angular emission distribution in the molecular frame. Panel A shows a symmetric distribution averaging over all emitted Auger electrons. The distribution in Panel B becomes asymmetric (depicting localization) as a gate on distinct Auger electron emission directions is applied (thus selecting a *gerade/ungerade*-superposition).

5.6   Efimov State of the He Trimer

Since more than hundred years, long-range van der Waal forces have attracted great interest in molecular physics. In their origin they differ from the Coulombic and the

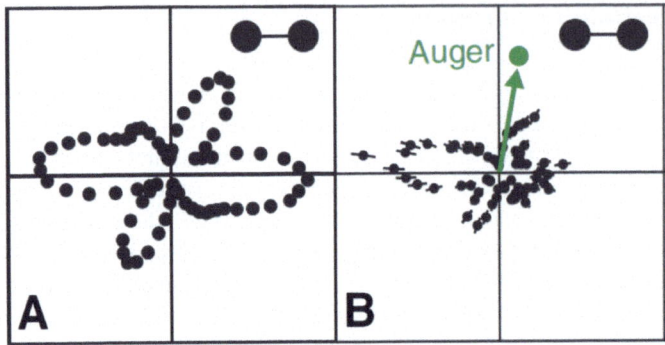

Fig. 43 Angular distribution for of 9 eV photo-electrons emitted from the K shell N_2. The circular polarized photons had an energy of 419 eV. The propagation of the photons is perpendicular to the plotted distributions. The molecule orientation is indicated by the bar-bell. A: Integrated over all Auger-electrons, B: Photon electron distribution coincident with a specific Auger electron emission direction (green arrow) [134]

covalent bonding force and are created by dynamical correlation (or better dynamical entanglement), Van der Waal forces can create bonding at huge inter-nuclear distance. Efimov [136] predicted in the late 60-ties of the last century a universal three-body state which exists as any dominating two-body force vanishes. Such bound three-body states have been termed since then "Efimov-states". It has been predicted that at very low temperature an excited He trimer molecule may form an Efimov state with several hundred Angström inter-nuclear distance between its atomic constituents. Already the He dimer is one of the largest, naturally occurring system (exceeding by far 100 Å) with a binding energy of only a few hundred neV. It was discovered in 1994 by Schöllkopf and Toennies [137] by matter-wave-diffraction and analysis of the observed interference structures.

Starting from initial work on He dimers [138], Kunitski et al. [139] succeeded in 2015 to produce, identify He trimers in an Efimov state and measure the vibrational wave-function of the ^{4}He Efimov-trimer. They prepared the excited state by employing the matter-wave diffraction technique of Schöllkopf and Toennies [137] and multiply-ionized the trimers with a short, highly intense Laser pulse. The rapid ionization yielded a Coulomb explosion of the trimer and—using the C-REMI approach—the momenta of the ionic fragments were measured in coincidence. From the measured momenta the spatial structure was determined, which is shown in Fig. 44. The agreement between experiment and theory is very good. The Efimov trimer consists in principle of a He dimer with the third He atom orbiting at even further distance. This experiment has proven that C-REMI is able to clearly identify even very rare events in the presence of other hugely dominating processes or background, due to the coincident detection of all fragments with the precise measurement of momenta.

excited state: theory   excited state: experiment   ground state: theory

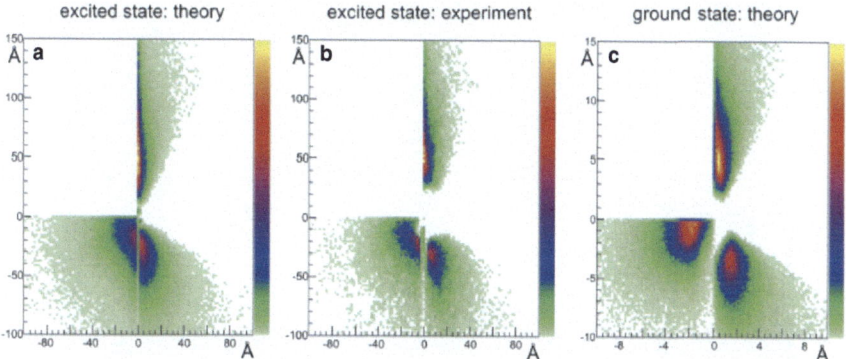

Fig. 44 Structure of the He trimer. **a** The structure predicted by theory and **b** the measured one for the excited Efimov state [139, 140]. **c** for comparison, the ground state structure as predicted by theory. Notice the factor 10 difference in the size

5.7 Imaging of Structural Chirality

Many pharmaceutical drugs have a chiral structure. Since the "Contergan" case [141] in 1961 it became clear that the purity of drugs is a crucial condition for their application. One handedness is constitutional and the opposite handedness can be noxious. Even a very small impurity of the wrong handedness can be very dangerous. Thus it would be of great help if one can recognize for each molecule whether it has the proper chirality. A C-REMI can analyze molecules in the gas phase (and in the future eventually drugs) and decide practically with 100% certainty which handedness is present. Martin Pitzer from the Frankfurt group together with the chemistry group of Robert Berger in Marburg investigated the single-photon (710 eV) and strong-field induced complete fragmentation process of chiral molecules as, for example, CHBrClF and detected the five ionic fragments in coincidence [142, 143]. The molecules are randomly oriented in the gas phase, but as pointed out before, the coincident detection of ionic fragments allows for a determination of their orientation on a single molecule basis. Moreover, when investigating larger molecules, even the molecular structure can be reconstructed from the momentum measurement. As an example, the distribution of the measured momentum vectors is shown (after multiple ionization of CHBrClF using a fs-Laser) in Fig. 45. The Carbon ion is marked by the black sphere, the H ion by the white dots, the F ion by the green dots, the Cl ion by the yellow dots and the Bromine ions by the red dots. The multiple coincidence condition of 4 or five fragments reduces the background nearly to zero and allows to distinguish molecules of different handedness from a racemat, i.e., the experimenter can extract for each ionization event the handedness of the molecule. In Fig. 46 this unambiguous identification of the handedness by using C-REMI becomes obvious. Here the data are plotted as function of the chirality parameter.

$$\cos \Theta_{F(Cl \times Br)} = p_F \cdot (p_{CL} \times p_{Br}) / (|p_F| \cdot |p_{CL} \times p_{Br}|) \quad [137].$$

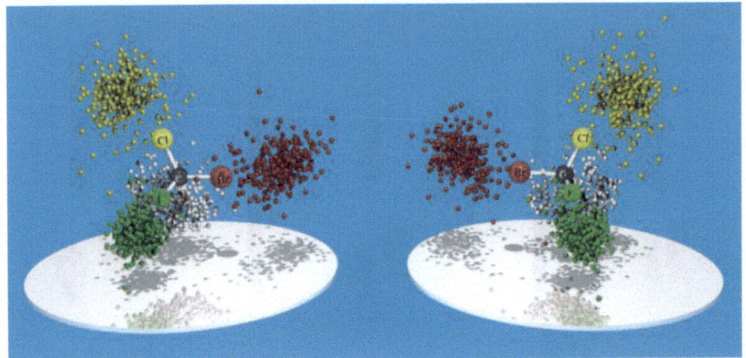

Fig. 45 Momentum vector distribution of ionic fragments of the chiral CHBrClF molecule after Laser ionization. Left: Left handedness, right: right handedness [142] (see text above)

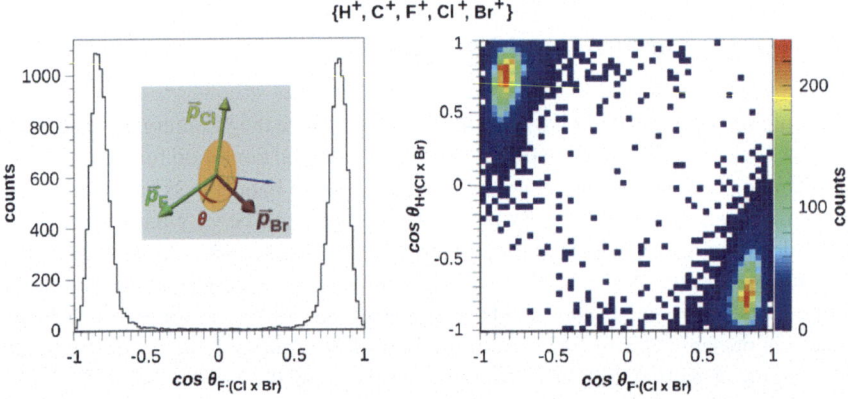

Fig. 46 Measured handedness distribution as function of the chirality parameter $\cos \Theta_{F(Cl \times Br)} = p_F \cdot (p_{CL} \times p_{Br})/(|p_F| \cdot |p_{CL} \times p_{Br}|)$ [142]

5.8 Spatial Imaging of the H_2 Vibrational Wave Function

Dependent on the gas temperature, molecules in the gas phase undergo repetitive collisions with neighboring molecules. This leads to excitation of vibrational or rotational states. The exploration of this intra-molecular motion with traditional x-ray or electron diffraction methods is complicated, as the method is not very sensitive to such features and yields the mean averaged spatial structure. Using Coulomb explosion imaging methods [143] with subsequent coincident measurement of all momenta of the ejected fragments, however, can yield information on this intra-molecular motion (i.e., the vibrational wave function of the nuclei) with high resolution. In Frankfurt Schmidt et al. [144] have investigated the vibrational states of excited H_2 molecules. 2.5 keV H_2^+ ions produced in a Penning ion source collided with a very cold super-sonic jet He beam and were neutralized by capturing one electron into the

Fig. 47 Calculated energy levels as a function of the inter-nuclear distance R for H_2 molecules depicting the concept of the "reflection approximation"

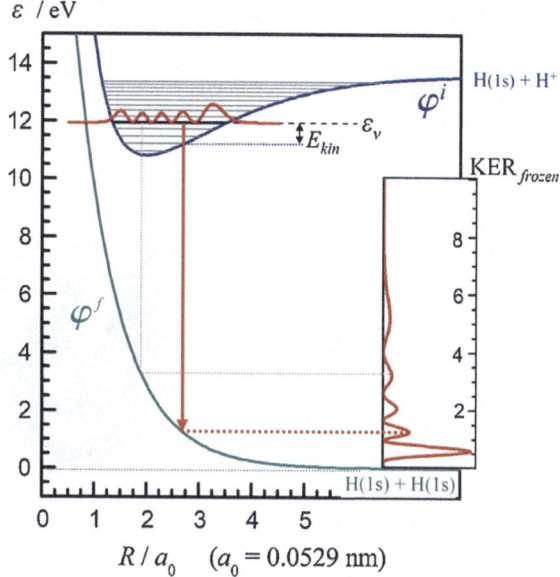

different vibrational states. Using the C-REMI approach the two neutralized H fragments were detected in forward direction by a multi-hit capable time- and position-sensitive detector and the He^+ ion was detected perpendicularly to the ion beam with a C-REMI system. Since the momenta of all three fragments were measured with high resolution (<0.04 a.u. => 3 micro eV) the kinetic energy release KER and the electronic excitation energy (different vibrational states) could be cleanly determined. From the measured H momenta, the H_2^+ inter-nuclear distance was inferred using the reflection approximation (see Fig. 47). The experimental density plot of vibrational states as function of the inter-nuclear distance and electronic excitation energy is shown in Fig. 48. The reflection methods yield slightly different results in case of approximating the nuclei as "frozen" or "moving". This difference becomes obvious from Fig. 49, when the data are analyzed for both reflection methods (green dots: frozen nuclei; red circles: moving nuclei). The solid line represents a mean value of both reflection methods.

5.9   Visualization of Directional Quantization of Quasi-Molecular Orbitals in Slow Ion-Atom Collisions

"Space quantization" or more appropriate "Directional quantization" (Richtungsquantelung) of atomic states in the presence of an outer magnetic field is known since 1916 when it was proposed by Debye and Sommerfeld [145] and its verification in 1922 in the Stern-Gerlach experiment [14]. The existence of such a directional

Fig. 48 Experimental density plot of vibrational states as function of the inter-nuclear distance and electronic excitation energy E_{vib}. The green line is the potential energy curve ($H_2^+(1s\sigma_g)$) calculated in the Born-Oppenheimer approximation [144]

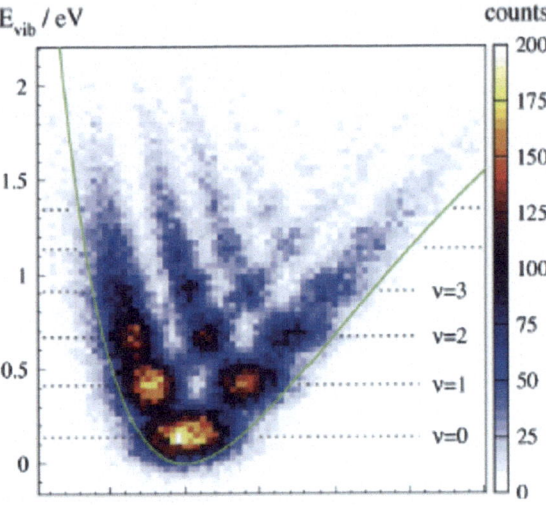

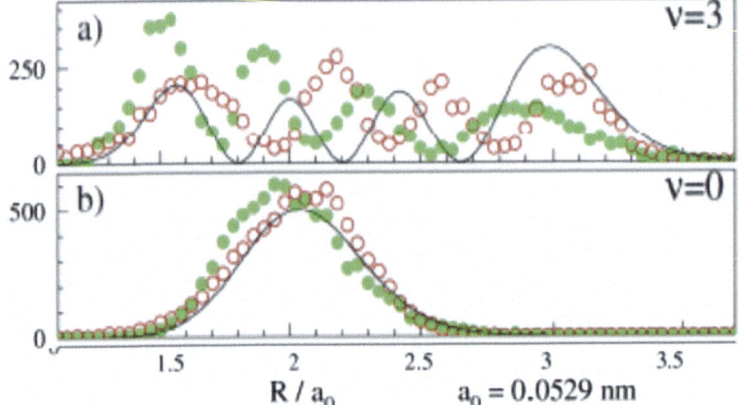

Fig. 49 Distribution of vibrational states as function of the inter-nuclear distance R, where R is calculated for the frozen and moving nuclei reflection methods (green dots: frozen nuclei; red circles: moving nuclei). The solid line represents a mean value of both reflection methods [144]

quantization also in electric field was already indirectly seen in the Stark-effect. The existence of a directional quantization of electronic quasi-molecular states was recently nicely explored by Lothar Schmidt in Frankfurt [146] where he measured the electron emission in slow 10 keV $He^{2+} + He \rightarrow He^+ + He^{2+} + e$ transfer ionization processes. By measuring all three emitted charged fragments in coincidence, the electron emission pattern with respect to the nuclear scattering plane were visualized. In this slow collision process inner-shell quasi-molecular orbitals are formed which are oriented in angular momentum (directional quantization) with respect to the nuclear collision system.

Averaging over all orientations of the nuclear scattering plane, the electron emission pattern does not show any sign of directional quantization, only when for each event the orientation of the nuclear plane is measured. In Fig. 50 the distribution of the emitted electron projected on the nuclear collision plane is shown (a experiment, c theory). The discrete structure corresponds to discrete angular momentum states. The abscissa and ordinate are given in units of the ion velocity v_p. A detailed discussion of this structure is given in [146]. In Fig. 50b, d the projections perpendicular to the nuclear scattering plane are presented. The comparison between experiment and theory shows perfect agreement and proves that in any quantum measurement where the experimenter is sensitive to angular momentum the quantum system reveals the principle existence of directional quantization, i.e. the ordering concept of dynamics in quantum systems.

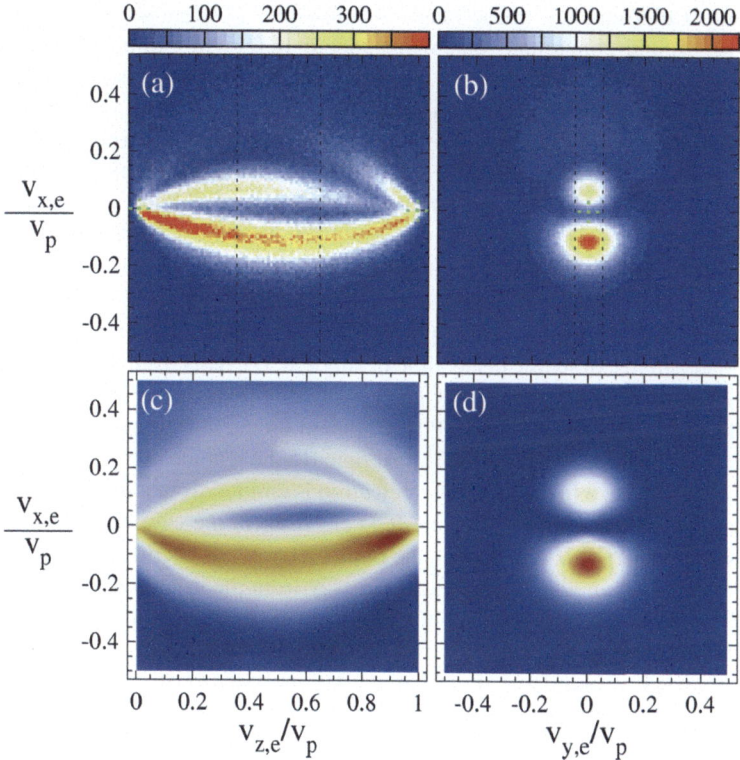

Fig. 50 Electron momentum plots in 10 keV $He^{2+} + He \rightarrow He^+ + He^{2+} + e$ transfer ionization processes [146]. The abscissa and the ordinate are in units of the projectile velocity v_p, i.e. the electron momenta, panels **a** and **b** are experimental data and **c** and **d** theoretical predictions. Panels **a** and **d** depict the projections on the nuclear scattering plane an **b** and **d** perpendicular to it

5.10 Time-Resolving Studies Employing Coincidence Detection Techniques

In the recent past it has been demonstrated, that time-resolving experiments are possible without having a projectile source with corresponding timing properties, as, for example, in a laser pump-probe scheme. In some cases, the temporal evolution on atomic or molecular time scales can be deduced from other information obtained from the coincident detection of ions and electrons. This subsection will provide three recent examples of such studies.

Interatomic Coulombic Decay (ICD) has been a subject of large interest, as described in section V.d. Its efficiency (and thus the lifetime of IC-decaying states) is strongly linked to the inter-nuclear distance between the participating entities. As typical ICD lifetimes are in the range of a few tens of femtoseconds to picoseconds, the excited compound, that will undergo ICD, will exhibit changes of its geometry prior to the decay. These nuclear dynamics triggered strong interest in performing time-resolved measurements of ICD during the last decade, because—as mentioned above—the nuclear motion alters dynamically the electronic decay probability, making ICD a prototype process for distinct non-exponential decay behavior. A molecular movie of the nuclear motion during ICD in helium dimers has been obtained in 2013 by Trinter and coworkers [147]. They used a synchrotron source for triggering ICD in He_2, which has obviously no timing properties, that allow for a direction determination of single event ICD lifetimes (typical synchrotron light pulses have a duration of approx. 100 ps). Accordingly, Trinter et al. introduced a novel approach to extract the decay time of single ICD events from their coincidence measurement. By the so-called "PCI-streaking" the decay time is encoded in the photoelectron kinetic energy. PCI (Post Collision Interaction) is an effect studied in detail already since the 1970-ties [148]. Adopted to the scheme of ICD, the following process takes place: a low energy photoelectron is emitted from a dimer creating the IC-decaying state. As ICD occurs, an ICD electron (in the case of He_2 of approx. 10 eV kinetic energy) is released. If the photoelectron has been chosen sufficiently slow (by selecting an appropriate photon energy from the synchrotron light source) the ICD electron will overtake the photoelectron which causes a change of the effective potential the photoelectron is emerging from, i.e., the potential changes from effectively singly charged to doubly charged. The more attractive potential will decelerate the photoelectron, and the amount of deceleration depends on the emission time of the ICD electron. Thus, by performing a high resolution measurement of the photoelectron momenta and the two ions created in process, the decay time can be inferred from the photoelectron energy and the inter-nuclear distance of the two atoms of the dimer from the ions' kinetic energy release. Employing this approach, Trinter et al. were able to create snap shots of the nuclear motion during ICD covering the first picosecond after the excitation.

Similarly, but using a different approach in detail, Sann et al. showed, how an electronic orbital transforms from being *molecular* to *atomic* upon dissociation of a molecule [110]. A resonant excitation of a HCl molecule triggered its (ultrafast) dissociation [149]. During the dissociation an Auger electron is emitted. Depending

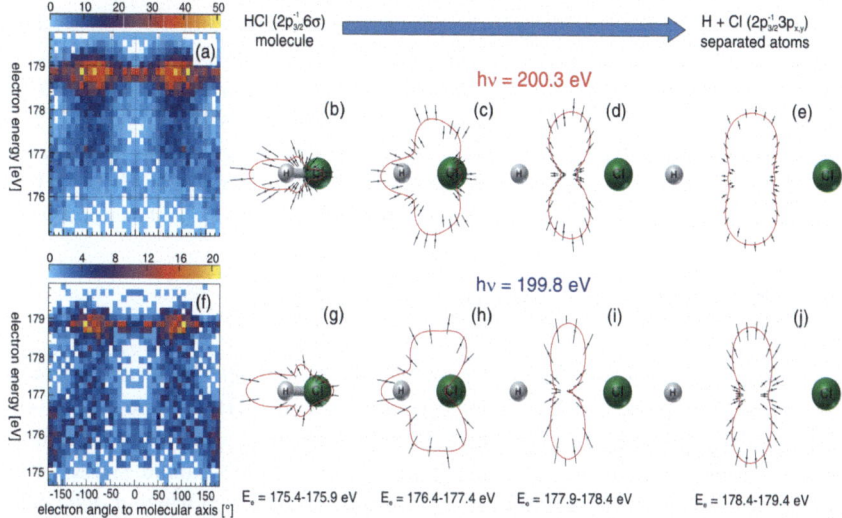

Fig. 51 **b–e** and **g–j** show the transformation of the molecular frame angular distributions of Auger electrons emitted during ultrafast dissociation of HCl. The MFPAD shows initially molecular features (left) and becomes atomic for larger inter-nuclear separations (right). The figure has been taken from [110]

on the emission time, the Auger electron is either emitted from the still intact HCl molecule, an intermediate state or—at later times—from the Cl atom. Sann and coworkers investigated the molecular frame angular distributions of the Auger electron for different inter-nuclear distances during the dissociation. The inter-nuclear distance has been inferred from the energy of the ion measured in coincidence with the Auger electron, providing (as the overall dissociation and decay process typically occurs within 5 fs) information on the timing, as well. The molecular frame angular distributions changed during the dissociation from showing signatures of a molecular orbital to an atomic distribution as shown in Fig. 51.

Very recently, Grundmann et al. investigated the following question employing a multi-particle coincidence approach [13]: Is an electron emitted simultaneously from all across a molecular orbital as it is released by photoionization, or is it first released from that portion of its orbital that is "illuminated first" by the photon? A H_2 molecule has been used as prototype test bench to answer this question. Photoemission from a homo-nuclear molecule can be—due to the two-center nature of the molecule—intuitively regarded as a microscopic analog to scattering at a classical double slit. The photoelectron wave is emitted as a superposition from the "left" and the "right" atom of the molecule, which, indeed, causes Young-type interference patterns in the molecular frame angular emission distribution of the electron. Grundmann and coworkers showed, that the molecular frame angular emission distribution changes subtly if the molecule is oriented along the photon propagation direction or perpendicular to it during the photoionization process. Within the double slit picture these changes are understandable: if the molecule is oriented perpendicular to the photon direction, the photon arrives at both nuclei of the molecule at the same time. However, if it oriented in parallel to the propagation direction, one of the atom is hit prior to the second one. This delay in the arrival time can be modelled as a phase shift of the one of the emerging photoelectron waves, or, in the double slit picture, a phase shift in one of the two slits, which causes a measureable displacement of the double slit interference pattern. From this displacement a birth time delay of approx. 250 zeptoseconds was resolved in the experiment, which nicely corresponds to the travel time of the photon along the molecule. The sensitivity of this approach is below a few 10 zeptoseconds, despite employing synchrotron light pulses of >100 ps duration.

5.11 Proposed Experiments in Neutrino Physics

In an article published in 1994 in Comments on Atomic and Molecular Physics Ullrich et al. [150] presented future perspectives of the C-REMI technique. One exotic one is worth mentioning here: With the C-REMI approach, i.e. measuring with ultra-high resolution the momenta, one can determine in principle from one single event also the mass of a particle. Therefor it was thoroughly discussed whether one could use a C-REMI to measure, in the decay process of Tritium, the neutrino momentum by performing a He^{1+} recoil ion-electron coincidence. To be sensitive to a very small neutrino mass of about one eV in this decay event the electron kinetic energy must be very close to the Q-value of the tritium decay. These events are extremely rare. Thus because of the huge number of random coincidences the required time of measurement would nearly approach the life-time of an experimenter. May be somebody will discover a way to handle such a high random rate?

6 Conclusion

The C-REMI technique can be considered as the "Bubble Chamber" or "Time Projection Chamber" in atomic and molecular physics. Using the multi-coincidence concept initially developed in nuclear and high energy particle physics, a C-REMI can image the whole momentum space in a single-event quantum process. Using ultra-cold targets in the gas phase and electro-magnetic spectrometer designs with focusing conditions an excellent sub-atomic momentum resolution and a large multi-hit coincidence efficiency are obtained. Thus visualizing the complete dynamics in a single event the dynamical entanglement in many particle systems can be explored. The C-REMI is now a standard detection system in many fields of physics and chemistry and is used by many groups around the world.

Acknowledgements Many of the scientists contributing to the C-REMI development are co-authors of this paper. But we are indebted to many other colleagues: Reinhold Schuch, Siegbert Hagmann, Amin Cassimi, Nora Berrah, Andre Staudte, Harald Bräuning, Angela Bräuning Demian, Paul Corkum, Kiyoshi Ueda, Tadashi Kambara, Yasu Yamazaki, Paul Mokler, Thomas Stöhlker, Klaus Blaum, Erhard Salzborn, Alfred Müller, Karl Ontjes Groeneveld, Hans Joachim Specht,

Bernd Sonntag, Jochen Schneider, Berthold Krässig, Timor Osipov, etc. for a long close cooperation and many theorists providing ideas for measurements like John Briggs, Burkhard Fricke, Hans Jürgen Lüdde, Jan Michael Rost, Steve Manson, etc. Additionally we want to thank BMBF (Dietrich Hartwig) at GSI, the Deutsche Forschungsgemeinschaft, the people at GSI, ALS, at Hasylab, Bessy, Grenoble, Soleil Paris, Spring8, the mechanics and electro technicians in Frankfurt, KSU, Heidelberg, GSI and Berkeley workshops for continuous support.

The early team laying the foundations of C-REMI. Painting by Jürgen Jaumann 2011

References

1. H. Schmidt-Böcking, S. Eckart, H. J. Lüdde, G. Gruber, T. Jahnke, The precision limits in a single-event quantum measurement of electron momentum and position, these proceedings
2. O. Stern, Zur Methode der Molekularstrahlen I. Z. Physik. **39**, 751–763 (1926); F. Knauer, O. Stern, Zur Methode der Molekularstrahlen II. Z. Physik. **39**, 764–779 (1926)
3. H. Schmidt-Böcking, K. Reich, Otto Stern-Physiker, Querdenker, Nobelpreisträger. Goethe-Universität Frankfurt, Herausgeber. Gründer, Gönner und Gelehrte. Societätsverlag, Reihe. ISBN 978-3-942921-23-7 (2011)
4. J. Ullrich, R. Dörner, V. Mergel, O. Jagutzki, L. Spielberger, H. Schmidt-Böcking, Cold-target recoil-ion momentum-spectroscopy: first results and future perspectives of a novel high resolution technique for the investigation of collision induced many-particle reactions. Comments Atomic Mol. Phys. **30**, 285 (1994)
5. J. Ullrich, R. Dörner, H. Schmidt-Böcking, A New "Momentum Microscope" Views Atomic Collision Dynamics, Physics News. American Institute of Physics (1996), p. 12

6. R. Moshammer, M. Unverzagt, W. Schmitt, J. Ullrich, H. Schmidt-Böcking, A 4 π recoil-ion electron momentum analyser: a high-resolution, "microscope" for the investigation of the dynamics of atomic, molecular and nuclear reactions. Nucl. Instrum. Meth. B **108**, 425 (1996)
7. J. Ullrich, R. Moshammer, R. Dörner, O. Jagutzki, V. Mergel, H. Schmidt-Böcking, L. Spielberger, Recoil ion momentum spectroscopy. J. Phys. B At. Mol. Opt. **30**, 2917 (1997); J. Ullrich, W. Schmitt, R. Dörner, O. Jagutzki, V. Mergel, R. Moshammer, H. Schmidt-Böcking, L. Spielberger, M. Unverzagt, R.E. Olson, in Recoil Ion Momentum Spectroscopy Photonic, ed. by F. Aumayr et al. Electr. Atomic Coll., World Scient. (1997), p. 421
8. R. Dörner, V. Mergel, H. Bräuning, M. Achler, T. Weber, Kh. Khayyat, O. Jagutzki, L. Spielberger, J. Ullrich, R. Moshammer, Y. Azuma, M. H. Prior, C.L. Cocke, H. Schmidt-Böcking, in Recoil ion momentum spectroscopy—a "momentum microscope" to view atomic collision dynamics. Proceedings of the AIP Conference—Atomic Processes in Plasmas, ed. by E. Oks, M. Pindzola (1998), p. 443
9. R. Dörner, V. Mergel, O. Jagutzki, L. Spielberger, J. Ullrich, R. Moshammer, H. Schmidt-Böcking, Cold target recoil ion momentum spectroscopy: a "momentum microscope" to view atomic collision dynamics. Phys. Rep. **330**, 95 (2000)
10. R. Dörner, Th. Weber, M. Weckenbrock, A. Staudte, M. Hattass, R. Moshammer, J. Ullrich, H. Schmidt-Böcking, Multiple Ionization in Strong Laser Fields Advances in Atomic and Molecular Physics, vol. 48, ed. by B. Bederson, H. Walther. Academic Press (2002), p. 1
11. J. Ullrich, R. Moshammer, A. Dorn, R. Dörner, L. Ph. H. Schmidt, H. Schmidt-Böcking Recoil-ion and electron momentum spectroscopy: reaction-microscopes. Rep. Prog. Phys. **66**, 1463 (2003)
12. O. Stern, Eine direkte Messung der thermischen Molekulargeschwindigkeit. Z. Physik **2**, 49–56 (1920); O. Stern, Nachtrag zu meiner Arbeit: „Eine direkte Messung der thermischen Molekulargeschwindigkeit, Z. Physik **3**, 417–421 (1920)
13. S. Grundmann, D. Trabert, K. Fehre, N. Strenger, A. Pier, L. Kaiser, M. Kircher, M. Weller, S. Eckart, L. Ph. H. Schmidt, F. Trinter, T. Jahnke, M. S. Schöffler, R. Dörner, Zeptosecond Birth Time Delay in Molecular Photoionization. Sci. **370**, 339–341 (2020)
14. L. Dunoyer, Le Radium **8**, 142 (1911)
15. W. Gerlach, O. Stern, Der experimentelle Nachweis der Richtungsquantelung im Magnetfeld. Z. Physik, **9**, 349–352 (1922); W. Gerlach, O. Stern, Über die Richtungsquantelung im Magnetfeld. Ann. Physik, **74**, 673–699 (1924)
16. H. Schmidt-Böcking, H. Reich, K. Templeton, W. Trageser, V. Vill (Hrsg.). Otto Sterns Veröffentlichungen—Band 1 bis V Sterns Veröffentlichungen von 1912 bis 1916 Springer Verlag. ISBN 978-3-662-46953-8 (2016)
17. O.R. Frisch, Experimenteller Nachweis des Einsteinschen Strahlungsrückstoßes. Z. Phys. **86**, 42–48 (1933)
18. Center for History of Science, The Royal Swedish Academy of Sciences, Box 50005, SE-104 05 Stockholm, Sweden
19. S. Kelbch, H. Schmidt-Böcking, J. Ullrich, R. Schuch, E. Justiniano, H. Ingwersen, C.L. Cocke, The contributions of K-electron capture for the production of highly charged Ne recoil ions by 156 MeV bromine impact. Z. Phys. A **317**, 9 (1984)
20. J. Ullrich, C.L. Cocke, S. Kelbch, R. Mann, P. Richard, H. Schmidt-Böcking, A parasite ion source for bare-ion production on a high energy heavy-ion accelerator. J. Phys. B **17**, L 785 (1984)
21. S. Kelbch, J. Ullrich, R. Mann, P. Richard, H. Schmidt-Böcking, Cross sections for the production of highly charged Argon and Xenon recoil ions in collisions with high velocity uranium projectile. J. Phys. B **18**, 323 (1985)
22. P. Richard, J. Ullrich, S. Kelbch, H. Schmidt-Böcking, R. Mann, C.L. Cocke, The production of highly charged Ar and Xe recoil ions by fast uranium impact. Nucl. Instr. Meth. A **240**, 532 (1985)
23. H. Schmidt-Böcking, C.L. Cocke, S. Kelbch, R. Mann, P. Richard, J. Ullrich, Multiple Ionization of Argon Atoms by Fast Uranium Impact and its Possible Application as an Ion Source for Highly Ionized Rare Gas Atoms, High Energy Ion Atom Collisions, eds. by D. Berenyi, G. Hock. Akademia Kiado, Budapest (1985)

24. S. Kelbch, J. Ullrich, W. Rauch, H. Schmidt-Böcking, M. Horbatsch, R. Dreizler, S. Hagmann, R. Anholt, A.S. Schlachter, A. Müller, P. Richard, Ch. Stoller, C. L. Cocke, R. Mann, W. E. Meyerhof, J.D. Rasmussen, Multiple Ionization of Ne, Ar, Kr and I by nearly Relativistic U Ions. J. Phys. B **19**, L 47 (1986)

25. R.E. Olson, J. Ullrich, H. Schmidt-Böcking. J. Phys. B20, L809 (1987); Grandin et al. Europhys. Lett. **6**, 683 (1988)

26. C.L. Cocke, R.E. Olson, Recoil ions. Phys. Rep. **205**, 155 (1991)

27. U. Buck, M. Düker, H. Pauly, D. Rust, Proceed. of the IV Int. Symp. Molecular beams (1974) 70; H. Haberland, U. Buck, M. Tolle. Rev. Sci. Instr. **56**, 1712 (1985)

28. M. van der Poel, C.V. Nielsen, M.A. Gearba, N. Andersen. Phys. Rev. Lett. **87**, 123–201 (2001); J.W. Turkstra, R. Hoekstra, S. Knoop, D. Meyer, R. Morgenstern, R.E. Olson, Phys. Rev. Lett. **87**, 123–202 (2001); X. Flechard, H. Nguyen, E. Wells, I. Ben-Itzhak, B.D. DePaola, Phys. Rev. Lett. **87**, 123–203 (2001)

29. R. Mann, F. Folkmann, K.O. Groeneveld, strong molecular effects in heavy-ion-induced carbon and nitrogen auger transitions. Phys. Rev. Lett. 1674 (1976); R. Mann, C.L. Cocke, A.S. Schlachter, M. Prior, R. Marrus, Selective final-state population in electron capture by low-energy highly charged projectiles studied by energy-gain spectroscopy. Phys. Rev. Lett. 1329 (1982)

30. G. Gaukler, H. Schmidt-Böcking, R. Schuch, R. Schulé, H.J. Specht, I. Tserruya, A position sensitive parallel plate avalanche detector for heavy-ion X-ray coincidence measurements. Nucl. Instr. Meth. **141**, 115 (1977)

31. J.L. Wiza, Microchannel plate detectors. Nucl. Instrum. Meth. **162**, 587–601 (1979)

32. C. Martin, P. Jelinsky, M. Lampton, R.F. Malina, H.O. Anger, Rev. Sci. Instrum. **52**, 1067 (1981)

33. O. Jagutzki, V. Mergel, K. Ullmann-Pfleger, L. Spielberger, U. Meyer, R. Dörner, H. Schmidt-Böcking, Fast position and time resolved read-out of micro-channelplates with the delay-line technique for single particle and photon detection, imaging spectroscopy IV. Proceedings of the International Symposium on Optimal Science Engineering & Instr., eds. by M.R. Descour, S.S. Shen, vol. 3438. Proc SPIE (1998), pp. 322–333; K. Fehre, D. Trojanowskaja, J. Gatzke, M. Kunitski, F. Trinter, S. Zeller, L. Ph. H. Schmidt, J. Stohner, R. Berger, A. Czasch, O. Jagutzki, T. Jahnke, R. Dörner, M.S. Schöffler, Absolute ion detection efficiencies of microchannel plates and funnel microchannel plates for multi-coincidence detection. Rev. Sci. Instrum. **89**, 045112 (2018)

34. S.E. Sobottka, M.B. Williams, IEEE Trans. Nucl. Sci. **35**, 348 (1988)

35. http://roentdek.com/detectors/ M. S. Schöffler, Grundzustandskorrelationen und dynamische Prozesse untersucht in Ion-Helium-Stößen, Dissertation(2006) Universität Frankfurt

36. https://aktuelles.uni-frankfurt.de/menschen/vom-selbstgebauten-pc-zum-supercomputer/; https://www.gsi.de/work/kurier/Ausgabe/19;2018.htm?no_cache=1&cHash=4fe9358d35b5 f37e110eefdd2d86e1b1

37. K. Ullmann, V. Mergel, L. Spielberger, T. Vogt, U. Meyer, R. Dörner, O. Jagutzki, M. Unverzagt, I. Ali, J. Ullrich, W. Schmitt, R. Moshammer, C.L. Cocke, T. Kambara, Y. Awaya, H. Schmidt-Böcking, Cold target recoil ion momentum spectroscopy, in Proceedings of the 4th US-Mexican Symposium on Atomic and Molecular Physics, eds. by I. Alvarez, C. Cisneros, T.J. Morgan. World Scientific (1995), p. 269

38. https://groups.nscl.msu.edu/nscl_library/manuals/eggortec/453.pdf

39. https://www.cronologic.de/

40. J. Ullrich, H. Schmidt-Böcking, S. Kelbch, C. Kelbch, V. Dangendorf, A. Visser, D. Weisinger, Stoßparameterabhängigkeit der Vielfachionisationswahrscheinlichkeit. Annual Report Institute for Nuclear Physics, University Frankfurt, (1984), p. 20

41. J. Ullrich, H. Schmidt-Böcking, Time of flight spectrometer for the determination of micro-radian projectile scattering angles in atomic collisions. Phys. Lett. A **125**, 193 (1987)

42. R.E. Olson, J. Ullrich, R. Dörner, H. Schmidt-Böcking, Single and double ionization cross sections for angular scattering of fast protons by helium. Phys. Rev. A **40**, R2843 (1989)

43. J. Ullrich, R. Olson, R. Dörner, V. Dangendorf, S. Kelbch, H. Berg, H. Schmidt-Böcking, Influence of ionized electrons on heavy nuclei angular differential scattering cross section. J. Phys. B-At. Mol. Opt. **22**, 627 (1989)

44. R. Dörner, J. Ullrich, R.E. Olson, H. Schmidt-Böcking, Three-body interactions in proton-helium angular scattering. Phys. Rev. Lett. **63**, 147 (1989)

45. C.E. Gonzales Lepra, M. Breining, J. Burgdörfer, R. DeSerio, S.B. Elston, J.P. Gibbons, H.P. Hülskötter, L. Liljeby, R.T. Short, C.R. Vane, Nucl. Instr. Meth. B24/25 (1987), p. 316

46. J.C. Levin, R.T. Short, C.S. O., H. Cederquist, S.B. Elston, J.P. Gibbons, I.A. Sellin, H. Schmidt-Böcking, Steep dependence of recoil-ion energy on coincident projectile and target ionization in swift ion-atom collisions. Phys. Rev. **A36**, 1649 (1987)

47. I.A. Sellin, J.C. Levin, O.C.-S., H. Cederquist, S.B. Elston, R.T. Short, H. Schmidt-Böcking, Cold highly ionized ions: comparison of energies of recoil ions produced by heavy ions and by synchrotron radiation x-rays. Physica Scripta **T22**, 178–182 (1988)

48. R. Ali, V. Frohne, C.L. Cocke, M. Stöckli, S. Cheng, M.L.A. Raphaelian, Phys. Rev. Lett. **69**, 2491 (1992)

49. V. Frohne, S. Cheng, R. Ali, M. Raphaeilien, C.L. Cocke, R.E. Olson, Phys. Rev. Lett. **71**, 696 (1993)

50. W.C. Wiley, I.H. McLaren, Time-of-flight mass spectrometer with improved resolution. Rev. Sci. Instrum. **26**, 1150 (1955)

51. R. Dörner, J. Ullrich, O. Jagutzki, S. Lencinas, A. Gensmantel, H. Schmidt-Böcking, in Electronic and Atomic Collisions, ed. by W.R. MacGillivray, I.E. McCarthy, and M.C. Standage (Adam Hilger, Bristol, 1992), p. 351

52. V. Mergel, R. Dörner, J. Ullrich, O. Jagutzki, S. Lencinas, S. Nüttgens, L. Spielberger, M. Unverzagt, C.L. Cocke, R. E. Olson, M. Schulz, U. Buck, E. Zanger, W. Theisinger, M. Isser, S. Geis, H. Schmidt- Böcking, State selective scattering angle dependent capture cross sections using cold target recoil ion momentum spectroscopy (COLTRIMS), Phys. Rev. Lett. **74**, 2200 (1995); Diplomarbeit (1994), Institute f. Nucl. Physics, University Frankfurt

53. A. Cassimi, S. Duponchel, X. Flechard, P. Jardin, P. Sortais, D. Hennecart, R.E. Olson, Phys. Rev. Lett. **76**, 20 3679 (1996)

54. J.P. Grandin, D. Hennecart, X. Husson, D. Lecler, I. Lesteven-Vaisse, D. Lisfi, Europhys. Lett. **6**, 683 (1988)

55. P. Jardin, J.P. Grandin, A. Cassimi, J.P. Lemoigne, A. Gosslin, X. Husson, D. Hennecart, A. Lepontre, in 5th. Conference on Atomic Physics of Highly Charged Ions (AIP Proceed. 274) (1990), p. 291

56. P. Jardin, A. Cassimi, J.P. Grandin, H. Rothard, J.P. Lemoigne, A. Gosslin, X. Husson, D. Hennecart, A. Lepontre, Nucl. Instr. Meth. **B 107** (1996), p. 41

57. Dominique Akoury. Photodoppelionisation von molekularem Wasserstoff bei hohen Photonenenergien, Diplomarbeit (2008) https://www.atom.uni-frankfurt.de/publications/

58. M.S. Schöffler, Grundzustandskorrelationen und dynamische Prozesse untersucht in Ion-Helium-Stößen, Dissertation, Universität Frankfurt, and private communication, 2006

59. N.V. Federenko, V.V. Afrosimov, Sov. Phys.-Tech. Phys. **1**, 1872 (1956) 1956

60. L.J. Puckett, D.W. Martin, Phys. Rev. A **5**, 1432 (1976)

61. W. Steckelmacher, R. Strong, M.W. Lucas, J. Phys. **B11** (1978), 1553; W. Steckelmacher, R. Strong, M.W. Lucas, A simple atomic or molecular beam as target for ion-atom collision studies. J. Phys. D: Appl. Phys. **11**, 1553 (1978)

62. B. Fastrup, Inelastic Energy-Loss Measurements in Single Collisions, in Methods of Experimental Physics, vol. 17. (Academic Press, 1980), p. 149

63. X. Flechard, C. Harel, H. Jouin, B. Pons, L. Adoui, F. Freemont, A. Cassimi, D. Hennecart, J. Phys. B. **34**, 2759 (2001)

64. Th Weber, Kh Khayyat, R. Dörner, V.D. Rodriguez, V. Mergel, O. Jagutzki, L. Schmidt, K.A. Müller, F. Afaneh, A. Gonzales, H. Schmidt-Böcking, Abrupt rise of the longitudinal recoil ion momentum distribution for ionizing collisions. Phys. Rev. Lett. **86**, 224 (2001)

65. D. Fischer, B. Feuerstein, R.D. Dubois, R. Moshammer, J.R. Crespo Lopèz-Urrutia, I. Draganic, H. Lörch, A.N. Perumal, J. Ullrich, J. Phys. B **35**, 1369 (2002); H.K. Kim, M.S.

Schöffler, S. Houamer, O. Chuluunbaatar, J. N. Titze, L.Ph.H. Schmidt, T. Jahnke, H. Schmidt-Böcking, A. Galstyan, Yu.V. Popov, R. Dörner, Electron transfer in fast proton-helium collisions. Phys. Rev. A, **85**, 022707 (2012)

66. R. Dörner, V. Mergel, R. Ali, U. Buck, C.L. Cocke, K. Froschauer, O. Jagutzki, S. Lencinas, W.E. Meyerhof, S. Nüttgens, R.E. Olson, H. Schmidt-Böcking, L. Spielberger, K. Tökesi, J. Ullrich, M. Unverzagt, W. Wu, Electron-electron interaction in projectile ionization investigated by high resolution recoil ion momentum spectroscopy. Phys. Rev. Lett. **72**, 3166 (1994)

67. W. Wu, K.L. Wong, R. Ali, C.Y. Chen, C.L. Cocke, V. Frohne, J.P. Giese, M. Raphaelian, B. Walch, R. Dörner, V. Mergel, H. Schmidt-Böcking, W.E. Meyerhof, Experimental separation of electron-electron and electron-nuclear contributions to ionisation of fast hydrogenlike ions colliding with He. Phys. Rev. Lett. **72**, 3170 (1994); W. Wu, K.L. Wong, E.C. Montenegro, R. Ali, C.Y. Chen, C.L. Cocke, R. Dörner, V. Frohne, J.P. Giese, V. Mergel, W.E. Meyerhof, M. Raphaelian, H. Schmidt-Böcking, B. Walch, Electron-electron interaction in the ionization of O^{7+} by He. Phys. Rev. A **55**, 2771 (1997)

68. R. Moshammer, J. Ullrich, M. Unverzagt, W. Schmidt, P. Jardin, R.E. Olson, R. Mann, R. Dörner, V. Mergel, U. Buck, H. Schmidt-Böcking, Low-energy electrons and their dynamical correlation with the recoil-ions for single ionization of helium by fast, heavy-ion impact. Phys. Rev. Lett. **73**, 3371 (1994)

69. R. Moshammer private communication, GSI report (1997)

70. J. Ullrich, R. Dörner, V. Mergel, O. Jagutzki, L. Spielberger, H. Schmidt-Böcking, Cold-target recoil-ion momentum-spectroscopy: first results and future perspectives of a novel high resolution technique for the investigation of collision induced many-particle reactions. Comments Atomic Mol. Phys. **30**, 285 (1994)

71. M. Unverzagt, R. Moshammer, W. Schmitt, R. E. Olson, P. Jardin, V. Mergel, H. Schmidt-Böcking, Collective behavior of electrons emitted in multiply ionizing of 5.9 MeV/m U^{65+} with Ne. Phys. Rev. Lett. **76**, 1043 (1996)

72. R. Moshammer, J. Ullrich, H. Kollmus, W. Schmitt, M. Unverzagt, O. Jagutzki, V. Mergel, H. Schmidt-Böcking, R. Mann, C. Woods, R.E. Olson, Double ionization of helium and neon for fast heavy-ion impact: correlated motion of electrons from bound in continuum states. Phys. Rev. Lett. **77**, 1242 (1996)

73. R. Moshammer, J. Ullrich, H. Kollmus, W. Schmitt, M. Unverzagt, H. Schmidt-Böcking, C.J. Wood, R.E. Olson, Complete momentum balance for single ionization of helium by fast ion impact: experiment. Phys. Rev. A **56**, 1351 (1997)

74. R. Moshammer, W. Schmitt, J. Ullrich, H. Kollmus, A. Cassimi, R. Dörner, O. Jagutzki, R. Mann, R.E. Olson, H.T. Prinz, H. Schmidt-Böcking, L. Spielberger, Ionization of helium in the attosecond equivalent light pulse of 1 GeV/Nucleon U^{92+} projectiles. Phys. Rev. Lett. **79**, 3621 (1997)

75. M. Schulz, R. Moshammer, D. Fischer, H. Kollmus, D.H. Madison, S. Jones, J. Ullrich, Three-dimensional imaging of atomic four-body processes. Nature **422**, 48–50 (2003)

76. R. Dörner, T. Vogt, V. Mergel, H. Khemliche, S. Kravis, C.L. Cocke, J. Ullrich, M. Unverzagt, L. Spielberger, M. Damrau, O. Jagutzki, I. Ali, B. Weaver, K. Ullmann, C.C. Hsu, M. Jung, E.P. Kanter, B. Sonntag, M.H. Prior, E. Rotenberg, J. Denlinger, T. Warwick, S.T. Manson, H. Schmidt-Böcking, Ratio of cross sections for double to single ionization of He by 85-400 eV photons. Phys. Rev. Lett. **76**, 2654 (1996)

77. H. Schmidt-Böcking, C.L. Cocke, R. Dörner, O. Jagutzki, T. Kambara, V. Mergel, R. Moshammer, M.H. Prior, L. Spielberger, W. Schmitt, K. Ullmann, M. Unverzagt, J. Ullrich, W. Wu, in Accelerator-based atomic physics techniques and application, ed. by S.M. Shafroth, J.C. Austin (AIP, Woodbury, New York, 1996), pp. 723–745

78. L. Spielberger, O. Jagutzki, R. Dörner, J. Ullrich, U. Meyer, V. Mergel, M. Unverzagt, M. Damrau, T. Vogt, I. Ali, Kh. Khayyat, D. Bahr, H. G. Schmidt, R. Frahm, H. Schmidt-Böcking, Separation of photoabsorption and compton scattering contributions to He single and double ionization. Phys. Rev. Lett. **74**, 4615 (1995); L. Spielberger, O. Jagutzki, B. Krässig, U. Meyer, Kh. Khayyat, V. Mergel, Th. Tschentscher, Th. Buslaps, H. Bräuning, R. Dörner, T.

Vogt, M. Achler, J. Ullrich, D.S. Gemmell, H. Schmidt-Böcking, Double and single ionization of helium by 58-keV X-Rays. Phys. Rev. Lett. **76**, 4685 (1996)

79. O. Schwarzkopf, B. Krässig, J. Elminger, V. Schmidt, Phys. Rev. Lett. **70**, 3008 (1993)
80. A. Huetz, P. Laplanque, L. Andric, P. Selles, J. Mazeau, J. Phys. B: At. Mol. Opt. Phys. **27**, L13 (1994)
81. U. Becker, D. Shirley, in: VUV and Soft X-ray Photoionization (Chapter 5). (Springer, 2012)
82. R. Dörner, J. Feagin, C.L. Cocke, H. Bräuning, O. Jagutzki, M. Jung, E.P. Kanter, H. Khemliche, S. Kravis, V. Mergel, M.H. Prior, H. Schmidt-Böcking, L. Spielberger, J. Ullrich, M. Unverzagt, T. Vogt, Fully differential cross sections for double photoionization of He measured by recoil ion momentum spectroscopy. Phys. Rev. Lett. **77**, 1024 (1996)
83. R. Dörner, H. Bräuning, O. Jagutzki, V. Mergel, M. Achler, R. Moshammer, J.M. Feagin, T. Osipov, A. Bräuning-Demian, L. Spielberger, J.H. McGuire, M.H. Prior, N. Berrah, J.D. Bozek, C.L. Cocke, H. Schmidt-Böcking, Double photoionization of spatially aligned D_2. Phys. Rev. Lett. **81**, 5776 (1998)
84. V. Mergel, M. Achler, R. Dörner, Kh. Khayyat, T. Kambara, Y. Awaya, V. Zoran, B. Nyström, L. Spielberger, J.H. McGuire, J. Feagin, J. Berakdar, Y. Azuma H. Schmidt-Böcking, Helicity dependence of the photon-induced three-body coulomb fragmentation of helium investigated by COLTRIMS. Phys. Rev. Lett. **80**, 5301 (1998)
85. R. Dörner, H. Khemliche, M.H. Prior, C.L. Cocke, J.A. Gary, R.E. Olson, V. Mergel, J. Ullrich, H. Schmidt-Böcking, Imaging of saddle point electron emission in slow p-He collisions. Phys. Rev. Lett. **77**, 4520 (1996)
86. E. Horsdal-Pedersen, C.L. Cocke, M. Stöckli, Phys. Rev. Lett. **50**, 1910 (1983)
87. H. Vogt, R. Schuch, E. Justiniano, M. Schulz, W. Schwab, Phys. Rev. Lett. **57**, 2250 (1986)
88. L.H. Thomas, Proc. R. Soc. London A **114**, 561 (1927)
89. E. Horsdal-Pedersen, B. Jensen, K.O. Nielsen, Phys. Rev. Lett. **57**, 1414 (1986)
90. R. Shakeshaft, L. Spruch, Rev.Mod.Phys. **51**, 369 (1979)
91. V. Mergel, R. Dörner, M. Achler, Kh. Khayyat, S. Lencinas, J. Euler, O. Jagutzki, S. Nüttgens, M. Unverzagt, L. Spielberger, W. Wu, R. Ali, J. Ullrich, H. Cederquist, A. Salin, C.J. Wood, R.E. Olson, Dz. Belkic, C.L. Cocke, H. Schmidt-Böcking, Intra-atomic electron-electron scattering in p-He collisions (Thomas process) investigated by cold target recoil ion momentum spectroscopy. Phys. Rev. Lett. **79**, 387 (1997); V. Mergel, Ph.D. thesis, University Frankfurt. ISBN 3-8265-2067-X (1996) unpublished
92. V. Mergel, R. Dörner, Kh Khayyat, M. Achler, T. Weber, H. Schmidt-Böcking, H.J. Lüdde, Strong correlations in the He ground state momentum wave function—observed in the fully differential momentum distributions for the four particle p + He transfer ionization process. Phys. Rev. Lett. **86**, 2257 (2001)
93. M. Schöffler, A.L. Godunov, C.T. Whelan, H.R.J. Walters, V.S. Schipakov, V. Mergel, R. Dörner, O. Jagutzki, L.Ph.H Schmidt, J. Titze, E. Weigold, H. Schmidt-Böcking, Revealing the effect of angular correlation in the ground-state He wavefunction: a coincidence study of the transfer ionization process. J. Phys. B-At. Mol. Opt. **38**, L123 (2005)
94. M.S. Schöffler, Grundzustandskorrelationen und dynamische Prozesse untersucht in Ion-Helium-Stößen, Dissertation (2006); M.S. Schöffler, O. Chuluunbaatar, Yu. V. Popov, S. Houamer, J. Titze, T. Jahnke, L. Ph. H. Schmidt, O. Jagutzki, A.G. Galstyan, A.A. Gusev, Transfer ionization and its sensitivity to the ground-state wave function. Phys. Rev. A, **87**, 032715 (2013); M.S. Schöffler, O. Chuluunbaatar, S. Houamer, A. Galstyan, J.N. Titze, L.Ph.H. Schmidt, T. Jahnke, H. Schmidt-Böcking, R. Dörner, Yu.V. Popov, A.A. Gusev, C. Dal Cappello, Two-dimensional electron-momentum distributions for transfer ionization in fast proton-helium collisions. Phys. Rev. A, **88**, 042710 (2013)
95. A.L. Godunov, C.T. Whelan, H.R.J. Walters, V.S. Schipakov, M. Schöffler, V. Mergel, R. Dörner, O. Jagutzki, LPhH Schmidt, J. Titze, H. Schmidt-Böcking, Transfer ionization process $p + He \rightarrow H + He^{2+} + e$ with the ejected electron detected in the plane perpendicular to the incident beam direction. Phys. Rev. A **71**, 052712 (2005)
96. P.B. Corkum, Phys. Rev. Lett. **71**, 1994 (1993); K.C. Kulander, J. Cooper, K.J. Schafer, Phys. Rev. A **51**, 561 (1995)

97. R. Moshammer, B. Feuerstein, W. Schmitt, A. Dorn, C. D. Schroeter, J. Ullrich, H. Rottke, C. Trump, M: Wittmann, G. Korn, K. Hoffmann, W. Sandner et al. Momentum distribution of Ne ions created by an intense utrashort laser pulse, Phys. Rev. Lett. 84, 447-450 (2000); R. Moshammer, J. Ullrich, B. Feuerstein, D. Fischer, A. Dorn, C.D. Schröter, J.R. Crespo López-Urrutia, C. Höhr, H. Rottke, C. Trump, M. Wittmann, G. Korn, W. Sandner, Rescattering of ultra-low energy electrons for single ionization of Ne in the tunneling regime. Phys. Rev. Lett. 91, 113002 (2003)

98. Th. Weber, M. Weckenbrock, A. Staudte, L. Spielberger, O. Jagutzki, V. Mergel, F. Afaneh, G. Urbasch, M. Vollmer, H. Giessen, R. Dörner, Recoil-ion momentum distributions for single and double ionization of helium in strong laser fields. Phys. Rev. Lett. 84, 443 (2000); Th. Weber, H. Giessen, M. Weckenbrock, G. Urbasch, A. Staudte, L. Spielberger, O. Jagutzki, V. Mergel, M. Vollmer, R. Dörner, Correlated electron emission in multiphoton double ionization. Nature, 405, 658 (2000)

99. R. Moshammer, J. Ullrich, B. Feuerstein, D. Fischer, A. Dorn, C.D. Schröter, J.R. Crespo López-Urrutia, C. Höhr, H. Rottke, C. Trump, M. Wittmann, G. Korn, W. Sandner, Strongly directed electron emission in non-sequential double ionization of Ne by intense Laser pulses, Journal of Physics B-Atomic, Molecular and Optical Physics 36 (2003) L113-119; M. Weckenbrock, D. Zeidler, A. Staudte, Th. Weber, M. Schöffler, M. Meckel, S. Kammer, M. Smolarski, O. Jagutzki, V.R. Bhardwaj, D.M. Rayner, D.M. Villeneuve, P.B. Corkum, R. Dörner, Fully differential rates for femtosecond multiphoton double ionization of neon. Phys. Rev. Lett. 92, 213002 (2004)

100. J. Burgdörfer, C. Lemmel, X. Tong, Invited lecture at ICPEAC 2019. arXiv:2001.02900v1 [quant-ph] 9 Jan 2020

101. D. Zeidler, A. Staudte, A.B. Bardon, D.M. Villeneuve, R. Dörner, P.B. Corkum, Controlling attosecond double ionization dynamics via molecular alignment. Phys. Rev. Lett. 95, 203003 (2005)

102. P. Eckle, A. Pfeiffer, C. Cirelli, A. Staudte, R. Dörner, H.G. Muller, M. Büttiker, U. Keller, Attosecond ionization and tunneling delay time measurements. Science 322, 1525 (2008)

103. M. Kress, T. Löffler, M.D. Thomson, R. Dörner, H. Gimpel, K. Zrost, T. Ergler, R. Moshammer, U. Morgner, J. Ullrich, H.G. Roskos, Determination of the carrier-envelope phase of few-cycle laser pulses with terahertz-emission spectroscopy. Nat. Phys. 2, 327 (2006)

104. M. Dürr, A. Dorn, J. Ullrich, S.P. Cao, A. Czasch, A.S. Kheifets, J.R. Götz, J.S. Briggs, (e,3e) on helium at low impact energy: the strongly correlated three-electron continuum. Phys. Rev. Lett. 98, 193201 (2007)

105. R. Moshammer, Y.H. Jiang, L. Foucar, A. Rudenko, Th Ergler, C.D. Schröter, S. Lüdemann, K. Zrost, D. Fischer, J. Titze, T. Jahnke, M. Schöffler, T. Weber, R. Dörner, T.J.M. Zouros, A. Dorn, T. Ferger, K.U. Kühnel, S. Düsterer, R. Treusch, P. Radcliffe, E. Plönjes, J. Ullrich, Few-photon multiple ionization of Ne and Ar by strong free-electron-laser pulses. Phys. Rev. Lett. 98, 203001 (2007)

106. A. Staudte, C. Ruiz, M. Schöffler, S. Schössler, D. Zeidler, Th Weber, M. Meckel, D.M. Villeneuve, P.B. Corkum, A. Becker, R. Dörner, Binary and recoil collisions in strong field double ionization of helium. Phys. Rev. Lett. 99, 263002 (2007)

107. X.-J. Liu, H. Fukuzawa, T. Teranishi, A. De Fanis, M. Takahashi, H. Yoshida, A. Cassimi, A. Czasch, L. Schmidt, R. Dörner, K. Wang, B. Zimmermann, V. McKoy, I. Koyano, N. Saito, K. Ueda, Breakdown of the two-step model in K-shell photoemission and subsequent decay probed by the molecular-frame photoelectron angular distributions of CO_2. Phys. Rev. Lett. 101, 083001 (2008)

108. A. Rudenko, L. Foucar, M. Kurka, Th Ergler, K.U. Kühnel, Y.H. Jiang, A. Voitkiv, B. Najjari, A. Kheifets, S. Lüdemann, T. Havermeier, M. Smolarski, S. Schössler, K. Cole, M. Schöffler, R. Dörner, S. Düsterer, W. Li, B. Keitel, R. Treusch, M. Gensch, C.D. Schröter, R. Moshammer, J. Ullrich, Recoil-ion momentum distributions for two-photon double ionization of He and Ne by 44 eV free-electron laser radiation. Phys. Rev. Lett. 101, 073003 (2008)

109. M. Meckel, D. Comtois, D. Zeidler, A. Staudte, D. Pavicic, H.C. Bandulet, H. Pépin, J.C. Kieffer, R. Dörner, D.M. Villeneuve, P.B. Corkum, Laser-induced electron tunneling and diffraction. Science 320, 1478 (2008)

110. H. Sann, T. Havermeier, C. Müller, H.-K. Kim, F. Trinter, M. Waitz, J. Voigtsberger, F. Sturm, T. Bauer, R. Wallauer, D. Schneider, M. Weller, C. Goihl, J. Tross, K. Cole, J. Wu, M.S. Schöffler, H. Schmidt-Böcking, T. Jahnke, M. Simon, R. Dörner, Imaging the temporal evolution of molecular orbitals during ultrafast dissociation. Phys. Rev. Lett. **117**, 243002 (2016)

111. H. Kang, K. Henrichs, M. Kunitski, Y. Wang, X. Hao, K. Fehre, A. Czasch, S. Eckart, L. Ph. H Schmidt, M. Schöffler, T. Jahnke, X. Liu, R. Dörner, Timing recollision in nonsequential double ionization by intense elliptically polarized laser pulses. Phys. Rev. Lett. **120**, 223204 (2018)

112. Y.H. Jiang, A. Rudenko, M. Kurka, K.U. Kühnel, Th Ergler, L. Foucar, M. Schöffler, S. Schössler, T. Havermeier, M. Smolarski, K. Cole, R. Dörner, S. Düsterer, R. Treusch, M. Gensch, C.D. Schröter, R. Moshammer, J. Ullrich, Few-photon multiple ionization of N2 by extreme ultraviolet free-electron laser radiation. Phys. Rev. Lett. **102**, 123002 (2009)

113. E. Shigemasa, J. Electron Spectrosc. Relat. Phenom. **88–91**, 9 (1998)

114. F. Heiser, O. Gessner, J. Viefhaus, K. Wieliczek, R. Hentges, U. Becker, Phys. Rev. Lett. **79**, 2435 (1997)

115. https://www.researchgate.net/scientific-contributions/31908446_P_M_Guyon ZEKE

116. A. Landers, Th Weber, I. Ali, A. Cassimi, M. Hattass, O. Jagutzki, A. Nauert, T. Osipov, A. Staudte, M.H. Prior, H. Schmidt-Böcking, C.L. Cocke, R. Dörner, Photoelectron diffraction mapping: molecules illuminated from within. Phys. Rev. Lett. **87**, 013002 (2001)

117. A. Lafosse, M. Lebech, J.C. Brenot, P.M. Guyon, O. Jagutzki, L. Spielberger, M. Vervloet, J.C. Houver, D. Dowek, Vector correlations in dissociative photoionization of diatomic molecules in the VUV range: strong anisotropies in electron emission from spatially oriented NO molecule. Phys. Rev. Lett. **84**, 5987 (2000)

118. T. Jahnke, Th. Weber, A.L. Landers, A. Knapp, S. Schössler, J. Nickles, S. Kammer, O. Jagutzki, L. Schmidt, A. Czasch, T. Osipov, E. Arenholz, A.T. Young, R. Diez Muino, D. Rolles, F.J. Garcia de Abajo, C.S. Fadley, M.A. Van Hove, S.K. Semenov, N.A. Cherepkov, J. Rösch, M.H. Prior, H. Schmidt-Böcking, C.L. Cocke, R. Dörner, Circular dichroism in K-shell ionization from fixed-in-space CO and N$_2$ molecules. Phys. Rev. Lett. **88**, 073002 (2002)

119. K. Ueda, A. De Fanis, N. Saito, M. Machida, K. Kubozuka, H. Chiba, Y. Muramatu, Y. Sato, A. Czasch, O. Jaguzki, R. Dörner, A. Cassimi, M. Kitajima, T. Furuta, H. Tanaka, S.L. Sorensen, K. Okada, S. Tanimoto, K. Ikejiri, Y. Tamenori, H. Ohashi, I. Koyano, Nuclear motion and symmetry breaking of the B 1S-exited BF3 molecule. Chem. Phys. **289**, 135 (2003)

120. T. Jahnke, L. Foucar, J. Titze, R. Wallauer, T. Osipov, E. P. Benis, A. Alnaser, O. Jagutzki, W. Arnold, S. K. Semenov, N. A. Cherepkov, L. Ph. H. Schmidt, A. Czasch, A. Staudte, M. Schöffler, C. L. Cocke, M.H. Prior, H. Schmidt-Böcking, R. Dörner, Vibrationally resolved K-shell photoionization of CO with circularly polarized light. Phys. Rev. Lett. **93**, 083002 (2004)

121. F. Trinter, L.Ph.H. Schmidt, T. Jahnke, M.S. Schöffler, O. Jagutzki, A. Czasch, J. Lower, T.A. Isaev, R. Berger, A.L. Landers, Th. Weber, R. Dörner, H. Schmidt-Böcking, Multi-fragment vector correlation imaging. A search for hidden dynamical symmetries in many-particle molecular fragmentation processes. Mol. Phys. **110**, 1863 (2012)

122. Th. Weber, M. Weckenbrock, M. Balser, L. Schmidt, O. Jagutzki, W. Arnold, O. Hohn, M. Schöffler, E. Arenholz, T. Young, T. Osipov, L. Foucar, A. De Fanis, R. Diez Muino, H. Schmidt-Böcking, C.L. Cocke, M.H. Prior, R. Dörner, Auger electron emission from fixed-in-space CO. Phys. Rev. Lett. **90**, 153003 (2003); T. Osipov, C.L. Cocke, M.H. Prior, A. Landers, T. Weber, O. Jagutzki, L. Ph. H. Schmidt, H. Schmidt-Böcking, R. Dörner, Photoelectron-photo-ion momentum spectroscopy as a clock for chemical rearrangements: isomerization of the dication of acetylene to the vinylidene configuration. Phys. Rev. Lett. **90**, 233002 (2003); T. Weber, A.O. Czasch, O. Jagutzki, A.K. Müller, V. Mergel, A. Kheifets, E. Rotenberg, G. Meigs, M.H. Prior, S. Daveau, A. Landers, C.L. Cocke, T. Osipov, R. Diez Muino, H. Schmidt-Böcking, R. Dörner, Complete photo-fragmentation of the deuterium molecule. Nature, **431**, 437 (2004); Th. Weber, A. Czasch, O. Jagutzki, A. Müller, V. Mergel, A. Kheifets, J. Feagin, E. Rotenberg, G. Meigs, M.H. Prior, S. Daveau, A.L. Landers, C.L. Cocke, T. Osipov, H.

Schmidt-Böcking, R. Dörner, Fully differential cross sections for photo-double-ionization of D2. Phys. Rev. Lett. **92**, 163001 (2004); F. Martín, J. Fernández, T. Havermeier, L. Foucar, Th. Weber, K. Kreidi, M. Schöffler, L. Schmidt, T. Jahnke, O. Jagutzki, A. Czasch, E.P. Benis, T. Osipov, A.L. Landers, A. Belkacem, M.H. Prior, H. Schmidt-Böcking, C.L. Cocke, R. Dörner, Single photon induced symmetry breaking of H2 dissociation. Science, **315**, 629 (2007); J.B. Williams, C.S. Trevisan, M.S. Schöffler, T. Jahnke, I. Bocharova, H. Kim, B. Ulrich, R. Wallauer, F. Sturm, T.N. Rescigno, A. Belkacem, R. Dörner, Th. Weber, C.W. McCurdy, A.L. Landers, Imaging polyatomic molecules in three dimensions using molecular frame photoelectron angular distributions. Phys. Rev. Lett. **108**, 233002 (2012)

123. L.S. Cederbaum, J. Zobeley, F. Tarantelli, Phys. Rev. Lett. **79**, 4778 (1997)
124. S. Marburger, O. Kugeler, U. Hergenhahn, T. Möller, Phys. Rev. Lett. **90**, 203401 (2003)
125. T. Jahnke, A. Czasch, M.S. Schöffler, S. Schössler, A. Knapp, M. Käsz, J. Titze, C. Wimmer, K. Kreidi, R.E. Grisenti, A. Staudte, O. Jagutzki, U. Hergenhahn, H. Schmidt-Böcking, R. Dörner, Experimental observation of interatomic coulombic decay in neon dimers. Phys. Rev. Lett. **93**, 163401 (2004)
126. T. Jahnke, A. Czasch, M. Schöffler, S. Schössler, M. Käsz, J. Titze, K. Kreidi, R. E. Grisenti, A. Staudte, O. Jagutzki, L. Ph. H. Schmidt, Th Weber, H. Schmidt-Böcking, K. Ueda, R. Dörner, Experimental separation of virtual photon exchange and electron transfer in interatomic coulombic decay of neon dimers. Phys. Rev. Lett. **99**, 153401 (2007)
127. K. Kreidi, Ph. V Demekhin, T. Jahnke, Th Weber, T. Havermeier, X.-J. Liu, Y. Morisita, S. Schössler, L. Ph. H. Schmidt, M. Schöffler, M. Odenweller, N. Neumann, L. Foucar, J. Titze, B. Ulrich, F. Sturm, C. Stuck, R. Wallauer, S. Voss, I. Lauter, H. K. Kim, M. Rudloff, H. Fukuzawa, G. Prümper, N. Saito, K. Ueda, A. Czasch, O. Jagutzki, H. Schmidt-Böcking, S. Scheit, L. S. Cederbaum, R. Dörner, Photo- and auger-electron recoil induced dynamics of interatomic coulombic decay. Phys. Rev. Lett. **103**, 033001 (2009)
128. N. Sisourat, H. Sann, N.V. Kryzhevoi, P. Kolorenč, T. Havermeier, F. Sturm, T. Jahnke, H.K. Kim, R. Dörner, L.S. Cederbaum, Interatomic electronic decay driven by nuclear motion. Phys. Rev. Lett. **105**, 173401 (2010)
129. T. Havermeier, T. Jahnke, K. Kreidi, R. Wallauer, S. Voss, M. Schöffler, S. Schössler, L. Foucar, N. Neumann, J. Titze, H. Sann, M. Kühnel, J. Voigtsberger, J.H. Morilla, W. Schöllkopf, H. Schmidt-Böcking, R.E. Grisenti, R. Dörner, Interatomic coulombic decay following photoionization of the helium dimer: observation of vibrational structure. Phys. Rev. Lett. **104**, 133401 (2010)
130. J. Titze, M. S. Schöffler, H.-K. Kim, F. Trinter, M. Waitz, J. Voigtsberger, N. Neumann, B. Ulrich, K. Kreidi, R. Wallauer, M. Odenweller, T. Havermeier, S. Schössler, M. Meckel, L. Foucar, T. Jahnke, A. Czasch, L. Ph. H. Schmidt, O. Jagutzki, R. E. Grisenti, H. Schmidt-Böcking, H. J. Lüdde, R. Dörner, Ionization dynamics of helium dimers in fast collisions with He^{2+}. Phys. Rev. Lett. **106**, 033201 (2011)
131. F. Trinter, J.B. Williams, M. Weller, M. Waitz, M. Pitzer, J. Voigtsberger, C. Schober, G. Kastirke, C. Müller, C. Goihl, P. Burzynski, F. Wiegandt, R. Wallauer, A. Kalinin, LPhH Schmidt, M.S. Schöffler, Y.-C. Chiang, K. Gokhberg, T. Jahnke, R. Dörner, Vibrationally resolved decay width of interatomic coulombic decay in HeNe. Phys. Rev. Lett. **111**, 233004 (2013)
132. T. Jahnke, U. Hergenhahn, B. Winter, R. Dörner, U. Frühling, Ph. V. Demekhin, K. Gokhberg, L. S. Cederbaum, A. Ehresmann, A. Knie, and A. Dreuw, Interatomic and Intermolecular Coulombic Decay, Chem. Rev. 120, 11295–11369 (2020)
133. B. Najjari, A.B. Voitkiv, C. Müller, Two-center resonant photoionization. Phys. Rev. Lett. **105**, 153002 (2010)
134. M.S. Schöffler, J. Titze, N. Petridis, T. Jahnke, K. Cole, LPhH Schmidt, A. Czasch, D. Akoury, O. Jagutzki, J.B. Williams, N.A. Cherepkov, S.K. Semenov, C.W. McCurdy, T.N. Rescigno, C.L. Cocke, T. Osipov, S. Lee, M.H. Prior, A. Belkacem, A.L. Landers, H. Schmidt-Böcking, Th Weber, R. Dörner, Ultrafast probing of core hole localization in N$_2$. Science **320**, 920 (2008)

135. P.S. Bagus, H.F. Schäfer, J. Chem. Phys. **56**, 224 (1972); L.C. Snyder, J. Chem. Phys. **55**, 95 (1971); J.F. Stanton, J. Gauss, R.J. Bartlett, J. Chem. Phys. **97**, 5554 (1992); J.A. Kintop, W.V.M. Machando, L.G. Ferreira, Phys. Rev. A **43**, 3348 (1991); D. Dill, S. Wallace, Phys. Rev. Lett. **41**, 1230 (1978); L.S. Cederbaum, W. Domcke, J. Chem. Phys. **66**, 5084 (1977)
136. V. Efimov, Phys. Lett. B **33**, 563–564 (1970)
137. R. E. Grisenti, W. Schöllkopf, J. P. Toennies, J. R. Manson, T. A. Savas, and Henry I. Smith, He-atom diffraction from nanostructure transmission gratings: The role of imperfections, Phys. Rev. A 61, 033608 (2000); W. Schöllkopf, J. P. Toennies, Nondestructive Mass Selection of Small van der Waals Clusters, Science **266**, 1345–1348 (1994)
138. S. Zeller, M. Kunitski, J. Voigtsberger, A. Kalinin, A. Schottelius, C. Schober, M. Waitz, H. Sann, A. Hartung, T. Bauer, M. Pitzer, F. Trinter, Ch. Goihl, Ch. Janke, Martin Richter, G. Kastirke, M. Weller, A. Czasch, M. Kitzler, M. Braune, R. E. Grisenti, Wieland Schöllkopf, L. Ph. H. Schmidt, M. Schöffler, J. B. Williams, T. Jahnke, R. Dörner, Imaging the He2 quantum halo state using a free electron laser, P. Natl. Acad. Sci. USA **113**, 14651 (2016)
139. M. Kunitski, S. Zeller, J. Voigtsberger, A. Kalinin, L. Ph. H. Schmidt, M. Schöffler, A. Czasch, W. Schöllkopf, R. E. Grisenti, T. Jahnke, D. Blume, R. Dörner, Observation of the Efimov state of the helium trimer. Science **348**, 551 (2015)
140. W. Cencek et al., J. Chem. Phys. **136**, 224303 (2012); D. Blume, C.H. Greene. B.D. Esry, J. Chem. **113**, 2145–2158 (2000); D. Blume, C.H. Greene. B.D. Esry, J. Chem. **141**, 069901 (E) (2014)
141. [https://de.wikipedia.org/wiki/Contergan-Skandal]
142. M. Pitzer, M. Kunitski, A.S. Johnson, T. Jahnke, H. Sann, F. Sturm, L.Ph.H. Schmidt, H. Schmidt-Böcking, R. Dörner, J. Stohner, J. Kiedrowski, M. Reggelin, S. Marquardt, A. Schießer, R. Berger, M.S. Schöffler, Direct determination of absolute molecular stereochemistry in gas phase by Coulomb explosion imaging. Science **341**, 1096 (2013); M. Pitzer, G. Kastirke, P. Burzynski, M. Weller, D. Metz, J. Neff, M. Waitz, F. Trinter, L.Ph.H Schmidt, J.B. Williams, T. Jahnke, H. Schmidt-Böcking, R. Berger, R. Dörner, M. Schöffler, Stereochemical configuration and selective excitation of the chiral molecule halothane. J. Phys. B-At. Mol. Opt. **49**, 234001 (2016); M. Pitzer, G. Kastirke, M. Kunitski, T. Jahnke, T. Bauer, C. Goihl, F. Trinter, C. Schober, K. Henrichs, J. Becht, S. Zeller, H. Gassert, M. Waitz, A. Kuhlins, H. Sann, F. Sturm, F. Wiegandt, R. Wallauer, L.Ph.H. Schmidt, A.S. Johnson, M. Mazenauer, B. Spenger, S. Marquardt, S. Marquardt, H. Schmidt-Böcking, J. Stohner, R. Dörner, M. Schöffler, R. Berger, Absolute configuration from different multifragmentation pathways in light-induced Coulomb explosion imaging. Chem. Phys. Chem. **17**, 2465 (2016); M. Pitzer, How to determine the handedness of single molecules using Coulomb explosion imaging. J. Phys. B-At. Mol. Opt. **50**, 153001 (2017); M. Pitzer, K. Fehre, M. Kunitski, T. Jahnke, L. Ph. H. Schmidt, H. Schmidt-Böcking, R. Dörner, M. Schöffler, Coulomb explosion imaging as a tool to distinguish between stereoisomers. JOVE - J. Vis Exp **126**, e56062 (2017); M. Pitzer, R. Berger, J. Stohner, R. Dörner, M. Schöffler, Investigating absolute stereochemical configuration with Coulomb explosion imaging, Chimia, **72**, 384 (2018). https://doi.org/10. 2533/chimia.2018.384; K. Fehre, S. Eckart, M. Kunitski, M. Pitzer, S. Zeller, C. Janke, D. Trabert, J. Rist, M. Weller, A. Hartung, L.Ph.H. Schmidt, T. Jahnke, R. Berger, R. Dörner, M. S. Schöffler, Enantioselective fragmentation of an achiral molecule in a strong laser field. Sci. Adv., **5**(3) (2019); K. Fehre, S. Eckart, M. Kunitski, M. Pitzer, S. Zeller, C. Janke, D. Trabert, J. Rist, M. Weller, A. Hartung, L.Ph.H. Schmidt, T. Jahnke, R. Berger, A. Dörner, M.S. Schöffler. Enantioselective fragmentation of an achiral molecule in a strong laser field. Sci. Adv. **5**(3) (2019)
143. Z. Vager, R. Naaman, E.P. Kanter, Science **244**, 426–431 (1989)
144. L. Ph. H. Schmidt, T. Jahnke, A. Czasch, M. Schöffler, H. Schmidt-Böcking, R. Dörner, Spatial imaging of the H_2 vibrational wave function at the quantum limit. Phys. Rev. Lett. **108**, 073202 (2012)
145. P. Debye, Göttinger Nachrichten 1916 and A. Sommerfeld, Physikalische Zeitschrift **17**, 491–507 (1916)

146. L.Ph.H. Schmidt, C. Goihl, D. Metz, H. Schmidt-Böcking, R. Dörner, SYu. Ovchinnikov, J.H. Macek, D.R. Schultz, Vortices associated with the wave function of a single electron emitted in slow ion-atom collisions. Phys. Rev. Lett. **112**, 083201 (2014)
147. F. Trinter, J.B. Williams, M. Weller, M. Waitz, M. Pitzer, J. Voigtsberger, C. Schober, G. Kastirke, C. Müller, C. Goihl, P. Burzynski, F. Wiegandt, T. Bauer, R. Wallauer, H. Sann, A. Kalinin, LPhH Schmidt, M. Schöffler, N. Sisourat, T. Jahnke, Evolution of interatomic coulombic decay in the time domain. Phys. Rev. Lett. **111**, 093401 (2013)
148. A. Niehaus. J. Phys. B: Atom. Molec. Phys. **10**, 1977
149. P. Morin, I. Nenner, Phys. Rev. Lett. **56**, 1913 (1986)
150. J. Ullrich, R. Dörner, V. Mergel, O. Jagutzki, L. Spielberger, H. Schmidt-Böcking, Cold-target recoil-ion momentum-spectroscopy: first results and future perspectives of a novel high resolution technique for the investigation of collision induced many-particle reactions. Comments Atomic and Mol. Phys. **30**, 285 (1994)

Part IV
Cold and Controlled Molecules

Chapter 19
STIRAP: A Historical Perspective and Some News

Klaas Bergmann

A very brief outline of what STIRAP is and does is followed by the presentation of the sequence of experiments, which started some 50 years ago, the visions developed and experimental efforts undertaken, that finally led to the development of STIRAP.

1 What Is STIRAP?

Stimulated Raman Adiabatic Passage (STIRAP, [1]) is a process which allows efficient and selective population transfer between discrete states of a quantum system, in its simplest form shown in Fig. 1. Level 1 is initially populated. The goal is to transfer all of that population to level 3. In most cases of interest, a direct one-photon dipole coupling between levels 1 and 3 is not possible. Therefore, one needs to invoke an intermediate level 2, often in a different electronic state. The characteristic and initially surprising feature of STIRAP is that the quantum system needs to be exposed first to the S-laser, which couples initially unpopulated levels. When the intensity of the S-laser is reduced, the intensity of the P-laser, which provides the coupling to the populated level, rises. If the switching-off of the S-laser and the switching-on of the P-laser is properly coordinated and the so-called adiabatic condition is fulfilled [2] nearly 100% of the initial population in level 1 will reach the target level 3 without ever establishing significant population in level 2. The underlying physics is interference of transition amplitudes, which—in the adiabatic limit—prevents population in level 2. Therefore, loss of population during the transfer process through spontaneous emission does not occur or is much reduced.

K. Bergmann (✉)
Fachbereich Physik der Technischen Universität Kaiserslautern, Erwin Schrödinger Str. 49, 67663 Kaiserslautern, Germany
e-mail: bergmann@rhrk.uni-kl.de

© The Author(s) 2021
B. Friedrich and H. Schmidt-Böcking (eds.), *Molecular Beams in Physics and Chemistry*,
https://doi.org/10.1007/978-3-030-63963-1_19

Fig. 1 The upper part shows
a three-level system with the
S- and P-lasers, coupling the
levels 2–3 and 1–2,
respectively. Level 4 stands
for all levels that can also be
reached by radiation from
level 2. The lower part shows
the geometric arrangement
(S-before-P) for population
transfer within particles of a
molecular beam

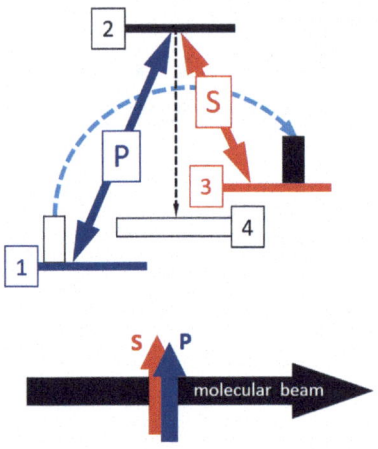

The S-before-P sequence, called "counter-intuitive pulse sequence" in the early days, can be implemented either by suitably delayed laser pulses when applied to molecules in a gas cell (Fig. 2), or by spatially shifting the parallel axes of continuous lasers when population transfer within particles of a molecular beam is to be realized, as shown in the lower part of Fig. 1. It is the directional flow, which guarantees that a given molecule experiences a time variation of the coupling between levels as shown in the upper part of Fig. 2.

Another important feature, namely the robustness of STIRAP, made the scheme popular in many laboratories for applications in a wide and diverse range of quantum systems (see Sect. 6). Robustness means that a small variation of the S- or P-laser intensities or their time-delay does not reduce the transfer efficiency.

The original publication, reporting the main features of STIRAP and its theoretical foundation [1], was followed over the years by a number of review articles, e.g. [3–6]. The wide range of applications is documented in [7].

Fig. 2 The upper part shows
the STIRAP-sequence of
laser interactions with the
quantum system
(S-before-P). The variation
of the Rabi frequencies,
which determine the
coupling strength between
levels, is shown. The lower
part shows the corresponding
flow of population P_x from
level 1 to level 3 (see Fig. 1).
In the adiabatic limit no
population is deposited in
level 2

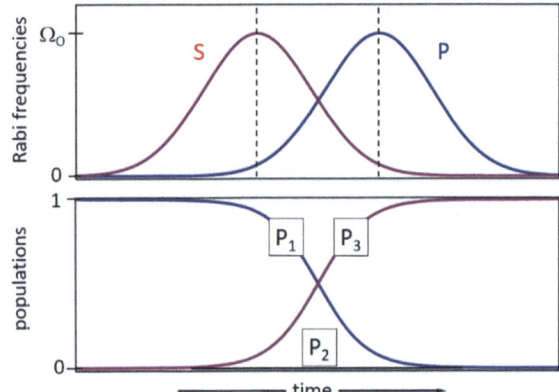

This article does not offer a detailed discussion of the physics of STIRAP. It describes, in the format of a memoir, the background, the vision, the various steps and the systematic plan followed, which finally led to realizing how a complete and robust population transfer between quantum states can be achieved. This work concludes with the presentation of a short list of topics or problems which benefited from the application of STIRAP. Although STIRAP has also been applied to many types of quantum systems, including a polyatomic molecule, the specific discussions and comments that follow relate mainly to diatomic molecules.

2   Background and Motivation

The deep roots of STIRAP reach back in time more than 50 years. The topic of my diploma thesis (submitted in early 1968) was the dynamics of photodissociation of some polyatomic molecules, using a classical pulsed high pressure discharge source [8]. After completion of that work my response to the question whether I wanted to continue this kind of experiments was a determined "no". I stated the reason: such work would very soon be done with lasers. In 1968 lasers were known for only 8 years.

Lasers did indeed play a central role in my PhD thesis, completed in early 1972. That work led to one of the very first applications of lasers to collision dynamics. The topic of the thesis emerged from spectroscopic work in sodium beams done by W. Demtröder while visiting R. N. Zare in Boulder [9]. In my work home-built Argon-ion lasers were used to excite a single rovibronic level (v', j') in the B-state of sodium molecules in a cell with rare gases added. (Here and below I use the traditional convention from spectroscopy: a single prime marks a level in an electronically excited state, while a double prime refers to a level in the electronic ground state.) Atom-molecule collisions induced transfer of population to neighboring rotational levels. That transfer was monitored by observing collision-induced spectral satellite lines. The pressure dependence of the intensity of those lines allowed the determination of rate constants. Of particular interest was the difference between rotational energy transfer to levels $(v', j' + \Delta j)$ and $(v', j' - \Delta j)$. The first paper on this topic [10] appeared in print only a few months after J. Steinfeld had published similar studies for I_2, also involving a laser [11]. Because the transferred energy was small compared to the mean kinetic energy, the observed difference of the rate constants for excitation and deexcitation processes with the same $|\Delta j|$ (called propensity) was unexpected. It was later explained through a detailed analysis of the wave functions involved [12].

While doing Ph.D. work, I learned about the then very popular molecular-beam technique through close contact with students working in a neighboring laboratory. The offer to continue academic work at the University of Kaiserslautern, founded in 1970, triggered the plan to combine molecular beams and lasers in future research. In early 1973, while carefully studying a paper by R. Drullinger and R. N. Zare on optical pumping of molecules [13], in particular their discussion of excitation and relaxation

Fig. 3 The level scheme relevant for optical pumping of molecules in a gas cell showing laser-excitation, spontaneous emission and relaxation via collisions

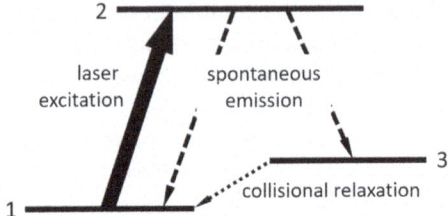

pathways (see Fig. 3), I realized that the relaxation path after laser excitation and spontaneous emission back to the initially pumped level would be missing in the collision-free environment of a molecular beam. Thus the entire population of a specific thermally populated rotational level j''_{pump} could be removed. Controlled by Franck-Condon factors and optical selection rules, only a very small fraction of the laser-excited molecules would return to levels near j''_{pump} by spontaneous emission. This consideration led to the crossed beams arrangement as shown in Fig. 4. Particles scattered under the angle ϑ into the level j''_{probe} were probed by laser-induced fluorescence (see Fig. 5) while the pump laser would periodically switch off the population in level j''_{pump}. Most of the experiments involved levels in $v'' = 0$.

With the pump laser turned off, all thermally populated levels may contribute to the scattering into the probed level. With the pump laser turned on, the contribution from the pumped level would be missing. The difference of the scattering signal with pump laser off and on isolates the scattering rate from the level j''_{pump} into the level $j''_{probe} = j''_{pump} + \Delta j$ ($\Delta j > 0$ and $\Delta j < 0$ possible) under the scattering angle ϑ which is determined by the position of the narrow entry slit of a rotatable detector.

The molecular-beam laboratory for doing such experiments in Kaiserslautern (built after my post-doctoral work in Berkeley with C. B. Moore on laser-induced

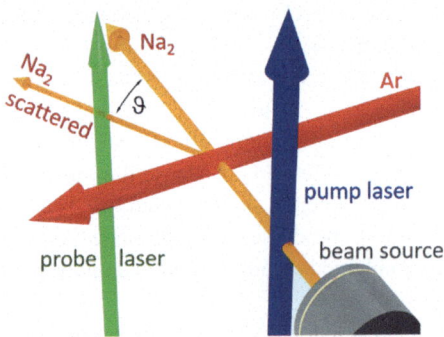

Fig. 4 The crossed beams arrangement for the study of state-to-state angle-resolved inelastic scattering, with a Na$_2$-beam, an Ar-beam, the pump-laser and a laser beam for monitoring the flux of particles scattered under the angle ϑ into the quantum state j''_{probe}. The device for collecting the fluorescence induced by the probe laser is not shown

Fig. 5 The level scheme relevant for the scattering experiment as shown in Fig. 4. For further discussion, see the text

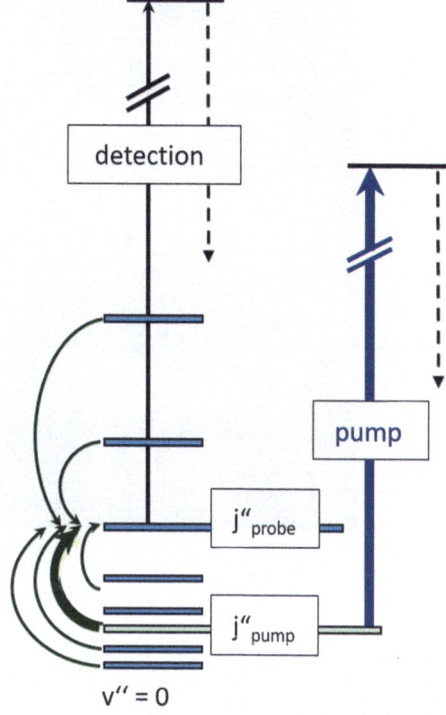

chemistry [14, 15]) had several innovative features. The entire apparatus was designed from the very beginning with the central role of lasers in mind. By design, some components, which were traditionally considered absolutely necessary, were not even included. In particular hot-wire detectors for detecting alkali atoms or molecules were replaced by lasers. Equally relevant: the mechanical flexibility of the detector, required for measuring angular distributions, was provided through the use of single-mode optical fibers in combination with a new design for efficient collection of laser-induced fluorescence [16]. A prerequisite of this work was a careful state-resolved characterization of the molecular beam [17].

Figure 6 shows what is most likely the very first AMO research laboratory with an optical-fiber network implemented, only a few years after Corning had manufactured the first Germanium doped single mode fibers. The photo shown was taken in 1977. Several lasers were connected to a number of experimental stations in different rooms with single-mode optical fibers donated by the fiber-research laboratory of Schott/Mainz. It was the late colleague Walter Heinlein from the electric engineering department of my university who introduced me to the relevant researchers in that laboratory. None of the optical components needed for coupling laser radiation into and out of fibers were commercially available. The photo was first shown in public at the ICPEAC conference 1979 in Kyoto. This photo triggered much more interest than the content of the related scientific presentation, namely the first laser-based

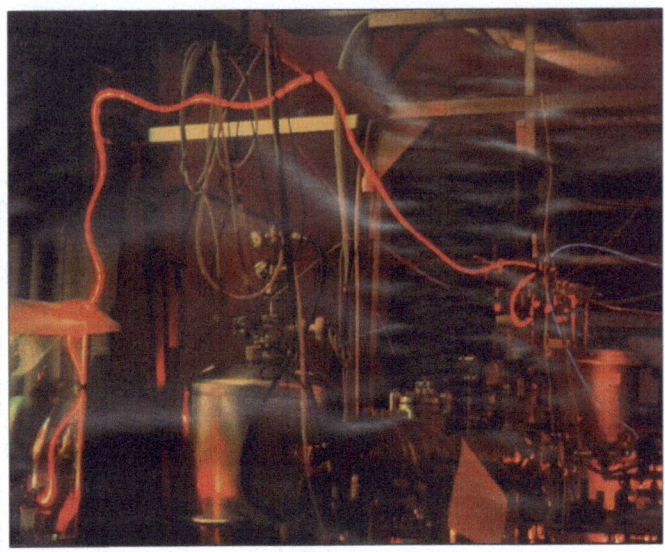

Fig. 6 The photo from 1977 shows what is probably the first laboratory with an optical fiber network installed, for flexibly connecting a number of lasers with various experimental stations

angularly resolved state-to-state energy transfer cross section [18]. The laser-based approach to molecular-beam scattering proved very successful. It led to a series of experiments yielding fully resolved state-to-state differential energy-transfer cross sections, with "rotational rainbows" being a prominent feature, see e.g. [19, 20], including even m-selectivity [21].

3   The Vision and the Challenge

Motivated by the success of the work mentioned in Sect. 2, I was considering in the late 1970s to use in my future work laser-based molecular state selection for laboratory studies of collision processes with relevance to atmospheric chemistry. It was known that chemical reactions and photodissociation processes in the higher atmosphere may lead to highly vibrationally excited molecules. However, little was known about how such vibrational excitation would change reaction-rates. The problem, though, was that the state-selection by optical pumping as described above works only for thermally populated levels. In a molecular-beam environment the levels v'' $\gg 1$ of interest are not populated. In order to use an approach similar to the one of Sect. 2 and shown in Fig. 4, one needs to efficiently and selectively populate a single rotational level in a highly vibrationally excited level. "Efficiency" was needed to realize a sufficiently high flux of excited molecules for scattering studies. "Selectivity" was crucial because reaction rates may sensitively depend on the vibrational level.

Further considerations, which are documented in my Habilitation Thesis submitted in late 1979 [22], led quickly to the conclusion that a direct one- or multi-photon excitation of a $v'' \gg 1$ level in the molecular electronic ground state would not be possible. The transition moments for high-overtone excitation are too small. Extremely strong laser pulses would be needed. At such high laser intensity many detrimental multi-photon excitation and ionization paths would make the approach inefficient. For homonuclear molecules the relevant transition moments are zero anyway.

Thus, it was straightforward to conclude that any efficient transfer scheme must invoke an auxiliary third level, most likely in an electronically excited state. On- or off-resonance Raman scattering, π-pulses or a sequential chirped adiabatic passage process were candidates. However, any process that drives the population through a level in an electronically excited state would suffer from unavoidable loss of population through spontaneous emission. Such processes would not only reduce the transfer efficiency to the target level; even worse, spontaneous emission would spoil selectivity because levels adjacent to the target level would also be populated. Off-resonance Raman scattering would reduce or avoid spontaneous emission, but the optical selection rules would prevent reaching levels $v'' \gg 1$ from thermally populated rotational levels in $v'' = 0$. Neither would the use of π-pulses allow reaching the goal, because the pulse area of the radiative interaction (roughly speaking: the product of the mean Rabi frequency $\Omega = \mu\, E/\hbar$ and the pulse duration, with μ being the transition dipole moment and E the electric field of the laser radiation) needs to be precisely controlled. Such transfer would not be robust. The transfer efficiency would depend very sensitively on small changes of relevant parameters. It would be different for different m-states within the rotational level. A sequence of two π-pulses would therefore allow the transfer of only a small fraction of the population of molecules in a given j" level, also because the condition for efficient transfer would be satisfied only for a tightly restricted number of trajectories of the molecule across the laser beam. A flux of molecules sufficiently high for scattering experiments would not be achievable.

Soon after reaching these conclusions the process of stimulated emission pumping (SEP) was proposed [23], which turned out to be very powerful for some collision dynamics experiments [24]. However, in most cases, the transfer efficiency did not exceed 10%. Thus, the problem was temporarily put aside with the hope that a new idea or a new inspiration would come along.

4 An Intermediate Step: The Molecular Beam Laser

The new inspiration came through discussions with B. Wellegehausen and after reading his papers on optically pumped lasers with alkali dimers in a heat pipe (length of the order of 10 cm) serving as gain medium [25]. In that work it was shown that high gain can be realized for transitions from levels in electronically excited states (populated by laser-pumping from thermally populated levels) to many high lying vibrational levels (see Fig. 7) in the electronic ground state. For many such transitions lasing was observed with a power of the pump laser as low as 1 mW. My quick back-of-the-envelope calculation showed that the gain in a molecular beam about 1 cm downstream from the nozzle would be even larger, despite the small extension of only a few mm. The reason is that the population distribution over low lying rovibronic levels in a supersonic beam is characterized by a temperature on the order of 10 K [17] rather than the ≈750 K in a heat pipe.

That conclusion quickly led to a preliminary design of a cavity around the vacuum chamber supporting the molecular beam. Figure 8 shows the second generation of such a cavity. An essential difference between a molecular-beam laser and a heat pipe relates to the relaxation process of population reaching the lower laser level (level 3 in Fig. 7). Such relaxation, removing continuously population from the lower laser level, is needed to allow continuous laser operation. In a heat pipe relaxation is dominated by collisions. In the molecular-beam, however, it is the directional flow that continuously transports new molecules in low-lying thermally populated levels (level 1 in Fig. 7) into the region of the laser cavity and at the same time removes molecules in level 3 from the cavity. These latter molecules do not experience further collisions. They remain in the highly vibrationally excited level and are available for collision experiments. Rotating the birefringent filter allows choosing which vibrational level is populated.

It was a very crucial moment when Uli Hefter and Pat Jones tried for the first time to get the molecular-beam laser going. It rarely happens that such experiments work at the first try. In this case it did happen on July 28, 1981. The cavity was aligned

Fig. 7 The level scheme relevant for an optically pumped dimer laser, using molecules either in a heat-pipe or in a molecular beam

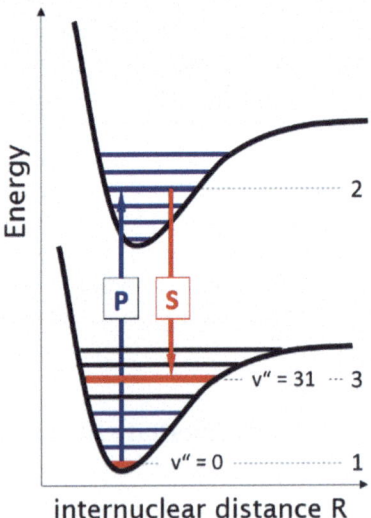

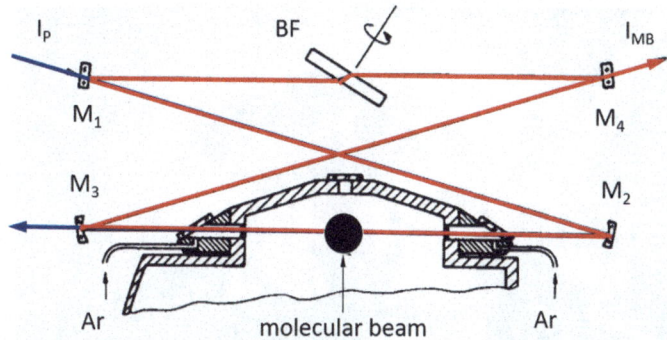

Fig. 8 The optical laser cavity built around the molecular beam in a vacuum chamber. I_P is the pump laser and I_{MB} the molecular-beam laser radiation. Only a small fraction of the pump laser is absorbed in the beam. Therefore, the pump laser exits again through the mirror M_3. BF is the birefringent (a tunable optical) filter that controls to which vibrational level v'' in the electronic ground state lasing is possible

and the molecular beam was operating. Upon turning on the pump laser lasing was immediately observed. The laser operated for about 15 min and then it went off. It took us 6 months (!!) to get it back into operation. This task was accomplished by Uli Gaubatz, who had joined the group shortly after the first operation of the laser was observed. Without these crucial 15 min, proving that the concept works, we would have probably given up after a few months and, most likely, the successful path towards realizing STIRAP would have been abandoned.

It turned out that one of the problems was the alkali deposit on the intra-cavity windows (see Figs. 8 and 9), separating the vacuum region from the outside. After several trial-and-error modifications, the problems were overcome. One of the measures was the heating of the windows to high temperatures what required the use of metal rings for vacuum-tight sealing. Furthermore, the installation of small pipes that directed a flow of Argon atoms away from the windows reduced the rate of deposit of alkali atoms and molecules on the windows. With these measures the molecular-beam laser could be routinely operated and in 1986 we set out to determine the properties of the transfer process, in particular its efficiency and selectivity [26].

The delay of a couple of years between the first demonstration of successful laser operation and the attempt to use such a laser for quantitative population transfer was in part due to the fact that no apparatus was available for such experiments. Students needed to first complete their ongoing experiments. The time was used to further explore the physics of the molecular-beam laser [27, 28]. Later we also built a molecular-beam laser with iodine molecules using a slit-nozzle expansion. For the latter system an optical pump-power threshold for starting the laser operation as low as 250 nW [29] was demonstrated.

Fig. 9 View through a window into the gain region of the molecular beam laser. The glowing red part is the molecular beam source with a cooler (darker) thermocouple attached. The thin yellow trace marks the pump laser beam which excites also background molecules. The thicker yellow region is seen because of radiation diffusion between particles along streamlines of the molecular beam. The gain region is the crossing between the two yellow traces

5   The Breakthrough

Figure 10 explains how the molecular-beam laser induced transfer efficiency was

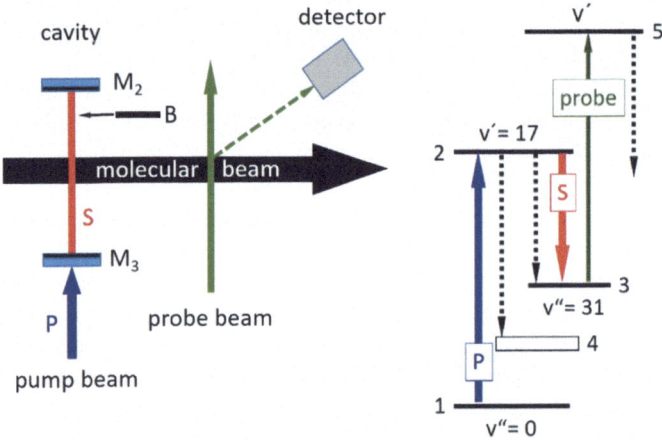

Fig. 10 Schematic of the set up (left part) for the calibration of the transfer efficiency induced by the molecular beam laser from level 1 ($v'' = 0$) to level 3 ($v'' = 31$). The element B is used to block the cavity. The related level scheme is shown in the right. Levels other than 1 and 3 that can also be reached from level 2 by spontaneous emission are summarized as level 4. The transfer into level 3 is probed by laser-induced fluorescence from level 5

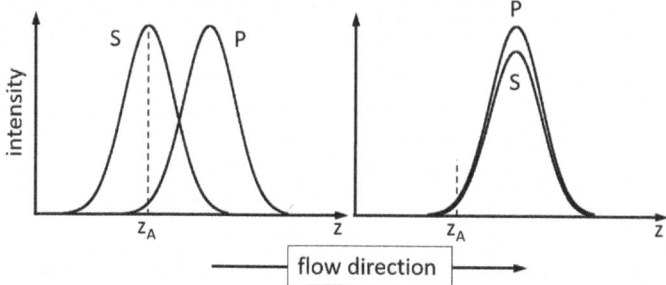

Fig. 11 The right part shows the P- and S- laser with coincident axes, as realized in the molecular-beam laser arrangement of Fig. 8. The left part shows the S-before-P arrangement of the lasers as also seen in the lower part of Fig. 1. The position z_A marks the location near which the pump laser would start pumping molecules out of level 1

determined. The pump laser excited molecules from the level $v'' = 0$ to $v' = 17$ in the B-state of Na_2, followed by spontaneous emission. The population in the target level of interest, e.g. $v'' = 31$, was probed by laser-induced fluorescence further downstream with the cavity blocked. Because the optical transition probabilities are known and after confirming that the population in level 1 was entirely depleted, the transfer efficiency of the population reaching level $v'' = 31$ by spontaneous emission could be determined. Unblocking the cavity allowed the molecular-beam laser to operate and the population in the target level increased. The increase of the population in relation to the known transfer efficiency by spontaneous emission yielded the beam-laser induced transfer efficiency. It was found [26] that the transfer efficiency was as large as 75% (larger than any other scheme would allow) but it was still far from the goal of $\approx 100\%$. The question thus was: what limits the transfer efficiency to about 75%?

The solution, which paved the final segment on the path to STIRAP, was surprisingly simple, as shown in Fig. 11. Results from earlier work on the consequences of optical pumping in two-step photoionization [30] led the way. The right part of Fig. 11 shows the profiles of the pump laser and the molecular-beam laser which appears after the pump laser is switched on. The axes of the two laser fields coincide. The molecules travel from left to right. As soon as the molecules reach the wings of the pump laser profile (P) they are efficiently pumped to the upper laser level. At the location z_A, however, the local molecular-beam laser intensity, which is supported by molecules that had already crossed the cavity, is still weak. Therefore, stimulated emission induced by the radiation field S that is supposed to populate level 3 (the target level) cannot yet compete with spontaneous emission. In fact, the transit time of the molecules across the cavity is about one order of magnitude longer than the radiative lifetime in level 2, the upper laser level. Therefore, only a fraction of the relevant molecules reaches the axis of the cavity where the beam laser is sufficiently strong to allow stimulated emission to compete successfully with spontaneous processes.

The final conclusion was again straightforward. Despite the significant effort invested to realize it, the molecular-beam laser approach had to be abandoned. In addition to the pump laser, an external laser for driving the stimulated emission process was needed, with its axis placed upstream of the axis of the pump laser, as shown in the left part of Fig. 11. As soon as the molecules enter the wings of the pump laser profile, they should be exposed to the maximum possible intensity of the S-Laser to optimize the chance for successful competition of stimulated emission with spontaneous processes. The informed reader realizes that the above argument, with optical pumping processes in mind, doesn't yet properly catch an essential part of STIRAP physics. It provided, however, the rationale for placing the axis of the S-beam upstream of the one for the P-beam, with a suitable overlap between the two.

It was very fortunate that Piotr Rudecki, a visiting scientist from Torun/Poland, arrived in the fall of 1987 a few days after I had come to the conclusion that a S-before-P arrangement was needed. The experiment which he wanted to join was not ready yet. However, he had some experience in modeling radiative processes. Therefore, I asked him to take our code for simulating the molecular-beam laser, which we had developed with the help of Wellegehausen, and modify it in accordance with the new geometry. The hope was to quantitatively understand the benefit of the S-before-P configuration from results of simulation studies. I certainly did expect a transfer efficiency of more than 75%.

About 10 days later, Rudecki presented his results: nearly 100% transfer. Expecting something near 90%, my reaction was: "hard to believe, please check for errors". A few days later, Piotr joined the group for the traditional after-lunch coffee-and-discussion meeting at a round table near the lab, presented the results of new calculations (see Fig. 12) and stated firmly "no errors—100% is correct". I

Fig. 12 The first numerical results for the transfer of the population of level 1 to level 3 (the final level) with a laser arrangement as shown in the lower part of Fig. 1. The lower part shows the transient population in the intermediate level 2. While, at late times, the final state population approaches unity, the maximal transient population of the intermediate level is 10^{-3}

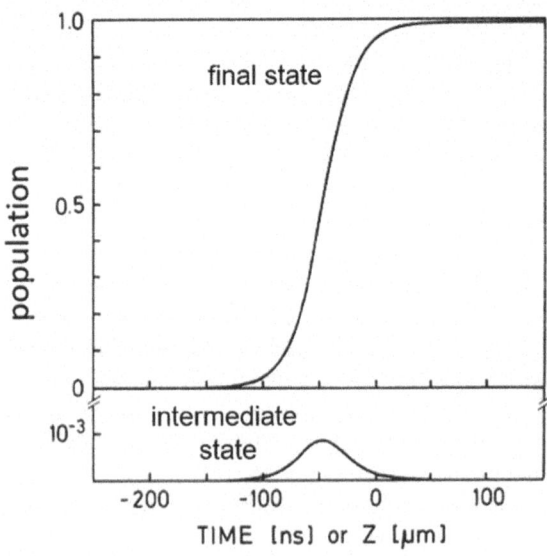

clearly recall my prompt reaction: "congratulation—this result will be a big bang in the community and will determine what we do in the lab for the next 10 years".

A few months later, in early 1988, we managed to demonstrate that a very high population transfer can indeed be realized in the S-before-P configuration. In a first short publication [31], we showed results, but the theoretical basis had not yet been clearly sorted out. This gap was closed by the May 1, 1990 publication in J. Chem. Phys. [1], with Uli Gaubatz being the leading graduate student who had noted the close connection of what we did with the work by Claude Cohen-Tannoudji [32]. Prior to submitting the manuscript of paper [1], a detailed discussion of the adiabatic condition was published [2]. The paper [1] remains a standard reference for STIRAP and indeed, all the basic features that make STIRAP unique are experimentally documented and theoretically properly analyzed in that work. In that paper, also the acronym STIRAP was introduced. Regarding the latter, I had learned earlier that it is important to give a name to a new technique, method or process before others will do it. After the first rough draft of the paper was ready, I told my students that instead of joining the after-lunch meeting, I will spend one hour or two in my home-office and come back with a suggestion for an acronym. At home, I wrote down all physical processes or phenomena that had some connection to the transfer process, looked at the initial letter or letters and wrote them down in different orders. The criteria were: the acronym should have no more than two syllables and pronunciation should be easy. I returned to the lab with the suggestion STIRAP, which was accepted by all involved.

We defined the publication date or ref [1], May 1st, 1990 the "birthday" of STIRAP and celebrated its 25th anniversary in September 2015 with a well-attended and well-received international conference in Kaiserslautern [33].

6   Some STIRAP Highlights that Followed

The most recent compilation of some highlights regarding STIRAP applications can be found under [7]. Here a few topics are listed, with only one or two references given:

- preparation of ultracold molecules, see e.g. [34, 35]
- reduction of the upper limit of the electric dipole moment of the electron [36]
- controlling the phase of superposition states [37, 38]
- new tools for matter wave optics [39, 40]
- population transfer in superconducting circuits with relevance to quantum information [41, 42]
- single photon generation by sending atoms through an optical cavity [43]
- control of the pathway of light in optical fiber networks [44]
- population transfer in a solid-state environment [45, 46]
- controlled modification of the quantum state in strings of ions bound in a trap [47]

- use of the concept for acoustic waves with the potential to improve hearing aids [48]
- control of the flow of spin-waves in a network of suitable wave-guides [49]

The application of the STIRAP-approach to acoustic waves [48] is a particularly nice example for how far reaching the concept is. It also underlines that STIRAP is not a purely quantum mechanical process.

There are also a number of proposals with a detailed analysis of the feasibility of STIRAP applications, such as

- implementing quantum gates, e.g. [50–52]
- cooling particles in an atomic beam [53]
- excitation of molecular Rydberg states, by-passing predissociation levels [54]
- preparation of highly polarized molecular quantum states [55]
- storage of energy using nuclear isomers [56]
- spatial adiabatic passage (SAP, transfer of particles between traps) [57]
- digital pulse sequences, optimized via a learning algorithm, to speed up the process [58]

The example (SAP, in earlier years also called CTAP—coherent transport by adiabatic passage) would be a particularly intriguing demonstration of the STIRAP-concept: Consider three traps A, B and C in a linear arrangement and close proximity. Each trap is able to hold a single particle. Assume that one particle is initially in trap A while B and C are empty. When the coupling between the traps is properly varied as required for STIRAP (e.g. by lowering the barriers between them while keeping the quantum states in the traps in resonance) the particle is removed from A and appears in C without establishing a significant transient population in B.

7 Final Remarks

Following the original publication in 1990, the STIRAP concept has been systematically developed, both experimentally and theoretically, in Kaiserslautern (with too many publications to be all listed), also for applications beyond the canonical three-level system. That work benefitted greatly from the contributions of the visiting scientists Bruce W. Shore, Leonid P. Yatsenko, Razmik Unanyan, Matthew Fewell, and Nikolay V. Vitanov. Experimental progress was achieved through the dedicated work of many excellent students: Axel Kuhn, Stefan Schiemann, Jürgen Martin, Thomas Halfmann, Heiko Theuer, and Frank Vewinger to name at least some. The post-docs George Coulston, Horst-Günter Rubahn, and Stéphane Guérin also contributed significantly to the successful developments.

At the occasion of my first public presentation of the concept in the colloquium at JILA/Boulder on March 1st, 1990 (i.e. prior to the publication of [1]) I had the chance to discuss STIRAP with Peter Zoller. It was the follow-up theoretical and experimental work from the groups of Peter Zoller and Bill Phillips [59, 60], respectively, on matter-wave mirrors and beam splitters and the experimental work in the group of Steve Chu on atom interferometry [39], that made STIRAP quickly known in the AMO community. Nevertheless, it took more than 10 years after the original publication [1] before STIRAP was used in many laboratories and in different areas of research.

Several proposals did appear in the literature discussing the prospects for applying STIRAP to poly-atomic molecules. However, nearly all of them are based on model systems that did not adequately include relevant properties, such as the realistically modelled (detrimental) high level density. The consequences of the inclusion of these properties are carefully analyzed in an extensive simulation study [61] involving the HCN molecule. To the best of my knowledge, SO_2 is still the largest, or most complex, molecule to which STIRAP has been successfully applied [62] in an experiment.

As explained in Sect. 3, the STIRAP-concept was developed with reaction dynamics experiments involving vibrationally excited molecules in mind. One early experiment of that kind has been completed in Kaiserslautern ($Na_2(v'')$ + Cl $\rightarrow$ NaCl + Na* [63]). Using STIRAP was also essential in the recent observation of bimolecular reactions at ultracold temperatures [64]. However, the initial motivation had reactions of relevance to atmospheric chemistry in mind. Such an application still awaits its realization. Because of recent developments of coherent radiation sources for the region $\lambda < 200$ nm, this situation may change soon. The related requirements for the molecules H_2, N_2, O_2, and OH are discussed in the appendix of [5].

References

1. U. Gaubatz, P. Rudecki, S. Schiemann, K. Bergmann, Population transfer between molecular vibrational levels by stimulated Raman Scattering with partially overlapping laser: a new concept and experimental results. J. Chem. Phys. **92**, 5363–5376 (1990)
2. J.K. Kuklinski, U. Gaubatz, F.T. Hioe, K. Bergmann, Adiabatic population transfer in a three level system driven by delayed laser pulses. Phys. Rev. A **40**, 6741–6744 (1989)
3. K. Bergmann, H. Theuer, B.W. Shore, Coherent population transfer among quantum states of atoms and molecules. Rev. Mod. Phys. **70**, 1003–1026 (1998)
4. N.V. Vitanov, M. Fleischhauer, B.W. Shore, K. Bergmann, *Coherent Manipulation of Atoms and Molecules by Sequential Pulses*, in Advances of Atomic, Molecular, and Optical Physics, eds. by B. Bederson, H. Walther, vol 46 (Academic Press, 2001), pp. 55–190
5. K. Bergmann, N.V. Vitanov, B.W. Shore, Perspective—stimulated Raman adiabatic passage: the status after 25 years. J. Chem. Phys. **142**, 170901 (1–20) (2015)
6. N.V. Vitanov, A. Rangelov, B.W. Shore, K. Bergmann, Stimulated raman adiabatic passage in physics, chemistry and beyond. Rev. Mod. Phys. **89**, 015006 (1–66) (2017)
7. K. Bergmann (guest editor), Roadmap on STIRAP applications. J. Phys. B At. Mol. Opt. Phys. **52**, 202001 (1–55) (2019)
8. K. Bergmann, W. Demtröder, Mass-spectrometric investigation of the primary processes in the photo dissociation of 1,3-butadiene. J. Chem. Phys. **48**, 18–22 (1968)
9. W. Demtröder, M. McClintock, R.N. Zare, Spectroscopy of Na_2 using laser-induced fluorescence. J. Chem. Phys. **51**, 5495–5508 (1969)

10. K. Bergmann, W. Demtröder, Inelastic collision cross section for excited molecules: I. rotational energy transfer within the $B^1\Pi_u$-state of Na_2 induced by collision with He. Z. Phys. **243**, 1–13 (1971)

11. R.B. Kurzel, J.I. Steinfeld, Energy-transfer processes in monochromatically excited iodine molecules III. Quenching and multiquantum transfer from $v' = 43$. J. Chem. Phys. **53**, 3293–3303 (1970)

12. K. Bergmann, H. Klar, W. Schlecht, Asymmetries in collision induced rotational transitions. Chem. Phys. Lett. **12**, 522–525 (1972)

13. R.E. Drullinger, R.N. Zare, Optical pumping of molecules. J. Chem. Phys. **51**, 5523–5542 (1969)

14. K. Bergmann, C.B. Moore, Energy dependence and isotope effect for the total reaction rate of Cl + HI and Cl + HBr. J. Chem. Phys. **63**, 643–649 (1975)

15. K. Bergmann, S.R. Leone, C.B. Moore, Effect of reagent electronic excitation on the chemical reaction $Br(^2P_{1/2,3/2})$ + HI. J. Chem. Phys. **63**, 4161–4166 (1975)

16. K. Bergmann, R. Engelhardt, U. Hefter, J. Witt, A detector for state-resolved molecular beam experiments using optical fibers. J. Phys. E **12**, 507–514 (1979)

17. K. Bergmann, U. Hefter, P. Hering, Molecular-beam diagnostic with internal state selection: velocity distribution and dimer formation in a supersonic Na/Na_2 beam. Chem. Phys. **32**, 329–348 (1978)

18. K. Bergmann, R. Engelhardt, U. Hefter, P. Hering, J. Witt, State resolved differential cross sections for rotational transitions in Na_2 + Ne(He) collisions. Phys. Rev. Lett. **40**, 1446–1450 (1978)

19. U. Hefter, P.L. Jones, A. Mattheus, J. Witt, K. Bergmann, R. Schinke, Resolution of supernumerary rotational rainbows in Na_2-Ne scattering. Phys. Rev. Lett. **46**, 915–918 (1981)

20. G. Ziegler, M. Rädle, O. Pütz, K. Jung, H. Ehrhardt, K. Bergmann, Rotational rainbows in electron-molecule scattering. Phys. Rev. Lett. **58**, 2642–2645 (1987)

21. A. Mattheus, A. Fischer, G. Ziegler, E. Gottwald, K. Bergmann, Experimental proof of $|\Delta m|<$ $< j$ propensity rule in rotationally inelastic differential scattering. Phys. Rev. Lett. **56**, 712–715 (1986)

22. K. Bergmann, Molecular-beam experiments with internal state selection by laser optical pumping. Habilitation thesis (University of Kaiserslautern, 1979)

23. C. Kittrell, E. Abramson, J.L. Kinsey, S.A. McDonald, D.E. Reisner, R.W. Field, D.H. Katayama, Selective vibrational excitation by stimulated emission pumping. J. Chem. Phys. **75**, 2056–2059 (1981)

24. J.M. Price, J.A. Mack, C.A. Rogaski, A.M. Wodtke, Vibrational state-specific self-relaxation rate constant: measurements of highly vibrationally excited O2 (v = 19–28). Chem. Phys. **176**, 83–98 (1993)

25. B. Wellegehausen, Optically pumped cw Dimer laser. IEEE-QE **15**, 1108–1130 (1979)

26. M. Becker, U. Gaubatz, P.L. Jones, K. Bergmann, Efficient and selective population of high vibrational levels by near resonance Raman Scattering. J. Chem. Phys. **87**, 5064–5076 (1987)

27. P.L. Jones, U. Gaubatz, U. Hefter, B. Wellegehausen, K. Bergmann, Optically pumped sodium-dimer supersonic beam laser. Appl. Phys. Lett. **42**, 222–224 (1983)

28. P.L. Jones, U. Gaubatz, H. Bissantz, U. Hefter, I. Colomb de Daunant, K. Bergmann, Optically-pumped supersonic beam lasers: basic concept and results. J. Opt. Soc. Am. B **6**, 1386–1400 (1989)

29. I.C.M. Littler, S. Balle, K. Bergmann, Molecular beam Raman Laser with a 250 nW threshold pump power. Opt. Commun. **77**, 390–394 (1990)

30. K. Bergmann, E. Gottwald, Effect of optical pumping in two step photoionization of Na_2 in molecular beams. Chem. Phys. Lett. **78**, 515–519 (1981)

31. U. Gaubatz, P. Rudecki, M. Becker, S. Schiemann, M. Külz, K. Bergmann, Population switching between vibrational levels in molecular beams. Chem. Phys. Lett. **149**, 463–468 (1988)

32. C. Cohen-Tannoudji, S. Reynaud, Dressed-atom description of resonance fluorescence and absorption spectra of a multi-level atom in an intense laser beam. J. Phys. B **10**, 345–363 (1977)

33. K. Bergmann (conference chair), *Stimulated Raman Adiabatic Passage in Physics, Chemistry and Technology*. https://www.physik.uni-kl.de/bergmann/stirap-symposium-2015/
34. J.G. Danzl, M.J. Mark, E. Haller, M. Gustavsson, R. Hart, J. Aldegunde, J.M. Hutson, H.-C. Naegerl, An ultracold high-density sample of rovibronic ground-state molecules in an optical lattice. Nat. Phys. **6**, 265–270 (2010)
35. K.K. Ni, S. Ospelkaus, M.H.G. de Miranda, A. Pe'er, B. Neyenhuis, J.J. Zirbel, S. Kotochigova, P.S. Julienne, D.S. Jin, J. Ye, A high phase-space-density gas of polar molecules. Science **322**, 231–235 (2008)
36. V. Andreev, D.G. Ang, D. DeMille, J.M. Doyle, G. Gabrielse, J. Haefner, N.R. Hutzler, Z. Lasner, C. Meisenhelder, B.R. O'Leary, C.D. Panda, A.D. West, E.P. West, X. Wu, Improved limit on the electric dipole moment of the electron. Nature **562**, 355–364 (2018)
37. R.G. Unanyan, M. Fleischhauer, K. Bergmann, B.W. Shore, Robust creation and phase-sensitive probing of superposition states via stimulated Raman adiabatic passage (STIRAP) with degenerate dark states. Opt. Commun. **155**, 144–154 (1998)
38. F. Vewinger, B.W. Shore, K. Bergmann, *Superposition of Degenerated Atomic Quantum States: Preparation and Detection in Atomic Beams*, eds. by E. Arimondo, P.R. Berman, C.C. Lin. Advances in Atomic, Molecular and Optical Physics, vol. 58 (Academic Press, USA, 2010), pp. 113–172
39. M. Weitz, B. Young, S. Chu, Atomic interferometer based on adiabatic population transfer. Phys. Rev. Lett. **73**, 2563–2566 (1994)
40. H. Theuer, R.G. Unanyan, C. Habscheid, K. ̀Klein, K. Bergmann, Novel laser-controlled variable matter wave beamsplitter. Opt. Expr. **4**, 77–83 (1999)
41. K.S. Kumar, A. Vepsäläinen, S. Danilin, G. S. Paraoanu, Stimulated Raman adiabatic passage in a three-level superconducting circuit. Nat. Commun. **7**, 10628 (1–6) (2016)
42. H.K. Zhu, C. Song, W.Y. Liu, G.M. Xue, F.F. Su, H. Deng, Ye Tian, D.N. Zhen, Siyuan Han, Y.P. Zhong, H. Wang, Yu-xi Liu, S.P. Zhao, *Coherent population transfer between uncoupled or weakly coupled states in ladder-type superconducting quitrits*. Nat. Commun. **7**, 11019 (1–6) (2016)
43. A.Kuhn, M. Hennrich, G. Rempe, Deterministic single-photon source for distributed quantum networking. Phys. Rev. Lett. **89**, 067901 (1–4) (2002)
44. S. Longhi, G. Della Valle, M. Ornigotti, P. Laporta, Coherent tunneling by adiabatic passage in an optical waveguide system. Phys. Rev. B. **76**, 201101 (1–4) 2007
45. J. Klein, F. Beil, T. Halfmann, Robust population transfer by stimulated Raman adiabatic passage in a Pr3 + :Y2SiO5 crystal. Phys. Rev. Lett. **99**, 113003 (1–4) (2007)
46. D.A. Golter, H. Wang, Optically driven rabi oscillations and adiabatic passage of single electron spins in diamond. Phys. Rev. Lett. **112**, (1–5) 116403 (2014)
47. J.L. Sørensen, D. Møller, T. Iversen, J.B. Thomsen, F. Jensen, P. Staanum, D. Voigt, M. Drewsen, Efficient coherent internal state transfer in trapped ions using stimulated Raman adiabatic passage. New J. Phys. **8**, 261 (1–10) (2006)
48. Y.X. Shen, Y.G. Peng, D.G. Zhao, X.C. Chen, J. Zhu, X.F. Zhu, One-way localized adiabatic passage in an acoustic system. Phys. Rev. Lett. **122**, 094501 (1–7) (2019)
49. P. Pirro, B. Hillebrands, The magnonic STIRAP process. J. Phys. B At. Mol. Opt. Phys. **52**, 202001 (22–23) (2019)
50. R.G. Unanyan, M. Fleischhauer, Geometric phase gate without dynamical phase. Phys. Rev. A **69**, 050302 (1–4) (2004)
51. D. Möller, J.L. Sørensen, J.B. Thomson, M. Drewsen, Efficient qubit detection using alkaline-earth-metal ions and a double stimulated Raman adiabatic passage. Phys. Rev. A **76**, 062321 (1–12) (2007)
52. B. Rousseaux, S. Guérin, N.V. Vitanov, Arbitrary qudit gates by adiabatic passage. Phys. Rev. A **87**, 032328 (1–4) (2013)
53. M.G. Raizen, D. Budker, S.M. Rochester, J. Narevicius, E. Narevicius, Magnetooptical cooling of atoms. Opt. Lett. **39**, 4502–4505 (2014)
54. T. J. Barnum, D. D. Grimes, R.W. Field, Populating Rydberg states of molecules by STIRAP. J. Phys. B At. Mol. Opt. Phys. **52**, 202001 (43–44) (2019)

55. S. Rochester, S. Pustelny, K. Szymanski, M. Raizen, M. Auzinsh, D. Budker, Efficient polarization of high-angular-momentum systems. Phys. Rev. A **94**, 043416 (1–12) (2016)
56. W.T. Liao, A. Pálffy, C. H. Keitel, A three-beam setup for coherently controlling nuclear state populations. Phys. Rev. C **87**, 054609 (1–12) (2013)
57. R. Menchon-Enrich, A. Benseny, V. Ahufinger, A.D. Greentree, T. Busch. J. Mompart, Spatial adiabatic passage: a review of recent progress. Rep. Prog. Phys. **79**, 074401 (1–31) (2016)
58. I. Poarelle, L. Moro, E. Prati, Digitally stimulated Raman passage by deep reinforcement learning. Phys. Lett. A **384**, 126266 (1–10) (2020)
59. P. Marte, P. Zoller, J.L. Hall, Coherent atomic mirrors and beam splitters by adiabatic passage in multilevel systems. Phys. Rev. A **44**, R4118–R4121 (1991)
60. L.S. Goldner, C. Gerz, R.J. Spreeuw, S.L. Rolston, C.I. Westbrook, W.D. Phillips, P. Marte, P. Zoller, Momentum transfer in laser-cooled cesium by adiabatic passage in a light field. Phys. Rev. Lett. **72**, 997–1000 (1994)
61. W. Jakubetz, Limitations of STIRAP-like population transfer in extended systems: The three-level system embedded in a web of background states. J. Chem. Phys. **137**, 224312 (1–16) (2012)
62. T. Halfmann, K. Bergmann, Coherent population transfer and dark resonances in SO_2. J. Chem. Phys. **104**, 7068–7072 (1996)
63. P. Dittmann, F.P. Pesl, J. Martin, G. Coulston, G.Z. He, K. Bergmann, The effect of vibrational excitation ($3 \leq v" \leq 19$) on the chemiluminescent channel od the reaction $Na2(v") + Cl \rightarrow NaCl + Na^*(3p)$. J. Chem. Phys. **97**, 9472–9475 (1992)
64. M.G. Hu, Y. Liu, D.D. Grimes, Y.-W. Lin, A.H. George, R. Vexlau, N. Bouloufa-Maafa, O. Dulieu, T. Rosenband, K.-K. Ni, Direct observation of bimolecular reactions of ultracold KRb molecules. Science **366**, 1111–1115 (2019)

Chapter 20
Manipulation and Control of Molecular Beams: The Development of the Stark-Decelerator

Gerard Meijer

Abstract State-selective manipulation of beams of atoms and molecules with electric and magnetic fields has been crucial for the success of the field of molecular beams. Originally, this manipulation only involved the *transverse* motion. In this Chapter, the development of the Stark-decelerator, that allows to also manipulate and control the *longitudinal* motion of molecules in a beam, is presented.

1 Introduction

"*Born in leaks, the original sin of vacuum technology, molecular beams are collimated wisps of molecules traversing the chambered void that is their theatre [...]. On stage for only milliseconds between their entrances and exits, they have captivated an ever growing audience by the variety and range of their repertoire*". This is how John B. Fenn affectionately phrased it over 30 years ago, when he reflected on the long and rich history of molecular beams in his foreword to one of the classic books on this subject [1]. He could not have foreseen the spectacular leap forward that the level of control over molecular beams would take. In particular, methods that have been developed since then to slow down and store molecular beams – thereby stretching the duration of their performance on stage by orders of magnitude – have made whole new classes of experiments possible.

The motion of neutral molecules in a beam can be manipulated and controlled with inhomogeneous electric and magnetic fields. Static fields can be used to deflect or focus molecules, whereas time-varying fields can be used to decelerate or accelerate beams of molecules to any desired velocity. In this paper we present an historical overview, emphasizing the important role of molecular beam deflection and focusing experiments in the development and testing of quantum mechanics. We describe the original attempts and the successful implementation of schemes to decelerate and accelerate molecular beams with electric fields, that is, the development of the Stark-decelerator. The various elements, using electric as well as magnetic fields, that have

G. Meijer (✉)
Fritz-Haber-Institut der Max-Planck-Gesellschaft, Faradayweg 4-6, 14195 Berlin, Germany
e-mail: meijer@fhi-berlin.mpg.de

© The Author(s) 2021
B. Friedrich and H. Schmidt-Böcking (eds.), *Molecular Beams in Physics and Chemistry*,
https://doi.org/10.1007/978-3-030-63963-1_20

463

been developed for the manipulation and control of molecular beams since the first successful demonstration of the Stark-decelerator in 1998, have resulted in setups in which the molecules can be stored in stationary traps or injected in a molecular storage ring or synchrotron, for instance. Novel crossed-beam scattering studies at low collision energies, high-resolution spectroscopy studies on trapped or slow molecules and lifetime measurements of trapped molecules have become possible [2].

2 Deflection and Focusing of Molecular Beams

Atomic and molecular beams have played central roles in many experiments in physics and chemistry – from seminal tests of fundamental aspects of quantum mechanics to molecular reaction dynamics – and have found a wide range of applications [1]. Nowadays, sophisticated laser-based methods exist to perform sensitive and quantum state selective detection of the atoms and molecules in the beams. In the early days, such detection methods were lacking and the particles in the beam were detected, for instance, by a "hot wire" (Langmuir-Taylor) detector, by electron-impact ionization or by deposition and *ex-situ* investigation of the particles on a substrate at the end of the beam-machine. To achieve quantum state selectivity in the overall detection process, these methods were combined with inhomogeneous electric and magnetic field sections to influence the trajectories of the particles on their way to the detector.

The first paper discussing the degree of deflection for a beam of polar molecules passing through an inhomogeneous electric field was submitted almost a century ago, at the end of July 1921, to the *Zeitschrift für Physik* [3]. The paper was written by Hartmut Kallmann and Fritz Reiche, coworkers of Fritz Haber at the Kaiser Wilhelm Institute for Physical Chemistry and Electrochemistry in Berlin, the present Fritz-Haber-Institut der Max-Planck-Gesellschaft. Kallmann and Reiche write in their paper that they performed their analysis in support of experiments that "are ongoing at the institute" to determine whether the dipole moment is a property of an individual molecule or whether this is only induced in the molecule when it is in close proximity to other molecules, an issue that was intensively debated at that time. By passing a molecular beam, that is, a dilute but highly collimated sample of molecules, through an inhomogeneous electric field, they argued, it should be possible to monitor the deflection of the molecules and to thus determine the value of their dipole moment – provided they have one. In the introduction of their paper, they discuss in general terms the forces on a moving dipolar molecule due to inhomogeneous electric fields as well as due to inhomogeneous magnetic fields. In their analysis, however, they restrict themselves to the electric field case, and assume that the dipole moment of the diatomic molecule is perpendicular to the angular momentum vector, i.e., that the component of angular momentum along the internuclear axis is zero.

Shortly after its submission, the paper by Kallmann and Reiche came to the eyes of Otto Stern, pressing him to write up the theory behind an experiment that – as he wrote – "he had been involved in for some time with his colleague Walther Gerlach". Stern submitted his paper only about one month later, at the end of August 1921, to the same journal. In a footnote to his paper, he explicitly acknowledges that the reason for publishing his paper is the upcoming paper of Kallmann and Reiche, and concludes that both papers are nicely complementary as his paper "discusses the case in which the dipole moment is parallel to the angular momentum vector, as is generally the case for magnetic atoms" [4]. In his paper, Stern describes a method to experimentally test space quantization via measuring the deflection of a beam of atoms with a magnetic moment when moving through an inhomogeneous magnetic field [4]. No further account on the early electric deflection experiment in Berlin is to be found in the literature or in the archives, while Stern and Gerlach performed their famous experiment within one year [5], in February of 1922. Electric deflection of a beam of polar molecules was first demonstrated by Erwin Wrede, a graduate student of Stern, several years later in Hamburg [6]. It is interesting to note that up to our re-discovery of the article by Kallmann and Reiche in 2009, this article went largely unnoticed, being cited only seven times; the important mentioning of the article by Otto Stern in a footnote in his paper could not include the final reference yet. Since then, it has been cited more than once per year.

All the original experimental geometries were devised to create strong magnetic or electric field gradients to efficiently deflect particles from the beam axis. In 1939, Isidor Rabi introduced the molecular beam magnetic resonance method by using two magnets in succession to produce inhomogeneous magnetic fields of oppositely directed gradients. In this set-up the deflection of particles caused by the first magnet is compensated by the second magnet, such that the particles are directed on a sigmoid path to the detector. A transition to "other states of space quantization" induced in between the magnet sections can be detected via the resulting reduction of the detector signal. This provided a new method to accurately measure nuclear or other magnetic moments [7]. Later, both magnetic [8, 9] and electric [10] field geometries were designed to focus particles in selected quantum states onto the detector. An electrostatic quadrupole focuser, i.e., an arrangement of four cylindrical electrodes with alternating positive and negative voltages, was used to couple a beam of ammonia molecules into a microwave cavity. Such an electrostatic quadrupole lens focuses ammonia molecules that are in the so-called low-field seeking, upper level of the inversion doublet while it simultaneously defocuses those that are in the lower, high-field seeking, level. The inverted population distribution of the ammonia molecules that is thus produced in the microwave cavity led to the invention of the maser by Gordon, Zeiger and Townes in 1954–1955 [11, 12]. Apart from the spectacular observation of the amplification of microwaves by stimulated emission, these focusing elements more generally enabled the recording, with high resolution and good sensitivity, of microwave spectra in a molecular beam. By using several multipole focusers in succession, with interaction regions with electromagnetic radiation in between, versatile set-ups to unravel the quantum structure of atoms and molecules were developed. In scattering experiments, multipole focusers were exploited to study steric effects, that is, to study how the orientation of an attacking molecule affects its reactivity [13]. Variants of the molecular beam resonance methods as well as scattering machines that employed state-selectors were implemented in many laboratories, and have yielded a wealth of detailed information on stable molecules, radicals and molecular complexes, thereby contributing enormously to our present understanding of intra- and inter-molecular forces.

3   Early Attempts to Decelerate or Accelerate Molecular Beams

The state-selective manipulation of beams of atoms and molecules with electric and magnetic fields is thus about as old as the field of atomic and molecular beams itself, and it actually has been crucial for the success of the latter field. In his autobiography, Norman Ramsey, who himself later invented the separated oscillatory fields method and wrote a very influential book on molecular beam methods [14], recalls that Rabi was rather discouraged about the future of molecular beam research when he arrived in Rabi's lab in 1937, and that this discouragement only vanished when Rabi invented the molecular beam magnetic resonance method [15]. However, even though the manipulation of beams of molecules with external fields has been used extensively and with great success in the past, this manipulation exclusively involved the *transverse* motion of the molecules.

When the velocity distribution in a molecular beam is rather broad, the state-selective deflection fields can be used to provide some velocity selection. Offset or angled molecular beam geometries have been used, in which deflection fields cause only the slow (fast) atoms or molecules to obtain sufficiently large (small) deflections to pass through the apparatus and to reach the detector, for instance. This approach has been attempted to selectively load slow ammonia molecules in a microwave cavity to produce a maser with an ultra-narrow linewidth [16]. At Bell telephone laboratories an electrostatic parabolic reflector was designed to selectively couple slow ammonia molecules in the microwave cavity, that is, after deflecting them by 180 degrees [17]. These approaches suffered from the deficiency that, as stated then, "it is generally known that the velocity distribution of molecules emanating from a hole in a box is not Maxwellian, but departs from it by having fewer low velocity molecules", possibly caused by collisions with fast molecules "from behind" [17].

At the end of the nineteen-fifties, alternative approaches, with electric fields designed such as to create "multiple retardation barriers" have been proposed to actively manipulate the *longitudinal* motion of molecules in a beam, that is, to slow down ammonia molecules [18]. A rather compact experimental setup consisting out of a source chamber, a deceleration chamber and a slow-molecule deflection and detection chamber was constructed for exactly this purpose in the physics department of MIT, under the supervision of John G. King [19]. The approximately 20 cm long decelerator consisted of a linear array of ten parallel plate capacitors, capable of maintaining a voltage difference of 30 kV across 1 mm plate separation. Ammonia molecules in the low-field seeking, upper level of the $J = K = 1$ inversion doublet lose kinetic energy when entering the high electric field region of the capacitor. When the electric field is slowly turned to zero while the molecules are in the homogeneous electric field inside the capacitor, the molecules do not regain the lost kinetic energy when exiting the capacitor and the process can be repeated. The experiments were performed using continuous beams and the same high voltage was applied to the whole array of electrodes in the form of a sine wave with a fixed frequency of 6 kHz. Therefore, the distances between the adjacent parallel plate capacitors as well as their lengths needed to be made such that it takes the ever slower molecules always exactly the same amount of time to reach the next stage, i.e., these distances and lengths needed to gradually decrease along the molecular beam. The machine was designed to slow down ammonia molecules from an initial speed of 200 m/s to a final speed of 35 m/s, selectively detecting the slowed molecules at the exit. The project suffered from the same deficiency as mentioned earlier, namely that there were not enough ammonia molecules at the initial speed of 200 m/s to yield a detectable signal. The work is described in detail in the Ph.D. thesis of Robert Golub [20], but no further publication has resulted from this work.

In the physical chemistry community the experimental efforts of Lennard Wharton, to demonstrate electric field *acceleration* of a molecular beam, are much better known. In the mid nineteen-sixties, at the University of Chicago, he constructed a molecular beam machine, containing a thirty-three foot long accelerator. The accelerator consisted of an array of 600 acceleration stages intended to increase the kinetic energy of LiF molecules in high-field seeking states from 0.2 eV to 2.0 eV, that is, speeding the LiF molecules up from 1200 m/s to 3800 m/s, with the aim to use these high energy beams for reactive scattering studies. Each acceleration stage consisted of two hemispherically ended rods with a diameter of 0.5 mm, spaced 0.5 mm apart. The beam was transversely kept together using additional alternate-gradient lenses. A popular scientific account of this work, together with a schematic drawing of the acceleration principle and a photograph of the "Chemical Accelerator", appeared in *Scientific American* in 1968 [21]. The photograph of the about eleven meter long machine is reproduced in Fig. 1. Also in this case, continuous molecular beams were used and the high voltage was applied as a sine wave with a fixed frequency to the array of electrodes, implying that the adjacent electric field stages needed to be put ever further apart to compensate for the molecules being ever faster, explaining the length of the molecular beam machine. An excellent paper, in which the focusing of beams of polar molecules in high-field seeking states was theoretically analyzed, resulted from this work [22]. The acceleration experiment was not successful, however, not only because the alignment of the array of electrodes is very critical in an alternate-gradient setup but also due to an overly optimistic view on the magnitude of the electric fields that could be stably obtained when designing the accelerator module. This is what the Ph.D. student who was working on this project, Edward A. Bromberg, concluded in his Ph.D. thesis in which he summarised that "it has not been possible to show either that particles have been accelerated, or that neutral particles cannot be accelerated" [23]. Both, the deceleration experiments of John King and the acceleration experiments of Lennard Wharton were not continued after the Ph.D. students completed their theses. Whereas interest in slow molecules as a maser

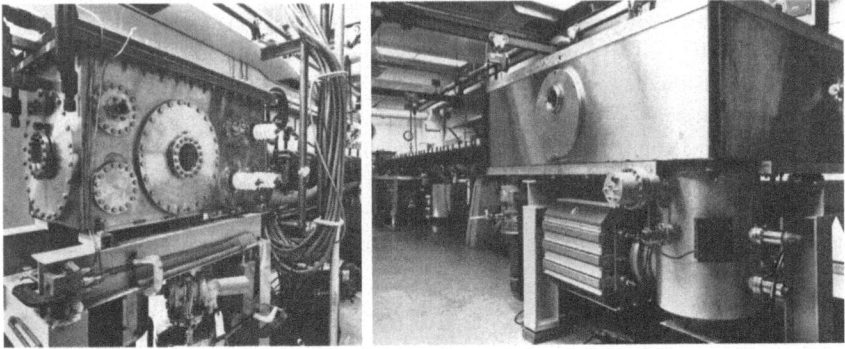

Fig. 1 Photograph of the "Chemical Accelerator" built at the University of Chicago by Lennard Wharton, to produce molecular beams with an energy of about two electron volts. The photographs show the long dipole accelerator from both ends. Molecules produced by heating in an oven (left) are accelerated down the long tube by electric fields to the reaction chamber (right). Reproduced from [21]

medium declined owing to the invention of the laser, the molecular beam accelerator was made obsolete by the seminal demonstration of John Fenn and co-workers of gas dynamic acceleration of heavy species in seeded supersonic He and H_2 beams [24]. Because of these unsuccessful early attempts, researchers who wanted to pursue molecular beam deceleration with electric or magnetic fields for high-resolution spectroscopy and metrology in the following decades, had difficulties getting this financed [25].

4   Deceleration of CO ($a^3 \Pi$) Molecules with Electric Fields

I studied physics at the Katholieke Universiteit Nijmegen in Nijmegen, The Netherlands – since 2004 renamed into Radboud University – and performed my undergraduate and Ph.D. research in the Atomic and Molecular Physics department, headed by Antoni Dymanus and Jörg Reuss. In this department, I was exposed from day one, that is, from February 1984 on, to a wide variety of molecular beam machines; electrostatic and magnetic state-selectors and focusing elements were used throughout. The magnetic properties and the molecular quadrupole tensor of the water molecule [26] as well as the electric and magnetic properties of carbon monoxide [27], for instance, were measured with those machines by Antoni Dymanus and his students already in the nineteen-seventies and are still the standards in the field. During my time as Ph.D. student, Jörg Reuss wrote the chapter entitled "State-selection by non-optical methods" as contribution from Nijmegen to the earlier mentioned classic book on "Atomic and Molecular beam methods" [1].

After post-doctoral periods abroad, I was offered the opportunity to start up my own research program in Nijmegen, in what was then just renamed the Molecular and Laser Physics department, and I became the successor of Jörg Reuss per January 1995. With my Ph.D. student Rienk Jongma, we performed two-dimensional imaging of metastable CO ($a^3\Pi$) molecules, to study with modern tools the phenomenon of mass-focusing in a seeded molecular beam as well as the performance of an electrostatic hexapole focuser [28]. We showed that metastable CO molecules, prepared with a pulsed laser at a well-defined time, at a well-defined position and in a single quantum state, are ideally suited to study velocity distributions and spatial distributions in molecular beams in general, and to study electric and magnetic field manipulation of molecular beams in particular. The metastable CO molecules live several milliseconds [29], long enough for molecular beam experiments, and with about 6 eV internal energy they can be efficiently detected – temporally and spatially resolved – via the Auger electrons that are released when they impinge on a surface. Whereas the Stark shift of the rotational levels of CO in its electronic ground state is very small, the shift of the $J = 1$ rotational level in the metastable $a^3\Pi_1$ state is on the order of 1 cm^{-1} in electric fields of about 100 kV/cm. When the CO molecules are state-selectively prepared in the metastable state with a pulsed laser *inside* a strong electric field, the interaction with the electric field is suddenly "switched on". This led us to propose a scheme for confining metastable CO molecules in stable "planetary" orbits in an appropriately shaped electrostatic trap [30]. As this confinement scheme would only work for CO molecules with speeds below about 22 m/s, we discussed how such slow molecules could be obtained, and in the final section of our paper we mentioned the possibility to slow down polar molecules with electric fields [30]. The latter was critically commented upon by the (anonymous) referee of our paper, which motivated us to try to experimentally demonstrate that when CO ($a^3\Pi$) molecules are prepared in a high-field seeking state *inside* a large electric field, the molecules will indeed slow down when leaving the high-field region, loosing an amount of kinetic energy that is identical to the Stark-shift of the levels. Even when seeded in Xe, however, the kinetic energy of the CO molecules in our beam was too large, to be able to unambiguously detect the deceleration effect of a single parallel-plate electric field stage of 140 kV/cm. It was clear that we would need either a considerably higher electric field or more electric field stages. When I informed Jörg Reuss in the spring of 1997 about these experiments, he told me that he vaguely remembered an experiment along similar lines by "Wharton in Chicago", but he did not know any further details. This was the first time I learned about Lennard Wharton's earlier experiments, which was comforting as I had been wondering for quite a while already why acceleration or deceleration of molecular beams with electric fields had never been demonstrated – at least now I knew it had been tried.

In December 1997, Rick Bethlem started in my department as Ph.D. student. He was hired on a project to study the physical properties of endohedral fullerenes, which would include the synthesis of small molecules like CO encapsulated in a fullerene cage, and the subsequent spectroscopic characterization of the motion of such a "dumbbell in a box". In one of our first meetings, prior to his start as Ph.D. student, I informed him about our proposed scheme to confine slow CO molecules in stable orbits and about our attempt to demonstrate molecular beam deceleration with a single electric field stage. Rick was fascinated by this topic, that actually fitted better to his background and interest, and we decided that he would switch research projects and start on the deceleration of metastable CO molecules with electric fields. At that time, I was also actively involved with other members in my department in experiments with an infrared free electron laser [31] and over lunch we had discussions about the operation principle of a linear accelerator (LINAC) for electrons, and whether and how the equivalent device for neutral, polar molecules could be constructed. We discussed in particular whether, as proposed by Basov [18], the "multiple retardation barriers" would need to be combined with electromagnetic radiation fields to be able to repeat the deceleration process, or that one could also rapidly switch the fields on and off; we concluded that both should be possible and decided to go for the approach with the time-varying electric fields. Searching for the term "slow molecules" on the internet gave us then – quite unexpectedly – as one of the first hits the one-page conference abstract by John King from 1959 in which he briefly described the ongoing ammonia deceleration experiments in his laboratory [19] (he there mentioned a 1 m long, 25-stage array of parallel-plate capacitors, using electric fields of 100 kV/cm, i.e., longer but with lower electric fields than what Robert Golub reported upon later [20]), to which we had found no reference in the later work from Lennard Wharton. Different from the earlier attempts by John King and Lennard Wharton, we used seeded pulsed beams and we did not have to rely on time-varying high voltages at a certain fixed frequency. Instead, we could make use of commercially available high voltage switches that can be rapidly (sub-μs) switched on and off in any pre-programmed time-sequence. This made it possible to design an array of – in our first experiment – 63 equidistant electric field stages with a center-to-center distance of 5.5 mm and to have complete flexibility in the input and output velocity that we would like to use. The electric field in each stage is formed by applying a high voltage to two parallel 3-mm-diameter cylindrical rods, centered 4.6 mm apart, leaving 1.6 mm opening for the molecular beam. The two opposing rods are simultaneously switched by two independent high-voltage switches from ground potential to voltages of +10 kV and −10 kV, yielding maximum electric fields of 125 kV/cm in a geometry that also provides transverse focusing. To obtain transverse focusing in two dimensions, adjacent electrode pairs are alternately positioned horizontally and vertically. All horizontal and all vertical electrode pairs are electrically connected and alternately switched, requiring a total of four independent high-voltage switches. A photograph of the prototype, 35 cm long, so-called "Stark-decelerator" as well as a scheme of the experimental set-up is shown in Fig. 2. The two electric field configurations between which switching takes place are schematically shown on the right hand side of Fig. 2.

In Fig. 3, the original measurements are shown that gave us the first hint of signal due to decelerated molecules, indicated by the blue dashed line as a "guide to the eye" – which is needed here. Rick Bethlem had been scheduled to give an oral presentation on the deceleration experiments at the annual meeting of the Dutch AMO community in Lunteren, The Netherlands, in November 1998, and there he presented these results, that had been obtained just a few days before that. In the weeks following this meeting, we managed to significantly improve upon these first results and we demonstrated deceleration of metastable CO ($a^3\Pi_1$, $J = 1$) molecules in low-

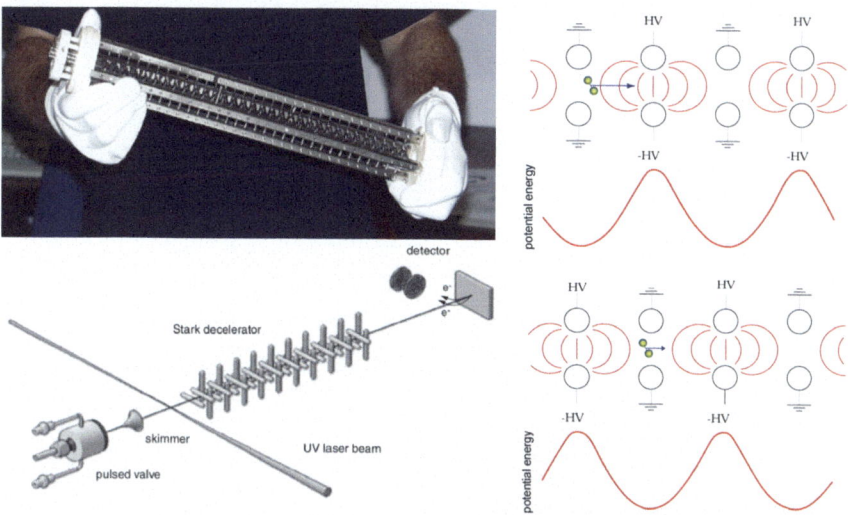

Fig. 2 Rick Bethlem, holding the prototype Stark-decelerator in his hands. The hexapole that is used to optimize the coupling of the beam into the Stark-decelerator can be seen sticking out on the left. A scheme of the experimental setup, omitting the hexapole, is shown underneath the photograph. A pulsed molecular beam is produced by expanding a mixture of 5% of CO seeded in Xe through a pulsed valve into vacuum. After passing through a skimmer, the CO molecules are excited with a pulsed laser to the low-field seeking, upper Λ-doublet component of the $J = 1$ level in the $a^3\Pi_1$ state. The metastable molecules then pass through the Stark-decelerator and are detected by monitoring the electrons that are emitted when they impinge on a clean gold surface. On the right, the two electric field configurations that are alternately used to slow down the metastable CO molecules are schematically shown.

field seeking states from 225 m/s to 98 m/s reducing their kinetic energy by almost 0.8 cm^{-1} per electric field stage [32]. Two years after this, we demonstrated phase-stable acceleration (and deceleration) of – again – CO ($a^3\Pi_1$, $J = 1$) molecules, but this time in high-field seeking states, using an array of twelve dipole lenses in alternate-gradient configuration [33]. These were the successful demonstrations of the experiments that John King and Lennard Wharton set out to perform almost forty years earlier, made possible by the choice of our system, that is, by the advantages that laser-prepared, metastable CO molecules offered, and by the advances in high-voltage switching technology.

When our first manuscript on the Stark-decelerator was still under review, I presented the main results and our future plans during a workshop at ITAMP, in July 1999. There, Daniel Kleppner informed me that although none of the attempted deceleration work of John King had been published, there should exist a Ph.D. thesis from one of his students, and we subsequently traced down the Ph.D. thesis of Robert Golub. Interestingly enough, Robert Golub mentions in his Ph.D. thesis from 1967 that "there has recently been a proposal to use a similar scheme to accelerate molecules by L. Wharton" [20]; it remains unclear in how far Lennard Wharton was

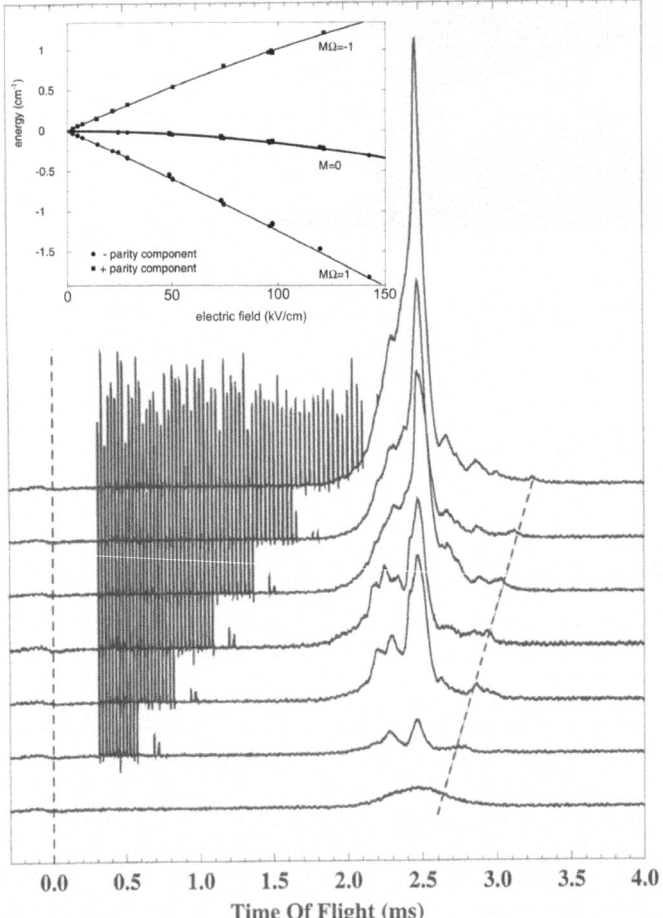

Fig. 3 Observed arrival time distribution of metastable CO molecules when an increasing number of deceleration stages is being used, from none at all (lowest curve) to all of them (uppermost curve). With no fields applied, the arrival time is centered around 2.5 ms after firing of the pulsed laser, i.e., 2.5 ms after preparation of the metastable CO molecules, with a more-or-less Gaussian distribution. The spikes in the signal prior to the arrival of the molecules on the detector are due to electrical noise from the high-voltage switches, and indicate how often the fields have been switched. When electric fields are applied, transverse focusing takes place, and more molecules are seen to arrive on the detector, with a highly structured temporal distribution. The blue, dashed line indicates the arrival time of the synchronous molecules, i.e., those CO molecules that experience the aimed-for, constant amount of deceleration per stage, and these are seen to arrive later when more deceleration stages are being used. In the inset, the measured (solid dots and squares) and calculated Stark shift of the components of the $J = 1$ level in the $a^3\Pi_1$ state of CO are shown in electric fields up to 150 kV/cm.

aware of John King's experiments. During the same workshop, Hossein Sadeghpour informed me on ongoing experiments in the group of Harvey Gould in Berkeley, aimed at decelerating molecules with electric fields. Shortly after our work was published, they published an article in which they presented data on the deceleration of Cs atoms with time-varying electric fields [34].

5   Concluding Remarks

The Stark-decelerator has made it possible to produce packets of state-selected molecules, oriented in space, with computer-controlled six-dimensional phase-space distributions. This level of control of molecular beams has first been demonstrated with electric fields, as outlined in this Chapter, but has also been obtained using time-varying magnetic fields and – to a lesser extent – using electro-magnetic radiation fields by now. Together, these methods have made a whole variety of new experiments possible. It would go too far, to (try to) list these here, and the interested reader is referred to the earlier mentioned Review from 2012 [2], and the references therein. As selected highlights that have appeared since then, I would like to mention the experimental realization of a molecular fountain by Rick Bethlem and co-workers [35], the demonstration of a cryogenic molecular centrifuge in the group of Gerhard Rempe [36], the magnetic deceleration and trapping of molecular oxygen in the group of Edvardas Narevicius [37] and the high-resolution collision experiments in the group of Bas van de Meerakker [38].

"If one extends the rules of two-dimensional focusing to three dimensions, one possesses all ingredients for particle trapping." This is how Wolfgang Paul stated it in his Nobel lecture [39], and as far as the underlying physics principles of particle traps are concerned, it is indeed as simple as that. To experimentally realize the trapping of neutral particles, however, the main challenge is to produce particles that are sufficiently slow that they can be trapped in the relatively shallow traps that can be made. When the particles are confined along a line, rather than around a point, the requirements on the kinetic energy of the particles are more relaxed, and storage of neutrons in a one meter diameter magnetic hexapole torus could thus be demonstrated first [40]. Trapping of atoms in a 3D trap only became feasible when Na atoms were laser cooled to sufficiently low temperatures that they could be confined in a quadrupole magnetic trap [41]. The Stark-decelerator enabled the first demonstration of 3D trapping of neutral ammonia molecules in a quadrupole electrostatic trap [42] even before it was used in the demonstration of an electrostatic storage ring for neutral molecules [43].

There obviously are large similarities between the manipulation of polar molecules and the manipulation of charged particles, and concepts used in the field of charged particle physics can and have been applied to neutral polar molecules, and *vice versa*. Both Hartmut Kallmann and Wolfgang Paul worked on the deflection and focusing of beams of neutral molecules before they turned their attention to controlling the motion of charged particles; it is interesting to realize that multipole fields were actually used

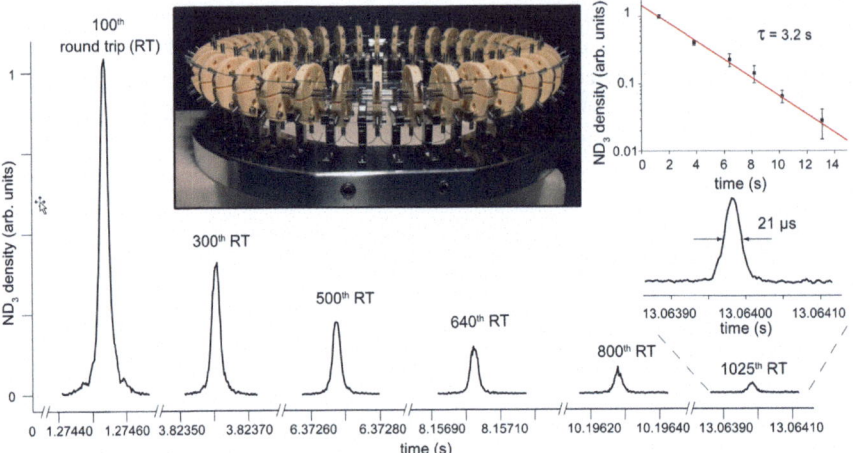

Fig. 4 Photograph of the 0.5 m diameter molecular synchrotron, consisting out of 40 straight hexapoles arranged in a circle. Deuterated ammonia molecules with a velocity of 125 m/s are injected in the synchrotron and stay confined to a 2.5 mm long packet, also after having made more than thousand round trips, i.e., after having travelled a distance of over one mile. The inset shows that the amount of trapped molecules decays with a $1/e$ time of 3.2 s, caused in equal parts by optical pumping due to black-body radiation and collisions with background gas.

in molecular beam physics first. Inspiration from charged particle physics has been instrumental for the development of the Stark-decelerator, the LINAC for neutral, polar molecules. It has also inspired the realization of a molecular synchrotron, shown in Fig. 4, in which state-selected, neutral molecules are kept together in a compact packet for a distance of over one mile, extending their duration on stage to many seconds [44].

References

1. G. Scoles, Ed., *Atomic and Molecular Beam Methods*, vol. 1 & 2 (Oxford University Press, New York, NY, USA, 1988 & 1992). ISBN 0195042808
2. S.Y.T. van de Meerakker, H.L. Bethlem, N. Vanhaecke, G. Meijer, Manipulation and control of molecular beams. Chem. Rev. **112**, 4828–4878 (2012)
3. H. Kallmann, F. Reiche, Über den Durchgang bewegter Moleküle durch inhomogene Kraftfelder. Zeitschrift für Physik **6**, 352–375 (1921)
4. O. Stern, Ein Weg zur experimentellen Prüfung der Richtungsquantelung im Magnetfeld. Zeitschrift für Physik **7**, 249–253 (1921)
5. W. Gerlach, O. Stern, Der experimentelle Nachweis der Richtungsquantelung im Magnetfeld. Zeitschrift für Physik **9**, 349–352 (1922)
6. E. Wrede, Über die Ablenkung von Molekularstrahlen elektrischer Dipolmoleküle im inhomogenen elektrischen Feld. Zeitschrift für Physik **44**, 261–268 (1927)
7. I.I. Rabi, S. Millman, P. Kusch, J.R. Zacharias, The molecular beam resonance method for measuring nuclear magnetic moments. Phys. Rev. **55**, 526–535 (1939)

8. H. Friedburg, W. Paul, Optische Abbildung mit neutralen Atomen. Die Naturwissenschaften. **38**, 159–160 (1951)
9. H.G. Bennewitz, W. Paul, Eine Methode zur Bestimmung von Kernmomenten mit fokussiertem Atomstrahl. Zeitschrift für Physik **139**, 489–497 (1954)
10. H.G. Bennewitz, W. Paul, C. Schlier, Fokussierung polarer Moleküle. Zeitschrift für Physik **141**, 6–15 (1955)
11. J.P. Gordon, H.J. Zeiger, C.H. Townes, Molecular microwave oscillator and new hyperfine structure in the microwave spectrum of NH_3. Phys. Rev. **95**, 282–284 (1954)
12. J.P. Gordon, H.J. Zeiger, C.H. Townes, The maser—new type of microwave amplifier, frequency standard, and spectrometer. Phys. Rev. **99**, 1264–1274 (1955)
13. R. Levine, R. Bernstein, *Molecular Reaction Dynamics and Chemical Reactivity* (Oxford University Press, New York, 1987)
14. N.F. Ramsey, *Molecular Beams*, in The International Series of Monographs on Physics (Oxford University Press, London, 1956)
15. N.F. Ramsey, http://nobelprize.org/nobel_prizes/physics/laureates/1989/ramsey-autobio.html
16. D.C. Lainé, Molecular beam masers. Rep. Prog. Phys. **33**, 1001–1067 (1970)
17. L.D. White, Ammonia maser work at Bell telephone laboratories, in *Proceedings of the 13th Annual Symposium on Frequency Control* (Fort Monmouth, U.S. Army Signal Research and Development Laboratory, 1959), pp. 596–602
18. N.G. Basov, A.N. Oraevskii, Use of slow molecules in a maser. Soviet Phys. JETP **37**, 761–763 (1960)
19. J.G. King, Experiments with slow molecules, in *Proceedings of the 13th Annual Symposium on Frequency Control* (Fort Monmouth, U.S. Army Signal Research and Development Laboratory, Asbury Park, 1959), p. 603
20. R. Golub, *On Decelerating Molecules* (Ph.D. thesis, MIT, Cambridge, USA, 1967)
21. R. Wolfgang, Chemical accelerators. Sci. Am. **219**(4), 44–52 (1968)
22. D. Auerbach, E.E.A. Bromberg, L. Wharton, Alternate-gradient focusing of molecular beams. J. Chem. Phys. **45**, 2160–2166 (1966)
23. E.E.A. Bromberg. *Acceleration and Alternate-Gradient Focusing of Neutral Polar Diatomic Molecules*. Ph.D. thesis, University of Chicago, USA (1972)
24. N. Abuaf, J.B.A.R.P. Andres, J.B. Fenn, D.G.H. Marsden, Molecular beams with energies above one volt. Science **155**, 997–999 (1967)
25. E.A. Hinds, Private communication (Heidelberg, 1999)
26. J. Verhoeven, A. Dymanus, Magnetic properties and molecular quadrupole tensor of the water molecule by beam-maser Zeeman spectroscopy. J. Chem. Phys. **52**, 3222–3233 (1970)
27. W.L. Meerts, A. Dymanus, Electric and magnetic properties of carbon monoxide by molecular-beam electric-resonance spectroscopy. Chem. Phys. **22**, 319–324 (1977)
28. R.T. Jongma, Th Rasing, G. Meijer, Two-dimensional imaging of metastable CO molecules. J. Chem. Phys. **102**, 1925–1933 (1995)
29. J.J. Gilijamse, S. Hoekstra, S.A. Meek, M.Metsälä, S.Y.T. van de Meerakker, G. Meijer, G.C. Groenenboom, The radiative lifetime of metastable CO ($a^3 \Pi$,v=0). J. Chem. Phys. **127**, 221102-1–221102-4 (2007)
30. R.T. Jongma, G. von Helden, G. Berden, G. Meijer, Confining CO molecules in stable orbits. Chem. Phys. Lett. **270**, 304–308 (1997)
31. G. von Helden, I. Holleman, G.M.H. Knippels, A.F.G. van der Meer, G. Meijer, Infrared resonance enhanced multiphoton ionization of fullerenes. Phys. Rev. Lett. **79**, 5234–5237 (1997)
32. H.L. Bethlem, G. Berden, G. Meijer, Decelerating neutral dipolar molecules. Phys. Rev. Lett. **83**, 1558–1561 (1999)
33. H.L. Bethlem, A.J.A. van Roij, R.T. Jongma, G. Meijer, Alternate gradient focusing and deceleration of a moleclar beam. Phys. Rev. Lett. **88**, 133003-1–133003-4 (2002)
34. J.A. Maddi, T.P. Dinneen, H. Gould, Slowing and cooling molecules and neutral atoms by time-varying electric-field gradients. Phys. Rev. A **60**, 3882–3891 (1999)

35. C. Feng, A.P.P. van der Poel, P. Jansen, M. Quintero-Pérez, T.E. Wall, W. Ubachs, H.L. Bethlem, Molecular fountain. Phys. Rev. Lett. **117**, 253201-1–253201-5 (2016)
36. X. Wu, T. Gantner, M. Koller, M. Zeppenfeld, S. Chervenkov, G. Rempe, A cryofuge for cold-collision experiments with slow polar molecules. Science **358**, 645–648 (2017)
37. Y. Segev, M. Pitzer, M. Karpov, N. Akerman, J. Narevicius, E. Narevicius, Collisions between cold molecules in a superconducting magnetic trap. Nature **572**, 189–193 (2019)
38. T. de Jongh, M. Besemer, Q. Shuai, T. Karman, A. van der Avoird, G.C. Groenenboom, S.Y.T. van de Meerakker, Imaging the onset of the resonance regime in low-energy NO-He collisions. Science **368**, 626–630 (2020)
39. W. Paul, Electromagnetic traps for charged and neutral particles. Angew. Chem. Int. Ed. Engl. **29**, 739–748 (1990)
40. K.-J. Kügler, W. Paul, U. Trinks, A magnetic storage ring for neutrons. Phys. Lett. B. **72**, 422–424 (1978)
41. A.L. Migdall, J.V. Prodan, W.D. Phillips, T.H. Bergeman, H.J. Metcalf, First observation of magnetically trapped neutral atoms. Phys. Rev. Lett. **54**, 2596–2599 (1985)
42. H.L. Bethlem, G. Berden, F.M.H. Crompvoets, R.T. Jongma, A.J.A. van Roij, G. Meijer, Electrostatic trapping of ammonia molecules. Nature **406**, 491–494 (2000)
43. F.M.H. Crompvoets, H.L. Bethlem, R.T. Jongma, G. Meijer, A prototype storage ring for neutral molecules. Nature **411**, 174 (2001)
44. P.C. Zieger, S.Y.T. van de Meerakker, C.E. Heiner, H.L. Bethlem, A.J.A. van Roij, G. Meijer, Multiple packets of neutral molecules revolving for over one mile. Phys. Rev. Lett. **105**, 173001-1–173001-4 (2010)

Chapter 21
Quantum Effects in Cold and Controlled Molecular Dynamics

Christiane P. Koch

Abstract This chapter discusses three examples of quantum effects that can be observed in state-of-the-art experiments with molecular beams—scattering resonances as a probe of interparticle interactions in cold collisions, the protection of Fano-Feshbach resonances against decay despite resonant coupling to a scattering continuum, and a circular dichroism in photoelectron angular distributions arising in the photoionization of randomly oriented chiral molecules. The molecular beam setup provides molecules in well-defined quantum states. This, together with a theoretical description based on first principles, allows for excellent agreement between theoretical prediction and experimental observation and thus a rigorous understanding of the observed quantum effects.

1 Introduction

When you ask young students entering a university physics course today for the term they associate most with quantum mechanics, many of them will respond with "entanglement". This reflects the rise of quantum information science out of an often ridiculed ivory tower to the decision-making levels of the big tech companies and to the headlines of well-respected media outlets. In the waves of excitement created by the "second quantum revolution" [1, 2], it may be overlooked that features more traditionally associated with quantum mechanics continue to fascinate and challenge our classically trained intuition.

This chapter reviews three examples of such quantum effects beyond entanglement from recent work of my group—tunneling resonances that emerge in cold collisions and that can be used to probe interparticle interactions [3], Fano-Feshbach resonances that can be protected against decay by a suitable phase condition [4], and quantum pathway interference in the circular dichroism of photoelectrons that is observed after the photoionization of chiral molecules [5, 6]. All three examples share a rigorous theoretical description based on first principles. More importantly

C. P. Koch (✉)
Arnimallee 14, 14195 Berlin, Germany
e-mail: christiane.koch@fu-berlin.de

© The Author(s) 2021
B. Friedrich and H. Schmidt-Böcking (eds.), *Molecular Beams in Physics and Chemistry*,
https://doi.org/10.1007/978-3-030-63963-1_21

for the present contribution, the quantum effects discussed here have been observed in experiments with molecular beams [3, 4, 7]. They thus testify to the topicality and continuing significance of the molecular beam technique developed by Otto Stern and colleagues.

2 Quantum Scattering Resonances in Cold Collisions

The wave nature of colliding particles emerges most prominently at low scattering energies [8, 9], where quantum resonances dominate the scattering cross section before threshold laws take over [10]. While quantum resonances are also present at higher scattering energies, this presence is hidden in the ensemble average over all quantum states that are populated at a given energy. In other words, collisions are "cold" when only a few partial waves contribute to the scattering cross section [8]. At the corresponding collision energies, the dynamics are often dominated by the long-range behavior of the interparticle interactions [10], easing the interpretation of the collision studies.

In order to observe quantum scattering resonances experimentally, one needs to ensure a sufficiently narrow velocity distribution—narrower than the resonance width—in addition to the capability to finely tune the relative kinetic energy of the colliding particles to very low values. Both requirements can be met with merged neutral beams [11, 12], the use of which has allowed for the observation of tunneling resonances in Penning ionization reactions [13]. Since then, scattering resonances have also been observed in inelastic low-energy collisions, see Ref. [14] and references therein.

Penning ionization reactions occur when the excitation energy of a particle prepared in a metastable quantum state is sufficient to ionize its collision partner [15]. Metastable nobel gas atoms, for example, feature excitation energies of more than 10 eV which is above the ionization potential of most molecules. The scattering resonances in this case are tunneling, or shape, resonances that occcur when the kinetic energy of the colliding particles matches the energy of a quasi-bound state that is trapped behind the rotational barrier of an otherwise barrier-less potential, cf. Fig. 1. Peaks in the ionization cross section indicate not only the presence of a tunneling resonance but also highlight the corresponding quantization of the intermolecular motion. In a more pictorial description, when the colliding particles hit a resonance upon tunneling through the rotational barrier, their amplitude gets trapped at short interparticle distances. This results in peaks in the cross section since the ionization probability grows inverse-exponentially with interparticle distance [15].

The experiments leading to the first observation of tunneling resonances in Penning ionization reactions were carried out with metastable helium colliding with argon atoms and dihydrogen molecules [13]. At the time, no qualitative difference was observed for the Penning ionization of an atom compared to that of a molecule. In general, however, one would expect the molecular degrees of freedom—rotations or vibrations—as well as the anisotropy of the interparticle interaction potential (the

Fig. 1 Tunneling, or shape, resonances (top) versus Fano-Feshbach resonances (bottom): Tunneling resonances form on a single potential curve displaying a barrier. In contrast, Fano-Feshbach resonances arise from the coupling of scattering states to a bound state belonging to another scattering channel which is energetically closed and asymptotically characterized by a different set of quantum numbers. In both cases, scattering amplitude gets trapped at short interparticle distance

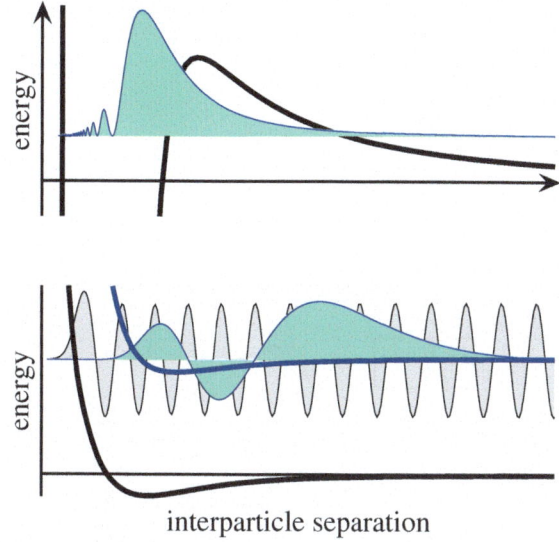

interparticle separation

difference between the molecule hitting the metastable atom head-on or in a T-shaped geometry) to come into play and modify the cross section, respectively the reaction rate. For low-energy collisions involving dihydrogen molecules, a change of the H_2 vibrational state is energetically not accessible. In order to assess the role of the anisotropy of the interparticle interaction, it is useful to expand the angular dependence of the interaction potential into Legendre polynomials,

$$V(R, \theta) = V_0(R) + V_2(R) \cos^2 \theta + \cdots .$$

Here, R and θ denote the interparticle separation and the angle between intermolecular axis and collision axis, respectively. A comparison of the magnitude of $V_2(R)$ to the rotational constant of the molecule yields, for H_2, energy equivalents of 1 K versus roughly 200 K. The collision is thus highly unlikely to induce changes of the rotational state. Nevertheless, the anisotropy of the interaction does affect the reaction rate! It determines the occurrence of quantum scattering resonances in slow barrier-less reactions [3] and dictates the scaling of the reaction rate coefficient with collision energy in fast barrier-less reactions [16].

The role of the anisotropy for the occurence of quantum scattering resonances is most readily understood using an adiabatic separation of interparticle motion and internal molecular rotation [17]. Denoting the eigenstates of the internal molecular rotation by $|j, m_j\rangle$, the potential energy operator is effectively diagonal in this basis (reflecting the absence of rotational state changing collisions), but the diagonal matrix elements differ for $j = 0$ and $j = 1$: While the matrix element for $j = 0$ contains only the isotropic part of the interaction, $V_0(R)$, those for $j = 1$ also depend on the anisotropy, $V_2(R)$. This allows one to directly probe the role of the anisotropy

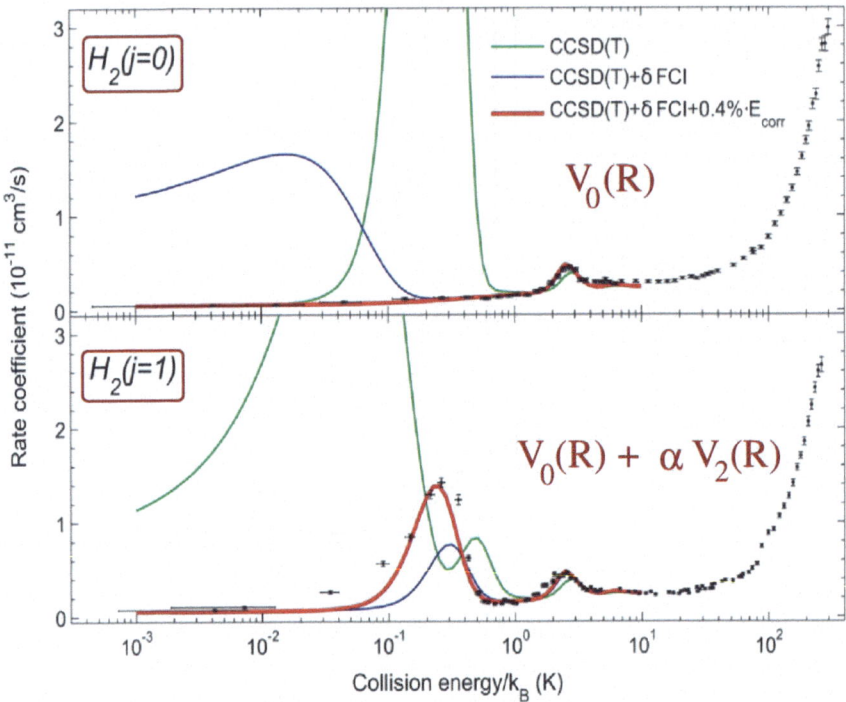

Fig. 2 Rate coefficient for Penning ionization of dihydrogen colliding with metastable helium: The tunneling resonance at a collision energy of $k_B \times 270$ mK appears only if the molecule is rotationally excited. This is due to the anisotropy of the interparticle interaction (equal to V_2 in leading order). Adapted from Ref. [3]

in atom-molecule collisions via quantum scattering resonances [3], provided the molecule can be selectively prepared in $j = 0$ or $j = 1$.

Rotational state selection in H_2 molecules is made possible by the large rotational splitting together with the nuclear spin symmetry. The latter implies that para-H_2 must have an even rotational wavefunction, thus $j = 0$, whereas the rotational wavefunctions for ortho-H_2 are odd, i.e., $j = 1$ for ortho-dihydrogen. Higher rotational levels are not populated in molecular beam experiments due to the large rotational splitting. Samples of almost pure para-H_2 are readily obtained by catalytic ortho-para conversion. It is thus comparatively straightforward to carry out molecular beam studies with rotational state selection [3, 16].

Repeating the Penning ionization experiments of Ref. [13] with dihydrogen molecules in a well defined rotational state led to the surprising observation, cf. Fig. 2, of the low-energy resonance (at a collision energy of $k_B \times 270$ mK) occuring only for ortho-H_2, i.e., only for rotationally excited molecules. A theoretical model capturing this phenomenon requires spectroscopic accuracy. Systematic corrections to high-level state-of-the-art quantum chemistry combined with coupled channels calculations allow for reaching this accuracy. This is illustrated by the green, blue and red curves in Fig. 2, depicting the results obtained with a potential energy surface derived by coupled cluster theory with single, double and non-iterative triple excitations (green), including the full CI correction (blue) and further improved by uniformly scaling the correlation energy by 0.4% (red) [3].

The excellent agreement between the theoretical results and the experimental data observed in Fig. 2 allows for a more indepth examination of the role of the anisotropy. In the theoretical model, we can easily modify the relative weight of the anisotropic part of the interaction potential, denoted by α in Fig. 2. Increasing α up to 50% does not introduce a noticeable change on the results obtained with para-H_2, i.e., molecules in their rotational ground state [3]. On the other hand, the low-energy resonance observed only for rotationally excited H_2 in Fig. 2 depends very sensitively on the scaling factor of the anisotropic potential, shifting to lower energies with decreasing anisotropy [3].

The reason underlying this behavior can be unveiled using the adiabatic approximation mentioned above [17]. As is often the case with perturbation theory, it does not yield a quantitative description but qualitatively provides the correct picture. When examining the adiabatically separated scattering channels, a bound state just below the dissociation threshold occurs for rotational ground state para-dihydrogen (with $\ell = 3$, $j = 0$, and $J = 3$). For rotationally excited ortho-dihydrogen, the anisotropic part of the interparticle interaction potential introduces an energy shift that is added to the effective potential. This pushes, what is a weakly bound state for para-dihydrogen, above the dissociation threshold for ortho-dihydrogen, turning the bound state into a shape resonance (with $\ell = 3$, $j = 1$, and $J = 3$) [3], thus solving the riddle of resonance (dis)appearance.

To conclude this section, quantum scattering resonances testify to the emergence of the wave nature of matter in "cold" collisions. Using dihydrogen molecules in merged beam studies allows for simple rotational state selection which in turn can be used to probe the anisotropic part of the interparticle interaction governing Penning ionization reactions [3, 16]. While an excellent agreement between theory and experiment can be reached when including appropriate corrections for electron correlations, the calculation of quantum resonances involving metastable states continues to present a significant challenge even for the highest available levels of first principles based theory.

3 Phase-Protection in Fano-Feshbach Resonances

After understanding the dramatic effects that very small shifts in energy may have on a shape resonance, in the present section, another type of quantum resonance will highlight the sensitivity of resonances to small changes in phase. A Fano-Feshbach resonance describes the decay of a bound quantum state due to coupling with a continuum of scattering states. In contrast to shape resonances, it involves two distinct scattering channels characterized by different quantum numbers, as sketched in Fig. 1. In the original theory, the coupling between bound and continuum states described an effective nucleon-nucleon interaction [18], resp. configuration interaction in autoionization [19]. In recent years, Fano-Feshbach resonances due to the hyperfine interaction between different nuclear spin states have attracted attention as key tool for control in ultracold gases [20]. The present example considers rovibrational predissociation resonances due to the spin-orbit interaction [21].

Predissociation resonances may be populated by associative ionization which accompanies a Penning ionization reaction, provided the dependence of the Penning ionization rate on interparticle separation matches the ionic potential energy curve [15]. The ionization then populates bound levels of noble gas diatomic molecules such as $HeAr^+$, $NeAr^+$, or $HeKr^+$ which may or may not be spin-excited. These seemingly simple molecules with very similar electronic structure possess (spin-excited) predissociation resonances with surprisingly different lifetimes.

In order to estimate lifetimes, one typically inspects the coupling responsible for the decay. When comparing predissociation in $HeAr^+$ and $NeAr^+$, the spin-orbit splitting Δ in the two cases is identical. The term dominating the coupling between spin-excited and ground states is radial coupling [4] which scales as Δ/μ, where μ is the reduced mass. The lifetime of the resonances then scales with μ^2. Since, for the two diatoms, the reduced masses differ by a factor of approximately four, the lifetime of $NeAr^+$ would be expected to be larger than that of $HeAr^+$ by about an order of magnitude. This expectation derived from scaling arguments ignores, however, a phase dependence of the lifetimes which may entirely alter the picture, as shown next.

A qualitative understanding of the resonance lifetimes can be obtained in first order perturbation theory, using Fermi's golden rule:

$$\tau_\varphi^{-1} \sim |\langle k_{\text{res}}|\mathbf{W}_{\text{coupl}}|\varphi\rangle|^2, \tag{1}$$

where $|\varphi\rangle$ denotes the quasi-bound state, $\mathbf{W}_{\text{coupl}}$ the coupling operator, and $|k_{\text{res}}\rangle$ the continuum state with scattering momentum k_{res}, in resonance with the quasi-bound state, cf. Fig. 1. Since the continuum state describes the scattering off a potential, a phase shift δ is associated with the position of the repulsive wall. It is this phase shift that determines the value of the complex overlap in Eq. (1). In particular, there exist combinations of k and δ for which the overlap vanishes such that $\tau_\varphi \rightarrow \infty$ despite non-zero coupling! We have termed this phenomenon "phase protection" [4]. In fact, it is straightforward to show that vanishing overlap is equivalent to

$$\arg(\tilde{\varphi}(k)) + \delta = m\pi \quad \text{with } m \in \mathbb{Z} \tag{2}$$

with $\tilde{\varphi}(k)$ the Fourier transform of the quasi-bound state, when neglecting the energy dependence of the phase shift δ and assuming s-wave scattering.

In a real molecule, the resonance lifetime will never strictly go to infinity since eventually other decay mechanisms will become relevant. However, the lifetime can indeed become very large. This is shown in Fig. 3 displaying the potential energy curves, derived from spectroscopic data, respectively coupled-cluster calculations,

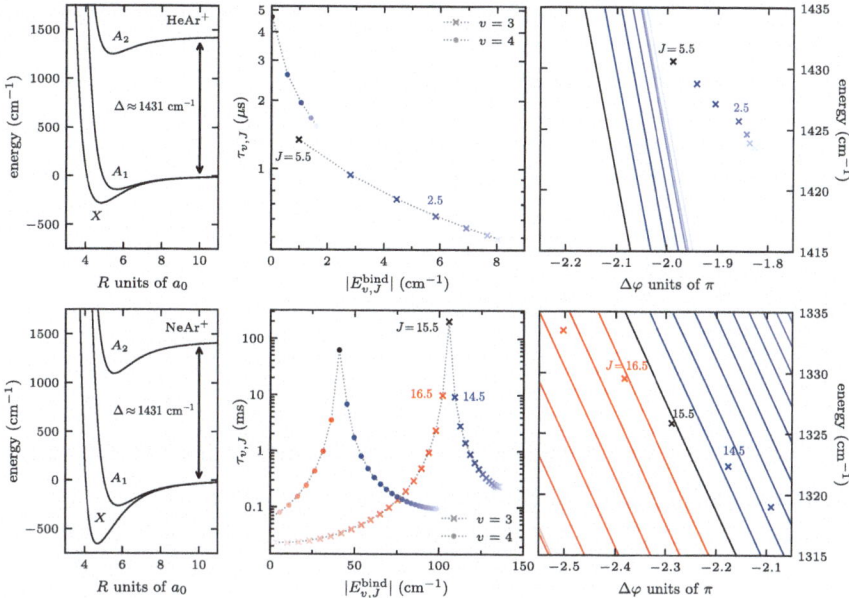

Fig. 3 Potential energy curves (left), lifetimes (center), and resonance positions (right) of predissociation resonances in spin-excited noble gas dimers. Adapted from Ref. [4]

for $HeAr^+$ and $NeAr^+$, together with the lifetimes obtained with a complex absorbing potential and including relativistic and angular couplings [4]. The potential energy curves shown in the left-hand part of Fig. 3 highlight the similarity of the potential energy curves and spin-orbit coupling Δ for the two molecules. In contrast, a striking difference is observed in the middle panel of Fig. 3, where the predissociation lifetimes of spin-excited $NeAr^+$ span more than four orders of magnitude whereas those of $HeAr^+$ differ by only a factor of 10. This difference is rationalized in terms of the condition for phase protection, cf. Eq. (2), visualized in the right part of Fig. 3. The lines indicate the combinations of k (respectively scattering energy) and δ for which the overlap in Eq. (1) vanishes. For $NeAr^+$, several resonance positions are located very close to those lines. The closer the resonance position to its corresponding phase protection line, the larger becomes the corresponding lifetime. In contrast, for $HeAr^+$, none of the resonances fulfills even approximately the condition for phase protection which is why the lifetimes span a much smaller range.

Experimental evidence for the lifetimes of spin-excited $NeAr^+$ spanning several orders of magnitude is provided by two experiments [4], using molecular beams. First, velocity-map imaging (VMI) of the reaction products of Penning and associative ionization of argon by metastable helium revealed predissociation for $HeAr^+$ to occur on a timescale smaller than the time of flight of the VMI setup, of the order of $10\,\mu s$. In contrast, the predissociation feature was missing in the VMI for $NeAr^+$, suggesting lifetimes significantly larger than the time of flight [4]. A second experiment, designed to assess the range of NeAr$^+$ lifetimes more directly, injected the molecular ions generated by associative ionization into an electrostatic ion beam trap [22] and recorded oscillations of the molecules between two electrostatic mirrors for several hundreds of milliseconds via trap loss of neutral particles. The experimental data was found to decay non-exponentially, with decay times ranging from below 50 μs up to 100 ms [4], in excellent agreement with the predictions of Fig. 3.

One may wonder whether there is a way to predict, for a given molecule, the chance of phase protection to occur. Assuming one can approximate the potential energy curve by a Morse potential, the lifetime of level v is determined by a phase and the Fourier transform of the corresponding eigenfunction, $\tilde{\psi}_v(k)$,

$$\tau_v^{-1} \propto \sin{(\delta + \phi_v(k))^2} \left| \tilde{\psi}_v(k) \right|^2,$$

where $\phi_v(k) = \arg[\tilde{\psi}_v(k)]$. On top of a smooth variation with k (or energy) due to $\left| \tilde{\psi}_v(k) \right|^2$, the lifetimes τ_v oscillate and vanish whenever the condition for phase protection, cf. Eq. (1), is fulfilled. In order to answer the question how often the latter will happen, one can exploit the fact that, for a Morse potential, $\tilde{\psi}_v(k)$ is known analytically, in terms of complex Γ-functions [23]. This allows for deriving the scaling of the lifetimes with the parameters of the Morse potential and the reduced mass: The number of times τ_v vanishes, and thus the chance for phase protection, increases with μ as well as the well depth and equilibrium distance and decreases with the potential width [4]. This agrees well with the observations for NeAr$^+$ and HeAr$^+$ in Fig. 3 since the reduced mass and the well depth are larger for NeAr$^+$ than for HeAr$^+$, while the position of the minimum and the potential width are very similar. Moreover, this scaling argument predicts an isotope effect on predissociation lifetimes which has indeed been observed experimentally for N$_2^+$ [24] and Ne$_2^+$ [25].

To summarize this section, phase protection refers to the fact that the lifetime of a quasi-bound quantum state can become very, very large despite non-zero coupling with a continuum of scattering states. The protection from decay occurs whenever the relative phase in the overlap of bound and scattering state becomes a multiple of π. Experiments with spin-orbit excited noble gas dimers have confirmed our theoretical prediction of phase protection. The probability of phase protection increases with reduced mass, well depth and equilibrium distance of the potential supporting the quasi-bound state, providing a blueprint to identify quantum states that are intrinsically protected against undesired decay.

4   Photoelectron Circular Dichroism and Its Coherent Control

The third example showcases a quantum effect that is observed in experiments with molecular beams when circularly polarized light ionizes molecules which are chiral [7]. Remarkably, when a molecule is chiral, i.e., when its nuclear scaffold exhibits a handedness, ionization of randomly oriented molecules with right circularly polarized light does not yield the same photoelectron angular distribution as that with left circularly polarized light [26]. The difference between the photoelectron angular distributions obtained with left and right circularly polarized light is called photoelectron circular dichroism (PECD). Instead of exchanging the polarization direction of the light, one can also exchange the handedness of the molecule to observe PECD [26]. Unlike other dichroic effects, PECD does not involve a magnetic dipole moment and is obtained merely within the electric dipole approximation of the light-matter interaction [26].

First order perturbation theory for the photoionization cross section reveals the mechanism underlying the dichroic effect [26]: For a chiral molecule, the photoionization matrix elements with opposite spatial orientation do not cancel when averaging over the Euler angles. This gives rise to terms in the cross section that are odd under inversion of the polar angle and thus to dichroism. More intuitively, the photoionization cross section involves two rotations—of the polarization axis into the molecular frame and of the photoelectron momentum into the lab frame—which are sufficient to be sensitive to the handedness of the molecular scaffold. This geometric picture can be made more rigorous by noticing that the angle-resolved photoionization cross section is a vector observable which can be expressed as a triple product [27]. For non-coplanar vectors forming the triple product, the observable becomes enantiosensitive [27].

While theoretically predicted in the mid-1970s [26] and first observed with synchrotron radiation in the early 2000s [28], femtosecond laser pulses driving multiphoton ionization have made PECD accessible in table-top experiments [7, 29–32]. Since multi-photon ionization probes intermediate electronically excited states [7, 33], a theoretical description beyond first order in the perturbation theory for the light-matter interaction is called for. At the same time, the bicyclic ketones with which the experiments have been carried out, for example, fenchone, camphor, or limonene, are amenable to a high-level treatment of their electronically excited states. In contrast, modeling the photoionization continuum from first principles is rather challenging.

A way to address this challenge, applied to the specific example of resonantly enhanced (2+1) multi-photon ionization (REMPI) of fenchone and camphor [5], separates the non-resonant two-photon excitation from the one-photon ionization. The former can be described with coupled cluster theory whereas for the latter a single active electron approach using hydrogenic orbitals captures the essential physics of the photoelectron moving in a Coulombic potential [5]. Neglecting any coherent effects during the excitation, the ionization probes an anisotropic distribution of electronically excited molecules. This is due to the anisotropy of the two-photon absorption tensors of the molecules [5]. The properties of the two-photon absorption alone are, however, not sufficient to determine which intermediate state is probed by the REMPI process. Based on energetic arguments, there are five possible candidate states, only one of which is ruled out when using the information of the two-photon absorption tensors in the calculation of the photoionization cross section [5]. When including also the properties of the probed electronically excited states, only one out of the five candidates is in agreement with the experimental data [5]. Interestingly, the theoretically predicted intermediate state differs for fenchone compared to camphor [5]. This might explain the different photoelectron angular distributions and different PECD observed for these two chemically very similar molecules [7].

There are many intriguing questions that arise in the context of PECD, for example whether the effect is determined by the initial, the final, or the intermediate state of the process. In order to answer such questions, a more rigorous description of PECD and related effects in the ionization of chiral molecules is required. In particular, a better description of the photoionization continuum including electron correlation effects and the ability to treat coupled electronic and vibrational motion are called for. At the same time, PECD provides a very convenient experimental handle to probe the chiroptical response of molecules, and it is natural to ask whether this response can be enhanced by suitably shaping the ionizing field, in the spirit of coherent control of photoinduced dynamics [34, 35].

This question can be answered by combining the time-dependent configuration interaction singles (CIS) method for the electronic structure [36], second order perturbation theory for the light-matter interaction, and parameter optimization for the electric field using sequential parametrization updates [37]. Figure 4 illustrates the ionization pathways that are accounted for in this description. Pathways ending at the same final kinetic energy of the photoelectron (within the spectral bandwidth of the pulse) can interfere with each other, cf. the red and green arrows in Fig. 4. This can be thought of as a spectral realization of a double-slit experiment, with the additional advantage that the relative phase between the two pathways can be adjusted by properly tuning the ionizing electric field in its amplitude and phase [35].

A bichromatic pulse driving the red and green ionization pathways in Fig. 4 is thus a natural starting point for optimizing the electric field. When ionizing a chiral methan derivate with such a pulse, assuming a flat spectral phase, a PECD of about 4% (relative to the isotropic ionization yield) is obtained [6]. When optimizing the

Fig. 4 One-photon and two-photon ionization pathways within the time-dependent CIS framework

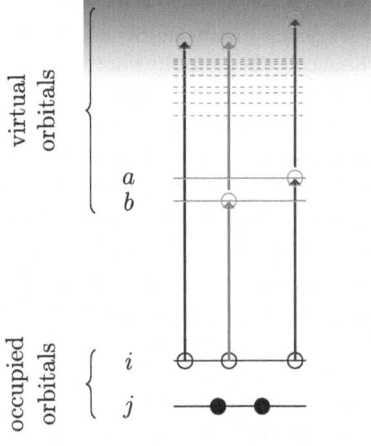

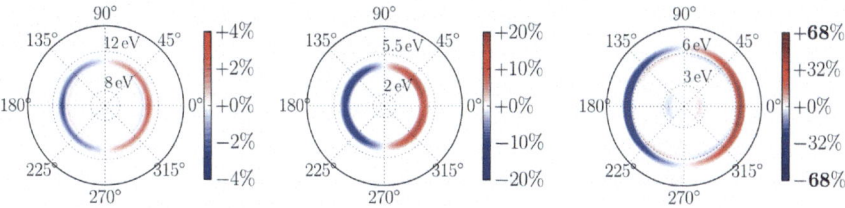

Fig. 5 PECD after ionization of CHBrClF with a bichromatic guess pulse with flat spectral phase (left), optimized bichromatic (center) and freely optimized (right) electric fields. The percentage is taken with respect to the isotropic yield, and the increase is due to interference between one-photon and two-photon pathways (center), respectively between different two-photon pathways probing different intermediate states (right). Adapted from Ref. [6]

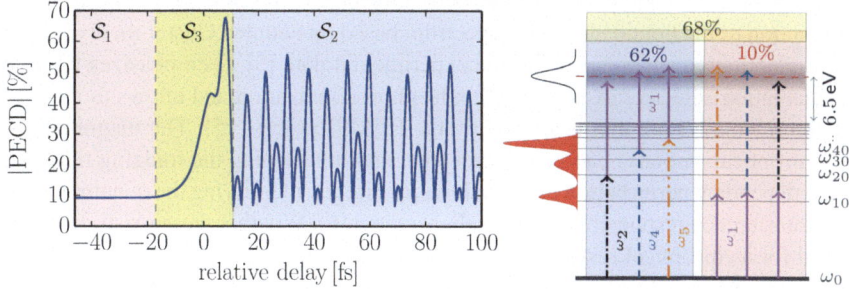

Fig. 6 The freely optimized field drives various two-photon ionization pathways that, depending on the linear chirp of the low-frequency component of the ionizing field, respectively the time-delay between low-frequency and high-frequency pulse compontents, interfere constructively or destructively. Adapted from Ref. [6]

field, constraining the pulse to be bichromatic, the PECD is pushed up to about 20%. This is shown in the center panel of Fig. 5. Compared to the initial guess pulse, the photon energies are lowered and, more importantly, the spectral phases of the two pulse components are adjusted to ensure that one-photon and two-photon ionization pathways interfere constructively. In contrast, with a flat spectral phase, the interference is mainly destructive. Indeed, constructive ionization of one-photon and two-photon pathway makes up for 17% out of the calculated 20% of PECD [6].

An even larger increase in PECD, up to almost 70%, is obtained when freely optimizing the ionizing electric field [6], cf. right panel of Fig. 5. This increase is driven by interference between various two-photon ionization pathways probing different intermediate, electronically excited states, illustrated in Fig. 6 (right). The optimized field turns out to have two spectral components, with the lower one exhibiting a linear spectral phase [6]. Such a "chirp" provides a very convenient handle to analyze the interference pattern in more detail, since changing the chirp rate, i.e., the slope of the spectral phase, amounts to changing the relative time delay between the low-frequency and the high-frequency component of the optimized pulse. If the low-frequency component precedes the high-frequency one, the PECD does not depend on the specific time delay and amounts to about 10%, cf. red-shaded parts in Fig. 6. If, in contrast, the high-frequency component arrives first, an electronic wavepacket is created which is then ionized by the low-frequency component of the pulse, cf. blue-shaded parts in Fig. 6. In this case, PECD sensitively depends on the time-delay, reflecting the time evolution of the electronic wave packet, oscillating between 6% and 62%. The largest PECD is obtained when the two spectral components of the optimized pulse overlap in time [6]. This allows for maximum interference between all the two-photon ionization pathways, highlighted by the yellow-shaded parts in Fig. 6.

In summary, photoelectron circular dichroism is a sensitive chiroptical probe of electron dynamics in a chiral potential, measured in experiments with molecular beams. Time-independent perturbation theory of resonantly enhanced (2+1) photoionization of camphor and fenchone molecules, combined with an *ab initio* description of the bound electronic spectrum based on coupled cluster theory, yields semi-quantitative agreement with the experimental data [5]. It emphasizes the role of orientation-selective excitation in multi-photon ionization and allows to identify the intermediate state that is probed by the REMPI process [5]. The magnitude of the chiroptical response can be enhanced by suitably shaping the ionizing field [6]. Time-dependent perturbation theory allows for directly identifying the quantum pathway interference responsible for the enhancement [6]. Whether there exists an upper bound for a chiroptical response such as photoelectron circular dichroism is one of the many open questions in the quantum control of molecular chirality.

5 Conclusions

This chapter reviewed three examples of quantum effects beyond entanglement — the (dis)appearance of a tunneling resonance in cold collisions, the protection of predissociation resonances by a phase, and the circular dichroism observed in the photoelectron spectrum of chiral molecules. In each case, a theoretical description based on first principles was key to elucidating the rather surprising observations made in experiments with molecular beams. The examples thus testify to the intriguing nature of quantum mechanics as well as to the topicality of Otto Stern's legacy.

Acknowledgements I would like to thank my students and postdocs, in particular Alexander Blech, Daniel Reich, Esteban Goetz and Wojtek Skomorowski, for their dedication and determination in carrying out the work that I have reviewed here. I am indebted to my colleagues Edvardas Narevicius and Robert Berger for many years of inspiring and fruitful collaboration. I am grateful for a Rosi and Max Varon Visiting Professorship to the Weizmann Institute of Science and for financial support from the German-Israeli Foundation (grant no. 1254), the State Hessen Initiative for the Development of Scientific and Economic Excellence (LOEWE) within the focus project Electron Dynamic of Chiral Systems (ELCH), and the Deutsche Forschungsgemeinschaft (Projektnummer 328961117—SFB 1319). This work was carried out while I was at the University of Kassel; and I would like to thank my Kassel colleagues for creating a friendly and stimulating research atmosphere.

References

1. J.P. Dowling, G.J. Milburn, Philos. Trans. R. Soc. A **361**, 1655 (2003)
2. A. Acín et al., New J. Phys. **20**, 080201 (2018)
3. A. Klein et al., Nat. Phys. **13**, 35 (2016)
4. A. Blech et al., Nat. Commun. **11**, 999 (2020)
5. R.E. Goetz, T.A. Isaev, B. Nikoobakht, R. Berger, C.P. Koch, J. Chem. Phys. **146**, 024306 (2017)
6. R.E. Goetz, C.P. Koch, L. Greenman, Phys. Rev. Lett. **122**, 013204 (2019)
7. C. Lux, M. Wollenhaupt, T. Bolze, Q. Liang, J. Köhler, C. Sarpe, T. Baumert, Angew. Chem. Int. Ed. **51**, 5001 (2012)
8. O. Dulieu, A. Osterwalder (eds.), *Cold Chemistry* (The Royal Society of Chemistry, Theoretical and Computational Chemistry Series, 2018)
9. R.V. Krems, *Molecules in Electromagnetic Fields* (Wiley, Hoboken, NJ, 2019)
10. H.R. Sadeghpour, J.L. Bohn, M.J. Cavagnero, B.D. Esry, I.I. Fabrikant, J.H. Macek, A.R.P. Rau, J. Phys. B **33**, R93 (2000)
11. Y. Shagam, E. Narevicius, J. Phys. Chem. C **117**, 22454 (2013)
12. A. Osterwalder, E.P.J. Techn, Instrum. **2**, 10 (2015)
13. A.B. Henson, S. Gersten, Y. Shagam, J. Narevicius, E. Narevicius, Science **338**, 234 (2012)
14. M. Costes, C. Naulin, Chem. Sci. **7**, 2462 (2016)
15. P.E. Siska, Rev. Mod. Phys. **65**, 337 (1993)
16. Y. Shagam, A. Klein, W. Skomorowski, R. Yun, V. Averbukh, C.P. Koch, E. Narevicius, Nat. Chem. **7**, 921 (2015)
17. M. Pawlak, Y. Shagam, E. Narevicius, N. Moiseyev, J. Chem. Phys. **143**, 074114 (2015)
18. H. Feshbach, Ann. Phys. **5**, 357 (1958)
19. U. Fano, Phys. Rev. **124**, 1866 (1961)
20. C. Chin, R. Grimm, P. Julienne, E. Tiesinga, Rev. Mod. Phys. **82**, 1225 (2010)
21. A. Carrington, T.P. Softley, Chem. Phys. **92**, 199 (1985)
22. D. Zajfman, O. Heber, L. Vejby-Christensen, I. Ben-Itzhak, M. Rappaport, R. Fishman, M. Dahan, Phys. Rev. A **55**, R1577 (1997)
23. M. Bancewicz, J. Phys. A **31**, 3461 (1998)
24. T.R. Govers, C.A. van de Runstraat, F.J. de Heer, J. Phys. B **6**, L73 (1973)
25. K. Gluch, J. Fedor, R. Parajuli, O. Echt, S. Matt-Leubner, P. Scheier, T.D. Märk, Eur. Phys. J. D **43**, 77 (2007)
26. B. Ritchie, Phys. Rev. A **13**, 1411 (1976)
27. A.F. Ordonez, O. Smirnova, Phys. Rev. A **98**, 063428 (2018)
28. N. Böwering, T. Lischke, B. Schmidtke, N. Müller, T. Khalil, U. Heinzmann, Phys. Rev. Lett. **86**, 1187 (2001)
29. C.S. Lehmann, N.B. Ram, I. Powis, M.H.M. Janssen, J. Chem. Phys. **139** (2013)
30. C. Lux, M. Wollenhaupt, C. Sarpe, T. Baumert, ChemPhysChem **16**, 115 (2015)
31. R. Cireasa et al., Nat. Phys. **11**, 654-658 (2015)
32. A. Comby et al., J. Phys. Chem. Lett. **7**, 4514 (2016)
33. A. Kastner, T. Ring, B. C. Krüger, G.B. Park, T. Schäfer, A. Senftleben, T. Baumert, J. Chem. Phys. **147**, 013926 (2017)
34. S.A. Rice, M. Zhao, *Optical Control of Molecular Dynamics* (Wiley, New York, 2000)
35. M. Shapiro, P. Brumer, *Quantum Control of Molecular Processes*, 2nd, revised and enlarged edition edition (Wiley Interscience, New York, 2012)
36. L. Greenman, P.J. Ho, S. Pabst, E. Kamarchik, D.A. Mazziotti, R. Santra, Phys. Rev. A **82**, 023406 (2010)
37. R.E. Goetz, M. Merkel, A. Karamatskou, R. Santra, C.P. Koch, Phys. Rev. A **94**, 023420 (2016)

Chapter 22
From Hot Beams to Trapped Ultracold Molecules: Motivations, Methods and Future Directions

N. J. Fitch and M. R. Tarbutt

Abstract Over the past century, the molecular beam methods pioneered by Otto Stern have advanced our knowledge and understanding of the world enormously. Stern and his colleagues used these new techniques to measure the magnetic dipole moments of fundamental particles with results that challenged the prevailing ideas in fundamental physics at that time. Similarly, recent measurements of fundamental electric dipole moments challenge our present day theories of what lies beyond the Standard Model of particle physics. Measurements of the electron's electric dipole moment (eEDM) rely on the techniques invented by Stern and later developed by Rabi and Ramsey. We give a brief review of this historical development and the current status of eEDM measurements. These experiments, and many others, are likely to benefit from ultracold molecules produced by laser cooling. We explain how laser cooling can be applied to molecules, review recent progress in this field, and outline some eagerly anticipated applications.

1 Introduction

It has been nearly a hundred years since Otto Stern and Walther Gerlach used an atomic beam to reveal one of the most striking aspects of the then-burgeoning quantum theory, space quantization [1]. Their work introduced new techniques that would later be used in countless experiments in physics and chemistry. Stern saw clearly the great promise of his new method, stating [2]

> The molecular beam method must be made so sensitive that in many instances it will become possible to measure effects and tackle new problems which presently are not accessible with known experimental methods.

He was right. The molecular beam method has been at the heart of atomic and molecular physics ever since and remains the method of choice for a huge number of experiments.

N. J. Fitch · M. R. Tarbutt (✉)
Centre for Cold Matter, Blackett Laboratory, Imperial College London, Prince Consort Road, London SW7 2AZ, UK
e-mail: m.tarbutt@imperial.ac.uk

© The Author(s) 2021
B. Friedrich and H. Schmidt-Böcking (eds.), *Molecular Beams in Physics and Chemistry*,
https://doi.org/10.1007/978-3-030-63963-1_22

A more recent development—laser cooling—can also be traced back to Stern, via Frisch who used the molecular beam method to measure the photon recoil momentum [3]. By controlling this recoil, modern atomic physics experiments routinely cool atoms and ions to μK temperatures. Until recently, experiments with molecules lagged behind, usually because of the difficulty of cooling them. Nevertheless, molecules have many useful properties that are increasingly being exploited for a variety of applications including tests of fundamental physics. An important example is the measurement of the electron's electric dipole moment (eEDM) where the precision of molecular experiments exceeds that achieved using atoms. The desire to improve these measurements provides strong motivation to extend cooling and trapping methods to molecules, and this has been achieved in the last decade. Laser cooling has been used to collimate and decelerate molecular beams, capture molecules in magneto-optical traps, and then cool them to ultracold temperature. These ultracold molecules can be used to address a wide variety of important problems - exploring what new forces lie beyond the Standard Model of particle physics [4, 5], studying collisions and reactions at the quantum level [6, 7], simulating the behaviour of many-body quantum systems [8–10], and processing quantum information [11, 12].

In this article, we review some of these past developments and future prospects. We begin in Sect. 2 with a brief review of molecular beam sources. In Sect. 3 we consider how the development of molecular beam methods for measuring magnetic dipole moments eventually enabled measurements of the electric dipole moments of fundamental particles. We briefly review the current status of these experiments in Sect. 4 and explain the importance of laser cooling to the future of this endeavour. In Sects. 5 and 6 we explain how laser cooling works for molecules and present recent achievements in this field. Finally, in Sect. 7, we give a brief overview of some applications of ultracold molecules and how they might be realized.

2 Molecular Beam Sources

The molecular beam method developed by Stern [13, 14] is the foundation for innumerable experiments in atomic and molecular physics. Here, we give a brief review of the three main types of atomic and molecular beam sources in use today: effusive beams, supersonic beams, and cryogenic buffer gas beams. Their velocity distributions and flux are compared in Fig. 1 for the case of YbF molecules, one of the few species for which all three types of beam source have been realized.

Effusive sources typically use heated ovens to generate a sufficient vapour pressure of the atoms or molecules of interest. They operate at low pressure so that there are no collisions in the vicinity of the exit aperture. As first shown experimentally by Stern [13, 14], these sources produce beams with a broad velocity distribution, whose mean and width both scale as $\sqrt{T/m_s}$, where T is the temperature of the source and m_s is the mass of the species. As a consequence of the high oven temperature, effusive beams are characterized by a wide velocity distribution and low flux in any single quantum state.

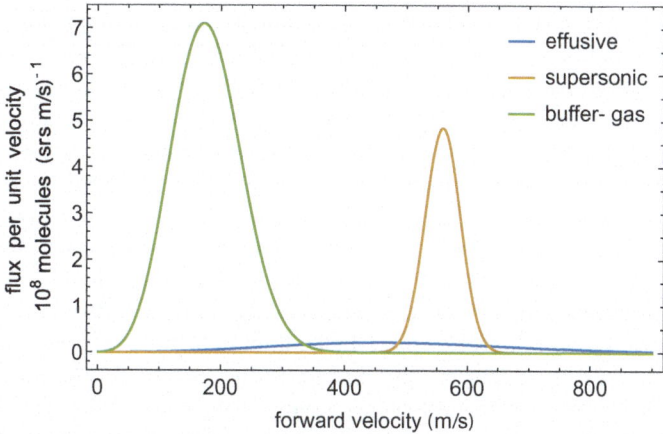

Fig. 1 Velocity distributions of effusive, supersonic, and buffer-gas beams. The vertical axis indicates the number of molecules per unit solid angle per unit time and per unit interval of velocity. The plots are for YbF molecules in their absolute ground state, chosen because, for this molecule, all three sources have been developed and characterised. Realistic operating conditions have been taken. For the effusive case, molecules are generated from an oven at 1500 K [15]. For the supersonic case, a carrier gas of argon with a reservoir temperature of 300 K is assumed, with an internal beam temperature of 4 K and a repetition rate of 25 Hz [16]. For the cryogenic buffer-gas case, a carrier gas of helium at 4 K creates a beam with moderate hydrodynamic boosting, operating at 10 Hz [17]

In supersonic sources [18, 19], a gas held at high pressure expands through a nozzle into a vacuum chamber. There are a large number of collisions in the vicinity of the nozzle. The slower particles are bumped from behind, while the faster ones bump into those ahead, so that all the particles end up travelling at nearly the same speed. In this way, the random thermal motion is converted into forward kinetic energy, producing a cold, fast beam—the mean velocity is high, but the velocity distribution is narrow, as illustrated in Fig. 1. The collisions also transfer the rotational and vibrational energy of the molecules to forward kinetic energy, resulting in a beam that is cold in all degrees of freedom. The first supersonic beams were continuous, but the method was soon extended by using pulsed valves with short opening times. In this way, intense pulses can be produced without an excessive gas load. A wide variety of methods have been developed to introduce atoms and molecules of interest into the supersonic expansion, including laser ablation, electric discharge, and photodissociation. Translational temperatures of 1 K are typical, and beam speeds are 1800 m/s when the carrier gas is room temperature helium, and 400 m/s for room temperature krypton.

The third type of molecular beam source is the cryogenic buffer gas source [20, 21]. Here, the molecules of interest are formed inside a cryogenically-cooled cell containing a cold buffer gas, often helium at 4 K. The molecules are commonly formed by laser ablation or introduced into the cell through a capillary. The internal and motional degrees of freedom of the molecules thermalize through collisions with the buffer gas, and then the molecules exit through a hole in the cell to form a beam. The density of buffer gas in the cell is determined by the gas flow rate and the size of the exit hole. When the density is low, there are few collisions near the aperture, so the beam tends towards the effusive regime. These beams are slow, especially for heavy species, for then the mass m_s is large whereas the temperature T is small, typically two orders of magnitude lower than a standard effusive source. Beam speeds as low as 40 m/s have been achieved this way [22]. However, the molecular flux tends to be low in this regime because most molecules diffuse to the cell walls, where they freeze, instead of passing through the exit aperture. As the buffer gas flow is increased it sweeps more molecules out of the cell, increasing the beam flux. However, collisions near the aperture boost the beam speed. In the limit of high density the speed of the molecules reaches the supersonic speed of the buffer gas which scales as $\sqrt{T/m_b}$, where m_b is the mass of the buffer gas atoms. Very often, cryogenic buffer gas sources are operated in an intermediate flow regime where the flow is high enough to extract a substantial fraction of the molecules from the cell, but low enough for a moderate beam speed. Speeds in the range 100–200 m/s are typical, as illustrated in Fig. 1. Due to the high flux and low relative beam speeds, these sources are becoming increasingly popular, especially for experiments on laser cooling of molecules and tests of fundamental physics.

3 Particle Dipole Moments

Stern's pioneering experiments established the reality of space quantization and determined the magnetic dipole moments of the electron and proton [1, 23–25]. It is interesting to consider whether elementary particles might also have *electric* dipole moments. Just like the magnetic moment, such an electric dipole would have to be oriented along the particle's spin. Furthermore, this orientation must be fixed, since the particle would otherwise have an additional degree of freedom that would, for example, change the filling of electron orbitals in the periodic table. A spin defines a direction of circulation, as does a magnetic moment, so it seems natural for the two to be associated. Far less natural is the association of an electric dipole – a charge separation – with this direction of circulation. Indeed, such an electric dipole moment (EDM) implies a difference between left- and right-handed coordinate systems, and implies a fundamental arrow of time. To see this, consider the Hamiltonian for a particle with magnetic moment μ and EDM d, both fixed relative to the spin, interacting with magnetic and electric fields B and E:

$$\mathcal{H} = -\boldsymbol{\mu} \cdot \boldsymbol{B} - \boldsymbol{d} \cdot \boldsymbol{E}. \tag{1}$$

Reflection in a mirror, equivalent to the parity operation, reverses E but does not reverse B, μ or d. Conversely, reversing the direction of time reverses B, μ and d, but not E. We see that while the first term in $\mathcal{H}$ is even under both the parity and time-reversal operations, the second term is odd and so the existence of an EDM violates both symmetries.

Prior to the 1950s, it was generally considered that nature did not distinguish between left and right, or between forwards and backwards in time, and this seemed to be a powerful argument against the existence of fundamental electric dipoles, implying that $|\boldsymbol{d}| = 0$. This idea was challenged by Purcell and Ramsey who insisted that it was *"a purely experimental matter"*, noted that existing evidence was weak, and declared their intention to measure the EDM of the neutron [26]. Regarding the need for experimental evidence to determine whether parity (P), time-reversal (T) and charge conjugation (C) are symmetries of nature, Hermann Weyl was similarly emphatic, writing that [27]

> a priori evidence is not sufficient to settle the question; the empirical facts have to be consulted.

Along this line of thought, in 1956 Lee and Yang [28] noted that, for the weak interaction,

> parity conservation is so far only an extrapolated hypothesis unsupported by experimental evidence.

Within a few months it was discovered that the weak interaction violates P symmetry [29–31]. In 1964 it was found that the weak interaction also violates CP symmetry, the combined symmetry of charge conjugation and parity [32]. CP violation is equivalent to T-violation in most theories, and so the last theoretical objection to the existence of fundamental EDMs was removed. Suddenly, and ever since, the question was not whether particles could have electric dipoles, but why those dipoles are so small.

In considering how the electric dipole moment of a particle might be measured, it is instructive to reflect on Stern's method for measuring magnetic moments, which is illustrated in Fig. 2a. The magnetic moment of an atom is proportional to its internal angular momentum, $\hbar \boldsymbol{F}$, so we often write $\boldsymbol{\mu} = -g\mu_B \boldsymbol{F}$, where μ_B is the Bohr magneton and g is a proportionality factor. Taking $\boldsymbol{B}$ in the z-direction, the interaction energy is $W_B = \langle -\boldsymbol{\mu} \cdot \boldsymbol{B} \rangle = g\mu_B B \langle F_z \rangle = g\mu_B B m_F$, where m_F is the projection of the angular momentum onto the z-axis. Space quantization is expressed by only discrete values being allowed for m_F. In the inhomogeneous field of the Stern-Gerlach magnets there is a force on the atoms $\boldsymbol{F}_B = -\nabla W_B = -g\mu_B \nabla B m_F$, leading to the deflection proportional to m_F observed by Stern and Gerlach. Let us now consider an atom with an unpaired electron that has an eEDM, d_e. If an electric field E is applied in the z-direction, there is an interaction energy $W_E = -Rd_e E m_F / |F|$, analogous to the magnetic interaction. The proportionality factor R depends on the choice of atom and is often called the enhancement factor because, for heavy atoms, $|R|$ can be considerably larger than 1 [33]. For example, for Cs, $R = 120$ [34]. Suppose we pass a beam of Cs atoms through a region of inhomogeneous electric field. Of course, the electric field produces an induced electric dipole in the atom, resulting in a force which is often used to deflect atoms and molecules. This force is the same for states of opposite m_F and is not the one of interest here. The force

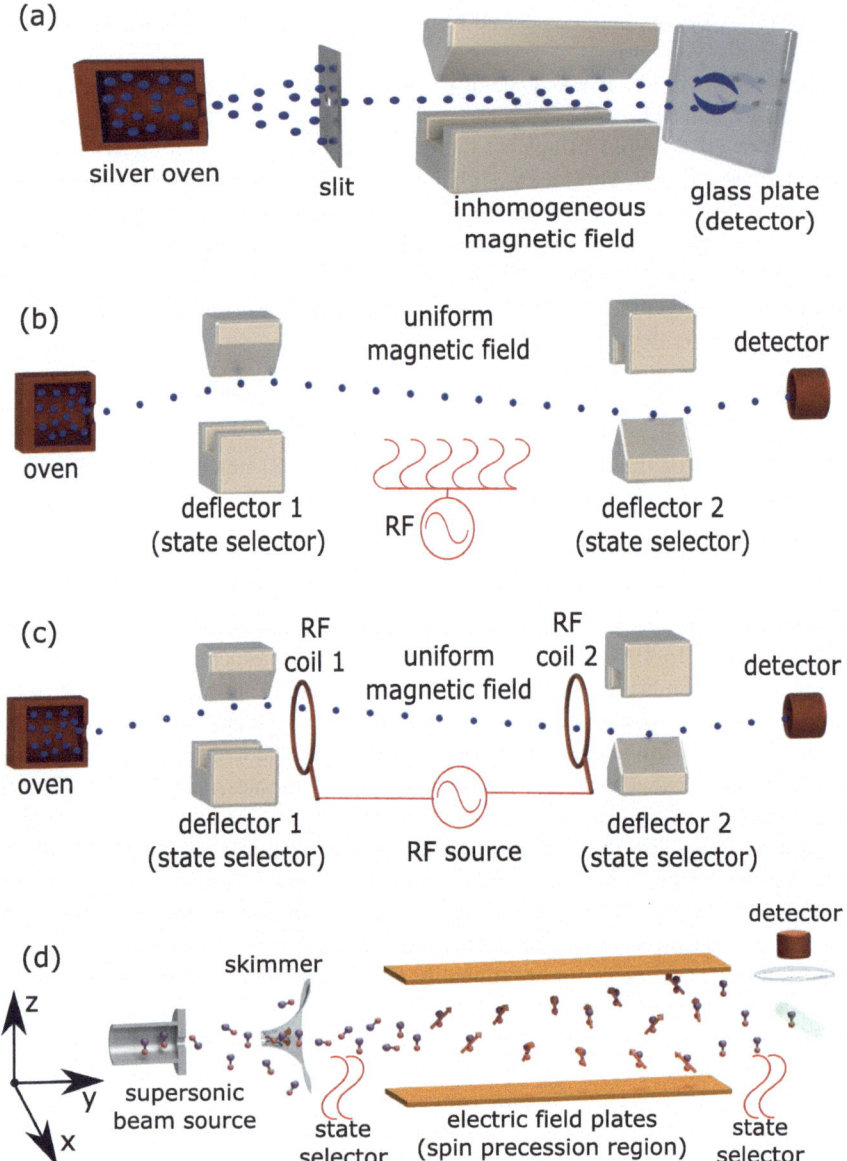

Fig. 2 The evolution of molecular beam methods for measuring dipole moments. (a) Stern and Gerlach's experiment for measuring magnetic moments and demonstrating space quantization. (b) Rabi's improved method for measuring magnetic moments. Between the two deflectors there is a uniform magnetic field and an oscillating field which, when resonant, changes the spin state. (c) A further improvement by Ramsey separates the oscillating field into two short regions. (d) Method for determining the electron's electric dipole moment by measuring the spin precession in an applied electric field. In all its key aspects, the technique is the same as Ramsey's method

due to the permanent EDM is $F_E = -\nabla W_E = Rd_e \nabla E \, m_F / |F|$, and deflects states of opposite m_F in opposite directions. Taking $d_e = 10^{-29}$ e cm, which is close to the current upper limit [35], a field gradient of 100 kV/cm^2, and a speed of 200 m/s, the deflection after propagating 1 m is about 10^{-19} m. Clearly, this is not a good way to measure d_e. Nevertheless, the subsequent development of molecular beam methods inspired by Stern's techniques became so sensitive that measurements of d_e soon became feasible.

In the late 1930s, Rabi introduced a new idea that greatly improved Stern's method of measuring magnetic moments [36]. Instead of using the magnetic deflector as a measuring device, he used a pair of them as state selectors, as illustrated in Fig. 2b. The deflectors are arranged such that molecules reach a detector provided they remain in the same state so that they have the same magnetic moment throughout. Between the two deflectors, Rabi produced a uniform magnetic field, B_0, so that neighbouring m_F states are separated by an energy $\hbar\omega_0 = g\mu B_0$. In this region, an rf field of frequency $\omega \approx \omega_0$ resonantly drives transitions from one m_F state to another. Molecules that change m_F are deflected by the second magnet and miss the detector, resulting in a dip in the detected signal at $\omega = \omega_0$. This measurement of the resonant frequency, together with a measurement of B_0, determines the magnetic moment. The precision of this measurement is proportional to the interaction time with the rf field, so it's desirable to make this as long as possible. In practice however, this time is limited by the difficulty of keeping the fields uniform enough.

By the 1950s, Ramsey had solved this problem by separating the rf region into two short sections driven by the same oscillator [37], as illustrated in Fig. 2c. The first deflecting magnet prepares the molecules in a chosen spin state, say spin up. The first rf region rotates the spin so that it is orthogonal to B_0. The spin then precesses in the uniform magnetic field with angular frequency ω_0 for a time T. When $\omega = \omega_0$ the rf oscillation is in phase with the spin precession, so the second rf region rotates the spin in the same direction as the first, producing the spin-down state which will miss the detector. If there is a frequency difference $\omega - \omega_0 = \pm\pi/T$, the extra half rotation means that the spin will be driven back to the spin-up state in the second rf region and will reach the detector. The signal at the detector oscillates as ω is scanned, allowing ω_0 to be determined with an uncertainty inversely proportional to the free precession time T. Ramsey's method has such high precision that it is suitable for measuring the tiny electric dipole moments of fundamental particles. All that is needed is to add to B_0 a uniform electric field E_0, and then measure the change in the precession frequency when the direction of E_0 is reversed. This change is proportional to the EDM. This was the method used by Smith, Purcell and Ramsey in their first measurement of the neutron EDM [38], and the one used for all subsequent measurements of particle electric dipole moments.

Figure 2d illustrates a typical electron EDM measurement that uses Ramsey's molecular beam method. A molecular beam passes through a state selector, which could be a magnetic or electric deflector but in modern experiments is often a laser beam that optically pumps molecules to the desired state. Next, a first region of oscillating field aligns the spin along x. This can be done using an rf field, or a laser field that drives a Raman transition or optical pumping process. The spin now precesses around z due to the combination of μ interacting with B_0 and d_e interacting with E_0. Finally the spin direction is measured, for example by mapping its direction to a pair of states that are easily distinguished spectroscopically. The change in the spin precession angle that correlates with the reversal of E_0 determines d_e.

4 Current Status and Future Directions of eEDM Experiments

In the Standard Model of particle physics, the eEDM is predicted to be $d_e \approx 10^{-38}$ e cm [39]. Theories that extend the Standard Model often introduce new CP-violating interactions, which are needed to explain the observed asymmetry between matter and antimatter in the universe [40], and these new interactions lead to much larger eEDM values. Thus, eEDM measurements can be excellent probes of these theories. Early measurements used heavy atoms and yielded results consistent with zero, eventually assigning an upper limit of $|d_e| < 1.6 \times 10^{-27}$ e cm [41, 42]. Although more difficult to produce and control, heavy polar molecules can be far more sensitive to the eEDM than atoms [43, 44]. The sensitivity is proportional to the degree of polarization and therefore to the electric-field-induced mixing of opposite parity states. In atoms, these are different electronic states whose spacing is typically ~ 1 eV, but in molecules they are neighbouring rotational states whose energy spacing is about four orders of magnitude smaller, or the opposite parity states of an Ω-doublet where the spacing is even smaller still. Because the levels are closely spaced, only a modest electric field is needed to fully polarize the molecule. In this case, it is common to define an effective electric field $E_{\text{eff}} = RE$ which saturates to a maximum value $E_{\text{eff}}^{\text{max}}$. The effective field is enormous for some species, and its maximum value is often easy to reach. For example, $E_{\text{eff}}^{\text{max}} \approx 26$ GV/cm for YbF and ≈ 78 GV/cm for ThO.

The enormous effective fields make eEDM experiments with molecules very attractive, and measurements have been made using beams of YbF [45, 46], beams of ThO [35, 47], a cell of PbO vapour [48], and trapped HfF$^+$ molecular ions [49]. The results of these measurements are all consistent with zero, and the best upper limit is currently $|d_e| < 1.1 \times 10^{-29}$ e cm at the 90% confidence level [35]. Remarkably, this experiment and ones like it test theories that extend the Standard Model at an energy scale similar to, and even exceeding, the maximum collision energy of the Large Hadron Collider.

Given the great significance of eEDM experiments, it is natural to consider how to make the next leap in sensitivity. The uncertainty in measuring d_e scales as $1/(T\sqrt{N})$ where T is the spin precession time and N is the number of molecules used in the measurement. In a molecular beam experiment of length L, where the spin precession region occupies most of the space and the beam has diverged sufficiently that it fills the detector, T is proportional to L but N falls as $1/L^2$ because of the divergence of the beam. Consequently, there is no benefit in increasing L. This can be circumvented by cooling the molecules to much lower temperatures. A beam that is cooled in the transverse directions is collimated and can travel for long distances without spreading, allowing T to increase without reducing N. Going further, molecules that are cold enough can be launched into a fountain [50] or stored in a trap [51], giving access to even longer spin precession times. These ideas require molecules cooled to μK temperatures. Such low temperatures can be reached either by associating ultracold atoms into molecules [52], or by direct laser cooling [53]. EDM experiments using laser-cooled YbF, BaF, YbOH and TlF are all currently being developed [54–57]. They will have unprecedented sensitivity and tremendously exciting potential for new discoveries in fundamental physics.

5 Laser Cooling of Molecules: Principles

5.1 Laser Cooling Scheme

Figure 3 illustrates the energy level structure of a typical diatomic molecule, showing the electronic, vibrational, rotational, and hyperfine structure and the notation used to label the levels. For molecules to be slowed, cooled and trapped by radiation pressure, they must scatter many photons from the light, typically 10^4 or more. This calls for a cooling scheme where an upper level decays to only a few lower levels, so that only a few transitions need to be addressed. The inset to Fig. 3 shows an example of such a scheme. The upper level is the lowest level of positive parity in the first electronically excited state, labelled here as A, $v = 0$, $R = 0$, $+$. Electric dipole transitions to the X state must change the parity and obey the selection rule $\Delta R = 0, \pm 1$, which means that only the $R = 1$ rotational state is accessible.[1] However, the molecule can decay to any vibrational state, since there is no selection rule dictating how v can change in an electronic transition. The branching ratio to each vibrational state is mainly given by the square of the overlap integral between the vibrational wavefunctions in the lower and upper electronic states, which is known as the Franck Condon factor. For molecules where the optically active electron is not involved in the bonding, the sets of vibrational wavefunctions for the two electronic states are very similar. In this case, the branching ratio is close to 1 for the $\Delta v = 0$ transition and diminishes rapidly with increasing Δv. These molecules are the ones best suited to laser cooling because only a few vibrational bands need to be addressed, requiring just a few lasers, as indicated in the figure. Hyperfine components of these transitions can usually be addressed by adding radio-frequency sidebands to each laser using acousto-optic or electro-optic modulators.

---

[1] Often, R is not a good quantum number because it is coupled strongly to other angular momenta in the molecule, such as the orbital angular momenta of the electrons. In this case, R may be replaced by the relevant coupled angular momentum and the same principles apply.

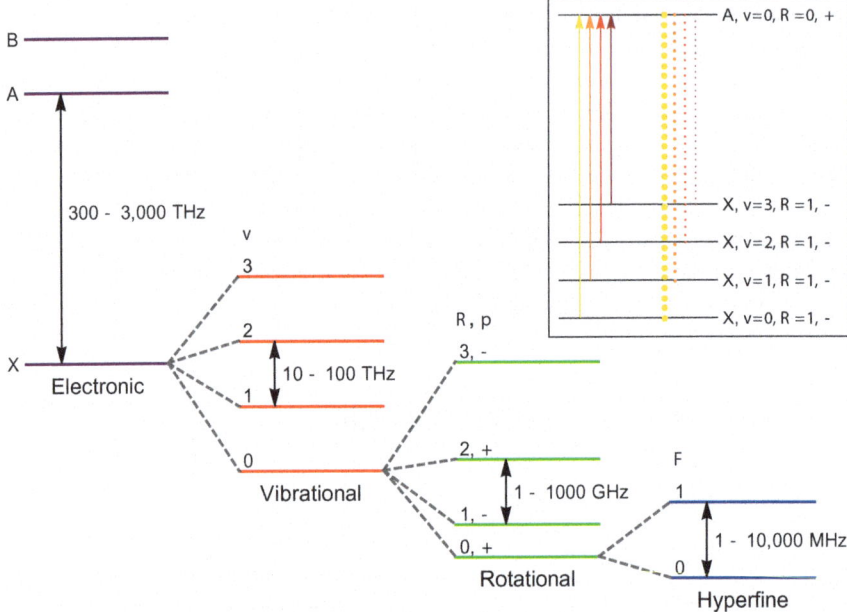

Fig. 3 Energy level structure of a typical diatomic molecule, with indicative transition frequencies. The ground electronic state is labelled X, and the excited states are A, B, etc. Transitions between electronic states are usually in, or near, the visible part of the electromagnetic spectrum. Each electronic state has a set of vibrational states, labelled by v. Vibrational transitions are usually in the mid-infrared. Each vibrational state has a set of rotational states, labelled by the rotational angular momentum, R, and the parity p. Rotational transitions are in the microwave regime. Each rotational state usually has a set of hyperfine states with total angular momentum F determined by the electronic angular momentum and the nuclear spin. In the example shown, both are 1/2. The inset shows the laser cooling scheme discussed in the text. Arrows show transitions driven by lasers, and the weights of the dotted lines indicate the relative branching ratios of the decay channels

5.2   Doppler and Sub-Doppler Cooling

Despite the complexity of the molecular structure outlined above, and the need to drive many transitions using several laser frequencies, the basic principles of laser cooling can be understood by focussing on just two or three levels. Figure 4a illustrates the principle of Doppler cooling applied in one dimension to a hypothetical two-level molecule. A molecule moving to the right with speed v interacts with a pair of identical, counter-propagating laser beams with wavevector k. The frequency of the light, ω, is slightly smaller than the molecular transition frequency, ω_0. The laser beam from the right is Doppler shifted closer to resonance, so the molecule scatters more photons from this beam and slows down as a result of this imbalanced radiation pressure. The force on the molecule due to each of the beams is

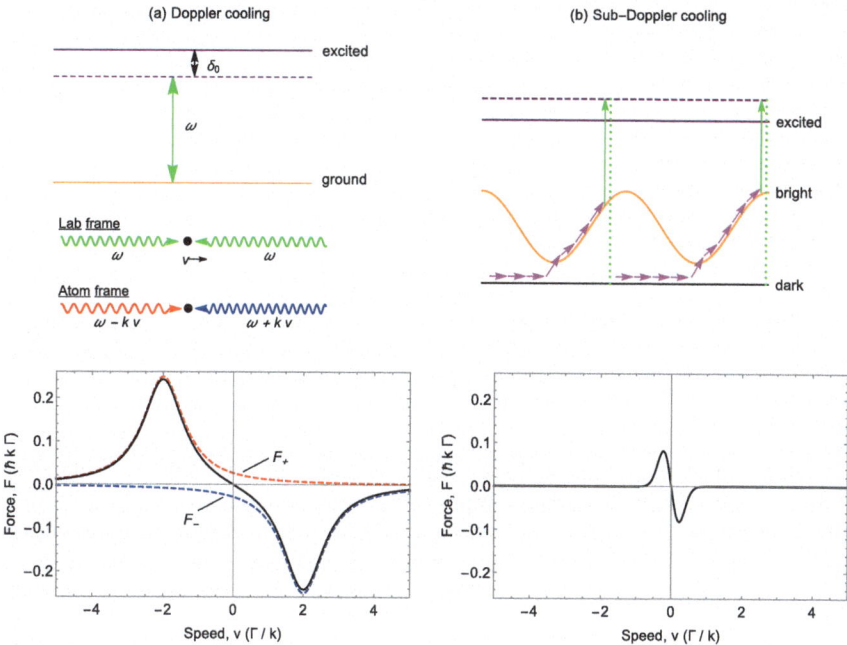

Fig. 4 Principles of **a** Doppler cooling, where a negative detuning is used and **b** sub-Doppler cooling, where a positive detuning is used

$$F_{\pm} = \pm \frac{\hbar k \Gamma}{2} \frac{I/I_{s}}{(1 + I/I_{s} + 4(\delta_0 \mp kv)^2/\Gamma^2)}, \qquad (2)$$

where $\delta_0 = \omega - \omega_0$ is the detuning, Γ is the natural linewidth of the transition, I is the laser intensity and I_s is a characteristic intensity known as the saturation intensity. The graph in Fig. 4a shows these two forces as a function of v in the case where $I = I_s$ and $\delta_0 = -2\Gamma$. The solid line is their sum and shows that there is a force driving the molecule towards zero velocity. In addition to this cooling force, there is heating due to the randomly-directed momentum kicks associated with the photon scattering events. When the heating and cooling rates are balanced, the molecule reaches its equilibrium temperature. The minimum temperature for Doppler cooling is known as the Doppler limit and is $T_{D,\min} = \hbar\Gamma/(2k_B)$.

Figure 4b illustrates a method of sub-Doppler cooling. Here, we distinguish two Zeeman sub-levels of the ground state. Due to the choice of states and the angular momentum selection rules, laser light of a given polarization cannot drive transitions from one of the ground states. We call this the dark state because it does not couple to the light. The other state is called the bright state. The bright state has an ac Stark shift which is positive when the detuning is positive. The dark state has no ac Stark shift because it does not couple to the light. If the two counter-propagating laser beams have different polarizations, neither parallel nor orthogonal, both the intensity and

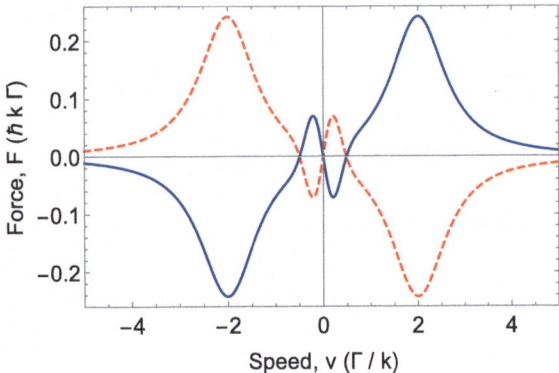

Fig. 5 Typical form of the total force as a function of speed due to the combination of Doppler and sub-Doppler processes illustrated in Fig. 4. The two lines are for equal and opposite detunings: dashed red for negative, and solid blue for positive

polarization of the light field will vary with position. This causes the ac Stark shift of the bright state to be modulated in position, setting up potential energy hills and valleys which the molecules move through. A molecule in the bright state will be excited by the light and optically pumped into the dark state. This is most likely to happen at positions where the intensity of the light field is high, which are also the positions where the ac Stark shift is largest, i.e. near the tops of the hills. A moving molecule can make a non-adiabatic transition back to the bright state because the polarization changes as it moves. This is most likely to happen where the energy gap between bright and dark states is smallest, i.e. near the bottom of the hills. As a result, molecules moving through the light field lose energy because they climb hills more often than they descend into valleys. The graph in Fig. 4b shows the typical force produced by this mechanism. It operates over a smaller range of velocities than Doppler cooling, and produces a smaller maximum force. Crucially however, the gradient near zero velocity, which is the damping constant, is substantially higher than for Doppler cooling. Furthermore, because the molecule spends much of its time in the dark state, there is less photon scattering, and thus a lower heating rate. Thus, for small velocities, the cooling rate is higher while the heating rate is lower, leading to much lower temperatures.

The Doppler cooling mechanism shown in Fig. 4a requires a negative detuning, while the sub-Doppler cooling mechanism requires a positive detuning.[2] The two mechanisms often appear together, resulting in the typical velocity-dependent force illustrated in Fig. 5. A negative detuning is useful for capturing molecules with a wide range of initial velocities and cooling them to lower velocity. However, the lowest temperatures are not reached because the total force has the wrong sign at low

---

[2]Note that there are other methods of sub-Doppler cooling, commonly used to cool atoms, that work for negative detunings. For molecules, it appears that sub-Doppler cooling always requires a positive detuning.

velocity. Once molecules are slow enough, the frequency of the light can be switched to a positive detuning so that the sub-Doppler mechanism cools them further. In this way, molecules have been cooled to temperatures far below the Doppler limit.

6   Laser Cooling of Molecules: Practice

Figure 6 illustrates an apparatus for laser cooling and trapping of molecules. The experiments begin with a molecular beam, a testament to the experimental power of the method developed by Stern and subsequent researchers. The cryogenic buffer gas sources described in Sect. 2 are ideal for this application because they deliver the critical combination of a high flux of molecules with a low initial speed. The illustration shows how this beam is laser cooled in the transverse directions, decelerated to low speed using radiation pressure, and then captured and cooled in a magneto-optical trap. We discuss each of these steps in turn.

6.1   Transverse Laser Cooling of a Molecular Beam

The density in a molecular beam drops with distance from the source because the beam spreads out as it propagates. Laser cooling can reduce the transverse temperature enormously, resulting in an intense, highly collimated molecular beam. The pioneering work on laser cooling of molecules was done at Yale [58]. They worked with a beam of SrF molecules and showed how to cool the beam in one transverse direction using both Doppler cooling and sub-Doppler cooling. Several other diatomic and polyatomic molecular species have since been cooled using similar

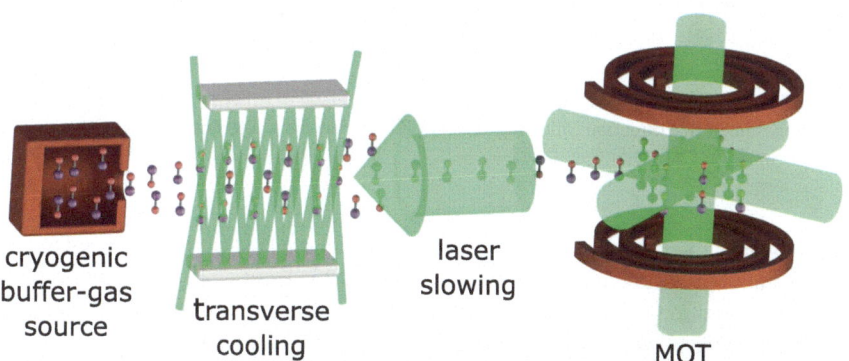

Fig. 6 An illustration of an apparatus for laser cooling and trapping of molecules. A beam of molecules from a cryogenic buffer gas source is cooled in the transverse directions, decelerated by the radiation pressure of a counter-propagating laser beam, and captured in a magneto-optical trap

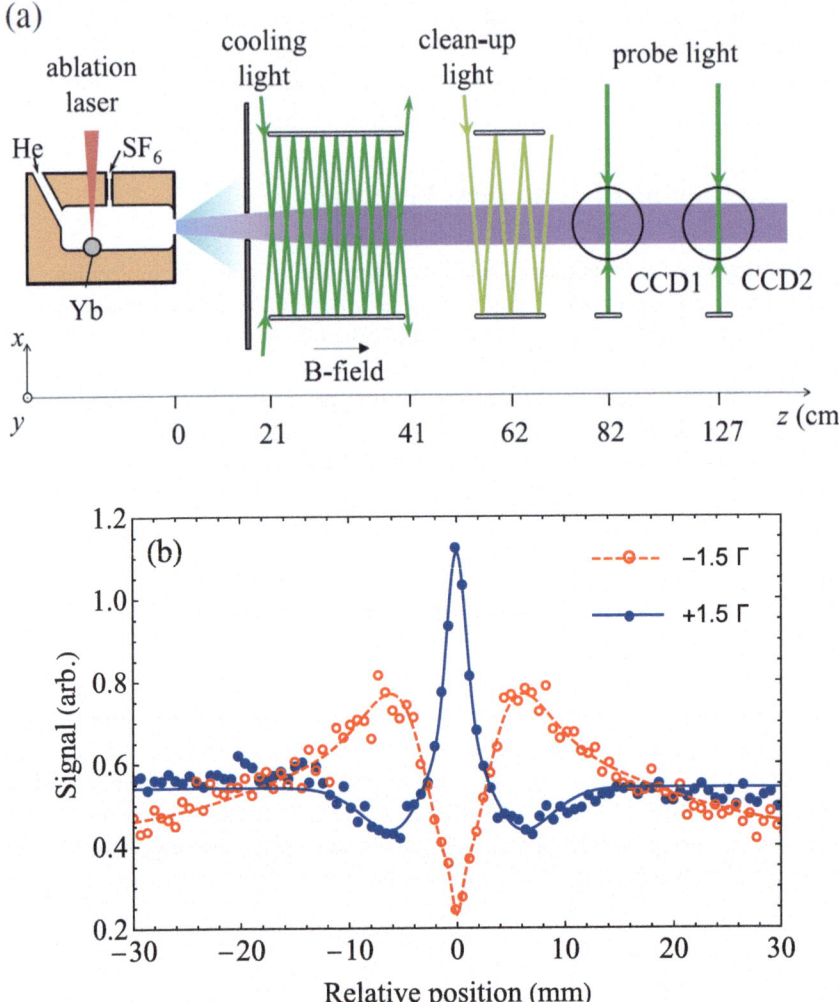

Fig. 7 Transverse laser cooling of YbF molecules. **a** Schematic of experiment. **b** Density distribution in one transverse direction following laser cooling in that direction. The lines are fits to a sum of four Gaussians. Adapted from [54]

methods [54, 59–64]. Figure 7a illustrates an experiment [54] to cool YbF molecules in one transverse direction, x. A beam of molecules from a cryogenic buffer gas source passes through a 20 cm long sheet of laser cooling light that forms a standing wave in the x direction. All molecules are then optically pumped to the lowest vibrational level by the clean-up light, and then detected by laser-induced fluorescence on a camera. Figure 7b shows the resulting density distribution of these molecules along x. When the detuning of the laser light is positive, a narrow peak appears at the centre of the distribution, corresponding to molecules that have been cooled to low temperature by sub-Doppler cooling. In this experiment, the transverse temperature of the cooled molecules was found to be below 100 μK. When the detuning is negative, a dip appears at the centre with broad wings on either side where the molecules have accumulated. The reason for this profile can be appreciated from the form of the force curve in Fig. 5. For negative detuning, the sub-Doppler mechanism forces molecules near zero velocity to a higher velocity, while Doppler cooling forces high velocity molecules towards lower velocities. As a result, molecules accumulate around the non-zero velocity where the force curve crosses zero.

6.2   Slowing a Molecular Beam with Radiation Pressure

Transverse laser cooling produces a highly collimated molecular beam, but the molecules still have a high forward speed. They can be decelerated using the radiation pressure of a laser beam propagating in the opposite direction. Here, it is essential to account for the changing Doppler shift as the molecules slow down. This can be done by chirping the frequency of the laser so that it follows the changing Doppler shift, or by broadening the frequency spectrum of the laser to cover the full range of Doppler shifts. Laser slowing of molecules was first demonstrated by the group at Yale using SrF [65], and similar methods have been applied to other molecules [66–69].

Figure 8 illustrates frequency-chirped laser slowing of a cryogenic buffer-gas beam of CaF molecules [69]. The black curves show the velocity distributions with no slowing applied, and the coloured curves show the distributions after slowing using various frequency chirps. The initial frequency of the laser is tuned to be resonant with molecules moving at about 180 m/s. When there is no chirp, the molecules are decelerated to about 100 m/s. They bunch up around this speed because the faster molecules are initially closer to resonance so are decelerated more than the slower ones. The distribution is shifted to lower velocities as the chirp increases, but the number of detected molecules drops at low velocities. This is because there is no transverse cooling in these experiments, so the beam diverges rapidly as it slows down, reducing the number of molecules that pass through the detector.

6.3   Trapping the Molecules

With the molecules slowed to low velocity, it becomes possible to trap them. Magneto-optical traps (MOTs) have been used to cool and trap atoms for decades [71]. In a MOT, counter-propagating laser beams result in a velocity-dependent force which cools the atoms, as described in Sect. 5.2. The detuning of the light is usually negative so that Doppler cooling, with its large capture velocity, is the dominant process. This alone does not trap the atoms because the force does not depend on position. To produce a position-dependent force, a magnetic field gradient is added,

Fig. 8 Radiation pressure slowing of a beam of CaF molecules. The laser light propagates in the opposite direction to the molecular beam and is frequency chirped. Black lines show the velocity distributions without slowing, and coloured lines are the distributions when the slowing is applied with various frequency chirps. The dashed lines show the resonant velocity at the beginning and end of the chirp. Adapted from [69]

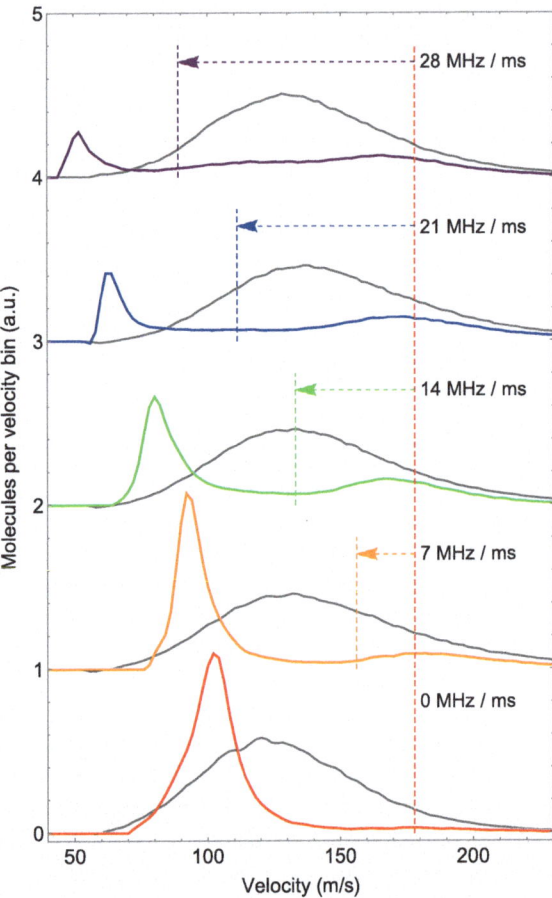

typically by using a pair of coils as illustrated in Fig. 6. The current flows in opposite directions in the two coils so that the magnitude of the field is zero at the centre and increases linearly in all directions away from this point. For a stationary atom at the field-zero, there is no net force because the atom is equally likely to scatter photons from any of the laser beams. When the atom is displaced, the Zeeman effect shifts the frequencies of transitions with $\Delta m_F = \pm 1$ in opposite directions, one closer to resonance and the other further away. These two transitions are driven by circularly polarized light of opposite handedness. By choosing the handedness of the beams in the correct way, the transition closest to resonance is always driven by the beam that pushes the atom back towards the field-zero. This traps the atoms.

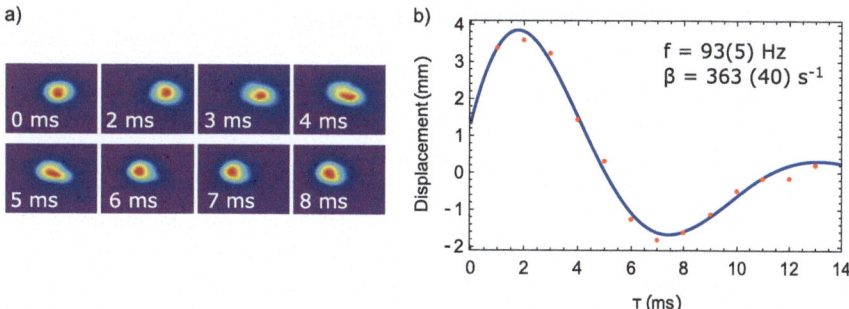

Fig. 9 Damped oscillations of CaF molecules in a magneto-optical trap. **a** Fluorescence images of the trapped molecules at various times after giving the molecules a radial push. **b** Red points: position of the centre of the cloud versus time. Blue line: fit to a damped harmonic oscillator model. Figure taken from [70]

The first three-dimensional magneto-optical trap of molecules was demonstrated in 2014 by the Yale group [72]. They captured a few hundred SrF molecules for about 50 ms at a density of 6×10^2 cm^{-3}, and cooled them to 2.3 mK. They went on to investigate several other trapping configurations [73–75] and were soon able to trap about 10^4 molecules for 500 ms at a density of 2.5×10^5 cm^{-3} and temperatures down to 250 μK. MOTs of CaF [70, 76, 77] and YO [78] molecules have also been produced, with steadily increasing number densities. Figure 9a is a sequence of pictures showing CaF molecules trapped in a MOT. Each picture is made by imaging the fluorescence of the trapped molecules onto a camera. The molecules were given a sudden push in the horizontal direction at time T=0 and the subsequent images show them oscillating in the trap. Figure 9b shows the displacement of the cloud versus T together with a fit to the motion of a damped harmonic oscillator. The results show that the MOT exhibits both a restoring force and a damping force, as expected.

Once molecules have been trapped in a MOT, they can be cooled to much lower temperature using the sub-Doppler cooling method described in Sect. 5.2. Typically, the magnetic field is turned off and the detuning of the lasers is switched from negative to positive. The molecules typically cool below the Doppler limit [76, 79, 80] in about 1 ms, and temperatures as low as 4 μK have been reached in this way [81–83]. At this point the cooling can be turned off and the molecules stored in conservative traps where their quantum states can be controlled and preserved for long periods. Ensembles of laser-cooled molecules have recently been confined in pure magnetic traps [80, 84] and in optical dipole traps [79], and single molecules have been held in tightly-focussed tweezer traps [85]. Coherent control of the hyperfine and rotational states of molecules has been studied [10, 84] and rotational superpositions with long coherence times have been demonstrated for trapped molecules [86].

7   Applications of Laser-Cooled Molecules

The ultracold molecules produced by direct laser cooling are well suited to a wide variety of exciting applications. Many of these applications require molecules trapped for long periods, long-lived coherences, and control over all degrees of freedom at the quantum level, often including the motional degree of freedom. Figure 10 illustrates four experimental approaches that satisfy some, or all, of these requirements. The molecules may be launched into a molecular fountain so that they are in free fall throughout a measurement [50, 87], or they could be trapped near the surface of a chip that integrates microscopic traps with superconducting microwave resonators [12]. Small, reconfigurable arrays of molecules can be produced using optical tweezer traps [85], while larger arrays can be made using optical lattices in one, two or three dimensions [88]. Here, we give an overview of future research directions using these platforms.

7.1   Testing Fundamental Physics

Ultracold molecules provide several avenues to constrain new theories or discover new physics [5]. As discussed in Sects. 3 and 4, the use of molecules to measure fundamental electric dipole moments is an amazingly powerful probe of symmetry-violating physics beyond the Standard Model. Other kinds of symmetry tests can also be done using molecules. For example, they can be used to explore parity violation in nuclei with unprecedented sensitivity [89]. Of particular interest is the measurement of nuclear anapole moments which arise from weak interactions within nuclei. A recent experiment with a beam of BaF has demonstrated the exceptional sensitivity achievable [90]. It is also of fundamental interest to measure the parity-violating energy difference between left- and right-handed chiral molecules, which is predicted but has not yet been observed [91]. The recent extension of laser cooling to polyatomic symmetric top molecules [63] shows that quite complex molecules can be cooled to sub-millikelvin temperature, making a parity-violation measurement using a laser cooled chiral species feasible in the future.

Precise measurements of molecular transition frequencies can also be used to test the idea that the fundamental constants may actually vary in time or space, or according to the local density of matter. Such variations are predicted by theories that aim to unify gravity with the other forces, and by some theories of dark energy [92, 93]. The frequencies of molecular transitions depend primarily on two fundamental constants, the fine-structure constant α and the proton-to-electron mass ratio $\mu = m_p/m_e$. The rotational and vibrational frequencies of molecules scale as μ^{-1} and $\mu^{-1/2}$ respectively, a direct dependence that an electronic transition in an atom does not have. Moreover, certain transitions have enhanced sensitivity to α or μ [94], sometimes because the transition energy results from a near cancellation between two large contributions of different origin. Astrophysical observations of atomic and molec-

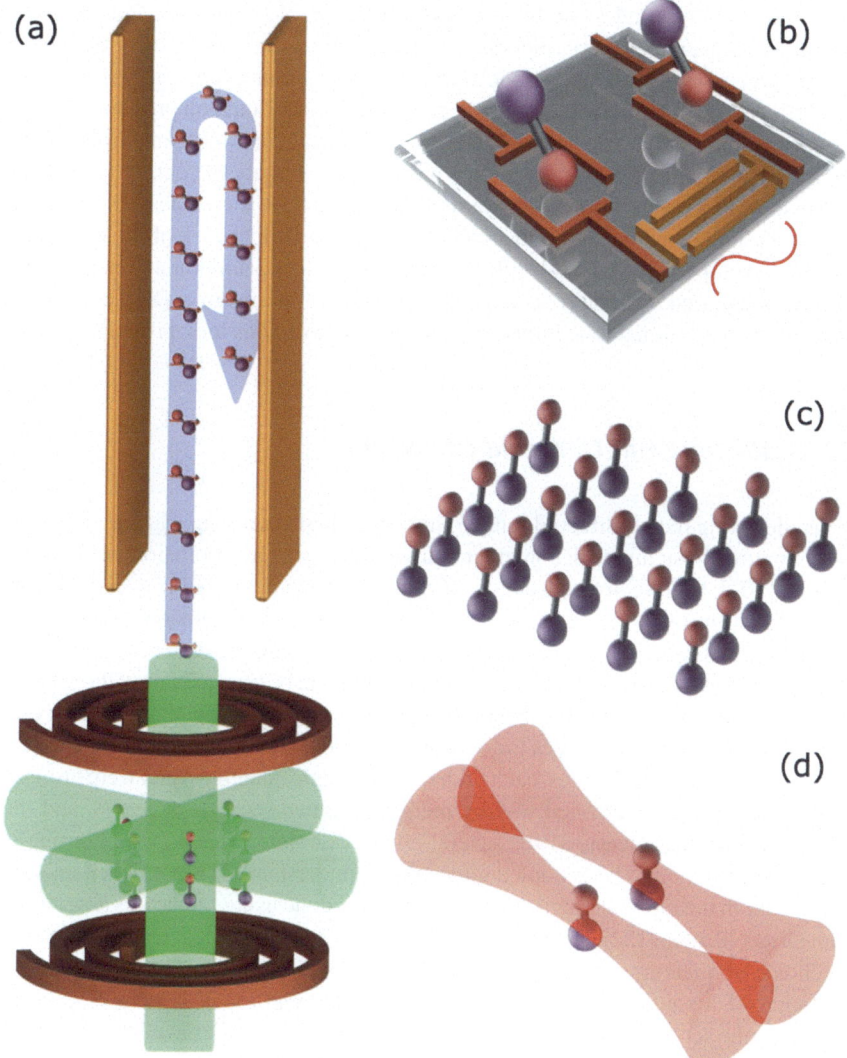

(a)

(b)

(c)

(d)

Fig. 10 Techniques for controlling ultracold molecules. **a** Molecules are launched into a fountain. **b** Molecules are stored on the surface of a chip using either electric or magnetic potentials created by planar electrodes or current-carrying wires. **c** Molecules are trapped in a 1D, 2D, or 3D lattice created by interfering counter-propagating lasers. **d** Molecules are trapped using an array of optical tweezers

ular spectra can be used to study variations on a cosmological timescale [5]. Here, laboratory measurements are important for establishing the present-day frequencies to high precision, as has been done using cold molecular beams of OH [95] and CH [96]. Atomic and molecular clocks can be used to set limits on present-day variations on a timescale of a few years. So far, the most stringent limits come from atomic clock measurements [97, 98], but molecular clocks are likely to contribute valuable information in the near future. For example, ultracold KRb molecules were recently used to set limits on the temporal variation of μ [99], a lattice clock of Sr_2 molecules is being developed [100], a molecular fountain of ultracold ammonia molecules has been demonstrated and could be used to search for variations in μ [87], and clocks based on the vibrational transitions of laser-cooled molecules look promising [101].

7.2 Collisions and Ultracold Chemistry

Molecules prepared at ultracold temperature in a single quantum state are ideal for studying how those molecules interact and what happens in a collision or chemical reaction [102]. With such a high degree of control, it becomes possible to explore how the rotational or hyperfine state influences the outcome of a collision, and to study collisions in a single partial wave regime. Electric or magnetic fields can be used to tune through collision resonances, and electric fields can be used to control the relative orientation of the colliding molecules. A fascinating recent advance in this direction is the study of collisions between individual laser-cooled molecules trapped in optical tweezers [103]. Two molecules, each prepared in a single quantum state, were brought together in a highly controlled way by merging the two separate tweezers, and the collisional loss rate measured for several choices of state. This experiment marks the first contribution of laser-cooled molecules to understanding ultracold chemistry. Some work in the ultracold regime has already been done using molecules assembled from ultracold atoms [52]. Examples include the control of chemical reactions though the choice of quantum state [104] or molecule orientation [105], and the controlled formation of a single molecule from a pair of atoms [106]. Direct laser cooling diversifies the set of ultracold molecules and molecular properties available for these studies, which is an exciting prospect for future research.

7.3 Quantum Simulation

It is important to understand the behaviour of systems consisting of many quantum particles all interacting with one another. These many-body quantum systems exhibit remarkable phenomena that are poorly understood at present, such as high temperature superconductivity, magnetism, the fractional Hall effect, and the structure of nuclei. We often use computer simulations to help understand complicated systems, but it is impossible for a (classical) computer to simulate more than a few tens of interacting quantum particles. Instead, we may try to engineer a well-controlled quantum system in such a way that it simulates a many-body problem that we wish to understand [107]. One system that has been developed with considerable success is an optical lattice of ultracold atoms. These lattices have been used to study important problems in condensed matter physics, such as the quantum phase transition between a superfluid and a Mott insulator [108], and models of antiferromagnetism [109]. However, the variety of systems that can be simulated is limited because atoms only have short-range interactions, meaning that they only interact appreciably when they are on the same site of the lattice. Complex many-body phenomena usually arise from long-range interactions among many particles, which atoms in a lattice struggle to emulate. A lattice of ultracold polar molecules solves this problem because the molecules interact through the long-range dipole-dipole interaction. This interaction has two main effects. First, the energy of the system depends on the configuration of the dipoles, and second, the interaction can mediate the transport of excitations from one site to another. Both effects have a long-range, anisotropic character, and can be controlled using a dc electric field or a microwave field resonant with a rotational transition. This makes for a tremendously rich environment for exploring the behaviour of interacting quantum systems [9]. Taking a first step in this direction, the effects of spin-exchange mediated by dipole-dipole interactions have been studied in a lattice of polar molecules [110]. Recent progress that will advance this field includes the formation of a Fermi degenerate gas of molecules [111], collisional cooling methods for molecules [112], the ever-increasing variety of polar molecules being brought into the ultracold regime, and the improvements in controlling their hyperfine and rotational states [10, 86, 113, 114]. It seems likely that, in the near future, lattices of ultracold polar molecules will significantly advance our understanding of strongly-interacting many-body quantum systems.

7.4   Quantum Information Processing

There are many proposals for using ultracold molecules for quantum information processing [11, 12, 115–117]. The hyperfine and rotational states of molecules have extremely long lifetimes and so can serve as stable qubits or qudits. By using microwave fields to drive rotational transitions, single qubit operations can be done rapidly and robustly using very mature microwave technology. The dipole-dipole interaction can be used to entangle pairs of molecules and perform two-qubit operations. Each molecule in an array can be addressed separately either by using a field gradient to shift the frequency of the qubit transition differently for each molecule, or by using an addressing laser to produce an ac Stark shift at a chosen site. One interesting approach for quantum information processing is an array of optical tweezer traps with a single molecule in each trap, as illustrated in Fig. 10(d) and recently demonstrated [85]. The molecules can be tightly confined and cooled to the motional ground state of the trap [118], and the array of molecules can be reconfigured as needed [119] in order to implement quantum gates between selected pairs. Another interesting approach is to trap the molecules near a surface using microscopic electric or magnetic traps, as illustrated in Fig. 10(b) and proposed in [12]. The same chip can support superconducting microwave resonators with small mode volumes and at frequencies that match the rotational frequency of the chosen molecule. By reaching the regime of strong coupling between trapped molecules and the resonator it becomes possible to transfer quantum information between a molecule and a microwave photon in the resonator, and to use the resonator to couple distant molecules to one another. This architecture is thus a hybrid quantum processor that combines the advantages of molecules for storing and processing quantum information with the advantages of photons for exchanging that information.

8 Summary

Over the past century, the humble molecular beam method has pushed forward the frontiers of knowledge in physics and chemistry. Today, molecules laser cooled to ultracold temperatures are an exciting and powerful platform for investigating the boundaries of modern scientific knowledge, including what might lie beyond the Standard Model of particle physics, how chemistry works at a fundamental level, and how quantum phenomena lead to emergent collective behaviors. This rich history and bright future has been strongly shaped by the visionary work of Otto Stern and his colleagues.

Acknowledgements We are grateful to Bretislav Friedrich and Horst Schmidt-Böcking for inviting us to contribute to this Festschrift, and to Luke Caldwell and Ed Hinds for their helpful feedback. The authors are supported by EPSRC (EP/P01058X/1), STFC (ST/S000011/1), the Royal Society, the John Templeton Foundation (grant 61104), the Gordon and Betty Moore Foundation (grant 8864), and the Alfred P. Sloan Foundation (grant G-2019-12505).

References

1. O. Stern, W. Gerlach, Z. Phys. **9**, 349 (1922)
2. O. Stern, Z. Phys **39**, 751 (1926)
3. R. Frisch, Z. Phys. **86**, 42 (1933)
4. D. DeMille, J.M. Doyle, A.O. Sushkov, Science **357**, 990 (2017)
5. M.S. Safronova, D. Budker, D. DeMille, D.F.J. Kimball, A. Derevianko, C.W. Clark, Rev. Mod. Phys. **90**, 025008 (2018)
6. D.S. Jin, J. Ye, Chem. Rev. **112**, 4801 (2012)
7. J. Toscano, H.J. Lewandowski, B.R. Heazlewood, Phys. Chem. Chem. Phys. **22**, 9180 (2020)
8. A. Micheli, G. Pupillo, H.P. Büchler, P. Zoller, Phys. Rev. A **76**, 043604 (2007)
9. A.V. Gorshkov, S.R. Manmana, G. Chen, J. Ye, E. Demler, M.D. Lukin, A.M. Rey, Phys. Rev. Lett. **107**, 115301 (2011)

10. J.A. Blackmore, L. Caldwell, P.D. Gregory, E.M. Bridge, R. Sawant, J. Aldegunde, J. Mur-Petit, D. Jaksch, J.M. Hutson, B.E. Sauer, M.R. Tarbutt, S.L. Cornish, Quantum. Sci. Technol. **4**, 014010 (2018)
11. D. DeMille, Phys. Rev. Lett. **88**, 067901 (2002)
12. A. André, D. DeMille, J.M. Doyle, M.D. Lukin, S.E. Maxwell, P. Rabl, R.J. Schoelkopf, P. Zoller, Nat. Phys. **2**, 636 (2006)
13. O. Stern, Z. Phys. **2**, 49 (1920)
14. O. Stern, Z. Phys. **3**, 417 (1921)
15. B.E. Sauer, J. Wang, E.A. Hinds, Phys. Rev. Lett. **74**, 1554 (1995)
16. M.R. Tarbutt, J.J. Hudson, B.E. Sauer, E.A. Hinds, V.A. Ryzhov, V.L. Ryabov, V.F. Ezhov, J. Phys. B **35**, 5013 (2002)
17. N.E. Bulleid, S.M. Skoff, R.J. Hendricks, B.E. Sauer, E.A. Hinds, M.R. Tarbutt, Phys. Chem. Chem. Phys. **15**(29), 12299 (2013)
18. G. Scoles (ed.), *Atomic and Molecular Beam Methods* (Oxford University Press, Oxford, 1988)
19. R. Campargue (ed.), *Atomic and Molecular Beams: The State of the Art 2000* (Springer, Berlin, 2001)
20. S.E. Maxwell, N. Brahms, R. deCarvalho, D.R. Glenn, J.S. Helton, S.V. Nguyen, D. Patterson, J. Petricka, D. DeMille, J.M. Doyle, Phys. Rev. Lett. **95**, 173201 (2005)
21. N.R. Hutzler, H.I. Lu, J.M. Doyle, Chem. Rev. **112**, 4803 (2012)
22. H.I. Lu, J. Rasmussen, M.J. Wright, D. Patterson, J.M. Doyle, Phys. Chem. Chem. Phys. **13**, 18986 (2011)
23. O.R. Frisch, O. Stern, Z. Phys. **85**, 4 (1933)
24. I. Estermann, O. Stern, Z. Phys. **86**, 132 (1933)
25. I. Estermann, O. Stern, Z. Phys. **86**, 135 (1933)
26. E.M. Purcell, N.F. Ramsey, Phys. Rev. **78**, 807 (1950)
27. H. Weyl, *Symmetry* (Princeton University Press, Princeton, New Jersey, 1952)
28. T.D. Lee, C.N. Yang, Phys. Rev. **104**, 254 (1956)
29. C.S. Wu, E. Ambler, R.W. Hayward, D.D. Hoppes, R.P. Hudson, Phys. Rev. **105**, 1413 (1957)
30. R.L. Garwin, L.M. Lederman, M. Weinrich, Phys. Rev. **105**, 1415 (1957)
31. J.I. Friedman, V.L. Telgedi, Phys. Rev **105**, 1681 (1957)
32. J.H. Christenson, J.W. Cronin, V.L. Fitch, R. Turlay, Phys. Rev. Lett. **13**, 138 (1964)
33. P.G.H. Sandars, Phys. Lett. **14**, 196 (1965)
34. H.S. Nataraj, B.K. Sahoo, B.P. Das, D. Mukherjee, Phys. Rev. Lett. **101**, 033002 (2008)
35. V. Andreev, D.G. Ang, D. DeMille, J.M. Doyle, G. Gabrielse, J. Haefner, N.R. Hutzler, Z. Lasner, C. Meisenhelder, B.R. O'Leary, C.D. Panda, A.D. West, E.P. West, X. Wu, Nature **562**, 355 (2018)
36. I.I. Rabi, S. Millman, P. Kusch, J.R. Zacharias, Phys. Rev. **55**, 526 (1939)
37. N.F. Ramsey, Phys. Rev. **78**, 695 (1950)
38. J.H. Smith, E.M. Purcell, N.F. Ramsey, Phys. Rev. **108**, 120 (1957)
39. M. Pospelov, A. Ritz, Phys. Rev. D **89**, 056006 (2014)
40. A.D. Sakharov, Phys.-Uspekhi **34**(5), 392 (1991)
41. S.A. Murthy, J.D. Krause, Z.L. Li, L.R. Hunter, Phys. Rev. Lett. **63**, 965 (1989)
42. B.C. Regan, E.D. Commins, C.J. Schmidt, D. DeMille, Phys. Rev. Lett. **88**, 071805 (2002)
43. P.G.H. Sandars, in *Atomic Physics 4*, ed. by G. zu Pulitz (Plenum, New York, 1975), p. 71
44. E.A. Hinds, Physica Scripta **T70**, 34 (1997)
45. J.J. Hudson, B.E. Sauer, M.R. Tarbutt, E.A. Hinds, Phys. Rev. Lett. **89**, 023003 (2002)
46. J.J. Hudson, D.M. Kara, I.J. Smallman, B.E. Sauer, M.R. Tarbutt, E.A. Hinds, Nature **473**, 493 (2011)
47. J. Baron, W.C. Campbell, D. DeMille, J.M. Doyle, G. Gabrielse, Y.V. Gurevich, P.W. Hess, N.R. Hutzler, E. Kirilov, I. Kozyryev, B.R. O'Leary, C.D. Panda, M.F. Parsons, E.S. Petrik, B. Spaun, A.C. Vutha, A.D. West, Science **343**, 269 (2014)
48. S. Eckel, P. Hamilton, E. Kirilov, H.W. Smith, D. DeMille, Phys. Rev. A **87**, 052130 (2013)

49. W.B. Cairncross, D.N. Gresh, M. Grau, K.C. Cossel, T.S. Roussy, Y. Ni, Y. Zhou, J. Ye, E.A. Cornell, Phys. Rev. Lett. **119**, 153001 (2017)
50. M.R. Tarbutt, B.E. Sauer, J.J. Hudson, E.A. Hinds, New J. Phys. **15**, 053034 (2013)
51. M.R. Tarbutt, J.J. Hudson, B.E. Sauer, E.A. Hinds, Faraday Discuss. **142**, 37 (2009)
52. K.K. Ni, S. Ospelkaus, M.H.G. de Miranda, A. Pe'er, B. Neyenhuis, J.J. Zirbel, S. Kotochigova, P.S. Julienne, D.S. Jin, J. Ye, Science **322**, 231 (2008)
53. M.R. Tarbutt, Contemp. Phys. **59**, 356 (2018)
54. J. Lim, J.R. Almond, M.A. Trigatzis, J.A. Devlin, N.J. Fitch, B.E. Sauer, M.R. Tarbutt, E.A. Hinds, Phys. Rev. Lett. **120**, 123201 (2018)
55. The NL-eEDM collaboration, P. Aggarwal, H.L. Bethlem, A. Borschevsky, M. Denis, K. Esajas, P.A.B. Haase, Y. Hao, S. Hoekstra, K. Jungmann, T.B. Meijknecht, M.C. Mooij, R.G.E. Timmermans, W. Ubachs, L. Willmann, A. Zapara, Eur. Phys. J. D **72**, 197 (2018)
56. I. Kozyryev, N.R. Hutzler, Phys. Rev. Lett. **119**, 133002 (2017)
57. E.B. Norrgard, E.R. Edwards, D.J. McCarron, M.H. Steinecker, D. DeMille, S.S. Alam, S.K. Peck, N.S. Wadia, L.R. Hunter, Phys. Rev. A **95**, 062506 (2017)
58. E.S. Shuman, J.F. Barry, D. DeMille, Nature **467**, 820 (2010)
59. M.T. Hummon, M. Yeo, B.K. Stuhl, A.L. Collopy, Y. Xia, J. Ye, Phys. Rev. Lett. **110**, 143001 (2013)
60. I. Kozyryev, L. Baum, K. Matsuda, B.L. Augenbraun, L. Anderegg, A.P. Sedlack, J.M. Doyle, Phys. Rev. Lett. **118**, 173201 (2017)
61. B.L. Augenbraun, Z.D. Lasner, A. Frenett, H. Sawaoka, C. Miller, T.C. Steimle, J.M. Doyle, New J. Phys. **22**, 022003 (2020)
62. L. Baum, N.B. Vilas, C. Hallas, B.L. Augenbraun, S. Raval, D. Mitra, J.M. Doyle, Phys. Rev. Lett **124**, 133201 (2020)
63. D. Mitra, N.B. Vilas, C. Hallas, L. Anderegg, B.L. Augenbraun, L. Baum, C. Miller, S. Raval, J.M. Doyle, Science **369**, 1366 (2020)
64. R.L. McNally, I. Kozyryev, S. Vazquez-Carson, K. Wenz, T. Wang, T. Zelevinsky, New J. Phys. **22**, 083047 (2020)
65. J.F. Barry, E.S. Shuman, E.B. Norrgard, D. DeMille, Phys. Rev. Lett. **108**, 103002 (2012)
66. V. Zhelyazkova, A. Cournol, T.E. Wall, A. Matsushima, J.J. Hudson, E.A. Hinds, M.R. Tarbutt, B.E. Sauer, Phys. Rev. A **89**, 053416 (2014)
67. M. Yeo, M.T. Hummon, A.L. Collopy, B. Yan, B. Hemmerling, E. Chae, J.M. Doyle, J. Ye, Phys. Rev. Lett. **114**(22), 223003 (2015)
68. B. Hemmerling, E. Chae, A. Ravi, L. Anderegg, G.K. Drayna, N.R. Hutzler, A.L. Collopy, J. Ye, W. Ketterle, J.M. Doyle, J. Phys. B **49**, 174001 (2016)
69. S. Truppe, H.J. Williams, N.J. Fitch, M. Hambach, T.E. Wall, E.A. Hinds, B.E. Sauer, M.R. Tarbutt, New J. Phys. **19**, 022001 (2017)
70. H.J. Williams, S. Truppe, M. Hambach, L. Caldwell, N.J. Fitch, E.A. Hinds, B.E. Sauer, M.R. Tarbutt, New J. Phys. **19**, 113035 (2017)
71. E. Raab, M. Prentiss, A. Cable, S. Chu, D. Pritchard, Phys. Rev. Lett. **59**, 2631 (1987)
72. J.F. Barry, D.J. McCarron, E.B. Norrgard, M.H. Steinecker, D. DeMille, Nature **512**, 286 (2014)
73. D.J. McCarron, E.B. Norrgard, M.H. Steinecker, D. DeMille, New J. Phys. **17**, 035014 (2015)
74. E.B. Norrgard, D.J. McCarron, M.H. Steinecker, M.R. Tarbutt, D. DeMille, Phys. Rev. Lett. **116**, 063004 (2016)
75. M.H. Steinecker, D.J. McCarron, Y. Zhu, D. DeMille, Chem. Phys. Chem. **17**, 3664 (2016)
76. S. Truppe, H.J. Williams, M. Hambach, L. Caldwell, N.J. Fitch, E.A. Hinds, B.E. Sauer, M.R. Tarbutt, Nat. Phys. **13**, 1173 (2017)
77. L. Anderegg, B.L. Augenbraun, E. Chae, B. Hemmerling, N.R. Hutzler, A. Ravi, A. Collopy, J. Ye, W. Ketterle, J.M. Doyle, Phys. Rev. Lett. **119**, 103201 (2017)
78. A.L. Collopy, S. Ding, Y. Wu, I.A. Finneran, L. Anderegg, B.L. Augenbraun, J.M. Doyle, J. Ye, Phys. Rev. Lett. **121**, 213201 (2018)
79. L. Anderegg, B.L. Augenbraun, Y. Bao, S. Burchesky, L.W. Cheuk, W. Ketterle, J.M. Doyle, Nat. Phys. **14**, 890 (2018)

80. D.J. McCarron, M.H. Steinecker, Y. Zhu, D. DeMille, Phys. Rev. Lett. **121**, 013202 (2018)
81. L.W. Cheuk, L. Anderegg, B.L. Augenbraun, Y. Bao, S. Burchesky, W. Ketterle, J.M. Doyle, Phys. Rev. Lett. **121**, 083201 (2018)
82. L. Caldwell, J.A. Devlin, H.J. Williams, N.J. Fitch, E.A. Hinds, B.E. Sauer, M.R. Tarbutt, Phys. Rev. Lett. **123**, 033202 (2019)
83. S. Ding, Y. Wu, I.A. Finneran, J.J. Burau, J. Ye, Phys. Rev. X **10**, 021049 (2020)
84. H.J. Williams, L. Caldwell, N.J. Fitch, S. Truppe, J. Rodewald, E.A. Hinds, B.E. Sauer, M.R. Tarbutt, Phys. Rev. Lett. **120**, 163201 (2018)
85. L. Anderegg, L.W. Cheuk, Y. Bao, S. Burchesky, W. Ketterle, K.K. Ni, J.M. Doyle, Science **365**, 1156 (2019)
86. L. Caldwell, H.J. Williams, N.J. Fitch, J. Aldegunde, J.M. Hutson, B.E. Sauer, M.R. Tarbutt, Phys. Rev. Lett. **124**, 063001 (2020)
87. C. Cheng, A.P.P. van der Poel, P. Jansen, M. Quintero-Pérez, T.E. Wall, W. Ubachs, H.L. Bethlem, Phys. Rev. Lett. **117**, 253201 (2016)
88. S.A. Moses, J.P. Covey, M.T. Miecnikowski, B. Yan, B. Gadway, J. Ye, D.S. Jin, Science **350**, 659 (2015)
89. S.B. Cahn, J. Ammon, E. Kirilov, Y.V. Gurevich, D. Murphree, R. Paolino, D.A. Rahmlow, M.G. Kozlov, D. DeMille, Phys. Rev. Lett. **112**, 163002 (2014)
90. E. Altuntaş, J. Ammon, S.B. Cahn, D. DeMille, Phys. Rev. Lett. **120**, 142501 (2018)
91. S. Tokunaga, C. Stoeffler, F. Auguste, A. Shelkovnikov, C. Daussy, A. Amy-Klein, C. Chardonnet, B. Darquié, Mol. Phys. **111**, 2363 (2013)
92. J.P. Uzan, Living Rev. Relativity **14**, 1 (2011)
93. K.A. Olive, M. Pospelov, Phys. Rev. D **77**, 043524 (2010)
94. C. Chin, V.V. Flambaum, M.G. Kozlov, New J. Phys. **11**, 055048 (2009)
95. E.R. Hudson, H.J. Lewandowski, B.C. Sawyer, J. Ye, Phys. Rev. Lett. **96**, 143004 (2006)
96. S. Truppe, R.J. Hendricks, S.K. Tokunaga, H.J. Lewandowski, M.G. Kozlov, C. Henkel, E.A. Hinds, M.R. Tarbutt, Nat. Commun. **4**, 2600 (2013)
97. N. Huntemann, B. Lipphardt, C. Tamm, V. Gerginov, S. Weyers, E. Peik, Phys. Rev. Lett. **113**, 210802 (2014)
98. R.M. Godun, P.B.R. Nisbet-Jones, J.M. Jones, S.A. King, L.A.M. Johnson, H.S. Margolis, K. Szymaniec, S.N. Lea, K. Bongs, P. Gill, Phys. Rev. Lett. **113**, 210801 (2014)
99. J. Kobayashi, A. Ogino, S. Inouye, Nat. Comms. **10**, 3771 (2019)
100. S.S. Kondov, C.H. Lee, K.H. Leung, C. Liedl, I. Majewska, R. Moszynski, T. Zelevinsky, Nat. Phys. **15**, 1118 (2019)
101. M. Kajita, J. Phys. Soc. Jpn. **87**, 104301 (2018)
102. J.L. Bohn, A.M. Rey, J. Ye, Science **357**, 1002 (2017)
103. L.W. Cheuk, L. Anderegg, Y. Bao, S. Burchesky, S. Yu, W. Ketterle, K.K. Ni, J.M. Doyle, Phys. Rev. Lett. **125**, 043401 (2020)
104. S. Ospelkaus, K.K. Ni, D. Wang, M.H.G. de Miranda, B. Neyenhuis, G. Quéméner, P.S. Julienne, J.L. Bohn, D.S. Jin, J. Ye, Science **327**, 853 (2010)
105. M.H.G. de Miranda, A. Chotia, B. Neyenhuis, D. Wang, G. Quéméner, S. Ospelkaus, J.L. Bohn, J. Ye, D.S. Jin, Nat. Phys. **7**, 502 (2011)
106. L.R. Liu, J.D. Hood, Y. Yu, J.T. Zhang, N.R. Hutzler, T. Rosenband, K.K. Ni, Science **360**, 900 (2018)
107. R.P. Feynman, Int. J. Theo. Phys. **21**, 467 (1982)
108. M. Greiner, O. Mandel, T. Esslinger, T.W. Hänsch, I. Bloch, Nature **415**, 39 (2002)
109. A. Mazurenko, C.S. Chiu, G. Ji, M.F. Parsons, M. Kanász-Nagy, R. Schmidt, F. Grusdt, E. Demler, D. Greif, M. Greiner, Nature **545**, 462 (2017)
110. B. Yan, S.A. Moses, B. Gadway, J.P. Covey, K.R.A. Hazzard, A.M. Rey, D.S. Jin, J. Ye, Nature **501**, 521 (2013)
111. L.D. Marco, G. Valtolina, K. Matsuda, W.G. Tobias, J.P. Covey, J. Ye, Science **363**, 853 (2019)
112. H. Son, J.J. Park, W. Ketterle, A.O. Jamison, Nature **580**, 197 (2020)
113. J.W. Park, Z.Z. Yan, H. Loh, S.A. Will, M.W. Zwierlein, Science **357**, 372 (2017)

114. F. Seeßelberg, X.Y. Luo, M. Li, R. Bause, S. Kotochigova, I. Bloch, C. Gohle, Phys. Rev. Lett. **121**, 253401 (2018)
115. S.F. Yelin, K. Kirby, R. Côté, Phys. Rev. A **74**, 050301 (2006)
116. K.K. Ni, T. Rosenband, D.D. Grimes, Chem. Sci. **9**, 6830 (2018)
117. R. Sawant, J.A. Blackmore, P.D. Gregory, J. Mur-Petit, D. Jaksch, J. Aldegunde, J.M. Hutson, M.R. Tarbutt, S.L. Cornish, New J. Phys. **22**, 013027 (2020)
118. L. Caldwell, M.R. Tarbutt, Phys. Rev. Res. **2**, 013251 (2020)
119. D. Barredo, S. de Léséleuc, V. Lienhard, T. Lahaye, A. Browaeys, Science **354**, 1021 (2016)

Part V
Matter Waves

Chapter 23
Otto Stern and Wave-Particle Duality

J. Peter Toennies

Abstract The contributions of Otto Stern to the discovery of wave-particle duality of matter particles predicted by de Broglie are reviewed. After a short introduction to the early matter-vs-wave ideas about light, the events are highlighted which lead to de Broglie's idea that all particles, also massive particles, should exhibit wave behavior with a wavelength inversely proportional to their mass. The first confirming experimental evidence came for electrons from the diffraction experiments of Davisson and Germer and those of Thomson. The first demonstration for atoms, with three orders of magnitude smaller wave lengths, came from Otto Stern's laboratory shortly afterwards in 1929 in a remarkable *tour de force* experiment. After Stern's forced departure from Hamburg in 1933 it took more than 40 years to reach a similar level of experimental perfection as achieved then in Stern's laboratory. Today He atom diffraction is a powerful tool for studying the atomic and electronic structure and dynamics of surfaces. With the advent of nanotechnology nanoscopic transmission gratings have led to many new applications of matter waves in chemistry and physics, which are illustrated with a few examples and described in more detail in the following chapters.

1 Introduction

On September 8, 1926 Otto Stern submitted the first of a projected series of 30 articles all of which were to appear in the *Zeitschrift für Physik* under the subtitle "Untersuchungen zur Molekularstrahlmethode (UzM) aus dem Institut für physikalische Chemie der Hamburgischen Universität" [1]. In this initial article he outlined his plans for future physics experiments based on the method of molecular beams. It is remarkable that almost all of the projected experiments were successfully carried out in the ensuing 7 years until 1933 when Stern was forced to leave Germany. At the end of the list he had two special projects: "Der Einsteinsche Strahlungsdruck"

J. Peter Toennies (✉)
Max-Planck-Institut für Dynamik und Selbstorganisation, Am Fassberg 17, 37077 Göttingen, Germany
e-mail: jtoenni@gwdg.de

© The Author(s) 2021
B. Friedrich and H. Schmidt-Böcking (eds.), *Molecular Beams in Physics and Chemistry*,
https://doi.org/10.1007/978-3-030-63963-1_23

and "Die De Broglie-Wellen". Two years earlier, on November 1924, a little known young French physicist, Louis de Broglie, had defended his Sorbonne thesis [2]. In his thesis he published his famous formula $\lambda = h/m\upsilon$ [1] which introduced, for the first time, the concept of wave-particle duality of massive particles. Thus in 1926 Otto Stern was among the first who realized the tremendous importance of confirming de Broglie's revolutionary ideas by experiment.

After several marginally successful experiments, in May 1929 Otto Stern reported in a short note in the journal *Naturwissenschaften* well-resolved diffraction peaks in the reflection of He atoms from a cleaved NaCl crystal surfaces [3]. Thereby he provided the first direct evidence that the incident atoms of helium as massive particles had wave properties. Earlier in 1927, at about the same time as Stern and colleagues were still optimizing their experiment, Davisson and Germer had reported the first experimental evidence for the wave nature of electrons in *Nature* [4]. In the same year the wave nature of electrons was also independently confirmed in another *Nature* article by Thomson and Reid [5].

In the following, the history of the genesis of wave-particle duality will be reviewed. Then the diffraction experiments of Otto Stern will be described in more detail. Recent He atom diffraction experiments following on the footsteps of Stern and based partly on new developments are reviewed. These will serve as an introduction to the following articles describing the current state of experiments made possible by nanotechnologic advances which rely on the wave-particle duality of atoms and molecules.

2   History of Wave-Particle Duality and the de Broglie Relation

The early Greek philosophers Leucippus (fifth century BC) and his pupil Democritus (c. 460–c. 370 BC) were probably the first to introduce the idea that matter is composed of atoms (Greek for indivisible). Some of the early Greek philosophers also supposed that the *seen* object was emitting particles that bombarded their eyes. A modern theory that light was made of particles was first formulated much later by Isaac Newton (1643–1727). Somewhat earlier the idea that light was made of waves was postulated by Huygens (1629–1695). Soon after Thomas Young (1773–1829), August Jean Fresnel (1778–1827) and Joseph von Fraunhofer (1778–1826) carried out experiments which confirmed the wave nature of light. The wave versus particle dichotomy of light persisted up to the twentieth century. At the turn of the century, in 1900, Max Planck (1858–1947) introduced his famous radiation law and introduced the concept of a quantum of light which led the way to the development of Quantum Mechanics in 1926.

The modern development of the concept of matter waves can be traced back to 1905 when Einstein (1879–1955) used Planck's constant to explain the photoelectric

---

[1] λ is the wave length of a particle with mass m moving at velocity υ. h is Planck's constant.

effect with the simple equation

$$KE_{el} = h\nu - \phi, \tag{1}$$

where KE_{el} is the kinetic energy of the ejected electron, ν is the frequency of the incident photons, and ϕ is the work function of the solid. Thereby it was established that photons act like particles with a fixed energy. Some years later, in 1923, Arthur Compton (1892–1962) reported that the X-ray photons that had been scattered from an electron in a solid, had a fixed momentum $P_{ph} = mc = \frac{h\nu}{c}$. Moreover, the momentum of the rebounding electrons could be explained by conservation of momentum and energy by assuming that the incident X-ray was a particle.

In 1919, these developments attracted the attention of Louis de Broglie (1892–1987) after he had been released from the army after World War I. Working in the laboratory of his physicist brother he became interested in the new concept of a "quanta of light". In 1922 he published two short articles on black-body radiation [6, 7]. Then, in 1923 he published three additional short two-page articles in *Comptes Rendus* [8] in which he developed his ideas on the wave nature of light. These he summarized in a half-page *Nature* article in 1923 [9] and in the Philosophical Magazine [10]. In the *Nature* article he concluded: "A radiation of frequency ν has to be considered as divided into atoms of light of very small internal mass $\left(< 10^{-50} \text{ gm} \right)$ which move with a velocity very nearly equal to c given by $\frac{m_0 c^2}{\sqrt{1-\beta^2}} = h\nu$. The atom of light slides slowly upon the non-material wave the frequency of which is ν and velocity c/β, very little higher than c. The phase wave has a very great importance in determining the motion of any moving body, and I have been able to show that the stability conditions of the trajectories in Bohr's atom express that the wave is tuned with the length of the closed path." With this he anticipated his wave hypothesis in the case of the electron orbits in an atom by showing that the circumference of the orbits would be an integral multiple of the wavelength of the electron.

De Broglie's *These de doctoral* appeared in the following year 1924 and later was published in 1925 in *Annales de Physique* [11]. In only one place in the final short chapter in one of the last sections entitled *The New Conception of Gas Equilibrium* he writes his famous formula

$$\lambda = \frac{h}{m\upsilon}. \tag{2}$$

The fact that the formula appears only once suggests that at the time he was apparently more interested in discussing wave motion in general as applied to X-rays and electrons and did not fully realize then the far-reaching significance of the equation for which he is presently known.

According to the excellent reviews of de Broglie's discovery by Medicus [12] and MacKinnan [13] de Broglie's work did not become widely known, partly because *Comptes Rendus* was not very popular and partly because the reputation of the little-known young theoretician was controversial. De Broglie's thesis only became widely known after Paul Langevin, who was a member of his examination committee, had sent it to Einstein. Einstein immediately appreciated the far reaching consequences of de Broglie's ideas and wrote back that de Broglie had "lifted a corner of the great veil" and incorporated the new concept in his article in the *Proceedings of the Prussian Academy* which appeared in 1925 [14].

3   1925: Experimental Confirmation for Electrons

About the time of de Broglie's theories Clinton J. Davisson at the American Telephone and Telegraph (now AT&T) and the Western Electric Company in New York City was experimenting on the effect of electron bombardment on metal surfaces. This research was carried out in connection with understanding the physics of vacuum tubes which were a major product of the two companies. In 1921, Davisson and Kunsmann had reported their initial results on the measurements of the angular distributions of scattered electrons with incident energies up to 1000 eV in *Science* [15] and later in *Physical Review* [16]. At energies below 125 eV upon scattering from platinum and magnesium metal surfaces, they observed an unexpected small lobe in an otherwise Gauss-shaped distribution (Fig. 1). The lobe, they thought, could be related to the Bohr-model electron orbits in the metal.

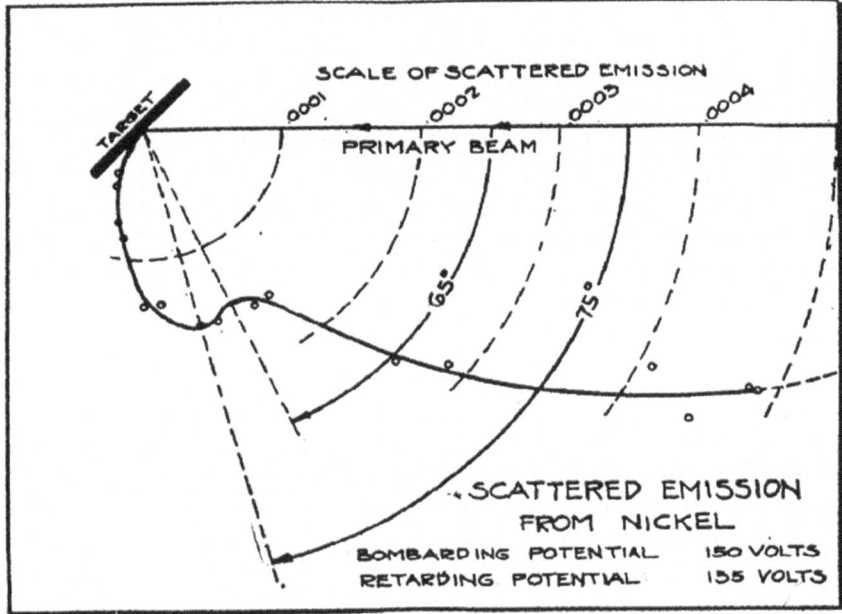

Fig. 1 The angular distribution of scattered electrons from nickel observed by Davisson and Kunsmann in 1921. The figure is taken from Ref. [16]

In the summer of 1925, in far away Göttingen, Friedrich Hund (1896–1997) gave a talk about the experiments of Davisson and Kunsmann in Prof. Max Born's seminar on *Die Struktur der Materie*. The Göttingen physicists were then also interested in electron scattering in connection with the 1924 Franck-Hertz experiment, which Franck and Hertz had carried out in Berlin before coming to Göttingen. In Göttingen it was one of the areas of research of the Born group. Walter Elsasser (1904–1942), a student attending the seminar, who had read about de Broglie's theory which had appeared in the same year, conjectured that the lobes reported by Davisson and Kunsman were in fact partly resolved diffraction peaks and could be the first experimental confirmation of de Broglie's theory. Since Elsasser was also attending the lecture course given by Prof. James Franck (1882–1964), he told Franck about his thoughts, whereupon Franck encouraged Elsasser to write a short article about his idea (Fig. 2). The half-page letter appeared in August 1925 in *Die Naturwissenschaften* [17].

The previous account is from the American National Academy of Sciences Biographical Memoir about Elsasser written by Rubin [18]. The important role of Elsasser is also supported by Hund [19] and also by Max Born in his article on the quantum mechanics of collisions [20]. Later, however, Max Born claimed that he had the idea and that he was the one to encourage Elsasser to write the note [21]. In the same article he does remark that he cannot fully remember the details. A somewhat

W. Elsasser
1904-1942

1925 Walter Elsasser, a student in Born's Seminar „Die Struktur der Materie" heard a talk by Frederick Hund about the experiments of Davisson and Kunsman

F. Hund
1896-1997

Walter came up with the idea that this could be evidence for de Broglie's idea of matter waves. James Franck suggested to Walter that he should submit an article to the journal *Die Naturwissenschaften*. The short note appeared on July 18,1925. Thus Elsasser was the first to point to experiments confirming de Broglie's ideas

J. Franck
1882-1964

Fig. 2 The story behind the first realization of experimental evidence for wave-particle duality according to the official National Academy of Science biography of Walter Elsasser [18]

different story can be found in Gehrenberg [22, 23]. It is also interesting to note that Davisson and his assistant Germer were not at all convinced by Elsasser's idea. In their 1927 article in *Physical Review* [24] they wrote "We would like to agree with Elsasser in his interpretation of the small lobe reported by Davisson and Kunsman in 1921 and 1923, but are unable to do so". At the time they were convinced that the curves seen in their initial experiments were "unrelated to crystal structure".

At about the same time, in April 1925, a momentous accident occurred in the laboratory of Clinton Davisson. The glass vacuum tube containing the electron scattering apparatus exploded while the metal target was at high temperatures. In an attempt to save the highly oxidized target it was subsequently baked out over an extended period of time in an effort to reduce the oxide coating. When the experiments were repeated, surprisingly, much sharper lobes were observed, which were especially apparent when the crystal was rotated in the azimuthal direction (Fig. 3). In their *Nature* article in 1927 [4] Davisson and Germer attributed the six sharp azimuthal features to matter-wave diffraction in agreement with the theory of de Broglie. In this same article they do finally give credit to Elsasser. In the same year they published a complete analysis in Physical Review [24]. There they attributed the appearance of the newly found diffraction peaks to the increased order of the sample, which had been partly crystallized by the annealing during the long bake out [24]. With serendipitous good fortune, they had finally succeeded in providing convincing confirmation of the

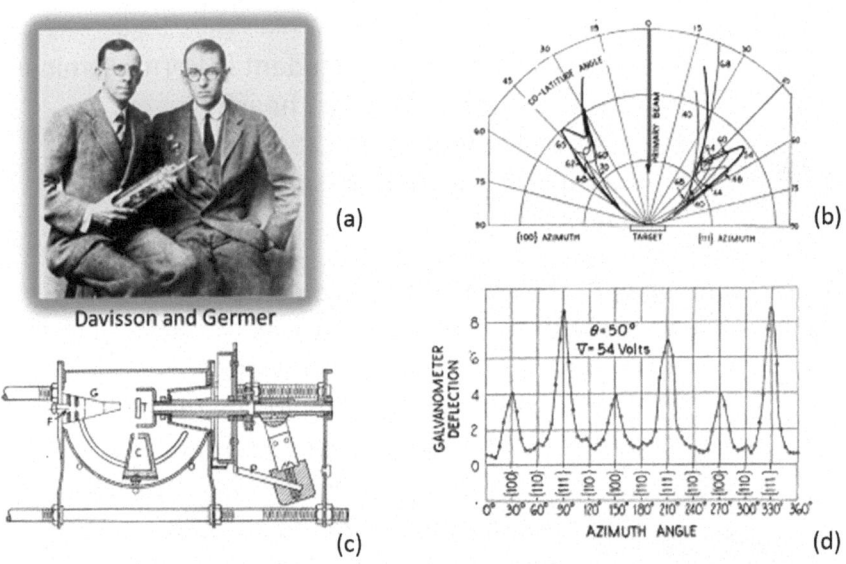

Fig. 3 The first unequivocal evidence for the de Broglie formula came from the 1927 electron diffraction experiments of Davisson and Germer (**a**). **b** The polar angle distribution shows diffraction peaks on both sides of the specularly reflected beam. **c** Schematic diagram of the electron scattering apparatus. **d** The sharp diffraction peaks observed on rotating the crystal. The latter 3 figures are from Ref. [24]

de Broglie relation. Of course this brief account does not in any way do justice to the many agonizing attempts that finally led to Davisson and Germer's accidental but successful experiment. The full story has been documented in detail by Gehrenberg [22].

This pioneering experiment was the forerunner of modern *Low Energy Electron Diffraction (LEED)* and *Electron Energy Loss Spectroscopy (EELS)* methods for surface analysis. Presently the former is widely used to determine the structure mostly of metal surfaces and the latter for measuring the surface phonons and the vibrations of clean and adsorbate-covered surfaces.

Then also in 1925, George P. Thomson (1892–1975), the son of the famous English physicist J. J. Thomson, reported on the diffraction of high energy electron beams (3,900–16,500 eV) upon passage through a 30 nm thick celluloid film. Their *Nature* article [8] appeared only two months after the *Nature* article of Davisson and Germer. With their simple but elegant experiment they were able to also verify the de Broglie relation (Fig. 4) [25]. In 1937, both Davisson and Thomson were awarded the Nobel Prize for *The Discovery of the Electron Waves*.

The significance of these developments has recently been highlighted in a thought provoking article by Steven Weinberg entitled *The Trouble with Quantum Mechanics* [27]. He begins his *critique* of quantum mechanics by noting "Then in the 1920s, according to theory of Louis de Broglie and Erwin Schrödinger, it appeared that

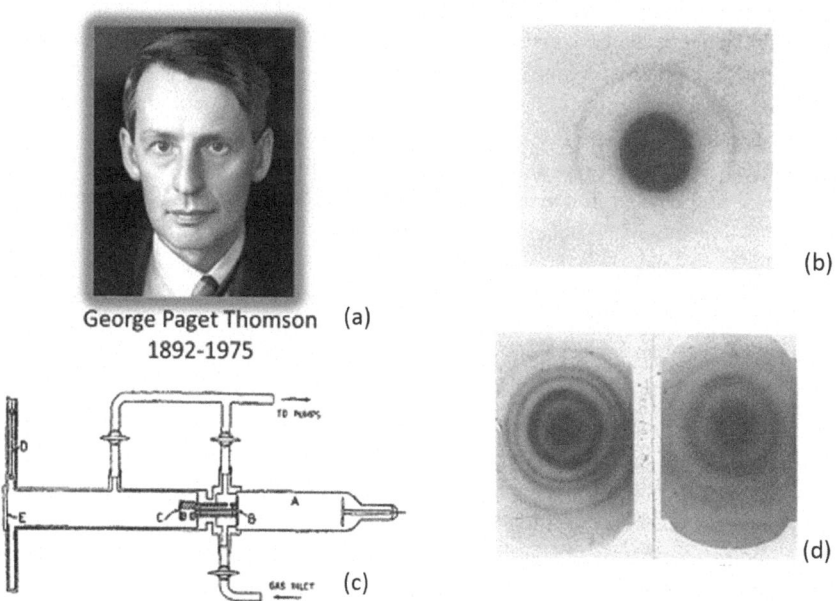

George Paget Thomson (a)
1892-1975

(b)

(c)

(d)

Fig. 4 The electron transmission experiment by G. P. Thomson which appeared a few months after the results of Davisson and Germer. **a** Photo of G. P. Thomson. **b** Diffraction rings on transmission through celluloid. **c** The electron transmission apparatus. **d** Diffraction rings seen on transmission through two different thin gold foils. The later 3 figures are from Ref. [26]

electrons which had always been recognized as particles, under some circumstances behaved as waves. In order to account for the energies of the stable states of atoms, physicists had to give up the notion that electrons in atoms are little Newtonian planets in orbit around the atomic nucleus. Electrons in atoms are better described as waves, fitting into an organ pipe. The world's categories had become all muddled."

4   Otto Stern's Experimental Confirmation for Atoms

Otto Stern's career as an experimentalist started in 1919 when he took up the position of assistant in Max Born's two-room theory group at the University of Frankfurt [28]. With a cleverly conceived apparatus he was able, for the first time, to measure the mean velocity in a molecular beam, which had been predicted by Clausius [29]. Then, in 1921 he embarked with Walter Gerlach on the famous Stern-Gerlach experiment [30], which led to the discovery of angular momenta and magnetic moments of atoms in magnetic fields. In the fall of the same year, Stern left Frankfurt to take up a new position as "Extraordinarius" (associate professor) at the University of Rostock. In the aftermath of World War I, the financial conditions in Rostock were such that he could not think of carrying out experimental research. Fortunately, Stern's Rostock period lasted only one year. In the following fall of 1922 he accepted an offer as an "Ordinarius" (full professor) of Physical Chemistry and Director of the Institute of Physical Chemistry at the University of Hamburg. On January 1, 1923, in the midst of the great inflation in Germany, he took up his research activities at Hamburg. Here Stern had the good fortune to be assigned four laboratory rooms in the basement of the Physics Institute. Now Stern was finally able to continue the experiments started in Frankfurt and to plan new molecular beam experiments.

Then, as already mentioned in the Introduction, on September 8, 1925 Otto Stern's manifesto appeared in *Zeitschrift für Physik* in which he outlined his plans for future molecular beam experiments including the confirmation of the de Broglie relation [19]. In this connection, he wrote (translated by the author) "...A question of great principle importance is the real existence of de Broglie waves, i.e. the question if with molecular beams, in analogy to light beams, diffraction and interference phenomena can be observed. Unfortunately, the wave lengths calculated with the formula of de Broglie $\left(\lambda = \frac{h}{m \cdot v}\right)$, even under the most favorable conditions (small mass and low temperatures), are less than 1 Å ($=10^{-1}$ nm). Nevertheless, the possibility in such an experiment to observe these phenomena cannot be excluded. Such experiments have so far not been successful." In a footnote he noted that at the time when they started their experiments in Hamburg he was not aware of the 1927 experiments of Davisson and Germer.

The report on the first experiments, which were judged publishable, was submitted on Christmas eve of 1928 as publication No. 11 in the series *Untersuchungen zur Molekularstrahlmethode* (UzM) with Friedrich Knauer [31]. In this initial experiment molecular beams of H_2 and He were scattered under grazing angles of only 10^{-3}

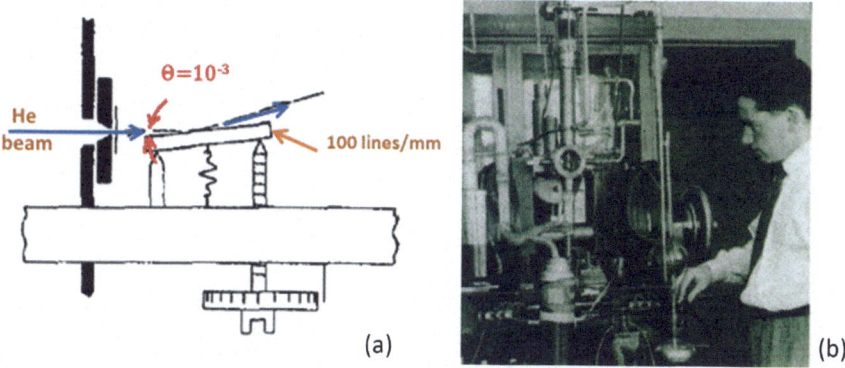

Fig. 5 The first experiment by Stern for a test of the de Broglie relation for massive particles. **a** A schematic diagram of the apparatus taken from the publication Ref. [31]. **b** A photo of a typical glass vacuum apparatus used by Stern and his colleagues

radians from a flat ruled optical grating made either of brass, glass or steel with up to 100 grooves per mm (Fig. 5). These experiments were only partially successful in the sense that only a sharp specular reflected beam but no diffracted beams were observed. The latter, it was concluded, were too close to the specular peak to be resolved. In the same article they describe a second apparatus which was designed for scattering at large angles. With this apparatus they scattered various atoms and molecules including H_2, He, Ne, Ar and CO_2 from a freshly cleaved and continually heated (100 C) NaCl crystal. At the time it was well established that the ionic alkali halide crystals were easily cleaved and that the resulting surfaces were relatively free of defects. Again, only for H_2 and He could relatively sharp specular scattering be observed. In the same UzM article they noted that parallel research was going on in the U.S. by Johnson [32] who had reported a specular peak with H atoms scattered from LiF and Ellet and Olsen [33] who scattered Cd and Hg atoms from NaCl, and that they had also been unsuccessful in observing diffraction.

Several months later on April 20, 1929, as mentioned in the Introduction, Otto Stern submitted a short note to the *Naturwissenschaften* reporting that with the second apparatus he had now found convincing evidence for the sought-after diffraction peaks with both He and H_2 in scattering from the surface of a single crystal of NaCl [6]. One year later, Immanuel Estermann and Otto Stern in UzM No. 15 reported that with an improved apparatus they were now able to observe well-resolved diffraction peaks both with H_2 and He from LiF, NaCl and KCl [34] (Fig. 6). The diffraction angles obeyed the de Broglie relation $\left(\lambda = \frac{h}{m \cdot v}\right)$ calculated from the lattice constants of the crystals and the masses of the scattering particles, both of which were well-known at the time.

Apparently, Otto Stern was still not completely satisfied judging by the fact that in the following year in 1930 he embarked on an ambitious project to rigorously and quantitatively check the de Broglie relation. For this it was necessary to use a beam with a well-defined velocity. In the following pioneering experiments two

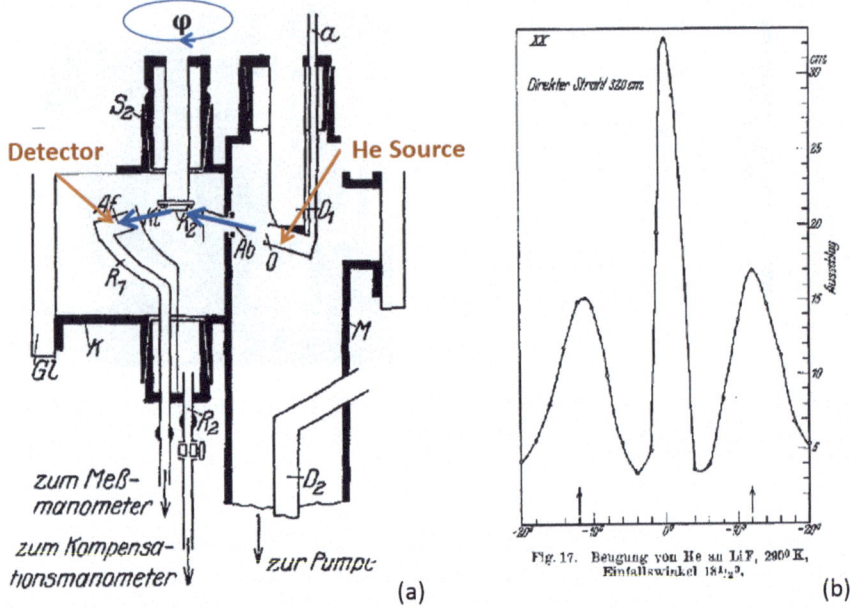

Fig. 17. Beugung von He an LiF, 290° K, Einfallswinkel 18½°,

Fig. 6 The first successful diffraction of a massive particle, He from crystalline LiF observed by Stern in 1929 [3]. **a** The glass encased vacuum apparatus in which the crystal was rotated. **b** The diffraction pattern showing two first order diffracted peaks on both sides of the specular peak [34]

new methods were used to select velocities from the broad Maxwell-Boltzmann distribution [34]. One method exploited the dependence of the first order diffraction angle on the incident beam velocity. In this apparatus, the He atom beam was first diffracted from one crystal surface and only those atoms diffracted into a chosen solid angle, corresponding to the desired velocity, were transmitted. These atoms were then directed at the second crystal surface under investigation. Figure 7 shows a schematic of the method and a more detailed view of the actual apparatus.

The second method was based on two rotating slotted discs on a common axis with the second downstream disc rotationally displaced. The discs were displaced in such a way that only atoms transmitted by the slots of the first disc with the desired velocity could pass through the slots of the second disc. For the motor they modified a Gaede-Siegbahn molecular turbo pump which was already available at the time.[2] The 1 cm thick discs with the pump veins were replaced by two 12 cm dia 1 mm thick discs 3.1 cm apart each with 408 slits with a width of 0.4 mm.[3] (Fig. 8).

---

[2]The molecular turbo pump was invented by Gaede in 1910; improved by Holweck in 1923 and by Manne Siegbahn in 1926.

[3]In the operation of the two disc velocity selector Stern and colleagues appear to have neglected the velocity side bands. For the transmitted velocity they assumed that only those atoms that passed a slit in the first disc could pass the next slit in the second disc-on the same axis-but displaced rotationally by one period with respect to the first disc. Thus the velocity is given by $v_0 = \ell \cdot v \cdot N$, where ℓ is the distance between discs, which is divided by the time for the disc to move by one slit, $\tau = 1/vN$,

Fig. 7 The apparatus used
by Estermann, Frisch and
Stern to measure diffraction
of a velocity selected He
atom beam [34]. **a** Schematic
of the apparatus. By
choosing the diffraction
angle only atoms with the
desired velocity are
transmitted to the second
crystal for diffraction. **b** Side
view of the apparatus

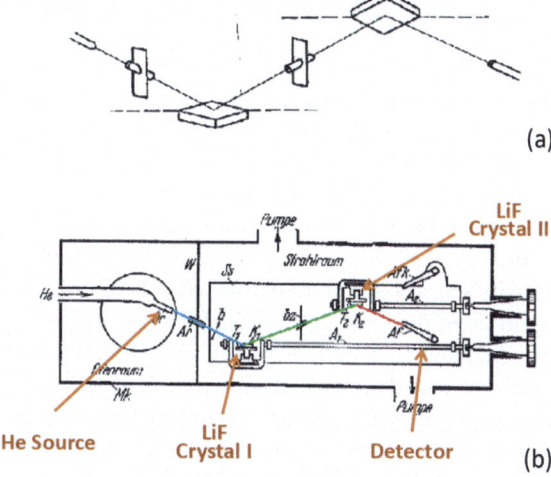

(a)

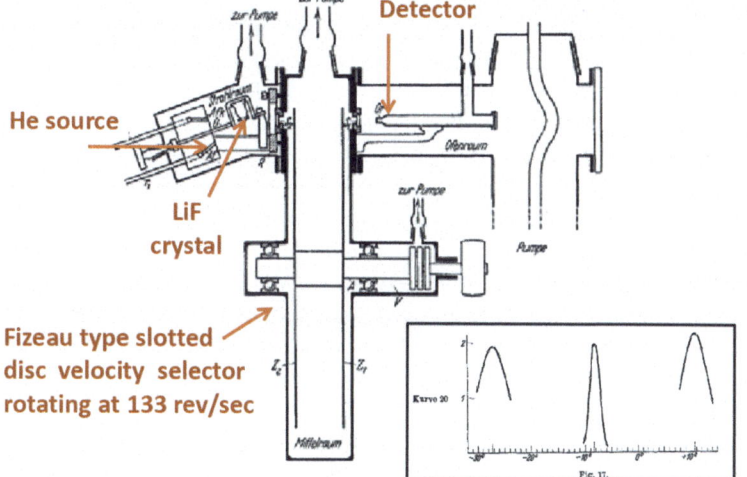

(b)

Fig. 8 The second apparatus used by the same authors as in Fig. 7 to measure the diffraction of
a velocity selected He atom beam [34]. **a** The velocity is selected by two rotating identical slotted
discs, rotational displaced. **b** The sharp diffraction peaks made it possible to confirm the de Broglie
wavelength within less than 1%

here v is the frequency of rotations and N is the number of slits. But atoms with slower velocities
will arrive at a later time when the second disc has rotated to the position of the second slit with
velocity $v_1 = \frac{1}{2}v_0$ and be transmitted. Also very fast atoms that pass the slit in the second disc in
the same position as in the first disc will be transmitted. For this reason, modern velocity selectors
have several additional discs at strategic distances along the axis to block out the unwanted velocity
side bands. The contribution of these side bands in the case of the Stern experiments is estimated
by the author to amount to about 25%.

With the second apparatus Immanuel Estermann, Robert Frisch, and Otto Stern in No. 18 of the UzM series reported highly resolved diffraction patterns of monochromatized He atoms from LiF [34] (Fig. 8). In a footnote of the same UzM article they mention that in the initial measurements the experimental wavelength was smaller by 3% than the predicted wavelength. This, they were convinced, was far too large a discrepancy to be possible. After a long search it was ultimately found that the commercial precision graduated disc used by the Hamburg machine shop to locate the slots to be milled in the discs had 408 positions instead of the 400 specified by the supplier! When this was accounted for and after a careful calibration of the velocity selector and extensive measurements at different rotational frequencies and on different days they established that the diffraction angle was 19.45 deg corresponding to a de Broglie wave length $\lambda = 0.600 \times 10^{-8}$ cm. The de Broglie wavelength calculated for the transmitted velocity was 0.604×10^{-8} cm with a deviation of less than 1% from the measured value. This was the first precision measurement of the de Broglie wavelength of a massive particle. Otto Stern's strong conviction that something must be wrong with the initial results illustrates once more his extraordinary acumen as an experimentalist.

After 4 years, Otto Stern had finally achieved one of the goals that he had laid out when he came to Hamburg. In fact, Otto Stern, as it later became clear, was still not satisfied with these experiments. In an interview with Res Jost in Zürich in 1961 [37] Otto Stern began his comments with reference to the above experiments: "*I especially like this experiment, it is not properly recognized. It is about the determination of the de Broglie wavelength. All the parts of the experiment were classical except for the lattice constant. All the parts came out of the shop. The atom velocity was specified with pulsed slotted discs. Hitler is to blame that we could not end these experiments in Hamburg. It was on the list of projects to be done.*" Here he implies that he had hoped to manufacture a regularly grooved grating in the shop as attempted in UZM, No. 11. Otto Stern would be happy to know that this experiment has recently been carried out in Berlin and is described in one of the following chapters by Wieland Schöllkopf.

In addition to the first observation of diffraction of massive particles and the confirmation of de Broglie's wave-particle duality, Otto Stern and his group made another important discovery. Already in the course of the first experiments, which led to clear diffraction peaks, Estermann and Stern in UzM No. 5 observed a series of four totally unexpected fairly sharp small minima upon azimuthal rotation of the target crystal [34]. They first considered that these could come from diffraction from unexpected structural modifications of the crystal. It was also speculated that they might be related to a layer of adsorbed molecules. To further investigate these anomalies, Frisch and Stern in UzM No. 23 constructed two new dedicated apparatus with additional degrees of freedom for moving the detector with respect to the crystal, but without velocity selection [35]. In one arrangement the diffraction angles could now be scanned out of the plane defined by the incoming beam and the crystal normal. In the other in-plane arrangement the incident angle could be varied and, in addition, the crystal could be rotated. In both arrangements, the distinct minima were confirmed and found to be even sharper. These they studied for He scattered from

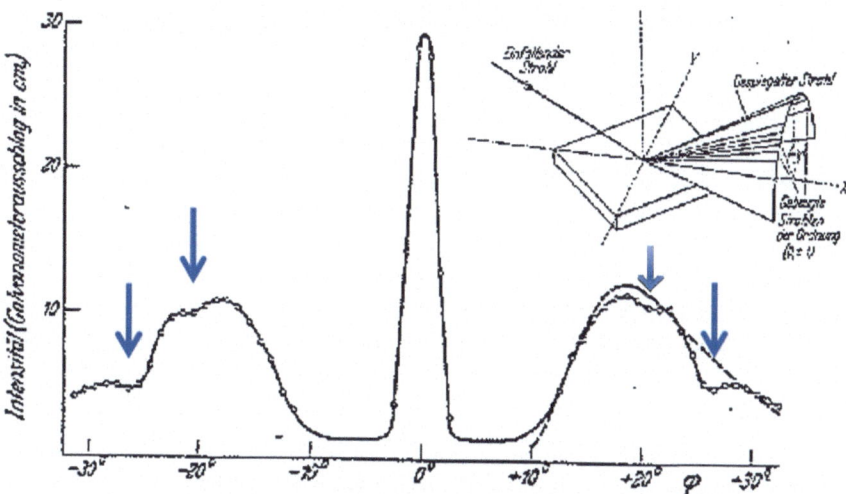

Fig. 9 Anomalous dips in the diffraction pattern, observed by Frisch in 1933 [35]. Three years later Lennard-Jones and Devonshire showed that the dips were due to atoms temporarily bound on the surface in a process called Selective Adsorption [36]

LiF over a wide range of angles in both arrangements. Minima were also found for H_2 on LiF and also for He scattered from NaF (Fig. 9). In summarizing their 1933 experiments in UzM No. 23 they write that they could not come up with a completely satisfactory explanation of these minima. They also report that the process depends in a specific way on the incident direction and energy of the particles. In 1933 in UzM No. 25, one of the last articles from the Hamburg group, Frisch was the only author since Stern, when the paper was submitted, left Germany three days later and had been busy with preparing his exodus. In this article Frisch further summarizes and characterizes the experimental conditions under which the minima appeared [38]. Here he correctly (see below) concludes that the adsorbed atoms are trapped in the two dimensional periodic force field of the surface from which they are diffusely reemitted. Since the perpendicular component of the motion of the atoms bound to the surface in the potential well must be quantized the incident atoms must initially have a specific incident direction and energy in order to be trapped.

Three years later in 1936 the careful and complete documentation of the experimental conditions under which the minima occurred enabled Lennard-Jones and Devonshire to develop the correct theory. The special angles and velocities at which the anomalies occurred were explained by the conditions of resonant trapping of the atoms by diffraction into the bound states of the atom-surface potential [36]. According to their theory: "*Atoms moving along the surface with the right energy and in the right direction may be diffracted as to leave the surface with positive energy and thus be evaporated. This is a new mechanism of evaporation which has not been previously expected.*" Since the minima only occur at special angles they named the new phenomenon Selective Adsorption (SA).

Since their discovery in 1933 and the explanation by Lennard-Jones and Devonshire, selective adsorption resonances have been extensively studied [39]. Presently, they provide the most sensitive and most direct probe of the bound states of the atom-surface potential and thereby the best method for determining the atom-surface potential. Since SA involves diffraction, the surface must have a sufficient corrugation for the effect to occur. Results are presently available for a wide variety of corrugated insulator surfaces and also for metal surfaces with sufficient corrugation as in the case of higher-index stepped surfaces [40]. Since 1981 several new types of resonances involving resonant *inelastic* processes, in which the bound states play an important role, have been found [39]. After a discussion of the elastic selective adsorption resonances the different types of inelastic resonances will be discussed below.

5 The Present Day Legacy of Otto Stern's Surface Scattering Experiments

On June 23, 1933 Stern and his colleagues Estermann, Frisch and Schnurmann were notified that they were discharged from the University, only Knauer, who was not Jewish, could remain. Stern was fortunate to soon after obtain a research professorship at Carnegie Institute of Technology in Pittsburgh. There he continued his molecular beam experiments, but did not continue the He atom surface diffraction experiments.[4] It appears that at the time he did not call attention or perhaps did not even realize the potential use of He atom diffraction for studying the structures of surfaces, which were largely unknown at the time. This we find surprising since in 1931 Thomas Johnson, who had reported diffraction of hydrogen atoms from LiF [42], wrote: "These experiments (diffraction of H_1, H_2 and He from LiF) are of interest not only because of their confirmation of the predictions of quantum mechanics, but also because they introduce the possibility of applying atom diffraction to investigations of the atomic constitution of surfaces. A beam of atomic hydrogen, ... has a range of wavelengths of the right magnitude ..., centering around 1 Å, and the complete absence of penetration of these waves will insure that the effects observed arise entirely from the outermost atomic layer".

Similarly, the significance and potential for surface physics of the 1927 electron scattering experiments of Davisson and Germer were not immediately realized. The experiments were only continued by Germer, who in 1929 reported on electron scattering to study gas adsorption [43]. Otherwise there were few immediate followers. One reason was that the apparatus used by Davisson and Germer were very fragile and prone to breakage. Also the preparation of metal surfaces was in the 1930s not well understood. Moreover, the depression in the 1930s and World War II halted

[4]In 1941, Bessey from the Carnegie Institute published a short note describing He and H_2 diffraction from LiF in which he thanks Otto Stern for suggesting the problem [41].

much of the fundamental research activities. One of the first to continue the experiments of Davisson and Germer was Harrison E. Farnsworth who was probably the first to use electron scattering to study the atomic structures of metal surfaces [44].

The remarkable experimental expertise of Otto Stern is well highlighted by the long time it took for others to repeat his He atom diffraction scattering experiments of 1929. The first post World War II attempt to repeat Stern's diffraction experiments was reported by J. Crews in 1962 [45] more than 30 years later. His angular distributions were not nearly as clearly resolved as in Stern's diffraction peaks. Also the 1968–1970 experiments of Okeefe et al [46, 47]. did not match up with the 1929 experiments. It was only after supersonic free jet expansion sources were introduced, with their inherent sharp velocity distributions, compared to the effusive atomic beam sources used by Stern, were comparable results achieved in 1973 [5]. Thus it took 44 years to arrive at a comparable technological-experimental level as achieved in the Hamburg group. This is even more surprising when it is realized that following the 1957 Sputnik shock the US embarked on a large program to compete scientifically with the Soviet Union. An important part of this program was to develop molecular beam research.

The outstanding experimental genius and foresight of Otto Stern was early on appreciated by the theoretician Max Born, who in 1919 had been Stern's colleague in Frankfurt. In his 1931 letter to the Nobel committee Born wrote: "According to my opinion Stern's achievement are far beyond those of other experimentalists through their conceptual boldness and also through the masterful overcoming of the experimental difficulties that I would like to propose no other physicist except him for the Nobel Prize."

The early history of atom and molecule surface scattering and diffraction experiments has been reviewed in the books by von Laue [48] and later in 1955 by Smith [49], and the more recent article by Comsa [50]. The first experimental studies of energy transfer in scattering from surfaces are described in the reviews by Beder [51], Stickney [52], and the author [53] and in 2018 in the monograph by Benedek and Toennies [39]. In 1969 Cabrera, Celli and Manson pointed out the possibility to observe single phonon excitations and their dispersion by inelastic diffraction of a He atom beam [53]. This stimulated several groups to carry out the corresponding scattering experiments. The first experiments to investigate the surface phonons of a crystal surface were performed by Brian Williams in Ottawa in 1971 [54]. He used essentially the same type of apparatus consisting of two diffraction surfaces that had been used by Estermann, Frisch and Stern for velocity selected diffraction in UzM No. 18. In the apparatus of Williams diffraction from the first crystal was used to select the velocity of the beam incident on the second crystal. The diffraction from the second crystal served to detect the inelastic change in velocity. Through an ingenious use of special kinematic conditions at certain incident and/or final directions additional small peaks in the angular distributions gave the first information on surface phonons on LiF(001) [54, 55]. Soon after in 1974 Boato and Cantini [56] also used high resolution angular distributions to study surface phonons with helium and neon scattered from LiF(001). Several groups also used time-of-flight inelastic scattering in further attempts to investigate the phonons on the surfaces of LiF [57] and on metals [58].

The first He atom inelastic time-of-flight experiments which were able to fully resolve single phonons and allowed the measurement of the dispersion curves of surface phonons out to the zone boundary were reported in 1981 by Brusdelylins, Doak and Toennies for the LiF surface [59] and in 1983 for a metal surface [60]. Similar measurements on metals using inelastic electron energy loss (EELS) were successfully carried out at about the same time [61, 62]. The He atom experiments were facilitated by the 1977 discovery that at high expansion pressures of about 100 bar free jet expansions of He have an inherent sharp velocity distribution corresponding to $\Delta \upsilon / \upsilon \approx 10^{-2}$[63]. It was no longer necessary to use diffraction and rotating discs to select the beam velocity.

The above experiments, continuing on the footsteps of Otto Stern, have established He atom scattering as a unique and indispensable surface science tool. A beam of He atoms in the thermal energy range (5–100 meV) is completely non-penetrating, chemically inert and produces no mechanical damage. Because of its matter-wave property it is the ideal method to project out of the chaotic vibrating surface by inelastic diffraction the phonon dispersion curves. Electron beams have also this property but since the electrons at the same wave length have much higher energies the electrons penetrate the surface and can damage the crystal. The modern experiments in which He atom diffraction is used to study the surface structures of insulating, semiconductor and metal surfaces have been reviewed by Rieder and Engel [64] and by Farias and Rieder [65]. The Helium Atom Scattering (HAS) studies of the phonon dispersion curves of clean surfaces and the vibrations of adsorbate-covered surfaces have very recently been surveyed in the 2018 book by Giorgio Benedek and J. P. Toennies entitled "Atomic Scale Dynamics at Surfaces: Theory and Experimental Studies with Helium Atom Scattering" [39]. There also the very recent understanding that He atoms interact with the electron densities at the surface and provide detailed information on the electron-phonon coupling constant is discussed.

6 New Applications of Matter-Wave Diffraction from Manufactured Nanoscopic Gratings

The advent of nanotechnology in the last 30–40 years has opened up new opportunities to utilize the wave-particle duality for investigations of the physical properties of atoms, molecules and clusters. Some very recent developments in this area are covered in the review articles in this book following this introductory historical review. Here only some early pioneering experiments are dealt with.

Free standing transmission gratings with a slit spacing commensurate with wave lengths in the soft X-ray and the extreme ultraviolet regime, where traditional optical elements are opaque, were first developed for spectroscopy. With decreasing structural dimensions, nanostructured transmission gratings have made it possible to carry over to massive particles with much smaller wave lengths the interference phenomena which had been exploited for light.

Keith, Schattenburg, Smith and Pritchard at MIT were the first in 1988 to report the diffraction of atoms from a fabricated transmission grating [66]. In their experiment Na atoms from an in Ar seeded beam with a de Broglie wave length of 0.017 nm (0.17 Å) were diffracted by an angle of about 70 μrad with an angular resolution of 25 μrad(!) after passing through a specially fabricated 0.2 μm period gold grating with slits and bars of 0.1 μm width. Previously, the same group had also shown that atoms could be diffracted from a standing wave of near resonant light [67]. Subsequently, in 1991, Carnal and Mlynek demonstrated a simple interferometer based on Young's double slit experiment using 2 μm wide slits [68]. Since these initial experiments transmission grating diffraction has been reported for molecules such as Na_2 [69], C_{60} [70, 71], $C_{60}F_{48}$ and $C_{44}H_{30}N_4$ and also for small helium clusters [72–74], with up to about 50 atoms [74–76].

The author's group in 1994 applied transmission grating diffraction as a type of mass spectrometer to establish the existence of the He atom dimer [72, 73]. The existence of the He dimer was long questioned since the long range attractive potential could not be predicted with sufficient accuracy to establish if the very weak attraction between the atoms would be strong enough to support a bound state. A 1993 claim based on mass spectrometer detection [77] was subsequently questioned [78, 79]. Figure 10 shows the diffraction apparatus used by our group [80]. The mass selection comes about since the de Broglie wave length and the diffraction angle are inversely proportional to the mass of the diffracted particles. Since only wavelets which pass through the slits coherently without any interaction with the grating bars can contribute to the diffraction peaks this type of mass spectrometer is completely non-destructive.

Figure 11 displays some diffraction patterns taken at different source temperatures and a high resolution measurement showing well resolved dimer and trimer

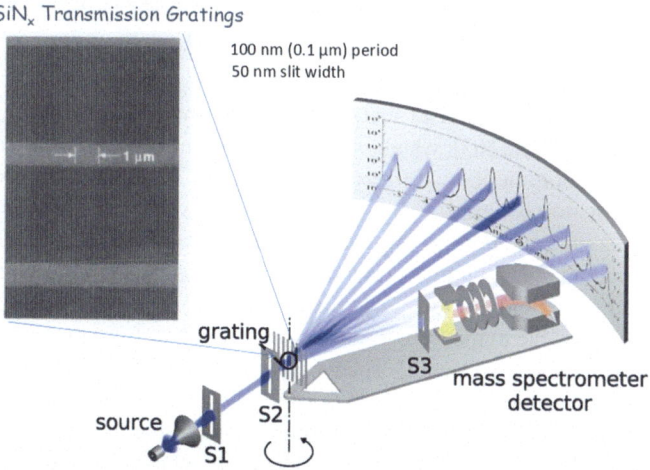

Fig. 10 The transmission grating diffraction apparatus used by the author's group to detect the helium dimer and other small He clusters [80]

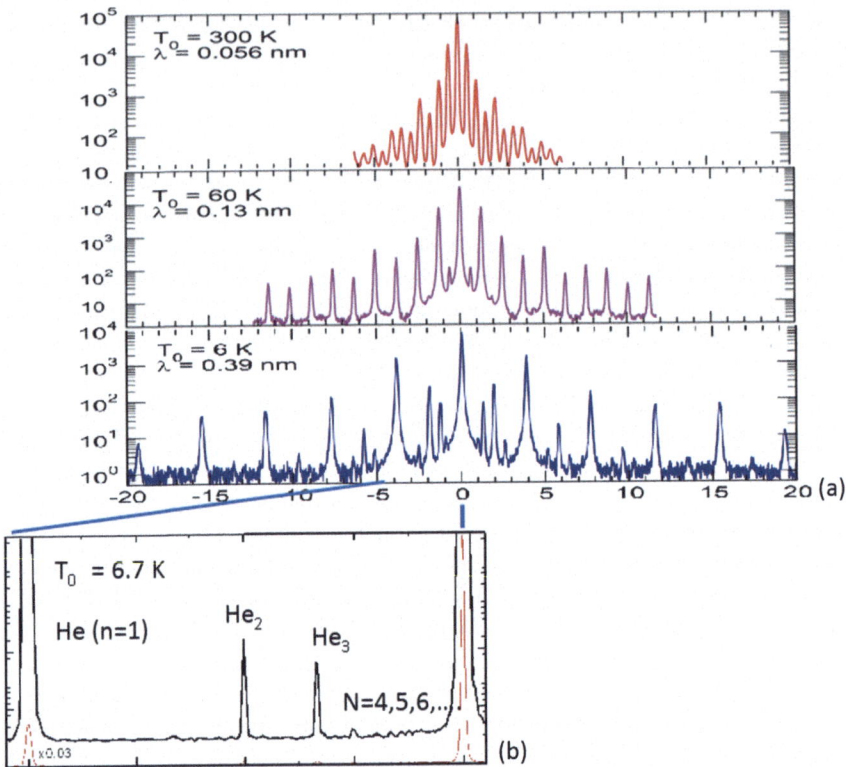

Fig. 11 **a** He atom diffraction patterns measured at three different decreasing source temperatures with increasingly large de Broglie wavelengths. At the lowest temperature between the central specular peak and the first order atom diffraction peaks at -4 and +4 degrees additional diffraction peaks appear which correspond to the dimer, trimer and tetramer of helium. **b** The same diffraction peaks measured with a much increased angular resolution [81]

first order diffraction peaks. The sharp velocity distributions of high-pressure free jet expansions of helium [63], which is also found for the clusters of helium, greatly facilitated the resolution of the diffraction experiments. The partly resolved anomalies in the diffraction patterns of larger helium clusters with up to 50 atoms revealed unexpected maxima and minima instead of a broad peak expected for liquid clusters. The corresponding magic numbers provided the first evidence for the quantum levels of these small superfluid clusters [74].

It was even possible to measure the size of the He-dimer from the intensity distribution of the diffraction peaks [82]. The weak bond of the dimer makes it sensitive to even the slightest interaction with the grating bars. Depending on the size of the dimer the effective slit and coherence width are reduced accordingly. The effective width was experimentally determined from the *slit function* which is the envelope over all the diffraction peaks out to high order. From extensive measurements of the diffraction patterns of the He atom and He dimer it was found that the dimer diffraction

Fig. 12 a The probability
amplitude of the He dimer
which is compatible with the
measured mean radius of the
dimer of 5.2 ± 0.4 nm [82].
b An expanded view of the
potential showing that the
highly quantum dimer
tunnels far beyond the
classical outer turning point
located at about 1.4 nm

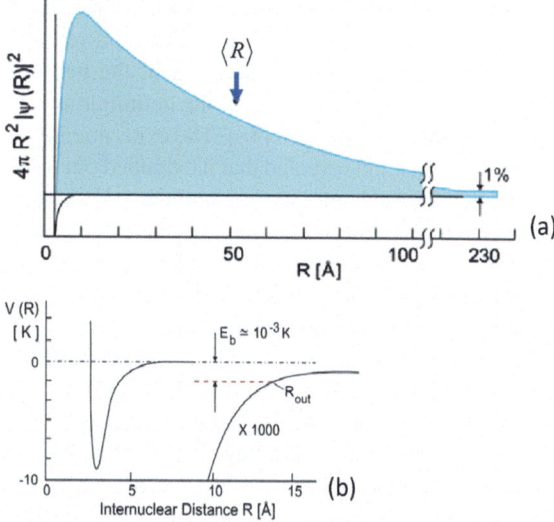

peaks were associated with a significantly smaller slit width, which corresponds to a
larger size, than that of the atoms. With the aid of a many-body quantum scattering
theory the difference in slit width could be referred to the mean internuclear distance
of the dimer $\langle R \rangle$ [83]. The extreme sensitivity is illustrated by the fact that $\langle R \rangle$ was
found to be 5.2 ± 0.4 nm. The uncertainty was only 0.4% when compared to the
100 nm overall slit width of the grating.[5] This experiment established that the dimer
is the largest of all ground state molecules and is about 70 times larger than the H_2
molecule (Fig. 12). It also provided the first measurement of the dimer van der Waals
bond which is still an important benchmark for quantum chemists. From the bond
distance the binding energy was calculated to be only $96^{+26}_{-17} \times 10^{-9}$ eV($1.1^{+0.3}_{-0.2}$ mK)
[82] which corresponds to a scattering length of about s = 100 Å.

In the case of other atoms and tightly bound molecules the effective slit and
coherence width depend on the long range van der Waals interaction with the solid
surface given by $V(l) = -C_3/l^3$, where l is the distance of the particle from the
slit surface as it passes through the slits of the grating. Both the magnitude of the
corresponding reduction in the effective slit width and its velocity dependence were
fitted with a scattering theory to provide values of the van der Waals constant C_3 [84].
The values obtained showed the expected linear increase of C_3 with the polarizability
of the particle increasing in the order of He, Ne, D_2, Ar und Kr. These experiments
represent the first quantitative measurements of C_3. The method was subsequently
used to determine the C_3 constants of the alkali atoms [85, 86] and of metastable
atoms [87] and to set limits on the strength of the non-Newtonian gravity at short
length scales [86].

---

[5]The method is so sensitive that the effect of a monolayer of adsorbed Xe in narrowing the slit
width by about 7 Å could be measured by comparing the slit width with a cold and a hot grating in
the presence of ambient Xe[(unpublished].

Quite recently, the non-destructive selection of small helium clusters by diffraction from a transmission gratings has facilitated a remarkable study. In this experiment the actual radial distribution function of the neutral atoms in the dimer could be measured from the distribution of the helium ions released in the femtosecond laser induced Coulomb explosion [88]. The experimental distribution confirmed the large size of the dimer and revealed that it extended out to distances of more than 23 nm far beyond the classical outer turning point at 1.4 nm [88]. From the exponential fall-off of the radial distribution the binding energy was determined to be $151.9 \pm 13.3 \times 10^{-9}$ eV(1.5 ± 0.13 mK) with the highest precision so far. With the same apparatus in another remarkable experiment the radial distributions of the three helium atoms in the first excited Efimov state of helium trimer could also be measured [89]. This is the first direct measurement of the size of an Efimov state. This unique quantum state was already predicted by Efimov in 1970 to occur for three bosons, of which each of the pairs are critically bound with a nearly vanishing binding energy [90]. As a result of the smaller binding energy the radial distribution of the Efimov trimer extends to even larger distances than in the dimer making the trimer even larger in size than the C_{60} molecule and many biological molecules and even viruses.

The de Broglie wavelength of atoms has also been the basis of using transmission Fresnel zone plates for focusing atomic beams. The focusing via Fresnel zone plates with many concentric slits was first demonstrated by Carnel et al. already in 1991 [68]. Using easily detected electronically excited He atoms they were able to focus down to a spot size of 15–20 μm with an intensity of 0.5 counts/s. In 1999, Doak et al. from our laboratory using a miniature 0.27 mm overall dia. zone plate with a smallest outermost slit width of 100 nm (this is also the predicted diffraction limited resolution) achieved with neutral He atoms a 2 μm dia spot with an intensity of 350 counts/s [91]. Recently, the spot size was reduced down to 1 μm but with an intensity of only 125 counts/s [92].

The first interferometers based on matter wave diffraction were with electrons [93–95] and neutrons [96, 97]. In both experiments the passage through a well-ordered crystal was used as the diffracting element. A Mach-Zehnder interferometer using transmission gratings with Na atoms was first reported by Keith et al. [98]. Subsequently, the same interferometer was used to demonstrate the loss of coherence by scattering photons from one of the paths through the interferometer [67]. Since these seminal demonstration experiments interferometry with atoms has become a wide field of activity stimulated by advances in laser and nanotechnology and laser cooling techniques [99, 100]. The recent advances are the subject of the article by Stefan Gerlich and colleagues in the next chapter.

The construction of a simple robust and compact three grating Mach-Zehnder interferometer developed in our group is shown in Fig. 13 [101]. A homogeneous electric field in one of the interfering arms was used to shift the phase in one of the two separate branches with respect to the other branch. Figure 14 shows the resulting interference fringes. In this experiment a novel extremely accurate lower limit on the velocity half-width of a He atom free jet beam could be demonstrated. In one branch of the interferometer an electric field was applied. With increasing field strength one half of the beam was shifted in phase with respect to the beam without an applied

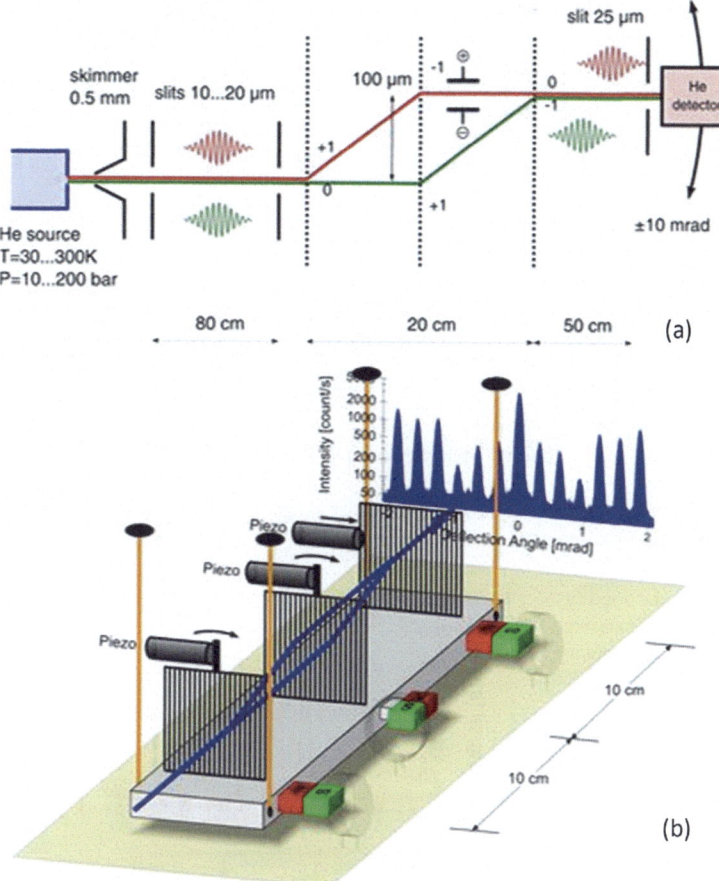

Fig. 13 a Schematic diagram of the compact Mach-Zehnder Interferometer constructed by the Göttingen group. **b** Perspective view of the interferometer showing the piezos to adjust the gratings and the magnetic vibration stabilization to reduce the vibrations to less than 1×10^{-6} m [101]

field. The interference pattern of Fig. 14 shows the interferences as one half of the cold 6 K beam with a de Broglie wave length of 0.4 nm (4 Å) had been shifted with respect to the other half by 25 wave lengths. The largest shift corresponds to a lower limit on the parallel coherence length of $25 \times 4\,\text{Å} = 100\,\text{Å}$, which corresponds to a velocity half width of only 10 m/sec.

A Mach-Zehnder interferometer also provides a precise method for measuring polarizabilities as was demonstrated for the highly polarizable sodium [102] and lithium [103]. Figure 15 shows the results of an interferometer experiment designed to measure the polarizability of the weakly bound He dimer. In the diffraction pattern at the outlet of the interferometer diffraction peaks due to the dimer could be identified between the more intense peaks of the He atom component. From the phase

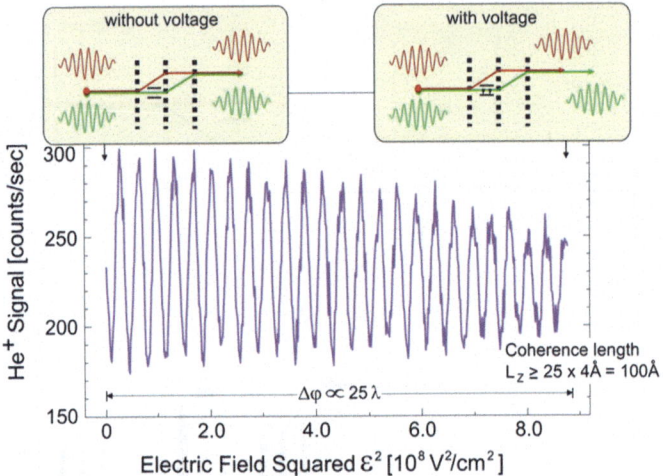

Fig. 14 Interference fringes obtained with a free jet expanded He atom beam with a de Broglie wavelength of 4 Å [101]

Fig. 15 The polarizability of the He dimer is measured in the same arrangement as Figs. 13 and 14 [101]. **a** From the diffraction pattern at the output of the interferometer, a small dimer peak could be identified at -1.4 mrad. **b** From the interference fringes measured at this angle with increasing voltage the polarizability of the He dimer was found to be somewhat smaller than that of two separate atoms

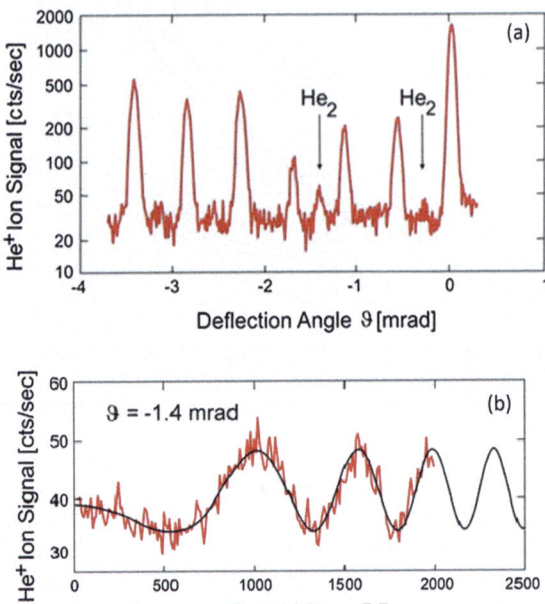

shift with voltage the polarizability of both the He atom and the dimer could be measured relative to each other. Since the atom polarizability is well-known from both metrological experiments and from theory the polarizability of the He dimer could be calibrated by comparison with the He atom beam component and found to be

$$\alpha \, (He_2) = 0.30884 \pm 0.001883 \, \text{Å}^3,$$

which is to be compared with the polarizability of two He atoms:

$$\alpha \, (2He) = 0.3113 \pm 0.0023 \, \text{Å}^3.$$

The smaller polarizability of the dimer is expected since the electrons in the dimer are very slightly more bound than in the fully separated atoms. This preliminary result serves to illustrate the remarkable sensitivity of the method. In a similar manner the comparatively large polarizabilities have been measured of the alkali atoms Li [104] and Na, K, and Rb [105].

7 Summary

In the 90 years since Otto Stern had demonstrated that wave-particle duality also applied to massive particles, the wave nature of atoms and molecules has found widespread applications in both physics and chemistry. The first experiments of He atom diffraction scattering from LiF crystals by Otto Stern in 1929 have since evolved into a gentle nondestructive and universal tool for the determination of the structure of clean and adsorbate-covered surfaces and for the measurement of the phonon dispersion curves at the surfaces of a wide range of different types of solids [39]. In the latter application, He atom scattering is the ideal complement to the scattering of neutrons, which since they pass through the crystal virtually unhindered, are only sensitive to the bulk phonons.

Since Otto Stern's days new nanotechnology advances have opened up an entire new field of experiments based on matter wave behavior of atoms and molecules. As discussed here these encompass on the one hand the non-destructive mass analysis of fragile clusters and on the other hand the precision interferometry of atomic and molecular properties. Certainly, Stern would have been happy to learn about the many wonderful and important applications of matter waves of atoms and molecules discussed here and described in the following chapters.

In this connection, the author, after having been occupied with the impact of Otto Stern's 1929 diffraction experiments, often wonders whether Stern at the time realized the many future important applications arising out of his pioneering experiments. It is interesting that he had attempted and had hoped to carry out diffraction experiments from man-made ruled gratings, only recently realized and described by Wieland Schöllkopf in the following chapter.

Otto Stern was definitely very much aware of the fundamental importance of his experiments in relation to the electron diffraction experiments and the implications for quantum theory as expressed in his 1943 Nobel lecture: *"With respect to the differences between the experiments with electrons and molecular rays, one can say that the molecular ray experiments go farther. Also the mass of the moving particle is varied (He, H_2). But the main point is again that we work in such a direct primitive manner with neutral particles. These experiments demonstrate clearly and directly the fundamental fact of the dual nature of rays of matter. It is no accident that in the development of the theory the molecular ray experiments played an important role. Not only the actual experiments were used, but also molecular ray experiments carried out only in thought. Bohr, Heisenberg, and Pauli used them in making clear their points on this direct simple example of an experiment..."* Here Otto Stern calls attention to the important impact of de Broglie's theory and his experiments on the development of quantum theory especially by Erwin Schrödinger. To do justice this aspect would be beyond the scope of this article.

Acknowledgements The author would like to thank Horst Schmidt-Böcking and Bretislav Friedrich for stimulating my interest in the life and scientific work of Otto Stern. I wish to also thank Katrin Glormann for her great care and enduring patience in preparing the manuscript and Bretslav Friedrich for a careful reading and corrections of the final manuscript.

References

1. O. Stern, Z. Physik **39**, 751 (1926)
2. L. de Broglie, *Thèse de doctoral*, Mason, Paris (1924); reprinted 1963; also Annales de Physique3, 22 (1925)
3. O. Stern, Naturwissenschaften **17**, 391 (1929)
4. C. Davisson, L.H. Germer, Nature **119**, 558 (1927)
5. G. Boato, P. Cantini, U. Garibaldi, A.C. Levi, L. Mattera, R. Spadacini, G.E. Tommei, J. Phys. C Solid State **6**, L394 (1973)
6. L. de Broglie, J. de Physique **3**, 422 (1922)
7. L. de Broglie, Comptes Rendus **175**, 881 (1922)
8. L. de Broglie, Comptes Rendus **177**, 507 (1923)
9. L. de Broglie, Nature **112**, 540 (1923)
10. L. de Broglie, Philos. Mag. **47**, 446 (1924)
11. L. de Broglie, Ann. Phys. **3**, 22 (1925)
12. H.A. Medicus, Phys. Today **27**, 38 (1974)
13. E. MacKinnon, Ann. J. Phys. **44**, 1047 (1976)
14. A. Einstein, Preuss. Akad. Wiss. Mathem.-Naturwisse. Kl. **23**, 3 (1925)
15. C. Davisson, C.H. Kunsman, Science **54**, 522 (1921)
16. E. Davisson, C.H. Kunsman, Phys. Rev. **22**, 242 (1923)
17. W. Elsassser, Naturwissenshaften 13, 711 (1925)
18. H. Rubin, *A Biographical Memoir of Walter M. Elsasser 1904–1991* (National Academy of Sciences, Washington D.C., 1995)
19. F. Hund, *Geschichte der Quantentheorie*, Bibliographisches Institut Mannheim (1967)
20. M. Born, Z. Phys. **38**, 863 (1926)
21. M. Born, *Experiments and Theory in Physics* (Dover, 1943)
22. R.K. Gehrenbeck, Phys. Today (January 1978), 34

23. R. Schlegel, R.K. Gehrenbeck, Phys. Today **31**, 9 (1978)
24. C. Davisson, L.H. Germer, Phys. Rev. A **30**, 705 (1927)
25. G.P. Thomson, A. Reid, Nature **119**, 890 (1927)
26. G.P. Thomson, A. Reid, Proc. Roy. Soc. (London) A 117 (1928) 601
27. S. Weinberg, *The Trouble with Quantum Mechanics,* The New York Times Review of Books (January 19, 2017)
28. J.P. Toennies, H. Schmidt-Bocking, B. Friedrich, J.C.A. Lower, Ann. Phys. **523**, 1045 (2011)
29. O. Stern, Phys. Z. **21**, 582 (1920)
30. W. Gerlach, O. Stern, Z. Phys. **9**, 349 (1922)
31. F. Knauer, O. Stern, Z. Phys. **53**, 779 (1929)
32. T.H. Johnson, J. Franklin Inst. **206**, 301 (1928)
33. A. Ellet, H.F. Olsen, Phys. Rev. **31**, 643 (1928)
34. I. Estermann, O. Stern, Z. Phys. **61**, 95 (1930)
35. R.O. Frisch, O. Stern, Z. Phys. **84**, 430 (1933)
36. J.E. Lennard-Jones, A.F. Devonshire, Nature **137**, 1069 (1936)
37. R. Jost, ETH-Bibliothek Zürich, Archive (1961)
38. R.O. Frisch, Z. Phys. **84**, 443 (1933)
39. G. Benedek, J.P. Toennies, *Atomic Scale Dynamics at Surfaces: Theory and Experimental Studies with Helium Atom Scattering* (Springer, 2018)
40. G. Vidali, G. Ihm, H.Y. Kim, M.W. Cole, Surf. Sci. Rep. **12**, 133 (1991)
41. W.H. Bessey, Phys. Rev. **59**, 459 (1941)
42. T.H. Johnson, Phys. Rev. **37**, 847 (1931)
43. L.H. Germer, Z. Phys. **54**, 408 (1929)
44. H.E. Farnsworth, Phys. Rev. **34**, 679 (1929)
45. J.C. Crews, J. Chem. Phys. **37**, 2004 (1962)
46. D.R. Okeefe, J.N. Smith, R.L. Palmer, H. Saltsburg, J. Chem. Phys. **52**, 4447 (1970)
47. R. Okeefe, R.L. Palmer, H. Saltsburg, J.N. Smith, J. Chem. Phys. **49**, 5194 (1968)
48. M. von Laue, *Materiewellen und ihr Interferenzen* (Akadem. Verl.-Ges. Becker & Erler, Geest und Portig, 1944)
49. K.F. Smith, *Molecular Beams* (Wiley, 1955)
50. G. Comsa, Surf. Sci. **299**, 77 (1994)
51. E.C. Beder, Advan. At. Mol. Phys. **3**, 205 (1968)
52. R.E. Stickney, Advan. At. Mol. Phys. **3**, 143 (1968)
53. N. Cabrera, V. Celli, R. Manson, Phys. Rev. Lett. **22**, 346 (1969)
54. B.R. Williams, J. Chem. Phys. **55**, 3220 (1971)
55. B.F. Mason, B.R. Williams, J. Chem. Phys. **56**, 1895 (1972)
56. G. Boato, P. Cantini, in *Dynamics Aspects of Surface Physics,* ed. by F.O. Goodman (1974), p. 707
57. S.S. Fisher, J.R. Bledsoe, J. Vac. Sci. Technol. **9**, 814 (1972)
58. S.C. Yerkes, D.R. Miller, J. Vac. Sci. Technol. **17**, 126 (1980)
59. G. Brusdeylins, R.B. Doak, J.P. Toennies, Phys. Rev. Lett. **46**, 437 (1981)
60. R.B. Doak, U. Harten, J.P. Toennies, Phys. Rev. Lett. **51**, 578 (1983)
61. S. Lehwald, J.M. Szeftel, H. Ibach, T.S. Rahman, D.L. Mills, Phys. Rev. Lett. **50**, 518 (1983)
62. J.M. Szeftel, S. Lehwald, H. Ibach, T.S. Rahman, J.E. Black, D.L. Mills, Phys. Rev. Lett. **51**, 268 (1983)
63. J.P. Toennies, K. Winkelmann, J. Chem. Phys. **66**, 3965 (1977)
64. T. Engel, K.-H. Rieder, *Structural Studies of Surfaces* (Springer-Verlag, Berlin-Heidelberg-New York, 1982)
65. D. Farias, K.H. Rieder, Rep. Prog. Phys. **61**, 1575 (1998)
66. D.W. Keith, M.L. Schattenburg, H.I. Smith, D.E. Pritchard, Phys. Rev. Lett. **61**, 1580 (1988)
67. M.S. Chapman, T.D. Hammond, A. Lenef, J. Schmiedmayer, R.A. Rubenstein, E. Smith, D.E. Pritchard, Phys. Rev. Lett. **75**, 3783 (1995)
68. O. Carnal, J. Mlynek, Phys. Rev. Lett. **66**, 2689 (1991)

69. M.S. Chapman, C.R. Ekstrom, T.D. Hammond, J. Schmiedmayer, S. Wehinger, D.E. Pritchard, Phys. Rev. Lett. **74**, 4783 (1995)
70. M. Arndt, O. Nairz, J. Vos-Andreae, C. Keller, G. van der Zouw, A. Zeilinger, Nature **401**, 680 (1999)
71. Y.Y. Fein, P. Geyer, F. Kialka, S. Gerlich, M. Arndt, Phys. Rev. Res. **1**, 033158 (2019)
72. W. Schöllkopf, J.P. Toennies, Science **266**, 1345 (1994)
73. W. Schollkopf, J.P. Toennies, J. Chem. Phys. **104**, 1155 (1996)
74. R. Guardiola, O. Kornilov, J. Navarro, J.P. Toennies, J. Chem. Phys. **124**, 084307 (2006)
75. R. Brühl, R. Guardiola, A. Kalinin, O. Kornilov, J. Navarro, T. Savas, J.P. Toennies, Phys. Rev. Lett. **92**, 185301 (2004)
76. L. Hackermuller, S. Uttenthaler, K. Hornberger, E. Reiger, B. Brezger, A. Zeilinger, M. Arndt, Phys. Rev. Lett. **91** (2003)
77. F. Luo, G.C. Mcbane, G.S. Kim, C.F. Giese, W.R. Gentry, J. Chem. Phys. **98**, 3564 (1993)
78. E.S. Meyer, J.C. Mester, I.F. Silvera, J. Chem. Phys. **100** (1994)
79. F. Luo, G.C. Mcbane, G. Kim, C.F. Giese, W.R. Gentry, J. Chem. Phys. **100**, 4023 (1994)
80. O. Kornilov, J.P. Toennies, Europhys. News **38**, 22 (2007)
81. J.P. Toennies, Mol. Phys. **111**, 1879 (2013)
82. R.E. Grisenti, W. Schöllkopf, J.P. Toennies, G.C. Hegerfeldt, T. Köhler, M. Stoll, Phys. Rev. Lett. **85**, 2284 (2000)
83. G.C. Hegerfeldt, T. Kohler, Phys. Rev. A **57**, 2021 (1998)
84. R.E. Grisenti, W. Schöllkopf, J.P. Toennies, C.C. Hegerfeldt, T. Köhler, Phys. Rev. Lett. **83**, 1755 (1999)
85. V.P.A. Lonij, C.E. Klauss, W.F. Holmgren, A.D. Cronin, J. Phys. Chem. A **115**, 7134 (2011)
86. S. Lepoutre, V.P.A. Lonij, H. Jelassi, G. Trenec, M. Buchner, A.D. Cronin, J. Vigue, Eur. Phys. J. D **62**, 309 (2011)
87. R. Brühl, P. Fouquet, R.E. Grisenti, J.P. Toennies, G.C. Hegerfeldt, T. Köhler, M. Stoll, C. Walter, Europhys. Lett. **59**, 357 (2002)
88. S. Zeller, M. Kunitski, J. Voigtsberger, A. Kalinin, A. Schottelius, C. Schober, M. Waitz, H. Sann, A. Hartung, T. Bauer, M. Pitzer, F. Trinter, C. Goihl, C. Janke, M. Richter, G. Kastirke, M. Weller, A. Czasch, M. Kitzler, M. Braune, R.E. Grisenti, W. Schöllkopf, L.P.H. Schmidt, M.S. Schöffler J.B. Williams, T. Jahnke, R. Dörner, P. Natl. Acad. Sci. USA **113**, 14651 (2016)
89. M. Kunitski, S. Zeller, J. Voigtsberger, A. Kalinin, L.P.H. Schmidt, M. Schöffler, A. Czasch, W. Schöllkopf, R.E. Grisenti, T. Jahnke, D. Blume, R. Dörner, Science **348**, 551 (2015)
90. V. Efimov, Sov J. Nucl. Phys. **12**, 1080 (1970)
91. R.B. Doak, R.E. Grisenti, S. Rehbein, G. Schmahl, J.P. Toennies, C. Wöll, Phys. Rev. Lett. **83**, 4229 (1999)
92. S.D. Eder, T. Reisinger, M.M. Greve, G. Bracco, B. Holst, New J. Phys. **14**, 073014 (2012)
93. L. Marton, J.A. Simpson, J.A. Suddeth, Phys. Rev. **90**, 490 (1953)
94. L. Marton, J.A. Simpson, J.A. Suddeth, Rev. Sci. Instrum. **25**, 1099 (1954)
95. G. Moellenstedt, H. Dueker, Die Naturwissenschaften **42**, 41 (1955)
96. H. Maier Leibnitz, T. Springer, Z. Phys. **167**, 386 (1962)
97. H. Rauch, A. Werner, *Neutron Interferometry: Lessons in Experimental Quantum Mechanics* (Clarendon Press, Oxford, 2000)
98. D.W. Keith, C.R. Ekstrom, Q.A. Turchette, D.E. Pritchard, Phys. Rev. Lett. **66**, 2693 (1991)
99. A.D. Cronin, J. Schmiedmayer, D.E. Pritchard, Rev. Mod. Phys. **81**, 1051 (2009)
100. K. Hornberger, S. Gerlich, P. Haslinger, S. Nimmrichter, M. Arndt, Rev. Mod. Phys. **84**, 157 (2012)
101. R. Brühl, R.B. Doak, J.P. Toennies, (unpublished)
102. C.R. Ekstrom, J. Schmiedmayer, M.S. Chapman, T.D. Hammond, D.E. Pritchard, Phys. Rev. A **51**, 3883 (1995)
103. A. Miffre, M. Jacquey, M. Buchner, G. Trenec, J. Vigue, Eur. Phys. J. D **38**, 353 (2006)
104. A. Miffre, M. Jacquey, M. Buchner, G. Trenec, J. Vigue, Phys. Rev. A **73**, 011603(R) (2006)
105. W.F. Holmgren, M.C. Revelle, V.P.A. Lonij, A.D. Cronin, Phys. Rev. A **81**, 053607 (2010)

Chapter 24
Otto Stern's Legacy in Quantum Optics: Matter Waves and Deflectometry

Stefan Gerlich, Yaakov Y. Fein, Armin Shayeghi, Valentin Köhler, Marcel Mayor, and Markus Arndt

Abstract Otto Stern became famous for molecular beam physics, matter-wave research and the discovery of the electron spin, with his work guiding several generations of physicists and chemists. Here we discuss how his legacy has inspired the realization of universal interferometers, which prepare matter waves from atomic, molecular, cluster or eventually nanoparticle beams. Such universal interferometers have proven to be sensitive tools for quantum-assisted force measurements, building on Stern's pioneering work on electric and magnetic deflectometry. The controlled shift and dephasing of interference fringes by external electric, magnetic or optical fields have been used to determine internal properties of a vast class of particles in a unified experimental framework.

1 From Otto Stern to Universal Molecule Interferometry

Our contribution honors the legacy of Otto Stern, who paved the path for 100 years of exciting research into atomic and molecular beams, spin physics and matter-wave interferometry. Many of his ideas and original methods are still implemented in our present-day experiments. On the one hand it is impressive how much progress these fields have made since Stern's time, but it is also humbling to realize how many of Stern's experimental challenges remain even in the most advanced experiments today.

S. Gerlich · Y. Y. Fein · A. Shayeghi · M. Arndt (✉)
Faculty of Physics, University of Vienna, Boltzmanngasse 5, 1090 Vienna, Austria
e-mail: markus.arndt@univie.ac.at

V. Köhler · M. Mayor
University of Basel, St. Johannsring 19, 4056 Basel, Switzerland
e-mail: marcel.mayor@unibas.ch

B. Friedrich and H. Schmidt-Böcking (eds.), *Molecular Beams in Physics and Chemistry*,
https://doi.org/10.1007/978-3-030-63963-1_24

1.1 Stern's Legacy in Molecular Beam Deflection

It is enlightening to look at one of Stern's early papers, *Zur Methode der Moleku-larstrahlen* [1], in which he describes the first applications of atomic and molecular beam deflectometry and formulates criteria for achieving the highest possible metro-logical sensitivity. His idea was straightforward and is sketched in Fig. 1a: a beam of atoms or molecules is launched into high vacuum, collimated, deflected by external fields and detected downstream with position resolution. Following Stern's notation, the deflection s of a particle of mass m after traveling a distance l with a velocity v in a uniform force field K is

$$s = \frac{1}{2}\frac{K}{m}\frac{l^2}{v^2}. \tag{1}$$

Knowing the beam velocity, geometry and fields involved, it is then straightforward to extract electronic or magnetic properties of the atoms or molecules, since they are contained within K. Stern formulated three criteria to achieve high sensitivity for deflectometry[1]:

1. **Narrow beam width**: *"make the beam as narrow as possible, because the nar-rower it is, the smaller the deflection s we can measure"*.
2. **Large deflection region**: *"make the path l through the field as long as possible, because s ∼ l²"*.
3. **Strong fields**: *"make the force K as big as possible, because s ∼ K"*.

Since the first two criteria reduce the flux of detected particles, Stern proposed to *"...increase the intensity to the required amount by placing 100 identical furrows next to each other on the pole shoe, all of them pointing to the same detector area, such that their images fall on top of each other."*

Stern envisioned far-reaching applications of his beam deflection apparatus, such as the measurement of nuclear magnetic moments,[2] induced moments, electric dipole moments, and higher-order moments. Many of our experiments with atoms, molecules and clusters are rooted in these ideas.

1.2 Stern's Legacy in Matter-Wave Research

Our experiments are also based on a second series of pioneering studies by Otto Stern: while textbooks correctly ascribe the first demonstration of matter-wave diffraction to the electron experiments by Davisson and Germer [3], it is noteworthy that Estermann and Stern were already working toward matter-wave experiments in 1926. In 1930

---

[1]Since Otto Stern's early papers were written in German, we use our own translation where a verbatim citation is indicated.
[2]Stern's Nobel Prize in 1943 was awarded for his measurement of the proton's magnetic moment.

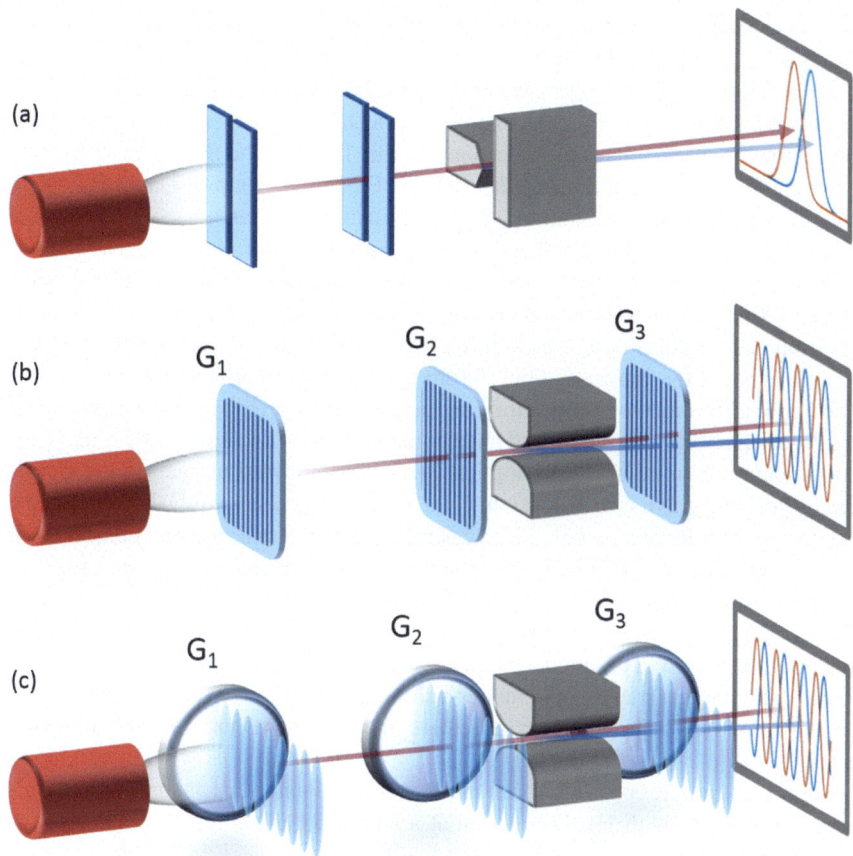

Fig. 1 **a** A molecular beam deflection experiment in the spirit of Otto Stern consists of an intense beam source, narrow collimators, an inhomogeneous deflection field and a position resolving detector. **b** Deflectometry in a near-field Talbot-Lau interferometer (TLI): the first grating, G_1, prepares transverse coherence from an initially incoherent beam, while the second grating imparts a superposition of momenta to the delocalized matter wave. Interference manifests itself as a particle density pattern with the same periodicity of as the gratings which can be detected by scanning the third grating and counting the transmitted particles. **c** Near-field interferometry requires either coherent sources or an absorptive first grating G_1. This can be realized by a material mask or photo-induced depletion of the molecular beam in a standing light wave grating [2]. In Sec. 2, different variants of Talbot-Lau interferometers are discussed, including the all-optical OTIMA experiment and the Kapitza-Dirac Talbot-Lau scheme, which constitutes a hybrid of (**b**) and (**c**) with G_1 and G_3 as material gratings, but an optical phase grating serving as G_2

they succeeded in demonstrating the first diffraction of atoms (He) and molecules (H_2) from a crystal surface [4].

Since then, huge progress has been made in atom interferometry based on improved beam sources, nanomechanical gratings, laser physics and optical beam-splitting techniques. Important milestones in this field are the first atom diffraction at optical [5, 6] and nanomechanical gratings [7] in the group of David Pritchard, who also realized the first atom interferometer using free-standing nanomechanical gratings [8]. Christian Bordé realized that single-photon absorption can act as a coherent beam splitter for atoms [9] and reinterpreted earlier spectroscopy experiments on SF_6 [10] as Ramsey-Bordé interferometry. A time-domain atom interferometer based on Raman transitions was built by Mark Kasevich and Steven Chu [11] and became a model for many atom interferometer realizations around the world.

Atom interferometry is now a thriving field of physics with applications ranging from precision tests of fundamental physics to quantum metrology, geodesy and inertial navigation. Some noteworthy applications include the measurement of the Earth's gravity [12], the gravitational constant G [13], tests of the weak equivalence principle and the universality of free-fall [14], measurements of the fine structure constant [15] and rotation sensing [16, 17]. Interferometry experiments have also been proposed for gravitational wave detection [18] and for dark matter [19] and dark energy [20, 21] searches. For reviews covering these topics see e.g. Refs. [22–24]. With the advent of ultra-cold quantum degenerate gases and Bose-Einstein condensates [25, 26], a wide range of mesoscopic matter-wave experiments have also become possible, with too many examples to be listed here; the same holds for molecular quantum gases [27, 28].

Significant progress has also been made in molecule interferometry since Stern's early experiments with H_2. Diffraction at a nanomechanical mask was key to the discovery of the extremely weakly bound helium dimer He_2 [29] and the basis for Mach-Zehnder interferometry with Na_2 [30]. The combination of four $\pi/2$-pulse beam splitters in a Ramsey-Bordé interferometer was demonstrated with I_2 [31]. We refer to contributions by Jan-Peter Toennies, Wieland Schoellkopf and David Pritchard in this book for more on these topics.

Stern's original idea was to exploit beam deflectometry as a tool to learn about the inner structure and physics of atoms and molecules, and the techniques of atom interferometry have enabled a number of such measurements. The nature of the bonding of He_2 [32] was measured via quantum reflection from a grating, since other techniques would have been too invasive to probe the extremely fragile bond. The static polarizability of sodium [33], lithium [34] and other alkali atoms [35] was measured using atom interferometry, and long-range potential properties [36], van der Waals coefficients [37], atomic tune-out wavelengths [38] and transition matrix elements [39, 40] as well as surface excitations [41] have all been studied using matter-wave diffraction.

2 Interferometer Concepts for Studying the 'Wave-Nature of Everything'

Building on the work with atoms and dimers and fueled by advances in lasers and nanotechnology, the investigation of the quantum nature of more massive objects became possible by the end of the 20th century.

The fullerene C_{60} was the first complex molecule for which de Broglie interference was demonstrated in **far-field diffraction** [42]. Fullerenes are particularly well-suited for beam experiments since they are thermally stable and can be evaporated in a simple furnace. Since a thermal source lacks both transverse and longitudinal coherence, the particles were sent through a pair of collimation slits to generate the required transverse coherence before being diffracted at a nanofabricated grating with a period of $d = 100$ nm. The far-field diffraction pattern is a convolution of the single-slit and multi-slit pattern as familiar from optics textbooks, with the relevant wavelength in this case $\lambda_{dB} = h/mv$. The molecular density pattern was detected by scanning a tightly focused green laser over the molecular beam to cause thermal ionization and create countable ions [43]. The experiment is illustrated in Fig. 2.

At first glance it is intriguing that high-contrast interference could be observed despite the fact that the molecules were heated to about 900 K, thus exciting many rotational and vibrational levels. While each individual molecule is distinguishable by virtue of its unique internal state, interference is still observed because each particle interferes with itself, and the evolution of its vibrational and rotational modes occurs simultaneously along each arm of the interferometer. The center-of-mass motion is the only relevant degree of freedom—as long as coupling to the environment can be suppressed.

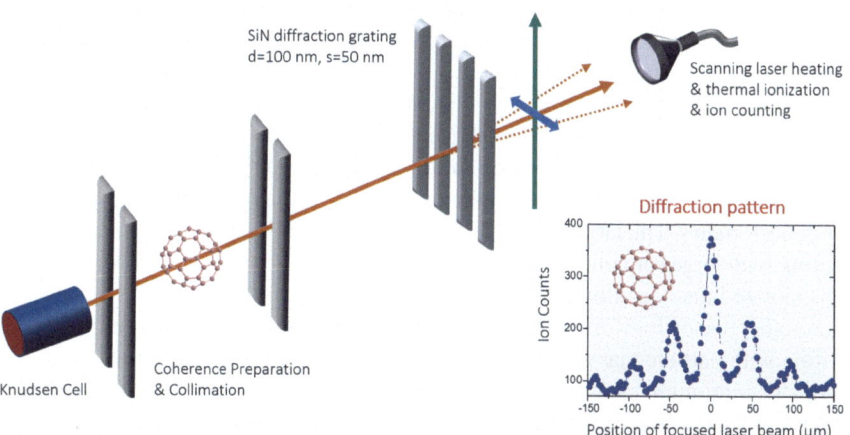

Fig. 2 Fullerene diffraction at a 100 nm period nanomechanical grating as realized in Vienna [42, 44]. For a central velocity of 136 m/s, the de Broglie wavelength is 4 pm and thus less than 200 times the molecular diameter. Two slits of 10 μm width separated by 104 cm prepare the transverse coherence required to illuminate several slits of the grating

A more visual way of revealing the molecular wave-particle duality is via fluorescence imaging. Starting from a micron sized laser emission source, molecules were diffracted at a nanomechanical mask and deposited on a quartz slide, where they were detected in real time by fluorescence microscopy with a spatial resolution of about 10 nm [45].

The de Broglie wavelength of a particle with $m \simeq 1000$ Da travelling at 100–300 m/s is of order $\lambda_{dB} \simeq 10^{-12}$ m. Typical gratings – both nanomechanical masks and standing light waves – have periods of $d \simeq 100$ nm or larger, which yield diffraction angles of 10 μrad. Resolving this small angle requires collimation and an angular resolution of the detector of a few μrad. Increasing the mass by a factor of ten thus requires improving the collimation and detector resolution by the same factor.

Textbook-style far-field diffraction thus becomes quickly impractical for beams of high-mass particles. Near-field optics, however, provides a viable solution. The phenomenon of lens-less grating self-imaging was first observed with light by Henry Fox Talbot in 1836 [46] and later extended by Ernst Lau to incoherent sources [47]. A **Talbot-Lau interferometer** (TLI) relies on this self-imaging phenomenon and can be formed with two gratings, as illustrated in Fig. 1b. In the symmetric configuration, two identical gratings are spaced apart by a multiple of the Talbot length, $L_T = d^2/\lambda_{dB}$, with d the grating period. In general, near field diffraction causes a complicated pattern, known as a Talbot carpet, to be imprinted into the light or matter-wave beam behind the second grating. At certain distances (in the symmetric case simply the same distance as the G_1–G_2 separation), the pattern is an exact self-image of the second grating, i.e., fringes with period d. A third grating with the same period can be employed to detect this self image: scanning it transversely to the beam will yield a sinusoidal modulation in the flux as detected by a spatially integrating detector.

The TLI scheme has several advantages for interferometry of massive particles. Compared to far-field schemes like the Mach-Zehnder interferometer, a TLI has relaxed requirements on transverse coherence, permitting the use of relatively uncollimated molecular beams. This is because each opening of the first grating acts as a coherent source for the second grating, with the symmetry of the setup ensuring that the many interferometer trajectories emanating from the first grating recombine in phase at the position of the third grating. In other words, the Talbot-Lau scheme benefits largely from multiplexing, a strategy already envisioned by Stern for deflectometry. The large gain in throughput is an important asset especially when dealing with large (organic) molecules, which are typically very fragile and difficult to volatilize intact, resulting in low beam intensities.

Another reason for working with near-field interferometers is due to their favorable scaling with particle mass, as pointed out by Clauser [48]. For a given length of the setup, the minimum resolvable de Broglie wavelength scales with d^2, in contrast to d in far-field diffraction. Increasing m by a factor of 100 – thereby reducing λ_{dB} by the same factor – would require gratings with a 10 times smaller period d in a near-field scheme, whereas they would have to be 100 times smaller to observe the diffraction of the same mass in the far-field. Talbot-Lau interferometry thus allows us to access a high mass range even with moderate interferometer baselines and grating periods.

There are a variety of ways to realize gratings in the lab, with the restriction that G_1 and G_3 must be transmission masks to fulfill their roles of spatially confining the beam and spatially filtering the beam, respectively.

Nanofabricated masks can be used for a large variety of particles because their action does not depend on any specific internal particle property or transition. Fullerenes once again served as the first species to be studied in a three-grating TLI[3] at the University of Vienna [49]. A TLI setup was also used to observe the wave nature of tetraphenylporphyrins (TPP) and the fluorofullerenes $C_{60}F_{48}$ [50] as well as to demonstrate the prospects of molecule lithography [51].

Material gratings, however, also induce strongly velocity-dependent (dispersive) Casimir-Polder phase shifts on matter waves, which limits the maximal fringe contrast particularly for highly polarizible, slow molecules [52]. This particle-wall interaction may also be enhanced by local charges deposited in the fabrication process. Surface effects can be partially mitigated by reducing the grating thickness, but even at the ultimate limit of an atomically thin diffraction grating made from single-layer graphene [53], a sizeable phase shift remains.

Optical beam splitters are particularly appealing for molecule interferometry, since they are free from effects of surface geometry and quality, contamination and charges, and they can be precisely defined both spatially and temporally. Diffraction at a standing light wave was originally proposed by Kapitza and Dirac for electrons in the Bragg regime [54]. It was first realized with atoms using off-resonant optical dipole force phase gratings in the Raman-Nath regime [5].[4] Kapitza-Dirac diffraction was demonstrated both for electrons [55] and fullerenes [56] in 2001.

While the polarizability of atoms varies by several orders of magnitude around an optical resonance, it does not typically vary by more than 50% across a large part of the optical spectrum for complex, warm molecules or nanoparticles. An optical phase grating of a fixed wavelength may therefore serve as a universal coherent beam splitter for a large variety of particles. This universality, however, comes at a price: without resonant enhancement, the polarizability remains moderate and the process requires high laser intensities. The wavelength should also be chosen to avoid photon absorption and re-emission processes, which tend to reduce interference contrast.

The benefits of both optical gratings and of near-field interferometry led to the proposal of a Talbot-Lau interferometer whose central element G_2 is an optical phase grating (see Fig. 1) [57]. In this scheme, the **Kapitza-Dirac-Talbot-Lau Interferometer** (KDTLI), the first and last grating remain material masks while the central grating is formed by a thin standing light wave with a period matching that of the outer two gratings. Such an interferometer was built and successfully employed with a variety of complex molecules with masses up to 10^4 Da [52, 58, 59].

---

[3]From here on, TLI will refer to Talbot-Lau interferometers implemented with three nanomechanical gratings.

[4]Modern matter-wave literature often associates the names of Kapitza and Dirac with diffraction at a thin dipole force phase grating, even though their original idea applied to non-polarizable electron diffracted at the ponderomotive potential created by interaction with the light field.

The use of nanomechanical gratings for G_1 and G_3 will eventually be limited when the Casimir Polder potential becomes sufficiently strong (for slow, highly-polarizable molecules) that the particles are completely deflected to the grating walls and cannot pass. It was therefore proposed to form transmission gratings using a standing light wave. This can work via photo-ionization, where particles passing the anti-nodes of a standing light wave are ionized and removed from the beam, leaving every node as an effective grating slit [2]. This process works best if the photon energy exceeds the ionization energy and the absorption cross section is sufficiently high to allow absorption of more than one photon in every anti-node, conditions which can be met, for example, by tryptophan-rich peptides or low work function metal clusters.

The **Optical TIme-domain MAtter-wave interferometer** (OTIMA) is a time-domain interferometer that employs pulsed absorptive optical gratings [60, 61]. Since all particles from the pulsed source interact with the same grating pulse at the same time, independent of their position, various dispersive phase shifts are eliminated, most prominently the shift related to Earth's gravitational acceleration: $\Delta \varphi \propto k g T^2$, where $k = 4\pi/\lambda_L$ is the wave number of the optical grating and T the pulse separation time. However, phase shifts with explicit velocity dependence remain, such as the Coriolis shift induced by the rotation of the Earth, $\Delta \varphi \propto k(2\vec{v} \times \vec{\Omega}_E)T^2$.

Photo-depletion gratings have been extensively studied in atom interferometry [62, 63]. For molecular quantum optics, we are developing mechanisms that can act on the widest class of particles possible. The fluorine (F_2) grating lasers in the OTIMA experiment have a wavelength of $\lambda_L = 157.6$ nm (7.9 eV). This suffices to ionize van der Waals clusters of aromatic anthracene and caffeine [61, 64] as well as tryptophan-rich polypeptides [65]. Combined with ultrafast desorption techniques [65], the first interference of a natural antibiotic polypeptide, Gramicidin A1, was recently demonstrated in this experiment [66] (see Fig. 3a).

Photo-fragmentation is another mechanism which can be used to achieve an optical transmission grating, as demonstrated with van der Waals clusters of hexafluorobenzene [64]. A major goal is to develop single-photon cleavage of tagged biopolymers for future interference experiments with proteins and DNA. The cleavage mechanism itself has recently been successfully demonstrated with functionalized insulin [68].

Since the grating periods of KDTLI and OTIMA are already at the lower limit of commercially available high power lasers, pushing high mass quantum experiments even further requires increasing the flight time between the gratings. This may be achieved by advanced particle cooling schemes or by increasing the interferometer length. The **Long-baseline Universal Matter-wave Interferometer** (LUMI) is a ten-fold stretched realization of the KDTLI experiment. This instrument accepts de Broglie wavelengths as small as 35 fm and has already demonstrated interference with a molecular library centered at 27,000 Da, with a typical molecule in the library consisting of nearly 2000 atoms and travelling at 260 m/s [67] (see Fig. 3b). These molecules were synthesized at the University of Basel for the purpose of these interference experiments, as described in Sect. 3.2. To date, they are the most massive and complex objects for which quantum interference has been demonstrated.

(a)   **(b)**

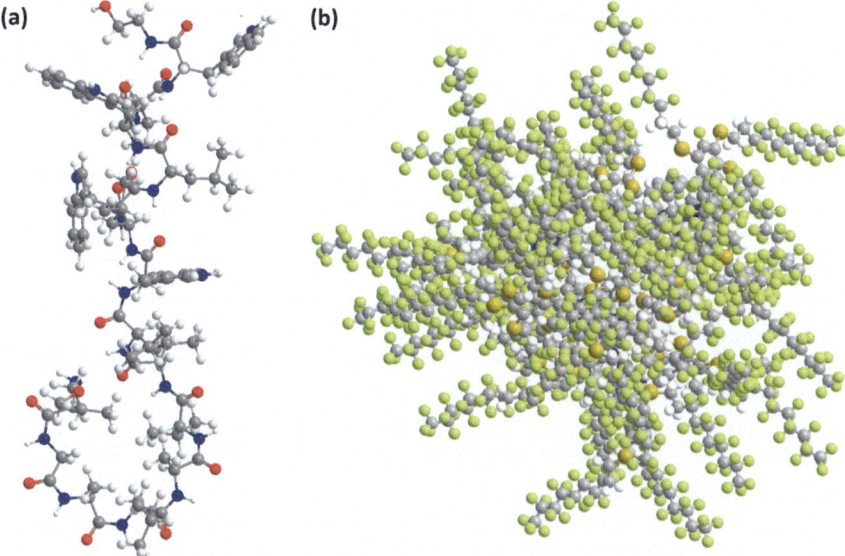

Fig. 3 **a** The antibiotic pentadecapeptide Gramicidin A1 is the most complex natural biomolecule in matter-wave experiments so far [66]. **b** A library of perfluoroalkyl-functionalized oligoporphyrins sets the current record for the most massive particles seen in matter-wave interference to date [67]

The LUMI design is modular: the central grating can be interchanged between a nanomechanical grating and an optical phase grating. Generalized versions of the scheme, including optical transmission gratings and surface detection have also been theoretically investigated [69]. The present LUMI experiment is compatible with both atoms and complex molecules, and an upgrade that is currently being implemented will allow it to also work with massive metal clusters. In this sense LUMI is a truly 'universal' interferometer as well as a powerful instrument for metrological studies.

3   Quantum-Assisted Deflectometry of Atoms and Complex Molecules

Otto Stern demonstrated that molecular beam methods can boost the precision in the measurement of atomic and molecular properties, external forces or fundamental constants. The Stern-Gerlach experiment has become standard textbook material, and his electric deflection experiments paved the way for a wide range of studies in physics and chemistry in the decades that followed.

Classical beam deflectometry measures the shift of a tightly collimated molecular beam as a result of its interaction with an external field (see Fig. 1 and Eq. (1)) [70–74]. The magnitude of the shift encodes information about the field and the molecular coupling to it. Beam deflectometry can also be used to separate isomers [72, 75, 76] or sort molecules by their quantum state [77]. The sensitivity of classical measurements is determined by the width of the beam and the detector resolution. Beam flux requirements typically constrain the beam width to $> 10\,\mu m$ while position-sensitive time-of-flight mass spectrometers can typically resolve shifts $> 50\,\mu m$ [78].

Here we review a method for improving the spatial resolution and flux in beam deflectometry: **Quantum-assisted deflectometry**. The technique is particularly fitting in a tribute to Otto Stern since it combines beam deflectometry and matter-wave diffraction, two fields he helped pioneer. Combining these techniques gains orders of magnitude in spatial resolution, allowing us to resolve nanometer, rather than micrometer, deflections.

As described in Sect. 2, the coherent evolution of molecules in a generalized Talbot-Lau interferometer (e.g. TLI, KDTLI, OTIMA, LUMI) manifests as density fringes imprinted into the molecular beam. In a symmetric setup these fringes have the same periodicity as the interferometer gratings, with $d = 79\,nm$ in OTIMA and $d = 266\,nm$ in KDTLI and LUMI. The capability to track fringe shifts on the nanometer level yields the high spatial resolution of this technique, enabling the measurement of tiny forces which would be nearly impossible to resolve with classical beam deflection methods. In analogy to Eq. (1), the fringe shift due to a uniform force acting transversely along the entire interferometer is

$$s = k\frac{K}{m}T^2 = k\frac{K}{m}\frac{L^2}{v^2} \qquad (2)$$

where $k = 2\pi/d$, T the time between gratings (for pulsed experiments like OTIMA) and L the inter-grating separation.

3.1 Quantum versus Classical Deflection

In the matter-wave deflectometry experiments described here, a force is applied to the molecules and the resulting phase shift of the interference fringes is detected. Consider a particle beam in a uniform electric field gradient. It will be broadened and/or shifted depending on whether the particles have a permanent electric dipole moment or are only polarizible. In quantum-assisted deflectometry this corresponds to contrast reduction and the deflection of the fringe pattern, where the fringes provide a ruler with high spatial resolution.

The Talbot-Lau deflectometry scheme can be compared to similar experiments using Mach-Zehnder interferometry [33] and classical Moiré deflectometry [79]. The three schemes are limiting cases of the same physical setup: a molecular beam traversing three gratings of period d each separated by a distance L. The relevant length scales are the Talbot length $L_T = d^2/\lambda_{dB}$ and the "aperture Talbot Length", $L_a = a^2/\lambda_{dB}$, where a is the width of the beam at the first grating.

1. The *far-field (Mach-Zehnder)* regime is reached when the grating separation satisfies $L \gg L_a$. In this limit, the diffraction orders emerge from G_1 as distinct partial beams which are diffracted back by G_2 and recombined by G_3. The isolated partial beams can be made to locally interact with potentials, which induce a measurable phase shift in the interference pattern. This setting was realized in the first atom interference [8] and metrology [33] experiments.

2. If the grating separation satisfies $L_a/N > L > L_T$, with N the number of illuminated grating slits, the apparatus realizes a *Talbot-Lau interferometer* [80, 81], which is the regime of our molecule interference experiments. Each individual molecule is still spatially delocalized across several, or in some cases up to 100, grating periods [82]. However, the molecular beam is so wide and the diffraction angles so small that all of these partial interferometers overlap. While this makes it impossible to address individual partial beams, the symmetry of the interferometer ensures that all partial waves converge at the position of the third grating, giving rise to high-contrast interference fringes. A potential gradient can be applied within the interferometer to induce an envelope phase shift of the interference fringes.

3. The third limit is that of classical Moiré deflectometry, in which $L_a, L_T \gg L$. In this setting, the wavelets originating in G_1 do not evolve fast enough to cover two slits in G_2 coherently. Even in this classical regime, sensitive force sensing is still possible [79]. This is the closest realization to the classical beam multiplexing proposed by Stern.

Among the three different regimes, only the far-field Mach-Zehnder features well-separated interferometer arms, and it still holds the record for the most sensitive polarizability measurements [33–35]. In this regime one can also explore topological and geometric phases, such as the Aharanov-Bohm [83], Aharanov-Casher [84], Berry [85] and He-McKellar-Wilkens phases [86].

The TLI regime, on the other hand, has the best mass scalability of the three limits and it is therefore currently the only setting compatible with quantum-enhanced measurements of large molecules. In both the far- and near-field limits (1 and 2 above), the sensitivity to external forces depends on the enclosed interferometer area and the detected signal-to-noise ratio. Compared to a classical Moiré deflectometer, the TLI employs smaller grating periods and/or longer machine length and therefore has intrinsically better sensitivity to small fringe displacements.

3.2 Molecules for Interferometry: Choice, Synthesis and Sources

3.2.1 General Strategies

Among all nanoscale particles, molecules are ideally suited for matter-wave experiments due to their monodisperse nature. Being virtually identical, they also exhibit a very narrow isotopomeric mass distribution, typically within a few Daltons. Over the years, we have explored a large variety of structures, from commercially available molecules to tailor-made model compounds with properties optimized for the particular experiment. The collaboration between experimental physicists and synthetic chemists has enabled access to higher mass regimes and the development of new diffraction mechanisms.

Three challenges need to be considered when selecting molecules for interference experiments: the preparation of neutral particle beams, novel diffraction mechanisms and detection schemes with high sensitivity and resolution. For different molecules different techniques may apply and one research goal is to find the most generic combinations that allow treating the largest class of particles.

While thermal evaporation from a Knudsen cell is a simple experimental technique, it requires molecules with sublimation temperatures below their degradation temperature to guarantee the launch of individual and intact molecules of known composition and mass. This calls for a molecular design of thermally stable molecules with minimal intermolecular attraction. An equally challenging criterion is to provide these substances in sufficient quantities (a few grams) to realize constant beams for a sufficient period of time to enable both alignment and interference experiments. While the requirement of 'large scale availability' constrains the variety of suitable structures, clever design and synthesis enabled us to push the mass limit to beyond 10 kDa [59]. Pulsed laser desorption of functionalized molecules from thin surfaces [87] has been shown to be more economical and applicable to an even larger variety of potential structures.

To minimize intermolecular attraction and to improve the volatility of molecules, their peripheral decoration with highly fluorinated alkyl chains is a successful strategy. The strong electron-withdrawing character of the fluorine atoms localizes the electron densities and decreases the electron mobility. The decoration with perfluorinated alkyl chains thus increases the mass of the target structure, while keeping the polarizability and induced dipole interaction low, a particularly appealing feature when 'heavy' particles are of interest. Furthermore, the stability of a C-F bond compares favorably with a C-H bond and the mass spectrum is kept clean, since fluorine is a monoisotopic element. The strategy has been successfully applied to a variety of model compounds ranging from simple dyes like azobenzenes [88], porphyrins [58, 89, 90], and phthalocyanines [45], to interlinked benzene subunits [91], and advanced molecular libraries pushing the limits of diffraction experiments [59, 67].

The feasibility of thermal peptide beams was studied using derivatives of a tryptophan-containing tripeptide. While an unprotected alanine-tryptophan-alanine (Ala-Trp-Ala) showed only fragments in the VUV-post-ionization mass spectrum ($\lambda = 157$ nm), the intact molecular ion could be observed after removing internal charges and hydrogen bond donors by acetylation and amidation of the termini and methylation of the peptidic amide protons [92, 93]. The introduction of fluoroalkyl chains at the N-terminus or both termini improved the relative intensity of the molecular ion substantially despite the increase in molecular mass. The best results with the least fragmentation were obtained when fluoroalkyl chains were introduced, and the

N-methylation was omitted. Considerably more massive peptidic constructs could be launched and VUV-ionized under femtosecond laser desorption even reaching beyond 20 kDa for a 50 amino acid Trp-Lys construct which was extensively decorated with fluoroalkyl chains [65].

The second crucial factor that must be considered in the molecular design is the detection method. The observation of neutral molecules is challenging at low beam densities, and fragmentation-free post-ionization becomes generally more challenging with higher molecular mass [94]. The detectability, however, can be improved substantially by molecular design. The presence of a suitable chromophore enables the observation of individual molecules by fluorescence, which allowed real-time single-molecule imaging in far-field diffraction experiments [45]. Large, electron-rich π-systems are also attractive for photo-induced post-ionization [87]. Oligopeptides with tryptophan units turned out to be particularly suited for photoionization mass spectrometry because of the high absorption cross section of the indole subunit, the only group in any natural amino acid that is susceptible to single-photon ionization at 157 nm [66, 93]. A high tryptophan density even permitted the VUV-ionization of a peptidic construct of more than 20 kDa, thereby exceeding the mass limit for VUV-ionization by one order of magnitude over the previous standard for biomolecules [65]. Detection by mass spectrometry is appealing, as it eases the requirement of monodispersivity and thereby allows a new approach based on molecular libraries. These ensembles of molecules have different numbers of identical subunits, have a broader mass range, but with well-defined and well-separated masses.

3.2.2   Specific Examples

Fullerenes were used in the first diffraction experiments with organic molecules [42] and in many calibration experiments ever since. Their high thermal stability facilitates the creation of intense thermal beams from simple Knudsen cells, and they can be detected by electron impact ionization or by thermal ionization in an intense laser field [43, 95] followed by ion counting. Pure fullerene powder is also readily available in bulk quantities. Vapor pressures of about 0.1 hPa can be reached by heating the powder to 900 K, which generates an intense molecular beam with velocities in the range of 100–200 m/s.

Vitamins and provitamins such as α-tocopherol (vitamin E), β-carotene (provitamin A), 7-dehydrocholesterol (converted to provitamin D3 upon absorption of UV light) and phylloquinone (vitamin K1) have been interfered and deflected in the KDTLI experiment. Thermal sublimation of such fragile biomolecules always competes with fragmentation. At 500 K, the beam would typically last for only about 30 minutes.

Natural peptides do not evaporate or sublimate intact in a continuous thermal source. However, they can be launched by nanosecond [96] or femtosecond pulsed laser desorption sources, if they are immediately entrained into an adiabatically expanding seed gas. This recently enabled interference of a polypeptide with 15 amino acids,

Gramicidin A, in the OTIMA experiment. The fragility of large peptides, limited photoionization cross sections (needed for optical gratings and post-ionization mass spectrometry) of complex peptides, and carrier gas velocity, are the reasons why Gramicidin A is the most complex natural peptide in matter-wave experiments to date [66] (Fig.3a).

Thermal beams of functionalized tripeptides: Peptides are very fragile compounds—even simple dipeptides hardly survive the temperature needed to build up the vapor pressure required for molecular beam experiments. Interestingly, per-fluoroalkly functionalization can facilitate the formation of thermal beams of even tripeptides [97] to a degree that matter wave interference became possible [93] (see Fig.6e).

Laser desorbed beams of large tailored polypeptides: The combination of fluoroalkyl decoration and ultrafast (femtosecond) laser desorption into a cold seed gas enabled launching even complex neutral peptides composed of up to 50 amino acids [65]. Their successful intact detection using single-photon ionization at 157 nm required optimizing the peptides for a tryptophan content as high as 50%.

Electrosprays of modified biopolymers: We have recently started investigating a novel approach to generating neutral biomolecular beams using bioconjugation techniques. Peptides with a photocleavable tag, introduced by amidation of surface amino groups with N-hydroxysuccinimide esters (NHS-esters) can be readily volatilized and ionized in an electrospray source. The emerging ions can be guided and manipulated using electric fields and they can even be cooled in a buffer gas. Subsequent neutralization can be achieved by selective photocleavage of the tag in an intense pulsed laser field. This mechanism has been demonstrated for various peptides [98] and even for human insulin [68].

Molecular libraries have been developed and optimized for high-mass interferometry. The concept is displayed in Fig.6f. A readily ionizable porphyrin architecture exposing numerous pentafluorophenyl groups was synthesized as a pure compound. In a subsequent aromatic nucleophilic substitution reaction fluorine atoms were substituted by highly fluorinated alkyl thiol chains. Since each reaction replaces exactly one fluorine atom by one fluorinated alkyl thiol chain, an entire molecular library emerges with precisely known masses, differing by the value of (M(fluorinated alkyl thiol chain)-M(FH)). Electron impact ionization mass spectrometry (EI-QMS) allowed the selective detection of a particular mass range.

Near-field interferometry tolerates a mass distribution even in excess of $\Delta m/m \simeq$ 10 % and neither isomers nor isotopes impair the experiment. Each molecule constitutes its own de Broglie wave and as long as the electromagnetic properties are similar the interference fringes will appear at the same position. Such a library was first built around a single tetrakispentafluorophenylporphyrin with 20 substitutable fluorine atoms [89] (see Fig. 6f), and successfully used in KDTL interferometry [59]. A dendritic porphyrin architecture with 60 substitutable fluorine atoms gave access to an even larger library that allowed us pushing the mass record in LUMI to beyond 25 kDa [67] (see Fig. 3b).

3.3   Molecule Interference Experiments

Numerous molecular properties have already been probed in quantum-assisted measurements. Here we restrict ourselves to experiments performed at the University of Vienna. They all rely on measurements of interference contrast and fringe deflection in generalized Talbot-Lau interferometers. Experimental results are divided into four categories; electronic, magnetic, and optical properties, as well as the measurement of inertial forces, as summarized in Table 1.

3.3.1   Electronic Properties

Electric deflection experiments require an electrode that provides a uniform force field. The transverse force on a polarizable particle (without permanent dipole moment) is then given by

$$F_x = -\nabla U_{ind} = -\nabla \frac{1}{2} \vec{d}_{ind} \cdot \vec{E} = \alpha_0 (E \cdot \nabla) E_x. \tag{3}$$

Here we include the possibility of a thermally induced dipole moment $d_{ind} = \alpha_0 E$. The factor of $1/2$ is due to the work done by the field inducing the dipole moment, and x is the direction transverse to both the molecular beam and to the grating bars.

Equation (3) shows that a field satisfying $(\vec{E} \cdot \nabla) E_x = $ const gives a constant transverse force proportional to the particles' static polarizability α_0. Electrodes have been designed and built with a tailored geometry to provide such a force, as described in Ref. [99] (see Fig. 4a).

Static polarizability: The first polarizability measurements in Talbot-Lau interferometry were made with fullerenes [101]. These measurements were repeated with improved precison and accuracy in the LUMI experiment [102]. Here, we took advantage of the ability to calibrate the setup in-situ with atomic cesium, the polarizability of which has been precisely measured with Mach-Zehnder interferomtery [35, 115] (see Fig. 5). Improving the precision even further, to better much than 1%, is only sensible for cold molecules, with improved control over the internal state.

Structural isomers have identical chemical composition and mass but different geometries and electronic properties (see Fig.6c). We consider two specially synthesized isomers which differ in their susceptibility by more than 20% because the molecule on the left of Fig.6c has a widely delocalized electron system, while electron delocalization is constrained to the phenyl rings in the molecule on the right. This can be easily distinguished in KDTLI deflectometry [91].

Dynamic dipole moment: While fullerenes are rigid, isotropic bodies, well-characterized by the scalar static polarizability α_0, this is not the case for floppy molecules such as the perfluoroalkyl-functionalized diazobenzenes [88] (see Fig. 6a). At a source temperature of 500 K these molecules undergo rapid conformational

Table 1 A list of quantum-assisted deflection experiments conducted in Vienna, sorted according to the type of measurement (electric, magnetic, optical, and inertial). Given are the particle properties measured, the interferometer in which the experiment was done, and the particle type with which it was done

Property	Interferometer	Particle
Electronic properties		
Static polarizability (*including dynamical electric dipole moment)	TLI, LUMI	C_{60}, C_{70} [101, 102]
	KDTLI	Perfluoroalkylated azobenzenes* [88]
	KDTLI	β-carotene*, vitamins E and K* [103]
	LUMI	Perfluoroalkylated tripeptides* [93]
Permanent dipole moment	KDTLI	Fe-TPP-Cl [104]
Fragment analysis	KDTLI	Perfluoroalkylated palladium complex and fragments [105]
Constitutional isomers	KDTLI	$C_{49}H_{16}F_{52}$ [91]
Magnetic properties		
Diamagnetic susceptibility, atoms	LUMI	Barium and strontium [82]
Diamagnetic susceptibility, molecules	LUMI	Anthracene and adamantane [106]
Optical properties		
Optical polarizability	KDTLI	C_{60}, C_{70} [107]
	KDTLI	C_{60}, C_{70}, $C_{60}F_{36}$, $C_{60}F_{48}$ [108]
	KDTLI	β-carotene, vitamins E and K [103]
Absorption cross section	KDTLI, grating interaction	C_{60}, C_{70}, $C_{60}F_{36}$, $C_{60}F_{48}$ [108, 109]
	KDTLI, recoil spectroscopy	C_{60} [110]
Conformer selection, proposed	Far-field, diffraction at dipole phase grating	Phenylethylamine [111]
Spectroscopy techniques, proposed	OTIMA, via recoil and depletion	[112]
Inertial forces		
Gravity/Weak equiv. princ.	OTIMA	TPP and derivatives [113]
Gravity-Coriolis compens.	LUMI	C_{60} [114]

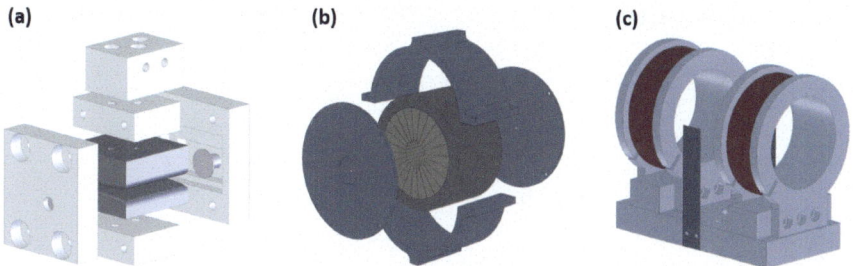

Fig. 4 Deflectors in our TLI, KDTLI and LUMI experiments. **a** A uniform $(\vec{E} \cdot \nabla)E_x$ field to measure polarizabilities and induced dipole moments [99]. **b** A modified Halbach array with a uniform $(\vec{B} \cdot \nabla)B_x$ to measure magnetic susceptibilities in LUMI [100]. **c** Anti-Helmholtz coils with a uniform ∇B_x for probing permanent magnetic moments in LUMI (image: S. Pedalino)

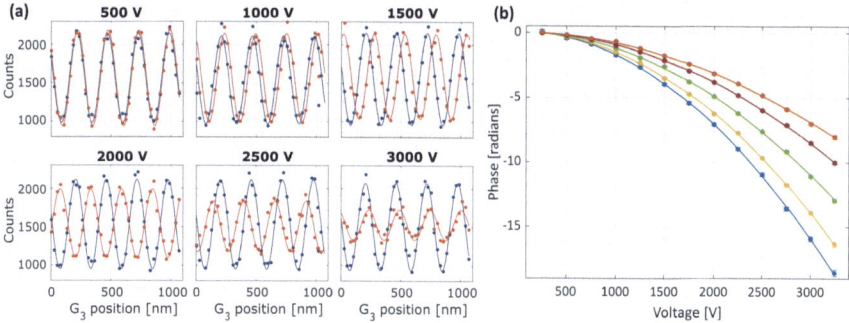

Fig. 5 **a** Electric deflection of C_{60} in the LUMI experiment. **b** The fringe shifts show the expected quadratic dependence on the electrode voltage for a selection of different molecular beam velocities

changes on the picosecond time scale, which leave the static and optical polarizability nearly constant, but may change the instantaneous electric dipole moment d_e by as much as 300%. This contributes to the net electronic susceptibility according to

$$\chi_{elec} = \alpha_0 + \frac{< d_e^2 >_T}{3k_B T}, \tag{4}$$

where the second term is due to the thermally averaged value of the dipole moment [116]. In the electric deflectometry experiments described here the total susceptibility is measured, and with the aid of ab initio molecular dynamics simulations the relative contributions of the static polarizability and the averaged thermally induced dipole moments can be extracted [88, 103]. It is interesting that de Broglie interferometry, which is primarily concerned with center-of-mass motion, can still reveal the influence of fast conformational changes through their influence on the molecules' response in an electric field. It is expected that molecular sequence isomers will exhibit different dynamic dipole moments and be separable in experiments with good signal-to-noise ratio [117].

(a) Time evolution of floppy azobenzenes

C₇F₁₅

t = 10 ns
d = 3.57 Debye

C₇F₁₅

t = 35 ns
d = 0.85 Debye

(b) Thermal fragmentation of a Pd catalyst precursor

(c) Structure isomers: same composition, different geometry

$\chi = 127 \times 4\pi\varepsilon_0$Å^3

$\chi = 102 \times 4\pi\varepsilon_0$Å^3

(d) Porphyrin derivatives with and without electric dipole moment

d = 2.7 Debye

d = 0 Debye

(e) Functionalized tripeptides

(f) Molecular library based on porphyrin derivatives with varying numbers of substituents

R = F → R = n x SC₂₀H₁₅F₂₆ and (20-n) x F

NaH
diglyme
microwave
220°C

n = 11 9360 Da
n = 12 10123 Da
n = 13 10885 Da
n = 14 10648 Da
n = 15 12410 Da

m/z →

6000 8000 10000 12000 14000

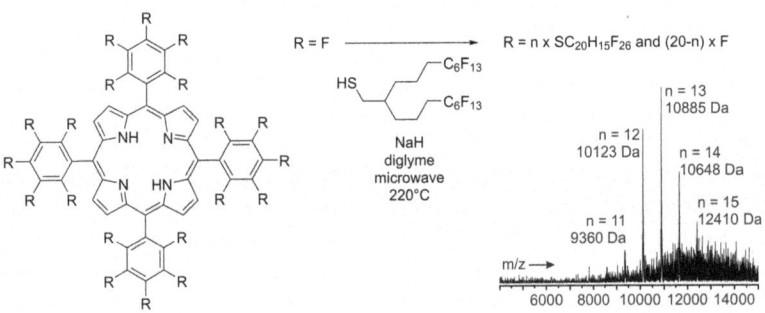

◀**Fig. 6** Interference-assisted deflectometry was used to elucidate the **dynamic dipole moments** of functionalized azobenzenes [88], **b** the **fragmentation** of a palladium catalyst precursor [105], **c** **electron delocalization** in constitutional isomers [91], and **d** permanent electric dipole moments in porpyhrin derivatives [104]. **e** Tripeptides optimized for both molecular beam formation and detection [93]. **f** Molecular libraries of members with well-defined molecular weight by random substitution of fluorine atoms with highly fluorinated alkyl chains [89]

Permanent dipole moment: Molecules with a permanent electric dipole moment experience a shift of their fringe pattern that depends upon the orientation of the molecule. Since most beam sources emit molecules with random initial orientation and in a mixture of thermally excited rotational states, each molecule experiences a different shift according to its orientation as it tumbles through the electric field. Molecules with a permanent electric dipole moment thus exhibit a reduced interference contrast, which can be used to distinguish them from polarizable particles with no permanent moment. This has been demonstrated with the porphyrin derivatives Fe-TPP and Fe-TPP-Cl (see Fig. 6d), which differ only by a single chlorine atom and a dipole moment of 2.7 D. Measuring the fringe deflection as a function of electrode voltage showed that both compounds have similar polarizabilities, while measuring the interference contrast revealed a much faster decay in contrast for the polar compound [104].

3.3.2   Magnetic Properties

Magnetic deflection, even more so than electric deflection, represents the huge impact of Otto Stern on the landscape of experimental physics, and has triggered a number of Stern-Gerlach type beam experiments [118–120]. However, only recently have similar experiments been performed in molecule interferometry.

The conceptual design of quantum-assisted magnetic deflectometry is identical to that of electric deflection. Here we aim to measure the magnetic susceptibility of particles subject to a uniform force which is introduced via a specially designed Halbach array of permanent magnets [100], as illustrated in Fig. 4b. The magnet can be translated in and out of the molecular beam, allowing us to take differential phase measurements referenced to a no-field situation.

In analogy to electric deflection, we require a region with

$$(\vec{B} \cdot \nabla)B_x = \text{const.} \tag{5}$$

such that magnetically susceptible particles with no permanent magnetic moment experience a uniform transverse force. Species with permanent magnetic moments, i.e. paramagnetic particles, will exhibit a reduced interference contrast, which is why this technique is best suited for measuring diamagnetic deflections or second order paramagnetic contributions (temperature-independent paramagnetism [116]).

The first measurements with the magnetic deflector described in Ref. [100] were performed in the LUMI experiment, taking advantage of the long interferometer baseline to observe the small diamagnetic deflection of thermal beams of the alkaline earth atoms barium and strontium [82]. The measured susceptibilities agreed well with the calculated values, and represent the first direct measurement of the ground-state diamagnetism of isolated particles. The sensitivity of the method was further illustrated by demonstrating the complete loss of interference contrast for the odd isotopes of barium and strontium which contain an unpaired nuclear spin, showing that even permanent moments on the order of a nuclear magneton are sufficient to completely dephase the interference fringes.

This work was recently extended to molecules [106], for which the situation is more complex than for atoms due to coupling with rotational states as well as alignment effects in the molecular beam. In this work, we measured the diamagnetic deflection of anthracene, a planar aromatic molecule, and adamantane, a tetrahedrally symmetric molecule. We observed the predicted isotropically averaged susceptibility for adamantane but a surprisingly large value for anthracene, which would be consistent with edge-on alignment of the planar molecules in the supersonic expansion. This alignment leads to the broadside orientation of anthracene being over-represented during its transit through the deflection region. Due to anthracene's aromaticity, the molecular plane has a significantly larger susceptibility tensor component than the other orientations, leading to the larger-than-isotropic observed deflection.

The difference to electric deflection experiments is apparent in the presence of a non-zero magnetic moment, as when there are unpaired nuclear or electron spins. In this case, as in the seminal Stern-Gerlach experiment, the quantization of the magnetic moment plays a role in the behavior of the particles in the magnetic field. When exposed to a $(\vec{B} \cdot \nabla)B_x$ field, the various projections will be deflected in different directions, leading to a reduction in interference contrast unless a spin state is selected and maintained in the interferometer (see Fig. 7). However, a constant ∇B_x field (such that the force on a permanent moment is uniform across the beam)

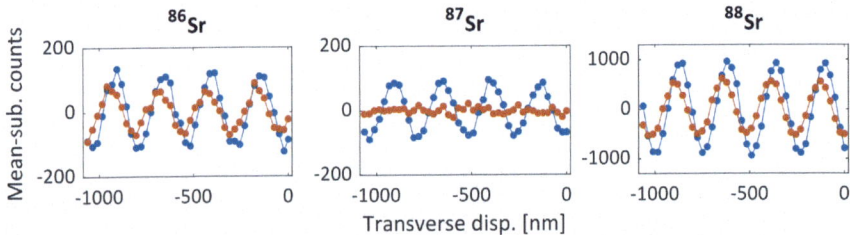

Fig. 7 In the LUMI experiment, we conducted the first beam deflection experiments to measure diamagnetic susceptibilities of atoms and molecules. In the first demonstration [82], the alkaline earth atoms barium and strontium were used in a TLI scheme. Here one can see both the diamagnetic deflection of the even isotopes of strontium (^{86}Sr and ^{88}Sr) and the complete washing out of the interference fringes of ^{87}Sr, which contains a small permanent magnetic moment due to an unpaired nuclear spin. Reference data is shown in blue and deflection data in red

can lead to revivals in interference visibility when the magnetic sub-levels are shifted by integer multiples of the grating period. This has been demonstrated in the LUMI experiment using anti-Helmholtz coils (illustrated in Fig. 4c) to observe the effect in atomic cesium [121], and experiments to observe the effect in triplet-excited fullerenes are in progress.

3.3.3 Optical Properties

There are several optical properties of molecules which can be extracted using quantum-assisted measurements. This can be accomplished by introducing an additional laser to the interferometer in analogy to the previously described deflectometry experiments and performing recoil spectroscopy, as proposed in Ref. [122]. The extraction of absolute absorption cross sections of dilute beams of C_{70} fullerenes was demonstrated using this technique in Ref. [110] in the KDTLI experiment.

Another approach to probe optical properties is to take advantage of the matter-light interactions which always occur in interferometer schemes with optical gratings. Since the contrast obtained in a KDTLI experiment depends on the AC polarizability of the molecule at the grating wavelength, this can be used as a measurement for optical polarizability, as done in Refs. [107, 108].

The sensitivity of the KDTLI scheme to optical polarizability can also be used to study molecular fragmentation [105]. The optical polarizability of a fragment of the palladium catalyst $C_{96}H_{48}C_{12}F_{102}P_2Pd$ was extracted by measuring the interference contrast as a function of the optical grating power. By comparing the measured polarizability to that of the intact particle versus the fragment, it could be determined whether the molecule fragmented already in the source, or only during the detection, after traversing the interferometer. Classical beam deflectometry could not have distinguished the origin of fragmentation, since the deflection depends only on the polarizability-to-mass ratio, which is nearly the same for the parent molecule and its fragments. Here, since the phase imprinted by the second grating depends only on the optical polarizability, it could be determined that the molecule fragmented in the source rather than the detector.

Measuring the optical polarizability is also useful for estimating the static polarizability of molecules, since for fullerenes and many other large organic molecules, we find that the static polarizability approximates the optical polarizability to within a few 10%. This is in agreement with the observation that most optical transitions for molecules in this complexity class are 30–50 nm wide.

There have been several proposals for other ways to utilize the sensitive dependence of molecule diffraction on optical interactions. Several spectroscopy setups have been proposed in the context of the OTIMA experiment, including multi-photon recoil and polarizability spectroscopy [112]. In the far-field diffraction experiment, a near-resonant ultraviolet optical grating could potentially be used for efficient sorting of conformers [111]. It has also been proposed to employ optical helicity fringes to create a diffraction grating that discriminates chiral enantiomers [123].

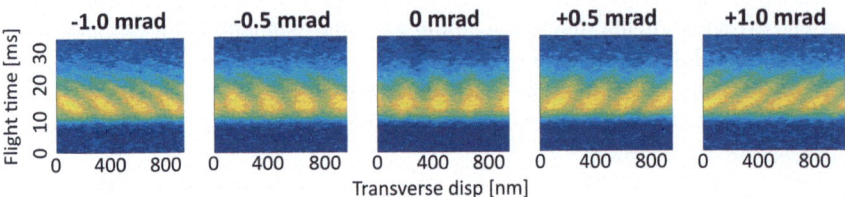

Fig. 8 Time-resolved interference scans showing the competing effects of gravitational and Coriolis phase shifts as a function of the interferometer roll angle. On the far left gravity is responsible for the large shearing, while on the far right the gratings are nearly aligned with gravity and the shearing is due to the Coriolis force. In Ref. [114] these two phase shifts were used to passively compensate one another, but a similar technique could be used for the purpose of measuring gravitational or Sagnac phases directly

3.3.4 Inertial Forces

Interferometers have long been used as inertial sensors, from Sagnac loop interferometers with light to sensitive gravity and gravity gradient sensors made with atom interferometers. Molecule interferometers do not compete in sensitivity due to the comparatively poor signal-to-noise ratios and smaller enclosed areas, but they are still sensitive to such effects. In Ref. [114], the competing phase shifts of fullerenes due to the Coriolis effect and gravity were mapped as a function of velocity for different roll angles of the interferometer setup (see Fig. 8). Molecule interferometry also enables weak equivalence principle measurements of a wider variety of species and internal energies and properties than atom interferometry experiments. This has been demonstrated in the OTIMA experiment by comparing the gravitational phase shift of various isotopomeres of tetraphenylporphyrin [113].

4 Outlook

Molecule interferometry and deflectometry have been inspired by work that was started by Otto Stern 100 years ago. Much of the research in the field since then can be seen as a very extended footnote to the ideas of Otto Stern. And yet we foresee years of exciting research in the attempt to push matter-wave interferometry to ever higher mass and complexity, and to further explore the interface to chemistry, biology and the classical world.

Otto Stern remarked in several of his writings on the particular challenge of preparing molecular beams. This is where quantum optics and chemistry have found a very fruitful overlap and where we still expect thrilling developments: the tailoring of molecules to the needs of quantum optics as well as the use of quantum optics to retrieve information about molecules is a new field of research that opens promising perspectives.

Acknowledgements We acknowledge funding by the European Research Council (Project No. 320694), the Austrian Science Funds (FWF P-30176, P-32543-N), the Swiss National Funds (Project No. 200020159730), the SNI PhD School (P1403) as well as the tireless contribution of many master and PhD students as well as postdocs who contributed over the years to various molecule interferometers and interferometer applications for molecular science.

References

1. O. Stern, Zeitschrift für Physik **39**, 751–763 (1926)
2. E. Reiger, L. Hackermüller, M. Berninger, M. Arndt, Opt. Commun. **264**(2), 326 (2006)
3. C. Davisson, L.H. Germer, Phys. Rev. **30**, 705 (1927)
4. I. Estermann, O. Stern, Z. Phys. **61**, 95 (1930)
5. P.E. Moskowitz, P.L. Gould, S.R. Atlas, D.E. Pritchard, Phys. Rev. Lett. **51**, 370 (1983)
6. P.L. Gould, G.A. Ruff, D.E. Pritchard, Phys. Rev. Lett. **56**, 827 (1986)
7. D.W. Keith, M.L. Schattenburg, H.I. Smith, D.E. Pritchard, Phys. Rev. Lett. **61**, 1580 (1988)
8. D.W. Keith, C.R. Ekstrom, Q.A. Turchette, D.E. Pritchard, Phys. Rev. Lett. **66**(21), 2693 (1991)
9. C.J. Bordé, Phys. Lett. A **140**, 10 (1989)
10. C.J. Bordé, S. Avrillier, A. Van Lerberghe, C. Salomon, D. Bassi, G. Scoles, J. Phys. Coll. **42**(C8), 15 (1981)
11. M. Kasevich, D.S. Weiss, E. Riis, K. Moler, S. Kasapi, S. Chu, Phys. Rev. Lett. **66**, 2297 (1991)
12. A. Peters, K. Yeow-Chung, S. Chu, Nature **400**, 849 (1999)
13. G. Lamporesi, A. Bertoldi, L. Cacciapuoti, M. Prevedelli, G. Tino, Phys. Rev. Lett. **100**, 5 (2008)
14. P. Asenbaum, C. Overstreet, M. Kim, J. Curti, M.A. Kasevich, arXiv:2005.11624v1 (2020)
15. R.H. Parker, C. Yu, W. Zhong, B. Estey, H. Müller, Science **360**(6385), 191 (2018)
16. I. Dutta, D. Savoie, B. Fang, B. Venon, C.L. Garrido Alzar, R. Geiger, A. Landragin, Phys. Rev. Lett. **116**(18), 183003 (2016)
17. D. Savoie, M. Altorio, B. Fang, L.A. Sidorenkov, R. Geiger, A. Landragin, Sci. Adv. **4**, 7948 (2018)
18. W. Chaibi, R. Geiger, B. Canuel, A. Bertoldi, A. Landragin, P. Bouyer, Phys. Rev. D **93**(2), 021101 (2016)
19. Y.A. El-Neaj, C. Alpigiani, S. Amairi-Pyka, H. Araújo, A. Balaž, A. Bassi, L. Bathe-Peters, B. Battelier, A. Belić, E. Bentine, J. Bernabeu, A. Bertoldi, R. Bingham, D. Blas, V. Bolpasi, K. Bongs, S. Bose, P. Bouyer, T. Bowcock, W. Bowden, O. Buchmueller, C. Burrage, X. Calmet, B. Canuel, L.I. Caramete, A. Carroll, G. Cella, V. Charmandaris, S. Chattopadhyay, X. Chen, M.L. Chiofalo, J. Coleman, J. Cotter, Y. Cui, A. Derevianko, A. De Roeck, G.S. Djordjevic, P. Dornan, M. Doser, I. Drougkakis, J. Dunningham, I. Dutan, S. Easo, G. Elertas, J. Ellis, M. El Sawy, F. Fassi, D. Felea, C.H. Feng, R. Flack, C. Foot, I. Fuentes, N. Gaaloul, A. Gauguet, R. Geiger, V. Gibson, G. Giudice, J. Goldwin, O. Grachov, P.W. Graham, D. Grasso, M. van der Grinten, M. Gündogan, M.G. Haehnelt, T. Harte, A. Hees, R. Hobson, J. Hogan, B. Holst, M. Holynski, M. Kasevich, B.J. Kavanagh, W. von Klitzing, T. Kovachy, B. Krikler, M. Krutzik, M. Lewicki, Y.H. Lien, M. Liu, G.G. Luciano, A. Magnon, M.A. Mahmoud, S. Malik, C. McCabe, J. Mitchell, J. Pahl, D. Pal, S. Pandey, D. Papazoglou, M. Paternostro, B. Penning, A. Peters, M. Prevedelli, V. Puthiya-Veettil, J. Quenby, E. Rasel, S. Ravenhall, J. Ringwood, A. Roura, D. Sabulsky et al., EPJ Quant. Technol. **7**, 1 (2020)
20. C. Burrage, E.J. Copeland, E.A. Hinds, J. Cosmol. Astroparticle Phys. **2015**(03), 042 (2015)

21. P. Hamilton, M. Jaffe, P. Haslinger, Q. Simmons, H. Müller, J.T. Khoury, Science **349**, 849 (2015)
22. P.R. Berman, B. Dubetsky, Phys. Rev. A **59**, 2269 (1999)
23. A.D. Cronin, J. Schmiedmayer, D.E. Pritchard, Rev. Mod. Phys. **81**(3), 1051 (2009)
24. G. Tino, M. Kasevich, Atom interferometry, in *Proceedings of the International School of Physics "Enrico Fermi"*, vol. 188 (IOS, Varenna, 2014)
25. M.H. Anderson, J.R. Ensher, M.R. Matthews, C.E. Wieman, E.A. Cornell, Science **269**, 198 (1995)
26. K.B. Davis, M.O. Mewes, M.R. Andrews, N.J. van Druten, D.S. Durfee, D.M. Kurn, W. Ketterle, Phys. Rev. Lett. **75**, 3969 (1995)
27. J. Herbig, T. Kraemer, M. Mark, T. Weber, C. Chin, H.C. Nagerl, R. Grimm, Science **301**, 1510 (2003)
28. C. Kohstall, S. Riedl, E.R. Sánchez Guajardo, L.A. Sidorenkov, J. Hecker Denschlag, R. Grimm, New. J. Phys. **13**(6), 065027 (2011)
29. W. Schöllkopf, J.P. Toennies, Science **266**, 1345 (1994)
30. M.S. Chapman, T.D. Hammond, A. Lenef, J. Schmiedmayer, R.A. Rubenstein, E. Smith, D.E. Pritchard, Phys. Rev. Lett. **75**, 3783 (1995)
31. C. Bordé, N. Courtier, F.D. Burck, A. Goncharov, M. Gorlicki, Phys. Lett. A **188**, 187 (1994)
32. B.S. Zhao, G. Meijer, W. Schoellkopf, Science **331**(6019), 892 (2011)
33. C. Ekstrom, J. Schmiedmayer, M. Chapman, T. Hammond, D. Pritchard, Phys. Rev. A **51**(5), 3883 (1995)
34. A. Miffre, M. Jacquey, M. Büchner, G. Trenec, J. Vigue, Phys. Rev. A **73**, 011603(R) (2006)
35. M.D. Gregoire, I. Hromada, W.F. Holmgren, R. Trubko, A.D. Cronin, Phys. Rev. A **92**, 5 (2015)
36. J. Schmiedmayer, M. Chapman, C. Ekstrom, T. Hammond, S. Wehinger, D. Pritchard, Phys. Rev. Lett. **74**(7), 1043 (1995)
37. V.P.A. Lonij, Atom optics, core electrons, and the van der Waals potential. Thesis (2011)
38. R. Trubko, J. Greenberg, M.T.S. Germaine, M.D. Gregoire, W.F. Holmgren, I. Hromada, A.D. Cronin, Phys. Rev. Lett. **114**, 14 (2015)
39. A. Fallon, C. Sackett, Atoms **4**, 2 (2016)
40. C. Lisdat, M. Frank, H. Knöckel, M.L. Almazor, E. Tiemann, Eur. Phys. J. D **12**, 235 (2000)
41. P. Rousseau, H. Khemliche, A.G. Borisov, P. Roncin, Phys. Rev. Lett. **98**, 1 (2007)
42. M. Arndt, O. Nairz, J. Voss-Andreae, C. Keller, G. van der Zouw, A. Zeilinger, Nature **401**, 680 (1999)
43. D. Ding, J. Huang, R. Compton, C. Klots, R. Haufler, Phys. Rev. Lett. **73**(8), 1084 (1994)
44. O. Nairz, M. Arndt, A. Zeilinger, Am. J. Phys. **71**(4), 319 (2003)
45. T. Juffmann, A. Milic, M. Müllneritsch, P. Asenbaum, A. Tsukernik, J. Tüxen, M. Mayor, O. Cheshnovsky, M. Arndt, Nature Nanotech. **7**, 297 (2012)
46. W.H.F. Talbot, Philos. Mag. **9**, 401 (1836)
47. E. Lau, Ann. Phys. **6**, 417 (1948)
48. J. Clauser, *De Broglie-Wave Interference of Small Rocks and Live Viruses* (Kluwer Academic, 1997), pp. 1–11
49. B. Brezger, L. Hackermüller, S. Uttenthaler, J. Petschinka, M. Arndt, A. Zeilinger, Phys. Rev. Lett. **88**, 100404 (2002)
50. L. Hackermüller, S. Uttenthaler, K. Hornberger, E. Reiger, B. Brezger, A. Zeilinger, M. Arndt, Phys. Rev. Lett. **91**(9), 090408 (2003)
51. T. Juffmann, S. Truppe, P. Geyer, A.G. Major, S. Deachapunya, H. Ulbricht, M. Arndt, Phys. Rev. Lett. **103**, 26 (2009)
52. S. Gerlich, L. Hackermüller, K. Hornberger, A. Stibor, H. Ulbricht, M. Gring, F. Goldfarb, T. Savas, M. Müri, M. Mayor, M. Arndt, Nat. Phys. **3**(10), 711 (2007)
53. C. Brand, M. Sclafani, C. Knobloch, Y. Lilach, T. Juffmann, J. Kotakoski, C. Mangler, A. Winter, A. Turchanin, J. Meyer, O. Cheshnovsky, M. Arndt, Nat. Nanotechnol. **10**, 845 (2015)
54. P.L. Kapitza, P.A.M. Dirac, Proc. Camb. Philos. Soc. **29**, 297 (1933)
55. D.L. Freimund, K. Aflatooni, H. Batelaan, Nature **413**, 142 (2001)

56. O. Nairz, B. Brezger, M. Arndt, A. Zeilinger, Phys. Rev. Lett. **87**, 160401 (2001)
57. B. Brezger, M. Arndt, A. Zeilinger, J. Opt. B. **5**, 82 (2003)
58. S. Gerlich, S. Eibenberger, M. Tomandl, S. Nimmrichter, K. Hornberger, P. Fagan, J. Tüxen, M. Mayor, M. Arndt, Nat. Commun. **2**, 263 (2011)
59. S. Eibenberger, S. Gerlich, M. Arndt, M. Mayor, J. Tüxen, Phys. Chem. Chem. Phys. **15**, 14696 (2013)
60. S. Nimmrichter, K. Hornberger, P. Haslinger, M. Arndt, Phys. Rev. A **83**, 043621 (2011)
61. P. Haslinger, N. Dörre, P. Geyer, J. Rodewald, S. Nimmrichter, M. Arndt, Nat. Phys. **9**, 144–148 (2013)
62. R. Abfalterer, S. Bernet, C. Keller, M. Oberthaler, J. Schmiedmayer, A. Zeilinger, Act. Phys. Slov. **47**(3/4), 165 (1997)
63. S. Fray, C.A. Diez, T.W. Hänsch, M. Weitz, Phys. Rev. Lett. **93**, 24 (2004)
64. N. Dörre, J. Rodewald, P. Geyer, B. von Issendorff, P. Haslinger, M. Arndt, Phys. Rev. Lett. **113**, 233001 (2014)
65. J. Schätti, P. Rieser, U. Sezer, G. Richter, P. Geyer, G.G. Rondina, D. Häussinger, M. Mayor, A. Shayeghi, V. Köhler, M. Arndt, Commun. Chem. **1**(1), 93 (2018)
66. A. Shayeghi, P. Rieser, G. Richter, U. Sezer, J. Rodewald, P. Geyer, T. Martinez, M. Arndt, Nat. Commun. **11**, 144 (2020)
67. Y.Y. Fein, P. Geyer, P. Zwick, F. Kiałka, S. Pedalino, M. Mayor, S. Gerlich, M. Arndt, Nat. Phys. (2019)
68. J. Schätti, M. Kriegleder, M. Debiossac, M. Kerschbaum, P. Geyer, M. Mayor, M. Arndt, V. Köhler, Chem. Commun. (Camb) **55**(83), 12507 (2019)
69. F. Kiałka, B. Stickler, K. Hornberger, Y.Y. Fein, P. Geyer, L. Mairhofer, S. Gerlich, M. Arndt, Physica Scripta (2018)
70. R. Schäfer, S. Schlecht, J. Woenckhaus, J.A. Becker, Phys. Rev. Lett. **76**(3), 471 (1996)
71. K. Bonin, V. Kresin, *Electric-Dipole Polarizabilities of Atoms, Molecules and Clusters* (World Scientific, 1997)
72. R. Antoine, I. Compagnon, D. Rayane, M. Broyer, P. Dugourd, N. Sommerer, M. Rossignol, D. Pippen, F.C. Hagemeister, M.F. Jarrold, Anal. Chem. **75**, 5512 (2003)
73. W.A. de Heer, V.V. Kresin, *Electric and Magnetic Dipole Moments of Free Nanoclusters* (CRC Press, 2011), pp. 10/1–13
74. T.M. Fuchs, R. Schäfer, Phys. Rev. A **98**, 6 (2018)
75. F. Filsinger, J. Kupper, G. Meijer, J.L. Hansen, J. Maurer, J.H. Nielsen, L. Holmegaard, H. Stapelfeldt, Angew Chem. Int. Ed. Engl. **48**(37), 6900 (2009)
76. Y.P. Chang, K. Dlugolecki, J. Küpper, D. Rösch, D. Wild, S. Willitsch, Science **342**(6154), 98 (2013)
77. E. Gershnabel, M. Shapiro, I. Averbukh, J. Chem. Phys. **135**(19), 194310 (2011)
78. M. Abd El Rahim, R. Antoine, L. Arnaud, M. Barbaire, M. Broyer, C. Clavier, I. Compagnon, P. Dugourd, J. Maurelli, D. Rayane, Rev. Sci. Instrum. **75**(12), 5221 (2004)
79. M.K. Oberthaler, S. Bernet, E.M. Rasel, J. Schmiedmayer, A. Zeilinger, Phys. Rev. A **54**, 3165 (1996)
80. J.F. Clauser, S. Li, Phys. Rev. A **49**, R2213 (1994)
81. K. Hornberger, S. Gerlich, P. Haslinger, S. Nimmrichter, M. Arndt, Rev. Mod. Phys. **84**, 157 (2012)
82. Y.Y. Fein, A. Shayeghi, L. Mairhofer, F. Kiałka, P. Rieser, P. Geyer, S. Gerlich, M. Arndt, Phys. Rev. X **10**, 011014 (2020)
83. M.A. Bouchiat, C. Bouchiat, Phys. Rev. A **83**, 5 (2011)
84. K. Zeiske, G. Zinner, F. Riehle, J. Helmcke, Appl. Phys. B **60**, 205 (1995)
85. E. Cohen, H. Larocque, F. Bouchard, F. Nejadsattari, Y. Gefen, E. Karimi, Nat. Rev. Phys. **1**(7), 437 (2019)
86. S. Lepoutre, A. Gauguet, M. Büchner, J. Vigué, Phys. Rev. A **88**, 4 (2013)
87. U. Sezer, L. Wörner, J. Horak, L. Felix, J. Tüxen, C. Götz, A. Vaziri, M. Mayor, M. Arndt, Anal. Chem. **87**, 5614–5619 (2015)

88. M. Gring, S. Gerlich, S. Eibenberger, S. Nimmrichter, T. Berrada, M. Arndt, H. Ulbricht, K. Hornberger, M. Müri, M. Mayor, M. Böckmann, N.L. Doltsinis, Phys. Rev. A **81**, 031604(R) (2010)

89. J. Tüxen, S. Eibenberger, S. Gerlich, M. Arndt, M. Mayor, Eur. J. Organ. Chem. (25), 4823 (2011)

90. P. Schmid, F. Stöhr, M. Arndt, J. Tüxen, M. Mayor, J. Am. Soc. Mass Spectrom. **24**(4), 602 (2013)

91. J. Tüxen, S. Gerlich, S. Eibenberger, M. Arndt, M. Mayor, Chem. Commun. **46**(23), 4145 (2010)

92. B.C. Das, S.D. Gero, E. Lederer, Biochem. Biophys. Res. Commun. **29**(2), 211 (1967)

93. J. Schätti, V. Köhler, M. Mayor, Y.Y. Fein, P. Geyer, L. Mairhofer, S. Gerlich, M. Arndt, J. Mass Spectrom. **55**(6), e4514 (2020). E4514 JMS-19-0196.R2

94. A. Akhmetov, J.F. Moore, G.L. Gasper, P.J. Koin, L. Hanley, J. Mass Spectrom. **45**(2), 137 (2010)

95. O. Nairz, M. Arndt, A. Zeilinger, J. Modern Opt. **47**(14–15), 2811 (2000)

96. M. Marksteiner, P. Haslinger, M. Sclafani, H. Ulbricht, M. Arndt, J. Phys. Chem. A **113**(37), 9952 (2009)

97. J. Schätti, U. Sezer, S. Pedalino, J.P. Cotter, M. Arndt, M. Mayor, V. Köhler, J. Mass Spectrom. **52**, 550 (2017)

98. M. Debiossac, J. Schätti, M. Kriegleder, P. Geyer, A. Shayeghi, M. Mayor, M. Arndt, V. Köhler, Phys. Chem. Chem. Phys. **20**, 11412 (2018)

99. A. Stefanov, M. Berninger, M. Arndt, Meas. Sci. Technol. **19**, 5 (2008)

100. L. Mairhofer, S. Eibenberger, A. Shayeghi, M. Arndt, Entropy **20**, 516 (2018)

101. M. Berninger, A. Stefanov, S. Deachapunya, M. Arndt, Phys. Rev. A **76**, 013607 (2007)

102. Y.Y. Fein, P. Geyer, F. Kiałka, S. Gerlich, M. Arndt, Phys. Rev. Res. **1**, 033158 (2019)

103. L. Mairhofer, S. Eibenberger, J.P. Cotter, M. Romirer, A. Shayeghi, M. Arndt, Angew. Chem. Int. Ed. **56**, 10947 (2017)

104. S. Eibenberger, S. Gerlich, M. Arndt, J. Tüxen, M. Mayor, New J. Phys. **13**(4), 043033 (2011)

105. S. Gerlich, M. Gring, H. Ulbricht, K. Hornberger, J. Tüxen, M. Mayor, M. Arndt, Angew Chem. Int. Ed. Engl. **47**(33), 6195 (2008)

106. Y.Y. Fein, A. Shayeghi, F. Kiałka, P. Geyer, S. Gerlich, M. Arndt, Phys. Chem. Chem. Phys. pp. 14,036–14,041 (2020)

107. L. Hackermüller, K. Hornberger, S. Gerlich, M. Gring, H. Ulbricht, M. Arndt, Appl. Phys. B **89**(4), 469 (2007)

108. K. Hornberger, S. Gerlich, H. Ulbricht, L. Hackermüller, S. Nimmrichter, I. Goldt, O. Boltalina, M. Arndt, New J. Phys. **11**, 043032 (2009)

109. J.P. Cotter, S. Eibenberger, L. Mairhofer, X. Cheng, P. Asenbaum, M. Arndt, K. Walter, S. Nimmrichter, K. Hornberger, Nat. Commun. **6**, 7336 (2015)

110. S. Eibenberger, X. Cheng, J.P. Cotter, M. Arndt, Phys. Rev. Lett. **112**, 250402 (2014)

111. C. Brand, B.A. Stickler, C. Knobloch, A. Shayegh, K. Hornberger, M. Arndt, Phys. Rev. Lett. **121**, 173002 (2018)

112. J. Rodewald, P. Haslinger, N. Dörre, B.A. Stickler, A. Shayeghi, K. Hornberger, M. Arndt, Appl. Phys. B **123**(1), 3 (2017)

113. J. Rodewald, N. Dörre, A. Grimaldi, P. Geyer, L. Felix, M. Mayor, A. Shayeghi, M. Arndt, New J. Phys. **20**, 033016 (2018)

114. Y.Y. Fein, F. Kiałka, P. Geyer, S. Gerlich, M. Arndt, New J. Phys. **22**, 033013 (2020)

115. M. Gregoire, N. Brooks, R. Trubko, A. Cronin, Atoms **4**(3), 21 (2016)

116. J.V. Vleck, *The Theory of Electric and Magnetic Susceptibilities* (Oxford University Press London, 1965)

117. H. Ulbricht, M. Berninger, S. Deachapunya, A. Stefanov, M. Arndt, Nanotechnology **19**, 045502 (2008)

118. W.D. Knight, R. Monot, E.R. Dietz, A.R. George, Phys. Rev. Lett. **40**, 1324 (1978)

119. U. Rohrmann, R. Schafer, Phys. Rev. Lett. **111**(13), 133401 (2013)

120. O. Amit, Y. Margalit, O. Dobkowski, Z. Zhou, Y. Japha, M. Zimmermann, M.A. Efremov, F.A. Narducci, E.M. Rasel, W.P. Schleich, R. Folman, Phys. Rev. Lett. **123**, 083601 (2019)
121. Y.Y. Fein, Long-baseline universal matter-wave interferometry. Thesis (2020)
122. S. Nimmrichter, K. Hornberger, Phys. Rev. A **78**, 023612 (2008)
123. R.P. Cameron, S.M. Barnett, A.M. Yao, New J. Phys. **16** (2014)

Chapter 25
Grating Diffraction of Molecular Beams: Present Day Implementations of Otto Stern's Concept

Wieland Schöllkopf

Abstract When Otto Stern embarked on molecular-beam experiments in his new lab at Hamburg University a century ago, one of his interests was to demonstrate the wave-nature of atoms and molecules that had been predicted shortly before by Louis de Broglie. As the effects of diffraction and interference provide conclusive evidence for wave-type behavior, Otto Stern and his coworkers conceived two *matter-wave* diffraction experiments employing their innovative molecular-beam method. The first concept assumed the molecular ray to coherently scatter off a plane ruled grating at grazing incidence conditions, while the second one was based on the coherent scattering from a cleaved crystal surface. The latter concept allowed Stern and his associates to demonstrate the wave behavior of atoms and molecules and to validate de Broglie's formula. The former experiment, however, fell short of providing evidence for diffraction of matter waves. It was not until 2007 that the grating diffraction experiment was retried with a modern molecular-beam apparatus. Fully resolved matter-wave diffraction patterns were observed, confirming the viability of Otto Stern's experimental concept. The correct explanation of the experiment accounts for *quantum reflection*, another wave effect incompatible with the particle picture, which was not foreseen by Stern and his contemporaries.

1   Introduction

The time when Otto Stern and his coworkers at the University of Hamburg were running their pioneering molecular-beam experiments almost a century ago, saw disruptive breakthroughs in quantum physics, experimental and theoretical alike. Among the latter was arguably the work of the french physicist Louis de Broglie on the wave nature of massive particles [1]. He came forward with a rather simple formula for the wavelength λ_{dB} of a *matter wave*, predicting that it equals the product

W. Schöllkopf (✉)
Fritz-Haber-Institut der Max-Planck-Gesellschaft, Faradayweg 4-6, 14195 Berlin, Germany
e-mail: wschoell@fhi-berlin.mpg.de

© The Author(s) 2021
B. Friedrich and H. Schmidt-Böcking (eds.), *Molecular Beams in Physics and Chemistry*,
https://doi.org/10.1007/978-3-030-63963-1_25

of Heisenberg's constant h and the inverse of the classical momentum $p = mv$ of a particle of mass m and velocity v.

$$\lambda_{\mathrm{dB}} = \frac{h}{mv} \tag{1}$$

Given the boldness of the concept of matter waves combined with the simplicity of de Broglie's formula it is not surprising that experimentalists—and a theorist-turned experimentalist such as Otto Stern in particular—must have felt challenged to seek experimental evidence for the existence of de Broglie's matter waves.

Diffraction and interference are unambiguous manifestation of wave-type behavior. As such, Otto Stern and his coworkers conceived two matter-wave diffraction experiments employing their molecular beam method. They had to cope with the fact that, according to Eq. (1), a typical de Broglie wavelength, even for lightweight atoms, at room temperature conditions is in the sub-nanometer regime. Observation of diffraction effects for a wavelength that small requires diffractive optical elements, such as gratings or grids, of a similarly small periodicity. In their article UzM[1] no. 11 *Über die Reflexion von Molekularstrahlen* (Fig. 1) that appeared in Zeitschrift für Physik in 1929 [2] Friedrich Knauer and Otto Stern outlined the two methods they considered promising and they pursued for observing diffraction of a molecular beam. The first one assumes the atoms or molecules to scatter off of a plane ruled (machined) grating at grazing incidence conditions, while the second one was based on the scattering from a cleaved crystal surface.

The latter method was essentially analogous to X-ray diffraction from a crystal lattice, a phenomenon already well known by the mid 1920s. Its first observation by Max von Laue, Walter Friedrich, and Paul Knipping in 1912 provided conclusive evidence for both the wave nature of X-rays and the periodic structure of a crystal lattice (for an historical account on von Laue's experiment see Ref. [3]). X-ray wavelengths are of the same order of magnitude as typical de Broglie wavelengths of light atoms at thermal energies. Crystal lattices, with sub-nanometer periodicity, present a natural match to these wavelengths resulting in comparatively large diffraction angles. Obviously, the main difference is that molecular beams, unlike X-rays, cannot penetrate a crystal. Thus, Otto Stern's approach relies on reflection of a molecular beam from a cleaved crystal surface, which needs to be clean and well-ordered on the atomic scale to allow for coherent scattering of atoms or molecules. While Knauer and Stern where able to observe specular reflection from a crystal surface, they could not present convincing evidence for diffraction of the molecular beam by the periodic crystal lattice. It was several months later, after some improvement of the experimental setup, that Otto Stern together with Immanuel Estermann was able to present unambiguous evidence for diffraction [4]. This work from Otto Stern's molecular beam lab provided the first definite evidence for matter-wave behavior of

---

[1]UzM stands for *Untersuchungen zur Molekularstrahlmethode*, the series of publications from Stern's molecular-beam lab in Hamburg termed *Investigations by the Molecular Ray Method*, c.f. *Otto Stern's Molecular Beam Method and its Impact on Quantum Physics* by Bretislav Friedrich and Horst Schmidt-Böcking in this volume.

(Untersuchungen zur Molekularstrahlmethode aus dem Institut für physikalische Chemie der Hamburgischen Universität. Nr. 11.)

Über die Reflexion von Molekularstrahlen *.

Von **F. Knauer** und **O. Stern** in Hamburg.

Mit 7 Abbildungen.   (Eingegangen am 24. Dezember 1928.)

Molekularstrahlen aus H_2 und He werden an hochpolierten Flächen bei nahezu streifendem Einfall spiegelnd reflektiert. Das Verhalten des Reflexionsvermögens ist in Übereinstimmung mit der de Broglieschen Wellentheorie. Die Versuche, Beugung an Strichgittern nachzuweisen, gaben noch kein Resultat. Auch bei Steinsalzspaltflächen wurde (bei steilerem Einfall) spiegelnde Reflexion gefunden. Die an Kristallspaltflächen beobachteten Erscheinungen sind wahrscheinlich als Beugung aufzufassen, wenngleich ihre vollständige Deutung noch aussteht.

Fig. 1 Front page of the UzM paper no. 11 *"On the reflection of molecular beams"* by Friedrich Knauer and Otto Stern as it appeared in Zeitschrift für Physik in 1929 (received by the journal on December 24, 1928) [2]. In English the abstract states: *Molecular beams of H_2 und He are specularly reflected from highly polished surfaces under near grazing incidence. The reflectivity behavior is in agreement with de Broglie's wave theory. Attempts to find evidence for diffraction from a ruled grating have not yet yielded results. For cleaved surfaces of rock salt (at steeper incidence) reflection was found as well. The phenomena observed with cleaved crystal surfaces are likely due to diffraction, albeit a complete interpretation is still missing*

atoms and molecules. In addition, Estermann and Stern were able to quantitatively check and validate Louis de Broglie's wavelength formula, Eq. (1), for atoms and molecules [4–6].

The success of Knauer's and Stern's second method leaves us with the question: What about the first concept they had conceived to demonstrate matter-wave diffraction with molecular beams; diffraction of a molecular beam reflected from a machined line grating under grazing incidence conditions? The remainder of this contribution will focus on this experiment. In Chap. 2 we will review how Knauer and Stern designed and implemented the grating diffraction experiment and see what results they got with their 1928 molecular beam apparatus. In Chap. 3 we will describe the modern implementation of the experiment, and we will discuss the explanation of the results accounting for quantum reflection. Quantum reflection from the attractive long-range branch of the atom–surface interaction potential is another quantum-wave phenomenon which is incompatible with a particle description. Quantum reflection was not foreseen by 1928, although it is a direct consequence of and evidence for quantum-wave behavior just as diffraction is.

2   The Grating Diffraction Experiment by Knauer
and Stern in 1928

In the first paragraph of the UzM no. 11 article Knauer and Stern describe their considerations regarding the two experimental methods they are pursuing with the aim to observe matter-wave diffraction. The original German text reads:

> Der Nachweis der Wellennatur schien uns am bequemsten mit einem Gitter zu führen zu sein. Am nächsten läge es, an die bei den Röntgenstrahlen mit so großem Erfolg benutzten Kristallgitter zu denken. Doch ist von vornherein schwer zu übersehen, ob hier Reflexion und Beugung auftreten werden, weil es im Gegensatz zu den Röntgenstrahlen bei den de Broglie-Wellen auf den Potentialverlauf an der äußeren Grenze der Kristalloberfläche ankommt. Wir beabsichtigen deshalb, optische Strichgitter zu benutzen. Hierfür ist die Voraussetzung, dass es zunächst gelingt, Molekularstrahlen spiegelnd zu reflektieren.

For convenience we are here providing an English translation to the unfortunate reader who is not proficient in German.

> Providing evidence of the wave nature appeared to be most straightforward by using a grating. The most obvious choice would be a crystal lattice which has been used to great success with X-rays. It is, however, difficult to predict if reflection and diffraction will occur, because for de Broglie waves, in contrast to X-rays, the shape of the interaction potential at the outer limit of the crystal surface matters. That is why we intend to use optical ruled gratings. It is prerequisite to first succeed in observing specular reflection of molecular beams.

Apparently, Knauer and Stern were not sure if and to what extent the interaction potential that an atom is exposed to at a crystal surface would possibly impede the observation of diffraction. Therefore, they intended to (also) employ optical ruled gratings in their quest for matter-wave diffraction. For this approach to work it is prerequisite to achieve mirror-like (specular) reflection.

In the following lines Knauer and Stern describe why grazing incidence represents a pivotal aspect of the experimental design. Firstly, the grazing incidence geometry allows for the use of a large (macroscopic) grating period length in the range of 0.01–0.1 mm to diffract wavelengths as small as 0.1 nm. At grazing incidence the effective period that determines the diffraction angles is given by the projection of the grating period along the direction of motion of the incoming molecular beam. Thus, for a grazing angle of 1 mrad the effective period of a 10-micron-period grating is as small as 10 nm. While this is still roughly a factor of 100 larger than the de Broglie wavelengths we are dealing with, it results in diffraction angles of several milliradian, which are well within the experimental resolution. Secondly, the surface roughness, which Knauer and Stern estimate to be on the order of 10–100 nm for their well polished surfaces, would prevent mirror-like reflection at steep incidence. However, as it is known from optics, even a rough surface becomes highly reflective at grazing incidence conditions. Knauer and Stern assume that the roughness times the sine of the incidence angle needs to be smaller than the de Broglie wavelengths if one wants to get good reflectivity. That is why they expect mirror-like reflection of molecular beams from their surfaces for milliradian incidence angles. Interestingly, in this consideration Knauer and Stern ignore a possible influence of the atom-surface interaction potential on the reflectivity.

2.1 Apparatus

The apparatus used by Knauer and Stern was already described in some detail in the preceding paper UzM no. 10 [7], which appeared back to back with their UzM no. 11 paper in volume 53 of Zeitschrift für Physik in 1929. The apparatus allowed them to generate molecular beams of various gases including He, H_2, Ne, Ar, CO_2. As can be seen in the original schematics replotted in Fig. 2, the main components included (i) the molecular beam source, (ii) the detector, (iii) the 3-slits beam collimation system, and (iv) the encasing vacuum system. It appears that all four components were state-of-the-art at that time representing significant improvements compared to previously used molecular beam setups.

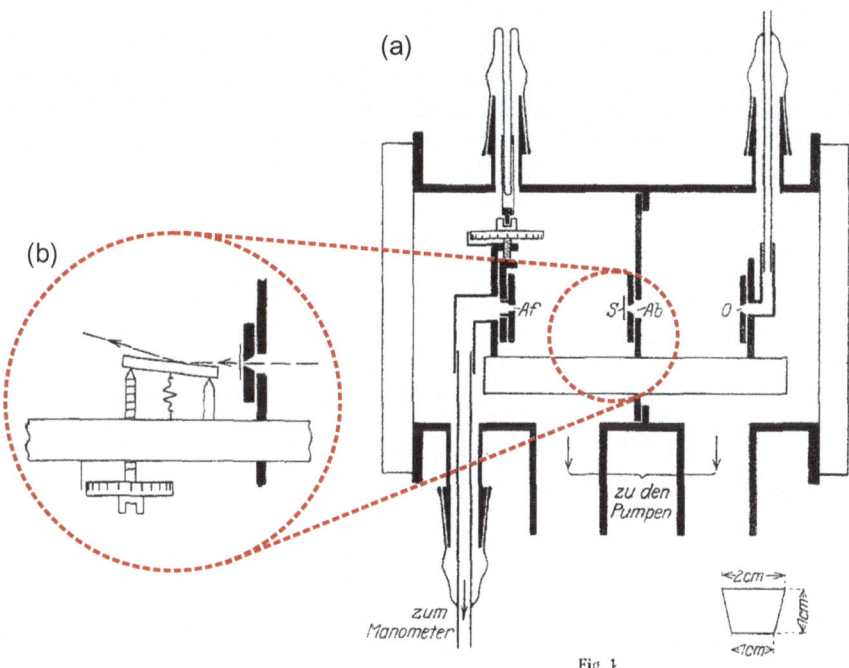

Fig. 2 Schematic of the experimental arrangement used by Knauer and Stern in 1928 copied from the original articles UzM no. 10 and 11. The dashed red circles indicate where the diffraction grating shown in the enlargement (**b**) (from Ref. [2]) was mounted in the molecular-beam apparatus shown in (**a**) (from Ref. [7]). The drawing in the lower right shows the trapezoidal cross section of the horizontal precision nickel-steel rod carrying the slits and the grating mount. The labels *Af*, *Ab*, and *O* denote the three collimating slits; *Auffängerspalt* (collector slit), *Abbildespalt* (imaging slit), and *Ofenspalt* (oven slit i.e. source slit). A shutter (*S*) is located downstream of the imaging slit. As described in the original text, a fourth slit *Vorspalt* (ante slit), not shown in the figure, was used in the grating diffraction experiments. It was located in between *O* and *Ab* and served as a differential pumping stage upstream of *Ab*

Molecular beam source: The gas was fed into a beam source from which it escaped through a narrow slit (10 to 20 μm wide) into the surrounding vacuum chamber. In today's jargon the source would be referred to as an effusive source, where the molecular beam inherits its velocity distribution from the Maxwell-Boltzmann distribution of the gas upstream the slit. Thus, unlike the supersonic free-jet expansion sources that have been available for the last few decades with their inherently narrow velocity distribution, the beams of Knauer and Stern were characterised by relatively large velocity spreads. This, via Eq. (1), corresponds to a wide distribution of de Broglie wavelengths. In other words, Knauer's and Stern's source did not generate monochromatic matter waves. Nonetheless, they were able to adjust the mean velocity and hence the mean de Broglie wavelength by as much as 50% by heating the source or by cooling it down to about 130 K.

Beam collimation and vacuum system: As in classical optics, matter-wave diffraction can only be observed, if spatial coherence is achieved in the experimental setup. This is done by three collimating slits: the first one defines the source (source slit described above); the second one limits the divergence of the beam; and the third one defines the angular resolution of the detector (detector slit described below). All three slits were 10–20 μm wide and 1 cm in height. From the schematic shown in Fig. 2b it can be seen that the second slit also served to effectively separate the vacuum of the right source chamber from the left beam chamber. The chambers were evacuated to a base vacuum pressure of 10^{-5} Torr by two mercury diffusion pumps made by Leybold [7].

Detector: In the Stern Gerlach experiment a few years earlier detection of the beam of silver atoms was accomplished by depositing the beam on a glass plate. After running the experiment for some time, one would check the thickness and location of the deposits. While this method allowed for the detection of the famous two-spot pattern in the Stern Gerlach experiment, it did not enable reliable quantitative measurements of beam intensities, not to mention the fact that the technique did not work with molecular beams of gases. Knauer's and Stern's detector, which is described in the UzM paper no. 10 [7], represents an enormous improvement. Their detector is essentially a Pitot tube with a narrow entrance slit (10–20 μm wide, just as the source slit). The stagnation pressure building up in the detector is measured by a modified Pirani-type vacuum gauge that Knauer and Stern were running at liquid nitrogen temperature. They were able to achieve an impressive absolute sensitivity on the order of 10^{-8} Torr [7]. With the base vacuum pressure in the 10^{-5} Torr regime, this translates into a relative sensitivity of 10^{-3} providing the required sensitivity for the diffraction experiment.

The optical elements—mirrors and gratings: Knauer and Stern were employing different materials for the mirrors they used in the reflectivity measurements: glass, steel, and speculum metal. The text states that each mirror was most thoroughly polished. The ruled gratings they employed in the diffraction experiments were all made from speculum metal with different periods of 10, 20 and 40 μm. No information is provided on how the ruling of the grating was done or what the groove shape might have been.

2.2 Results

Knauer and Stern observed specular reflection of He and H_2 beams from flat solid surfaces made out of any of the three materials. They found the reflectivity to slightly decrease from speculum metal to glass to steel, an observation they explained as a result of the somewhat different surface qualities achievable by polishing these materials. In addition, for each mirror they observed a strong decrease of reflectivity with increasing incidence angle. As an example they provide a reflectivity table for a beam of H_2 reflected from a speculum metal mirror including four data points decreasing from 5% at an incidence angle of 1 mrad to 0.75% at 2.25 mrad. Furthermore, they were able to increase the mean de Broglie wavelength of the molecular beam by cooling the beam source to −150 °C. They state that the observed reflectivity increased by a factor of 1.5 when the mean de Broglie wavelength of the molecular beam was increased by 1.5.

Despite the promising observation of mirror-like specular reflection of the molecular beam it was not possible for Knauer and Stern to observe diffraction when they were employing ruled gratings made out of speculum metal. They searched for diffraction signal in the incidence angle range from 0.5 to 3 mrad, but did not find reliable evidence:

> Wiewohl wir mehrmals Andeutungen eines Maximums gefunden zu haben glauben, gelang es uns nicht, sein Vorhandensein sicherzustellen.

> Although we believe to have seen hints of a [diffraction] peak several times, we were not able to verify its occurrence.

They make this clear in the article's abstract where they summarise *Die Versuche, Beugung an Strichgittern nachzuweisen, gaben noch kein Resultat*, which translates to English as *Attempts to find evidence for diffraction from a ruled grating have not yet yielded results*.

Apparently, Otto Stern planned to try matter-wave diffraction from ruled gratings again with an improved apparatus. As it is described on page 782 of UzM no. 11, a pump of "extremely high pumping speed" was already under construction for Stern's lab at the Leybold company. Stern hoped that the more powerful pump could further boost the molecular beam intensity. We do not know if this apparatus upgrade was implemented. By the time Stern had to leave Hamburg, diffraction from a ruled grating had not been observed in his lab. It would have been described, for instance, in the chapter that Robert Frisch and Otto Stern contributed to the Handbuch der Physik in 1933 [6]. In Peter Toennies's contribution *Otto Stern and Wave-Particle Duality* in this volume the reader will find an account of an interview Otto Stern gave in 1961, where he emphasised his interest in the grating diffraction experiment. As we will see in Sect. 3, Otto Stern's experimental approach was conceptually perfectly viable.

2.3 Historical Note on Friedrich Knauer

From a historical point of view it is intriguing to have a look at Friedrich Knauer, Otto Stern's assistant and only coauthor of the article UzM no. 11 on reflection from ruled gratings. The following basic biographical information is available at Knauer's *Wikipedia* entry: Born in Göttingen in 1897, Knauer studied in Göttingen and Hannover after the First Wold War. He was assistant to Robert Wichard Pohl in Göttingen in 1924, when he moved to Hamburg, where he worked as an associate of Otto Stern's until Stern left in 1933. In that year Knauer completed his Habilitation. He stayed at Hamburg University where he was appointed to the rank of associate professor in 1939. During the Second World War he contributed to the German nuclear project, the so-called *Uranverein*. After the war he continued working at the institute at Hamburg University till 1963. He died in 1979.

Unlike Otto Stern and Immanuel Estermann, Friedrich Knauer was not of Jewish decent. Thus, he was not forced to leave his position at the University in 1933. However, it is somewhat surprising that Friedrich Knauer signed the *Vow of Allegiance of the Professors of the German Universities and High-Schools to Adolf Hitler and the National Socialistic State*[2] that was presented to the public in Leipzig on November 11, 1933. Also Wilhelm Lenz was among the signatories. Lenz was director of the theoretical physics institute in Hamburg to which Wolfgang Pauli was associated. According to Isidor Rabi's recollections [8], Lenz must have had close ties to Stern's group (see also Ref. [9]). Why did Knauer and Lenz sign the Nazi allegiance? Was it out of conviction, fear or opportunism? And what does it mean with regard to Knauer's relation to Otto Stern and other group members like Estermann, who were forced to leave their positions just a few months earlier? After about nine years of fruitful collaboration on pioneering molecular beam experiments, wouldn't one expect some solidarity, or empathy at least, towards the former colleagues? While seeking answers to these questions is beyond the scope of this contribution, it appears doubtful that the few publicly available documents could provide hints to what the answers might be.

3 The Modern Implementation of the Knauer-Stern Experiment

From the two novel experimental methods introduced in UzM no. 11 only scattering from a crystal surface lead to the observation of matter-wave diffraction, while the scattering from a ruled grating did not. The former methods was used by Otto Stern an his associates in follow-up experiments and allowed them to observe, for instance, the appearance of anomalous dips in diffraction patterns [10]. This phenomenon was

---

[2]*Bekenntnis der Professoren an den Universitäten und Hochschulen zu Adolf Hitler und dem nationalsozialistischen Staat* überreicht vom Nationalsozialistischen Lehrerbund Deutschland, Gau Sachsen, 1933, Dresden-A. 1, Zinzendorfstr. 2.

later explained in terms of selective adsorption by Lennard-Jones and Devonshire [11] (see Peter Toennies's account on *Otto Stern and Wave-Particle Duality* in this volume), and it was developed into an important method of measuring atom-surface interaction potentials. Following the pioneering experiments in Stern's lab, helium atom scattering from crystal surfaces became a tool in surface science that has found widespread application in many labs around the word [12].

In contrast to the success story of scattering molecular beams off of crystal surfaces, the grating diffraction method was not further pursuit. Otto Stern might have tried the experiment again, had he been able to continue his work in Hamburg. But other groups working with molecular beams might not have considered the grating diffraction experiment worthwhile a try, because the de Broglie wave of atoms and molecules was now well established. As a result, Knauer's and Stern's grating diffraction experiment was pretty much forgotten with time.

The situation started to change in the 1990s when the new field of *atom optics and atom interferometry* [13, 14] emerged. The group of David Pritchard at MIT (Cambridge, MA, USA) first demonstrated diffraction of a beam of sodium atoms from a transmission grating [15]. The grating was a free-standing nanoscale structure with a period of only 200 nm. Fabrication of structures that small had become possible by enormous advances in micro-fanbrication and lithography techniques. Unlike in Knauer's and Stern's experiment, with a free-standing grating the molecular beam could pass through the sub-micron slits with no scattering and (almost) no interaction with the grating material. Thus, reflection from a solid surface was no longer prerequisite to observe the grating diffraction pattern. Subsequently, nanoscale transmission gratings were used in a variety of diffraction experiments with molecular beams of atoms, molecules and clusters including; Na_2 [16], metastable He^* [17], ground-state He, He_2, and He_3 [18, 19], rare-gas atoms [20], CH_3F and CHF_3 molecules [21]. In addition, Markus Arndt and his coworkers at Vienna University were able to observe diffraction patterns of massive and complex particles starting with C_{60} fullerenes [22] (see Markus Arndt's contribution in this volume). A comprehensive review of transmission-grating diffraction experiments as well as the use of transmission gratings in atom and molecular interferometers can be found in the literature [14].

While diffraction of molecular beams from nanoscale transmission gratings was done by several groups, the original Knauer and Stern experiment was not tried until 2007. In the following we describe the setup of the modern-day implementation of the experiment, the results observed and their interpretation.

3.1 Experimental Setup

The molecular-beam apparatus we used in 2007 is shown schematically in Fig. 3 [23]. The legacy of Otto Stern's molecular-beam experiments is still apparent although, of course, various technical implementations of the beam source, detector, vacuum

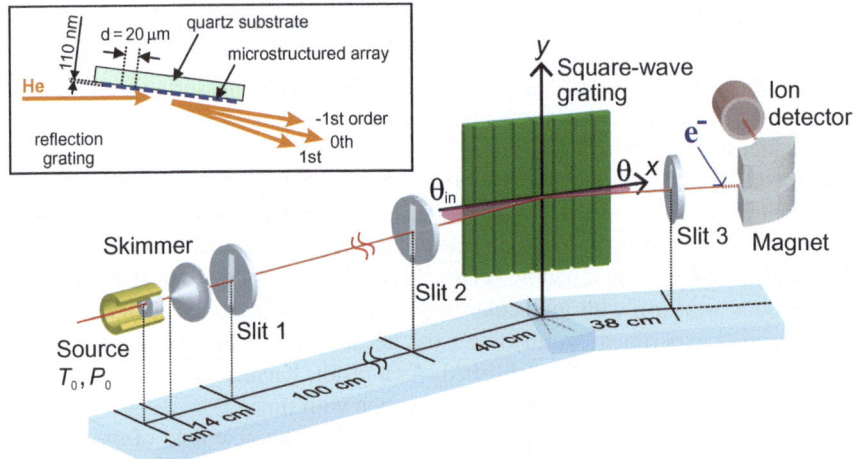

Fig. 3 Schematic of the experimental arrangement used by Zhao et al. in 2007 to observe diffraction of a molecular beam scattering coherently from a reflection grating under grazing incidence conditions [24]. Conceptually, the modern apparatus is essentially analogous to the setup used by Knauer and Stern in 1928, while the technical implementations are different due to, mainly, advances in vacuum and micro-fabrication technologies and in electronics. The inset in the upper left of the figure shows a sketch of the diffraction beams generated at the grating. We use the convention of negative diffraction orders being closer to the grating surface than the specular beam. The incidence angle θ_{in} and detection angle θ are defined with respect to the grating surface plane

system, and the grating are different, not to mention the computerised data acquisition method that was not available to Knauer and Stern.

The beam is formed by free-jet expansion of pure ^{4}He gas from a source cell (stagnation temperature T_0 and pressure P_0) through a 5-μm-diameter orifice into high vacuum. As indicated in Fig. 3, the beam is collimated by two narrow slits, each 20 μm wide, located 15 cm and 115 cm downstream from the source. A third 25-μm-wide detector-entrance slit, located 38 cm downstream from the grating, limits the angular width of the atomic beam to a full width at half maximum of $\approx$ 120 μrad. The detector, which is an electron-impact ionization mass spectrometer, can be rotated precisely around the angle θ indicated in Fig. 3. The reflection grating is positioned at the intersection of the horizontal atom beam axis and the vertical detector pivot axis such that the incident beam approaches under grazing incidence (incident grazing angle $\theta_{in} \leq 20$ mrad), with the grating lines oriented parallel to the pivot axis. Diffraction patterns are measured for fixed incidence angle by rotating the detector around θ and measuring the signal at each angular position. In addition, the grating can be removed from the beam path all together making it possible to measure the direct beam profile, i.e. the undisturbed incident beam, as a function of θ.

Comparison with Stern's apparatus reveals two essential differences beyond mere technological advancements. The first one is the modern molecular beam source. As a consequence of the high stagnation pressure in the source cell combined with the very small 5-μm-diameter aperture, the collision rate of the He atoms in the expanding gas is very large. As a result, the expansion is adiabatic and isentropic leading to a rapid cool down of the gas [25]. A low temperature in the gas is equivalent to a narrow velocity spread in the beam's velocity distribution. The narrowness is often quantified by the *speed ratio* which represents the ratio of mean beam velocity to velocity width. For helium, temperatures below 1 mK and speed ratios of several hundreds have been observed [26, 27]. The expansion efficiently transfers the kinetic energy of the helium gas in the source cell into uniform, directional motion of the molecular beam. As a consequence of de Broglie's formula, Eq. (1), a narrow velocity distribution implies a narrow wavelengths distribution. In other words, the helium beam generated by the modern source is, effectively, monochromatic. In contrast, the effusive beams in the 1920s were characterised by broad velocity and wavelength distributions.

A second qualitative improvement common in most if not all modern-day molecular-beam apparatus is the ionization detector. The neutral He atoms entering the detector are first ionized in collision with electrons of $\simeq 100$ eV kinetic energy. The ions are accelerated by high voltage, mass selected in a magnetic field and finally efficiently detected using a multiplier. The bottle neck of this detection scheme is the inefficient ionization step. It is assumed that only 10^{-6} to 10^{-5} of the neutral He atoms are ionized. However, this poor detection efficiency allows for a far better sensitivity than the Pitot-tube detection scheme that was available to Knauer and Stern in 1928. Interestingly, in an interview given to John L. Heilbron in December 1962 Immanuel Estermann describes that he was trying to build an electron-impact ionization detector in Stern's lab in the early 1920s [28]:

> One of the big problems in molecular beam technology was the (detector) end; we couldn't detect very well. (...) Then, I think, the great step forward was the Langmuir-Taylor detector which was worked out when Taylor was a visiting scientist in Hamburg. But that works only for a limited number of elements or substances. Then we tried all kinds of things, and one of the things that I tried is what is now known as the cross-fire method; it means to bombard the neutral atoms with electrons, thus ionizing them, and then collect the ions. Ions are far easier to detect than the neutral particles. But I did not succeed; I did not get it to work. It's a method which is now used in a number of places quite successfully, but it required much better vacuum technology and much better electronic technology than was available in those days. This must have been about 1923 or '24 when I tried this.

The reflection grating used in 2007 consists of a 56-mm-long micro-structured array of 110-nm-thick, 10-μm-wide, and 5-mm-long parallel chromium strips on a flat quartz substrate. It was made from a commercial chromium mask blank by e-beam lithography. As shown in the inset of Fig. 3 the center-to-center distance of the strips, and thereby the period d, is 20 μm. Given this geometry the quartz surface between the strips is completely shadowed by the strips for all the incidence angles used. We expect a chromium oxide surface to have formed while the grating was exposed to air before mounting it in the apparatus where the ambient vacuum is about 8×10^{-7} mbar. No in-situ surface preparation was done.

3.2   *Results*

Figure 4 shows a series of diffraction patterns measured with the source kept at $T_0 = 20$ K corresponding to a de Broglie wavelength of $\lambda_{dB} = 2.2$ Å. The incident grazing angle θ_{in} was varied from 3 up to 15 mrad. In each diffraction pattern the specular reflection (0th diffraction-order peak) appears as the strongest peak, for which the detection angle is equal to the incident grazing angle. The intensity of the specular peak decreases continuously from about 600 counts/s at $\theta_{in} = 3.1$ mrad to only 13 counts/s at $\theta_{in} = 15.2$ mrad. At $\theta_{in} = 3.1$ mrad at least seven positive-order diffraction peaks can be seen at angles larger than the specular angle (diffraction 'away from' the surface).

It is straightforward to calculate the nth-order diffraction angle θ_n for given incidence angle θ_{in}, grating period d, and de Broglie wavelength λ_{dB} from the grating equation $\cos(\theta_{in}) - \cos(\theta_n) = n\frac{\lambda_{dB}}{d}$ well known from classical optics [29]. The calculated diffraction angles agree with the observed ones within the experimental error confirming the interpretation of the peaks as grating-diffraction peaks [24]. Note that with increasing incidence angle the negative-order diffraction peaks appear succes-

Fig. 4 Diffraction patterns observed for He atom beams of $\lambda_{dB} = 2.2$ Å de Broglie wavelength scattered from a plane grating with 20 μm period at various incidence angles from 3 to 15 mrad (from Ref. [24]). Numbers indicate the diffraction-order assigned to the peaks. In each spectrum the specular peak is most intense. Its peak height decays rapidly with increasing incidence angle

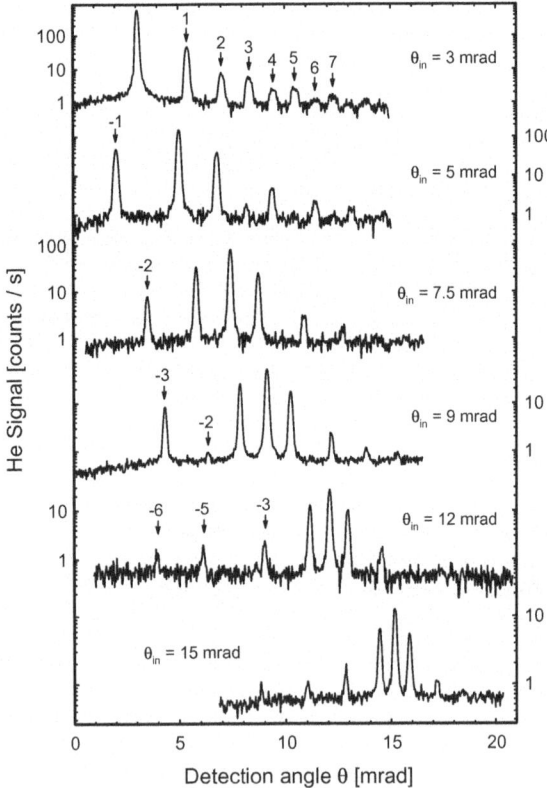

sively, emerging from the grating surface. Emergence of a new diffraction beam comes along with an abrupt redistribution of the flux among the diffraction peaks. These emerging-beam resonances have been studied for He atom beams diffracted by a blazed ruled grating [30].

The relative diffraction peak intensities change significantly with incident grazing angle. For instance, for $\theta_{in} = 3.1$ mrad even and odd order peaks have similar heights falling off almost monotonously with increasing diffraction order. With increasing incident grazing angle, however, the positive even-order diffraction peaks tend to disappear. Moreover, a distinct peak-height variation can be seen for the -2nd-order peak which decreases sharply when θ_{in} is increased from 7.4 to 9.1 mrad.

A diffraction pattern as the ones in Fig. 4 must have been exactly what Otto Stern was longing to see. The data shown here demonstrates that Stern's concept was perfectly viable. What prevented Knauer and Stern from observing diffraction with a machined grating was the limited experimental technology of their time that did not provide the required high beam flux, efficient detection, and high vacuum conditions.

3.3 Quantum Reflection

The peak heights decaying rapidly with increasing incidence angle confirm Otto Stern's conjecture that the highest reflectivity is to be found for the most grazing incidence conditions. A quantitative analysis of the reflectivity dependence on incidence angle is shown in Fig. 5. The reflectivity was determined by summing up the areas of all the peaks in a diffraction pattern. The sum was divided by the area of the direct beam to give the reflectivity. The direct beam (not shown in the plots) is measured when the grating is completely removed from the beam path. The reflectivity at various incidence angles and source temperatures is plotted in Fig. 5 in a semi-logarithmic plot as a function of the incident atoms' wave-vector component perpendicular to the surface, k_{perp}. It is apparent that all the data points fall on a single curve. At small perpendicular wave-vector up to about 0.11 nm^{-1} the curve decays rapidly, while at values larger than about 0.13 nm^{-1} it decays at reduced rate.

The same characteristic reflectivity behavior is also found when the diffraction grating is replaced by a plane surface [31]. This behavior cannot be understood by considering a wave scattering off of a rough surface, if one does not include the subtle effects of the atom–surface interaction. The expected dependence in the absence of a surface potential is also plotted in Fig. 5. While the reflectivity in the non-interaction model also tends to unity in the limit of vanishing perpendicular wave vector, this model obviously fails to describe the reflectivity dependence on k_{perp}.

A decent agreement with the observed data is found when quantum reflection from the atom–surface interaction potential is considered. Quantum reflection, just as diffraction, is a wave effect not compatible with the particle picture. The basic idea behind quantum reflection is illustrated in Fig. 6a, b, where the atom–surface interaction potential $V(z)$ is approximated by a square well potential, i.e., $V(z) = \infty$

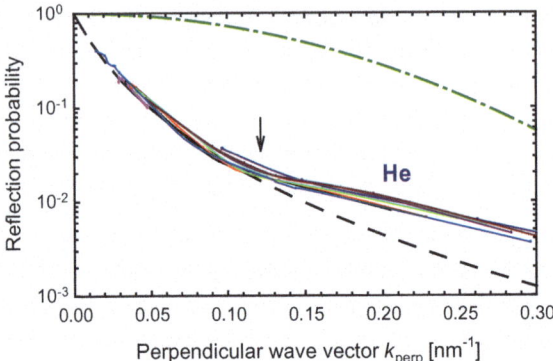

Fig. 5 Reflection probability of He atoms of various de Broglie wavelengths scattering from a plane grating with 20 μm period at a variety of grazing incidence angles plotted as a function of the normal component of the incident atom's wave vector [24]. Each color corresponds to a different wavelength. The dashed line presents a 1-dimensional quantum reflection calculation fitted to the data points at small perpendicular wave vector (left of the arrow). The dash-dotted line represents the reflection probability calculated for a wave scattering from a rough surface (4 nm root-mean-square roughness) in the absence of an atom–surface interaction potential

for $z < a$, $V(z) = 0$ for $z > b$, and $V(z) = V_0$, where V_0 is negative, for $a \leq z \leq b$. Here, the variable z denotes the atom to surface distance, and a and b are positive constants.

Within the classical particle description, Fig. 6a, an incident He atom approaching the well region from positive z with initial kinetic energy E_{kin} will gain energy upon entering the well at $z = b$; its kinetic energy is increased by exactly the well depth V_0. With correspondingly larger velocity the particle slams onto the steep repulsive wall where it is scattered back at the classical turning point. Upon leaving the well its kinetic energy is reduced to its initial value.

The description looks different in the quantum-wave picture, Fig. 6b, if the initial kinetic energy E_{kin} is sufficiently small such that quantum effects become observable. We then have to deal with a wave approaching a step in the potential at $z = b$. As quantum mechanics teaches us, in this situation there is a non-vanishing probability for reflection at the step (even for a "step down" as in our system). This quantum-wave reflection probability increases with increasing de Broglie wavelength of the incident atom. For a discontinuous step, as the one shown in Fig. 6b, it even approaches unity in the limit of vanishing kinetic energy.

The square-well model is a simplistic approximation to an atom–surface potential. In a more realistic model the steps will necessarily be smoothed out, as indicated in the depiction shown in Fig. 6c. Even then, there will be an appreciable reflection probability at the attractive branch of the potential as long as the incident energy of the atom is sufficiently small. This reflection mechanism of matter waves, referred to as quantum reflection, occurs in absence of a classical turning point and, paradoxically,

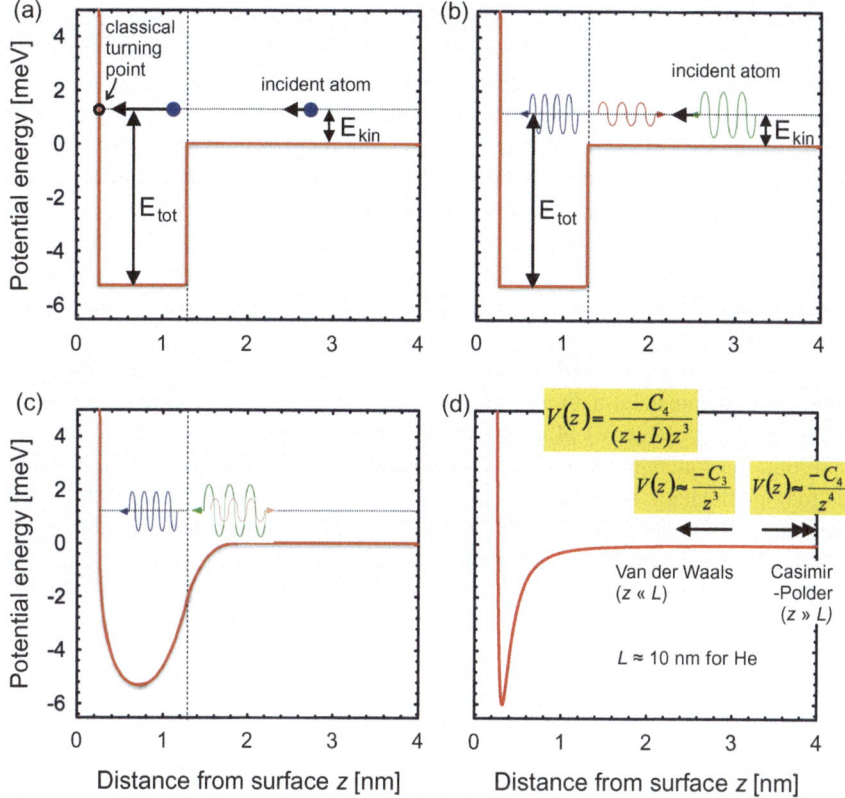

Fig. 6 Illustration of quantum reflection of an atom from a solid surface. In **a** and **b** the atom–surface interaction potential is approximated by a square-well model combined with the classical particle picture (**a**) and the quantum wave picture (**b**). In **c** the square well is smoothened, while **d** shows the potential between a He atom and a silver surface as an actual example. The long-range attractive branch of the potential exhibits a transition from the van der Waals regime at intermedium z to the Casimir–Polder regime at very large z

in the absence of a repulsive force acting on the incoming atom. The latter aspect in particular is counter-intuitive and incompatible with the classical description.

A realistic atom–surface potential model is displayed in Fig. 6d. The example shown describes the interaction between a He atom and a silver surface with a well depth of about 6 meV. The attractive potential branch is modelled by the function $V(z) = \frac{-C_4}{(z+L)z^3}$ describing a Van der Waals–Casimir interaction. The atom specific length L ($L \approx 10$ nm for He) marks the transition from the Van der Waals regime ($V(z) \propto z^{-3}$) at $z \ll L$ to the Casimir–Polder regime ($V(z) \propto z^{-4}$) at $z \gg L$. Although this potential looks smooth, it can be shown that its quantum reflection probability will approach unity in the zero-energy limit [32].

Quantum reflection of atoms from a solid surface was described by theory [32–38]. It was first observed in experiments with ultracold metastable atoms [39] and, later, also with a Bose-Einstein condensate [40, 41]. In these experiments extremely small atomic velocities needed to observe quantum reflection are achieved by cooling a dilute atomic gas in a trap to ultracold temperatures. An alternative approach to achieve those velocities is to scatter an atomic or molecular beam from a solid at grazing incidence [24, 31, 42], Due to the grazing incidence geometry the relevant velocity z-component, perpendicular to the surface, can approach extremely small values allowing observation of quantum reflection. The comparatively large parallel velocity component does not affect the quantum reflection process as long as the surface is (at least locally) homogenous. Quantum reflection of helium beams from plane surfaces [31, 42] as well as laminar [24] and blazed ruled [30] gratings has been reported.

In addition, reflection and diffraction from a grating was also observed for weakly bound ground-state helium dimers (He_2 binding energy ≈ 0.1 μeV) and trimers (He_3 binding energy ≈ 10 μeV) [43, 44]. Following the above description of classical scattering, the forces in the molecule–surface potential well region will inevitably lead to bond breakup, because the well depth (order of magnitude 10 meV) is $\approx 10^5$ times and $\approx 10^3$ times larger than the binding energy of helium dimers and trimers, respectively. Therefore, the observation of reflection of dimers and trimers provides direct evidence for quantum reflection. Furthermore, the fact that diffraction patterns are found indicates that quantum reflection leads to coherent reflection of matter waves.

4   Conclusion

The experiment described by Friedrich Knauer and Otto Stern in the UzM no. 11 article was designed to observe matter-wave diffraction of He and H_2 beams scattering off of a ruled diffraction grating at grazing incidence conditions. They were able to observe mirror-like, specular reflection with the reflectivity increasing with decreasing incidence angle and with increasing de Broglie wavelength. But, they fell short of detecting diffraction peaks. The modern reincarnation of the experiment was performed in 2007 in the Fritz-Haber-Institut der Max-Planck-Gesellschaft in Berlin, Germany. Conceptually, the modern apparatus is in line with the setup used in Otto Stern's lab. Yet, its intense molecular-beam source and sensitive electron-impact ionization detector out-perform their 1920s counterparts. In combination with better vacuum systems and modern electronics this made it possible to observe fully resolved diffraction patterns of He atom beams including peaks up to the seventh diffraction order. It can be assumed that the diffraction pattern observed in 2007 was exactly what Knauer and Stern were hoping to observe.

The results gained with the modern equipment make it clear that Knauer's and Stern's experiment was well conceived; it was bound to work, as soon as sufficient signal was available. This holds despite the fact that Otto Stern and his coworkers did not know that the atom–surface interactions can lead to quantum reflection and, thus, play a crucial role in the coherent scattering of atoms and molecules from surfaces.

While they did consider atom–surface interactions in the context of diffraction from a cleaved crystal surface, they ignored it for the gratings. It might well be that Friedrich Knauer and Otto Stern were the first to observe, unknowingly, quantum reflection of atoms from a solid surface more than 80 years before the first conclusive demonstration of this effect by Shimizu [39]. Ironically, quantum reflection itself is a wave phenomenon that cannot be explained with particles. As such, observation of quantum reflection of atoms provides evidence for de Broglie waves of an atom, exactly what Otto Stern was aiming to observe.

Acknowledgements I am indebted to my long-term collaborator, Prof. Bum Suk Zhao (UNIST, Ulsan, South Korea). His hard and skilful work was crucial to succeed with our joint diffraction experiments described in this contribution. I also want to thank Prof. Peter Tonnies (Max-Planck-Institut für Dynamik und Selbstorganisation, Göttingen, Germany). As my thesis adviser he taught me the art of running molecular-beam experiments; the legacy from Otto Stern a century ago. Furthermore, I thank Prof. Gerard Meijer (Fritz-Haber-Institut (FHI) der Max-Planck-Gesellschaft, Berlin, Germany) for making it possible for me to pursue molecular-beam diffraction experiments at the FHI. Last not least, I thank Prof. Bretislav Friedrich (FHI, Berlin, Germany) for numerous revealing discussions on Otto Stern and his legacy, and I want to thank him as well as Prof. Horst Schmidt-Böcking (Goethe Universität, Frankfurt, Germany) for their effort of organising this one-of-a-kind Otto-Stern tribute.

References

1. L. de Broglie, Recherches sur la théorie des quanta. Annales de Physique **10**, 22–128 (1925)
2. F. Knauer, O. Stern, The Reflection of Molecular Beams. Zeitschrift für Physik **53**, 779–791 (1929)
3. M. Eckert, Max von Laue and the discovery of X-ray diffraction in 1912. Ann. Phys. **524**, A83–A85 (2012)
4. I. Estermann, O. Stern, Diffraction of molecular beams. Zeitschrift für Physik **61**, 95–125 (1930)
5. I. Estermann, R. Frisch, O. Stern, Molecular ray problems. Experiments with monochromatic de Broglie waves of molecular beams. Physikalische Zeitschrift **32**, 670–674 (1931)
6. O.R. Frisch, O. Stern, *Handbuch der Physik*, ed. by H. Geiger, K. Scheel, Vol. XXII, II. Teil (Negative und positive Strahlen), 2 edn. (Springer, Berlin, Germany, 1933) , pp. 313–354
7. F. Knauer, O. Stern, Intensity measurements on molecular beams of gases. Zeitschrift für Physik **53**, 766–778 (1929)
8. T.S. Kuhn, Interview of I.I. Rabi (1963). Available online at www.aip.org/history-programs/niels-bohr-library/oral-histories/4836
9. J.P. Toennies, H. Schmidt-Böcking, B. Friedrich, J.C.A. Lower, Otto Stern (1888–1969): the founding father of experimental atomic physics. Ann. Phys. (Berlin) **523**, 1045–1070 (2011)
10. R. Frisch, O. Stern, Abnormality in the specular reflection and diffraction of molecular beams of crystal cleavage planes. I. Zeitschrift für Physik **84**, 430–442 (1933)
11. J.E. Lennard-Jones, A.F. Devonshire, Diffraction and selective adsorption of atoms at crystal surfaces. Nature **137**, 1069–1070 (1936)
12. G. Benedek, J.P. Toennies, *Atomic Scale Dynamics at Surfaces: Theory and Experimental Studies with Helium Atom Scattering* (Springer, Berlin, Germany, 2018)
13. P.R. Berman (ed.), *Atom Interferometry* (Academic Press, New York, 1997)
14. A.D. Cronin, J. Schmiedmayer, D.E. Pritchard, Optics and interferometry with atoms and molecules. Rev. Mod. Phys. **81**, 1051–1129 (2009)

15. D.W. Keith, M.L. Schattenburg, H.I. Smith, D.E. Pritchard, Diffraction of atoms by a transmission grating. Phys. Rev. Lett. **61**, 1580 (1988)
16. M.S. Chapman, et al., Optics and interferometry with Na_2 molecules. Phys. Rev. Lett. **74**, 4783 (1995)
17. O. Carnal, A. Faulstich, J. Mlynek, Diffraction of metastable helium atoms by a transmission grating. Appl. Phys. B **53**, 88 (1991)
18. W. Schöllkopf, J.P. Toennies, Nondestructive mass selection of small van der Waals clusters. Science **266**, 1345 (1994)
19. W. Schöllkopf, J.P. Toennies, The nondestructive detection of the helium dimer and trimer. J. Chem. Phys. **104**, 1155 (1996)
20. R.E. Grisenti, et al., Determination of atom-surface van der Waals potentials from transmission-grating diffraction intensities. Phys. Rev. Lett. **83**, 1755–1758 (1999)
21. W. Schöllkopf, R.E. Grisenti, J.P. Toennies, Time-of-flight resolved transmission-grating diffraction of molecular beams. Eur. Phys. J. D **28**, 125 (2004)
22. M. Arndt, O. Nairz, J. Vos-Andreae, C. Keller, G. van der Zouw, A. Zeilinger, Wave-particle duality of C_{60} molecules. Nature **401**, 680 (1999)
23. This apparatus was built in the 1990s in Prof. J.P. Toennies's Dept. of Molecular Interactions in the Max-Planck-Institut für Strömungsforschung in Göttingen, Germany, and was relocated to the Fritz-Haber-Institut der Max-Planck-Gesellschaft in Berlin, Germany in 2005 where it has been in use since then
24. B.S. Zhao, S.A. Schulz, S.A. Meek, G. Meijer, W. Schöllkopf, Quantum reflection of helium atom beams from a microstructured grating. Phys. Rev. A **78**, 010902(R) (2008)
25. H. Buchenau, E.L. Knuth, J. Northby, J.P. Toennies, C. Winkler, Mass spectra and time-of-flight distributions of helium cluster beams. J. Chem. Phys. **92**, 6875 (1990)
26. J. Wang, V.A. Shamamian, B.R. Thomas, J.M. Wilkinson, J. Riley, C.F. Giese, W.R. Gentry, Speed ratios greater than 1000 and temperatures less than 1 mk in a pulsed He beam. Phys. Rev. Lett. **60**, 696–699 (1988)
27. L.W. Bruch, W. Schöllkopf, J.P. Toennies, The formation of dimers and trimers in free jet ^{4}He cryogenic expansions. J. Chem. Phys. **117**, 1544–1566 (2002)
28. J.L. Heilbron, Interview of Immanuel Estermann (1962). Available online at www.aip.org/ history-programs/niels-bohr-library/oral-histories/4593
29. M. Born, E. Wolf, *Principles of Optics*, 6th edn. (Cambridge University Press, Cambridge, 1997)
30. B.S. Zhao, G. Meijer, W. Schöllkopf, Emerging beam resonances in atom diffraction from a reflection grating. Phys. Rev. Lett. **104**, 240404 (2010)
31. B.S. Zhao, H.C. Schewe, G. Meijer, W. Schöllkopf, Coherent reflection of He atom beams from rough surfaces at grazing incidence. Phys. Rev. Lett. **105**, 133203 (2010)
32. H. Friedrich, G. Jacoby, C.G. Meister, Quantum reflection by Casimir-van der Waals potential tails. Phys. Rev. A **65**, 032902 (2002)
33. R.B. Doak, A.V.G. Chizmeshya, Sufficiency conditions for quantum reflection. Europhys. Lett. **51**, 381–387 (2000)
34. A. Mody, M. Haggerty, J.M. Doyle, E.J. Heller, No-sticking effect and quantum reflection in ultracold collisions. Phys. Rev. B **64**, 085418 (2001)
35. S. Miret-Artés, E. Pollak, Scattering of He atoms from a microstructured grating: quantum reflection probabilities and diffraction patterns. J. Phys. Chem. Lett. **8**, 1009–1013 (2017)
36. J. Petersen, E. Pollak, S. Miret-Artés, Quantum threshold reflection is not a consequence of a region of the long-range attractive potential with rapidly varying de Broglie wavelength. Phys. Rev. A **97**, 042102 (2018)
37. G. Rojas-Lorenzo, J. Rubayo-Soneira, S. Miret-Artés, E. Pollak, Quantum reflection of rare-gas atoms and clusters from a grating. Phys. Rev. A **98**, 063604 (2018)
38. G. Rojas-Lorenzo, J. Rubayo-Soneira, S. Miret-Artés, E. Pollak, Quantum threshold reflection of He-atom beams from rough surfaces. Phys. Rev. A **101**, 022506 (2020)
39. F. Shimizu, Specular reflection of very slow metastable neon atoms from a solid surface. Phys. Rev. Lett. **86**, 987–990 (2001)

40. T.A. Pasquini, et al., Quantum reflection from a solid surface at normal incidence. Phys. Rev. Lett. **93**, 223201 (2004)
41. T.A. Pasquini, et al., Low velocity quantum reflection of Bose-Einstein condensates. Phys. Rev. Lett. **97**, 093201 (2006)
42. V. Druzhinina, M. DeKieviet, Experimental observation of quantum reflection far from threshold. Phys. Rev. Lett. **91**, 193202 (2003)
43. B.S. Zhao, G. Meijer, W. Schöllkopf, Quantum reflection of He_2 several nanometers above a grating surface. Science **331**, 892–894 (2011)
44. B.S. Zhao, W. Zhang, W. Schöllkopf, Non-destructive quantum reflection of helium dimers and trimers from a plane ruled grating. Mol. Phys. **111**, 1772–1780 (2013)

Part VI
Exotic Beams

Chapter 26
Liquid Micro Jet Studies of the Vacuum Surface of Water and of Chemical Solutions by Molecular Beams and by Soft X-Ray Photoelectron Spectroscopy

Manfred Faubel

Liquid water, with a vapor pressure of 6.1 mbar at freezing point, is rapidly evaporating in high vacuum, rapidly cooling off by the evaporational cooling, and is freezing to ice almost instantly.

Historically, this was noted already in the very earliest experiments with vacuum by Otto Guericke, the inventor of a first vacuum pump, and by Robert Boyle. In a contemporary, serendipitous report, published in 1664 in Nurnberg, G. Scotus in his "Technica curiosa" [1] is listing a number of noteworthy, miraculous and curious observations within the newly created vacuum space inside a bell jar, like: light is penetrating vacuum, the sound of a ringing bell can not penetrate to the outside, a candle flame is extinguishing, but, gun powder can be ignited inside a vacuum; small animals such as a mouse are dying quickly in a evacuated bell jar vacuum while certain insects can survive for a while. And eventually, as an item in his volume 2, Chap. 15, "Experimentum XXXVIII. Aqua intra evacuatum Recipientum congelatur", (experiment no. 38. Water inside the evacuated recipient is freezing to ice). With scholar diligence G. Scotus has annotated that with the "new" mechanically improved vacuum pump of Robert Boyle the water is freezing in vacuum while it was not observed to freeze in the equipment provided by Otto Guericke to the laboratory of G. Scotus. From our modern view we can conclude that Boyle's pump did reach a vacuum better than 6.1 mbar, the vapor pressure of ice at freezing, while Guericke's original pump, known to have been operated with water lubricated seals and not yet with fat or oil, did stop pumping slightly above the freezing point vapor pressure of water.

Thus, for centuries to come—including the first 80 years of modern age "Otto Stern type experiments"—liquid aqueous solutions were not considered as suitable

M. Faubel (✉)
Max-Planck-Institut für Dynamik und Selbstorganisation, Göttingen, Germany
e-mail: manfred.faubel@ds.mpg.de; Manfred.Faubel@mpibpc.mpg.de

B. Friedrich and H. Schmidt-Böcking (eds.), *Molecular Beams in Physics and Chemistry*,
https://doi.org/10.1007/978-3-030-63963-1_26

for research employing vacuum or any molecular beams technology or ultrahigh vacuum surface diagnostics.

1   The Free Vacuum Surface of Water Microjets

With this unfavorable veredictum for water in vacuum in mind, for advancing experiments successfully it is helpful to have a very close look at the dynamical processes on the surface of liquid water [2–4]. Here three principal problems dominate: (1) due to the high vapor pressure and the rapid evaporation a liquid water surface is always overcast by a dense gas cloud associated with the steady vapor stream into the adjacent vacuum space. At the minimum pressure of 6.1 mbar for liquid water, just above freezing, the molecular mean free path in this vapor is only $l = 10\ \mu$m. At a first glance this is preventing molecular beam type experiments at the surface, unless the geometrical extension of a water surface experiment could be shrunk to total dimensions smaller than 10 μm. Furthermore, (2) the un-obstructed unilateral free flow of molecules leaving the liquid surface with a Maxwellian velocity distribution at an average velocity of water molecules of approximately 1000 m/s is resulting in a massive gas flux of 600 mbar L/s/cm^2 for each square-centimeter of liquid surface, requiring extremely large pumping capacities in order to sustain a moderately decent vacuum. This vapor flow density, equivalent to 0.027 Mol/(s cm^2), that is 0.48 g s^{-1} cm^{-2} H$_2$O, corresponds on the liquid side of the separating surface to a liquid water surface ablation rate of 4.8 mm/s. (3) Considering the water heat of vaporization of 40 kJ/mol the surface cooling rate by evaporation is of the order of 1 kW/cm^2 for liquid water in free vacuum, inducing a very rapid freezing to ice of the vacuum exposed liquid surface.

In a practicable, efficient way these three challenges can be resolved with the use of a fast flowing very thin cylindrical liquid jet in high vacuum, a method developed since the 1980's (Faubel, Schlemmer, Toennies) and, gratefully, first published in a Zeitschrift für Physik "Otto Stern Centenial Issue", 1988 [2].

For a thin liquid jet of water with diameter D smaller than the vapor mean free path $l_{H_2O} = 10\ \mu$m: (1) Knudsen conditions, $D/l_{H_2O} < 0.5$, are met for collision-free "high vacuum" gas flow; (2) the total liquid surface is small and, thus, the vapor flow into the vacuum system is restricted to values lower than 0.5 mbar L/s, while (3) the rapid outflow of liquid from the thermostat controlled nozzle is continuously replacing the evaporation-cooled water surface before it can freeze to ice. One further benefit of the microjet technique is in providing an ultra-clean liquid surface due to the rapid replacement of the vacuum exposed surface section. A 1 mm long microjet surface section, flowing with a speed of 100 m/s, is replaced within 10 μsec. Usually, in surface science, for the handling of solid state surfaces, technically demanding ultra-high vacuum conditions with pressures lower than $p_{Vac} = 10^{-10}$ mbar are required to prevent surface coverage by impacting, and sticking, gas molecules during a typical experimental observation session of $t_{expos} \geq 1000$ s, according to Langmuir's surface coverage relation of: $p_{Vac} \cdot t_{expos} = 10^{-6}$ mbar s for coverage with 1 monolayer. On fast microjets, however, with short lived, continuously replacing surface the exposure time scales are $t_{expos}' = 10$ μs, only. Therefore, a background gas pressure of p_{Vac}' $= 10^{-3}$ mbar, with $p_{Vac}' \cdot t_{expos}' = 10^{-8}$ mbar s implying 1% surface coverage, is already the equivalent to perfectly clean, "ultrahigh vacuum" conditions from a point of view of surface coverage times.

In Fig. 1 the vacuum microjet setup scheme is depicted, showing a high-speed photography of a real jet, at 10 ns exposure time, and a sketched-in stream-tube along a vapor expansion line. A representative ensemble of gas molecules moving outward on a streamline progressively is increasing its available volume (blunt cone shape enclosures), indicated here for two different time instances, in order to illustrate the decrease in local gas density $n(r) \sim n_0 (r_0/r)^2$ in proportion to the quadratic radial distance r from the microjet. This results in a rapid increase in the local molecular mean free path $l = 1/(n(r) \cdot \sigma_{collision})$ in proportion to r^2. At the liquid surface position, at $r_0 = D/2$ given by the microjet diameter, the equilibrium vapor pressure density n_0 is assumed. Relevant for describing vacuum conditions (M. Knudsen 1905) is the number of molecular collisions on their path through a vacuum space: $dN =$ $dr/l_{H_2O}(r)$ which is determined by the ratio between the thickness dr of a gas layer in relation to the mean free path. Integration yields the total number of collisions on the gas path taken from the liquid surface at r_0 to infinity and shows the condition of one molecular collision encounter on this way out is $l_{H_2O}(r_0)/D > 1$, the well-known Knudsen condition for a free molecular beam source [2, 4] This simple model also explains, that for near surface conditions $l_{H_2O} \ll D$, i.e. at larger jet diameters or at higher temperatures of the liquid water and correspondingly higher jet vapor pressure, many molecular collisions occur in the outward streaming vapor and cause the onset of a supersonic jet hydrodynamic expansion into the vacuum.

The liquid jet photography shown in Fig. 1 was taken with a comparably large nominal nozzle aperture diameter $D = 18$ μm for reasons of optical resolution in this near diffraction limit of optical photography. The optically smooth, cylindrical jet propagates for a distance of 2 mm downstream from the nozzle exit before it starts to build up spontaneous surface oscillations and decays rapidly into a stream of fine droplets. The evaporation cooling of the jet filament amounts to ≈ 1 °C in 10 μs [2, 3]. At typical jet speeds between 30 and 100 m/s the jet surface temperature changes by only a few degrees centigrade on the first two millimeters of the smooth jet section to be employed in free vacuum experiments.

2   Liquid Jet Flow and Decay Into Droplets

In clearer detail liquid microjet flow contours are redrawn in Fig. 2 [3]. After leaving the nozzle channel the liquid jet free surface diameter contracts by 10–20%, depending somewhat on the nozzle shape, on the viscosity and on the surface tension of the liquid. The jet liquid flow stays strictly laminar up to very high flow velocities well above $v_{jet} \geq 100$ m/s. As a consequence of the small geometrical size of the micro-jet the Reynolds numbers for flow of low-viscosity, water-like,

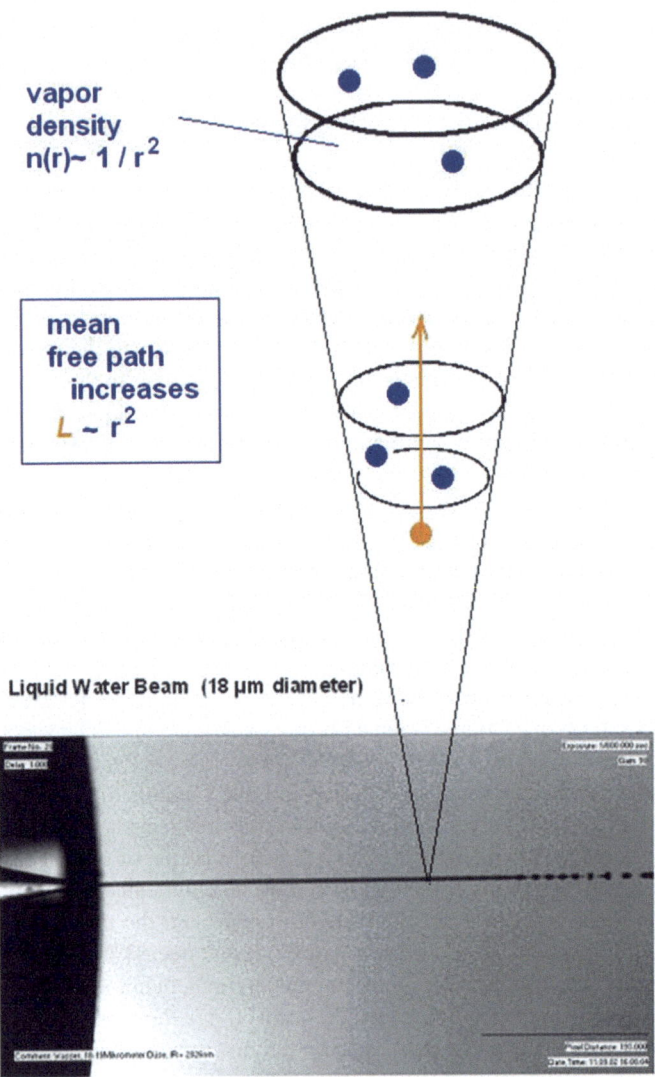

Fig. 1 A liquid microjet of water in vacuum. The photo insert shows a liquid microjet of 18 μm diameter, taken at 10 ns exposure time. The jet emerges from a quartz nozzle, and extends for several millimeters with a smooth cylindrical envelope, before decaying into a stream of droplets. Water molecules evaporating from the liquid surface rapidly are reducing in gas density with increasing distance from the liquid filament, as is indicated by the expansion stream tube cone with two control volume disks at different distances r. With decreasing densities the molecular mean free path increases. Fewer than one gas collisions occur in the emerging vapor beam when the jet diameter D is made smaller than the vapor mean free path in equilibrium, $l_{H_2O} \approx 10\,\mu m$ at T = 273 °C. This was Otto Stern's 'Knudsen condition' for operating molecular beam source ovens. By the intense vacuum evaporation the liquid jet cooling rate is ~1 °C in 10 μs, equivalent to 3 °C temperature drop on 1 mm length of flowing surface, for jets streaming with a velocity of 30 m/s

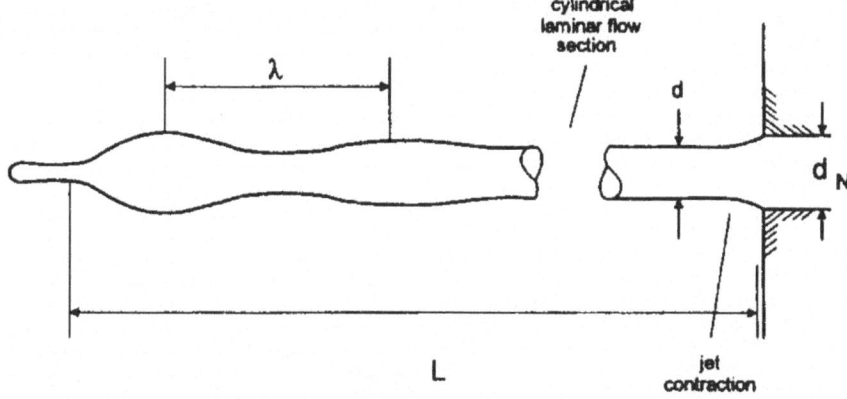

cylindrical
laminar flow
section

λ

d

d_N

L

jet
contraction

Fig. 2 Liquid jet flow and decay into droplets (figure reproduced from Ref. [3], Fig. 3)

liquids in microcapillaries are remaining well below the critical Re-values > 1500, the limit for the onset of turbulent flow [3]. This allows the production of very fast jets with smooth surfaces and long intact cylindrical sections, with moderate jet surface cooling by the powerful vacuum evaporation. Driven by surface tension (and delayed by higher liquid viscosity) the free jet filament in some distance from the nozzle begins to form contraction ripples, spontaneously, with a wavelength λ ≈ 6–8 D_{jet} and then decays rapidly into a stream of approximately uniform, equidistant droplets [3]. This decay is known as the Rayleigh spontaneous decay mode for free boundary liquid filaments. Lord Rayleigh's theory shows that the decay time to form droplets for a given liquid depends on the liquids' physical properties and on the jet diameter, only (similar to the droplet dripping time from a pipette mouth). Therefore, the contiguous length L of the smooth cylindrical section can be extended or shrunk at will, just by changing the flow speed of the jet.

A set of experimentally determined jet length values L as a function of the velocity of a liquid water jet is plotted in Fig. 3 for three different jet diameters, confirming the just discussed linear relationship between jet length L and jet velocity v_{jet} due to constant jet decay time [3]. In these measurements the nozzle was illuminated from the rear side by a laser. The light beam entered the jet like a light conducting fiber; and at the breakup point at the "end" of the liquid filament the red laser light was dispersed in all directions, creating a visible red spot which could be observed easily by a remote telescope as well for a vacuum jet as in atmosphere. The jet velocity, by Bernoulli's law, is related to the square root of the nozzle pressure $v_{jet} = $ Sqrt $(2 \, p_N/\rho)$ for a low viscosity liquid of density ρ. In the diagram the jet operation pressure is shown also, in the upper ordinate scale of the plot. For the smallest jets with approximately 6 μm actual jet diameter, emerging from a nozzle with $d_N = $ 10 μm aperture, the jet length increases linearly up to a maximum length L = 4 mm, at an approximate jet velocity near 150 m/s. At higher velocity the jet appears to decay into a diffuse, turbulent spray and the contiguous jet length begins to shorten with further increase in nozzle pressure, indicating the onset of turbulent flow and of

Fig. 3 Jet decay length for different jet velocities and different diameters (liquid water) (figure reproduced from Ref. [3], Fig. 4)

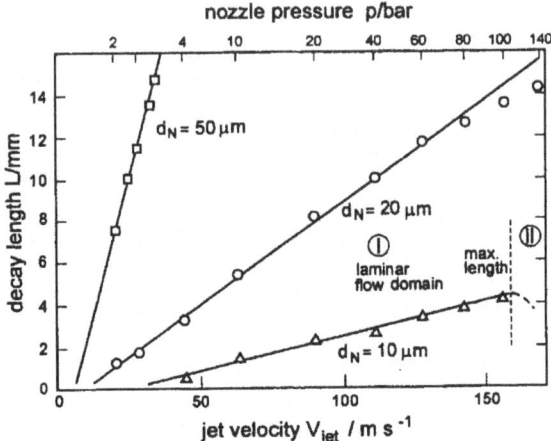

disruption instabilities by shearing forces. We can thus distinguish a laminar, smooth jet flow domain (I) for low velocities, which is well suited for surface experiments, from an unstable flow region (II) appearing above a certain critical speed. For the larger nozzle diameters $d_N = 20\ \mu m$ and $d_N = 50\ \mu m$ substantially longer, intact, cylindrical jet sections are obtained at much lower jet speed, as is to be expected from Rayleigh theory and its extensions, as discussed elsewhere [3, 4].

For the design of actual experiments in vacuum, microjet diameter and microjet operation conditions can be optimized and interpolated from these rough survey data in Fig. 3. From extrapolation of the jet length data shown here, it may be noted, also, that for very narrow, low viscosity water jets in the range of 1 μm diameter and smaller, the maximum decay length is decreasing dramatically to less than 0.1 mm and shorter, limiting their value for surface probing experiments with standard size microprobe devices.

3 Nascent Velocity Distribution of Evaporating Molecules

The first experiment for exploring the water microjet concept in vacuum were measurements of the velocity distribution of water molecules evaporating from the liquid microjet surface, in a set-up illustrated by Fig. 4a. [2, 3]. The liquid jet here is passing in front of a skimmer collimator entrance of a molecular beam time-of-flight spectrometer. Through a 5 mm diameter aperture further downstream the waste water jet enters in 2 cm distance from the nozzle into a separate beam catcher vacuum chamber where the by then supercooled liquid droplet stream freezes as slowly growing ice-needles onto a liquid nitrogen cooled cold trap, placed in 60 cm distance. With vacuum pumps of several 1000 L/s pumping speed and additional support by 10 000L/s cryotraps a vacuum of 10^{-6} mbar is sustained in the main

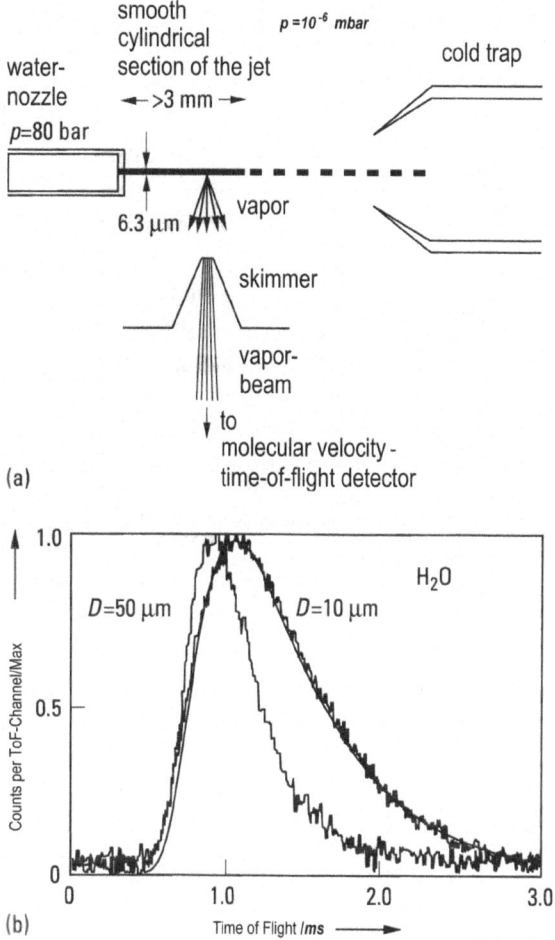

Fig. 4 Molecular beam time of flight spectroscopy of liquid microjet evaporation. **a** The microbeam setup shows the high-pressure nozzle, the free jet, and a beam-catcher cold trap for removal of the liquid jet. Part of the emerging vapor is extracted through an aperture in a conically shaped skimmer device. For velocity analysis a rotating disk with narrow slits is chopping the sampled molecular beam into short bursts, and molecules are detected in the ionizer of a mass spectrometer located at the end of a 0.81 m drift tube. **b** Two time of flight spectra of the velocity distribution of evaporated water from a thin 10 μm liquid jet and from a wider 50 μm jet, respectively, show a narrowed supersonic velocity distribution for the jet diameter larger than the vapor mean free path ($D = 50$ μm $> l_{H_2O}$) and, a broader Maxwell distribution (fitted by the smooth line) for collision free vapor expansion from 10 μm wide liquid surface into vacuum (figure reproduced from Ref. [3], Fig. 2)

chamber, surrounding the experimental surface probing region on the initial section of the intact, contiguous, 3 mm long liquid filament. For the molecular beam analysis of the vapor a sharp cone "skimmer" with ~ 50 μm entrance aperture, in two to five millimeters distance at right angle from the liquid jet propagation direction, is sampling a very small fraction of the radially evaporating, intense gas stream of water molecules. Not shown here in detail, these pass subsequently through a narrow slit in a rapidly rotating chopper wheel for producing short molecular beam bunches of 20 μs duration, and spread out in time in a drift tube on a vacuum flight path with 80 cm length. At the end of the drift region the dispersed molecules reach the electron bombardment ionizer region of a mass spectrometer placed in ultrahigh vacuum at pressures near 10^{-11} mbar for lowering the diffuse spurious gases background. By single ion counting and recording the time of ion arrival, time-of-flight spectra for the vapor molecule velocity distribution are accumulated, shown in Fig. 4b.

The two TOF spectra shown in Fig. 4b were obtained for two different sizes of liquid microjet nozzle diameter, one for $d_N = 50$ μm (which is $\gg l_{H_2O}$) and the other for $d_N = 10$ μm, respectively. The jet originating from the 10 μm pinhole nozzle is contracting to $D = 6.3$ μm diameter in the free surface flow region. The velocity distribution of the evaporated water from this narrower jet surface with free evaporation Knudsen conditions $D < l_{H_2O}$ is very close to the theoretical Maxwell distribution which is plotted also in Fig. 4b, as a smooth thin line superimposed to the "$d_N = 10$ μm" experimental spectrum. In contrast, the jet with the larger 50 μm nozzle, as an example for $D \gg l_{H_2O}$, shows a considerably narrower velocity distribution characteristic to ongoing supersonic jet expansion, driven by multiple molecular collisions in the early phase of vapor expansion. Thus, Fig. 4b gives the experimental proof that the free vacuum surface of liquid water with temperature above 0 °C can be prepared as high vacuum microjet surfaces when the jet diameter is smaller than the mean free path of the nascent vapor [2].

Numerical fitting of a Maxwell velocity distribution function to the measured distribution, in addition, yields the source temperature for the water vapor which is found to be a few degrees lower than the temperature of the nozzle and within expectations for a jet evaporational cooling model [2]. Finding a Maxwell distribution for the nascent vapor, at large, is in agreement with expectations for molecules which have to overcome a binding energy potential step barrier, i.e. the heat of evaporation, when moving from the liquid into the free vacuum space. Considering a thermal energy Boltzmann distribution in three independent cartesian coordinates, only the distribution component in the one coordinate perpendicular to the surface will be affected. Molecules will loose here the binding energy E_B in transit through the surface dividing plane, leaving a Boltzmann distribution with reduced intensity, however, with identical temperature T, as is easily seen by considering the mathematical separation formula: $\exp[(\frac{1}{2}mv^2 - E_B)/kT] = \exp(-E_B/kT)\exp(\frac{1}{2}mv^2/kT)$. This noteworthy finding that the vapor temperature, in spite of the evaporation energy loss of molecules, is identical to the liquid temperature, was an often disputed, surprising fact, although it had been published as early as in 1920 in a reply of Otto Stern to comments on his earliest Molecular Beam velocity distribution measurement [5].

In an additional upscaling experiment with the apparatus described in Fig. 4 we tried, in vain, to observe a direct evaporation of water dimer clusters from the surface of a microjet. Dimers are well known to occur in water vapor nozzle beam jets. In a mass spectrometer ionization source the dimers fractionize into $H–H_2O^+$ ions and are expected to appear at mass 19, next to the by far dominant water mass peak at mass 18. Actually, a faint, distinct signal peak could be observed at mass 19, 10^{-3} times smaller than the water monomer mass peak. However, when signal averaging the time-of flight spectrum over tens of hours at this purported peak the velocity distribution was exactly identical to the monomer peak. So it was identified to have come from the small fraction of deuterium and O^{17} atoms in HDO-water, and not from $(H_2O)_2$ water dimers with a mass of being twice the mass of monomer water that would result in average Maxwell distribution velocity smaller by a factor of $\sqrt{2}$. In conclusion, the water dimer fraction in evaporation was found to be smaller as least by a factor of five than the natural deuterium plus O^{17} abundance in hydrogen atom [2, 3, 6].

In continuing the search for direct dimer evaporation from liquid surfaces, carboxylic acids were studied which are known to form strongly bound dimers in a double hydrogen bridge structure COOH⊃HOOC-R, with binding energies in the order of 0.3 eV [6]. This is several times stronger than the water dimer hydrogen bond, and acetic acid microjets were found to emit large fractions, of ≥30%, of the vapor as dimers. The vapor velocity distribution measurements of the monomer species and of the dimer species, shown in the TOF spectra Fig. 5a, b, respectively, for liquid acetic acid, CH_3COOH, however, show yet another, very unexpected, phenomenon: When evaluating the measured velocity distribution by fitting a theoretical distribution function, the monomer distribution, Fig. 5a is very well fitted by a Maxwellian distribution function with a source temperature of ~252 K, well in the expected range of cooling for a vacuum microjet surface temperature. The acetic acid dimer velocity distribution, Fig. 5b, in contrast, can be fitted only by a slightly supersonic "floating Maxwellian" function representing the narrower half-width-spread by a Mach number of 3–4, and yielding a total dimer molecular beam enthalpy equivalent to a dimer molecules component apparent source temperature of 365 K for the liquid surface. This is 100 K higher than the apparent monomer source surface temperature and clearly above any error bar margins [6]. This anomaly in dimer source enthalpy of liquid surface vapor sources is further confirmed by vapor velocity measurements on a liquid jet of a mixture of 20% ethanoic acid in water, shown in Fig. 5c, d. The monomer distributions of the H_2O vapor component and of the CH_3COOH monomers, both displayed in Fig. 5c, are well fitted by simple Maxwellian function, with practically identical liquid surface source temperatures, 281 versus 275 K, although the average velocities of the two components differ by a factor of two, as to be expected for the mass ratio difference of 60:18 in molecular weight. This fitting result is also a proof of completely interaction-free, collisionless, vapor propagation of the two distinct mass components which show not any onset of dragging by the second component, known as the familiar seeded beams effect in more dense gas jet expansions. The dimer velocity component in this evaporating liquid mixture, in Fig. 5d, shows a slightly narrowed supersonic-like floating Maxwellian distribution

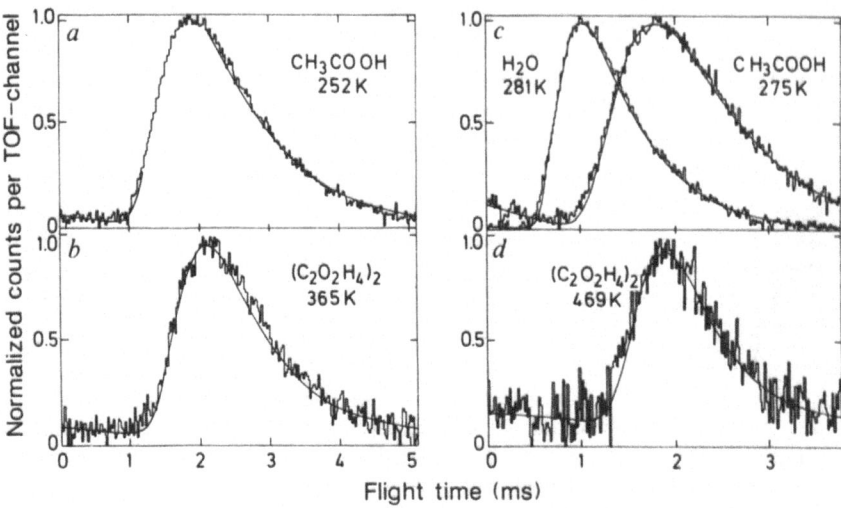

Fig. 5 Observation of non-equilibrium evaporation of dimer molecules of pure carboxylic acid and for a mixture of 20% CHCOOH in water. Temperatures shown are the (surface) source temperature calculated from the measured molecular beams time-of-flight velocity distributions. Dimers of acetic acid (Figs. 5b, d) show 100–200 K higher apparent source temperatures than the simultaneously evaporating monomers acetic acid (a) and for, both, H_2O and acetic acid monomers of the mixed solution (c) (figure reproduced from Ref. [6], Fig. 2)

with much higher apparent source temperature of 469 K, i.e., here the dimer source temperature is 200 K above the liquid surface temperature measured by the monomer vapor components emerging from the same liquid jet surface [6]. This astonishing anomaly is rising very interesting, far ranging questions about detailed microscopic balance of the liquid-evaporation/gas-phase-condensation process which requires the numbers of gas molecules evaporating per second and the number of molecules condensing must be in stationary equilibrium, as are to be their respective temperatures. As far as is known in statistical mechanics, or at least as far as is found in standard text books [7, 8], both, the distributions of evaporating molecules and the distribution of condensing molecules minus the distribution of molecules reflected from the liquid surface are all Maxwellian distributions with identical temperatures.

Although molecular liquid evaporation simulations for this phenomenon are not yet available, it was possible to give a semi-microscopic, intuitive model explanation for the observed velocity anomaly in dimer evaporation [6]: When the carboxylic dimers have formed in the liquid phase the two hydrogen bonding sites (at the O and at HO) of the, say, acetic acid COOH groups are crosswise saturated toward the adjacent dimer molecule. Then it may be reasonable to assume, that the two outward-pointing hydrophobic CH_3 groups act like a non-wetting nearly spherical inclusion in a water bubble or within a bubble of single hydrogen-bond, water molecules and monomerically dissolved acetic acid. When these preformed dimer inclusions approach the surface during the liquid evaporation process the bubble in this model

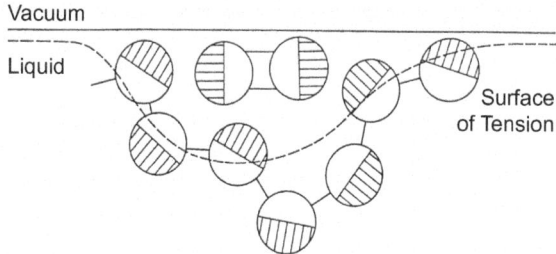

Fig. 6 Tentative surface tension model for the extra energy observed in dimer evaporation. For acetic acid CH_3COOH–$HOOCCH_3$ double hydrogen bonds form dimer inclusions in a liquid. The two CH_3 groups are pointing outwards and produce a hydrophobic dimer surface. The hydrophobic dimers are ejected by surface tension with an energy proportional to the released surface tension energy of the bubble hemisphere: $E \approx \sigma \pi R_2$. Different surface tensions of acetic acid ($\sigma = 1.7$ meV/Å_2) and of 20% acetic acid in water ($\sigma = 3.1$ meV/Å_2) explain readily the observed excess energies between dimer apparent source temperatures for the two liquids (figure reproduced from Ref. [6], Fig. 3)

picture will burst and the surface tension energy of the bubble will be released when the cavity is stretching out straight at the surface, transferring its freed energy like a stretched 'trampoline' to the outward moving dimer. This process is depicted in a cartoon drawing in Fig. 6. Assuming the hydrophobic dimer is sitting in a hemispherical cavity with radius of the average radius of the dimer molecule the change in surface area from hemispheric to plane, flat surface is just equal to the cross section area of the dimer molecule. With the macroscopically known surface tension value σ the energy gain for straightening out the Surface of Tension can then be calculated as $E_{excess} \approx \sigma \pi R^2$. For tabulated [9] macroscopic surface tension values of $\sigma_{Acetic} = 1.7$ meV Å^{-2} for pure acetic acid at 0 °C and of $\sigma_{20 \%Ac} = 3.1$ meV Å^{-2} for a mixture of water with 20% acetic acid, it is obvious to expect a factor of two larger excess energy in the dimer evaporation from the water/acetic acid mixture, in agreement with the actually observed change of "surface excess temperature" from a value of 100 K in Fig. 5b to 200 K excess in Fig. 5d. Further measurements of dimer evaporation from formic acid ($\sigma = 2.7$ meV Å^{-2}, $T_{excess} = 150$ K) and for propionic acid confirm this model prediction of linear correlation of experimental excess dimer temperatures with the respective surface tension values [6]. The formal change of surface area calculated from the actually observed experimental value of 100 K for absolute cavity energy release in this crude model gives a value of $\pi R^2 \approx 7$ Å^2 model-effective geometrical cavity cross section area of the dimer $(CH_3COOH)^2$ molecule.

There are clearly many further unanswered phenomena left to study in evaporation of liquids, and examples of actually ongoing work on liquid surface molecular collision processes will be presented in the here following contribution, by G. Nathanson, to this Otto-Stern-Fest collection of talks.

4   Photoelectron Spectroscopy of Liquid Water

The free vacuum microjet surface, equally well, is suited for electron scattering investigations and for photoelectron emission studies of liquid water, and for a majority
of volatile liquid solvents in use in chemistry [10]. In particular, it has shown to be of
appreciable value for providing an experimental data base of electronic structure of
liquid aqueous solutions with widespread use for basic chemistry, electrochemistry,
and for some studies on biological substances.

In addition to providing vacuum compatibility, here, the fast-flowing liquid jet is
alleviating some of the notorious surface charging problems associated with photoemission from non-conducting materials. Charged speckles of the insulating liquid
surface are just washed away, instantly, with the microjet streaming speed of several
ten to hundreds of meters per second. Also, radiation heating effects and surface
damage by intense photon beams are reduced by orders of magnitude by the rapid
target replacement in the quasi-stationary surface of the microjet. Drawbacks and
newly arising problems of the liquid jet method are: electrochemical double layer
potentials build up readily near the liquid surface such as the "Stern Layer", for
example, describing in greater detail electrical polarization of molecules in the
surface layers of electrolytes; the moving liquid may be charged up dramatically
by electrokinetic phenomena related to the internal Zeta-potential value and of the
herewith associated Debye Layer thickness of the investigated liquid; in modestly
well conducting electrolytic liquids external superimposed electric fields can charge
up the liquid tip by current flow through the liquid filament column, leading to time
dependent surface charges interdependent with the droplet decay times [3, 10]. Thus,
additional measures had to be worked out for stabilizing, over the time of a photoelectron spectrum record, the electrical surface potential of poorly conducting liquid
jets in order to get meaningful absolute reference potential values for measurements
of photoelectron orbital binding energies from microjet photoemission spectra of
aqueous solutions.

The principal construction scheme of the microjet photoelectron spectroscopy
apparatus [11], shown in Fig. 7, resembles the earlier liquid jet free evaporation
molecular beam sampling apparatus that was given in Fig. 1. With the extension
for a photon beam directed onto the microjet on a third experimental axis, and,
after replacement of the previously used molecular beam time-of-flight detector by a
photoelectron hemispherical analyzer for the energy analysis, emitted photoelectrons
are detected at right angle with respect to the incoming photon beam and at right
angle with respect to the liquid direction. The UV and soft X-ray radiation used in
photoelectron spectroscopy are strongly absorbed by gases and need the microjet
vacuum for being able to penetrate the vapor shroud and to reach the liquid water
surface. The mean free path of electrons in gases is one order of magnitude larger
than the previously discussed free path for molecules.

For exploratory development and proofs of the technology the equipment was
tested initially with a Helium-I lamp laboratory radiation source for photon energy
$h\nu = 21.22$ eV (lambda $= 58.43$ nm). And after optimization for jet and for charging

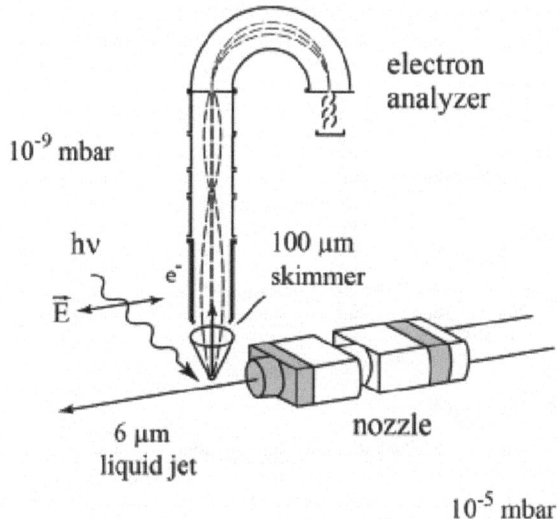

Fig. 7 Photoelectron spectroscopy on a free vacuum surface of a liquid microjet of water. A focused beam of photons with defined energy $E_{ph} = h\nu$, ranging from vacuum ultraviolet radiation to soft X-rays, is directed onto the microjet surface. Emitted photoelectrons are transferred into an electron energy analyzer with hemispherical electric deflection field for photoelectron spectra recording. The electron "vertical" binding energy E_B is determining the measured electron kinetic energy $E_{kin} = h\nu - E_B$ (figure reproduced from Ref. [11], Fig. 18)

stability, it was providing new photoelectron binding energy spectra for the three outer valence band electrons of pure liquid water as well as for some solvents such as ethanol or gasoline, and early data for solvated halogen ions in water solutions [3, 11]. Soon after, the PES microjet apparatus could be moved to 3rd generation Synchrotron tunable radiation sources (such as Bessy II) becoming available at the end of the 1990's. These yielded for microjets photoelectron spectroscopy studies sufficiently high radiation intensities of larger than 10^{16} monochromatized photons per square centimeter, focused onto a perfectly microjet-suited tiny spot of size <100 to <10 μm [10]. The Synchrotron radiation beam outlet port, not shown here in detail in Fig. 7, is protected by a series of several narrow collimator plus vacuum pump stages, needed to separate and to protect the Synchrotron storage ring ultrahigh vacuum region, at 10^{-10} mbar, from the water vapor loaded microjet surface intersection region.

A representative, typical set of energy resolved photoelectron spectra is shown in Fig. 8. It was obtained with Synchrotron radiation photons at $h\nu = 100$ eV for salt solutions of the diatomic alkali-halide salt series CsI, KI, NaI and LiI in water [13]. The photoelectron spectrum, in a simplified point of view, is imaging the electron density of states populations in the liquid electrolyte into an energy distribution spectrum of emitted electrons. The electron orbital binding energies are determined by the difference of the incident photon energy and the measured kinetic energy of emitted photoelectrons: $E_{bind} = h\nu - E_{kin}$. For the microjet spectra in Fig. 8, in molar concentrations of 2 m to 3 m, one salt molecule is dissolved in 20 molecules

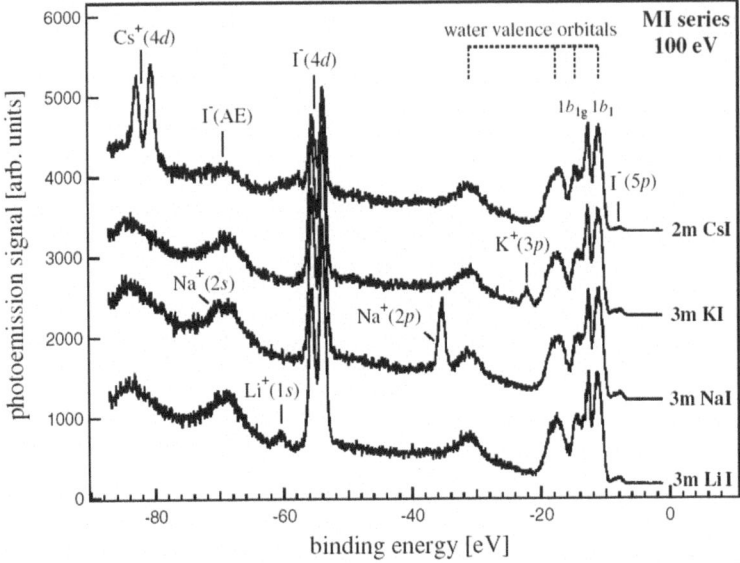

Fig. 8 Photoelectron spectra of alkali-halide salts in aqueous solution showing the valence electron states of liquid water, of the I⁻ ion and of the anions Li⁺, Na⁺, K⁺, and Cs⁺ (figure reproduced from Ref. [13], Fig. 2)

of water. At binding energies higher than the ionization onset threshold for water, near 10 eV at the right hand side the spectra of Fig. 8, a progression of three peaks is visible which are resulting from ionization of the outermost valence orbitals of liquid water, the states designated $1b_1$, $3a_1$, $1b_2$ of the water molecule and a fourth, broader peak, the inner valence orbital peak $2a_1$ at 32 eV binding energy. Energy level assignments and orbital energies for the isolated water molecule are depicted in Fig. 9, together with the correlation diagram for the origin of the H_2O hybridized orbitals from states of the separate oxygen and hydrogen atoms [11]. In addition to the 4 valence orbitals, the water molecule has one K-shell electron state, $1a_1$, at ~540 eV binding energy, to be seen later on in spectra obtained at higher photon energies. The experimental liquid water valence orbital peak energy position assignments are indicated above the valence spectrum by the horizontal scale bar in Fig. 8. In addition, in between the liquid water valence spectrum peaks, narrower faint water gas phase photoelectron peaks are visible which are caused by the vapor cloud surrounding the microjet. The most prominent gas phase spike here is designated by its gas-phase peak assignment $1b_{1g}$ on the upper spectrum in Fig. 8 and it is shifted by approximately 1.5 eV "gas-liquid shift" with respect to the liquid phase $1b_1$ feature. During measurements, the water gas phase peak is a very helpful reference calibration point in undergoing microjet spectra evaluations. It is averaging over the electric field in the immediate surroundings of the liquid jet and, thereby, also can be used as an indicator of unintentional or unnoticed jet charging. When the potential at the jet surface is differing from the grounded chamber wall the resulting electric field

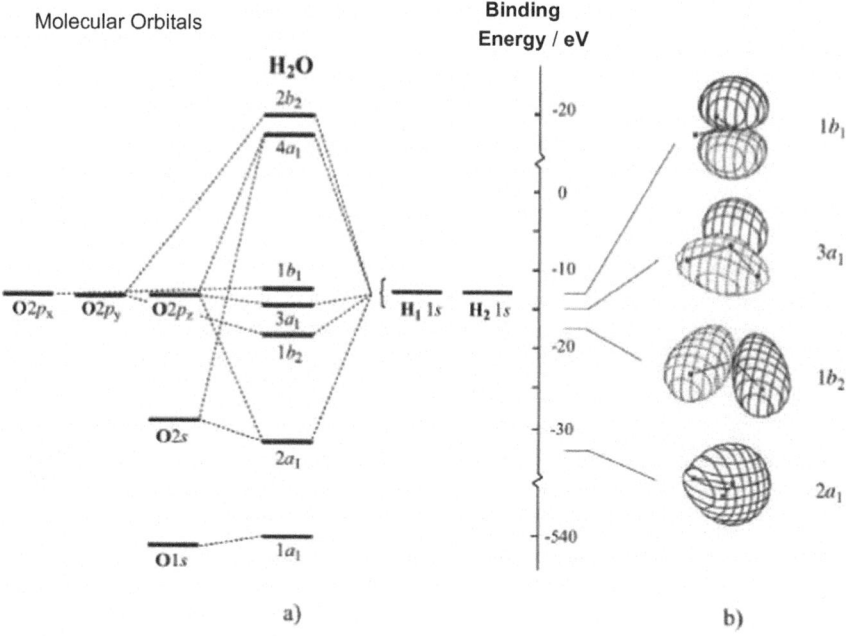

Fig. 9 a Molecular orbitals energy level diagram of the gas-phase water molecule. Intermittent diagonal lines indicate contributions of atomic orbital states of the separate H-atoms and O atom to the H_2O molecular orbitals. **b** Electron density contours for the highest occupied valence shell orbitals $1b_1$, $3a_1$, $1b_1$, and the "inner valence" orbital $2a_1$ (figure reproduced from Ref. [11], Fig. 20)

gradient is causing a clearly noticeable, field induced broadening of the gas phase $1b_{1g}$ peak and in a change of relative energy difference between liquid peak maximum and the gas phase peak.

The remarkable peak broadening of all liquid phase water valence electron energy levels with respect to the gas phase photoelectron spectrum, by amounts of the order of 3 eV peak half width, was surprising when found in the first measurements of liquid water photoelectron spectra [10]. It appears to be associated with the multiple and rapidly varying hydrogen bonding and thus seems to reflect the heterogeneous environment of liquid water molecules [11]. However, no simple broadening explanation is visible from water simulation calculations. The numerous salt ion peaks appearing in the electrolyte spectra are notably narrower, of an order of 1.1–1.5 eV, what is probably caused by the more rigid, stable hydration shells forming in ion solvation. The I^- ion in aqueous solution, for example, shows two marked photoemission peak doublets, a weak peak at 7.7/8.8 eV originating from ionization of the outermost electrons in the I^- (5p) orbital of iodine and a strong (here resonance enhanced) feature for the photoelectron emission from a lower orbital, I^- ($4d_{5/2,3/2}$) at ionization energy 53.8/55.5 eV. A weaker emission structure, I^-(AE), appearing in the spectra at the position near 70 eV on the binding energy scale, was identified as Auger electron emission occurring when an electron hole created in the I^-(4d) shell by direct photoemission is filled up shortly after by one $I^-(5p)$ outer orbital electron, at the simultaneous emission of a second 5p electron with the excess energy difference between the $I^-(5d)$ binding energy and the 5d → 4d transition energy. Constant emission energy at variations of the photon energy is the experimentally easy to confirm signature for this process. The alkali-anion peak structures in the spectra allow the experimental determination of the vertical ionization energies for the outer shell electrons of the series $Li^+(1 s)$, $Na^+(2p)$ and $K^+(3p)$ with values of $E_{aq}^{PES} = 60.4$ eV for Li^+, 35.4 eV for Na^+, 22.2 eV for K^+, respectively [11, 13]. For Cs^+ the vertical ionization energy peak, estimated at about 15 eV, is buried in the intense liquid water valence structure peaks. The vertical binding energy values, thus obtained, are of great relevance for theoretical description of chemical processes in aqueous solution. They differ from the chemical solvation enthalpy G^0 of equilibrium caloric experiments by the relaxation energies which arise by the rapid formation and realignment of the solvation shell of solvent molecules around a newly formed solvate ion species, and therefore allow an experimental determination of the important reorganization energies in electrochemical processes in polar solvents. Noteworthy, a comparison with the already known and tabulated gas phase ionization energies of the alkali and halide ions [9] shows the solvated states are changing by remarkably large amounts of energies: for I^- gas phase ions the ionization energy is 3.1 eV and shows a gas-liquid shift to 7.7 eV in the aqueous solution experiment shown in Fig. 8. This shift is opposite in sign to the gas-liquid shift of about 12 eV for Na^+ ions, decreasing from the gas phase ionization energy of 47.3 eV to lower electron binding energy of 35.4 eV in the liquid.

Whereas direct simulation of the peak solvation energy shifts is proofing to be demanding and time expensive, Max Born in 1922 had already proposed an elegant, very descriptive dielectric cavity (DCS) model for ion solvation which is sometimes still in use in phenomenological description of electrolytic fluids. Here the ion is thought to sit inside an empty spherical cavity inside a continuous dielectric medium with dielectric constant $\varepsilon \, \varepsilon_0$ different from the vacuum dielectric value ε_0. Using the electrostatic Maxwell equation for calculating the energy of a sphere with charge Z and radius R in an infinite dielectric medium and comparing versus vacuum with dielectric constant $\varepsilon = 1$ he obtains the formula: $\Delta E_{solvation} = (Z^2 \, e^2/8 \, \pi \, \varepsilon_0 \, R) \, 1/(1 - 1/\varepsilon)$ for the solvation energy of an ion with charge Z and ion radius R, giving reasonably good numerical results for positive monatomic ions in water [10]. In the case of I^- photodetachment Born-model results are far less accurate, simply because at final state with zero charge, the reorganization time of the dielectric water is difficult to include appropriately. The comparably small gas-liquid shift of 1.5 eV, only, in the ionization energy of the neutral water orbitals may also be rationalized using educated guess estimates for time-dependent dielectric constants in the DCS model. For water the static value of the dielectric constant is $\varepsilon \approx 80$. This constant is decreasing for time varying electric fields. On extremely short experimental time scales, such as the time period of light, the dielectric constant ε approaches the square value of the optical diffraction index n of water ($n = \sqrt{\varepsilon} \approx 1.4$) according to Maxwell's relation between the velocity of light and dielectric constant and the fact that diffraction indices are deriving from ratios of velocities of light. Therefore, for water dielectric reorganization on photoelectron emission with time spans of 10^{-15} s before the electron is distanced, the more decent estimate for the effective, dynamic dielectric response constant to the suddenly formed ion is likely to be closer to a value of $\varepsilon_{fs} \approx 2$. In accord with this consideration, using the "optical" value $\varepsilon_{fs} \approx 2$ in Born's DCS formula yields a gas-liquid shift of the order of 1.5 eV for newly formed H_2O^+ ions vertical ionization energies, in fairly good agreement with the observed shifts in the present water PES spectrum measurements, Fig. 8. The neutral liquid water had no time yet to respond to the appearance of the photoionized H_2O^+ ion in the neutral water medium. The DCM asymptotic model is thus giving at hands an intuitive, and roughly predictive formula which should be useful when discussing the far more accurate, but, by far less transparent results of realistic computer simulations of liquids.

The measurements in a wide range of ion concentrations and for different counterions show a remarkable independence of the ionization energies which are not changing within the current experimental accuracy of 30 meV (equivalent 3 kJ/mol in more familiar caloric units). These vertical ionization energies, also, are not varying for different locations in the electrolyte, independent of whether the ionization takes place near the surface or in greater depth in bulk phase environment, as is demonstrated by a series of measurements on 2 m solutions of NaI with photon energies being changed between 200 and 1000 eV, in Fig. 10. Here is made use of the fact

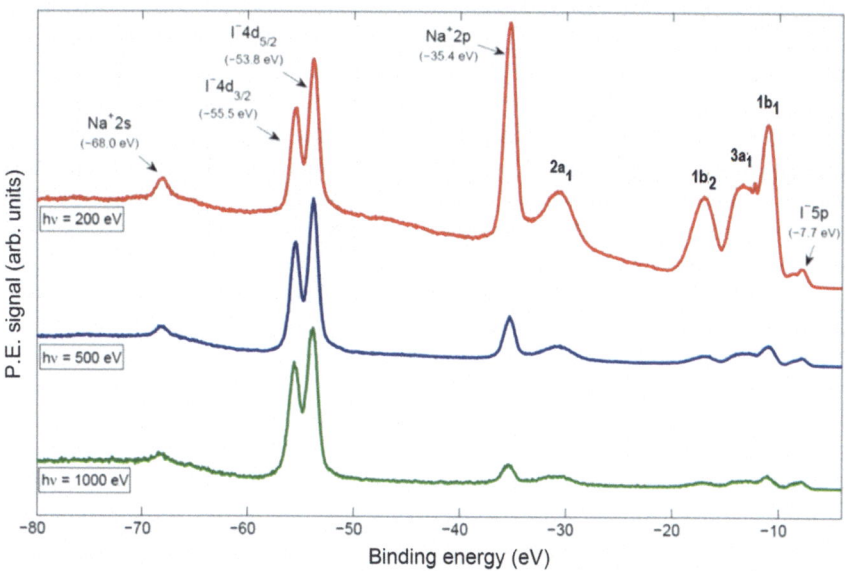

Fig. 10 Photoelectron spectrum of 4m NaI aqueous solution, taken at three different photon energies 200, 500, and 1000 eV. With increasing kinetic energy of photoelectrons deeper bulk regions of the aqueous solution are probed. The minimum escape depth is 2 or 3 water molecule diameters at 200 eV electron kinetic energy, and the probing depth increases to a layer thickness of 10–12 H_2O-diameters for 1000 eV photoelectron energy (figure reproduced from Ref. [14], Fig. 8)

that for photoelectrons with different kinetic energies the escape depth for electrons increases with increasing energy [14]. The thickness of the water layer probed increases, approximately, from 3 molecular layers at 150 eV ($\approx$12 Å) to a layer thickness of 12–15 molecular diameters of water at 950 eV kinetic energy. The observed photoelectron signal for a given depth and given initial energy is exponentially attenuated when the photoelectrons penetrate to the vacuum space over larger distances below the surface, whereas the photoelectron energy is not altered much by electronic stopping power on these comparably short average escape path lengths. This is confirmed by the observation, that the NaI photoelectron peak shapes are not broadening with increasing photon energy, although, due to the associated higher kinetic electron energies, photoelectrons are probed for significantly greater sampling depth [14].

It may be noticed that the photoelectron spectra recorded in Fig. 10 are very smooth and show almost negligibly small statistical counting noise in comparison to the earlier measurements of Fig. 8. This is the result of continuing experimental improvements of the synchrotron radiation photon beamline and of the water photoelectron spectroscopy apparatus. The photon intensity is higher and it is focused to a smaller spot size, illuminating fully a 15 μm liquid jet. In the liquid water valence peak features, therefore, the gas phase peak $1b_{1g}$ has vanished almost completely, and is visible only as the very weak spike residue between the $1b_1$ and the $3a_1$ liquid water photoelectron peaks. The three spectra are shown here with intensities normalized to the strong I^- 4d peak feature. The wide variation of all other relative peak intensities in between the three different spectrum records is caused, primarily, by changes in photoemission cross sections for different incident photon energies.

Basically, however, the photoelectron peak intensities are proportional to absolute concentrations of molecules in the liquid target probe. Depth profile probing with different photon energies, hence, allows, also, quantitative studies of concentration changes near the liquid-gas surface of solutions. Listed ionization cross section data of reasonable computational accuracy are available by NIST [12]. Using these and the experimental apparatus functions for collection and transmission of photoelectrons it is straightforward to evaluate absolute ion concentrations for Na^+ and for I^- from the PES measurements [14]. The thus obtained I^-/Na^+ ratios over the photoelectron kinetic energy range from $E_{kin} = 100$–1000 eV, are plotted versus the electron kinetic energy, in Fig. 11. They show a clear enhancement by almost a factor of two in favor of I^- anions near 200 eV kinetic energy, sampling the composition within the first two or perhaps three water molecule diameters in the uppermost layers of the liquid. With increasing photoelectron energy the evaluated ion ratios decrease and, above 400 eV, are approaching asymptotically the value of one, expected for the bulk stoichiometry of the sodium iodide salt solution. In a computational molecular dynamics modelling study of 1.2 m alkali halide salt aqueous solutions (with 18 Na^+ and 18 halogen-anion molecules in a slab of 864 H_2O molecules in the numerical sample calculation) the surface enhancement for halogen anions concentrations with increasing anion radii had been studied theoretically, one decade earlier [15], and can here be compared with detailed experiments. Snapshots of ion and water molecules distributions from this molecular modelling study are reproduced as a side insert in Fig. 11, adjacent to

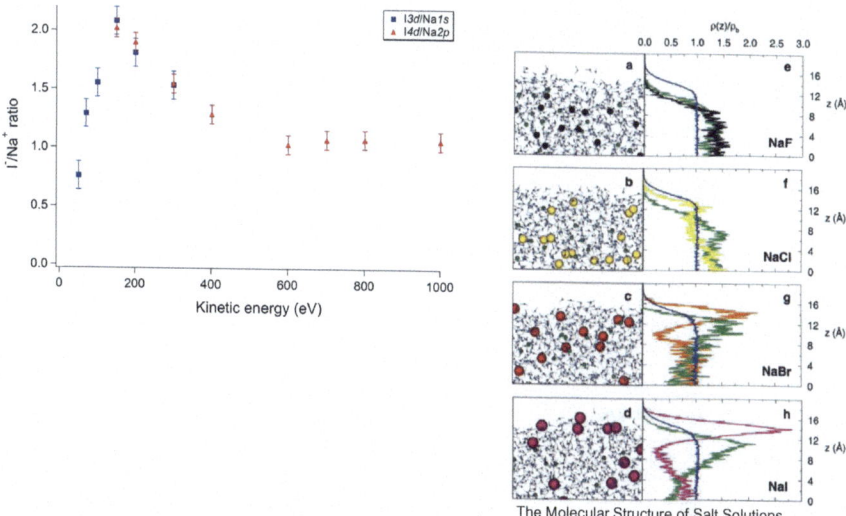

Fig. 11 (At left) evaluated anion/cation intensity ratios at different photoelectron kinetic energies show a significant propensity of I^- ions over Na+ near the surface of the 4 m NaI aqueous solution. At electron kinetic energies larger 500 eV the measured ratio approaches the bulk solution ratio 1:1. (At right) snapshots of computer simulations of the molecular structure of alkali-halide salts in aqueous solution show a preference for Iodine ions to the liquid-gas surface. Evaluated anion-, cation- and, water molecule-densities plotted versus the z-position coordinate of the simulation slab (Fig. 10h) are showing in quantitative detail an enhancement of the (magenta) I^- anion concentration at the water surface and a tendency for immersion of Na+ cations (red) into the bulk aqueous solution (O-atoms:blue) (figure 10 adapted from Ref. 12, Fig. 14 and Ref. 15, Fig. 1)

the measurement, and illustrate the prevalence of I^- anions over the Na^+ cation near the model water surface [15].

5   Core Electron Spectroscopy of Protonation/Deprotanation in Aqueus Solution

High resolution photoelectron spectroscopy at here available energies up to 1000 eV provides, also, a useful, sensitive, tool for chemical-environment sensitive K-shell core electron spectroscopy in a number of low-Z atoms, such as C atoms, N, or O-atoms which are prevailing in solute organic molecules. Chemical environment induced shifts in K-shell photoelectron spectroscopy are well known since the earliest studies of Kai Siegbahn's group in the 1960's (on solid state probes), then coined as ESCA, the electron spectroscopy chemical analysis. A great advantage of core shell spectra is in identifying chemical changes near a single, specific, atom in a chemical compound.

In Figs. 12 and 13 it is illustrated how the pH value induced change by protonation

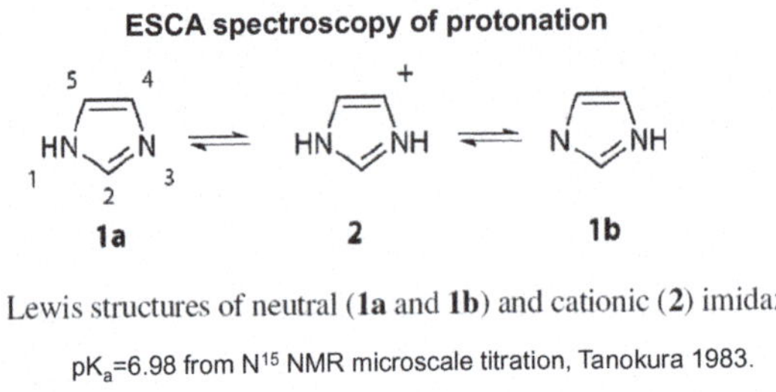

Fig. 12 Lewis structures of neutral (1a and 1b) and of cationic (2) imidazole. Known pKₐ = 6.98 from N¹⁵ NMR- microscale titration (Tanokura 1983) (figure reproduced from Ref. [16], Fig. 1)

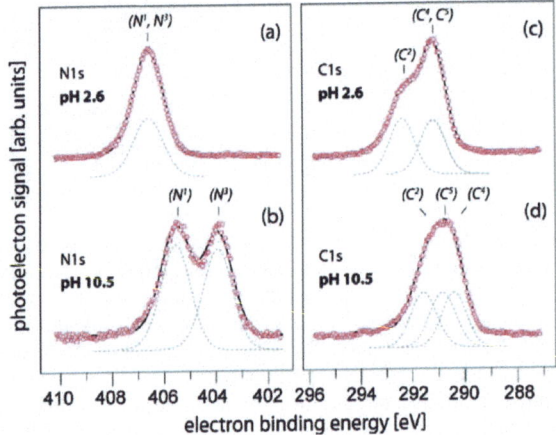

ESCA spectroscopy of protonation

Lewis structures of neutral (**1a** and **1b**) and cationic (**2**) imidazole.

pKₐ=6.98 from N¹⁵ NMR microscale titration, Tanokura 1983.

Fig. 13 Inner shell Nitrogen 1s (**a, b**) and Carbon 1s (**c, d**) photoelectron spectra show energy shifts for protonation and deprotonation of 2m imidazole aqueous solutions measured at pH 2.6 (cationic structure 2), and at pH 10.5 (neutral molecule 1a, 1b). The photon energy is 480 eV for the N1s measurement and 380 eV for C1s. Smooth intermittent lines show fitting results to the experimental spectra (red circles). In the protonated state the two nitrogen atoms in imidazole are indistinguishable (**a**) (figure reproduced from Ref. 16, Fig. 2)

can be traced on individual atoms of a solvated organic molecule [16]. Imidazole is a five atom ring compound, made of two N-atoms and three C-atoms, shown in the Lewis-structure representation of the molecule in Fig. 12. Protonation of the neutral dissolved imidazole molecule leads to two identical NH groups appear in the molecule, Fig. 12 structure (2), while the charge of the added proton is delocalized over the whole molecule. The pK value for imidazole protonation of pKₐ = 6.98 had been determined by, isotopically enriched, N¹⁵ NMR microscale titration by Tanokura, in 1983 [16]. The 1 s core electron photoelectron spectra for the N1s and for the C1s core states in 2 m imidazole solutions for two pH values, in Fig. 13, show the effects of the transition from neutral to protonated molecules. At pH 10.5, in the lower row of spectra, the imidazole molecule is neutral and, as expected from the structure formulas 1a and 1b for the neutral compound, two separate N1 s energy levels are observed for the two different nitrogen atoms in the ring, one peak for the nitrogen atom N^1 where the hydrogen atom is attached and, separated from the first peak by 1.5 eV, a different peak N^3 for the second nitrogen in position 3 of the ring (indicated in the structure drawing in Fig. 12). When changing the pH value both peaks are decreasing and a third peak, in the position of the N1s single peak labeled (N^1, N^3) in the upper spectrum is growing up in intensity. In subsequent measurement with solutions of different pH values, not shown here for shortness, after crossing pK = 7 value the first double peaked structure is shrinking further in amplitude until the fully protonated imidazole solution at pH 2.6 shows only one single photoelectron peak for nitrogen, confirming theoretical chemistry results that here two pseudo-equivalent NH groups exist with the positive charge/electron hole-orbital distributed equally over the location of both N-atoms, and not a NH group and a distinct NH^+ is formed after the proton attachment. This transition of the binding sites of the two distinct N atoms in the neutral state to two identically bound nitrogens in the protonized ion is further reflected in the C1s peak structures of the three carbon atoms bound in the molecule's ring structure. For the neutral imidazole, Fig. 13d, at pH 10.5 a broadened carbon C1s photoelectron spectrum is observed which can be deconvoluted into 3 nearby lying C1s states with similar amplitudes, corresponding to the three different C/N neighborhood bonding configurations of the three carbon atoms in the positions 2, 5 and 4, respectively, indicated in the, Fig. 12, structure scheme drawings 1a and 1b. In the charged state, Fig. 13c at pH 2.6, however, these carbon C1s levels contract to two overlapping states in new positions on the energy scale, as shows the deconvolution of the peak structure into two standard width peaks. The stronger peak is attributed to, both, the C^4 and the C^5 atoms, which are now in identical neighborhoods, as expected for the protonated imidazole species. The deconvoluted core shell C1s peak amplitudes, individually image the stoichiometric ratios for different atoms in the photoelectron spectrum and, accordingly, the joint (C^4, C^5) carbon 1s peak shows twice the amplitude of the separate C^2 peak originating from bottom C-atom C^2 connecting the two identical NH groups in the protonated imidazole. In summary this shows, here titration of charged/neutral molecular states can be performed quantitatively, in stoichiometric precision.

In further detail, this 1s core level PES titration demonstrates, in addition, the very distinct methodical advantage of the exceptionally high intrinsic time resolution on the order of sub-femtoseconds, given for the photoemission process by the time scale for the removal of the fast electron from the parent atom. The protonation/deprotonation bond-making and bond-breaking processes in solution take place on time scales of 10^{-12} s. Thus, in the 1 s photoelectron spectra of solutions near the pK_a point always two distinct peaks for protonated and deprotonated species populations appear simultaneously. In contrast in the classical NMR microtitration procedure the averaging time is limited by the period of the absorbed resonance frequency, in an order of 10^{-8}–10^{-9} s. Therefore, averaging over many proton bond making-and-breaking cycles occurs in the NMR method which results in a frequency shift with weighted averaging over the two distinct states, only, without simultaneous separation of both levels.

A somewhat more complex case of chemical adsorption and reaction is investigated by photoelectron spectroscopy studies shown, in Fig. 14, for analysis of the details of carbon dioxide capture in industrially used solutions of Monoethanolamine ($HOC_2H_4NH_2$) for washing CO_2 from flue gas [17]. Known for more than a century, the chemical steps involved in the gas capture process have been extensively studied and characterized in great detail. 30% monoethanolamine (MEA) in aqueous solution has a CO_2 load capacity of 0.25 mol/L. The principal capture reaction is:

$$2\,MEA + CO_2 \rightarrow MEA - COO^- + MEA - H^+$$

Acid/base equilibria are:

$$MEA + H_2O \leftrightarrow MEA - H^+ + OH^- \quad pK_a = 9.55$$

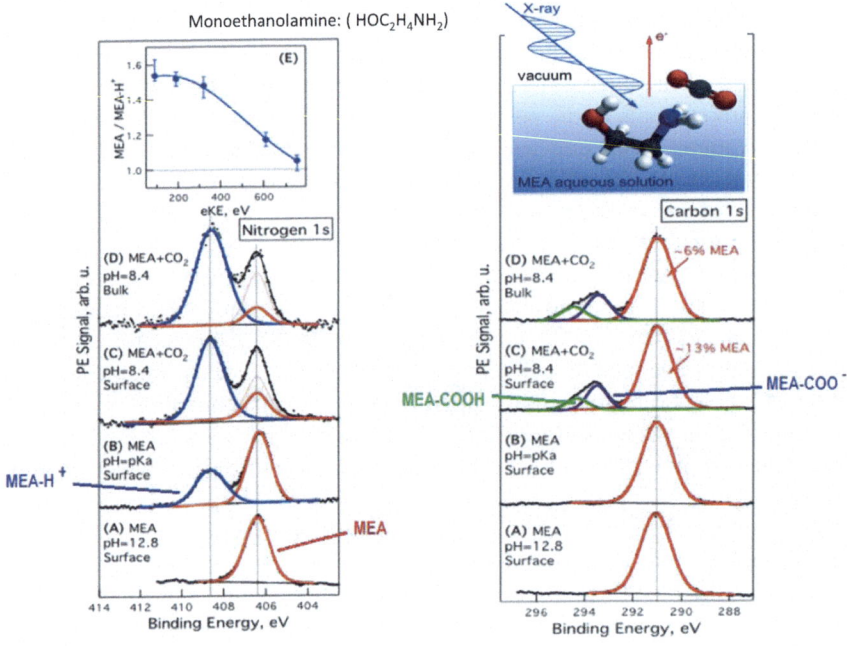

Fig. 14 CO_2 capture in MEA solution (Left column) nitrogen 1s photoelectron spectra for 4.9 m MEA (A, B) and for CO_2 treated MEA solutions (C, D) show varying contributions from MEA in its neutral form (red, 406.3 eV BE) and from protonated MEA (blue, 408.6 eV). Red intermittent lines represent carbamate reaction product contributions (see text). The ratio (E) of neutral and protonated MEA molecules changes as a function of depth in a solution of CO_2 loaded MEA at pH 8.4 (Right column) carbon 1s PES spectra for MEA (A, B) and CO_2 loaded MEA (C, D). In CO_2 treated MEA separate peaks from carbamate (low BE, purple) and carbamic acid (green, high BE) appear. Red labels on the peaks for BE=291 eV indicate the percentage of neutral MEA contribution (figure from Ref. [17], Figs. 1 and 2)

$$MEA - COOH + H_2O \leftrightarrow MEA - COO^- + H_3O^+ \quad pK_a \text{ unknown } (7 - 9?)$$

Photoelectron spectra for C1s and for N1s were taken for MEA-CO_2, gas loaded solutions (Fig. 14c, d) and for 30% MEA solution without gas load (Fig. 14a, b). pH values are adjusted to the technical working point, or to other values, when needed for analysis of details. At pH 12.8 the MEA in solution is completely neutral, and the N1s binding energy spectrum (A), at the bottom line of Fig. 14, shows a single nitrogen atom peak centered at $E_B = 406.4$ eV with half width 1.3 eV. The spectrum is a surface spectrum, taken at electron kinetic excess energy 90 eV. In (B) the pH value was adjusted to the (bulk) $pK_a = 9.5$. In this surface spectrum a second peak, smaller by a ratio 1.6, appears at 408.8 eV BE and is identified as the N1s signature of the protonated NH_3^+ group in the MEA-H^+ fraction. The ratio of the two peak amplitudes represents the quantitative ratio of the two MEA species in the locally probed region of the solution. The ratio was determined in a series of additional measurements with progressively higher photoelectron excess kinetic energies in a range up to 750 eV and shows, in plot (E) in the upper row of Fig. 14, the ratio is reducing from an excess 1.6:1 of neutral MEA-molecules near the surface to a 1:1 ratio for 750 eV electrons which originate in greater depth from the bulk phase of the liquid.

The two remaining N1s spectra (C) and (D) are taken for gas saturated MEA solution with a CO_2 load of 0.24 mol/L. These solutions change the original, equilibrium pH value from pK_a to pH = 8.4 in the loaded state. The transformation of MEA to carbamate MEA-COO^- and to carbamic acid MEA-COOH changes little in the N1s binding energy of the NH_2 group on the opposite end of the molecule. The peak appearing at the position of "neutral MEA" in these spectra is the superposition of unknown fractions of contributions from loaded and non-loaded MEA with two almost identical peak shapes. More can be learned from a consideration of C1s spectra shown, adjacent to respective N1s results, on the right-hand side of Fig. 14. For neutral MEA at pH 12.8, and for MEA at pH = 9.5 a single narrow C1s peak structure for all carbon atoms in the compound is observed at 291 eV binding energy in the spectra shown in (A) and (B), respectively. In the CO_2 saturated MEA solution at the surface (C) and for bulk solution (D) two new, distinctly visible C1s peaks arise from carbamate (MEA-COO^-) and for carbamic acid (MEA-COOH), with energies shifted to higher binding energy. From the intensity ratios of the carbamate and of the carbic acid peaks, at known pH-value of the solution, the first experimental determination of the previously uncertain pK_a equilibrium with a $pK_a = 8.2$ is here obtained. Further evaluation of the C1s peak ratios, in combination with measured relations between neutral and protonated species from the MEA/MEA-H^+ ratios of the simultaneously recorded N1s peak spectra, eventually, yields an absolute concentration ratio of (MEA) over (MEA-COO^- + MEA-COOH) of ~0.22 when probing the surface and of ~0.09 for probing the bulk. The carbamate products have a preference for moving into the bulk and MEA a tendency to be enriched on the surface. This provides a perfectly cooperative cycling support for the CO_2 trapping process

at the interface and for the subsequent removal of carbamates into the bulk of the
washing fluid [17].

6 Excited States of Water, Resonant Auger Spectroscopy of H_2O_{aq} and OH^-_{aq}

Liquid water is dissociating spontaneously into a very small fraction of H_3O^+ and
OH^- ions, with far ranging consequences on the properties of aqueous solutions.
The ions with concentrations of 10^{-7} mol/L in pure water, by far, are too small
to be observed in photoelectron spectra. Thus, solutions of strong acids (1 m HCl,
corresponding to pH $= -1$) and of strong bases ($LiOH_{aq}$), instead, are to be used for
photoelectron spectroscopy of the self-ionization products of water [18]. In spectra,
shown in Fig. 15, valence orbitals for H_3O^+ and for OH^- have been studied and show
a weak perturbation of the dominant H_2O valence photoelectron structure, superimposed by faint photoelectron emission from H3O+, in Fig. 15a, and a small peak
localized at the water ionization threshold with an OH^- ionization energy of 9.2 eV,

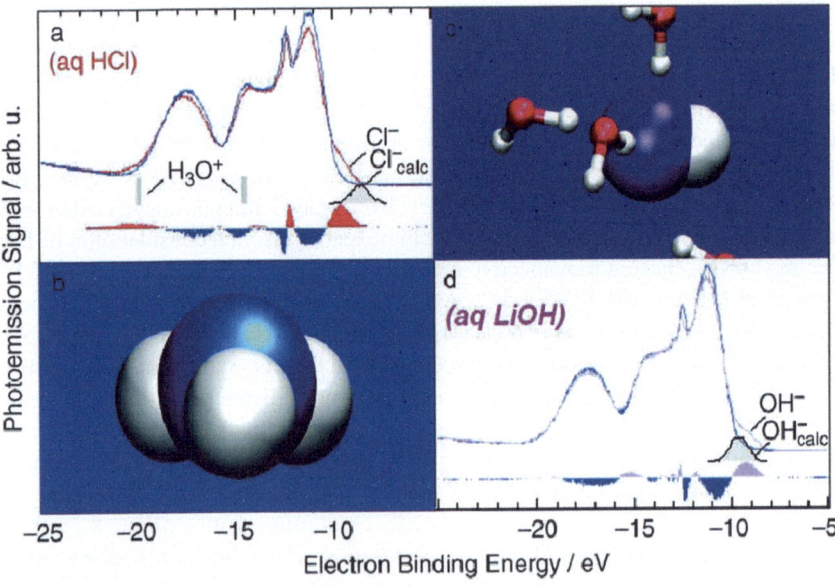

Fig. 15 $H_3O^+_{aq}$ and OH^-_{aq} valence photoelectron spectra obtained by difference measurements of
3 m HCl aqueous solution with neat water (**a**) and of a 2 m LiOH solution (d). The $H_3O^+_{aq}$ ion signature
appears as a weak peak at 20.8 eV BE, in the tail region of the $2a_1$ inner valence orbital of water. The
OH^- electron binding energy in solution is comparable to other halogen ions. Also, for comparison
with the experiment, theoretically assigned ionization energy positions for $H_3O^+_{aq}$ and OH^-_{aq} are
indicated by grey shaded features (figure reproduced from Ref. [18], Fig. 2)

for the LiOH solution in Fig. 15d, similar in magnitude to the previously considered photoionization energies of halogen anions. Separately taken, photoelectron spectra of pure water are here compared to the photoelectron spectra of the H^+ and OH^- in solutions and are used for extracting difference spectra shown in the lower part of Fig. 15a, d. Simulated spectra for calculated values for the OH^- and the Cl^- ionization potential are also shown, and are found to be in reasonable agreement with the measurement [18]. For the solvated proton different hydronium configurations were considered in the model calculation. The "Eigen"-like aqueous cation structure H_3O^+ was judged to agree best with the experimental ionization energies derived from the HCl solution spectra, as indicated by vertical bars for theory results, in Fig. (15a).

OH^- in the gas phase exists for a single negative ion ground state, only, and similarly to the anions of halogen atoms the attractive potential well is too shallow to support any exited electronic state. In contrast, in aqueous solutions the anions are embedded in liquid water in the additional polarization potential described by the dielectric-cavity/Born-model which increases the well depth and binding energy to about 10 eV, a value large enough to allow for the existence of an excited electron state, at a binding energy of the order of 1 or 2 eV below the ionization threshold. The optical transition to this excited state, a very strong s $\rightarrow$ p absorption line, was first observed in the 1930's in UV absorption spectra of I_{aq}^-. The phenomenon was called charge transfer to solvent (CTTS) and has long drawn the attention of spectroscopists because the CTTS states existed in liquids, only.

With the availability of narrow band, tunable soft X-ray synchrotron radiation it became possible, also, to explore this CTTS band by resonant excitation from an inner core level, and monitoring the resonance in the Auger electron emission spectrum. This has the advantage that the limitations of "classical" UV/VUV spectroscopy by the onset of the strong absorption of liquid water above 9 or 10 eV can be offset in the Auger method. Before testing with OH_{aq}^- in liquid water, we explored the technology first on Cl^- ions where the process is simpler and better known from previous optical VUV spectroscopy work [19]. An illustration of the different possible Auger excitation processes is shown by the three schemes drawn, in Fig. 16, for the Cl^- anion. Direct Auger electrons (1) are emitted following an excitation of a core hole vacancy by photoelectron emission. In a rapidly following step, within a few fs, the hole in the inner shell 2p-level is filled by an outer valence 3p electron and the gained energy is transferred to a second outer valence shell 3p electron which is emitted as the (LMM-) Auger electron. Spectator Auger electrons (2) are emitted after resonant excitation of a 2p electron into unoccupied levels "e1, e2, ..." of the solvated Cl_{aq}^- ion. The energy of the spectator-Auger electron is higher than for normal Auger electrons because the originally photon-excited electron is still present in the ion and increases the coulomb forces acting on the outgoing Auger electron. A third process may occur (3) in a shake down, transferring electron-energy of the resonantly populated levels to other internal states before the Auger electron is emitted, with the result of an additional change in the kinetic energy of the emitted spectator-Auger electron.

In a series of photoelectron spectra records at closely spaced energies of the incoming synchrotron radiation, tuned over the region of interest for expected 2p to CTTS transition, shown in Fig. 17, these discussed Auger phenomena can be

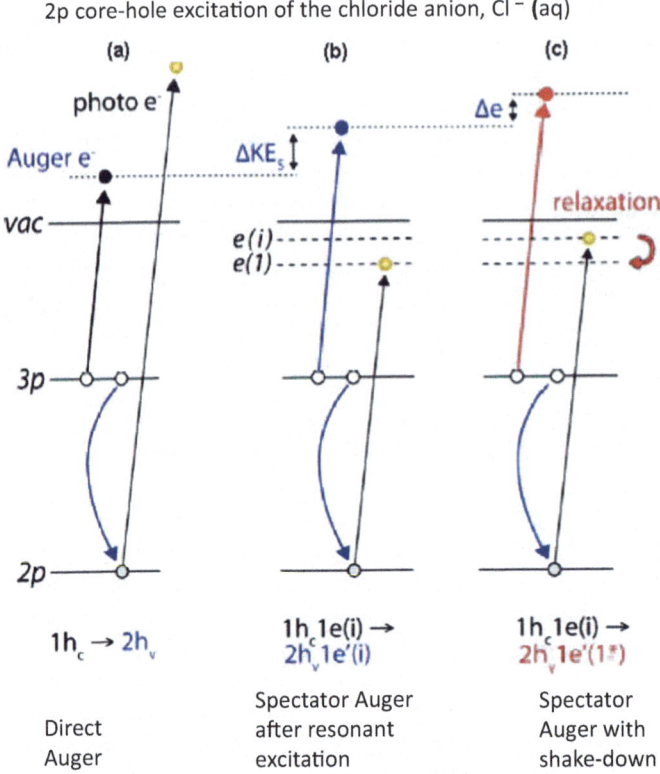

2p core-hole excitation of the chloride anion, Cl⁻ (aq)

Fig. 16 Auger processes associated with 2p core-hole excitation of the chloride anion, Cl_{aq}^-, in aqueous solution. Negative halogen ions have no excited states in the gas phase. In aqueous solution one or several new excited states appear, called charge transfer to solvent states (CTTS). **a** Direct Auger process with emission of a photoelectron followed by emission of the Auger electron. **b**, **c** Spectator Auger emission after resonant excitation of one of the CTTS states $e(1) - e(i)$. The spectator Auger electron kinetic energies are shifted due to electrostatic interaction by the presence of the resonance electron. **c** Occasionally, energy of the resonantly excited state is transferred to nearby internal states before the Auger electron is emitted (figure reproduced from Ref. [19], Fig. 5)

actually observed [19]. At the lowest photon energy, at 200 eV, the electron kinetic energy spectrum shows only the familiar peak structures of the liquid water valence bond states, and in addition, the small Cl_{aq}^- peak shoulder adjacent to the right-hand side of the water $1b_1$ peak. At the highest photon energy 204.8, the electron kinetic energy spectrum of the water valence structure photoelectron spectrum is shifted toward 4.8 eV higher kinetic energies in accordance with the higher photon energy. In addition, the 204.8 eV spectrum shows the strong, fully developed, regular Auger peak for Cl_{aq}^- LMM Auger emission with indicated doublet splitting n, n̲. The LMM Auger peak starts to develop at photon energies larger than the $2p_{1/2,3/2}$ level ionization energy of 201 eV. Its signature is the constant kinetic electron energy, independent from the incident photon energy. Most importantly, however, in the intermediate

Fig. 17 Photoelectron and Auger electron spectra for Cl_{aq}^- measured at photon energies near the detachment energy for 2p orbital electrons. A 3m LiCl aqueous solution liquid jet is used. Normal Auger peaks for the Cl_{aq}^- $2p_{3/2}$ (**n**) and $2p_{1/2}$ (**n_**) LMM process (see Fig. 16a) occur at constant kinetic energy. Emitted photoelectron peak positions for water move with increasing photon energy. Resonant absorption transitions into unoccupied CTTS states of Cl_{aq}^- are readily observable at the resonant Auger line positions **1_, 2, 2_, 3_, 4**. Resonance features **2, 4** are attributed to transitions originating from $p_{3/2}$, **1_, 2_** and **3_** from $p_{1/2}$ Auger resonances in CTTS state (figure reproduced from Ref. 19, Fig. 2)

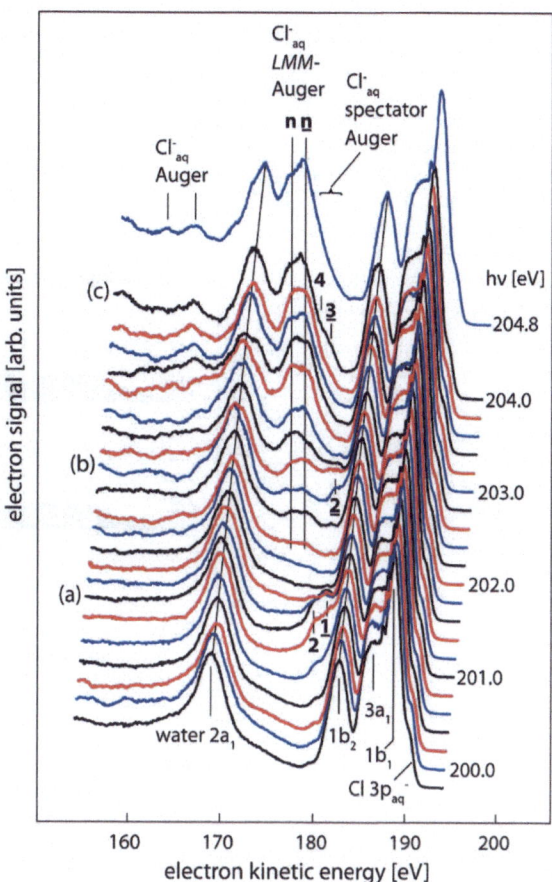

region of scanned photon excitation energies some photoelectron spectra show small intensity bumps appearing only in very narrow excitation energy ranges ≤ 0.3 eV full half width, which originate from the searched-for resonant excitation of Cl^- unoccupied level states. The resonance Auger features are identified in Fig. 17 by the numbers 1, 2, 3 and 2 or 4, pointing to the observed features at three different excitation photon energies where resonances could be detected. The region of the (2, 1) Cl_{aq}^- spectator-Auger peaks group is drawn enlarged in Fig. 18, showing the signal evolution as a function of photon energy. The blue peaks are Gaussian fits of 2 and 1 to the experimental kinetic energy spectra envelopes. After analogous evaluation of all other observed spectator-Auger resonances an energy level diagram of these newly identified unoccupied electronic levels can be constructed and is shown in Fig. 19 together with the also determined absolute energy level values for the occupied electron orbital states 3p and $2p_{1/2}$, $2p_{3/2}$ which were obtained simultaneously from the measured photoelectron spectra of Cl_{aq}^-. The Cl^- excited states orbitals in aqueous solutions are here found at binding energies of 2.5 and

Fig. 18 Enlarged region of Cl^-_{aq} photoelectron spectra in Fig. 17, showing the signal evolution of spectator Auger-electron peaks **2** and **1_** at photon energies between 200.4 eV and 201.4 eV. Blue peaks are Gaussian fits to structures **2** and **1_** in the experimental photoelectron spectra (red) (figure reproduced from Ref. 19, Fig. 3)

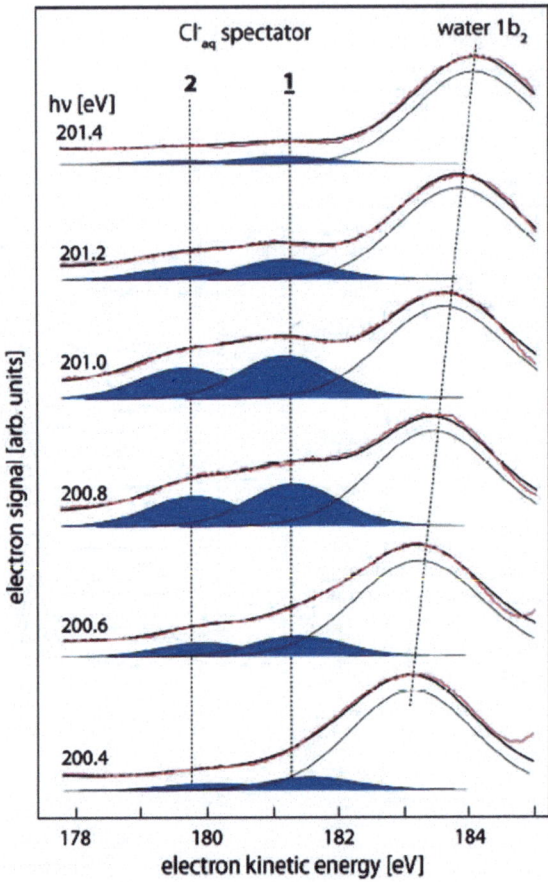

Fig. 19 Experimental CTTS states of Cl-**aq** and electron binding energies for the 3p and 2p states of the negative chlorine ion in 3 m LiCl aqueous solution (figure reproduced from Ref. 19, Fig. 6)

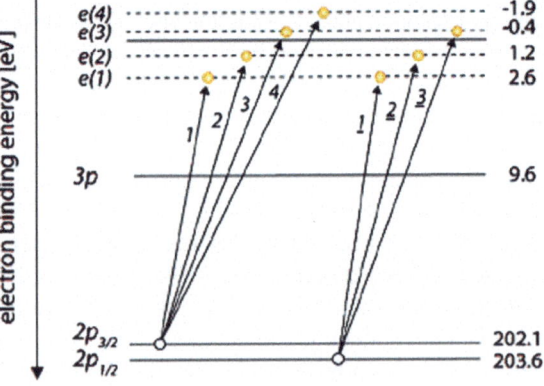

of 1.3 eV, and two further, antibound state resonances were identified, appearing slightly above the vacuum level, at -0.4 and -1.9 eV, respectively [19].

With the expertise acquired in resonant spectroscopy of CTTS states in the simple, spherical Cl^-_{aq} anion, we return to the OH^-_{aq} with a study to find the CTTS state here in resonant excitation from the 1 s core state of the O-atom in OH^-_{aq} which is the only inner shell state available in this anion entity [20]. Fig. 20 shows electron kinetic energy spectra obtained in a narrow scan range of photon energies near the expected O1s to CTTS transition. In a series of preceding photoelectron experiments the OH^-_{aq} O1s binding energy had been determined to be 536.0 eV (shown in Fig. 21, in a combined experimental energy level diagram for liquid H_2O and aqueous OH^-). In search for the 1s resonance of OH^-_{aq} near the continuum threshold, kinetic energy electron spectra are shown, in Fig. 20, for 4 m NaOH solutions (red lines) and for

Fig. 20 4 m NaOH aqueous solution photoelectron spectra, showing resonant Auger-electron spectra (**b**, **c**) at 532.2 and 532.8 eV, and (ultrafast) intermolecular coulombic decay of OH^-_{aq}. The photon energy is scanned near the ionization threshold for the oxygen O1s inner shell orbital. In addition, photoelectron spectra of pure water are recorded (blue) for reference. Resonance Auger electron peaks 2, 3, and 4 reveal the existence of a very fast energy transfer process from excited OH^- to the H_2O solvent. The small peak at highest kinetic energy, far right, arises from O1s ionization by spurious, second harmonics photons with energy 2 hν, and, provides a method for highly accurate absolute energy calibration (figure reproduced from Ref. [20], Fig. 1)

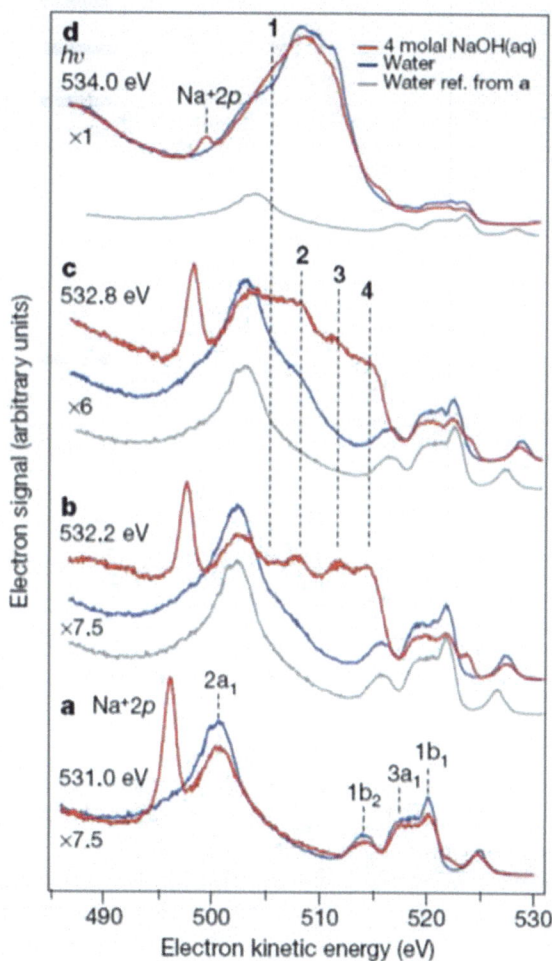

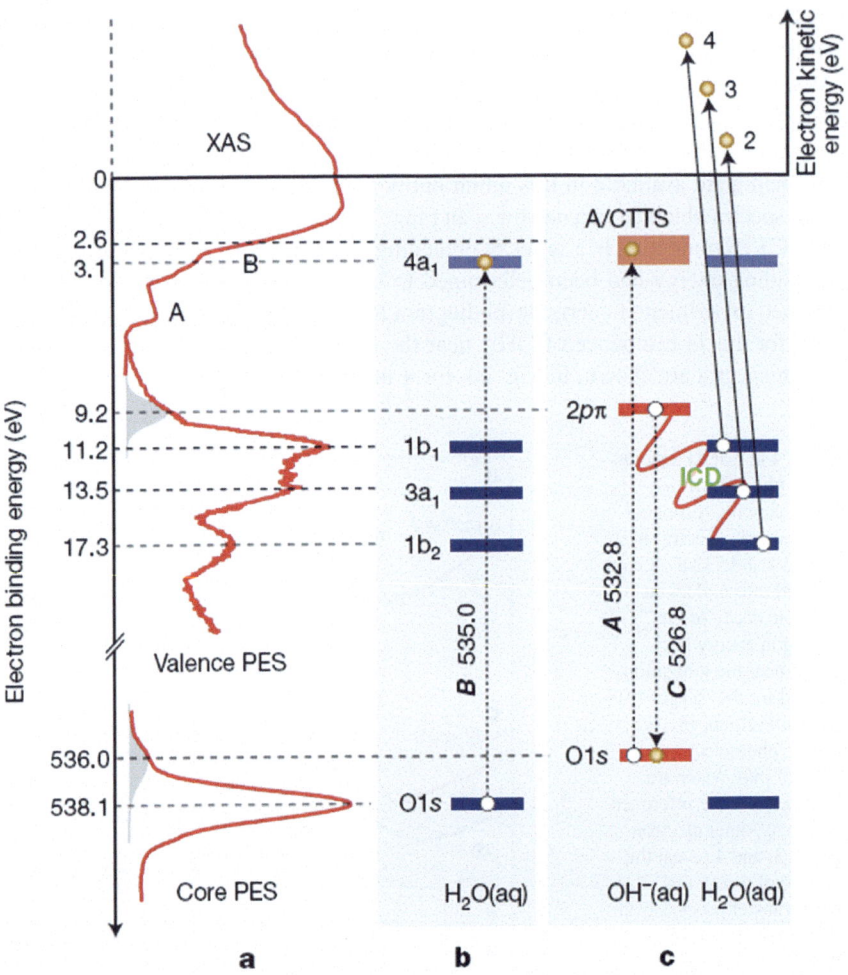

Fig. 21 Complete, experimental energy level diagram of OH^-_{aq} and $H_2O^-_{aq}$ obtained from liquid microjet photoelectron spectroscopy of a 4m NaOH solution. Shown are bound states and, in addition, un-occupied energy levels ($4a_1$ of H_2O and a CTTS of OH^-_{aq}) identified by resonant Auger spectroscopy. Plotted on the left side (**a**) are experimental PES spectra used for the energy level determination; shaded Gaussian peaks (light grey) are the deconvoluted contribution of OH^-_{aq} states (figure reproduced from Ref. [20], Fig. 2)

neat water (blue line) at the 4 photon energies 531.0, 532.2, 532.8, and 534.0 eV. Figure 20a, at 531.0 eV photon energy, shows a reference spectrum taken at an energy slightly below the onset of resonances. In the (blue line) neat water valence spectrum the well known four peaks $2a_1$, $1b_2$, $3a_1$, $1b_1$ are observed and the (red line) NaOH$_{aq}$ spectrum is showing water plus the additional Na$^+_{aq}$ 2p lines as well as the small valence OH$^-_{aq}$ peak with $E_{kin} = 521.8$ eV. At the subsequent photon energy values at 532.2 eV and at the close by value 532.8 eV the NaOH brine spectra (red) have changed dramatically from the, respective, neat water photoelectron spectra (blue), showing O1s resonance spectator-Auger features in a remarkably high intensity and with very broad additional structure. Further scans for a slightly higher photon energy 534.0 eV, in Fig. 20d, show this is off resonance, already. Except for the OH⁻ peak and the Na⁺, there are no big differences left between the recorded neat water and the NaOH spectrum. A broad and intense peak structure beginning to appear in, both, neat water and in NaOH solution in Fig. 20d arises from a superposition of the water valence state $2a_1$ photoelectron emission peak and an onset of strong water Auger peak structure resulting from excitation of the $4a_1$ unoccupied state of H_2O_{aq} from O1s with excitation centerline energy 535.0 eV. It is designated by the label 1 on top of the spectrum. Within the spectacular OH⁻ spectator resonance spectra, Fig. 20b, c, three adjacent peak features are recognized with spacings closely matching the energy differences and overall structure between the $1b_2$, $1a_1$ and $1b_1$ valence photoelectron spectra of H_2O. They are marked by the letters 2, 3 and 4 and three additional vertical lines, for underlining the fact that the kinetic energies are constant for Auger electron emission. The apparent, unexpected mixing of valence states of adjacent H_2O molecules into the Auger decay process of core hole excited OH⁻ is an obviously new phenomenon, difficult to understand [20]. Although it is well known, that the high electrical mobility of the OH⁻ ion in liquid water has to be attributed to a charge migration process, rather than to molecular diffusion, the time scale is 10^{-12} s for the established charge migration model where a local $OH^-(H_2O)_n$ hydration cluster is rearranging bonds and hopping the charge to a newly formed OH⁻ center. This mechanism, therefore, must be disputed as being very unlikely because the charge migration step takes orders of magnitudes longer than the decay time of ~ 7 fs, available for the here observed O1s Auger processes. A different, more recent theory attributes the mixing of adjacent molecules' molecular states in an Auger process to a very rapid Intermolecular Coulombic Decay (ICD) phenomenon, with quantitative details for OH_{aq}^- resonance here also still open to discussion.

With the tentative assignments of the OH_{aq}^- resonance emission peak structures number 2, 3 and 4 to Auger emission of water valence electrons $1b_1$, $2a_1$, $1b_2$, in Fig. 20b, c, the picture emerges that one electron from the OH_{aq}^- valence state $2p\pi$ is filling the O1s hole in OH_{aq}^- and, simultaneously, in undisclosed dynamical detail, one of the neighbor-H_2O valence electrons is emitted and observed in the recorded kinetic energy electron spectra. This Auger process interpretation is symbolically depicted in the energy level diagrams at the right-hand side in Fig. 21c. In this energy level scheme are compiled all experimentally determined energies for OH_{aq}^- and, also, all levels for liquid H_2O (in Fig. 21b) which have been obtained from the here described photoelectron spectroscopy measurements. On the left-hand side of the level diagram, Fig. 21a, characteristic photoelectron spectra traces for core and valence states of water and of OH⁻ in solution are displayed for illustrating the relationship between measured PES features and the herewith determined electron orbital energies in aqueous solution. The also shown energetic positions of the un-occupied excited state $4a_1$ of water in aqueous solution and the new CTTS state for solute OH⁻ ions are determined with the resonant Auger spectroscopy method, just discussed.

For comparison with more traditional and longer established XAS methods, in the upper part of Fig. 21a is shown, furthermore, a measurement of electronic structure in liquids by X-ray absorption spectroscopy on the K-edge, in the region of O1s excitation to the ionization vacuum level of an aqueous NaOH solution. In the X-ray near edge absorption scan with high resolution tunable x-ray radiation, here the total absorption increases in a large step at the K-edge of water. In addition, pre-edge absorption XAS structures appear at slightly lower photon energies, assigned with the letters A and B and originating from resonance absorptions from unoccupied bound states. The pre-edge structure peak B is related to the excitation of the $4a_1$ unoccupied state of water at 3.1 eV electron binding energy according to the photoelectron energy absolute calibration in Fig. 21b. The peak A observed for OH⁻ in the XAS spectrum appears at lower excitation energy and its position can be interpreted correctly, only, after it is known from PES results in Fig. 21c that the vertical ionization energy of O1s electrons of the OH_{aq}^- with 536.0 eV BE is smaller by 2.1 eV than the liquid water O1s ionization energy value. Not visible at all in XAS, of course, is the Auger decay transfer from OH⁻ to vicinal H_2O molecules, following excitation of CTTS state.

7 Concluding Remarks

In going through this retrospective on the development of liquid water microjets for molecular beam studies of evaporating nascent molecular velocity distributions, and then extended, for X-ray photoelectron spectroscopy as diverse as simple determination of valence energies in electrolytes, concentration measurements of surface versus bulk abundance, diagnostics of pH sensitivity of protonation-deprotonation in 1s K shell states individual molecular group atoms, and the detailed spectroscopy of unoccupied near vacuum level states of solvent and solute molecules by resonant Auger spectroscopy, in summary, I feel strongly compelled to thank very many coauthoring colleagues who were with me on this journey at different times over more than three decades. For names I can refer here, only, to the shorter list of coauthors given in the cited references, although it was many more people who have lent their hands, discussed ideas and kept the projects going by their support. Also, I gratefully acknowledge the continuing support by my home institution, Max-Planck-Institut für Strömungsforschung/MPI Dynamics and Self-Organization/, by the Deutsche Forschungsgemeinschaft, by the BESSY synchrotron radiation facility, and by the Max-Born-Institut.

References

1. Gasparus Scotus "Technica Curiosa" p. 172, experimentum XXXVIII; Nuremberg 1664 (Digital Edition: Herzog August Library, Wolfenbüttel)
2. M. Faubel, S. Schlemmer, J.P. Toennies, A molecular beam study of the evaporation of water from a liquid jet. Z Phys D Atoms Molecul Clust **10**, 269–277 (1988)
3. M. Faubel, Photoelectron spectroscopy at liquid surfaces, Chap. 12, vol. I, in *Photoionization and photodetachment*, vol. 101A, ed. by C.Y. Ng (World Scientific Publishing, Singapore, 2000), pp. 634–690
4. S. Hess, M. Faubel "Gase und Molekularstrahlen" Fig. 1.70, in: *Bergmann-Schäfer Lehrbuch der Experimentalphysik* Bd. 5, ed. by K. Kleinermanns, Walter de Gruyter (Berlin, New York 2006), p. 119
5. O. Stern, Nachtrag zu meiner Arbeit: Eine direkte Messung der thermischen Moleku-larstrahlgeschwindigkeit (Comment to my paper: A direct measurement of the thermal molecular beams velocity). Z. f. Phys. **3**, 417–421 (1920) "Zusammenfassung: Es wird die Frage der Geschwindigkeitsverteilung der von einer Flüssigkeitsoberfläche ausgehenden Moleküle diskutiert. Ferner werden die Resultate einiger neuerer Messungen der mittleren Geschwindigkeit von Silberatomen mitgeteilt" 'Summary: I discuss the question of the velocity distribution of molecules emerging from the surface of a liquid. In addition, results of some new measurements of the mean velocity of silver atoms are communicated."
6. M. Faubel, T. Kisters, Non-equilibrium molecular evaporation of carboxylic acid dimers. Nature **339**, 527–529 (1989)
7. D. Chandler, *Introduction to Modern Statistical Mechanics* (Oxford University Press, New York, 1987)
8. M.P. Allen, D.J. Tildesley, *Computer Simulation of Liquids* (Oxford University Press, Oxford, 1987)
9. J.R. Rumble (ed.), *CRC handbook of chemistry and physics*, 98th edn. (CRC Press, Taylor & Francis Ltd, Boca Raton, London, 2017)
10. M. Faubel, B. Steiner, J.P. Toennies, Photoelectron spectroscopy of liquid water, some alcohols, and pure nonane in free micro jets. J. Chem. Phys. **106**, 9013 (1997)
11. M. Faubel, B. Winter, Photoemission from liquid aqueous solutions. Chem. Rev. **106**(4), 1176–1211 (2006)
12. W.S.M. Werner, W. Smekal, C.J. Powell, *NIST database for simulation of electron spectra for surface analysis* (U.S. Department of Commerce, National Institute of Standards and Technology, Gaithersburgh MD, 2005)
13. R. Weber, B. Winter, P.M. Schmidt, W. Widdra, I.V. Hertel, M. Dittmar, M. Faubel, Photoe-mission from aqueous alkali-metal—iodide salt solutions using EUV synchrotron radiation. J. Phys. Chem. B **108**(15), 4729–4736 (2004)
14. N. Ottosson, M. Faubel, S.E. Bradforth, P. Jungwirth, B. Winter, Photoelectron spectroscopy of liquid water and aqueous solution: Electron effective attenuation length and emission-angle anisotropy. J. El. Rel. Phen. **177**, 60–70 (2010)
15. P. Jungwirth, D.J. Tobias, The molecular structure of salt solutions. J. Phys. Chem. B **105**, 10468–10472 (2001)
16. D. Nolting, N. Ottosson, M. Faubel, I.V. Hertel, B. Winter, Pseudoequivalent nitrogen atoms in aqueous imidazole distinguished by chemical shifts in photoelectron spectroscopy. J. Am. Chem. Soc. **130**(26), 8150–8151 (2008)
17. T. Lewis, M. Faubel, B. Winter, J.C. Hemminger, CO2 Capture in amine-based aqueous solution: role of the gas–solution interface. Angew. Chem. Int. Ed. **50**, 10178 (2011)
18. B. Winter, M. Faubel, I.V. Hertel, C. Pettenkofer, S.E. Bradforth, B. Jagoda-Cwiklik, L. Cwiklic, P. Jungwirth, Electron binding energies of hydrated H3O+ and OH−: photoelectron spec-troscopy of aqueous acid and base solutions combined with electronic structure calculations. J. Am. Chem. Soc. **128**(12), 3864–3865 (2005)

19. B. Winter, E.F. Aziz, N. Ottosson, M. Faubel, N. Kosugi, I.V. Hertel, Electron dynamics in charge-transfer-to-solvent states of aqueous chloride revealed by Cl⁻ 2p resonant auger-electron spectroscopy. J. Am. Chem. Soc. **130**(22), 7130–7138 (2008)
20. E.F. Aziz, N. Ottosson, M. Faubel, I.V. Hertel, B. Winter, Interaction between liquid water and hydroxide revealed by core-hole de-excitation. Nature **455**, 89–91 (2008)

Chapter 27
When Liquid Rays Become Gas Rays: Can Evaporation Ever Be Non-Maxwellian?

Gilbert M. Nathanson

Abstract A rare mistake by Otto Stern led to a confusion between density and flux in his first measurement of a Maxwellian speed distribution. This error reveals the key role of speed itself in Stern's development of "the method of molecular rays". What if the gas-phase speed distributions are not Maxwellian to begin with? The molecular beam technique so beautifully advanced by Stern can also be used to explore the speed distribution of gases evaporating from liquid microjets, a tool developed by Manfred Faubel. We employ liquid water and alkane microjets containing dissolved helium atoms to monitor the speed of evaporating He atoms into vacuum. While most dissolved gases evaporate in Maxwellian speed distributions, the He evaporation flux is *super*-Maxwellian, with energies up to 70% higher than the flux-weighted average energy of $2\,RT_{liq}$. The explanation of this high-energy evaporation involves two beautiful concepts in physical chemistry: detailed balancing between He atom evaporation and condensation (starting with gas-surface collisions) and the potential of mean force on the He atom (starting with He atoms just below the surface). We hope that these measurements continue to fulfill Stern's dream of the "directness and simplicity of the molecular ray method."

1 Introduction: J. C. Maxwell and Otto Stern

Otto Stern's first publication, in 1920, described an ingenious Coriolis measurement of the root-mean-square (rms) speed of a Maxwellian distribution of silver atoms emitted from a hot oven ("gas rays") [1]. It is remarkable that this distribution had not been measured before, but even more remarkable was the correction to Stern's article later in 1920. Stern's postdoctoral advisor, Albert Einstein, pointed out to Stern that he had calculated the rms speed, $\langle c^2 \rangle_{density}^{1/2} = (3RT/m)^{1/2}$ using the density weighting $n(c)$ instead of the flux (velocity) weighted average $\langle c^2 \rangle_{flux}^{1/2} = (4RT/m)^{1/2}$, where the

G. M. Nathanson (✉)

Department of Chemistry, University of Wisconsin-Madison, 1101 University Avenue, Madison, WI 53706, USA

e-mail: nathanson@chem.wisc.edu

© The Author(s) 2021

B. Friedrich and H. Schmidt-Böcking (eds.), *Molecular Beams in Physics and Chemistry*,

https://doi.org/10.1007/978-3-030-63963-1_27

flux $J(c) = c \cdot n(c)$. Stern immediately published a correction that agreed more closely with his measured value [2], and 27 years later published a measurement of the full distribution [3]. It is heartening to know that even the great Otto Stern made mistakes, although it took someone of the stature of Einstein to correct him! (See Chap. 5 for more history.) In a sense, this chapter starts with Stern's mistake by exploring the nature of speed distributions, but with a focus on the speeds of evaporating gases dissolved in liquid microjets in vacuum ("liquid rays"). Our discussion of non-Maxwellian evaporation weaves a tale that involves two beautiful concepts in physical chemistry, namely detailed balancing between condensation and evaporation and the potential of mean force for a dissolved gas in solution.

The Maxwellian properties of number-density and flux distributions are thoroughly summarized by David and Comsa, a review article I highly recommend [4]. These two distributions can be imagined using the fingers on one hand. Cup the air within your fist: the molecules trapped inside have a Maxwellian speed distribution given by $n(c) \sim c^2 e^{-mc^2/2RT} n_{gas}$. Here $n(c)dc$ is the number of molecules per unit volume in a narrow speed interval dc. Now make an "O" with your thumb and forefinger: the speed distribution of molecules passing through the "O" is instead the flux (speed-weighted) distribution, $J(c, \theta) \sim c^3 e^{-mc^2/2RT} \cos\theta \, n_{gas}$, where θ is the polar angle. In this case, $J(c, \theta)\sin\theta \, d\theta \, d\phi \, dc$ is the number of molecules passing through a unit area per second per unit speed and solid angle interval. This distribution is shifted toward higher speeds (c^3 vs. c^2) because faster molecules traverse the area of the "O" more frequently than do slower molecules. $J(c, \theta)$ is also weighted by $\cos\theta$ because the normal velocity, $c_z = c \cdot \cos\theta$, is the component that transports the gas molecule to the surface formed by the "O", such that the integrated gas-surface collision frequency with a unit area is given by $(RT/2\pi m)^{1/2} n_{gas}$ [5]. Next situate your "O" over the surface of a glass of water: the flux of water vapor or other gas molecules striking the surface, as pictured as in Fig. 1, is just the same $J(c, \theta)$. This review addresses how the probabilities of dissolution and evaporation vary with the translational energy of the gas molecule, and what this dependence tells us about the mechanisms of solvation.

Maxwell's seminal 1860 article derived the number-density speed distribution of molecules that bears his name and often that of Boltzmann [6]. In a later 1879 article, Maxwell included comments on collisions of molecules with surfaces [7]. He categorized gas molecules striking a surface in two distinct ways: adsorption, which refers to the trapping of molecules at the surface (bound in a physisorption or chemisorption well), and reflection, which corresponds to an immediate, direct bounce from the surface. The fact that not all gases stick upon collision with a surface was in fact proved by Estermann and Stern in their celebrated study of the diffraction of helium atoms from the surface of crystalline lithium fluoride in 1930 [8]. Maxwell's and Stern's paths intersected more than once!

Fig. 1 Condensation and evaporation are reverse processes. Water molecules strike the surface in a cosine angular distribution and velocity(flux)-weighted Maxwellian distribution of translational energies. When every approaching water molecule sticks, the evaporation distribution is also cosine and Maxwellian. The simulation snapshot of the surface of water is adapted with permission from P. Jungwirth, Water's wafer-thin surface, Nature, **474**, 168–169 (2011)

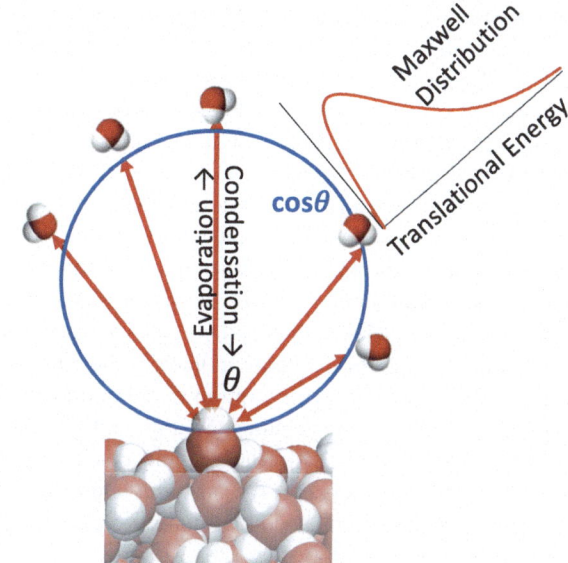

2 Condensation and Evaporation as Reverse Processes

We now know that molecules colliding with a surface interact in numerous ways, as summarized in recommended reviews [9–18]. During a single or multi-bounce nonreactive collision, these pathways include not only translational energy exchange but also vibrational, rotational, and electronic transitions (including spin-orbit) in the gas-phase molecule and in the surface and subsurface molecules within the collision zone. The range of energy exchange can vary from zero (elastic collisions such as occurs in diffraction) through production of "hot" adsorbed species to complete energy equilibration at the substrate temperature (also called thermalization) and momentary trapping within the gas-surface potential (often called sticking if the species remains on the surface for long times, often longer than the measurement). It is often said that the trapped molecule "loses memory" of its initial trajectory after its microscopic motions are scrambled through numerous interactions with surface atoms [19, 20]. These adsorbed molecules may subsequently desorb back into the gas phase (trapping-desorption [21, 22]) at rates that are determined by the surface temperature but by not its initial trajectory or internal states.

We also know that, when the gas-solid or gas-liquid system has come to equilibrium, the outgoing and incoming fluxes of each species must be equal. Langmuir stated this criterion in 1916 with extraordinary prescience: "Since evaporation and condensation are in general thermodynamically reversible phenomena, *the mechanism of evaporation must be the exact reverse of that of condensation,* even down to the smallest detail." [23] In modern terms, a molecular dynamics simulation of gas-solid or gas-liquid collisions can be run backward to simulate the reverse process for every internal [20, 24–26] and velocity component [4, 20] (see water movie at nathanson.chem.wisc.edu by Varilly and Chandler [27]). This microscopic reversibility, a detailed balancing of every molecular process, has an astonishing implication at equilibrium: because the flux of molecules arriving at a surface is Maxwellian and cosine, the flux of molecules leaving the surface must be Maxwellian and cosine too. If the trapping probability depends on incident energy or angle, then the flux of just the desorbing molecules will be non-Maxwellian and non-cosine, with the difference made up by the molecules that directly scatter from the surface! Only the sum of all scattering and desorbing molecules must be Maxwellian and cosine. Thus, if one could observe just the desorbing molecules, one might measure a distribution that is non-Maxwellian and non-cosine, and then infer from it the energy and angular distribution of incoming molecules that undergo trapping and solvation.

A measurement of the desorption distribution can indeed be made in a vacuum experiment (where there is almost no impinging flux) if one assumes that the distribution out of equilibrium in the vacuum chamber is the same as at equilibrium. The history of these concepts for gas-solid interactions is told with great clarity and suspense by Comsa and David [4] and by Kolasinski [20]. I have also learned much from several original references [28–30].

3   Rules of Thumb for Gas-Surface Energy Transfer and Trapping

Three key concepts and examples from gas-surface scattering can be used to appreciate the implications of detailed balancing, as summarized below.

1. The kinematics of the collision govern energy transfer: light gas atoms or molecules bounce off heavy surface atoms or molecules, transferring just a fraction of their translational energy upon collision [31]. Conversely, gas species that are heavier than the surface species (often the case for liquid water) will undergo multiple collisions that lead to efficient energy transfer. For an incoming sphere colliding head-on with an initially stationary sphere (zero impact parameter), the energy transfer is given by $\Delta E/E_{inc} = 4\mu/(1 + \mu)^2$, where $\mu = m_{gas}/m_{surf}$ and $E_{inc} = 1/2\, m_{gas}\, c_{inc}^2$. This equation also models an atom striking a flat cube in a perpendicular direction. When $\mu = 1/4$, 64% of the incident kinetic energy of the gas atom is transferred to the surface atom, while it rises to 89% for $\mu = 1/2$. Numerous experiments verify that energy transfer indeed increases with heavier gas and lighter surface molecules [13, 32–34]. Further studies show that grazing collisions (large impact parameter) transfer less energy, and thermal motions generally decrease the overall energy transfer as well. Sophisticated models of energy transfer have been developed that take into account the shape of molecules [35] and surface and their internal excitation, including the development of a "surface Newton diagram" [18, 36].

Fig. 2 Two-step mechanism for the dissolution of a gas atom or molecule. In general, high translational energies and grazing collisions lead to direct scattering from the surface, while lower incident energies and more perpendicular collisions lead to energy loss and momentary trapping. This trapping is typically followed by desorption back into the gas phase or diffusion and solvation in the bulk

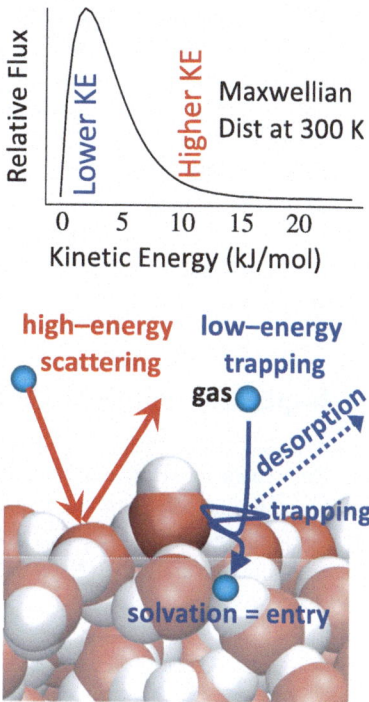

2. Attractive forces create gas-surface potential energy wells that can momentarily trap the incoming molecule once it has dissipated its excess energy after one or several bounces, as pictured in Fig. 2. For the simple model above, the minimum initial translational energy required to escape the potential energy well is

$$E_{\text{min}} = 4\mu/(1 - \mu)^2 \cdot \varepsilon \tag{1}$$

where ε is the well depth [11, 30]. This expression neatly separates into kinematic (mass) and potential energy terms. For $\mu = 1/4$ and $\varepsilon = 20$ kJ/mol (a hydrogen bond between gas and liquid), E_{min} is 36 kJ/mol or 14 RT_{liq} at 300 K—only gases with higher energy will escape thermalization and momentary trapping. Again, experiments verify that heavy gas atoms/light surface atoms and strong attractive forces enhance trapping via the strength of the reagent or product desorption signal [13, 33, 37]. We note the inherent distributions of attractive forces and impact parameters arising from bumpy surfaces, molecular orientation, varying approach angles, and multiple collisions, along with thermal motions of the surface atoms, will broaden the sharp cutoff imposed by Eq. 1. The value of E_{min} might then be taken as midway along the trapping probability curve [20, 30].

3. The Maxwellian flux distribution in terms of translational energy E is given by $J(E) = E/(RT)^2 e^{-E/RT}$. This function peaks at $E = RT = 2.5$ kJ/mol at

300 K and has an average value of $2\,RT = 5.0$ kJ/mol (not $3/2\,RT$, which is the average energy of the number-density distribution). In the example above, only 1 in 120,000 molecules at 300 K have translational energies greater than 36 kJ/mol (only 1 in 160 have energies greater than 18 kJ/mol and 1 in 8 have energies greater than 9 kJ/mol). This is a general result: while even heavy gases will often scatter directly from a surface at high collision energies of many 10s to 100s of kJ/mol, these energies have vanishingly low probabilities in a room temperature Maxwellian distribution. Full energy dissipation and trapping (adsorption) is the rule rather than the exception for most molecules on most surfaces near room temperature.

4 Implications of Detailed Balance

The three rules above have immediate implications for the evaporation of gases from solids and liquids. By detailed balancing, the desorption flux J_{des} is equal to the flux of impinging molecules J_{trap} that are momentarily trapped in the interfacial region. The trapping probability $\beta(E, \theta)$ then connects the desorbing and impinging fluxes via [30]

$$
\begin{aligned}
J_{des}(E, \theta) = J_{trap}(E, \theta) &= \beta(E, \theta) \cdot J_{inc}(E, \theta) \\
&= \beta(E, \theta) \cdot \frac{E}{(RT)^2} e^{-E/RT} \cdot \frac{\cos\theta}{4\pi} \cdot n_{gas}
\end{aligned}
\tag{2}
$$

such that $\beta(E, \theta)$ may be considered both the trapping probability (J_{trap}/J_{inc}) and the evaporation probability (J_{des}/J_{inc}). The rules of thumb above suggest that $\beta(E, \theta)$ will be constant and close to one for most gases, especially when the liquids are made of light molecules such as water (where μ often exceeds one) and where dispersion, dipolar, and hydrogen bonding interactions occur. Thus, most gas molecules should evaporate in a distribution that is close to Maxwellian and cosine at room temperature from water and organic liquids, but perhaps not from solid or liquid metals [30, 38].

Deviations from the typical rules for trapping can reveal underlying mechanisms. One deviation occurs when $\beta(E, \theta)$ changes significantly over the energies in a Maxwellian distribution (0 to ~7 RT_{liq}) or at grazing angles, most likely because of light gas/heavy surface masses (small μ) and weak attractions ε. In these cases, collisions at low energy should lead to trapping while the molecules will scatter at higher energies (as predicted by Eq. 1). Detailed balancing then requires that the adsorbate will desorb in a speed distribution tilted toward lower translational energies because $\beta(E, \theta)$ steadily declines from high to low values as E increases. Rettner and coworkers indeed show this behavior for argon atoms desorbing from hydrogen-covered tungsten, whose sub-Maxwellian desorption matches the distribution of incoming Ar atoms that are momentarily trapped at the surface [30]. This study is mandatory reading for its clarity and precision.

Conversely, imagine an H_2 molecule dissociating upon collision with a metal surface, such as copper. It must have enough translational energy to overcome the ~20 kJ/mol barrier in order to break the H–H bond and form surface Cu–H bonds. High energies along the surface normal facilitate this dissociative adsorption. In the reverse associative desorption, the two adsorbed H atoms come together to generate an H_2 molecule that suddenly finds itself repulsively close to the surface and leaves the surface at high energies, preferentially along the surface normal, that match the incoming energies that lead to dissociation. This detailed balancing of dissociative adsorption and recombinative desorption is observed in pioneering experiments by Cardillo and coworkers and by others [4, 28].

5 Maxwellian Evaporation and a Two-Step Model for Solvation

Now we come to the question in this chapter. Are there also deviations in Maxwellian evaporation from liquids and solutes dissolved in them? During 30 years of observation, we have monitored the vacuum evaporation of liquids such as glycerol, ethylene glycol, alkanes and aromatics, fluorinated ethers, and water from sulfuric acid and pure and salty water itself [39–43]. We have also recorded the evaporation of solute atoms and molecules such as Ar, N_2, O_2, HCl, HBr, HI, Cl_2, Br_2, BrCl, N_2O_5, HNO_3, CO_2, SO_2, $HC(O)OCH_3$, CH_3OCH_3, CH_3NHCH_3, and butanol from one or more of the solvents listed above and others [39, 40, 42, 44–48]. We observed Maxwellian speed distributions in every case (except when the vapor pressure is so high that the gas expands supersonically [41, 49]). This observation is in accord with the arguments above, so we were not surprised. For solvent evaporation and condensation, the mass of the evaporating solvent is necessarily equal to the mass of the surface molecules ($\mu = 1$). In this case, there is very efficient energy transfer (just like billiard balls) and the attractive forces that cohere the molecules into a liquid also trap the gaseous solvent molecule upon collision with the surface. For hydrogen-bonding gases, the attractive forces are also very strong and lead to significant trapping.

We also find, however, that even Ar and N_2 evaporating from salty water evaporate in Maxwellian distributions (within our signal to noise) [42, 43, 50]. By detailed balancing, this Maxwellian evaporation implies that collisions of Ar and N_2 at energies populated in a Maxwellian distribution must thermally equilibrate upon collision. This nearly complete thermalization is likely promoted by the soft nature of surfaces composed of water and organic molecules (it is not true for liquid metals) and weak attractions of a few RT_{liq}.

Separate studies provide insights into the mechanism of dissolution and reaction. Reactive scattering experiments probe HCl $\rightarrow$ DCl exchange in collisions of HCl with liquid D_2O/D_2SO_4 and of $Cl_2 \rightarrow Br_2$ exchange and $N_2O_5 \rightarrow Br_2$ oxidation in NaBr/glycerol [39, 45, 46]. In all three cases, the product DCl or Br_2 evaporates in a Maxwellian distribution at the temperature of the liquid. The measurements reveal that the ratio of the desorbing product to trapping-desorption (TD) component of the reactant (J(product)/J_{TD}(reactant)) is independent of reactant collision energy from near-thermal to hyperthermal energies. These observations suggest a two-step process for dissolution and reaction (as illustrated in Fig. 2): [39] (1) incoming molecules either directly scatter from the surface or dissipate their excess translational energy, becoming momentarily trapped within the gas-surface potential energy well and losing memory of their initial trajectory, and then (2) these trapped molecules either evaporate or dissolve into the bulk at rates that are determined by gas and liquid properties and temperature, but not by their initial trajectory. In this two-step process, reaction occurs after thermalization within the interfacial region or deeper in the bulk. For the reversible solvation of a non-reacting gas (such as Ar or CH_3OCH_3), evaporation occurs along the reverse pathways, starting with the solute molecule diffusing from the bulk to the surface and then being jettisoned in a Maxwellian distribution into the vacuum by numerous energy-exchanging encounters with surface molecules. This two-step mechanism likely applies to the dissolution and evaporation of most gaseous solutes, but are there exceptions?

6 Non-Maxwellian Evaporation Discovered!

Faubel and Kisters first observed the non-Maxwellian evaporation of acetic acid dimers from a water microjet in 1988, which they attributed to repulsive ejection of the hydrophobic dimer at the surface of water [51]. This single observation persisted until we recorded the non-Maxwellian evaporation of helium atoms in 2014 [50]. Our measurements came about by accident: we were generating microjets [42, 50, 52] of alkane solutions to mimic evaporation of jet fuel in vacuum, which were created by pressurizing a sealed reservoir of the liquid with Ar or N_2 (as first developed by Manfred Faubel [53, 54] and described in Chap. 26). As shown in Fig. 3, the pressurized liquid then emerges from a glass tube with a tapered hole as narrow as 10 μm in diameter. We found that, by vigorously shaking the reservoir, gas can be dissolved into the liquid, which then evaporates as the thin liquid stream exits the nozzle and passes through the vacuum chamber. We have exploited the Maxwellian evaporation of dissolved Ar atoms as "argon jet thermometry" because the Ar speed distribution yields the instantaneous temperature of the jet.

Helium may also be used as a pressurizing gas to create microjets. To our astonishment, we found that He evaporation is non-Maxwellian for every solvent tested, including octane, dodecane, squalane, jet fuel, ethylene glycol, and pure and salty water (as shown in Fig. 4 for 7 M LiBr and 7 M LiCl in water) [42, 43, 50]. Importantly, its behavior is opposite to expectations: instead of evaporating in a slower, sub-Maxwellian distribution, as predicted by argon desorbing from tungsten mentioned above, the He atoms evaporate in a distinctly faster, *super-Maxwellian* distribution! The extent of non-Maxwellian behavior can be gauged by the average translational energy of the exiting He atoms: $1.14 \cdot (2RT_{liq})$ for dodecane at 295 K, $1.37 \cdot (2RT_{liq})$

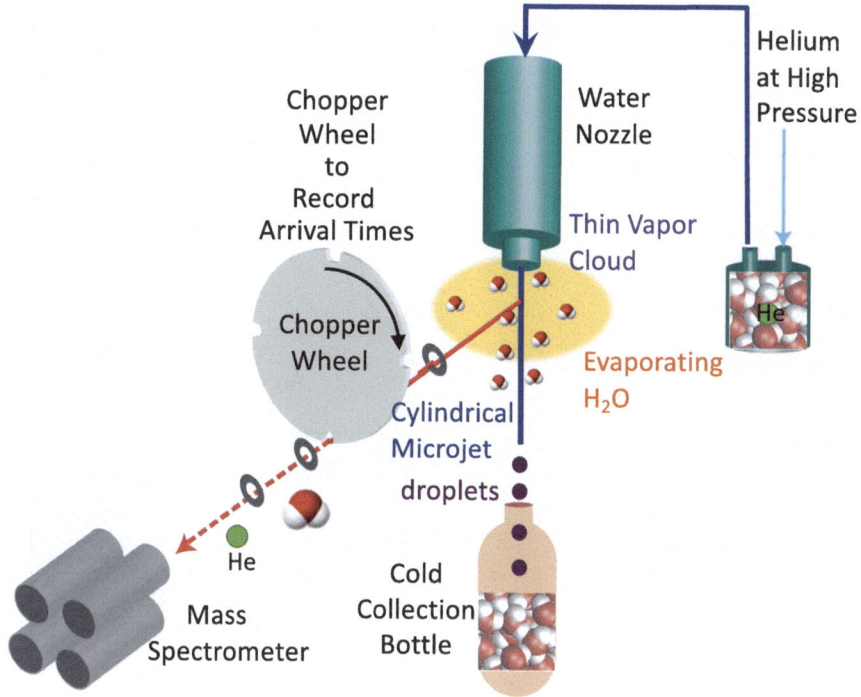

Fig. 3 Vacuum evaporation of helium from pure and salty water microjets. The microjet is a fast-moving thin stream of solvent typically thinner than a strand of hair. When the jet radius is significantly smaller than the He-water mean free path, nearly all He atoms avoid collisions with evaporating water molecules in the vapor cloud surrounding the jet. The jet diameters range from 10 to 35 μm and travel at ~20 m/s. The breakup lengths vary from less than 1 mm for pure water at 252 K to 7 mm for 7 M LiBr/H_2O at 235 K

for pure supercooled water at 252 K, and $1.70 \cdot (2RT_{\text{liq}})$ for 7 M LiBr/H_2O at 255 K, which are 14, 37, and 70% higher than expected [42, 43].

Detailed balancing provides a fascinating interpretation: the super-Maxwellian evaporation of He atoms implies that the reverse process of He dissolution must also be super-Maxwellian.[1] The translational energies of He atoms that dissolve are shifted to higher values, such that the solvation probability, the analog of the trapping probability, increases with increasing collision energy. This result may be interpreted to mean that some He atoms dissolve by "ballistic penetration", pushing water molecules slightly aside as they pass through the interfacial region and enter the liquid! The measured, relative evaporation probabilities $\beta(E)$ for 7 M LiCl and LiBr in water are shown in Fig. 4, which by detailed balance are also the relative solvation

---

[1] We note that the reverse scattering experiments of He from liquids is complex, involving at least four pathways: direct recoil, trapping-desorption, trapping-dissolution-evaporation, and "ballistic" entry and evaporation. Our attempts to separate the processes have not been successful, as the TOF spectra are dominated by direct recoil. Dissolution appears to be a rare event.

Fig. 4 Examples of helium evaporation from salty water. (**a**) TOF spectra of He atoms evaporating from 8 molal (7 M) LiBr (232 K) and LiCl (237 K), which peak at significantly shorter arrival times (higher speeds and kinetic energies) than the dashed Maxwellian distributions at each temperature. (**b**) The corresponding translational energy distributions of the He atoms, again in comparison to Maxwellian distributions (dashed lines, here called P_{MB}). The relative solvation probabilities $\beta(E)$ (dot-dash) each rise steadily with kinetic energy (see Footnote 2). Panel c shows the excellent agreement between the Skinner/Kann simulations and measurements. This figure is reproduced from Ref. [43]

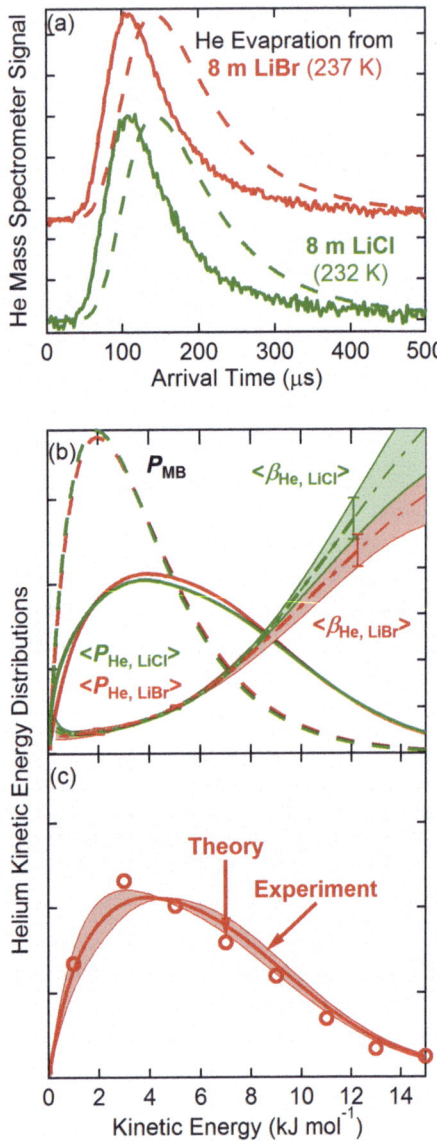

probabilities for He atoms.[2] Both curves rise steadily with increasing evaporation energy (which is also the collision energy for the reversed trajectories). Many of these He atoms therefore circumvent the two-step trapping-dissolution mechanism

---

[2]Because we measure only relative He fluxes in the experiments, not absolute fluxes, $J_{des}(E)$ and $J_{inc}(E)$ are each area-normalized in Fig. 4, only ratios of $\beta(E)$ at different E are meaningful. The angular average over $\beta(E, \theta)$ for the cylindrical microjet is described in Refs. [42, 43].

described above as they pass through the interfacial region. But why? Helium atoms have the lowest polarizability of any atom in the periodic Table (0.2 Å^3). In turn, the He-surface potential energy may be so shallow (less than RT_{liq}) that the attractive forces cannot capture He atoms at the surface for the time needed for He atoms to dissolve—thermal motions of the surface molecules instead immediately kick most of the He atoms back into the gas phase. In this case, a substantial fraction of the He atoms cannot enter the liquid via the adsorbed state because the shallow well cannot trap them. We know that the high-energy evaporation of He does not originate from its low mass because the opposite behavior is observed for the evaporation of H_2 from water. For this even lighter gas, evaporation is indeed sub-Maxwellian, as predicted kinematically: energy transfer between H_2 and water is inefficient ($\mu = 0.11$), and only the low energy H_2 molecules lose enough energy to be trapped in the H_2-water potential energy well (H_2 is 4 times more polarizable than He). We also note that the only other gas we have observed that displays super-Maxwellian behavior is neon, which is also weakly polarizable (0.4 Å^3).

7   A View from the Interior

Detailed balancing arguments are beautiful and rigorous and in accord with experiments, but they leave us yearning to know more. How do the dissolved He atoms "know" to evaporate in the same super-Maxwellian distribution that leads to dissolution? It must be so because a single (but complex) potential energy function for all He-water and water-water interactions governs the reverse evaporation and condensation processes [29]. Comsa and David [4] quote an early pioneer, Peter Clausing, who described the detailed balancing requirement of the cosine angular distribution for condensation and evaporation as an "incomprehensible wonder machine", but this statement could apply to the speed distribution as well. My theory colleague Jim Skinner and his student Zak Kann set out to make helium evaporation from water comprehensible, but their explanation is still full of wonder.

Skinner and Kann first performed classical molecular dynamics simulations of He atoms dissolved in pure liquid water [43]. Their simulations indeed show that dissolved He atoms possess a Maxwellian speed distribution right up to the top one to two layers of water, where the He atom is then accelerated into vacuum during the final few collisions of He with H_2O molecules moving outward. My students and I had hoped that the measured He speed distributions would reveal new features of gas-water interactions, but the agreement between simulation and measurement was excellent! Skinner and Kann then went a step further, calculating the Potential of Mean Force (PMF) on the He atom. This potential is equal to the free energy of the He atom as it is dragged infinitely slowly through the interface and into the bulk, sampling all configurations of the water molecules along the way. The free energy curve (PMF) of He in pure water calculated at 255 K is shown in Fig. 5. It starts high in bulk water and decreases to the gas phase value: the difference between the asymptotes is equal to the (very positive) free energy of solvation of approximately

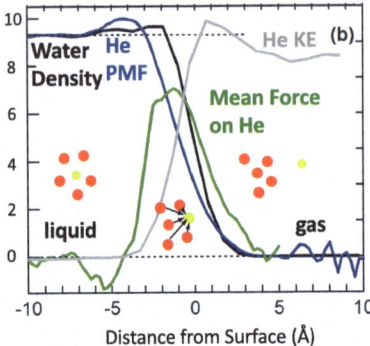

Fig. 5 Potential of Mean Force (PMF) description of an He atom being expelled from water at 255 K. The black curve is the liquid water density, for which the 0 distance is the Gibbs dividing surface. The blue curve is the calculated helium PMF (free energy of solvation), and the green curve is the mean force (negative derivative of the PMF), which spikes in the interfacial region. The grey curve is the resulting helium atom kinetic energy. For these curves, the PMF spans 0 to 10 kJ/mol, the He density-averaged KE spans $(3/2)RT = 3.2$ kJ/mol to $\sim1.5\times(3/2)RT = 4.7$ kJ/mol, and the mean force spans -0.5 to $+2.3$ kJ/mol/Å. The small drop in He kinetic energy after 1 Å reflects the weak attractive force between He and surface water molecules decelerating the He atom as it leaves. This figure is adapted from Ref. [43]

9.5 kJ/mol at the supercooled 252 K temperature of the 10 μm diameter microjet. Accordingly, helium has the lowest solubility of any gas in water, equal to n_{water}/n_{gas} ~ 1/100 at 252 K. The free energy curve may possess a small barrier (<0.5 kJ/mol) between the surface and bulk regions, but displays at most a very shallow minimum at the surface. This is unlike even N_2 or O_2, which are also weakly soluble but whose attraction to H_2O generate weak adsorption wells (>2 kJ/mol) [55].

Why then do He atoms emerge at higher than Maxwellian translational energies? The negative derivative of the free energy curve is just the "mean force" associated with the PMF—it is the repulsive force acting on the He atom itself as it moves infinitely slowly through the liquid! Figure 5 shows this mean force spikes right at the interface where the He atoms are accelerated [43]. Here is the key point: if the He atom indeed moved slowly through the interfacial region, it would undergo enough energy-exchanging collisions with water molecules at each point to maintain a Maxwellian distribution. But the He atoms do not move slowly, and at some point they stop equilibrating as the interfacial density becomes sparser (as in Fig. 5) and there are insufficient He–H_2O collisions to absorb the extra He atom energy. In this case, the He atom "detaches" from the PMF and exits into vacuum, carrying its excess energy with it imparted by the repulsive forces. In a sense, the water molecules "squeeze" the interloping He atom into vacuum as the water-water hydrogen bonds "heal" to their native structure.

We note that the PMF only describes the force perpendicular to the surface. Skinner and Kann have also investigated the angular distributions of evaporating atoms, and deduce that the perpendicular component is even more super-Maxwellian but is partially canceled by sub-Maxwellian parallel components [56]. This study includes a wide-ranging investigation of the effects of solute mass and solute-solvent attractive forces on solute evaporation, including confirmation that H_2 is sub-Maxwellian and Ne is super-Maxwellian. Parallel simulations by Williams, Patel, and Koehler of He evaporation from dodecane lead to a fascinating "cone and crater" mechanism by which He atoms are expelled in an exposed cone at the surface whose walls may crater inward, accelerating the He atom from the cone [57, 58].

One rule of thumb emerges from these investigations: the more insoluble the gas, the steeper the PMF, the greater the force on the evaporating gas atoms, and the more likely that the He atom will emerge in a non-Maxwellian distribution. Thus, higher He atom exit energies should accompany lower solubilities in different solvents, a trend that we observe experimentally [43]. This correlation is not quantitative, however, because the PMF describes a slowly moving solute atom that fully equilibrates as it moves through solution and samples all configurations of the water molecules—it is the breakdown of this picture arising from insufficient He-water collisions in the outermost region that gives rise to an excess kinetic energy. A focus on the mean force and interfacial collisions instead provides an exquisite statistical framework that can guide future investigations.

8   Future Non-Maxwellian Adventures

What are some potential new directions for helium evaporation experiments? The demonstration of super-Maxwellian He evaporation is the closest we have come to He atom diffraction from periodic solid surfaces. The question of what can be learned from He scattering from liquids was one my students and I asked when we began in 1988, and it took until now to address it: super-Maxwellian He evaporation from liquids reflects the forces acting on the He atom in the outermost layers of the liquid. Skinner's and Kann's successful simulations [43] suggest that He evaporation from pure and salty water may not contribute to a refined picture of gas-water interactions because they were already so successful in replicating the energy distributions. But water is almost never pure or even just salty. Oceans, lakes, aerosol particles, and tap water contain numerous organic species, many of which are surface active [59–61]. We hope in future studies to investigate surfactant-coated microjets prepared with soluble ionic species such as tetrabutylammonium bromide and neutral ones such as butanol or pentanoic acid [47]. Helium evaporation from these surfactant solutions may reveal how gases move through loosely to tightly packet alkyl chains, depending on their bulk-phase concentration, and thus provide information on the mechanisms of gas transport through monolayers [62]. It will also be intriguing to mimic the seminal studies of the Cardillo and Comsa groups [4, 28], who investigated H_2 permeation and desorption through metals. We can monitor the parallel evaporation of He atoms through thin polymer films of functionalized organic polymers and even self-assembled monolayers over a wide range of exit angles. It is inspiring to imagine that Stern might have enjoyed these studies, an extension of his "method of molecular rays" to liquids in vacuum, "for which I [Stern] consider the directness and simplicity as the distinguishing property."

Acknowledgements This work was supported by the National Science Foundation (CHE-1152737) and the Air Force Office of Scientific Research. I thank the many contributions of my students working on the projects described here, including Alexis Johnson, Diane Lancaster, Jennifer Faust, Christine Hahn, and Tom Sobyra, who all became masters of microjets and committed untold hours to making these experiments work. I am also indebted to Zak Kann and Jim Skinner for a bountiful collaboration and for teaching us the extraordinary insights that statistical mechanics can provide. I am grateful to Dudley Herschbach for introducing me to the life of Otto Stern and to Dudley and Peter Toennies for being Honorary Chairpersons of the Otto Stern conference. My special thanks go to Bretislav Friedrich and Horst Schmidt-Böcking for organizing and directing every aspect of the conference, which spanned science, history, philosophy, music, science outreach, and a dazzling excursion on the Rhein River. Otto Stern leapt off the page through loving tributes, history lessons, and the astonishingly manifold applications of the molecular ray technique that he founded. We all glimpsed a multidimensional human being—a fantastically insightful and fearless scientist and outstandingly moral mentor in dark times. His words and deeds have become a constant companion.

References

1. O. Stern, A direct measurement of thermal molecular speed. Z. Phys. **2**, 49–56 (1920)
2. O. Stern, Addition to my work "a direct measurement of thermal molecular speed". Z. Phys. **3**, 417–421 (1920)
3. I. Estermann, O.C. Simpson, O. Stern, The free fall of atoms and the measurement of the velocity distribution in a molecular beam of cesium atoms. Phys. Rev. **71**, 238–249 (1947)
4. G. Comsa, R. David, Dynamical parameters of desorbing molecules. Surf. Sci. Rep. **5**, 145–198 (1985)
5. A.H. Persad, C.A. Ward, Expressions for the evaporation and condensation coefficients in the Hertz-Knudsen relation. Chem. Rev. **116**, 7727–7767 (2016)
6. J.C. Maxwell, Illustrations of the dynamical theory of gases—Part I. On the motions and collisions of perfectly elastic spheres. London, Edinburgh Dublin Phil. Mag. J. Sci. **19**, 19–32 (1860)
7. J.C. Maxwell, VII. On stresses in rarified gases arising from inequalities of temperature. Phil. Trans. **170**, 231–256 (1879)
8. I. Estermann, O. Stern, Diffraction of molecular beams. Z. Phys. **61**, 95–125 (1930)
9. J.A. Barker, D.J. Auerbach, Gas-surface interactions and dynamics; thermal energy atomic and molecular beam studies. Surf. Sci. Rep. **4**, 1–99 (1985)
10. S.T. Ceyer, New mechanisms for chemical at surfaces. Science **249**, 133–139 (1990)
11. C.T. Rettner, D.J. Auerbach, J.C. Tully, A.W. Kleyn, Chemical dynamics at the gas-surface interface. J. Phys. Chem. **100**, 13021–13033 (1996)
12. G.M. Nathanson, Molecular beam studies of gas-liquid interfaces. Ann. Rev. Phys. Chem. **55**, 231–255 (2004)
13. J.W. Lu, B.S. Day, L.R. Fiegland, E.D. Davis, W.A. Alexander, D. Troya, J.R. Morris, Interfacial energy exchange and reaction dynamics in collisions of gases on model organic surfaces. Prog. Surf. Sci. **87**, 221–252 (2012)
14. H. Chadwick, R.D. Beck, Quantum state resolved gas-surface reaction dynamics experiments: a tutorial review. Chem. Soc. Rev. **45**, 3576–3594 (2016)
15. M.A. Tesa-Serrate, E.J. Smoll, T.K. Minton, K.G. McKendrick, Atomic and molecular collisions at liquid surfaces, in *Ann. Rev. Phys. Chem.*, vol. 67, ed. by M.A. Johnson, T.J. Martinez (2016), pp. 515–540

16. C.H. Hoffman, D.J. Nesbitt, Quantum state resolved 3D velocity map imaging of surface scattered molecules: incident energy effects in HCl plus self-assembled monolayer collisions. J. Phys. Chem. C **120**, 16687–16698 (2016)
17. F. Zaera, Use of molecular beams for kinetic measurements of chemical reactions on solid surfaces. Surf. Sci. Rep. **72**, 59–104 (2017)
18. W.A. Alexander, Particle beam scattering from the vacuum-liquid interface, in *Physical Chemistry of Gas-Liquid Interfaces*, ed. by J.A. Faust, J.E. House (Elsevier, The Netherlands, 2018), pp. 195–234
19. C.T. Rettner, D.J. Auerbach, Distinguishing the direct and indirect products of a gas-surface reaction. Science **263**, 365–367 (1994)
20. K.W. Kolasinski, *Surface Science: Foundations of Catalysis and Nanoscience*, 3rd edn. (Wiley, United Kingdom, 2012). Ch. 3
21. J.E. Hurst, C.A. Becker, J.P. Cowin, K.C. Janda, L. Wharton, D.J. Auerbach, Observation of direct inelastic scattering in the presence of trapping-desorption scattering—XE ON Pt(111). Phys. Rev. Lett. **43**, 1175–1177 (1979)
22. M.E. Saecker, G.M. Nathanson, Collisions of protic and aprotic gases with hydrogen-bonding and hydrocarbon liquids. J. Chem. Phys. **99**, 7056–7075 (1993)
23. I. Langmuir, The constitution and fundamental properties of solids and liquids part I solids. J. Am. Chem. Soc. **38**, 2221–2295 (1916)
24. D.F. Padowitz, S.J. Sibener, Sublimation of nitric oxide films—rotation and angular distributions of desorbing molecules. Surf. Sci. **217**, 233–246 (1989)
25. H.A. Michelsen, C.T. Rettner, D.J. Auerbach, R.N. Zare, Effect of rotation on the translational and vibrational energy dependence of the dissociative adsorption of D_2 on Cu(111). J. Chem. Phys. **98**, 8294–8307 (1993)
26. M.J. Weida, J.M. Sperhac, D.J. Nesbitt, Sublimation dynamics of CO_2 thin films: a high resolution diode laser study of quantum state resolved sticking coefficients. J. Chem. Phys. **105**, 749–766 (1996)
27. P. Varilly, D. Chandler, Water evaporation: a transition path sampling study. J. Phys. Chem. B **117**, 1419–1428 (2013)
28. M.J. Cardillo, M. Balooch, R.E. Stickney, Detailed balancing and quasi-equilibrium in adsorption of hydrogen on copper. Surf. Sci. **50**, 263–278 (1975)
29. J.C. Tully, Dynamics of gas-surface interactions: thermal desorption of Ar and Xe from platinum. Surf. Sci. **111**, 461–478 (1981)
30. C.T. Rettner, E.K. Schweizer, C.B. Mullins, Desorption and trapping of argon at a 2H–W(100) surface and a test of the applicability of detailed balance to a nonequilibrium system. J. Chem. Phys. **90**, 3800–3813 (1989)
31. E.K. Grimmelmann, J.C. Tully, M.J. Cardillo, Hard-cube model analysis of gas-surface energy accommodation. J. Chem. Phys. **72**, 1039–1043 (1980)
32. M.E. King, G.M. Nathanson, M.A. Hanninglee, T.K. Minton, Probing the microscopic corrugation of liquid surfaces with gas-liquid collisions. Phys. Rev. Lett. **70**, 1026–1029 (1993)
33. M.E. Saecker, G.M. Nathanson, Collisions of protic and aprotic gases with a perfluorinated liquid. J. Chem. Phys. **100**, 3999–4005 (1994)
34. B.G. Perkins, D.J. Nesbitt, Quantum-state-resolved CO_2 scattering dynamics at the gas-liquid interface: Incident collision energy and liquid dependence. J. Phys. Chem. B **110**, 17126–17137 (2006)
35. T.Y. Yan, W.L. Hase, J.C. Tully, A washboard with moment of inertia model of gas-surface scattering. J. Chem. Phys. **120**, 1031–1043 (2004)
36. W.A. Alexander, J.M. Zhang, V.J. Murray, G.M. Nathanson, T.K. Minton, Kinematics and dynamics of atomic-beam scattering on liquid and self-assembled monolayer surfaces. Faraday Discuss. **157**, 355–374 (2012)
37. T.B. Sobyra, M.P. Melvin, G.M. Nathanson, Liquid microjet measurements of the entry of organic acids and bases into salty water. J. Phys. Chem. C **121**, 20911–20924 (2017)
38. M. Manning, J.A. Morgan, D.J. Castro, G.M. Nathanson, Examination of liquid metal surfaces through angular and energy measurements of inert gas collisions with liquid Ga, In, and Bi. J. Chem. Phys. **119**, 12593–12604 (2003)

39. J.R. Morris, P. Behr, M.D. Antman, B.R. Ringeisen, J. Splan, G.M. Nathanson, Molecular beam scattering from supercooled sulfuric acid: collisions of HCl, HBr, and HNO$_3$ with 70 wt % D$_2$SO$_4$. J. Phys. Chem. A **104**, 6738–6751 (2000)
40. A.H. Muenter, J.L. DeZwaan, G.M. Nathanson, Collisions of DCl with pure and salty glycerol: enhancement of interfacial D → H exchange by dissolved NaI. J. Phys. Chem. B **110**, 4881–4891 (2006)
41. S.M. Brastad, G.M. Nathanson, Molecular beam studies of HCl dissolution and dissociation in cold salty water. Phys. Chem. Chem. Phys. **13**, 8284–8295 (2011)
42. D.K. Lancaster, A.M. Johnson, K. Kappes, G.M. Nathanson, Probing gas–liquid interfacial dynamics by helium evaporation from hydrocarbon liquids and jet fuels. J. Phys. Chem. A 14613–14623 (2015)
43. C. Hahn, Z.R. Kann, J.A. Faust, J.L. Skinner, G.M. Nathanson, Super-Maxwellian Helium evaporation from pure and salty water. J. Chem. Phys. **144** (2016)
44. B.R. Ringeisen, A.H. Muenter, G.M. Nathanson, Collisions of HCl, DCl, and HBr with liquid glycerol: gas uptake, D → H exchange, and solution thermodynamics. J. Phys. Chem. B **106**, 4988–4998 (2002)
45. L.P. Dempsey, J.A. Faust, G.M. Nathanson, Near-interfacial halogen atom exchange in collisions of Cl$_2$ with 2.7 M NaBr-glycerol. J. Phys. Chem. B. **116**, 12306–12318 (2012)
46. M.A. Shaloski, J.R. Gord, S. Staudt, S.L. Quinn, T.H. Bertram, G.M. Nathanson, Reactions of N$_2$O$_5$ with salty and surfactant-coated glycerol: interfacial conversion of Br- to Br 2 mediated by alkylammonium cations. J. Phys. Chem. A **121**, 3708–3719 (2017)
47. T.B. Sobyra, H. Pliszka, T.H. Bertram, G.M. Nathanson, Production of Br$_2$ from N$_2$O$_5$ and Br⁻ in salty and surfactant-coated water microjets. J. Phys. Chem. A **123**, 8942–8953 (2019)
48. T. Krebs, G.M. Nathanson, Reactive collisions of sulfur dioxide with molten carbonates. Proc. Natl. Acad. Sci. U.S.A. **107**, 6622–6627 (2010)
49. D.K. Lancaster, A.M. Johnson, D.K. Burden, J.P. Wiens, G.M. Nathanson, Inert gas scattering from liquid hydrocarbon microjets. J. Chem. Phys. Lett. **4**, 3045–3049 (2013)
50. A.M. Johnson, D.K. Lancaster, J.A. Faust, C. Hahn, A. Reznickova, G.M. Nathanson, Ballistic evaporation and solvation of helium atoms at the surfaces of protic and hydrocarbon liquids. J. Phys. Chem. Lett. **5**, 3914–3918 (2014)
51. M. Faubel, T. Kisters, Non-equilibrium molecular evaporation of carboxylic acid dimers. Nature **339**, 527–529 (1989)
52. J.A. Faust, G.M. Nathanson, Microjets and coated wheels: versatile tools for exploring collisions and reactions at gas-liquid interfaces. Chem. Soc. Rev. **45**, 3609–3620 (2016)
53. M. Faubel, S. Schlemmer, J.P. Toennies, A molecular beam study of the evaporation of water from a liquid jet. Z. Phys. D. **10**, 269–277 (1988)
54. M. Faubel, Photoelectron spectroscopy at liquid surfaces, in *Photoionization and Photodetachment: Part I*, vol. 10A, ed. by C.-Y. Ng (World Scientific, Singapore, 2000), pp. 634–690
55. R. Vácha, P. Slavíček, M. Mucha, B.J. Finlayson-Pitts, P. Jungwirth, Adsorption of atmospherically relevant gases at the air/water interface: free energy profiles of aqueous solvation of N$_2$, O$_2$, O$_3$, OH, H$_2$O, HO$_2$, and H$_2$O$_2$. J. Phys. Chem. A **108**, 11573–11579 (2004)
56. Z.R. Kann, J.L. Skinner, Sub- and super-Maxwellian evaporation of simple gases from liquid water. J. Chem, Phys **144**, 154701 (2016)
57. S.V.P. Koehler, M.A. Williams, MD simulations of he evaporting from dodecane. Chem. Phys. Lett. **629**, 53–57 (2015)
58. E.H. Patel, M.A. Williams, S.P.K. Koehler, Kinetic energy and angular distributions of He and Ar atoms evaporating from liquid dodecane. J. Phys. Chem. B **121**, 233–239 (2017)
59. R.E. Cochran, O. Laskina, J.V. Trueblood, A.D. Estillore, H.S. Morris, T. Jayarathne, C.M. Sultana, C. Lee, P. Lin, J. Laskin, A. Laskin, J.A. Dowling, Z. Qin, C.D. Cappa, T.H. Bertram, A.V. Tivanski, E.A. Stone, K.A. Prather, V.H. Grassian, Molecular diversity of sea spray aerosol particles: impact of ocean biology on particle composition and hygroscopicity. Chem **2**, 655–667 (2017)
60. T.H. Bertram, R.E. Cochran, V.H. Grassian, E.A. Stone, Sea spray aerosol chemical composition: elemental and molecular mimics for laboratory studies of heterogeneous and multiphase reactions. Chem. Soc. Rev. **47**, 2374–2400 (2018)

61. K. Jardak, P. Drogui, R. Daghrir, Surfactants in aquatic and terrestrial environment: occurrence, behavior, and treatment processes. Environ. Sci. Pollut. Res. **23**, 3195–3216 (2016)
62. G.T. Barnes, Permeation through Monolayers. Colloids Surf. A **126**, 149–158 (1997)

The manufacturer's authorised representative in the EU is Springer
Nature Customer Service Centre GmbH, Europaplatz 3, 69115 Heidelberg,
Germany. If you have any concerns regarding our products, please
contact ProductSafety@springernature.com

Printed and bound by CPI Group (UK) Ltd, Croydon, CR0 4YY
29/04/2026
02099459-0010